HOW TO REMOVE POLLUTANTS AND TOXIC MATERIALS FROM AIR AND WATER

How to Remove Pollutants and Toxic Materials from Air and Water

A Practical Guide

Marshall Sittig

NOYES DATA CORPORATION
Park Ridge, New Jersey, U.S.A.
1977

Library of Congress Catalog Card Number: 77-71309
ISBN: 0-8155-0654-6
Printed in the United States

Published in the United States of America by
Noyes Data Corporation
Noyes Building, Park Ridge, New Jersey 07656

FOREWORD

This, the thirty-second volume in our Pollution Technology Review series is a practical book dealing with the removal of pollutants and toxic materials from air and water. Subject entries are arranged in alphabetic sequence. Because of the encyclopedic nature of the book it should prove to be very valuable to all those concerned with the problems of pollution abatement and engineering.

As explained in the introduction, this book is based almost exclusively on U.S. patents with coverage extending to a very late date. Because it supplies technical information in detail, it can also be used as a guide to the U.S. patent literature in this field. By indicating all the information that is significant, and eliminating legal jargon and juristic phraseology, this book presents an advanced, technically oriented review of modern pollutant removal practices as depicted in U.S. patents.

The U.S. patent literature is the largest and most comprehensive collection of technical information in the world. There is more practical, commercial, timely process information assembled here than is available from any other source. The technical information obtained from a patent is extremely reliable and comprehensive; sufficient information must be included to avoid rejection for "insufficient disclosure."

The patent literature covers a substantial amount of information not available in the journal literature. The patent literature is a prime source of basic commercially useful information. This information is overlooked by those who rely primarily on the periodical journal literature. It is realized that there is a lag between a patent application on a new process development and the granting of a patent, but it is felt that this may roughly parallel or even anticipate the lag in putting that development into commercial practice.

Many of these patents are being utilized commercially. Whether used or not, they offer opportunities for technological transfer. Also, a major purpose of this book is to describe the number of technical possibilities available, which may open up profitable areas of research and development. The information contained in this book will allow you to establish a sound background before launching into research in this field.

Advanced composition and production methods developed by Noyes Data are employed to bring our new durably bound books to you in a minimum of time. Special techniques are used to close the gap between "manuscript" and "completed book." Industrial technology is progressing so rapidly that time-honored, conventional typesetting, binding and shipping methods are no longer suitable. We have bypassed the delays in the conventional book publishing cycle and provide the user with an effective and convenient means of reviewing up-to-date information in depth.

15 Reasons Why the U.S. Patent Office Literature Is Important to You —

1. The U.S. patent literature is the largest and most comprehensive collection of technical information in the world. There is more practical commercial process information assembled here than is available from any other source.

2. The technical information obtained from the patent literature is extremely comprehensive; sufficient information must be included to avoid rejection for "insufficient disclosure."

3. The patent literature is a prime source of basic commercially utilizable information. This information is overlooked by those who rely primarily on the periodical journal literature.

4. An important feature of the patent literature is that it can serve to avoid duplication of research and development.

5. Patents, unlike periodical literature, are bound by definition to contain new information, data and ideas.

6. It can serve as a source of new ideas in a different but related field, and may be outside the patent protection offered the original invention.

7. Since claims are narrowly defined, much valuable information is included that may be outside the legal protection afforded by the claims.

8. Patents discuss the difficulties associated with previous research, development or production techniques, and offer a specific method of overcoming problems. This gives clues to current process information that has not been published in periodicals or books.

9. Can aid in process design by providing a selection of alternate techniques. A powerful research and engineering tool.

10. Obtain licenses — many U.S. chemical patents have not been developed commercially.

11. Patents provide an excellent starting point for the next investigator.

12. Frequently, innovations derived from research are first disclosed in the patent literature, prior to coverage in the periodical literature.

13. Patents offer a most valuable method of keeping abreast of latest technologies, serving an individual's own "current awareness" program.

14. Copies of U.S. patents are easily obtained from the U.S. Patent Office at 50¢ a copy.

15. It is a creative source of ideas for those with imagination.

TABLE OF CONTENTS

Table of Contents

INTRODUCTION

This book is designed to provide a one-volume ready reference for the handling of toxic materials and other pollutants emerging into the air and water from industrial processes.

The developments are coming so fast that getting this information between the covers of one book is virtually impossible. Hence the present volume supplements and updates the earlier (1973) *Pollutant Removal Handbook* from this same publisher.

This book is based almost entirely on material from U.S. patents dealing with practical environmental control systems. It surveys some 500 patents in the 1973 to 1976 period with exhaustive coverage up to November 1, 1976. Since environmental patents are given priority handling by the U.S. Patent Office, many of these were applied for in late 1975 and even early 1976.

So in this volume we offer broad coverage of timely, practical information. This book is addressed to:

—the industrialist who wants to and must keep abreast of the latest control techniques

—the environmental protection or public health official

—legislators who are contemplating control measures and their advisory staffs

—the conservationist who is interested in exactly what can be done about the effluents of local factories

—the manufacturer of pollution control equipment

Since, in the interest of combining comprehensiveness with compactness, all literature references cannot be cited here, the reader is referred to the following companion volumes, also published by Noyes Data Corp, for more detail on specific topics.

Mercury Pollution Control (1971)
Sulfuric Acid Manufacture and Effluent Control (1971)
Environmental Control in the Organic and Petrochemical Industries (1972)
Environmental Control in the Inorganic Chemical Industry (1972)
Detergents and Pollution (1972)
Fine Dust and Particulates Removal (1972)
Pollution Control in the Nonferrous Metals Industry (1972)
Waste Disposal Control in the Fruit and Vegetable Industry (1973)
Pollution Control in the Textile Industry (1973)
Pollution Control and Chemical Recovery in the Pulp and Paper Industry (1973)
Pollutant Removal Handbook (1973)
Pollution Control in the Petroleum Industry (1973)
Pollution Control in the Metal Finishing Industry (1973)
Pollution Control in Meat, Poultry and Seafood Processing (1974)
Pollution Control in the Dairy Industry (1974)
Pollution Detection and Monitoring Handbook (1974)
Pollution Control in the Organic Chemical Industry (1974)
Oil Spill Prevention and Removal Handbook (1974)
Environmental Sources and Emissions Handbook (1975)
Pollution Control in the Plastics and Rubber Industry (1975)
Pollution Control in the Asbestos, Cement, Glass and Mineral Industries (1975)
Resource Recovery and Recycling Handbook of Industrial Wastes (1975)
How to Dispose of Toxic Substances and Industrial Wastes (1976)

ACETONE CYANOHYDRIN

Removal from Water

A process developed by *W. Fries; U.S. Patent 3,984,314; October 5, 1976; assigned to Rohm and Haas Company* relates to the purification of industrial effluents containing cyanide ions, and cyanide precursors like acetone-cyanohydrin. The purification of such effluents is carried out by utilizing a complexing compound followed by treatment with an anion exchange resin and optionally cation exchange resin to remove the cyanide complexes. The cyanide values are recovered from the resin by acid regeneration.

ACID MINE WATERS

See Mining Effluents (U.S. Patent 3,823,081)
See Mining Effluents (U.S. Patent 3,717,073)
See Mining Effluents (U.S. Patent 3,795,609)
See Iron Oxides (U.S. Patent 3,537,966)

ACROLEIN PROCESS EFFLUENTS

The toxicity of alpha,beta-ethylenically unsaturated aldehydes and ketones even in low concentrations to biological treatment systems has been recognized by those skilled in the art. A review article on this problem is presented by V.T. Stack, Jr. in *Industrial and Engineering Chemistry,* Volume 49, No. 5, page 913 (1957). In the manufacture of such compounds, the wastewaters containing these organic substances must be processed at very low concentrations if they are further treated by a biological system. If not, the biomass is in danger of being killed or inhibited to a very low level of activity. This problem is particularly acute in the treatment of wastewaters from the manufacture of acrolein, acrylic acid and acrylic acid esters.

Removal from Water

Although disposal of toxic wastes by injection into a deep well has been used, this is a method of questionable efficiency and is not a long-term solution to the problem. Incineration of toxic wastes is also not economical because such methods are more expensive than biological oxidation and can have adverse environmental consequences. Wastewaters containing alpha,beta-ethylenically unsaturated aldehydes or ketones have been processed in the past by slowly adding the contaminated waters to a waste stream effluent so that the level of toxic component is diluted below that which is detrimental to the biological system. This method suffers from the disadvantage of being slow and inconvenient and is subject to the danger of at times exceeding the toxic limit of the contaminant, because of metering problems, thereby upsetting the operation of the treatment plant. The required dilution may also involve recycle of previously treated wastewater adding significantly to the hydraulic flow and thus to the cost of the biological treatment plant.

A process developed by *E.R. Lashley, Jr.; U.S. Patent 3,923,648; December 2, 1975; assigned to Union Carbide Corporation* for the disposal of wastewaters containing alpha,beta-ethylenically unsaturated aldehydes or ketones containing 3 to 10 carbons comprises contacting the wastewaters with sufficient base to render the pH of the wastewaters alkaline, maintaining the alkaline wastewaters at a temperature of about 25° to 100°C for at least about 15 minutes and then degrading the wastewater in a biological system containing active biomass.

The method outlined above is particularly preferred for use with wastewaters containing acrolein as the alpha,beta-ethylenically unsaturated aldehyde. It is equally useful for other aldehydes such as methacrylaldehyde, crotonaldehyde, 2,4-hexadienal, 2-ethylcrotonaldehyde, and the like as well as alpha,beta-ethylenically unsaturated ketones such as methyl vinyl ketone, and the like.

ACRYLIC RESIN PROCESS EMISSIONS

Water having organic residues such as acrylics, where the acrylic emulsion constitutes only about 2% or so of the total solution by weight, have been difficult to treat because such a solution does not lend itself to conventional chemical treatment or to conventional drying procedures. There is thus a long felt and extremely urgent need for an efficient and high capacity process for purification of such polluted liquids.

Removal from Water

A process developed by *J. Greenberg; U.S. Patent 3,766,087; October 16, 1973; assigned to Anti-Pollution Systems, Inc.* is one in which the polluted water is fed onto the surface of a brine solution maintained at the boiling point, the emulsion being broken instantly at the surface of the brine and accumulated as an organic residue, while the water is boiled off. The organic residue is skimmed off for separate handling.

Such a process is shown in Figure 1. Referring to the drawing, there is shown a container **10** in which is maintained a brine solution **11** having a level **12**. The solution is preferably a concentrated brine solution of water and salt. The process is nonspecific as to the salt, and may be worked with good results with ordinary NaCl, chlorates, nitrates and carbonates. In a preferred example, the brine is a saturated solution of NaCl (39.8% by weight), with some small amount of salt settling at the bottom of container **10**.

A gas fired immersion tube **13** is positioned within container **10** below the brine surface **12**, and provides a source of heating energy to maintain the brine at or just above the boiling level. In the preferred example, the brine solution is maintained at the boiling point, i.e., above 100°C. It is, of course, understood that any suitable means of heating may be employed. The polluted water solution is fed from a holding tank **15** through a spigot feed **16**, having a plurality of spigots **17** from which the polluted solution is entrained or dropped onto the surface **12** of the brine. A pump **17** is used for transporting the solution into the spigot feed. A second pump **18** takes brine collected from an inlet **19** near the surface **12**, and pumps it through pipe **29** to an ejecting orifice **20** positioned so that the emitted brine pushes the accumulated organic residue (noted at **25**) on the surface **12** from left to right as seen in the drawing.

FIGURE 1: APPARATUS FOR TREATING ACRYLIC RESIN PROCESS WASTEWATERS WITH LOW TEMPERATURE BRINE SOLUTION

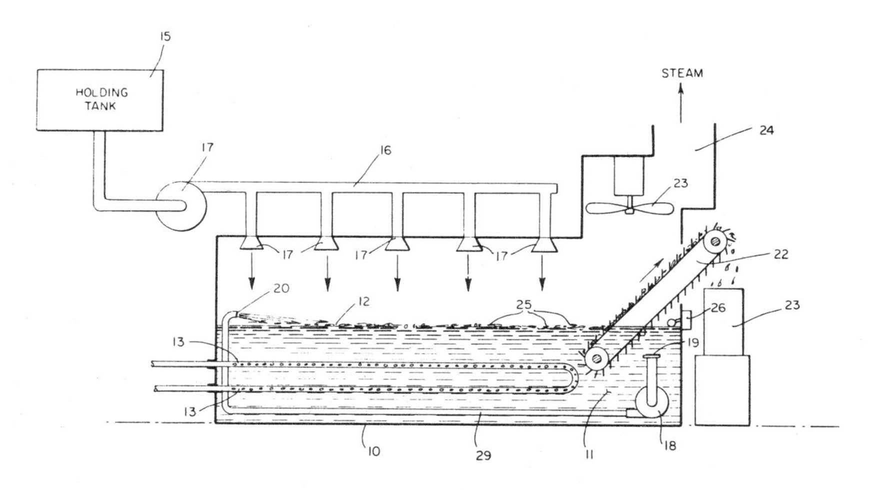

Source: U.S. Patent 3,766,087

There is thus, in the figure, a clockwise circulation of the brine, transporting the surface accumulation of organic residue, or scum, toward a conventional skimmer **22** which removes the scum and deposits it into a collection drum **23**. This skimming system is exemplary only, and any type of conventional skimming apparatus and/or system may be utilized which has the function of transporting the scum on the surface to a point where it may be skimmed or otherwise removed into a suitable collection instrument.

In operation, the polluted solution is spread from spigot **17** and distributed substantially uniformly onto the surface **12** of the brine. The brine, which is maintained at the boiling point, causes the emulsion to be broken instantly at the surface, such that it accumulates as organic residue. This effect in breaking the emulsion is also known as salting-out. At the same time that the emulsion is being broken, the water is being boiled off, and is drawn by exhaust fan **23** through port **24** where it is passed into the atmosphere in a safe and substantially pure form. The remaining organic residue, or scum, is pushed and/or drawn toward the skimmer and removed into the collection drum. Since there is about a 5% carryout of the salt with the curdled scum, or organic matter, the salt must be periodically replenished. In practice, the pump **18** and skimmer **22** operate continuously, drawing off the unburned residue. The system has been operated successfully at a rate of pressing 400 gal/hr of polluted water having a 2% emulsion content. It is, of course, understood that the system capacity may be enlarged by appropriate engineering design, and the method as claimed is not limited by the system rate. A water balance is maintained in the system, by adapting the rate of feed from the holding tank to the vapor exhaust rate, such that the amount of water passing through port **24** substantially equals that being fed in through spigot feed **16**. The surface **12** may be monitored, by conventional level detecting device **26**, with a signal derived to control the operation of pump **17** so as to provide a closed loop servo-type system for maintaining surface **12** within any prescribed range.

ACRYLONITRILE PROCESS EFFLUENTS

In the process for producing acrylonitrile from propylene, ammonia and air, a waste gas is exhausted after the absorption of acrylonitrile. The waste gas contains hydrocarbons, carbon monoxide, hydrogen cyanide, acrylonitrile and nitrogen oxides which are harmful or release a bad smell and sometimes cause photochemical smog, and therefore, it is unfavorable to discharge the waste gas into the atmosphere without removal of the harmful materials.

Removal from Air

There is a method for removing harmful materials from such a waste gas comprising subjecting the waste gas containing such harmful materials to an oxidative or reductive combustion at a high temperature in the presence of a catalyst and thereby changing the harmful materials to harmless materials.

A process developed by *T. Ohrui, Y. Sakakibara, T. Hoshikuma, O. Imai and M. Iwasa; U.S. Patent 3,988,423; October 26, 1976; assigned to Sumitomo Chemical Company, Limited, Japan* involves dividing the waste gas exhausted after the absorption of acrylonitrile into two flows and mixing the first flow gas with air in an amount necessary for combustion of the combustible gas contained in the whole waste gas. The process then involves subjecting the main flow gas of the mixed gas thus obtained to heat exchange with the gas of the outlet of a first catalyst layer. It then involves combining the main flow gas with the remainder of the mixed gas, thereby preheating the mixed gas at a temperature of 200° to 450°C.

The preheated mixed gas is passed through a first catalyst layer wherein at least one noble metal is dispersed on an alumina carrier, and therein burning the gas at a temperature of 650° to 750°C in the presence of an excess amount of oxygen to give a combustion gas. The combustion gas is combined with the second flow gas which is preheated by subjecting the main flow gas of the second flow gas to heat-exchange with the gas of the outlet of the second catalyst layer and combining the main flow gas with the remainder of the second flow gas, and thereby controlling the temperature of the combustion gas to 250° to 450°C. The combustion gas combined with the second flow gas is then passed through the second catalyst layer, and therein burning the gas at a temperature of the outlet of 600° to 750°C in a concentration of the remaining oxygen of 0 to 0.5% by volume.

In this process, the efficient removal of combustible gas and the nitrogen oxides contained in the waste gas exhausted from the absorption tower in an acrylonitrile producing plant can be achieved economically and in simple apparatus and operation by minimizing as much as possible the amount of the supplemented air and further by recovering efficiently the waste heat produced by the combustion of the waste gas.

The reaction effluent from the manufacture of unsaturated nitriles and more particularly from an acrylonitrile or methacrylonitrile plant will contain an appreciable amount of unreacted ammonia, some unreacted feed materials, e.g., propylene or isobutylene, oxygen, nitrogen (particularly if air is used as the source of molecular oxygen) and also one or more other reaction by-products such as water, hydrogen cyanide, acrolein, methacrolein, by-product saturated aliphatic nitriles such as acetonitrile and propionitrile, etc. The desired reaction product is generally recovered in a refrigerated absorber by absorption in a suitable solvent such as water, during which step some additional relatively heavy organic compounds may be formed.

The absorption is usually done after the reaction effluent is countercurrently scrubbed with a dilute mineral acid in a quench tower. The acid, typically sulfuric, reacts with the ammonia to form an ammonium salt and thereby makes the ammonia unavailable for the formation of undesired by-products, e.g., the formation of aminopropionitrile or aminodipropionitrile resulting from the direct reaction of ammonia and acrylonitrile. However, some cyanoethylation still takes place and these by-products tend to react with other constituents of the reactor effluent to form various polymers.

Some of these polymers are relatively heavy and almost all of them are characteristically soluble in water. For example, in the manufacture of acrylonitrile, the dilute water solution of the ammonia salt issuing from the bottom of the quench tower typically contains some acrylonitrile and other reaction products but is contaminated with organic "heavies" such as carbonyl polymers, hydrolyzed polyacrylonitrile, polyacrylamide, and cyanoethylated side reaction products. The overhead from the quench tower is lead into the absorber while the bottoms from the quench tower is split into two streams, one of which is cooled and recycled to the quench tower and the other of which is a wastewater stream. Dissolved acrylonitrile and usually one or more other compounds such as acetonitrile, hydrogen cyanide, propionitrile, etc., are recovered from this waste stream by conventional means, e.g., steam stripping.

One of the most pressing problems associated with plants for producing unsaturated aliphatic- and aromatic-nitriles is the ultimate disposal of waste by-products. For example, typically in an acrylonitrile plant the waste treatment facilities may include a vent stack for off-gases, a flare stack for relief valves, an incinerator for HCN, acetonitrile, and organic waste streams, an ammonium salt (e.g., ammonium sulfate) recovery plant, and an additional incinerator or biological waste treatment unit for wastewater streams. However, such facilities are expensive and do not provide a satisfactory solution to the problem of disposing of waste materials in a manner that meets governmental standards. A further problem is that certain of the by-products have only limited sales appeal.

A process developed by *H.R. Sheely; U.S. Patent 3,895,050; July 15, 1975; assigned to The Badger Company, Inc.* is one in which wastewater, unreacted ammonia and by-products such as HCN and acetonitrile are not condensed but remain with the absorber off-gas for ultimate disposal by incineration. The method employs a hydrocarbon solvent to adiabatically quench the reactor effluent and, after removal of polymer by-products, the partially quenched effluent is passed to a hot absorber column where the nitrile product but no ammonia and only some of the HCN are absorbed by the hydrocarbon solvent. The nitrile-solvent mixture is distilled to separately recover the solvent and nitrile product. The solvent is recycled. The hot absorber off-gases are cooled to recover water and then incinerated, with ammonia, HCN, acetonitrile and some vaporized solvent furnishing the necessary fuel values. In a preferred example, solvent in the absorber overhead vapors is recovered by scrubbing with a high boiling oil.

Removal from Water

A process developed by *W.O. Fitzgibbons, E.M. Schwerko and A.H. Brainard; U.S. Patent 3,734,943; May 22, 1973; assigned to The Standard Oil Company* is one in which a portion of the bottom effluent from the wet acetonitrile distillation column is removed and disposed of by pumping into a deep well after it has first been treated with a mixture of acrolein and ammonium sulfate, acrolein per se, or an aqueous solution containing these materials. Failure to treat the wet acetonitrile bottoms in this manner results in plugging of the disposal well when the material is pumped therein over a period of time.

The mechanism whereby the addition of acrolein and ammonium sulfate to the wet acetonitrile bottoms stream eliminates plugging of the deep disposal well is not known. However, plugging is believed to be caused by suspended particles in the waste stream in the particle size range of about 0.45 to 5.0 μ which cause a decrease in the permeability of the rock core and a corresponding increase in the injection pressure at the well. Plugging of the well is virtually eliminated when the wet acetonitrile bottoms stream is filtered through sand after it is treated with the acrolein or the acrolein-ammonium sulfate mixture. The acrylonitrile process is shown in Figure 2 together with the streams going to waste treatment.

In the process, approximately stoichiometric quantities of propylene, ammonia and oxygen (as air) are introduced via spargers and are reacted at elevated temperatures and essentially atmospheric pressures in catalytic reactor **1**. The effluent from the reactor is introduced into a quench tower **2** where the unreacted ammonia in the reactor effluent is neutralized with sulfuric acid and the resulting ammonium sulfate, water and heavy impurities are separated from the stream in the wastewater stripper column **3**. The overhead from the quench tower is taken to a recovery tower **4** where it is scrubbed with water in order to remove soluble organic products as an aqueous solution.

The aqueous solution of organic products is removed from the product scrubber by means of line **5**. This solution is then taken to a product separator **6** into which is also added an aqueous stream containing minor amounts of heavy impurities by means of line **15**. This column provides a wet acrylonitrile overhead **7** and a wet acetonitrile bottoms stream **12**. The wet acrylonitrile overhead is dried and purified by azeotropic distillation in column **8**, and conventional distillation in column **9**. The overhead **10** from distillation column **9** consists of product acrylonitrile, and the residue removed through line **11** comprises a mixture of small amounts of by-products and heavier impurities. The wet acetonitrile is transferred from the product separator **6** via line **12** to the acetonitrile distillation column **13**. The bottoms stream **14** from the acetonitrile distillation column is divided and a portion of the stream is recycled to the product separator **6** and the remainder is treated and disposed of according to the process. The overhead acetonitrile from the acetonitrile distillation column **13** is further concentrated and dried.

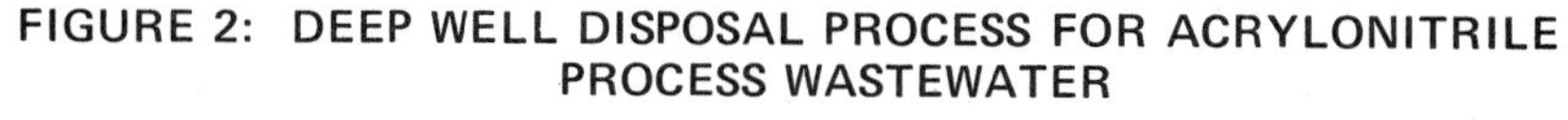

FIGURE 2: DEEP WELL DISPOSAL PROCESS FOR ACRYLONITRILE PROCESS WASTEWATER

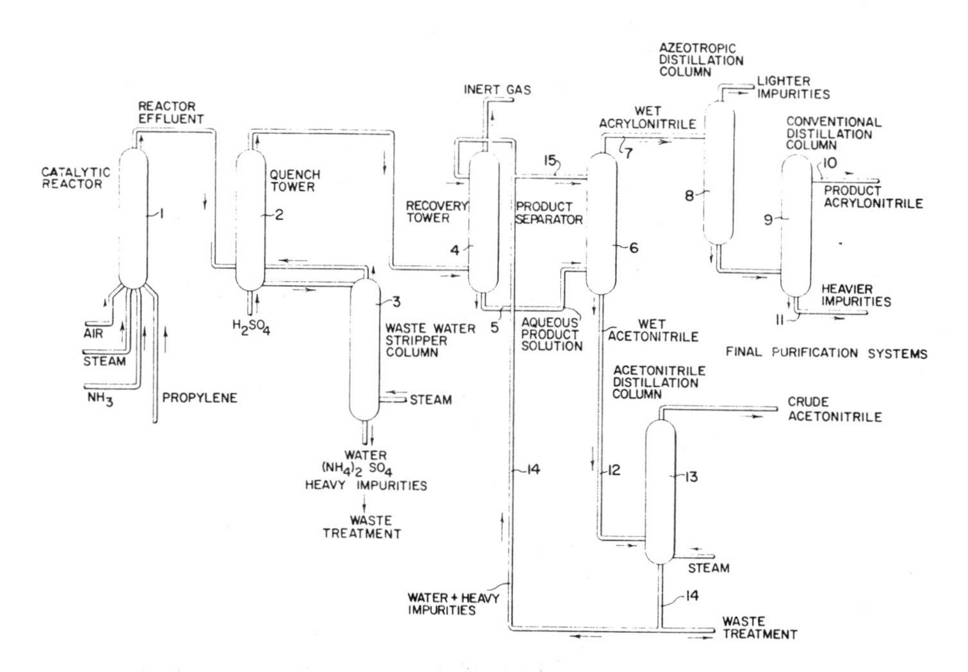

Source: U.S. Patent 3,734,943

The wet acetonitrile bottoms stream **14**, obtained from the acetonitrile distillation column **13** contains various by-products and heavy organic material resulting from the ammoxidation reaction. These products have been characterized as polynitriles, partially hydrolyzed polynitriles and cyanoethylated products of ammonia which in turn react with other constituents to form polymers, and polymers of hydrogen cyanide, unsaturated aldehydes, ketones, cyanohydrins, and the like. Chemical analysis of a typical wet acetonitrile bottoms stream indicates the following composition:

	% Concentration
Water	98.2
Ammonium sulfate	0.02
Heavy organic material	1.74

In the treatment of the aqueous stream of wet acetonitrile bottoms a synergistic effect on flow rate of the wastewater into the disposal well is observed for a mixture of acrolein and ammonium sulfate. However, satisfactory flow rates can also be maintained by treating the wastewater stream with small amounts of acrolein per se. Conveniently some waste streams from the acrylonitrile process, as for example, the bottoms stream from the wastewater column **3** contain sufficient amounts of acrolein and ammonium sulfate, which are present as a reaction product of acrolein and ammonium sulfate, so that when this stream is mixed with the waste stream from the acetonitrile distillation column, no plugging is observed in the well as evidenced by the lack of increase in injection pressure. Ammonium sulfate by itself, however, has little effect on the injectability when added to the wet acetonitrile bottoms waste stream.

A process developed by *H.C. Wu; U.S. Patent 3,936,360; February 3, 1976; assigned to The Standard Oil Company* is one in which substantial capital and operating cost savings and improved recovery of acrylonitrile and methacrylonitrile are realized by the recycle of the product column bottoms to the quench liquid of the reactor effluent quench system. Such a process variation is shown in Figure 3.

FIGURE 3: ACRYLONITRILE REACTOR EFFLUENT PURIFICATION SYSTEM SHOWING WASTE DISPOSAL

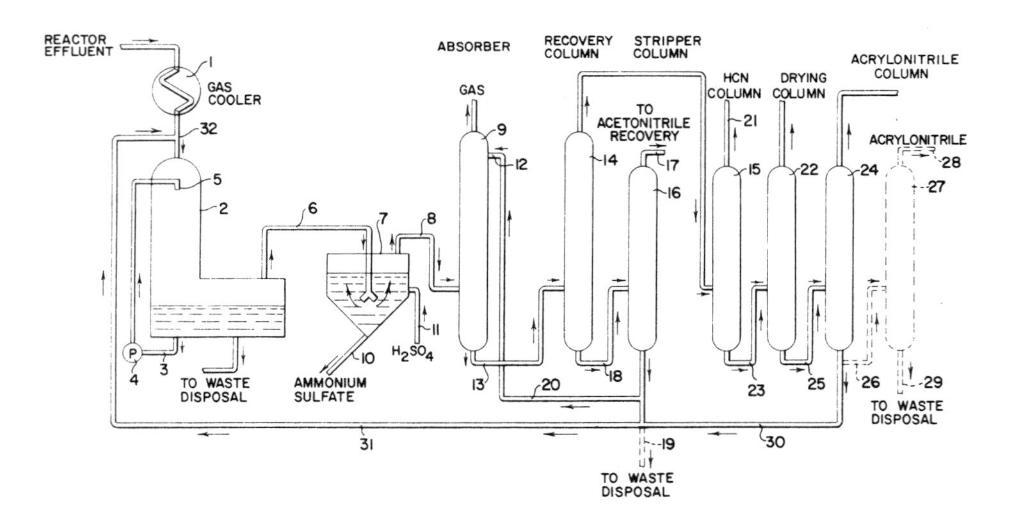

Source: U.S. Patent 3,936,360

The effluent from the ammoxidation reactor enters a gas cooler **1** in which it is cooled from a temperature of about 800° to 450°F. The effluent from the gas cooler **1** enters a quench system which consists of a recycle stream (to be described later) sprayed into the gas at point **32** and a gas washer **2**. The gas washer **2** contains an aqueous liquid in the bottom thereof which is held at a temperature of about 180°F. The aqueous liquid is continuously cycled through passage **3** by means of pump **4** to a spray nozzle **5** near the top of the washer **2** where the aqueous liquid spray contacts the incoming reactor effluent. The recycle stream could be added to this aqueous liquid between pump **4** and spray nozzle

5 by a line. The high-boiling material from the reactor effluent is deposited in the aqueous liquid in the bottom of the gas washer **2** and the volatile materials in the effluent are passed as vapor through a line **6** into a bubble chamber **7** containing aqueous sulfuric acid wherein most of the ammonia in the effluent is converted to ammonium sulfate.

The aqueous sulfuric acid in the chamber **7** is maintained at a temperature of about 180°F and the volatile products other than ammonia continue through line **8** into an absorber **9**. The level of aqueous sulfuric acid in the bubble chamber **7** is maintained substantially constant by the continuous withdrawal of ammonium sulfate solution at the bottom **10** and addition of fresh sulfuric acid through line **11**. In the absorber **9** a stream of water at **12** contacts the gaseous stream from **8** and the water-soluble material is taken off at the bottom of the absorber column **9** through line **13** and is conducted to the recovery column **14** wherein a light overhead is collected and transferred to the HCN column **15** and the bottoms from the recovery column **14** go to the stripper column **16** through line **18**.

From the stripper column **16** there is withdrawn an overhead **17** to by-product acetonitrile recovery. The major portion of the bottoms from the stripper column is transmitted by line **20** back to the absorber column **9** at point **12** previously described. The remaining portion, which was formerly taken to waste disposal **19** is taken to point **32** through line **31** and sprayed into the reactor effluent. In the further processing of the acrylonitrile, the HCN is removed as overhead, **21**, in the HCN column **15** and the bottoms from this column are transmitted to a drying column **22** through line **23**. In the drying column **22** water is removed as the overhead and the bottoms go to the acrylonitrile column **24** through line **25**. Acrylonitrile is recovered as the overhead from the acrylonitrile column **24**.

According to the prior art procedure, the bottoms from the acrylonitrile column were transferred by line **26** to a second distillation column **27** wherein acrylonitrile was taken as overhead **28** and the bottoms went to waste disposal through line **29**. In this process, however, the bottoms from column **24** which are normally slightly acidic (pH about 3 to 5) are conducted through line **30** into the line **31** and back to point **32** where they are mixed with the reactor effluent which has just been cooled in the gas cooler **1** which then enters the gas washer **2** and completes the recovery cycle. Alternatively, this recycle stream could be sprayed through nozzle **5** rather than at point **32**.

A. Kato and K. Yamamura; U.S. Patent 3,940,332; February 24, 1976; assigned to Sumitomo Heavy Industries, Ltd., Japan developed a process wherein a wastewater effluent containing nitriles and cyanides is treated by passing it through an acclimated, activated sludge containing a microorganism capable of degrading nitriles and cyanides, such as one selected from the genus Nocardia, e.g., *Nocardia rubropertincta* ATCC 21930. The wastewater effluent containing 50 to 250 ppm of nitriles and 10 to 50 ppm of cyanide and 500 to 2,000 ppm of COD (potassium dichromate method) can be purified with a high efficiency.

ADIPIC ACID PROCESS EFFLUENTS

In general, in the process for oxidizing cyclohexanone or cyclohexanol with an aqueous solution of a concentrated nitric acid in the presence of a metal oxide catalyst, a mother liquid from which adipic acid has been separated in the form of crystals normally contains various types of carboxylic acids produced by oxidative decomposition of the starting material, in addition to excess nitric acid and metal ions used as a catalyst. These carboxylic acids referred to above include monobasic acids such as formic acid and acetic acid; and dibasic acids such as oxalic, succinic, glutaric and adipic acids.

Removal from Water

A process developed by *M. Seko, A. Yomiyama, T. Miyake and H. Iwashita; U.S. Patent 3,673,068; June 27, 1972; assigned to Asahi Kasei Kogyo K.K., Japan* involves recovering a substantial portion of nitric acid and metal ions from the acidic waste liquid produced in the process by supplying the waste liquid to an electrodialysis apparatus including one or more electrodializers to recover the nitric acid and metal ions in a recovering liquid selected from water and a diluted aqueous nitric acid. This enables the use of the recovered solution in the subsequent oxidation reaction of cyclohexanone or cyclohexanol with nitric acid.

A process developed by *C.R. Campbell, D.E. Danly and M.J. Mathews, III; U.S. Patent 3,267,029; August 16, 1966; assigned to Monsanto Company* provides for the removal of monobasic and dibasic acids, mineral acids, and other organic and inorganic material from aqueous adipic acid process mixtures prior to the discharge thereof to waste.

In this process, the aqueous mixture is contacted either in single or multiple stages with a liquid ion exchange material dissolved in a suitable solvent to remove the chemical-oxygen-demand-causing components in the solvent phase, thereby leaving purified water suitable for safe, nonchemical-polluting waste discharge to rivers or other natural waste areas. The resulting solvent phase may be contacted with an inorganic base, anhydrous ammonia, an aqueous solution of an inorganic base or ammonia, or mixtures thereof to regenerate the solvent solution of ion exchange material for recycle or other use, and the remaining concentrated aqueous solution of salts of the organic acids and other chemical-oxygen-demand-causing components which remain in much reduced volume may be disposed of by incineration or other suitable nonpolluting means or recovered for profitable use.

The results of a typical operation of a single-stage process on a continuous basis according to the preferred example are summarized below; all numbers are in pounds per hour unless otherwise specified.

	Feed	Product
Water	100,000	97,137
Nitric acid	287	0
Organic acids	729	176
Ammonium salts	0	22
Solvent	0	9
Cyclohexyl nitrate	49	5

As can be seen clearly from the above results, all of the inorganic acid and approximately 76% of the organic acids can be removed simply by the single-stage process. It is clear, also, that multistage purification by the use of this process can be accomplished easily if it is desired to reduce the chemical oxygen demand of the product below that which is possible with the use of a single stage.

ALDEHYDES

Formaldehyde is neither thoroughly recovered nor utilized in either the process for producing formaldehyde as the raw material, or in the industries for utilizing the formaldehyde, and consequently the formaldehyde is usually discharged, in most cases, as a waste gas or as a dilute solution. The formaldehyde is a very toxic substance, and therefore it is necessary to pay great attention to its disposal from both the social sanitary viewpoint and from the viewpoint of preventing environmental pollution.

As to the other aldehydes—for example, acetaldehyde which is an important substance as a raw material for preparing acetic acid or peracetic acid and acrolein which is also an important substance as an intermediate material for preparing acrylic acid or as a raw material for preparing synthetic resins as such—since their toxicities are equivalent to that of formaldehyde, the same attention must also be paid to the handling and disposal of these aldehydes as that for formaldehyde.

Removal from Air or Water

Heretofore, water absorption, catalytic oxidation-decomposition based on the use of platinum catalyst, or ammonia absorption have been known as methods for treating a waste gas containing formaldehyde. The catalytic oxidation-decomposition method is based on decomposition of formaldehyde to harmless carbon dioxide and water and can be said to be an ideal method for treating the formaldehyde, but it is not always an economical method because of the use of the expensive catalyst or the use of auxiliary fuel for heating.

On the other hand, the water absorption method is generally used widely as a method for removing most of the formaldehyde from the waste gas at a relatively low cost, but a large amount of water and a large scrubbing apparatus are necessary for completely removing the formaldehyde from the waste gas by water scrubbing from the viewpoint of vapor-liquid equilibrium of the aqueous formaldehyde solution; therefore the water absorption method is not always advantageous. Furthermore, a large amount of dilute formaldehyde solution from water scrubbing cannot be effluent from the viewpoint of environmental pollution. The ammonia absorption method also has various difficult problems in the disposal of by-products and residual ammonia.

A process developed by *S. Ishida, N. Oshima, K. Kurita, I. Suzuki and H. Ohno; U.S. Patent 3,909,408; September 30, 1975; assigned to Asahi Kasei Kogyo K.K., Japan* involves the effective removal of aldehyde from a mixture containing aldehyde which is either a gas or a solution which comprises treating the mixture at a pH of 6 to 11 with a mixed sulfite-bisulfite treating agent. The aldehyde of the mixture is ecologically-efficiently eliminated.

Removal from Water

Few industrially viable processes are known for treating effluents containing aldehydes or derivatives thereof. Biological treatment is a possibility if concentrations are low; and absorption on activated carbon is another possibility but it is a slow process and regeneration of the carbon restores the formaldehyde. Incineration may solve the problem but is only economical for high effluent concentrations, being very costly for low effluent concentrations because of the large amounts of water which have to be evaporated. A process developed by *J.P. Zumbrunn; U.S. Patent 3,929,636; December 30, 1975; assigned to L'Air Liquide S.A. pour l'Etude et l'Exploitation des Procedes, France* is a process for the treatment of industrial effluent containing toxic impurities wherein aldehydes are oxidized by a peroxy compound containing for example, the anion $SO_5^=$ in the form of Caro's acid (H_2SO_5) or in the form of a salt thereof.

See Acrolein Process Effluents (U.S. Patent 3,923,648)

ALKALIS

Removal from Air

A process developed by *T.S. Dean; U.S. Patent 3,647,395; March 7, 1972* involves the recovery of substantially pure alkali metal salts from exhaust gases and dust from the kilns of a cement-producing operation. As is well-known in the art, most cement raw materials contain alkali metal compounds in some amounts. As these raw materials enter the clinkering zone of a cement kiln, a portion of the alkalis are volatilized and are carried by the exhaust gases toward the exit door of the kiln and removed from the kiln itself.

In many of these installations, these exhaust gases and dust are vented into the atmosphere where they may cause a pollution problem. During the clinkering or burning of the cement raw materials, these alkali metals are converted into alkali metal sulfates or chlorides and these compounds if vented into the atmosphere carry along with them the sulfate or chloride radicals themselves. Both of these materials will cause air pollution problems. Additionally, it is well known to dissolve these alkali metal sulfates or chlorides in water after they leave the kiln or blast furnace and then once in solution they are discharged into any open stream. This also causes pollution problems in the water of that area.

The pH of the stream may be increased to the point where fish and water life will be killed. In addition, the volatilized parts of the raw material used in producing cement which has been carried out of the kiln with discrete particles will form dissolved solids in the water which, if allowed in too concentrated form are toxic to and will kill fish and water life. Also, the fine discrete particles of the raw materials themselves are not easily removed from the discharge water and therefore there is an increase in the suspended solids which may kill fish and other water life if allowed to increase in concentration.

These exhaust gases from a cement manufacturing plant kiln will carry with them finely divided particles of raw materials themselves. Thus, there is an economic loss of the raw materials if these discrete particles are exhausted into the atmosphere or discharged into an open stream or lake. This is costly in that there is a loss of part of the raw materials used in the production of concrete clinkers.

Along with the discrete particles of the raw materials carried from the kiln as previously pointed out, alkali metal salts, usually sulfates, are also moved from the kiln. It is of importance that these alkali metal sulfates which are soluble in water, be removed from the water which may be discharged into an open stream or sewer system. If they can be recovered in a substantially pure condition, they will bring a reasonable price on the open market and thus the cost of producing one barrel of cement clinkers will be substantially reduced. Further, if an attempt is made to return the discrete particles of raw materials to the kiln itself, without removal of these alkali metal salts there will be a buildup of these salts in the cement clinkers themselves which is most objectionable.

ALKALI CYANIDES

Removal from Air

In the process for producing ferromanganese in a conventional blast furnace higher temperatures are required than in iron ore reduction. At the high temperatures present in the ferromanganese blast furnace cyanides are synthesized from the nitrogen gas and the alkali metal carbides which are also present. The cyanides formed, particularly potassium cyanides are condensed higher in the furnace and carried out in the gas stream as a fine sublimate.

A process developed by *J.S. Mackay; U.S. Patent 2,877,086; March 10, 1959; assigned to Pittsburgh Coke & Chemical Company* relates to the removal of potassium cyanide and other solid cyanides present in the gases from ferromanganese blast furnaces.

In the past, the solid cyanides have conventionally been disposed of in two ways. First, the solids are removed from the gas by a Cottrell precipitator and then the powders are either burned to destroy the cyanides or alternatively where contamination is not a problem, the solids are washed out with water and disposed of in streams. Both of these procedures are relatively expensive and in areas where the streams must contain potable water the second procedure cannot be used at all. It is possible to destroy the cyanides in aqueous solution by chlorination in an alkali medium and the effluent then passed into the stream, but the cost of such a procedure is very appreciable.

It has been found that a large part of the cyanide can be removed by recycling the wash water liquor from the thickener back to the gas washer. Ideally, 100% of the water is recycled. As a practical matter, normally only about 95% of the water from the overflow is recycled. Normally up to 3% of the total water is removed with the sludge and it has been found desirable to recycle 80 to 97% of the total water or to recycle 80 to 99% of the clarified water after sludge removal.

ALKYL IODIDES

Removal from Air

Prior art processes have attempted the decontamination of air streams containing radioactive iodine by many techniques. One such technique employed the use of a mercuric nitrate-nitric acid solution at molarities of about 0.1 and 0.1, respectively. Although the mercury scrubbing process is known to remove elemental iodine effectively from air streams, removal of alkyl iodides, such as methyl iodide, has been shown to be inefficient in practical contacting equipment except at unpractically low gas flow rates. Since a significant fraction of the iodine in off-gases from fuel reprocessing plants is invariably in the form of alkyl iodides, efficient removal of these species is needed to provide a high overall iodine decontamination efficiency. Another hindrance to the use of a mercury scrubbing process has been the toxic nature of mercury.

Another prior art method for removing alkyl iodides from air streams, U.S. Patent 3,752,876, involves contacting the air stream with boiling, concentrated nitric acid in a reflux system wherein the organic iodides are decomposed and the iodine is oxidized and held in solution in the iodate form. The reflux system is controlled to permit the escape of nitrite-forming oxides of nitrogen which would otherwise reduce some iodate to free iodine and cause loss by volatilization. A disadvantage in that method lies in the critical control of the reflux system to prevent these losses of elemental iodine. Also, iodine in this system is highly corrosive and the ordinary materials of construction (stainless steels) cannot be used.

A process developed by *J.M. Schmitt, D.J. Crouse, Jr. and W.B. Howerton; U.S. Patent 3,852,407; December 3, 1974; assigned to U.S. Atomic Energy Commission* is an improved process for contacting and removing iodine and particularly alkyl iodides from an air stream comprising the use of a relatively concentrated nitric acid solution containing mercuric ions recycled through a gas contactor. Concurrently, a minor fraction of the solution is passed through an evaporator to oxidize and concentrate the accumulated iodine causing it to precipitate as mercuric iodate, thereby allowing essentially complete recycle of the mercury and nitric acid. Such a process is shown in Figure 4.

Air contaminated with I_2 and alkyl iodides is contacted countercurrently with a solution of nitric acid at a concentration of from about 6 to 14 but preferably from about 8 to 10 molar and at least 0.1 but preferably 0.2 to 0.6 molar in $Hg(NO_3)_2$ at a temperature of from 25° to 80°C and preferably about 25°C in a packed vertical column or in a bubble-cap or sieve-plate column. Column packing can be either ceramic berl saddles or preferably protruded, perforated stainless steel. The aqueous effluent from contacting column **1** accumulates in a surge tank **2**, where a portion of the effluent is pumped back to the top of column **1**. To remove accumulated iodine, a fraction of from about 1 to 15% of the effluent stream is evaporated in evaporator **3** to oxidize and concentrate the iodine, thereby causing precipitation of most of the iodine as mercuric iodate.

The mercuric iodate is somewhat slow in precipitating. In practice, it is satisfactory to allow the evaporated solution to remain in decanter **4** for about 3 hours at ambient temperature to allow precipitation to approach completion. After precipitation the mixture is fed to filter **6**. Under these conditions, the concentration of iodine in the supernatant solution is typically about 0.2 g/l. Fumes from evaporator **3** are fed to condenser **5**. The nitric acid from condenser **5** and the supernate (mercuric nitrate-nitric acid solution) from the filter **6** are recycled to surge tank **2**. The mercuric iodide precipitate can be stored as such or can be reacted with caustic in reactor **7** to form sodium iodate (for storage) and mercuric hydroxide. The latter can be dissolved in nitric acid in container **8** to provide complete mercury recycle. The rate at which solution is fed to the evaporator is dependent on the rate at which iodine enters the system in the incoming gas at source **9** and the desired iodine concentration in the solution being recycled to the top of the column. In practice the aqueous feed rate to the evaporator is usually set so that, at steady-state, the iodine concentration in the recycle solution is in the range of 0.1 to 0.3 g/l. The flow of solution through the column relative to the gas flow is not critical. Usually the solution pumping rate, on a volume basis, is maintained at 0.3 to 2% of the gas flow.

The system as shown provides very efficient removal of various iodine species including elemental iodine, methyl iodide, and other organic iodides, such as butyl iodide from gas streams issuing at **12**. Typically iodine decontamination factors when operating at gas flow rates (face velocity to the column) of 25 to 50 ft/min are higher than 10^3. As an alternative, it has been found desirable to use a second column in series with the first since there is some tendency for a very small fraction of the accumulated iodine in the solution pumped to the top of the first column to bleed off, i.e., decompose, into the gas stream leaving the column.

FIGURE 4: APPARATUS FOR REMOVING ALKYL IODIDES FROM AIR USING MERCURIC NITRATE SOLUTION

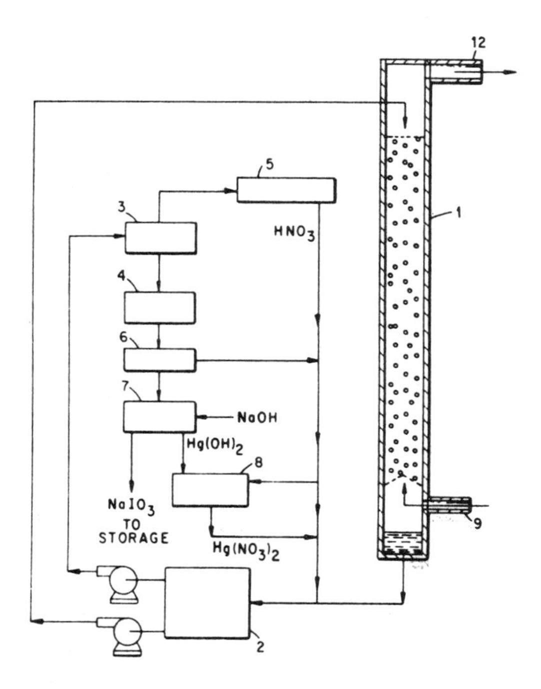

Source: U.S. Patent 3,852,407

See Iodine (U.S. Patent 3,630,942)
See Iodine (U.S. Patent 3,658,467)

ALKYLATION PROCESS EFFLUENTS

Removal from Air

Pollution of the air with HF acid vapor and light hydrocarbons occurs when noncondensible gas is vented from a depropanizer accumulator in an alkylation process. The venting is continuous. The vent gas heretofore has been neutralized in a closed system with sodium or potassium hydroxide or otherwise chemically contacted with an agent of which a suitable disposition must be made. In any event, however the HF contaminating the vent gas has been recovered, there has been a disposal problem as in disposing of spent caustic combined with fluoride and a concomitant waste of the HF.

A process developed by *C.O. Carter; U.S. Patent 3,972,956; August 3, 1976; assigned to Phillips Petroleum Company* utilizes at least two strippers, a first to remove substantially all readily vaporizable components, e.g., light hydrocarbons and HF, and to separate acid-soluble oils as a liquid residue, and a second to produce rerun HF vapors and a stream of water containing HF. In hydrocarbon alkylation operation, e.g., an isoparaffin by an olefin, the stream is used by reaction with an olefin to produce alkyl fluoride which can be fed to the alkylation reaction. Either before or after such reaction the stream is used to remove from noncondensible gas, as in a vent gas absorber, HF vapors therein contained and the stream is returned to the HF rerun unit.

ALUMINUM

The two most common chemical processing solutions that attain high concentrations of aluminum in aluminum finishing systems are: (a) A so-called basic aluminum etch containing free caustic soda that is continuously and slowly being enriched by aluminum which is dissolved from the surface of the metal workpiece and which is maintained until the aluminum concentration reaches a relatively high level at which time the solution may be dumped and a new one made up. Such a solution will contain aluminum in a range of 50 to 120 g/l at the time it is being dumped, with a free caustic soda content of about 30 to 50 g/l (NaOH). The aluminum is in the form of sodium aluminate, the basic hydroxide of aluminum.

(b) The second most common solution is an acid anodizing solution made up of about 250 to 300 g/l of sulfuric acid. After the aluminum content has increased, the solution has to either be dumped and renewed or some portion dumped and a new acid solution added to hold the aluminum concentration within a range of about 30 to 45 g/l.

Wash waters applied to a workpiece following either of the processes (a) or (b) contain the film that is dragged out from the process solution on surfaces of the workpeice which has been washed-off by the operation. The wash water usually contains considerably less and never more than about 1% concentration of the process solution. On neutralization of such used dissolved aluminum containing wash water, the aluminum hydroxide precipitates as a flocculent precipitate which is not easily compacted, compressed or thickened in any known manner.

Removal from Water

A process developed by *L.E. Lancy; U.S. Patent 3,738,868; June 12, 1973; assigned to Lancy Laboratories, Inc.* is one in which the yield of dry solid content of sludge is greatly increased in proportion to the quantity of flocculent material by taking off dissolved aluminum containing wash water solutions from an aluminum processing line, reconditioning and returning the solutions as aqueous wash solutions to the line and reusing them in the line, and treating and neutralizing the taken off solutions to settle and periodically remove sludge therefrom. The economy of the operation and yield of the dry solids content in the sludge is further enhanced by providing a relatively low acid pH before neutralization of alkaline waste, by combining acid and caustic solutions for treatment, and by the application of heat. Such a system is shown in Figure 5.

FIGURE 5: PROCESS FOR TREATMENT OF ALUMINUM-CONTAINING WASTEWATERS

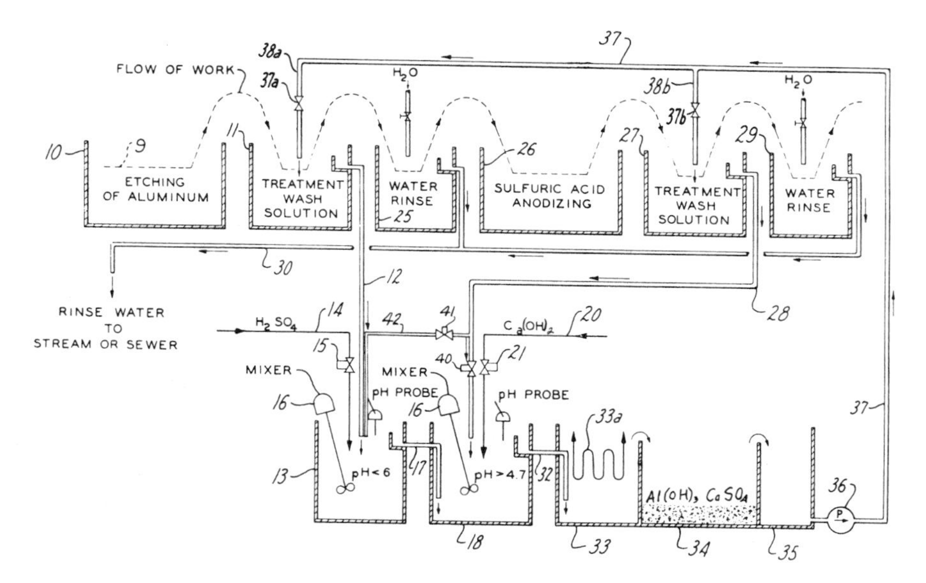

Source: U.S. Patent 3,738,868

In the figure, a workpiece **9** is shown adapted to move in line from left to right through a caustic (or acid) etching solution bath, tank or zone **10** and then into and through and out of a treatment wash solution bath, tank or zone **11**. It will be noted that overflow from the zone **11** is carried by return line **12** into a servicing or treatment tank or bath **13** of a reconditioning zone at which an inorganic acid, such as sulfuric acid, may be supplied through line **14** and control valve **15** to first convert the overflow representing the used aqueous waste wash solution from the zone **11** to an acid pH of less than 6, for better results, of less than 4.7, and for optimum results, to a pH of about 4. An electric mixer **16** is shown extending into the tank or bath **13** and a pH probe is also shown for checking its attained pH. Solution overflow from servicing bath or tank **13** moves through line **17** into a second servicing or treatment tank or bath **18** at which complete neutralization may be accomplished, although from a waste treatment standpoint, a pH of 4.7 allowing a full precipitation of aluminum may be sufficient. Since the second tank **18** is receiving acid wash water from wash zone **27** through return line **28**, an alkaline earth metal hydroxide, such as calcium hydroxide, may be supplied through line **20** and control valve **21**. The tank **18** also has a mixer **16** and a pH probe and is shown adapted to receive and mix overflow from the servicing tank **13** and from the wash treatment zone **27**.

Following washing treatment in the tank, bath or zone **11**, the workpiece **9** then moves through water rinse tank, bath or zone **25**, and then into and through a sulfuric acid anodizing tank, bath or zone **26**. On leaving the zone **26**, the workpiece is moved into and through treatment wash tank, bath or zone **27** which, like the zone **11** may be continuously supplied with reconstituted used aqueous washing solution, as moved along main return flow line **37** by pump **36**. Branch line **38a** supplies the used aqueous wash solution through control valve **37a** to zone **11**, and branch line **38b** and control valve **37b** supply the used aqueous solution to the zone **27**. The workpiece may then be moved from the zone **27** into a water rinse tank, bath or zone **29**. Since the water from the rinse zones **25** and **29** is relatively low in contaminants, it may as shown, be discharged from overflows through pipeline **30** to a stream or sewer.

Treated solution from the second tank **18** moves through overflow pipe **32** into a heating tank **33** which is shown provided with a steam coil **33a** to heat it to a temperature of at least 100°F. Then, the heated solution, after a period of about 10 minutes heating in tank or bath **33**, passes over an overflow into settling tank or bath **34** where it may be permitted to cool and where it may be fully reconstituted from the standpoint of the settling and removal of the aluminum containing sludge. The reconstituted used wash solution then moves into reservoir or holding tank **35** from which it may be supplied to the processing line.

A process developed by *Y. Aoyama; U.S. Patent 3,909,405; September 30, 1975; assigned to Dai-Doh Plant Engineering Corporation, Japan* involves treating an acidic or alkaline waste liquid containing aluminum dissolved therein to convert it to a neutral liquid free from colloidal aluminum hydroxide. The aluminum present in the hydrolysis system is hydrolyzed in the presence of 1 to 5 kg/kg of aluminum, of crystalline aluminum oxide having an average particle diameter of about 0.5 to 500 μ.

A process developed by *Y. Hanami and Y. Fukuyama; U.S. Patent 3,890,226; June 17, 1975; assigned to Kurita Water Industries, Ltd., Japan* aims to obtain dischargeable treated waters and also to recover aluminum hydroxide sludge having a moisture content of less than 60%, especially less than 50%, by subjecting various kinds of wastewaters from the aluminum surface treating process of aluminum manufacturing factories to a particular neutralization treatment.

This process involves treating aluminum-containing alkaline wastewater by neutralizing the wastewater with acid water. The neutralization reaction is carried out by adding the acid water to the wastewater slowly after aluminum hydroxide resultant from the neutralization begins to precipitate.

ALUMINUM CELL EFFLUENTS

The removal of the fluorides from gases from the smelting furnace is important for two basic reasons. First, the fluorides are a dangerous pollutant which must be removed from the gases in order to conform with pollution control laws and provide a safe operation. A second reason for removing the fluorides from the gases is that the fluorides are used in aluminum smelting operations. These fluorides have become expensive and their recapture from the exhaust gases is an economical saving to the aluminum smelter.

Removal from Air

A process developed by *T.B. Nix; U.S. Patent 3,876,394; April 8, 1975; assigned to Fuller Company* is one in which the fluorides are removed by achieving intimate contact between finely divided particles of alumina and the fluoride-containing gases. This is achieved by feeding the particles of alumina into a stream of gases collected from the aluminum smelting furnaces countercurrent to the flow of gases. The particles of alumina adsorb the fluorides. The particles of alumina and adsorbed fluorides are then separated in a gas-solids separator, such as a cyclone and the alumina may then be fed to the smelting furnaces.

Preferably, two stages are used. The gases supplied to the second or last cyclone in the direction of gas flow are supplied with new alumina countercurrent to the flow of gases. The alumina particles removed in this second cyclone are conveyed to the gas stream supplied to the first cyclone in the direction of gas flow. In a two-stage apparatus, these gases will be those evolved from the smelting furnace. The alumina separated from the first cyclone may be either conveyed to the smelting furnace or recirculated or a portion may be recirculated and a portion conveyed to the smelting furnace. The gases separated in the first cyclone are conveyed to the second cyclone.

A process developed by *M. Bahri and K. Carlsson; U.S. Patent 3,827,955; August 6, 1974; assigned to A/B Svenska Flaktfabriken, Sweden* is one in which alumina is injected into the waste gases in an amount controlled according to the concentration of hydrogen fluoride therein. The gases are caused to flow turbulently through a reactor and then through separators so that the alumina (and adsorbed hydrogen fluoride) may be directed into the furances and the cleansed gases may be discharged. Such a process is shown in Figure 6.

The figure shows schematically the method and apparatus for the cleaning of waste gases discharged from an aluminum main furnace in which so-called Soderberg electrodes are applied and an alternate furnace in which so-called prebaked electrodes are utilized. In the Soderberg process the crude gases discharged from the furnace **21** contain a large percentage of tar, 10 to 100 mg/Nm^3, which advantageously can be separated in an electrostatic precipitator, as shown at **1** in the figure. In view of the tar consistency, preferably a pelletizer is to be applied which converts the dust, prior to dumping or use, to pellets. In the electrostatic precipitator also dust and iron are separated from the gases. The gases from the electrostatic precipitator which contain a small percentage of tar, particulate fluorides and gaseous fluorides thereafter pass through a duct system **2**, in which with a certain dosage a quantity of alumina is injected at **22**. The quantity injected at **22** is controlled to insure entrainment of the alumina in the gas stream in a proper quantity to obtain the optimum relation of the alumina to the concentration of fluorides in the gases. To obtain proper injection, a metering device may be provided at **6** and a sampling and analyzing device **23** may be incorporated in the duct system **2** in advance of the injection point **22** so as to afford accurate control of the feeder **6**.

FIGURE 6: APPARATUS FOR CLEANING WASTE GASES CONTAINING HF FROM AN ALUMINUM PRODUCING FURNACE

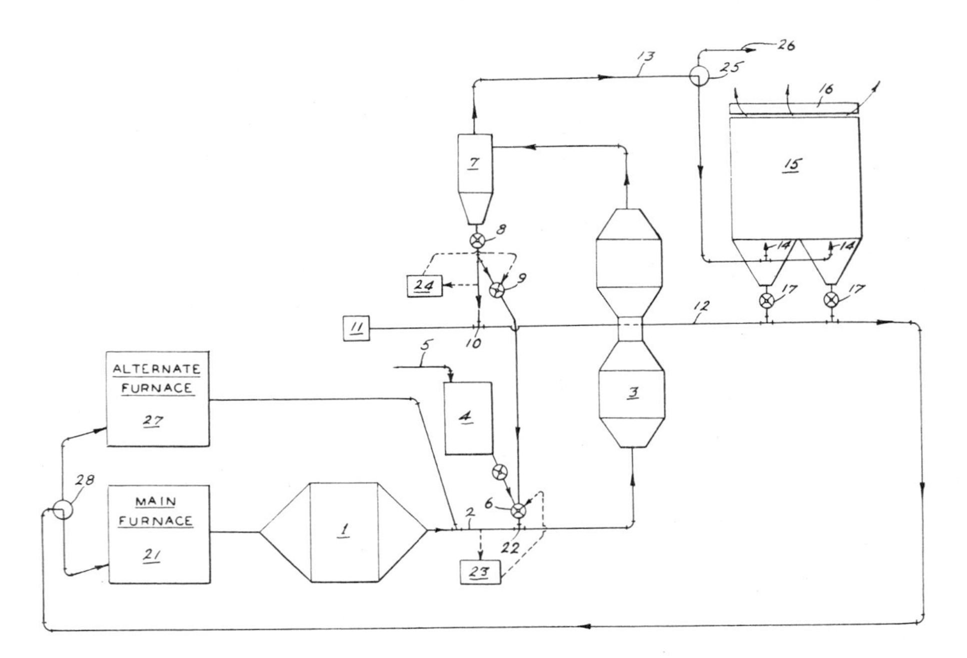

Source: U.S. Patent 3,827,955

The aluminum oxide then reacts with the gaseous fluorides in a reactor **3** of the design described below. The gases pass upwards from below and depending on the amount of aluminum oxide in relation to the fluoride content in the gas, temperature, fluoride concentration and residence time, an adsorption of the gas fluorides in the reactor and in the cyclone connected after the reactor at **7** in the drawing takes place whereby efficiency becomes of the magnitude up to 97%, for example at 50% alumina consumption.

The gases and the reactions products, as shown, thereafter pass through a cyclone separator **7**, in which 95 to 97% of the carried dust is separated. After having left the cyclone, the gases pass through a duct system **13** to a bag filter **15**. The degree of separation in the cyclone is chosen such that an amount of alumina controlled in advance is carried with the gases to the bag filter where a secondary reaction is effected between the gas fluorides and aluminum oxide. The dust discharged from the cyclone **7** through the feeder **8** may partially be recirculated through the reactor, for example by a sluice **9** to the injection point **22**, and the remaining part may be injected at **10**, and together with the dust obtained from the bag filter, may be transported back to the aluminum furnaces pneumatically by a blower **11** and conveyor **12** or by some other known method. The dust from the cyclone may also in its entirety be transported away to the furnaces, depending on the effective utilization of the alumina. A sampling device or analyzer may be provided at **24** to control the proportion of recycled alumina withdrawn by the sluice **9**.

The bag filter **15** may be either of so-called low-load or of high-load type. When using low-load filters, the vibration intervals (cleaning cycle) are to be optimized in order thereby to obtain the highest reaction degree. The total gas fluoride adsorption in this method, i.e., by means of reactor, cyclone and bag filter, was measured in an actual case to be 99 to 99.8% at 50% alumina consumption. The real mechanism for the adsorption of the gaseous impurities is not yet fully known because a conversion to the aluminum fluoride cannot be expected at these low temperatures. It is, thus, presupposed that the gas fluorides, which substantially contain hydrogen fluoride only influence the alumina surface layer, which in its turn is converted to aluminum fluoride without being split off from the surface and with no crystal change.

The electrostatic precipitator separates substantially particulate impurities, i.e., tar and dust with fluoride components and iron. Alumina is injected into the duct **2** through a metering device. Alumina is presupposed to be stored in a silo **4**, whch is filled with aluminum oxide by pneumatic transport, designated by **5**, from a large central silo. In the example shown, the reactor **3** is provided with a vertical pipe having different sections with different cross-sectional areas. The reactor may comprise more sections than the two shown in the figure. In a practical operation case, the gas rate at the narrowest reactor section was about 5 m/sec, and at the widest section the gas rate measured was about 0.8 m/sec. The height of each section was adjusted to the gas rate, and the residence time in the experimental plant was 6 to 7 seconds (volume divided by the gas amount per second).

Following the reactor **3**, the gases together with the aluminum oxide which has been diluted with hydrogen fluoride gas, pass through the cyclone separator **7** constructed for a definite degree of separation, in which separator 95 to 97% of the dust is to be separated. The amount of aluminum oxide, which according to above is injected at the point **22**, is mixed with oxide having been diluted with hydrogen fluoride gas in the reactor. The amount of aluminum fluoride recirculated is controlled by the device **9** acting as a variator as set forth above. The oxide from the cyclone **7** as mentioned above, may be transported away in its entirety to the aluminum furnaces by the duct system **12**.

The aluminum oxide, is evacuated from the bag filters **15** through the sluice apparatus **17** and in the direction of the arrow is to be supplied to the aluminum furnace **21**. In the schematic drawing shown, the bag filter was assumed to be a so-called low-load filter, into which the gases are introduced through funnels **14**, thereafter filtered through the bags and finally as cleaned gases evacuated to the atmosphere through a skylight designated by **16**. The bag filter preferably comprises several chambers equipped with automatic vibration means. The vibration intervals are determined by a suitable reaction time between aluminum oxide and hydrogen fluoride. In a test installation the optimum difference in time between the vibrations was of the magnitude 2 to 3 hours. The gas load on the bag filter was between 36 and 44 $m^3/hr/m^2$ cloth surface. An alternate furnace may be provided at **27**. In this case the furnace **27** utilized prebaked electrodes, eliminating the need for a preliminary electrofilter such as shown at **1**. A diverter is provided at **28** to direct the used alumina to either furnace **21** or **28**. At the same time, it would be possible to replace the above described bag filter by filters of so-called high-load type.

A process developed by *E. Böhm, L. Reh, E. Weckesser, G. Wilde and G. Winkhaus; U.S. Patent 3,907,971; September 23, 1975; assigned to Metallgesellschaft, AG and Vereinigte Aluminiumwerke AG, Germany* is one in which hydrogen fluoride is removed from a gas stream, e.g., the waste gas of an electrolysis cell for the production of aluminum, by using the gas as the fluidizing medium for an expanded fluid bed of particulate solids capable of taking up hydrogen fluoride. The solids content of the fluid bed chamber and the gas-flow parameters are adjusted to maintain a continuous solids concentration gradient throughout the entire vertically elongated chamber such that concentration decreases from the bottom to the top thereof.

In addition, a major proportion of the solids removed from the chamber are withdrawn at the top of the bed in entrainment with gases, are separated from the effluent gas stream, and are returned to the chamber. The solids may be fed to the electrolysis bath. Figure 7 shows the process and comprises a Hall-type electrolysis furnace **8** for the production of aluminum, above which is provided a hood **9** for collecting the exhaust gases. A blower **10** draws the exhaust gases from the hood **9** and forces them into the wind box **1a** of the fluid bed shaft **2**. The fluid bed, which consists of fine-grained alumina, absorbs hydrogen fluoride from the gas by a process or reaction which involves physical adsorption, chemisorption and chemical reaction. The discharged particles are separated from the gas stream by a cyclone **4a** and **4b** as previously described and recycled at **5a** and **5b** into the shaft **2**. For example, 20% by weight of the alumina required for the electrolytic cell of the furnace is recovered at **7** and delivered to the furnace **8** via line **15** through a dispenser or hopper **16**.

FIGURE 7: ALTERNATIVE SCHEME FOR REMOVING HYDROGEN FLUORIDE FROM GASES FROM ELECTROLYTIC FURNACE FOR ALUMINUM PRODUCTION

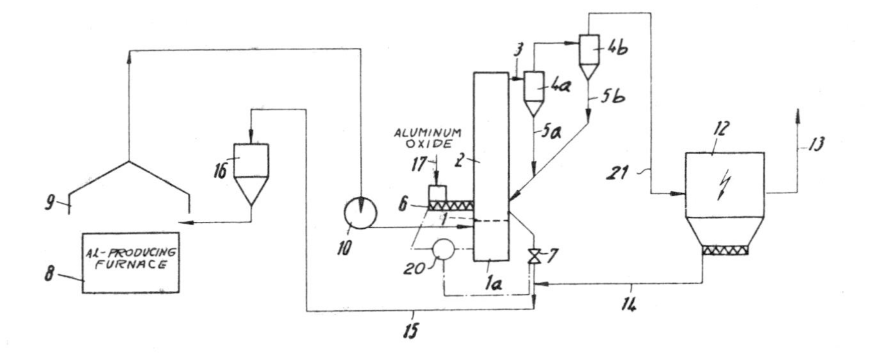

Source: U.S. Patent 3,907,971

Fresh aluminum oxide is supplied by conduit **17** to the feeder **6** which continuously supplies the alumina to the expanded fluid bed. A pressure sensor **20** responds to the pressure in the wind box **1a** and synchronously controls the feed device **6** and the discharge apparatus **7** to maintain the desired concentration of solids in the expanded layer substantially constant. The exhaust gases from the cyclones **4a, 4b** are supplied at **21** to an electrostatic precipitator **12** in which entrained dust is removed, the dust being combined with the alumina supplied via line **15** to the furnace **8** by a conduit **14**. The gases may then be discharged at **13** to the atmosphere.

A process for treating a gas steam, such as that from the incinerator stack of an aluminum electrolysis furnace containing hydrogen fluoride and sulfur dioxide, has been developed by *A.J. Teller; U.S. Patent 3, 919,392; November 11, 1975; assigned to Teller Environmental Systems, Inc.*

The gas stream is passed through two crossflow sorption zones in series and is contacted therein with separate adequate solutions of a basic material. In the first sorption zone the pH of the solution and the temperature of the gas stream are so maintained that only the hydrogen fluoride is removed. The soluble salt which forms is then reacted with calcium hydroxide to form insoluble calcium fluoride which is removed from the system; and the hydroxide solution is recycled to the first sorption zone.

In the second sorption zone the gas stream from which the hydrogen fluoride has been removed is contacted with a second aqueous solution of a basic material under conditions conducive to the formation of the corresponding soluble sulfite salt. The solution of sulfite is then reacted with calcium hydroxide to form insoluble calcium sulfite which is recovered. The remaining aqueous solution of hydroxide is recycled to the second sorption zone.

In the industrial production of aluminum it is important to maintain a clean working atmosphere in the furnace houses, and for this purpose expensive apparatus may be necessary, as the gases must be collected and cleaned before they can be discharged into the atmosphere. There are various systems which can be used. In one system, used in furnace houses in which the furnaces are in the open so that the exhaust gases escape into the house, the whole atmosphere in the furnace house is subjected to suction and drawn away through the roof. This requires powerful and expensive fans, as well as numerous gas purifiers extending over the whole length of the house and generally mounted in the roof. With such a system about 2,000 Nm^3 gas must be drawn off for each kilogram of aluminum produced.

In another system the individual furnaces are wholly enclosed and the exhaust gases are removed directly from each furnace. In this case the amount of air to be removed is only about 10% of that when the whole atmosphere in the house is subjected to suction, so less powerful fans are required, and this constitutes a substantial advantage. In addition, the gases can be led through a suction duct to a central purifying system, so that there is no need for the numerous purifiers in the roof extending over the whole length of the house. This second system, on the other hand, with its encased furnaces presents the considerable disadvantage that the servicing of the furnaces, that is to say the discharge of metal, the supply of alumina and so forth is complicated and thus the process is rendered more expensive. In fact, the complete removal of the exhaust gas is partially illusory, as the casings must be removed in order to service the furnaces. Not only does this involve additional work, but while the casing is removed the exhaust gases escape into the furnace house so that the atmosphere in it is no longer maintained completely pure. A further system involves the application of suction hoods over the furnaces. In this construction also a large part of the exhaust gases escapes into the surrounding atmosphere, particularly during the servicing of a furnace. The gas-air mixture to be removed is larger in volume than in the system last described.

In a scheme developed by *M. Laube; U.S. Patent 3,708,414; January 2, 1973; assigned to Swiss Aluminium Ltd., Switzerland* an air curtain directed towards a source of suction is set up around the furnace to prevent the escape of the exhaust gases into the atmosphere inside the furnace house.

Since the exhaust gases are removed as a comparatively concentrated gas mixture, the whole gas conveying and purifying plant can be smaller, thus reducing the capital cost of the installation. A further advantage is that there is no hindrance to the servicing of the furnace, and vice versa the furnace servicing does not have any adverse effect on the removal of the exhaust gases.

A process developed by *S.C. Jacobs and R.C. Schoener; U.S. Patent 3,904,494; September 9, 1975; assigned to Aluminum Company of America* involves a recycle system for the recovery and reuse of the effluent gas and additional values emitted in production of aluminum by electrolysis of aluminum chloride. The process scheme is shown in Figure 8.

FIGURE 8: EFFLUENT GAS RECYCLING AND RECOVERY SYSTEM

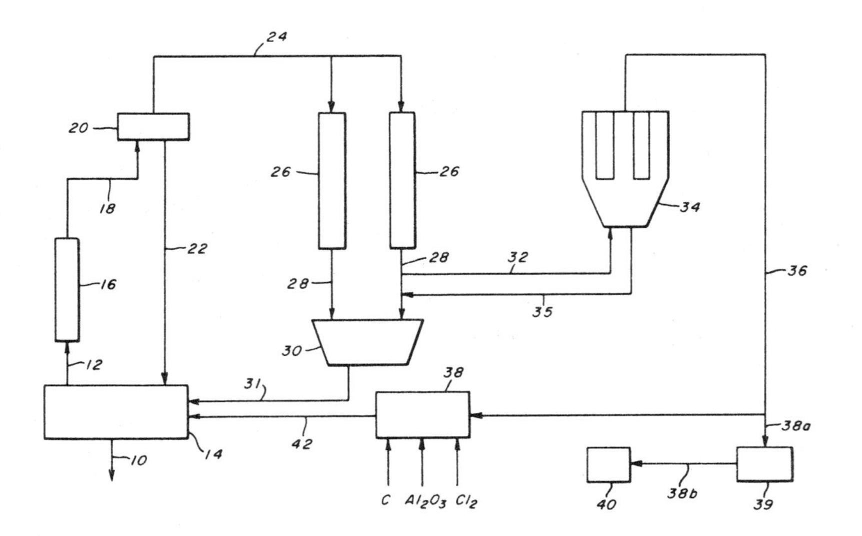

Source: U.S. Patent 3,904,494

Referring to the drawing, which schematically illustrates the sequence of operations in the practice of the improved recycle and recovery process, aluminum is produced in cell **14** by electrolysis of aluminum chloride dissolved in molten alkali or alkaline earth metal chlorides or mixtures thereof, the temperature within the cell **14** normally being in the neighborhood of 700°C. In operation of such a cell, aluminum is removed in molten form as indicated by the arrow **10**, and a gaseous effluent, comprising principally chlorine, together with a small amount of nitrogen and containing alkali metal chloride and/or alkaline earth metal chloride from the electrolyte and aluminum chloride, is evolved therefrom as indicated by the arrow **12**. Such a gaseous effluent may be exemplarily constituted of about 91.5% chlorine, 1.8% nitrogen, 4.8% alkali and/or alkaline earth metal chlorides, 1.9% aluminum chloride and traces of oxygen, and with the alkali and/or alkaline earth metal chlorides being in the form of a gas or in such finely divided form as to constitute a fume.

After evolvement and removal from the cell **14**, the effluent gas as above constituted, and at a temperature of approximately 700°C, is cooled, preferably by passage through a heat exchanger **16**, to reduce its temperature to a first predetermined value range that is sufficiently low as to effect selective condensation of all of the composite alkali and/or alkaline earth metal-aluminum chloride values without appreciable condensation or desublimation of the aluminum chloride values contained therein. The effluent gas may be reduced by the heat exchanger to a temperature range of from about 150° to 200°C, thereby effectively condensing such values into small liquid droplets or a mist.

Such cooled gaseous effluent is then conducted via line **18** to a coalescing zone such as that illustrated by the demister **20**, wherein the condensed liquid droplets of alkali and/or alkaline earth metal chloride-aluminum chloride combinations are coalesced and separated from the effluent gas. The coalesced liquids, including any aluminum chloride values dissolved therein, and, for example, amounting, in total, to about 0.01 to 0.2 lb/scf of effluent gas under standard conditions of 1 atmosphere and 25°C, are then directly returned through the line **22** to the electrolytic cell **14**, to thus continually replace the melt constituents being lost therefrom.

This depleted effluent gas, which still contains in the neighborhood of 0.004 lb/scf effluent gas of aluminum chloride values therein in gaseous form, is then passed via line **24** to a condensation zone or condensing device **26**, which may be, for example, a shell-and-tube heat exchanger or a fluidized bed of aluminum chloride operated at a second and lower predetermined temperature range that is often less than about 100°C to desublime the remaining aluminum chloride values thereof and effect a preliminary separation and transfer thereof via lines **28** to an appropriate collecting device, such as **30**, in crystalline form. Such collected aluminum chloride values may then be reintroduced via line **31** into the same or another electrolytic cell for electrolytic decomposition thereof.

This further depleted and relatively pure gaseous effluent is then conducted via line **32** to one or more bag filter units or assemblies **34**, by which any remaining or entrained solid impurities, and particularly those of small size, are removed. The solids collected by bag filter unit **34** will essentially comprise aluminum chloride and, if of sufficient purity, may be returned via line **35** to the collector **30** for eventual reintroduction, if desired, into an electrolytic cell **14**. The residual gaseous effluent from the filter **34**, in the form of relatively pure chlorine gas and some nitrogen, may then be conducted via line **36** to a locus of utilization thereof. Such chlorine gas, possibly along with other chlorine values, may be introduced into an aluminum chloride preparation zone **38** in which it may be reacted with alumina-bearing material in the presence of a reducing agent, for example, carbon. At least some of the chlorine values and carbon in such reaction may be employed in combined form, for example, as carbon tetrachloride or carbonyl chloride. Alternatively, such chlorine may be conducted via line **38a** to a condenser **39** and the resultant liquid chlorine transferred via line **38b** to storage as at **40**. As indicated at **42**, the aluminum chloride produced in zone **38** may be utilized in the same, or in another electrolytic cell **14**.

In aluminum production molten aluminum is formed in a reduction cell. The aluminum produced settles to the lower portion of the cell, which usually lies below floor level. In this process, the molten aluminum is removed from the base of the reduction cell by suction. An apparatus commonly referred to as a siphon spout is used in the suction collection of newly separated molten aluminum. The siphon spout is attached to a large container or collection pot, herein referred to as a crucible. High pressure air is used to create a partial vacuum in the crucible and siphon spout. The air travels through the straight legs of a T shaped valve located on the crucible lid, creating a partial vacuum in the right angle leg which has an opening into the crucible. The air exits the exhaust end of the T valve, hereinafter referred to as the exhaust pipe.

The high pressure air used to create a partial vacuum in the crucible requires a noise reducer, hereinafter referred to as a muffler, which usually is attached at a point where air exhausts from the straight leg (exhaust pipe) of the T shaped valve.

Prior art noise reducers generally have been commercially available diesel truck mufflers or mufflers of similar design. These mufflers have many disadvantages such as: excessive cost; large size which causes difficulty in handling and which necessitates removal from the crucible after each tapping operation; the interior frequently becomes stopped with small aluminum particles and the muffler must be cleaned or replaced, and excessive wear of muffler interior materials which necessitates frequent replacement.

An apparatus developed by *C.M. Benton and H.E. Niehaus; U.S. Patent 3,732,948; May 15, 1973; assigned to National-Southwire Aluminum Company* is one for reducing the noise level of a stream of air exiting a tapping crucible in aluminum reduction operations. The air is used to create a vacuum in the tapping crucible, thereby promoting siphon removal of molten aluminum from the reduction cell. This apparatus, commonly referred to as a muffler, comprises an outer conduit having entrance and exit openings, an inner conduit centrally attached within the outer conduit, the inner conduit having an entrance opening, an exit opening enclosed with a cup welded to the inner conduit, and exhaust slots equally spaced around the inner conduit's circumference.

Figure 9 shows, in turn, counterclockwise, a longitudinal section of the muffler, a transverse section, and a view of an aluminum crucible with the muffler attached. As shown in the figure, a reducer **11** is connected to the exit end of the exhaust pipe. One end of a bell reducer, **12**, is connected to reducer **11**. The other end of bell reducer **12** is connected to bushing **13**. The threads of bushing **13** are used to connect conduit **14** to bell reducer **12**. Conduit **14** can be a constant diameter collar or a bell reducer. Conduit **15** directs the air flow and is attached at the center of bushing **13**. The conduit **15** has a metal cap, **16**, attached to its end, preferably by welding, with a layer of absorbent material **17** glued to the inside of the cap. Absorbent material **17** preferably is composed of asbestos cloth or foam rubber. Advantageously the thickness of the absorbent material is approximately one-tenth inch for asbestos cloth and approximately one-half inch for foam rubber. Suitable glue is any commercially available glue which will bond cloth or rubber to metal and retain an effective bond at high temperatures. Preferably Elmer's glue is used.

Conduit **15** is encircled by rectangular grooves **18** sometimes hereinafter referred to as exhaust slots, which allow air to escape. The rectangular grooves **18** are equally spaced around the circumference of conduit **15**. These grooves are cut in such a manner that they begin on the inner wall of the conduit at a more narrow width than the groove opening on the outside wall of the conduit. A 30 degree slant of thickness of the grooves from the inner wall to the outer wall of the conduit causes air flow to be dispersed more evenly onto an absorbent material **19** on the interior walls of the muffler.

The walls of the main body **20** of the muffler are attached by threads to conduit **14**. These walls comprise a conduit with an open end. Absorbent material **19** glued to the inner walls of conduit **20** absorbs some of the exiting air's energy and transfers some to the walls of the muffler, thereby greatly lessening the decibel rating and lowering the pitch of sound emitted during tapping operations. Coating **19** preferably is composed of asbestos cloth or foam rubber. Advantageously the thickness of the coating is approximately one-tenth inch for asbestos cloth and one-half inch for foam rubber. Suitable glue is any commercially available glue which will bond cloth or rubber to metal and retain an effective bond at high temperatures.

Advantageously the muffler has an overall length of approximately 14 inches and can easily and quickly be attached to the crucible exhaust pipe by one person. The exhaust pipe for air exiting the crucible-siphon apparatus generally is about 2½ inches in diameter. Reducer **11** increases the size of the opening diameter from 2½ to 3 inches. Bell reducer **12** further increases the opening diameter from 3 inches to from 4 to 5 inches (a range of diameter increase of from 33 to 67%). Bushing **13** centers conduit **15** within conduits **14** and **20**, and reduces the opening diameter from 4 to 5 inches to 2 to 2½ inches, (range of diameter reduction of from 40 to 62.5%).

Advantageously conduit **14**, attached to bushing **13**, is a collar of constant diameter of 4 to 5 inches or is a bell reducer which increases the opening diameter from 4 inches to not more than 5 inches (an increase of up to 20%). Conduit **14** is attached to conduit **20** which has a diameter equal to the diameter of conduit **14**, i.e., from 4 to 5 inches.

Preferably conduit **15** is about 6 inches in length, approximately 43% the preferred exterior length of the muffler. Rectangular grooves **18** are approximately 3 inches in length, about 50% the length of conduit **15**, and located midway between the ends of conduit **15**, i.e., 1½ inches from either end. Grooves **18** are equally spaced around the diameter of conduit **15** and have a 5/16 inch space between the grooves on the interior wall. The grooves are flared out at a 30 degree slant from the interior wall to the exterior wall of conduit **15** facilitating a more even air flow from conduit **15**. The air flow which begins at the exhaust pipe on the crucible lid and travels toward the conduit walls of the muffler is substantially quieted as it passes from the muffler.

FIGURE 9: MUFFLER INSTALLATION FOR ALUMINUM CELL TAPPING CRUCIBLE

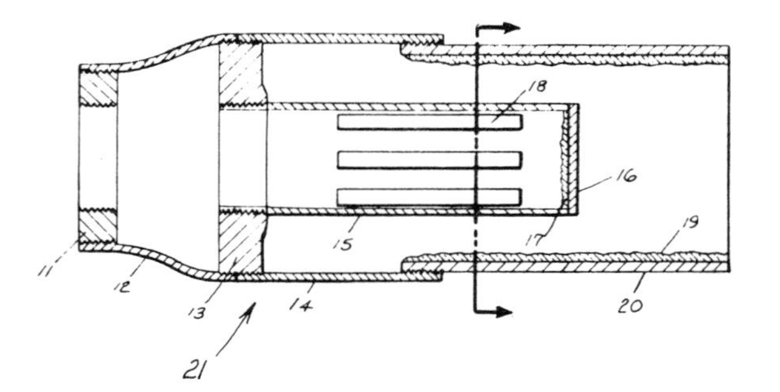

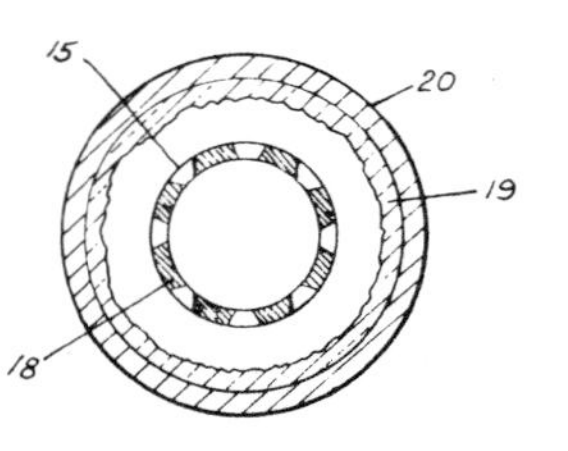

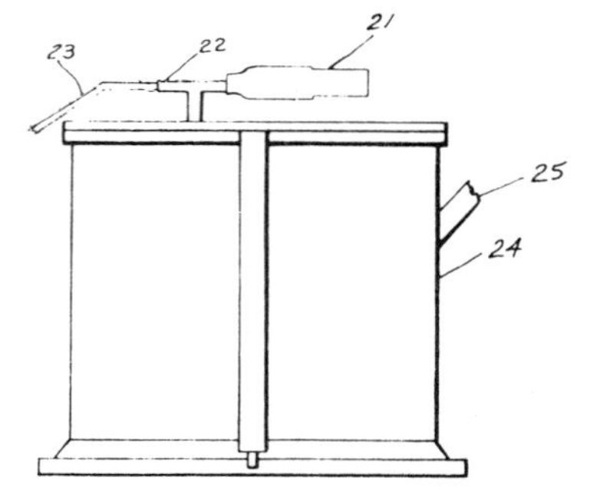

Source: U.S. Patent 3,732,948

The muffler **21** is attached to crucible **24** by a T shaped valve, **22**. During the siphon operation air hose **23** is connected to valve **22**. High pressure air passes from hose **23** through valve **22** and exits muffler **21**, thereby creating a vacuum or reduced pressure in crucible **24** and siphon spout **25**.

This muffer does not need to be removed after each tapping operation as prior art mufflers do. The cost of manufacturing the muffler is approximately one-fifth that of prior art mufflers, and prior art mufflers have a useful life markedly lower than this design. Aluminum particles do not readily collect in this muffler and the particles which do collect are readily expelled simply by removing the muffler from the exhaust pipe of the crucible and rapping it against the crucible or a similar stationary object. The efficiency of this design is superior to prior art mufflers in that it reduces the speed of air entering to approximately one-half upon exiting, reduces the decibel level from approximately 115 upon entering to approximately 90 upon exiting, and lowers the pitch of sound produced to reduce or eliminate the more piercing sounds.

ALUMINUM CHLORIDE

See Aluminum Cell Effluents (U.S. Patent 3,904,494)
See Aluminum Refining Effluents (U.S. Patent 3,900,298)

ALUMINUM CHLORIDE PRODUCTION EFFLUENTS

In aluminum chloride production, the metal is contacted with chlorine gas at an elevated temperature in a reaction chamber. Vaporous metal chloride is produced which is then fed to a condenser chamber so that it may be solidified and collected. The vaporous metal chloride solidifies mainly by crystallization on the cooler internal surfaces of this condenser chamber. This condenser chamber is operated at atmospheric pressure and is thus usually open at some point to the atmosphere. In many installations, any vaporous chlorine containing fumes are allowed to vent directly into the atmosphere. However, with more stringent antipollution ordinances being adopted, such procedures are not acceptable.

Removal from Air

The use of available commercial scrubbers in combination with aluminum chloride condensers has failed to solve this problem and yield acceptable results. The prime disadvantage of scrubbers is that such devices in operation develop a suction to pull vapors and gases into the scrubbing unit. When in combination with an aluminum chloride condenser, this suction acts also to pull amounts of vaporous chlorides into the condenser, thus decreasing the yield of aluminum chlorides, and producing deposits in the scrubbing unit.

A device developed by *T.O. Tongue; U.S. Patent 3,744,976; July 10, 1973; assigned to W.R. Grace & Company* is a vent gas purifying device which may be used in combination with aluminum chloride condenser chambers. The purifying device consists of a secondary condenser chamber in combination with a container having at least one layer of glass wool and at least one layer of a reactive or adsorbent material with the adsorbent or reactive layer being preferably a flake caustic. The device effectively vents aluminum chloride primary condenser chambers to the atmosphere without causing any deleterious back pressure or suction forces in the condenser chambers. Any noxious chlorine or other gases, and any particulate aluminum or other compounds are removed by this device precluding any venting into the atmosphere.

ALUMINUM ETCHING LIQUORS

Commercially, aluminum objects are etched or chemically milled by alkaline solutions, primarily hot aqueous solutions containing free alkali metal hydroxide, e.g., aqueous sodium hydroxide solutions, to shape or alter the dimensions of such objects. Such solutions are referred to hereinafter as caustic etching solutions, caustic etch liquors or caustic etchants. In the etching operation, ordinarily aluminum is contacted with an aqueous caustic (NaOH) solution at a temperature of from about 100°F up to the boiling point of the solution. Commonly, a solution containing about 10 weight percent NaOH and at a temperature of from about 190°F up to the boiling point of the solution is used as the etchant. During the etching process, aluminum goes into solution, probably in the form of an aluminate anion which in turn with the sodium values in the bath forms a complex hydrated sodium oxide·aluminum oxide. As the etching of the metal proceeds, the free alkali, i.e., the hydroxyl ions not combined with the aluminum in salt formation, decreases and the concentration of the aluminum salt in solution increases. These changes in bath composition in turn lead to a decrease in the rate of etching. Also, as the aluminate concentration builds up in the etching bath, this reaches a point where the solubility limit is exceeded and precipitation of objectionable quantities of an alumina-containing sludge can occur in the bath. Such etching baths, therefore, heretofore have had to be dumped and replaced in order that satisfactory etching of aluminum can be carried out.

Removal from Water

A process developed by *U.W. Weissenberg; U.S. Patent 3,712,838; January 23, 1973; assigned to The Dow Chemical Company* is a process for regenerating spent caustic etching solutions which provides a substantially aqueous insoluble innocuous product which does not contain appreciable hydroxyl ions and which, if not utilized, can be disposed of by conventional disposal means, e.g., dumping or piling, since it does not have a detrimental effect on the pH of streams, ground water and the like water sources. In this process, a spent caustic etching liquor, ordinarily at the etching bath temperature, is contacted with calcium oxide to precipitate aluminum values present in the spent liquor and regenerate hydroxyl ions. The precipitate is removed from the regenerated solution.

See Aluminum (U.S. Patent 3,738,868)

ALUMINUM REFINING EFFLUENTS

One source of particulate contaminants is the gaseous effluent discharged into the atmosphere from aluminum processing operations. Both primary and secondary aluminum processing operations use gaseous chlorine to purify the aluminum. The chlorine combines with the impurities in aluminum to form a slag which is skimmed from the top. During chlorination, gases and particulates are released from the molten metal and may pass into the flue stack and from there into the atmosphere. Periodic fluxing of the molten metal also contributes to the release of gases and particulates. These particulates are potential air pollutants and include inorganic metal halides such as aluminum chloride, sodium chloride, potassium chloride, magnesium chloride, and aluminum fluoride. These salts of chlorine and fluorine are typically in the form of finely divided solid particles which produce a white smoke or plume of varying density at the stack, and which may generate corrosive hydrogen chloride or hydrogen fluoride when they contact humid air.

Most of the particulates in the plume vary in size down to submicron levels, many being of an aerosol or colloid particle size, i.e., less than 0.5 micron in diameter. These aerosol-sized particles are very prone to produce fog or white smoke by nucleation of water droplets when such particles are introduced into a humid atmosphere. Prevention of such fogs or smokes requires very high efficiency scrubbing equipment in order to remove the aerosol-sized particles.

Removal from Air

Various methods have been adopted in the past in an attempt to remove particulates from gaseous effluents. Conventional bag filters of cotton or other textile materials have been used, but are limited by their inability to trap aerosol-sized particles. Scrubbing with caustic liquor has also been used with some success. However, such a caustic liquor scrubbing requires very high energy inputs to obtain an adequate collection efficiency for certain of the particulates, and has not been effective in removing a significant amount of the aerosol-sized metal halide particulates. Other methods which have been suggested for the removal of entrained solids in gaseous effluents include raining large amounts of an inert contact material such as coke downwardly through the gaseous effluent.

Similarly, a method of separating gaseous mixtures of metal chlorides using active carbon has been suggested. However, neither coke nor active carbon has been effective for removing the metal halide particulates of less than 0.5 micron in size from the gaseous effluent of aluminum processing plants. Calcium oxide (lime) is similarly ineffective.

It has also been proposed to contact gaseous effluent from a primary aluminum cell with a combined system of a bed of finely divided alumina pellets followed by a conventional bag filter. Since the bag filter itself is limited by an inability to trap aerosol-sized particles, the alumina bed which is the first portion of the system contacted by the effluent, acts as a gross collector of solids necessarily of a relatively large (e.g., greater than 0.6 micron) size.

A process developed by *L.A. St. Cyr and L.H. Young; U.S. Patent 3,900,298; August 19, 1975; assigned to Vulcan Materials Company* is one in which particulates less than 0.5 micron in size are removed from flue gases by passing the flue gases into intimate contact with a solid adsorbent of activated alumina. The solid adsorbent is particularly useful in removing metal halide solids, e.g., $AlCl_3$, KCl, NaCl, and AlF_3, of less than 0.1 micron in size from the gaseous effluent of aluminum processing operations which utilize a gaseous chlorine treatment of molten aluminum. Such a process is shown in Figure 10.

The fumes consisting of hot gases and particulate matter issue from molten aluminum **1** in a secondary aluminum melting furnace **2.** The aluminum melting furnace **2** comprises a combustion chamber **3** and an open hearth **4** where aluminum scrap or the like can be placed into the furnace. A burner **5** in the combustion chamber **3** is used to maintain the metal above its melting point and to maintain a reducing (nonoxidizing) atmosphere over the molten metal. A layer **6** of flux material such as a NaCl-KCl eutectic composition may be provided to prevent the formation of oxides and to absorb or gather impurities released from the metal during processing. A gaseous chlorine distributing pipe **7** may be immersed into the melt **1** at the open hearth section **4.** The gaseous chlorine is supplied to the pipe **7** from a pressure vessel (not shown) through a flexible conduit **8.** A suitable valve (not shown) in line **8** may permit metering of the amount of gaseous chlorine fed into the melt **1.**

A hood **9** is provided to direct fumes from the open hearth into a flue duct **10.** Similarly, fumes from the combustion chamber **3** are directed through a duct **11** into flue duct **10.** The fumes may be passed via ducts **10** and **11** through a preliminary filter unit **14** which may be composed of baffled foraminous trays **15** of, for example, 100 mesh silica sand, 1 to 50 cm deep, whereby soot, organic matter, and other coarse particulates generally having a particle size greater than about 1 micron may be removed. Thereafter, the filtered flue gas may be passed through a solid adsorbent unit **16** which may be composed of baffled foraminous trays **17** containing particulate highly adsorbent activated alumina at a bed depth of, for example, 5 to 50 cm. Thereafter, the flue gas may be passed into a natural draft stack **20**. A blower **19** may be provided to maintain a flue gas flow rate of, for example, about 1,000 to 1,000,000 liters per minute.

FIGURE 10: ACTIVATED ALUMINA ADSORBENT SYSTEM FOR REMOVING PARTICULATES FROM ALUMINUM REFINING EFFLUENTS

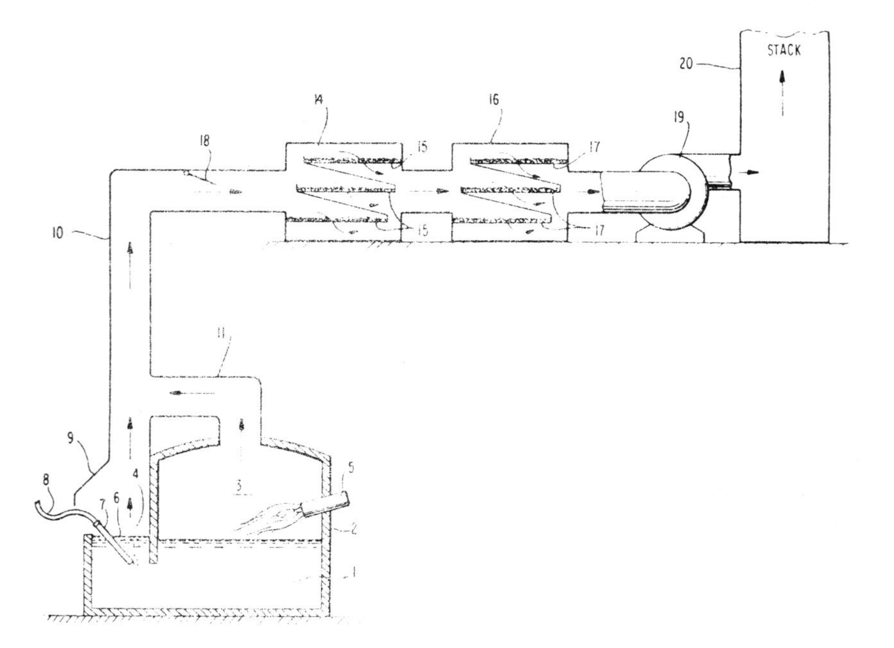

Source: U.S. Patent 3,900,298

Generally, the temperatures of fumes issuing from the aluminum furnace **2** will range from about 80°C in the open hearth section **4** to about 800°C in the combustion chamber **3**. Therefore, if the flue gas temperature immediately upstream of the preliminary filter unit **14** is still relatively high, e.g., 500°C, a damper **18** may be provided in the flue duct **10** to mix or bleed-in a gas, such as air, at a lower temperature, conveniently ambient temperature, e.g., 25°C.

An aluminum refining process has as its objects to degas the metal, to remove hydrogen from it, and to remove the undissolved inclusions, which generally are finely granular aluminum oxide. This purification is performed on the primary aluminum which will not be melted anew, for instance upon aluminum for slabs, plates or billets. It is also effected upon reclaimed aluminum and aluminum alloys and before casting these metals. The conventional method of purification is to inject chlorine into the molten metal. Chlorine injection suffers severe drawbacks since this gas is very corrosive and its lethal concentration in air for a sojourn of a few minutes is only about 5 ppm. The foundries receive liquid chlorine in tank cars, which gives rise to security problems during transportation, storage and handling. Chlorine is then fed to injection lances, by tubes and valves which are corroded unless special steels are used. Casual leaks are quite dangerous. The fumes from the metal bath into which chlorine is injected contains elemental chlorine and volatile chlorides which also are poisonous and corrosive. This necessitates costly and cumbersome devices for collecting and neutralizing the fumes. The neutralization itself pollutes the water in which is dissolved the reactant and the unavoidable leaks pollute the ambient air.

An improved process developed by *J. Foulard and J. Galey; U.S. Patent 3,743,500; July 3, 1973; assigned to L'Air Liquide S.A., France* is one in which, for the degasification of aluminum and aluminum alloys, a ladle is provided with two compartments interconnected by a bottom passage. Inert gas is insufflated in very fine bubbles into the lower portion of both compartments. The molten metal flows down the upstream compartment, then flows up the downstream compartment. Thus, chlorine pollution problems are avoided by replacement of the chlorine with an inert gas, such as nitrogen.

AMINES

See Foundry Casting Operation Emissions (U.S. Patent 3,795,726)
See Foundry Casting Operation Emissions (U.S. Patent 3,941,868)
See Foundry Casting Operation Emissions (U.S. Patent 3,937,272)

AMMONIA

Removal from Water

The removal of ammonia from gaseous mixtures is a highly developed art and known procedures for effecting such removal include absorption by porous absorbents which may be treated to improve the process, and complexing the ammonia to form amines. When efforts were made to treat an ammonia-containing liquid with absorbents which have been chemically or physically treated it was found that the liquid rapidly washed out the treatment from the absorbent and rendered the same ineffective for the removal of ammonia from liquids. Attempts to utilize conventional ion exchange resins for the removal of ammonia have been equally unsuccessful because such resins are not selective toward ammonia. In addition, chemical regeneration of the resin is required to restore it for reuse.

A process developed by *R.A. Dobbs; U.S. Patent 3,948,769; April 6, 1976; assigned to the U.S. Environmental Protection Agency* is one in which the selective removal of ammonia from wastewater is achieved by the use of a ligand exchanger which has been conditioned with a salt of a metal which is capable of forming a complex with ammonia. The ligand exchanger is regenerated by contacting it with low pressure steam.

J.W. Schroeder and A.C. Naso; U.S. Patent 3,920,419; November 18, 1975; assigned to Republic Steel Corporation use a method where free and fixed ammonia values are removed from ammonia liquor by a process which in the preferred embodiment includes the characterizing steps of: (1) automatically adjusting the pH of the liquor by the continuously controlled addition of caustic soda solution in an amount sufficient to maintain a minimum pH of 10.5; (2) stripping ammonia from the liquor by counterflowing air through a packed column at a temperature of from 140° to 180°F; and (3) controlling the flow of the air in direct relation to the amount of feed liquor so that at least 99% of the ammonia is removed from the liquor, the most preferred flow rate of the air being in the range of from about 50 to 100 cubic feet per gallon of feed liquor. The ammonia containing air is withdrawn from the column and the ammonia is extracted from the air in an acid absorber, preferably using a sulfuric acid solution having a pH of from 0.5 to 1.5. The air from the acid absorber is recycled to the column in a closed system which substantially eliminates problems of air pollution.

The process is particularly applicable to the stripping of ammonia from the ammonia liquor resulting from cooking operations. Ammonia liquor constitutes the condensed water vapor separated from the tar in a tar plant along with the flushing liquor and contains free and fixed ammonia in typical amounts of 5,000 parts or more per million. The process is effective in reducing this ammonia content by more than 99% to less than 50 parts per million. Figure 11 illustrates such a process.

FIGURE 11: PROCESS FOR STRIPPING AMMONIA FROM WASTEWATER

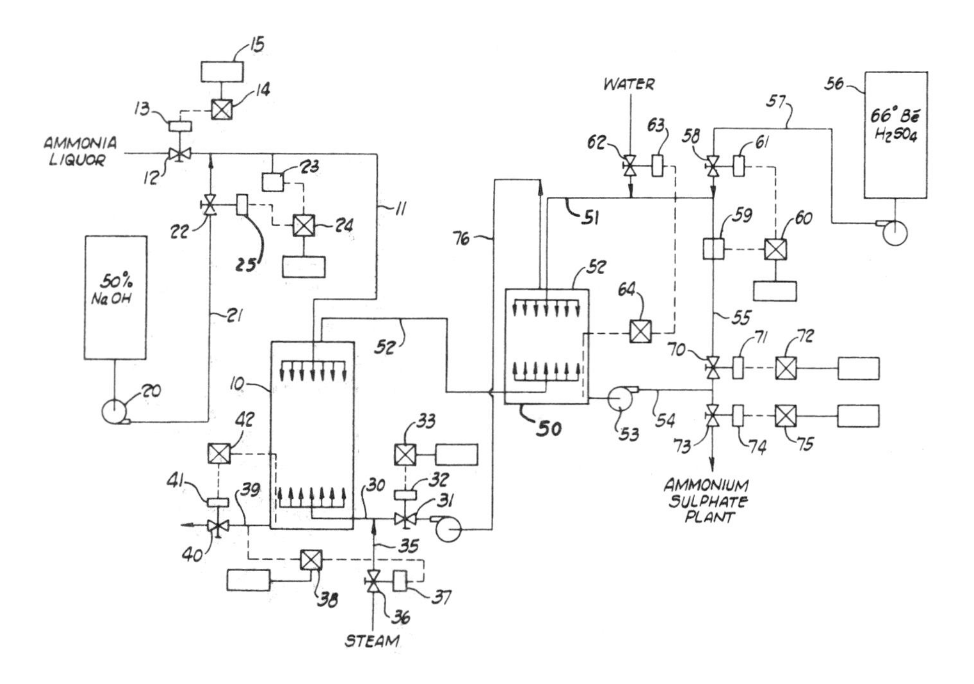

Source: U.S. Patent 3,920,419

The ammonia liquor is introduced into the top of a packed column **10** through line **11**. The process is such that the column **10** can be constructed of ordinary fiber glass reinforced plastic, thereby reducing the capital costs compared to conventional lime stills. The packing in the column **10** may be of any conventional type, such as plastic rings, saddles, etc. The flow of the ammonia liquor through the line **11** to the column **10** is automatically controlled by a flow control process valve **12**. The flow control valve **12** includes a pneumatic actuator **13** which is operated by a signal from a flow indicator control and recorder **14**.

In order to convert the fixed ammonia to the free state, the pH of the feed liquor is adjusted by the addition of a 50% by weight caustic soda solution so that the pH is at least 10.5 and is preferably in the range of from 10.5 to 11.5. The caustic soda solution is pumped by a pump **20** through a line **21** which connects with the line **11**. The flow of caustic soda to achieve the desired pH is automatically adjusted by a flow control process valve **22**. The automatic controls for the caustic soda solution include electrodes (not shown) which are contacted by the ammonia liquor. The electrodes produce a potential corresponding to the pH of the ammonia liquor and this potential is changed to a pressure impulse by suitable transducers **23**.

The signal from the transducers **23** is connected to a pressure indicator control and recorder **24** which operates a pneumatic actuator **25** for the valve **22**. This arrangement is such that the flow of caustic solution into the feed ammonia liquor is continuously adjusted to maintain the desired pH which is necessary to convert the fixed ammonia to the free state and to minimize an excess of caustic. The system can be precisely controlled so that the excess caustic in the effluent from the column **10** averages 2% by weight or less.

The feed ammonia liquor flows downwardly through the packed column **10** and the ammonia is stripped by concurrently counterflowing air at a preferred temperature of from 140° to 180°F. The air is introduced into the bottom of the column through a line **30**. The flow of air is controlled by a flow control valve **31** operated by a pneumatic actuator **32** and a flow indicator control and recorder **33**. The temperature of the air in the column is adjusted to be within the desired range of from 140° to 180°F by controllably admitting steam through a line **35** to the air line **30**. The flow of steam through the line **35** is automatically controlled by a process valve **36** operated by a pneumatic actuator **37** and a temperature indicator control and recorder **38**.

The temperature indicator control and recorder **38** is connected to the effluent line **39** so that the device senses the temperature of the effluent in the line **39** and operates the valve **36** to admit the necessary amount of steam to maintain the desired operating temperature in the column. The flow of air through the column is controlled in direct relation to the amount of feed liquor so as to reduce the ammonia content by at least 99%. A practical operating range is from about 40 to 180 cubic feet per gallon of feed liquor, and the preferred range is from about 50 to 100 cubic feet per gallon. Higher flow rates are possible, but it has been found that the amount of ammonia removed from the liquor is not significantly affected by increasing the flow rate above the preferred maximum limit of about 100 cubic feet per gallon.

The effluent from the column **10** is underflowed through the line **39** to exhaust. As shown, the flow of effluent in the line **39** is automatically controlled by a valve **40** operated by a pneumatic actuator **41** and a liquid level control **42** which senses the liquid level in the column. A simple U-seal can replace the liquid level control instruments where applicable.

The ammonia containing gas from the top of the column **10** is passed through a suitable acid absorbing system for extraction of the ammonia. In the illustrated example of the process, the acid absorbing system comprises a sulfuric acid absorber **50**. A sulfuric acid solution is introduced into the top of the absorber **50** through a line **51**, while the ammonia containing gas is introduced into the bottom of the absorber through a line **52** and flows upwardly in contact with the acid solution. The effluent from the absorber **50** is extracted by a pump **53** and is recirculated through a line **54** which is connected by a line **55** to the inlet line **51**.

The pH of the recirculated solution is maintained in the range of from 0.5 to 1.5 by the addition of concentrated (66°Bé) sulfuric acid supplied from a source **56** through a line **57** to the line **51**. The flow of the sulfuric acid into the line **51** is regulated by a flow control valve **58**. Suitable electrodes and transducers **59** are connected to the line **55**, and the signal from the transducers is transmitted to a pressure controller and regulator **60** which in turn operates a pneumatic actuator **61** for the valve **58**. Water is automatically added to the recirculated acid solution through a valve **62** operated by a pneumatic actuator **63**. The pneumatic actuator **63** is automatically controlled by a liquid level control **64** which senses the level of liquid in the absorber **50**.

The flow of the recirculated acid solution in the lines **51, 54, 55** is regulated by a flow control valve **70**. This flow control valve **70** is operated by a pneumatic actuator **71** under the control of a flow indicator control and recorder **72**. The acid solution is recirculated through the absorber so that the concentration of ammonium sulfate in the solution is in the range of from 20 to 35% by weight. The blowdown from the absorber system is pumped to an ammonium sulfate plant or the like through a flow control valve **73**. This valve is similar to valve **70** and is operated by pneumatic actuator **74** which is controlled by flow indicator control and recorder **75**.

As shown in the drawing, the air from the absorber **50** is recycled to the bottom of the column **10** through the line **76** in a closed system. This closed system in which the air from the absorber is recirculated to the column further reduces the amount of energy required to maintain the column at its operating temperature.

See Coke Oven Emissions (U.S. Patent 3,822,237)
See Coke Oven Emissions (U.S. Patent 3,915,655)
See Duplicating Machine Effluents (U.S. Patent 3,679,369)
See Melamine Process Effluents (U.S. Patent 3,555,784)
See Sewage Treatment Effluents (U.S. Patent 3,984,313)

AMMONIA-SODA PLANT EFFLUENTS

It is well known that, in ammonia-soda plants, the ammonium chloride liquors separated from the crude sodium bicarbonate precipitated in the carbonating towers are heated with steam in the presence of milk of lime, in order to regenerate and recover the ammonia. This operation, named distillation, is carried out in columns called distillers. An equivalent amount of calcium chloride is formed at the same time and remains in solution together with the sodium chloride that has passed unchanged through carbonating. Mixed with and suspended in this solution of calcium chloride and sodium chloride are various insoluble salts, among others calcium aluminate, silicate, carbonate and sulfate as well as silica, iron oxides and other insoluble materials introduced via the lime.

Removal from Water

The soda plants located in the vicinity of the sea or other large volumes of naturally-occurring water can dispose of waste liquors and solids in these water masses. On the other hand, inland soda plants are faced with the problem of waste disposal. It is possible, in some degree, to separate the solids in suspension by decantation and to set them aside in dams or embankments, the clear liquor being optionally concentrated by multiple effect evaporation in order to recover in sequence solid refined sodium chloride and liquid or crystallized calcium chloride. However, the need for calcium chloride in the market place may be below the amount obtained as by-product in the manufacture of sodium carbonate and the manufacturers are then bound to find other ways of disposal.

A process developed by *A. Bietlot; U.S. Patent 3,914,945; October 28, 1975; assigned to Solvay & Cie, Belgium* involves disposing of the effluents from the distillers of ammonia-soda plants at the bottom of subterranean cavities of disused salt boreholes, the cavities being filled with a sodium chloride brine to prevent subsidence and an equivalent volume of the brine being recovered. The effluent introduced at the bottom of the borehole comprises a calcium chloride solution which has been concentrated to a density significantly higher than that of the saturated solution of sodium chloride present in the borehole.

See Pickle Liquors (U.S. Patent 3,468,797)

AMMONIA SYNTHESIS PROCESS EFFLUENTS

In a process for manufacturing gases for use in ammonia synthesis by reforming hydrocarbons with steam, excess steam is condensed and discharged out of the system as process wastewaters which contain by-product ammonia and organic materials, e.g., mainly, methanol and a small amount of amines, and process gas components, e.g., mainly, carbon dioxide.

The quantity and properties of the process wastewaters, although varying depending upon the kind of raw materials to be used, operating conditions and the like, are usually as follows: 0.8 to 2.2 cubic meters of the process wastewaters per ton of product ammonia; temperature, 70° to 100°C; pressure, 10 to 30 atmospheric pressures; and the concentration of dissolved ingredients: ammonia, 500 to 1,500 mg/l; carbon dioxide, 1,500 to 4,000 mg/l; and organic materials, 700 to 2,500 mg/l.

Ammonium nitrogen present in wastewaters is considered to be one of the major causes of problems associated with nutritive fresh and seawaters, i.e., the abnormal growth of duckweeds or seaweeds, and organic materials present therein may increase the chemical oxygen demand (COD), so that removal of these materials is an urgent matter of today from the point of view of preventing water contamination.

Removal from Water

Processes for removing ammonium nitrogens and organic materials from process wastewaters have heretofore been proposed: the stripping method by means of air or steam; the ion exchange method; the adsorption method by means of activated carbon or other adsorbents; the biochemical methods; and the like. However, these methods are unsatisfactory in meeting technical and economical demands. For instance, the stripping of the process wastewaters by air or steam produces gases containing ammonia and organic materials which, when discharged directly into the atmosphere, will cause atmospheric pollution and be brought back again with rainfalls to the ground, so that this method does not give a fundamental solution to environmental contamination and pollution.

If ammonia is recovered from the stripped gases by recondensation or absorption in water, such an ammonia solution is obtained in an extremely low concentration of about 1% by weight or less at the most and also contains organic materials to some extent. Thus such a recovered solution is not suitable for industrial reuse. In the ion exchange and adsorption methods, it is difficult to remove concurrently both ammonium nitrogens and organic materials present in the process wastewaters in an effective way. Those methods also present problems associated with the regeneration of the used ion exchange resin or adsorbents.

The biochemical methods involve decomposition of the organic materials and nitrification of the ammonia with aerobic microorganisms (bio-oxidation) or denitrification thereof with anaerobic microorganisms (bio-reduction). However, these methods have the disadvantages that sufficient degrees of conversion are not always achieved and large equipment is required for industrial application of these methods.

In a procedure for manufacturing gases for use in synthesizing ammonia by reforming hydrocarbons with steam, both the heat and the power required for the reforming reaction are generated by combusting the hydrocarbon fuels which may optionally be combined with a gas purged from the ammonia synthesis system. The flue gases from the procedure contain oxides of nitrogen such as nitrogen monoxide, nitrogen dioxide, and the like. The quantity and properties of the flue gases, although they are varied depending upon the kind of fuel and operating conditions are usually as follows: flue gases, 3,000 to 7,000 Nm^3 per ton of product ammonia; temperature, 150° to 700°C; and the components: 70 to 72% of nitrogen; 7 to 12% of carbon dioxide; 2 to 4% of oxygen; 0.7 to 1% of argon; 12 to 19% of steam; 200 to 500 ppm of nitrogen oxides; and 0 to 500 ppm of sulfur oxides, all values measured on the basis of the volume of the flue gases.

The nitrogen oxides present in the flue gases not only are toxic to the human body but are considered to be one of the substances causing photochemical oxidants, so that the control and removal of such nocuous nitrogen oxides are an urgent matter. As a measure taken with respect to the reduction of nitrogen oxides there is known a method for suppressing the formation of the nitrogen oxides derived from the reaction of nitrogen and oxygen, by lowering the combustion temperature or the concentration of oxygen. However, this method is disadvantageous because of the reduction of nitrogen oxides to the extent of 50% at most. Some absorption methods by which nitrogen oxides are removed have also been found unsatisfactory because of a low absorption efficiency and the requirement of large equipment where nitrogen oxides are present in great dilution and the gases are employed in large quantities. These methods also present a problem associated with the treatment of the nitrogen oxides absorbed, so that they involve a technical difficulty and lack an economical advantage.

In addition thereto, there are methods which involve the catalytic reduction of nitrogen oxides by adding reducible gases such as hydrocarbons, carbon monoxide, hydrogen or ammonia to the flue gases which contain nitrogen oxides and by bringing the mixed gases into contact with a catalyst. Of these, the process wherein hydrocarbons, carbon monoxide or hydrogen are used permits these reducible gases to react preferentially with oxygen present in the flue gases and then to react with the nitrogen oxides after the oxygen present therein in large quantities was fully consumed. For this reason, a large quantity of the reducible gases is required. Furthermore, the heat of the reaction is so high that the temperature rise of a catalyst bed becomes remarkably high to a degree that the control of the temperature is rendered difficult. Such a high temperature also affects the catalyst life adversely.

A process developed by *T. Shiraishi, H. Fukusen, S. Oishi, S. Shimizu, H. Nishikawa and T. Wakabayashi; U.S. Patent 3,970,739; July 20, 1976; assigned to Sumitomo Chemical Company, Ltd., Japan* involves combining a step of removing ammonia from the process wastewaters with a step of removing nitrogen oxides from the flue gases, thereby rendering both of them concurrently innocuous.

More specifically, the process involves stripping ammonium nitrogens and organic materials, as gases, which are present in process wastewaters to be discharged from plants wherein gases for use in ammonia synthesis are manufactured by reforming hydrocarbons with steam and decomposing the organic materials selectively in the presence of a catalyst at a temperature of about 120° to 400°C.

The process then involves mixing the remaining gases with the flue gases from the plants so as to provide a gaseous mixture having ammonia therein in an amount of about 0.3 to 10 mols per mol of nitrogen oxides present in the flue gases, reacting the gaseous mixture over a catalyst at a temperature of about 150° to 700°C and oxidizing the unreacted ammonia, if any, in the presence of a catalyst at a temperature of about 150° to 700°C to render the nocuous substances innocuous.

AMMONIUM PHOSPHATE PLANT EFFLUENTS

Removal from Air

A process developed by *H.J. Clausen; U.S. Patent 3,687,618; August 29, 1972; assigned to Cities Service Company* is one whereby ammonia can be recovered from the exit gases of an ammonium polyphosphate plant by scrubbing the reactor exit gases with partially ammoniated superphosphoric acid having a pH above about 3, and recycling the product obtained thereby to the reactor.

A process developed by *W.J. Sackett, Sr.; U.S. Patent 3,499,731; March 10, 1970* provides a fully implemented control system for preventing the discharge of obnoxious wastes from plants such as those of the ammonium phosphate fertilizer types. The problem that the wastes are of different vapors and dusts in differing combinations is solved by the use of regenerating combinations of dry and wet cyclones and scrubbers taking into account that certain effluents are at elevated temperature.

AMMONIUM SULFATE

Removal from Air

A process developed by *W. Dieters; U.S. Patent 3,410,054; November 12, 1968; assigned to Inventa A.G.* involves the removal of solid polar particles, particularly of ammonium salts such as ammonium sulfate, from streaming gases in which they are suspended, by making the particles descend while rotating without eddy formation, whereby the particles agglomerate with each other and are then thrust out continuously and collected, while the gases escape in a pure state.

AMMONIUM SULFIDE

Removal from Water

A process developed by *K.B. Brown, et al; U.S. Patent 3,029,201; April 10, 1962; assigned to Universal Oil Products Company* is one in which wastewater is treated in such a manner as to convert the sulfide impurities to a form having an oxygen demand which is considerably reduced and in some cases is practically nil. The impurities in wastewater from petroleum refineries include ammonium sulfide, sodium sulfide, potassium sulfide, and in some cases hydrogen sulfide, as well as mercaptans, phenols, etc.

Although these impurities comprise a minute portion of a large volume of water, the sulfides, for example, consume oxygen when disposed in neighboring streams and rob aquatic life of necessary oxygen. This method of treating water containing such a sulfur impurity in a concentration of less than 5% by weight of the water, comprises reacting the sulfur impurity with an oxidizing agent in the presence of a phthalocyanic catalyst. The following is one specific example of the operation of this process.

A composite of cobalt phthalocyanine sulfonate on activated carbon was prepared by dissolving cobalt phthalocyanine sulfonate in water to which a trace of ammonium hydroxide (28%) solution was added. Activated carbon granules of 30 to 40 mesh were added to the solution with stirring. The mixture was allowed to stand overnight and then was filtered to separate excess water. The catalyst was then dried and was calculated to contain 1% by weight of the phthalocyanine catalyst. 10 cc of the composite catalyst prepared in the above manner were mixed in a separatory funnel with 100 ml of water containing 0.0112% by weight of ammonium sulfide. The mixture was shaken at room temperature and analyzed periodically by titration with silver nitrate to determine the disappearance of the sulfide ions. The air contained in the separatory funnel was sufficient for the desired purpose. After 13 minutes of contact in the above manner, the sulfide concentration was reduced to 0.00032% by weight. From the above data, it will be seen that the ammonium sulfide was reduced from 0.0112% by weight to 0.00032% by weight within 13 minutes.

AMMONIUM SULFITE

Removal from Water

A process developed by *P. Urban; U.S. Patent 3,574,097; April 6, 1971; assigned to Universal Oil Products Company* is one in which a water stream containing a water-soluble sulfite compound is treated in order to reduce its total sulfur content while minimizing the formation of sulfate by-products by the steps of: (a) converting the sulfite compound contained in the water stream to the corresponding thiosulfate compounds; (b) reacting the resulting thiosulfate compound with carbon monoxide at reduction conditions selected to produce the corresponding sulfide compound; and thereafter (c) stripping hydrogen sulfide from the effluent stream from step (b) to form a substantially sulfate-free treated water stream which is substantially reduced in total sulfur content relative to the input water stream.

Principal utility of this treatment procedure is associated with the regeneration of a sulfite-containing absorbent stream which is commonly produced by contacting a flue gas stream containing sulfur dioxide with a suitable aqueous absorbent stream containing an alkaline reagent. The treated water stream produced by this method can then be reused in the absorption process or discharged into a suitable sewer without causing pollution problems.

ANTIFOULING SHIP PAINT RESIDUES

A well-known maintenance procedure in ship overhaul work requires the removal or cleaning, and then the replacement at frequent intervals, of ships' hull bottoms antifouling (AF) paints which are used to prevent attachment and progressive growth of sea life. Operating requirements of today's high speed Navy have imposed severe demands on ships' coating systems, especially on antifouling paints. Fuel conservation, high speed capabilities, and extended periods between ship drydockings are naval objectives which depend to a great extent on the performance of antifouling (AF) paints.

The best known prior art antifouling paints were the old stand-by cuprous oxide paints which operate on the principle of leaching out, at a controlled rate, a toxic solution to kill or discourage sea life from attaching to the ships' bottoms. The cuprous oxide paints leached at a high rate in order to perform their function, and therefore had to be mechanically removed and renewed at frequent intervals. For example, one could expect an effective life of only 6 to 18 months, thus, one had to accept the low efficiency of a dirty bottom or the down time of drydocking. To overcome these difficulties, and to achieve the desired objectives discussed above, the old cuprous oxide AF paints are being rapidly replaced by improved AF paints and coatings containing organometallic compounds, such as for example, tributyltin oxide (TBTO), tributyltin fluoride (TBTF), tripropyltin oxide (TPrTO), and tripropyltin fluoride (TPrTF), etc. Also, toxic organometallic compounds of lead and other heavy metals, such as for example, triphenyllead acetate, may be used.

The advantage of these organometallic AF paints over the previous cuprous oxide type AF paints is that they are far more toxic to sea life and can be designed with very low leach rates to perform their AF function. Their antifouling life may thus be prolonged to a projected five year period. However, ships' hulls bearing these organometallic AF coatings do eventually require abrasive blasting to facilitate repainting. Since these organometallic compounds, particularly the commonly used organotins, are not biodegradable, remain toxic for long periods, are approximately 20 times more toxic than cuprous oxide, they therefore cannot be allowed to contaminate the water environment, i.e., harbor, and disposal of the spent abrasive material containing these paint residues has become a serious problem of growing proportions.

Removal from Air and Water

The current shipyard practice for disposal of spent abrasive materials containing organotin AF paint residues involves shoveling or otherwise collecting the material from the floor of the drydock into 55 gallon metal drums which are then sealed and transported to designated class 1 landfull sites for burial. A class 1 landfill offers minimal seepage risks. This procedure is unsatisfactory because firstly a class 1 landfill is not always available. For example, Hawaii does not have a class 1 landfill. Tons of the contaminated spent abrasive have therefore accumulated at Pearl Harbor awaiting shipment to the mainland states for disposal.

Secondly, the period over which the material remains toxic while underground has not been established. It is known that organotin compounds degrade under the influence of ultraviolet light from sunlight and by the action of some soil bacteria. However, packaging the material in metal drums effectively shields the organotin compounds from both the ultraviolet and the soil bacteria. Thirdly, the process does not really solve the pollution problem. It only transfers the pollutant from one environment to another. And fourthly, the process serves to concentrate the pollutant into discrete areas or pockets beneath the ground. This could cause problems in the future.

A technique developed by *A. Ticker; U.S. Patent 3,981,252; September 21, 1976* provides a process for neutralizing the toxic nature of organometallic antifouling (AF) paint particles, in intimate mixture with spent abrasive particles, derived from the abrasive blasting of ships' hulls. The spent abrasive containing the organometallic paint residue is collected and heated in a vapor-tight furnace which is fitted with a safety pressure release valve. When the ignition temperature of the organometallic paint is exceeded, the organometallic paint particles are oxidized to a harmless, nontoxic metal oxide, which may be safely disposed of. Volatile organometallic paint vapors are drawn through an afterburner so that exhaust residue consist only of nonpolluting carbon dioxide and water vapor. Metallic elements of commercial value, such as tin, may be recovered from the processed abrasive, which may also be reusable. To ensure complete combustion, air is fed into the combustion furnace from a blower fan, compressor, or other external source.

ASBESTOS

Removal from Air

Practical techniques for removal of asbestos from the air in the mining and milling of asbestos and in the manufacture of asbestos products have been reviewed by the U.S. Environmental Protection Agency in *Control Techniques for Asbestos Air Pollutants,* Research Triangle Park, N.C. (February 1973).

Removal from Water

Industries which employ asbestos materials are frequently faced with the problem of disposing of aqueous waste streams which contain relatively small percentages of asbestos. Disposal of such asbestos must be done in such a manner that the environment is protected and health and safety of workers and other people are not endangered. Waste streams containing asbestos should not be permitted to flow into bodies of water which serve or affect lives of animals or humans. The disposal of waste asbestos in such places as open pits or in landfill operations can pose an environmental problem in that it can become eluted into nearby waters or dry asbestos particles can easily become airborne.

There exists a need, then, for a safe, ecologically beneficial manner of disposing of waste asbestos which cannot be efficiently and economically recovered and recycled.

A technique has been developed by *T.F. Lagess and V.H. Maudlin; U.S. Patent 3,887,462; June 3, 1975; assigned to The Dow Chemical Company* whereby aqueous waste streams containing asbestos are pumped into underground cavities, such as brine wells, and the asbestos is caused to sink to the bottom of the cavities by being contacted with a relatively dense, finely divided insoluble settling agent, such as limestone. This provides safe disposal of waste asbestos, which otherwise could create a health or environmental problem.

ASPHALT VAPORS

Removal from Air

A device developed by *W.F. Scheetz; U.S. Patent 3,833,014; September 3, 1974; assigned to Hy-Way Heat Systems, Inc.* for a heated storage tank for liquid asphalt incorporates means for introducing and retaining low pressure inert gas as a sealing medium to provide emission free heated asphalt storage.

AUTOMOTIVE EXHAUST GASES

Removal from Air

A process developed by *G.E. Dolbear; U.S. Patent 3,767,764; October 23, 1973; assigned to W.R. Grace & Company* involves removing carbon monoxide, hydrocarbons and nitrogen oxides from the exhaust stream of internal combustion engines by passing the exhaust gas through a catalyst for conversion of CO to carbon dioxide and nitrogen oxides to ammonia, followed by admixing the effluent from the catalyst with a second exhaust gas stream diluted with air, and passing this combined effluent through a second catalytic converter, wherein the ammonia and nitrogen oxides react to form nitrogen, and the CO and hydrocarbons are converted to innocuous entities. Such a device is shown in Figure 12.

FIGURE 12: SYSTEM FOR REMOVING NITROGEN OXIDE FROM AUTOMOTIVE EXHAUST

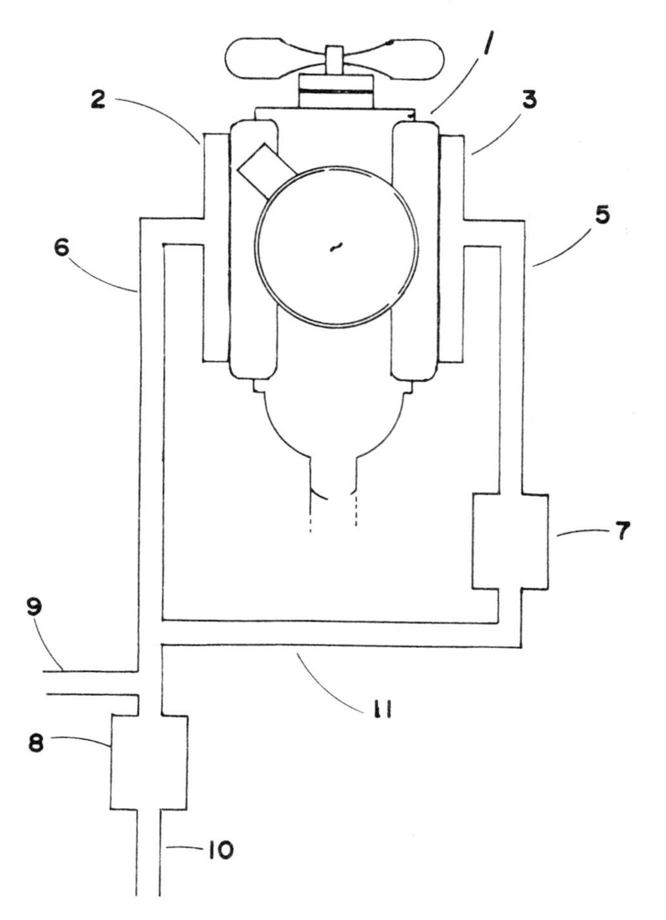

Source: U.S. Patent 3,767,764

The engine **1** has exhaust manifolds **2** and **3** attached to either side in the conventional manner. Each of the exhaust manifolds is connected through the pipe systems **5** and **6** to catalytic converters **7** and **8**. A provision for adding air to the system is provided from a pump (not shown) through air inlet **9**. The effluent from the converter **8** is passed through the collection tube **10** to the tail pipe (not shown) and is exhausted to the atmosphere.

In operation, the exhaust from the manifold **3** is conducted through the pipe **5** into the converter **7** which contains the catalysts for the conversion of nitrogen oxides to ammonia. The effluent from the reactor **7** is collected in the tube **11** and is admixed with the effluent from exhaust manifold **2** in the pipe **6**. Air is introduced through line **9** and the mixture is passed into the converter **8** where it is contacted with an oxidation catalyst for oxidizing the carbon monoxide hydrocarbons and nitrogen oxide to innocuous entities. The effluent from the catalytic converter **8** is passed to the atmosphere.

A process developed by *D.C. Gehri; U.S. Patent 3,718,733; February 27, 1973; assigned to North American Rockwell Corporation* is one in which nitrogen oxides are removed from an automotive exhaust gas by reaction with carbon monoxide in the presence of a wire mesh selected from copper-containing alloys. Where the exhaust gas contains lead species or other impurities which tend to poison the catalyst, the exhaust gas contacts a molten alkali metal carbonate mixture which removes the lead species prior to the gas contacting the catalyst. Other poisonous impurities such as sulfur dioxide are also removed by the carbonate mixture.

A process developed by *M.L. Unland; U.S. Patent 3,886,260; May 27, 1975; assigned to Monsanto Company* is one in which a catalyst containing a very small amount of rhodium or iridium is used under reducing conditions to remove the nitrogen oxides in automobile exhaust. The process involves a catalyst having a support containing thereon rhodium or iridium in amounts up to 0.005 part by weight per 100 parts by weight of support. A transition alumina is a suitable support. The reduction process can advantageously be used in conjunction with an oxidation process in a separate catalyst bed to oxidize other gaseous components.

The topic of automotive exhaust gas treatment is enough to fill one or more books in itself. Hence, in the interests of holding this volume to a reasonable size, the reader is referred to: *Catalytic Conversion of Automobile Exhaust,* by J. McDermott, Park Ridge, N.J., Noyes Data Corp. (1971) and *Noncatalytic Auto Exhaust Reduction,* by D. Post, Park Ridge, N.J., Noyes Data Corp. (1972).

BATTERY CHARGING EFFLUENTS

In the charging of lead accumulators the electrolyte in the form of sulfuric acid and distilled water often becomes violently agitated by the evolution of gas towards the end of the charge. On boost charging it often happens that the electrolyte boils and in both cases large quantities of gases and fumes are produced. Having regard both for the working personnel–the sulfuric acid in the gas evolved both destroys clothing and injures the skin by contact and causes unhygienic conditions for lungs and eyes in the event of a lengthy stay in the battery charging room—and also for environmental pollution—attack by the evacuated air on vegetable and metallic surfaces, especially aluminum surfaces, causes discoloration and often gradually and unnoticeably corrodes at places where ventilation pipes for evacuated air containing sulfuric acid emerge—it is an urgent matter to eliminate and render such sulfuric acid fumes innocuous.

Removal from Air

An apparatus developed by *B.G.O. Filén; U.S. Patent 3,926,598; December 16, 1975; assigned to AB Essve Produkter, Sweden* enables the separation of sulfuric acid fumes from ventilating air, for example the air discharged from battery-charging cabinets. The apparatus uses as an agglomerating-separating medium, a bed of long, unbroken, sharp-edged turnings of PVC (polyvinyl chloride) such as resulting from machining of PVC parts. The air is exhausted through the bed turbulently and the sulfuric acid agglomerates, forms liquid drops, and is drained from the bed. The battery-charging cabinets include a triangular chamber above each battery. Each chamber is filled with the PVC turnings and is arranged to drain the accumulated sulfuric acid from the chamber without entrainment in the ventilating air passing through the chamber.

Figure 13 shows such an apparatus. In the diagram, **1** denotes an evacuation channel from a space in which are generated sulfuric acid fumes accompanying the evacuation air. The channel is furnished with a container **2** for turnings **3** (not illustrated in detail) of the aforementioned kind, i.e., long, unbroken, sharp-edged turnings of PVC (polyvinyl chloride). **4** denotes an exhaust fan, **5** an electric driving motor for the fan.

The numeral **7** denotes a charging cabinet for lead accumulators **15**. The cabinet contains shelves **9** on a number of decks situated one above the other, carrying rows of the lead accumulators which rest on gratings **14** of wave-shaped cross section, and sloping towards the rear of the cabinet, so preventing the accumulators from standing in spilt battery acid. Above the accumulators, on every deck, there are obliquely downward sloping baffle-plates **8** with holes **8a** and with flanges **8b**. The baffle-plates form a triangular space **11** for reception of turnings. **10** denotes a slot along the rear of the battery charging cabinet. Through the slot, each deck communicates upwards with the evacuation channel **1**, which in this case as well is assumed to be equipped with a motor-driven exhaust fan **4, 5.** **12** denotes a drainage pipe on each deck **9** for spilt sulfuric acid. **13** is a pipe for collection of separated sulfuric acid in a replaceable flask **16** on the bottom of the charging cabinet.

FIGURE 13: APPARATUS FOR REMOVING SULFURIC ACID FUMES FROM BATTERY-CHARGING CABINETS

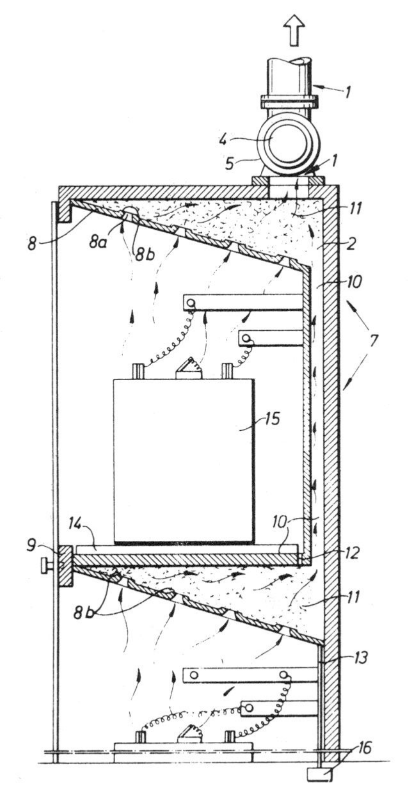

Source: U.S. Patent 3,926,598

BAYER ALUMINA PROCESS EFFLUENTS

The production of alumina by the Bayer process may be broadly divided into two steps. In the first step, bauxite is worked with the aqueous solution of caustic soda to dissolve alumina and obtain a slurry which contains the aqueous solution of sodium aluminate and insolubles. In the last step, solids are removed from this slurry and then hydrated alumina is precipitated from the aqueous solution of sodium aluminate. Bauxite containing monohydrated alumina is dissolved with the aqueous solution of caustic soda at an elevated temperature under an increased pressure in a reactor called a "digester."

Removal from Water

The slurry of high temperature and pressure formed in the digester is subjected to a plurality of stages of flashing so as to be successively lowered in both temperature and pressure. It is sent through a process designed to separate the sand, namely, coarse particles predominantly of quartz, from the slurry. Then it is forwarded to the red mud thickener. Steam is generated as the slurry is freed of high temperature and pressure in the flashing operation. The heat of this steam is recovered for heating the aqueous solution of caustic soda called "spent liquor" which is used for dissolving bauxite. As the slurry is passed through the successive stages of flashing, both temperature and pressure continue to fall by degrees. The steam generated in the last stage of flashing has a pressure equal to the atmospheric pressure and, therefore, does not have sufficiently high temperature. Unable to find any effective use, this steam has to date been blown off and wasted. Generally, the separation of sand is carried out subsequent to the blow-off. For the separation of sand, there are generally employed methods which utilize a liquid cyclone and other types of classifiers.

According to the conventional method, the discharge of waste steam and the separation of sand are carried out independently of each other. Lacking effective use of the steam, this method inevitably involves heavy consumption of steam. It also requires a special device designed exclusively for the removal of sand. This, accordingly, necessitates installation of pumps to effect the transfer of the slurry to these separate devices. Slurry pumps are highly liable to develop mechanical trouble and demand troublesome maintenance. It is, therefore, desirable to minimize use of slurry pumps.

The slurry which has been deprived of sand is washed in a multistage washing thickener to separate solids called "red mud" and produce an alumina-dissolved liquid called "pregnant liquor", which is forwarded to the process designed to precipitate hydrated alumina.

A process developed by *C. Sato, Y. Yamada and Y. Takenaka; U.S. Patent 3,869,537; March 4, 1975; assigned to Showa Denko Kabushiki Kaisha, Japan* permits heightening the ratio of heat recovery from the waste steam and improving the operation of sand removal. To be specific, the overflow from the red mud washing thickener is introduced into a column in which the flashing of slurry down to the atmospheric pressure is effected. Inside this column, the overflow liquid is brought into counterflow contact with the steam generated by the flashing so that the overflow liquid is heated by the heat recovered from the steam. The heated liquid is then refluxed to be used as the wash water in the washing thickener. In the meantime, the speed at which the slurry is discharged the bottom of the column is relatively below the speed at which the sand in the slurry settles so as to separate the sand from the slurry. The sand thus separated is discharged through the outlet provided at the bottom.

BLAST FURNACE EFFLUENTS

In the production of iron, iron ore is fed together with additional ingredients such as dolomite through the top of a blast furnace fired by coke. An air stream is blown upward from the bottom of the furnace through the subsequent molten materials. The carbon of the coke reduces the iron ore (Fe_2O_3) to iron metal. The molten iron is tapped from the bottom of the furnace while the slag is tapped from the middle of the furnace. The by-product of the carbon reduction is, of course, a combination of carbon dioxide and carbon monoxide which reacts with the calcium present to form the troublesome scaleformer, calcium carbonate and other solids: clay, slag, fines, etc.

As can be appreciated, the air stream blown upward contributes significantly to the impurity content of the flue gas, thus putting an extreme burden on the scrubbing system. The particulate load in the scrubbing medium ranges from about 1,000 to 2,000 parts per million because of the particulate load of the flue gas.

Removal from Air

The scrubbers that are used in blast furnace clean-up are often of the Venturi design and treat the off-gases from the furnace. These gases contain significant quantities of iron oxide, whose fine particle size allows it to be carried off in the gas stream. Also present may be coke fines, to a lesser extent and to some extent, particulate slag materials used, such as silicates and unused dolomites. The iron oxide has been subjected to high temperatures within the furnace and may therefore be in a sintered form of low surface activity. However, its fine particle size presents deposition problems in scrubbers and delivery lines.

Since HCN and $(CN)_2$ are also present in the gases, it has been the practice to maintain scrubber pH above 5 so as to maximize the formation of $(CN)_2$ in the water or minimize undissociated, volatile HCN by high pH control. CO_2 is also present in high concentrations and serves to maintain the pH in the desired range by acting as an acid buffer. CO is a definite problem in this application. It is generally economical to use the CO as a fuel in a waste heat boiler, but significant amounts can remain dissolved in the scrubber effluent. Its toxicity and vapor density make it hazardous as a chemical, while its combustibility may present an explosion and fire hazard. Although each of the various particulates contributed to the overall problem, the most significant factor is believed to be the iron oxide (Fe_2O_3) content of the flue gas.

A process developed by *K.R. Lange and Y.T. Hsu; U.S. Patent 3,880,620; April 29, 1975; assigned to Betz Laboratories, Inc.* is based on the belief that if the iron oxide deposition problems could be minimized, the overall effectiveness and, of course, the economics of the scrubbing system could be greatly enhanced. It was discovered that if a combination comprising a weight ratio basis of from about 1:10 to about 10:1 and preferably 1:5 to 5:1, of a water-soluble acrylic acid polymer having an average number molecular weight of from about 750 to 100,000 or water-soluble salt thereof (Na, K, NH_4) and an organophosphonate were added to the scrubbing medium in an amount of from 0.5 to about 300 preferably from about 1 to about 75 ppm, that the deposition of iron oxide (Fe_2O_3) and calcium carbonate could be effectively controlled so as to permit a more effective scrubbing operation. The acrylic acid polymers found to be most effective were those having an average number molecular weight of from about 1,000 to 10,000. The phosphonates which possess enhanced activity are those having shorter chains.

BLAST FURNACE SLAG QUENCHING EFFLUENTS

The cooling of sulfur-bearing slag such as blast furnace slag is conventionally accomplished by the use of water, for example by pouring it onto the slag in pits, by pouring molten slag into a reservoir of water, or dampening the ground on which thin layers of molten slag are poured. The slag is cooled by the water in order to permit its handling within a reasonable time after pouring. However, this cooling has heretofore been accompanied by the release of hydrogen sulfide which is malodorous, noxious and therefore considered an undesirable air pollutant.

Removal from Air

A process developed by *K.V. Hauser and L.A. Paulsen; U.S. Patent 3,941,585; March 2, 1976; assigned to Edward C. Levy Company* is one in which a readily reactive water soluble oxidant such as sodium hypochlorite (NaOCl) is mixed with the cooling water, and the sulfur-bearing slag is brought into contact with this solution so as to cool it sufficiently for digging or other handling. It has been found that during this cooling operation the oxidant inhibits the release of hydrogen sulfide from the slag to a major degree satisfying antipollution requirements. There is also no release of other contaminants such as sulfur dioxide.

A process developed by *F.H. Rehmus; U.S. Patent 3,900,304; August 19, 1975; assigned to Jones & Laughlin Steel Corporation* is one in which H_2S emissions during slag quenching are significantly minimized by a process involving the pouring of slag into holding pits in incremental layers and time intervals which lead to given slag thermal arrest temperatures. The incrementally poured slag may be air cooled for a 2 to 4 day period or quenched with oxidant containing water to further minimize H_2S emissions.

BORON

Over the last few years there has been a rapid growth in the use of boron-containing chemicals in households. Due to aggressive marketing tactics, there is every indication that the volume of such detergents consumed in the United States will continue to increase, resulting in a new and significant pollution problem. Municipal sewage treatment facilities currently do not remove boron and concentrations of boron in some effluents have already become appreciable.

Removal from Water

Control of boron contamination is important due to its toxicity to both plant and animal life. The U.S. Public Health Service recommends a boron concentration of 1.0 ppm for potable water (5.0 ppm maximum) and 0.7 ppm is recommended for most agricultural uses. Many water sources in the world exceed these limits. In the nuclear field even trace amounts of boron must be removed from aqueous coolants to avoid reactor poisoning.

Conventional methods for the removal of boron from water are unsatisfactory for handling the growing municipal boron pollution problem, or the existing agricultural and industrial problem. Reverse osmosis systems do not exclude borate ions and the ion exchange resins which have been proposed for such use are too expensive, or not applicable at such low concentrations, or not specific enough in the presence of interfering ionic species and/or organic contaminants. The art is searching for a method which will selectively permit removal of boron from aqueous solutions containing small but highly undesirable amounts of boron.

A process developed by *W.D. Peterson; U.S. Patent 3,856,670; December 24, 1974; assigned to Occidental Petroleum Corporation* involves contacting the aqueous solution containing borate ions with a water insoluble solid organic ion exchange resin, which resin is produced by the cocondensation of an aromatic ortho hydroxy carboxylic acid, a phenolic compound and an aldehyde to selectively absorb the boron value onto the resin in the presence of at least one cationic species selected from the group consisting of alkali, alkaline earth, nitrogen organic bases and ammonium cations. By the terms in the presence of at least one species selected from the group consisting of alkali, alkaline earth, nitrogen organic bases and ammonium cations is meant that such cations must be present either in the resin itself or in the system such as in the water containing the boron values. The cationic species can be incorporated into the resin during resin formation, or they can be incorporated into the resin by contacting the resin with a solution containing the desired species prior to utilizing the resin in this process.

BORON TRIFLUORIDE

Removal from Air

Boron fluoride has high water solubility like hydrogen fluoride. Therefore, it is assumed that when a waste gas containing a boron fluoride is washed with water, a boron fluoride would be easily absorbed by the water. Contrary to expectation, it has been found that a boron fluoride is not efficiently absorbed by water. When a boron fluoride diluted with an inert gas, such as air is washed with water by gas-liquid contact in a gas scrubber, such as a packed tower, a tower having multiplates, a fluidized bed and the like, absorption efficiency of the boron fluoride is very low. The absorption efficiency by means of a gas scrubber with a fluidized bed using an aqueous calcium hydroxide solution was measured in an amount of more than 95% for hydrogen fluoride, whereas it was about 30% for a boron fluoride.

In a process developed by *Y. Tsukamoto and S. Kondo; U.S. Patent 3,907,522; September 23, 1975; assigned to Mitsubishi Gas Chemical Co., Inc., Japan*, the fluorine compound(s)-containing gas is passed through a gas scrubber of gas-liquid contacting type, such as a packed tower, a tower having multiplates, a fluidized bed or the like to perform a gas-liquid contact, thereby removing a portion of the compound(s) from the gas. The gas so treated is passed through a mist-catching means comprising a porous filter medium in a filmy state having pores with a mean size of not more than 60μ and a nozzle or nozzles positioned above the surface of the filter medium, which can continuously or intermittently spray water or an aqueous alkaline solution on the surface of the filter medium into substantially the same direction as that of the stream of the gas, thereby removing a substantial amount of fluorine compound(s) from the gas. The boron fluoride which cannot be efficiently adsorbed by water, as well as hydrogen fluoride diluted with a large amount of an inert gas, are efficiently fixed or caught.

BREWERY EFFLUENTS

Removal from Water

A process developed by *C.B. Murchison; U.S. Patent 3,901,806; August 26, 1975; assigned to The Dow Chemical Company* is one in which organic compounds present in an aqueous liquor are oxidized by mixing gaseous oxygen with liquor at a pH of about 3 to about 4.5 in the presence of a catalytic quantity of uranyl ions under irradiation by light ranging in wavelength from about 5800 to about 2000 A to oxidize organic materials into CO_2, water, and lower weight organics. Cupric ions may be added to prevent the formation of a uranium-containing precipitate, which may occur during the oxidation of certain organic compounds.

The effectiveness of the process for treating a brewery plant waste was investigated on a sample containing carbohydrates (sugars, starch), ethanol, aliphatic acids (e.g., acetic, propionic), yeasts and sanitary sewage. To 750 ml of waste was added 1,050 ml H_2O, 9.4 grams $UO_2(NO_3)_2 \cdot 3H_2O$ (0.01 mol UO_2^{++}), and 4.4 grams $Cu(NO_3)_2 \cdot 3H_2O$ (0.01 mol Cu^{++}). The liquor was sparged with O_2 to saturation and was maintained at a temperature of 28°C and a pH of 3.5. The light source was a 100 watt Hanovia mercury vapor lamp placed in a quartz light well which furnished light ranging in wavelength from about 2000 to about 4000 A at a rate of 0.033 einstein/hour. The TOC removal obtained is shown as follows:

Time (min)	TOC (ppm)
0	308
64	271
147	263
266	257
615	187
1,302	123

A process developed by *E.H. Pavia; U.S. Patent 3,520,802; July 21, 1970* involves treating high protein liquid wastes under anaerobic conditions to effect a substantial reduction in the biochemical oxygen demand thereof, in a relatively short treatment time. With a relatively short treatment period, the BOD is substantially reduced, so that the effluent is in a more manageable condition for passage to municipal sewage disposal lines or to further treatment or to a stream. Further, the sludge represents a substantial buildup in protein, which may be recovered by suitable procedures known in the art, for commercial usage.

A process developed by *E. Krabbe; U.S. Patent 3,650,947; March 21, 1972* is a two-stage biological process for purification of brewery effluent where fermentation of sugars by means of aeration with yeast precedes treatment with activated sludge, thereby enhancing flocculation and separation of biota from the purified effluent. The yeast may be recovered.

As used here, the term "brewery effluent" refers to aqueous waste from the production of beer. Brewing is normally a batch operation, but it includes aqueous discharges from processing: such as the last sparges or tailings from lautering; press-water from spent grains and hop sparge; as well as various process precipitates such as trub, settlings and yeast. These discharges are flushed into the brewery sewers during tank cleaning.

The original sources of the organic load in brewery effluent are wort, beer and yeast, each having a BOD of 100,000 ppm or more; but, as a result of blending with rinse-water, cleaning solutions and cooling water, they are diluted 50 to 100 times as they are discharged into the sewers in the brewery.

The BOD or biological oxygen demand of the untreated brewery effluent may vary from 600 to 4,000 ppm. The average BOD load of the raw effluent from a modern brewery is frequently 2,000 ppm as compared to 200 to 250 ppm for domestic sewage. The discharge of waste from a brewery is about 10 times greater than the volume of the beer production. The total discharge of waste-water from all the breweries in the United States is of the order of 100,000,000 gpd.

See Glycolic Acid (U.S. Patent 3,819,516)

CADMIUM

See TV Tube Manufacturing Effluents (U.S. Patent 3,915,691)

CALCIUM OXIDE

Removal from Air

In the scrubbing of converter gases containing calcium oxide particles and carbon dioxide with water it is a standard practice to recirculate the scrubbing water and then to settle out the particles, decanting and recycling the liquid. Fresh water is added to the wash water to make up losses. A particular problem with such arrangements is that dust particles and calcium deposit in the water recirculating system. These deposits are a result of the fact that the calcium oxide particles picked up by the wash water form soluble calcium hydroxide which dissolves. The calcium ion combines with the carbon dioxide in the water to form insoluble calcium carbonate ($CaCO_3$).

It has been attempted to avoid these deposits by adding products to the water which ensure the precipitation of the dissolved calcium in the form of insoluble calcium compounds prior to recirculation. The additives necessary to accomplish this are relatively expensive and must be used in relatively large quantities, i.e., at least stoichiometrically equivalent to the calcium content of the gas treated. Approximately 75% of the wash water must be replaced with fresh water in conventional systems. Obviously such a system leads to further difficulties in the disposal of the contaminated wash water and the supply of the necessary large quantities of fresh water.

A process developed by *B. Krüger; U.S. Patent 3,988,422; October 26, 1976; assigned to Gottfried Bischoff Bau Kompl. Gasreinigungs- und Wasserruckkuhlungsanlagen KG, Germany* is one in which sodium carbonate is added to water used to scrub converter gases so that calcium oxide particles in the gas react with the aqueous sodium carbonate and a thin inactive film of calcium carbonate is formed on the calcium oxide particles, using carbon dioxide in the gas for the reaction. The coated particles are then separated from the wash water by settling and sedimentation and the wash water is recycled with the addition of more sodium carbonate and water.

CARBON BLACK PROCESS EFFLUENTS

Large tonnages of carbon black are used in many of the arts. Carbon black is used in all kinds of rubber including tire treads and other similar and widely differing applications. In the preparation of carbon black the bulk density as obtained is fairly low and it is customary to densify the carbon black. One way of doing this is to pelletize the carbon black. Carbon black is pelletized using pelletizing liquid which can be water and also this water can contain nitric acid which imparts desirable properties for certain use purposes to the pelleted black. Once the black has been pelleted, the wet pellets are passed to a dryer wherein the pellets are heated resulting in a dryer purge gas.

Removal from Air

A process developed by *E.W. Henderson; U.S. Patent 3,835,622; September 17, 1974; assigned to Phillips Petroleum Company* involves a combination of steps for removing nitrogen oxides from carbon black pellet dryer purge gas which, upon filtration, is scrubbed first with aqueous nitric acid solution and then with process water, the liquids being mixed, preferably, with aid of a jet venturi mixer. In one modification, the purge gas is heat interchanged with vent gas from at least one of the scrubbers to cool the purge gas before it is contacted with the scrubbing medium. Figure 14 is a flow diagram of the process showing the equipment used.

FIGURE 14: REMOVAL OF NO_x FROM CARBON BLACK PELLET DRYER PURGE GAS

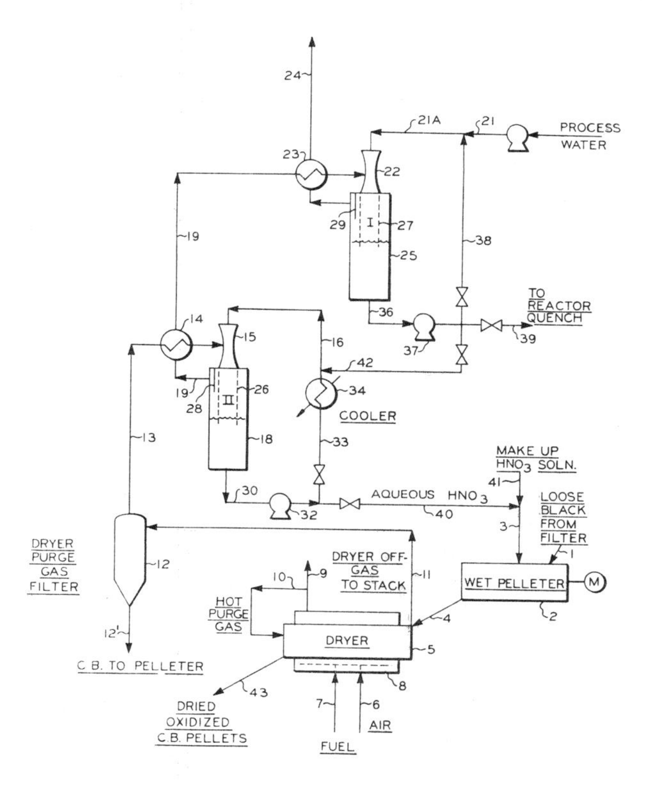

Source: U.S. Patent 3,835,622

Referring to the drawing, carbon black from filters operating upon smoke from carbon black reactors enters at **1** into wet pelleter **2** driven by motor M; to the wet pelleter there is also added at **3** a nitric acid solution and any other desired pelleting aid. The wet pellets produced in the pelleter pass by conduit **4** into dryer **5** which is heated in usual manner by air introduced at **6** and fuel introduced at **7** into combustion box **8** from which combustion gases leave at **9**.

At least a portion of these gases is passed by **10** into the dryer resulting in a dryer purge gas **11** which is passed to dryer purge gas filter **12** and from there by **13** through heat exchanger **14** into jet venturi mixer **15** wherein the gas is mixed with dilute nitric acid solution **16**. The details of the jet venturi mixing operation are well-known in the art and need not be here summarized. Suffice to say that there is immediate intimate mixture of the gases and the solution. The mixed purge gas and solution in this embodiment are directed downwardly with considerable force into scrubber **18** in which the gases disengage from liquid, the liquid collecting at the bottom and the gases leaving at the top portion of the vessel by **19**, and the heat exchanger for further treatment as later described.

The gases leaving at **19** which are considerably cooler than the gases moving in **13** on their way to the jet venturi mixer are heat interchanged at **14** and thus the gases in **13** are considerably cooled for more efficient mixing and more efficient nitrogen oxide removal. This results in considerably reducing the cost of the operation. The process water is introduced at **21** to jet venturi mixer **22** for mixture with the once scrubbed gases in **19**. Depending upon conditions, it may be advantageous to provide a second heat exchanger **23** to cool further the gases in **19** before these mix in jet venturi mixer **22**. The purge gas now substantially freed of its nitrogen oxide content is vented to the atmosphere at **24**.

In a variation of the embodiment described, downcomer pipes **26** and **27** are provided to guide the mixed fluids down into the surface and perhaps somewhat below the surface of the liquid settling in the bottom portions of the vessels **18** and **25**. In the alternative, there can be provided baffles or mist extractors as at **28** and **29** to remove any entrained liquid from the gases before these leave the respective vessels. Or, if desired, mist extractors alone can be used. Whether downcomer pipes and/or mist extractors or baffles are used, or for that matter any other equivalent device, will depend upon the positioning of the jet venturi mixers at **15** and **22**.

Although these mixers have been shown positioned directly atop these vessels, they can be positioned elsewhere as will be understood by one skilled in the art in possession of this disclosure having studied the same. Specifically, the jet venturi mixers can be positioned at one side of the top of the vessels or even along the side thereof or at the bottom. The design of the vessels and positioning of the jet venturi mixers involve subordinate concepts of the process. Their use in most economical and efficient manner is of prime importance.

Liquid is removed from the foot of scrubber vessel **18** at **30** and pumped by means of pump **32** and line **33** through an optionally provided fin cooler **34** and **16** to venturi mixer **15** for use as described. Liquid from the bottom of scrubber **25** is pumped from **36** by pump **37** and **38** to **21** and then to jet venturi mixer **22** for use as previously described. A portion of the used process water in **38** is passed by **39** to the carbon black reactor quench (not shown).

Carbon black removed from dryer purge gas **11** in filter **12** is recovered at **12'**. Finally, aqueous nitric acid is passed by **40** into mixture with makeup nitric acid solution introduced at **41** and passes together therewith by **3** to the wet pelleter. The used process water from the foot of vessel **25** is passed from pump **37** by **42** to **16** for use as already described. The dried oxidized carbon black pellets as known in the art are removed from the dryer **5** at **43**. A portion of liquid **36** can be recycled to scrubber **25** via **38** to build up a desired level of nitrogen oxides therein (reported as HNO_3); a portion of solution **30** can be recycled via **33** to build up a desired level of nitrogen oxides therein (reported also as HNO_3).

CARBON ELECTRODE MANUFACTURING EFFLUENTS

With respect to the manufacture of baked carbon articles, particularly anode blocks of the type used in aluminum reduction cells, such articles are prepared by forming a mixture of carbonaceous aggregate (such as coke) and binder (such as pitch) into a suitable shape. The resultant green carbon body is then baked to carbonize the binder. During the baking step, hydrocarbon volatiles are released from the green carbon bodies. In addition to hydrocarbon volatiles, the furnace effluent also contains entrained solid particulate matter. It is obviously desirable to remove these materials from the furnace effluent prior to discharging the residual effluent gas into the atmosphere.

Removal from Air

Various methods and apparatus directed to removing hydrocarbon volatiles and particulate matter have been proposed by the prior art. However, these methods and apparatus have various disadvantages. For example, probably the most successful prior art apparatus for removing undesirable components from furnace effluents are electrostatic precipitators. Electrostatic precipitators create electrostatic fields that attract, and thereby remove, undesirable components from furnace effluents. The precipitator plates are then cleaned by water sprays, which wash the attracted materials from the plates into suitable receptacles.

One of the major problems with such devices is the requirement that the resulting dirty water be cleaned, subsequent to its being used to clean the precipitator plates, but before it is returned to a stream, river or other ultimate disposal region. Another problem with electrostatic precipitators relates to their power requirements. Specifically, electric power must be applied to the precipitator plates and to the electric pumps used to spray the water over the plates. Obviously, the consumption of electric energy makes the use of these devices undesirable, particularly when compared to filter systems that do not consume large amounts of electric energy. Moreover, electrostatic precipitators are somewhat expensive to manufacture and install.

In the past, it has been known that certain carbonaceous materials, such as coke and the like, can be utilized as a filter medium. The problem with the use of such materials in a furnace effluent environment is that they rapidly become clogged and, thus, must be continuously replaced. In order to alleviate this problem, it has been proposed to provide filters wherein such filter materials are continuously moved through the filter. In this manner, clean carbonaceous

FIGURE 15: BLOCK FLOW SHEET FOR HYDROCARBON AND PARTICULATE REMOVAL FROM CARBON BAKING FURNACE EFFLUENTS

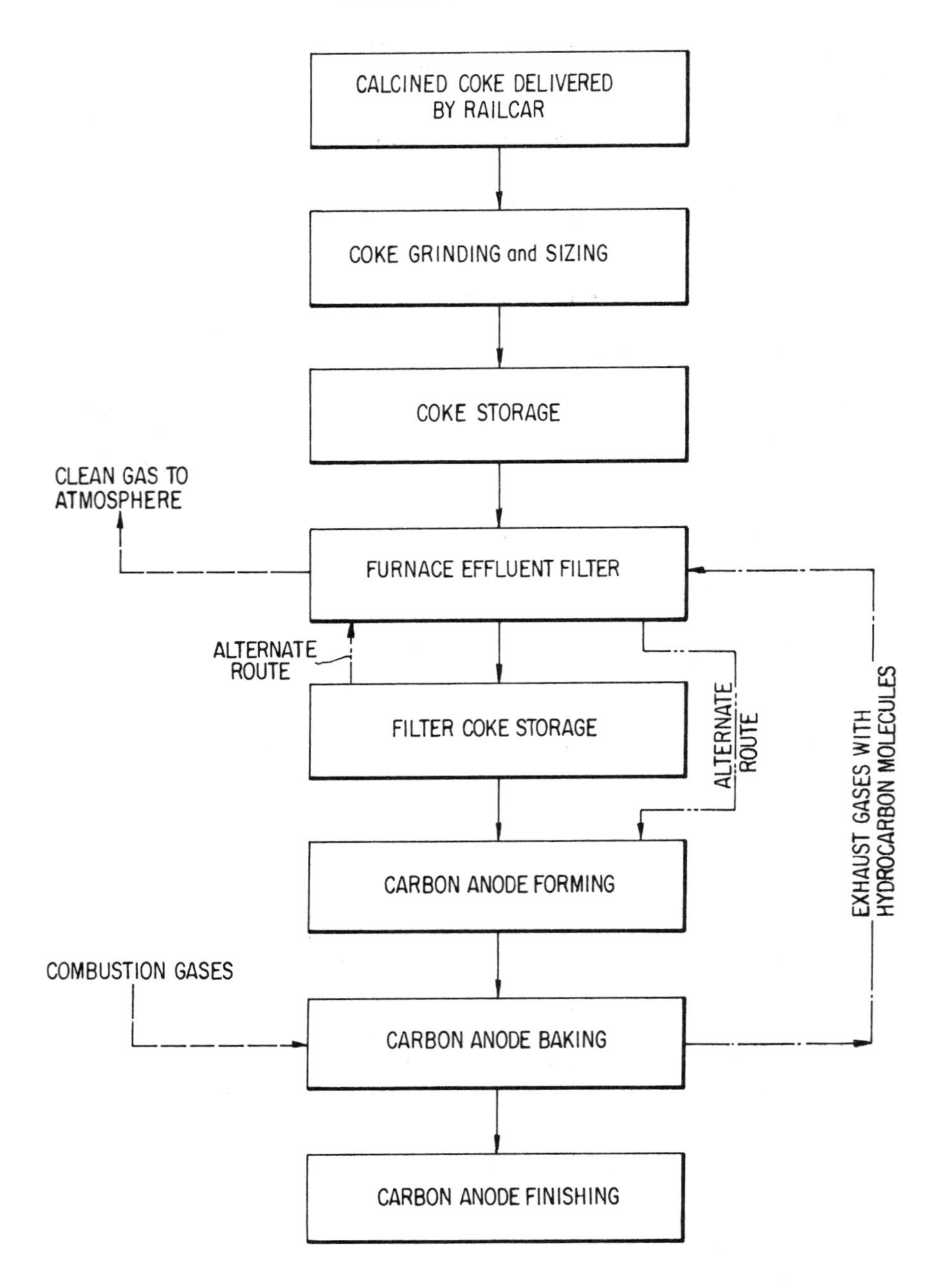

Source: U.S. Patent 3,940,237

material is continuously presented to the effluent stream for filtering purposes. While a variety of devices that function in this manner have been proposed, none of them have been proven to be entirely satisfactory. In some such devices, the depth or thickness of the portion of the filter material through which the effluent gases pass has varied, with the result that one portion of the filter medium becomes clogged while another portion remains relatively clean. This causes high pressure regions in which filter material blowouts occur, i.e., a hole is created in the filter material. In some such devices, expensive mechanical systems, adapted to move the filter material through a housing, have been required. In other such devices, complicated, and often unsatisfactory, discharge control mechanisms have been required.

A device developed by *J.M. Gonzalez and W.V. Nichols, Jr.; U.S. Patent 3,940,237; February 24, 1976; assigned to Reynolds Metals Company* is a carbon baking furnace effluent filter using as the filter medium a carbonaceous material of the type employed in making green carbon bodies to be baked in the furnace, and the used filter material is recovered in a form suitable for making such bodies. The filter unit includes a filter medium supply tube extending vertically downwardly into an enclosed funnel-shaped housing. An effluent inlet tube is coaxially arrayed within the filter medium supply tube.

Fresh filter material is added to renew a bed thereof in the funnel-shaped housing, through the filter medium supply tube, and used filter material is withdrawn therefrom via a valve located at the base of the housing. Effluent from the furnace enters the filter via the effluent inlet tube, rises upwardly through the bed of filter material, disposed generally in the region where the filter material leaves the filter medium supply tube, and exits via an outlet tube located atop the funnel-shaped housing. Subsequent to performing its filtering function, the used filter material is recovered for use, preferably for products of the type processed in the furnace, or recycled through the filter and reused therein.

Figure 15 is a block flow diagram of the electrode manufacturing process showing the pollution control scheme.

A device developed by *A.S. Russell, N. Jarrett, M.J. Bruno, J.A. Remper and L.K. King; U.S. Patent 3,977,846; August 31, 1976; assigned to Aluminum Company of America* uses a temperature controlled heat exchanger comprising fluidized particles to remove hydrocarbons from gas contaminated therewith and prevent hydrocarbon pollution of the atmosphere. Such a device is shown in Figure 16.

Referring to the hydrocarbon removal apparatus **10** of the drawing, sufficient solid particulate is introduced via line **12** to a fluidized bed **14** to maintain a desired depth, for example, from about 15" to about 30". A separate gaseous fluidizing medium such as air is introduced via line **16**. The bed may be cooled to hydrocarbon condensation temperature by a water-cooled heat exchange unit **18**. Gaseous effluent from the electrodes being baked is introduced to the bed at openings **20**, which, according to the process, may be insulated or specially heated, via line **22**. Exit line **24**, which leads to a dust collector or like unit (not shown), is for discharge of gases leaving unit **10** from which the hydrocarbons have been removed by deposition on the fluidized solid particles.

Drain **26**, or a side overflow arrangement (not shown), is for removal of solid particulate from bed **14**. If desired, for cooling the bed, instead of using a cooling unit such as **18**, coolant such as water may be introduced directly into the bed, for example, by means of jets. By introducing a hydrocarbon-contaminated gas at a temperature above the hydrocarbon condensation temperature into the fluidized bed of solid particles while maintaining the bed at a temperature not exceeding this condensation temperature, and by introducing the hydrocarbon-contaminated gas at a point spaced from the walls of the bed as well as from the entry of the fluidizing media, the objectionable hydrocarbons in the gas are not condensed until after entry into the fluidized bed, but after entry thereinto are deposited on the particles, the gas thereby being decontaminated or purified by separation from the particles containing the hydrocarbons.

FIGURE 16: FLUIDIZED BED DEVICE FOR HYDROCARBON REMOVAL FROM CARBON ANODE BAKING FURNACE EFFLUENTS

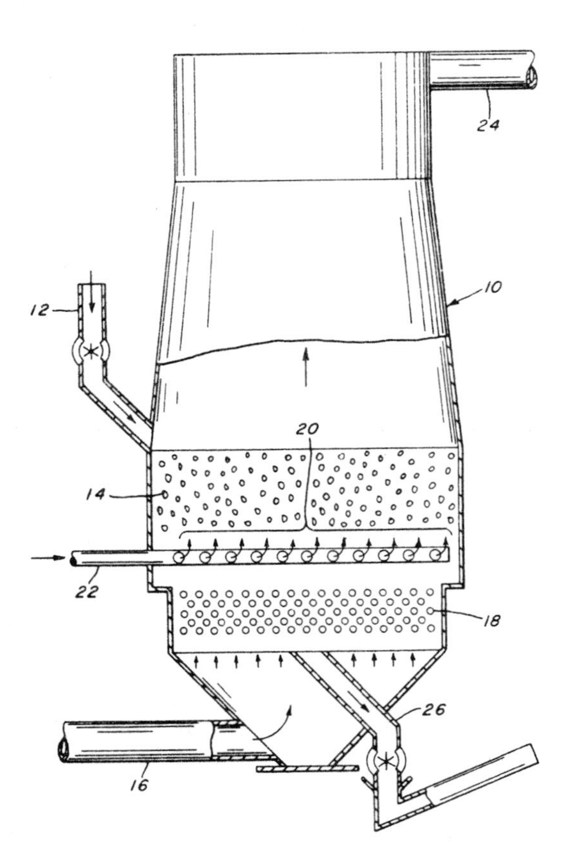

Source: U.S. Patent 3,977,846

Best results are obtained by maintaining the particles at a temperature not exceeding about 125°C and preferably from about 35° to about 80°C, the optimum not exceeding about 60°C, the temperature of entrance to the deposition zone always being at least slightly above the temperature of the particles thus preventing condensation and possible plugging of the entrance, as explained above.

Preferred material for use as the fluidized particles includes alumina, carbon, sand, calcium carbonate, silica and sorptive, pigment or particulate-type substances capable of collecting hydrocarbons on the surface thereof. Preferably, the fluidizing medium is a relatively clean gas such as filtered ambient air or nitrogen.

CARWASH EFFLUENTS

With the high cost of labor and the unreliability of low paid personnel frequently employed by vehicle wash installations, automatic vehicle washing mechanisms have gained in popularity. Such mechanisms frequently require use of high volumes of water and the addition of detergents and the like to such water in order to provide for fully effective washing with only a minimum of labor.

With the advent of pollution control legislation and water shortages throughout the country, it is becoming necessary for such vehicle washing installations to minimize the quantity of water taken from municipal water lines and to closely control the quality of wastewater returned to the city sewers.

Further, with the high cost of water and sewage disposal, vehicle washing installations have found it economically desirable to reclaim the used water employed in washing vehicles and to process such water for reuse. This necessitates separation of grit and sludge-like material from the water used in the initial wash cycle to clean such water for reuse in the wash cycle. By leaving certain quantities of detergent in such wash water, the necessity of adding additional detergent is minimized or, even, eliminated. Further, it is desirable to reclaim the water used in the final rinse of the vehicle and to filter and treat such water sufficiently to enable reuse thereof without leaving residue and stains on subsequent vehicles being washed by the reclaimed water.

Removal from Water

Many efforts have been made to provide effective and economical water reclaim systems for reclaiming substantially all or part of the water used in vehicle washing facilities. Reclaim systems of this type have been provided which include a wash sump separated into a collector compartment receiving water drained from a wash bay and a clean water compartment, the water being drawn from the collector compartment and recirculated direct to the wash arch or passed through a sludge separator to the arch or for return to the collection sump.

Some systems also incorporate a rinse water collection sump separated into a collector compartment and clean water compartment, the water from the rinse bay being drained into the collector compartment and then drawn therefrom for passage through rinse filters and returned to the clean water compartment from where it was subsequently withdrawn to pressurize rinse spray means for spraying rinse water on vehicles as they pass through the rinse bay.

Systems of this type suffer shortcomings in that they are not fully automatic, fresh water is required to periodically backwash the filter tanks thus resulting in unnecessary consumption of relatively large amounts of fresh water, and have short filter life, due to low efficiency cleaning, resulting in high operational costs.

A wash water reclaim system developed by *D.E. Wiltrout; U.S. Patent 3,911,938; October 14, 1975; assigned to The Allen Group Inc.* is characterized by automated means for withdrawing water from a wash water collection sump for separating the sludge therefrom and at the same time chemically treating the used water recirculated back to the sump or returning same to a wash water storage compartment in such sump and/or to pressurize a wash spray means in the wash bay.

Automated means are also provided for withdrawing used water from a rinse water collection sump, chemically treating the used water and directing it through a filter arrangement to a rinse arch, or storage, such filter arrangement including a plurality of pairs of filter tanks with pairs thereof being connected in series. A valving arrangement is provided for connecting the outlets of at least a pair of such filters in parallel with the outlet of a single remaining filter tank for back washing such single filter with the entire volume of water passed through such pair of filters.

A carwash water reclaim system developed by *D.E. Wiltrout; U.S. Patent 3,774,625; November 27, 1973; assigned to Ultradynamics Corporation* is shown schematically in Figure 17.

As the vehicle arrives over the wash pit, a switch is actuated as, for example, by the bumper of a car, and upon closure of this switch, a wash pump **12** commences to operate. The wash pump pumps water from the storage section **10a** of the pit, and into the line **14** leading directly to spray heads from which the wash water from the pit emerges directly onto the article. The water passing out of the spray heads **16** contains soap and detergents, but is free from grit and other solid particles which might be classified as sludge and which would ordinarily scratch the surface of the article, as a car for example, if it were to be present within the spray water.

After having been applied over the article, the wash water runs down off the article and drains into the area **10b** of the wash pit, which acts as a skimmer. The wash water draining into the section **10b** of the wash pit contains the grit and solid particles picked up as a result of washing the article which was conveyed over the wash pit. This wash water is then passed from the section **10b** of the wash pit to section **10c**.

To remove the grit and solids from the wash water, the latter is pumped from the section **10c** into a cyclone **18** by a pump **20**. The cyclone is a conventional device used to separate solids or sludge particles from a fluid water. Thus, the cyclone serves as a separator of the sludge and the water. The sludge, after separation, passes out of the end of the cyclone, and into a sludge cart **22**. The cleaned water, on the other hand, passes out of the cyclone at the end from which the largest proportion of the water is passed into the line **24** leading directly into section **10a** of the wash storage pit or tank until full.

FIGURE 17: FLOW DIAGRAM OF CARWASH WATER RECLAIM SYSTEM

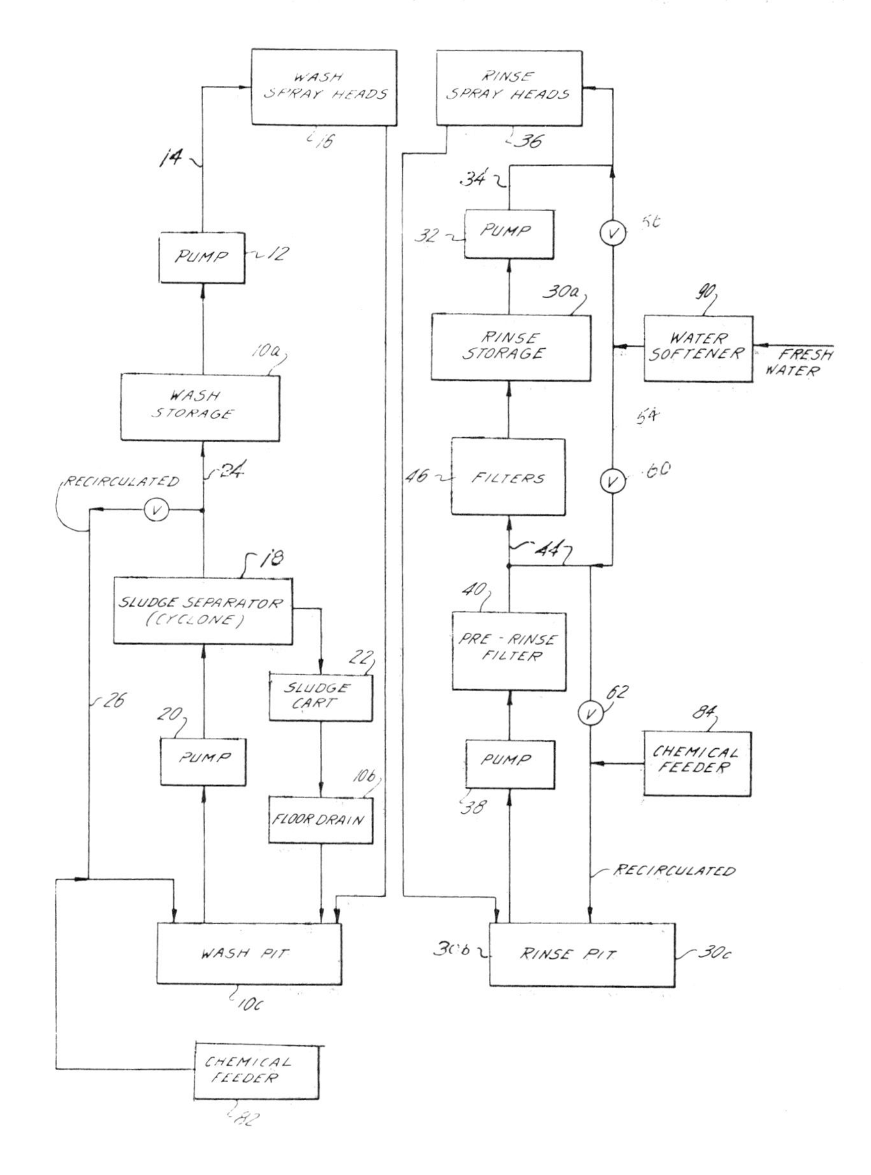

Source: U.S. Patent 3,774,625

A proportion of the cleaned water passing out of the cyclone is transmitted into the branch line **26** for recirculation into section **10c** of the wash pit. Thus, a proportion of the water treated by the cyclone **18** is continually returned into the wash pit for further retreating, rather than applying it directly to the car. Such an arrangement, in which a portion of the cyclone-treated water is returned for retreatment by the cyclone, results in an improved quality of the wash water applied to the surface. This amount of water returned to the wash pit section **10c** for further treatment, through line **26**, may be, for example, approximately 25% of the water leaving the cyclone when section **10a** is low, and 100% when section **10a** is full. This flow is controlled by a level control and shut-off valve.

The valve remains open until sufficient water treated by the cyclone fills the section **10a** of the wash pit to the extent that water overflows from section **10a** into **10c**. When water overflows from section **10a** of the wash pit, in this manner, sufficient treated water is always available to the wash operation through pump **12**. At that point, the valve closes and the entire flow is directed to line **26**. The water flowing through the line **14** and into the spray head **16**, is substantially light-weight water which includes soap and detergents. This water is reused only for washing purposes, and not for rinsing the car afterwards. Detergent is applied to the wash water as needed.

After the washed vehicle is moved from over the wash pit by a conveyor, for example, the pump ceases to deliver wash water to the wash spray heads. As the article is moved further over the rinse pit storage section a rinsing procedure is applied for the purpose of rinsing off the wash water remaining on the surface, and to leave the surface in a clean and waxed condition.

The actuation of a switch by, for example, the bumper of a car, as in the case where a car arrives over the wash pit, causes the pump **32** to commence operating and to draw rinse water from the storage section **30a** of the rinse pit. The water pumped from this section of the rinse pit, is passed into line **34** leading directly to the rinse spray heads. After the rinse water has been applied to the car surface and it flows off the car, the water drains into section **30b** of the rinse pit, which acts as a skimmer. From the latter section, the water is then passed into the section **30c**.

To cleanse the rinse water present in the section **30c** of the rinse water pit, the pump **38** draws water from section **30c** and passes the water into a prerinse filter **40**. The prerinse filter functions as a dirt or mud filter to remove whatever grit and dirt particles remained on the surface and were removed by the rinse water. After being pumped into the prerinse filter the water emerges from this filtering tank **40** through line **44** and passes through filtering tanks **46**. This filter **46** serves to remove soap, detergent, color and other impurities from the rinse water so as to leave clear reusable water. In order to achieve such filtering of the water, the filtering tank **46** contains special minerals, as, for example, activated carbon.

The processing capacity of a filtering tank **46** is substantially limited, and only a predetermined flow rate in gallons per minute, for example, can be accommodated by a single tank **46**. In order to process a larger quantity of rinse water, therefore, a series of tanks **46** are connected in parallel, so that each tank can carry a proportion of the water load to be treated or filtered. Thus the water from the supply line is distributed among three tanks **46** and these three tanks are able to process a substantially larger flow rate than would be possible with only a single tank. With the three tanks connected in parallel, substantially three times the flow rate of one tank can be accommodated. After passing through the filtering tanks **46**, the cleansed rinse water flows through the line **48** and into the storage section **30a** of the rinse pit.

The pump **38** continues to operate for the purpose of cleansing the rinse water until the level of the storage pit section **30a** is sufficiently high so as to overflow into section **30c** within the rinse pit. When this condition is reached, a liquid level sensor within the pit section switches off the pump **38**. The rinse spray heads **36** also cease to apply water to the car, after the pump **32** ceases to operate when the car has been moved from over the rinse pit **30**. To leave a polished and waxed surface on the car after the rinsing action, wax is passed into the line **34** leading to the rinse spray heads **36**.

After the filters **46** have been used repeatedly for reclaiming the rinse water, the filtering agents within the tank incur reduced filtering efficiency to the extent that they cannot perform, after a period of time, satisfactory filtering action. When that point is reached, it is essential to rejuvenate or reactivate the filtering agents, so that they can again perform their assigned duties. To reactivate or rejuvenate the filtering agents, the filtering tanks are back-washed.

As a result of evaporation and stray water losses, it is essential to replenish the lost water with fresh water. This replenishing action is provided by the line which passes fresh water directly into line **34** leading to the rinse spray heads upon opening of the solenoid valve **56**. This solenoid valve is actuated by a fluid level indicator within the rinse pit. When the water level section **30c** of the rinse pit drops below a predetermined height, the fluid level indicator transmits an electrical signal to the solenoid valve to open this valve and thereby permit fresh water to flow into line **34**. The water losses are, thereby, replenished and the added fresh water becomes finally accumulated within the rinse pit.

The fresh water is also passed into line **44**, when the filtering tanks are subjected to the back-wash process. Thus, fresh water is used for the purpose of back-washing the filtering tanks for rejuvenating the filtering or chemical agents therein. A valve **60** is used for this purpose of controlling a flow of fresh water from line **54** into line **44**, leading into the filtering tanks. A predetermined fraction or proportion of the water emerging from prerinse filter **40** is immediately recirculated and returned back to the rinse pit, rather than being transmitted to the filtering tanks **46** for final passage through the spray heads **36**. This procedure for recirculating a portion or fraction of the water treated by the prerinse filtering tank serves to improve the quality of the rinse water.

The car sensor, when actuated by the bumper of the car, for example, turns on both pump **32** and valve **76** so as to initiate rinsing of the car after the latter has been positioned over the rinse pit and has actuated thereby the sensor **74**. A similar such sensor commences the action of the wash spray heads **16**, turns on pump **12** and opens valve **13** when the car has become positioned over the wash pit and has actuated with its bumper the switch or sensor.

Chemical feeders **82** and **84** introduce chemical agents into the reclaimed wash water and rinse water for the purpose of retarding bacteria growth and chemically balancing the water while, at the same time, enabling the detergent to remain in the wash sump and decompose in the rinse sump.

Reverse osmosis may be further used to treat the filtered rinse water from filters

40 and/or **46**, if dissolved salts are excessive, as might be caused by the carry-in on vehicles in winter areas. The reverse osmosis process functions substantially as a demineralizer, when needed due to high total dissolved solids. In the reverse osmosis process, metal shells or permeators are packed with hollow, semipermeable fibers. A large number of fibers are used to create a large semipermeable surface area in a shell. In osmosis, pure water diffuses through the semipermeable membrane to dilute the salt solution. The effective driving force is called osmotic pressure. Reverse osmosis occurs when the pressure on the salt solution is greater than the osmotic pressure. Fresh water then diffuses through the membrane in the opposite direction and is collected.

A technique developed by *R.J. Adams; U.S. Patent 3,904,115; September 9, 1975; assigned to The Standard Oil Company, Ohio* is one whereby environmentally objectionable waste streams caused by rinsing automobiles can be eliminated by using a small volume of rinse water sprayed onto the automobile surface at a nozzle outlet pressure of at least about 120 psi.

CATALYTIC CRACKING PROCESS EFFLUENTS

Removal from Air

A process developed by *R.E. Evans and C.O. McKinney; U.S. Patent 3,817,872; June 18, 1974; assigned to Standard Oil Company, Indiana* provides an economical and more efficient system for reducing dust emissions from a gas-solid separation system which conventionally includes a series of cyclone separators located within a regeneration chamber. In particular, this system is suitable for reducing the emission of fines from the regenerator of a fluid catalytic cracking unit. The final two stages of the cyclone separators are provided with necked-in gas-discharge tube entrances, and the final-stage cyclone is provided with a solids-discharge line which is connected to a low-pressure and low-cost external recovery system.

Figure 18 shows the conventional prior art apparatus at the top and the improved arrangement at the bottom. In the prior art apparatus, spent catalyst enters the recovery chamber or regenerator **1** through line **2** into catalyst dense bed **3** and oxygen containing gas, usually air, enters the regenerator in the lower portion of the dense bed area through line **4**. Cyclone separator stages **5, 6** and **7** are located within the regeneration vessel and are supported by structure not shown. Each of the cyclone stages is provided with a solids-discharge line or dip-leg **15, 16** and **17**; inlets **25, 26** and **27**; and gas-discharge tubes **35, 36** and **37**, respectively.

The dip-legs may be provided with suitable valving not shown which serves to maintain the respective catalyst height **45, 46** and **47** at a level which will counterbalance the pressure drop in the cyclone system. If valves are not used, dependence on dip-leg sealing by the dense bed is involved. In operation, spent catalyst enters through line **2**. Catalyst is carried via the oxygen containing gas entering through line **4** and/or through other lines. Catalyst is swept upward and passes through inlet **25** into cyclone **5**. A portion of the solids is separated in first-stage cyclone **5**, the solids passing downward through dip-leg **15**, and the remaining gas and entrained solids leaving the cyclone through gas-discharge tube

35 and passing to second-stage cyclone 6.

FIGURE 18: PRIOR ART (ABOVE) AND IMPROVED DESIGN (BELOW) CYCLONE SEPARATION FOR CATALYTIC CRACKING REGENERATORS

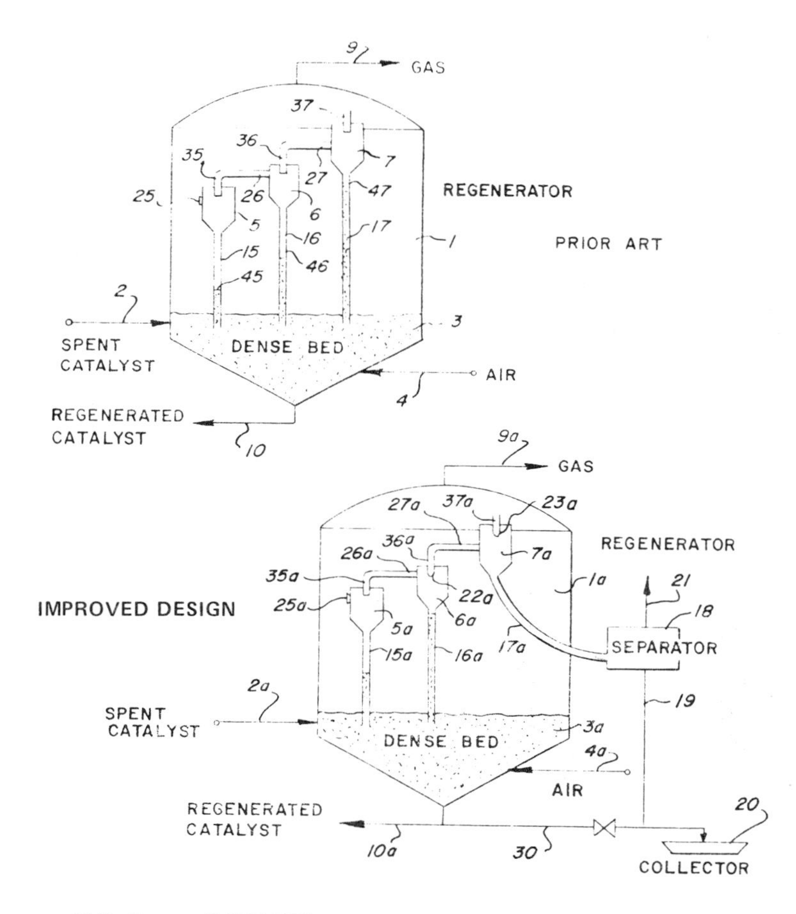

Source: U.S. Patent 3,817,872

Further solids separation takes place in the second-stage cyclone and the gas is passed to the third-stage cyclone where the final solids removal occurs. The exiting gas and the remaining dust particles leave the regenerator via line 9 which either exhausts the gas and dust particles to the atmosphere or to external recovery facilities. Regenerated catalyst is withdrawn from the regenerator via line 10 and returned to the reactor. As depicted, the height of the catalyst which must be maintained in the cyclone dip-leg becomes greater in each successive cyclone stage as a result of the cumulative pressure drop so that the overall pressure drop which may be utilized is limited by the maximum length of the dip-leg which can be used at the final cyclone stage. This in turn is limited by the height of the regenerator vessel and the height of catalyst therein required for combustion of the coke-on-catalyst.

In the improved design, spent catalyst enters the recovery chamber or regenerator **1a**, through line **2a** in the catalyst dense bed **3a** and oxygen containing gas enters the regenerator in the dense bed via line **4a**. The regenerator is equipped with a three-stage cyclone system containing cyclone stages **5a, 6a** and **7a**, having inlets **25a, 26a** and **27a** and gas-discharge tubes **35a, 36a** and **37a**, respectively. Cyclone stages **5a** and **6a** have dip-legs **15a** and **16a** which extend downward into the regenerator. Solids-discharge line **17a** of cyclone **7a** does not extend to the regenerator, but extends outside of the regenerator vessel and is connected to an external separator **18** which separates the solid dust particles from the associated gas.

The flow rate in line **17a** is adjusted to be small. This removes all the dust but little gas and makes final solid separation in separator **18** easy and inexpensive. The dust particles pass via line **19** to a dust collector **20**. Separated gases leave the separator via line **21**. Gas-discharge tubes **36a** and **37a** of second and third-stage cyclones **6a** and **7a** are provided with necked-in entrances **22a** and **23a**. As in the prior art design, exhaust gases are removed via line **9a** and regenerator catalyst is withdrawn from the regenerator via line **10a**.

See Petroleum Refining Effluents (U.S. Patent 3,671,422)
See Petroleum Refining Effluents (U.S. Patent 3,968,306)
See Sour Water (U.S. Patent 3,853,744)

CAUSTIC-CHLORINE PROCESS EFFLUENTS

See Mercury (U.S. Patent 3,725,730)
See Mercury (U.S. Patent 3,764,495)
See Mercury (U.S. Patent 3,736,253)

CELLULOSE FIBERS

See Papermill White Water (U.S. Patent 3,778,349)

CEMENT KILN DUSTS

A typical cement kiln discharges approximately 6,000,000 ft^3 of high temperature, dust-laden gases per hour, there being many tons of dust carried in these gases during each 24 hours of daily operation. Formerly the problem of attempting to separate such quantities of extremely fine dust from this huge volume of gas presented grave problems.

Removal from Air

To safequard against the escape of this dust during recent years, there has been proposed a great variety of dust collectors including baghouses, dust chambers, cyclones and electrostatic precipitators. Owing to the huge volume of hot gases with variable moisture content and of dust to be handled, these various types of collectors, of necessity, are bulky and costly. An ever present problem encountered in all types is the provision of effective means for removing separated dust in a manner maintaining as nearly uniform as possible the efficiency of the separating operation. The problems are aggravated if the wet process of cement making is employed since the gases discharging from the kiln contain large quantities of water vapor which condense and cake dust filtering equipment if the gas temperature falls below the dew point.

A process developed by *G.W. Barr; U.S. Patent 3,266,225; August 16, 1966; assigned to Southwestern Portland Cement Company* relates to baghouses and more particularly to an improved construction designed to operate continuously at high efficiency and capable of filtering large quantities of undesirable contaminants from the gases discharged by industrial processing equipment as for example, a continuously operating cement kiln.

A process developed by *S. Masuda; U.S. Patent 3,444,668; May 20, 1969; assigned to Onoda Cement Co., Ltd.* relates to an apparatus for the electrical precipitation of cement dust in a dust-containing gas.

A process developed by *H. Duessner; U.S. Patent 3,485,012; December 23, 1969; assigned to Klockner-Humboldt-Deutz AG* is a process for removing dust from the exhaust gases of a cement manufacturing installation in which the dust-laden exhaust gases, after leaving a raw-powder-preheater, are conducted into the lower portion of a vertically extending moistening compartment. Water in an excess amount is sprayed into the upwardly moving gas stream and the excess water loaded with dust collects in a sump at the bottom of the compartment.

A pump withdraws the muddy water from the sump and discharges it firstly into nozzles tangentially extending into the sump to circulate its content, and secondly into the upper portion of the compartment to wet the inner wall of the same. Another pump conveys a portion of the muddy water from the sump to the raw-powder-preheater at a point before the one where the raw-powder is introduced.

A process developed by *G. Deynat; U.S. Patent 3,503,187; March 31, 1970; assigned to Societe des Forges et Ateliers du Creusot* is one which extracts alkali from exhaust gases of a cement kiln and which includes a duct including a curtain of endless chain elements with chain cleaning arrangements for the chains.

A process developed by *L. Kraszewski et al; U.S. Patent 3,507,482; April 21, 1970* is one in which the alkali content of cement clinker produced in an apparatus having a suspension preheater which delivers preheated raw meal to a rotary kiln is reduced by the steps whereby:

(a) a portion of the hot exhaust kiln gases are withdrawn from a point adjacent the gas exhaust of the rotary kiln;

(b) the portion is passed through a wet scrubber to remove alkali and solid material therefrom and produce a thin slurry;

(c) the gaseous portion obtained from step (b) being then passed to an electrostatic precipitator; and

(d) the slurry is discharged from the scrubber to waste.

A process developed by *J.G. Hoad; U.S. Patent 3,577,709; May 4, 1971; assigned to John G. Hoad and Assoc., Inc.* involves a combined wet scrubber and clarifier or settling basin for removing dust and other particulate matter from exhaust gases associated with various metallurgical and chemical processes, such as cement manufacture.

CHLORINATED HYDROCARBONS

The disposal of halogenated organic materials formed as waste products in industrial halogenation processes give rise to difficult pollution problems. Since halogenated organic materials possess varying degrees of toxicity to plant and animal life, it is not desirable to dispose of them on land or by dumping at sea, and since their combustion products contain halogens and halogen acids which pollute the air, they are not suitably disposed of by using normal incineration techniques.

Removal from Air

A process developed by *D.F. Winnen; U.S. Patent 3,984,206; October 5, 1976; assigned to Shell Oil Company* is one in which halogenated organic materials are combusted at a temperature of at least 600°C in a refractory-lined furnace. The combustion gases formed are cooled and the acid constituent(s) thereof are subsequently stripped by contact with an aqueous liquid, the furnace being externally cooled such that the temperature of its metal casing lies between 140°C and 375°C thereby minimizing corrosion of the casing.

An apparatus suitable for the conduct of such a process is shown in Figure 19. The chlorinated organic material is passed via a line **1** and a burner **2** into a furnace **3** in which the material is combusted in the presence of fuel, air and steam which are introduced to the furnace via lines **4, 5** and **6** respectively. The furnace is supported by legs **7** resting on the ground **8**. The metal furnace casing **9** is lined with a refractory material **10** and surrounded with a metal cover **11** forming an air jacket **12** around the furnace casing.

Cooling air enters the air jacket at the bottom of the cover via a duct which runs along the entire length of the base of the cover, flows through the air jacket and passes out of the air jacket through stacks **13**. Each stack has a valve **14** for controlling the flow of air through the air jacket and also is provided with a weather protection cover **15**. The temperature of the metal furnace casing is monitored by thermocouples attached to the outside of the casing (not shown in the diagram).

The hot combustion gases pass out of the furnace via a line **16** and into the bottom of a scrubber **17**. The gases are cooled by passing them up the scrubber in countercurrent flow to water which is introduced via a line **18**. Hot halogen acid solution passes out of the scrubber via a line **19** and is recovered. A part of this acid solution if desired is recycled to the top of the scrubber after being cooled (not shown in the figure). Cooled gases having a substantially reduced halogen acid content pass out of the scrubber via a line **20** and pass into an alkaline scrubber **21** in order to remove the remaining halogen acid constituents. Alkali solution passes into this scrubber via a line **22** and, after absorbing substantially all the acidic constituents of the gases to which it passes countercurrently, passes out via a line **23**. Part of the alkali solution leaving the scrubber via line **23** is recycled to the top of the scrubber via line **24**. Cool substantially acid-free gases pass via line **25** into the atmosphere.

FIGURE 19: APPARATUS FOR THE COMBUSTION OF HALOGENATED HYDROCARBONS

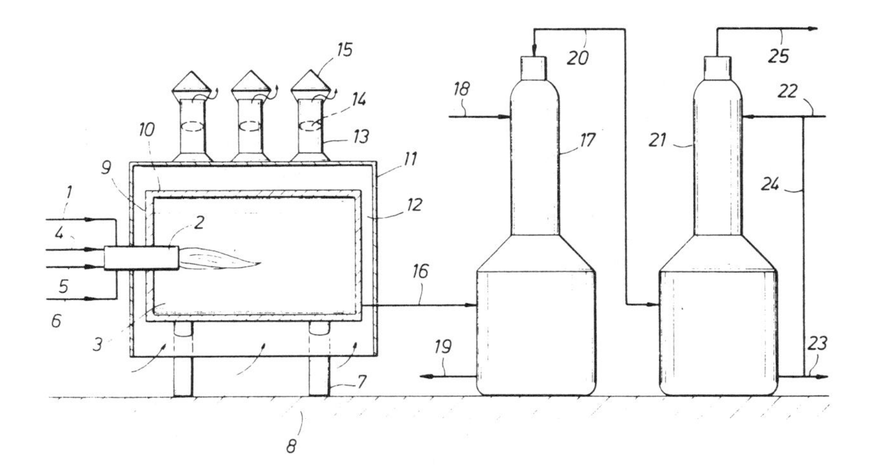

Source: U.S. Patent 3,984,206

A process developed by *Y. Kageyama; U.S. Patent 3,972,979; August 3, 1976; assigned to Mitsubishi Chemical Industries Ltd., Japan* is one in which an exhaust gas containing a halohydrocarbon having 1 to 10 carbons and molecular oxygen is contacted with chromium oxide or a boehmite supported platinum at an elevated temperature to decompose the halohydrocarbon to carbon dioxide, water, hydrogen halide and free halogen.

See Hydrogen Chloride (U.S. Patent 3,980,758)
See Photoresist Process Effluents (U.S. Patent 3,666,633)
See Polychlorinated Biphenyls (U.S. Patent 3,779,866)

CHLORINATED PHENOLS

Removal from Water

A process developed by *W. Hild, H. Krause, K. Scheffler, H. Gusten, E. Gilbert and R. Köster; U.S. Patent 3,971,717; July 27, 1976; assigned to Gesellschaft fur Kernforschung mbH, Germany* involves conditioning highly radioactive, solidified waste incorporated in a molded glass, ceramic or basaltic body prior to nonpolluting ultimate storage, the method comprising introducing the molded body into wastewater, containing a biologically-resistant organic chlorophenol contaminant, from a community or industrial system, introducing a stream of air into the wastewater to produce peroxide radicals which attack the chlorophenol and bring about decomposition of the chlorophenol, and maintaining the body in the wastewater until the heat output produced by radioactive irradiation from the body is reduced to about one-tenth of the amount of heat initially produced by the molded body thereby purifying the wastewater.

CHLORINATION PROCESS EFFLUENTS

Aqueous effluents containing caustic soda and oxychlorine compounds such as chlorites, hypochlorites and chlorates are produced in many industrial processes. One example of such processes is the manufacture of chlorocarbons. These effluents also contain considerable quantities of chloride. The chlorine and caustic soda contained in the effluents are products which can be recycled into the process from which the compounds originate, if they can be recovered by economical methods.

Removal from Water

A process developed by *C.M. Stander; U.S. Patent 3,875,031; April 1, 1975; assigned to AE & CI Limited, South Africa* involves treating aqueous effluent containing caustic soda and oxychlorine compounds and comprises filling the anode and cathode compartments of an electrolytic diaphragm cell with effluent to provide an electrolyte, electrolyzing the effluent and reducing the oxychlorine compounds to chloride ions and hydroxyl ions at the cathode. It further involves recovering chlorine from the anode compartment and sodium hydroxide solution from the cathode compartment while allowing the effluent to flow through the cell by introducing effluent into the cathode compartment.

See Hydrogen Chloride (U.S. Patent 2,730,194)
See Hydrogen Chloride (U.S. Patent 3,036,418)

CHLORINE

Removal from Air and Water

Scrubbing of chlorine-containing gases with alkali or alkaline earth metal hydroxide solution eliminates discharge to the atmosphere of most of the chlorine. However, the principal product of such scrubbing, hypochlorite, is often present in sufficiently high concentration to contaminate or pollute and create an objectionable odor in the streams or ponds of water receiving it.

A process developed by *D.L. Kinosz; U.S. Patent 3,965,249; June 22, 1976; assigned to Aluminum Company of America* involves the catalytic decomposition of such hypochlorites into basically nonpolluting products such as chlorides and oxygen using as catalyst a material containing one or more of the elements cobalt, nickel, copper and calcium, while operating in a pH range of 7 to 13. The catalyst concentration is at least 9 ppm and most advantageously is between 9 to 1,000 ppm. Representative materials for supplying the catalyst, which appears to be converted to the oxide form in the course of the decomposition of the hypochlorite, include (1) salts (nonoxides) such as the nitrates and chlorides, for example, the hydrated form $Co(NO_3)_2 \cdot 6H_2O$ for cobalt (the most effective catalyst), (2) the fused metal, and (3) the metal powder, although decomposition rates are generally slower for the catalyst in elemental form.

Figure 20 shows a suitable form of apparatus for the conduct of the process. In the drawing, chlorine-containing gas enters scrubber **10** via line **12** and is scrubbed by sodium hydroxide solution, which enters at line **14**. Effluent from scrubber **10** containing sodium hydroxide, sodium carbonate and sodium hypochlorite is conducted at a pH of 8.5 via line **16** to baffled decomposition tank **18**, which is maintained at a temperature of 60°C.

FIGURE 20: APPARATUS FOR SCRUBBING CHLORINE FROM AIR AND DECOMPOSING HYPOCHLORITE IN RESULTANT SCRUBBER SOLUTION

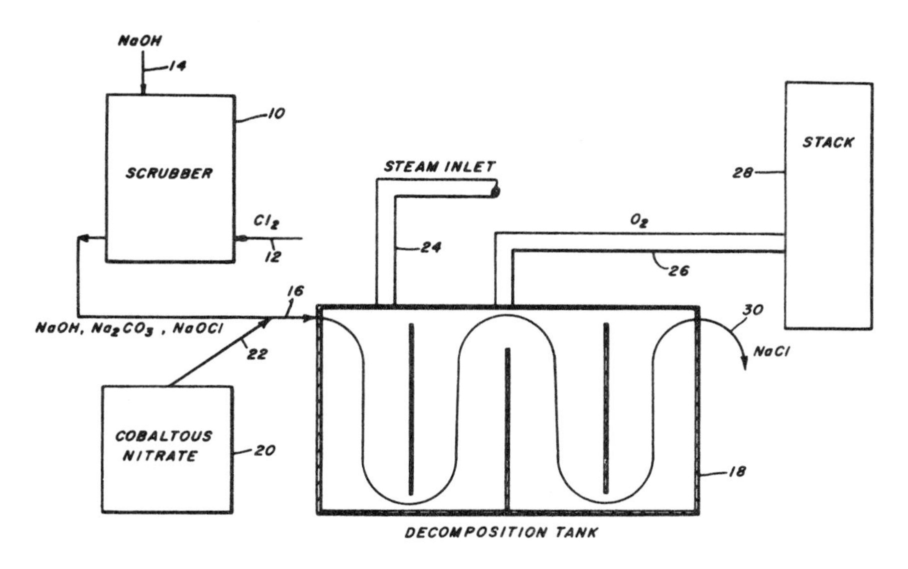

Source: U.S. Patent 3,965,249

Catalytic cobaltous nitrate hexahydrate supplied from source **20** enters the scrubber effluent via line **22**. Steam enters the tank at line **24**. During about 6 hours of residence time in the decomposition tank, the sodium hypochlorite, with the aid of the cobalt catalyst, is broken down into oxygen, which exits at line **26** and is discharged to stack **28**, and sodium chloride, which exits at line **30** in an effluent also containing sodium carbonate, the excess sodium hydroxide from the scrubbing operation and insoluble cobalt oxide (CoO).

See Aluminum Cell Effluents (U.S. Patent 3,904,494)

CHLOROISOCYANURIC ACID PROCESS EFFLUENTS

Chlorinated isocyanuric acids and their alkali metal salts are familiar chemical entities which are useful as a source of active chlorine. Especially important members are sodium dichloroisocyanurate and trichloroisocyanuric acid. These are high-purity, white crystalline solids, available in a variety of mesh sizes. Although active oxidizers, they can be handled and transported with relative ease and safety.

One of the important commercial applications of these products is in the area of water treatment where they have proved effective and convenient for controlling algae and pathogenic bacteria. The water in swimming pools, for example, is readily maintained in a clean and sanitary condition by the addition of chlorinated cyanuric acid derivatives. Other volume uses are as a dry bleach in cleansing, laundering and sanitizing compositions and the like. Alkali metal dichloroisocyanurates and trichloroisocyanuric acid are produced commercially by the chlorination in aqueous media of alkali metal cyanurates.

Removal from Air

A process developed by *L.C. Hirdler; U.S. Patent 3,896,213; July 22, 1975; assigned to Olin Corporation* is a process for disposing of off-gases comprised of a gaseous mixture of chlorine and carbon dioxide such as off-gases produced in hypochlorous acid generators. The off-gases are reacted with an aqueous solution of an alkaline compound and a reducing agent to produce a substantially neutral aqueous solution containing alkali metal salts which can be safely discarded.

This disposal process can be applied, for example, to a cyclic process for the production of chloroisocyanuric acids by the reaction of hypochlorous acid with cyanuric acid, an alkali metal cyanurate or an alkaline earth metal cyanurate, where chlorine containing carbon dioxide is used to generate the hypochlorous acid.

Removal from Water

The manufacture of sodium dichloroisocyanurate and trichloroisocyanuric acid is accompanied by the production of aqueous waste streams containing dissolved cyanurates, the pollution-free disposal of which presents a special problem. A practical solution to this difficulty is a requirement for a commercially successful operation.

One approach to the problem is to treat the aqueous waste streams with active carbon powders. The active carbon was found to exhibit a singularly strong affinity for dissolved cyanurates so that they were removed from the cyanurate waste streams after relatively brief contact times with the active carbon. Although the process effectively cleans up the waste streams, the need to purchase and handle large quantities of active carbon powder increases operating costs. The disadvantage is partially offset by recycling cyanurate values recovered from the exhausted carbon. Even so, the process is not as economically attractive as might be desired and further improvements in the treatment of chlorinated isocyanurate waste streams are being actively pursued.

In a process developed by *R.H. Carlson, R.N. Mesiah and H.R. Chancey; U.S. Patent 3,878,208; April 15, 1975; assigned to FMC Corporation,* aqueous waste streams containing dissolved chlorinated isocyanurates and sodium chloride are contacted with hydrogen peroxide at a pH of about 0.5 to 12.0 whereby the chlorinated isocyanurates are converted to substantially insoluble cyanurate values which are separated by filtration or otherwise removed. The cyanurate values consist mainly of cyanuric acid at the lower pH range and sodium cyanurates at the upper pH range. The resulting filtrate is much lower in organics than the original waste streams. The treatment constitutes an effective pollution control measure which can operate in conjunction with existing processes of manufacturing chlorinated isocyanurate chemicals.

Another process developed by *R.H. Carlson and R.N. Mesiah; U.S. Patent 3,907,794; September 23, 1975; assigned to FMC Corporation* is one in which aqueous waste streams containing dissolved chlorinated isocyanurate values are treated with a compound selected from the group consisting of sulfur dioxide and an alkali metal sulfur-containing reducing compound to dechlorinate and precipitate the isocyanurate values therefrom. The treatment effects recovery of isocyanurate values, and is an effective pollution control means which can operate with existing commercial facilities.

CHLOROMETHYL ETHERS

Chloromethyl ethers are used, among other things, for the manufacture of ion-exchange resins. They are also undesirable air pollutants.

Removal from Air

A process developed by *R.F. Black, C.P. Kurtz and R.J. Pasek; U.S. Patent 3,980,755; September 14, 1976; assigned to Rohm and Haas Company* is one whereby chloromethyl methyl ether and di(chloromethyl) ether contaminants are removed from air in laboratories, industrial plants and other locales wherein chloromethyl ether (more accurately called chloromethyl methyl ether) and bis-chloromethyl ether are used for various purposes. Silica gel and/or activated alumina are employed for removing the chloromethyl ethers either with or without other contaminants from the gaseous atmosphere wherein they occur.

The absorption of the ethers on the silica gel or activated alumina is carried out under conditions of temperature that have been found to result in the hydrolytic decomposition of these chloromethyl ethers, apparently by virtue of catalytic action occurring on the surface of the silica gel or activated alumina. The decomposition of these ethers on contact with the silica gel or activated alumina results in the production of hydrogen chloride, as well as methylal, methanol, formaldehyde or mixtures thereof depending on the particular ether (CME, BCME or both) present in the contaminated stream.

CHLOROPRENE MANUFACTURING EFFLUENTS

Chloroprene may be produced by the dehydrochlorination of 3,4-dichlorobutene-1 in an aqueous alkaline medium. The chloroprene is removed by alkaline medium. The chloroprene is removed by distillation and the spent aqueous medium from the dehydrochlorination is usually subjected to distillation and/or decantation to remove any major amounts of organic materials such as chloroprene and 3,4-dichlorobutene-1. The resulting aqueous effluent from this stage is essentially an aqueous solution of sodium chloride containing some sodium hydroxide and dissolved organic carbon compounds in quantities of up to a few thousand parts per million (ppm).

This effluent is usually discharged to waste since the presence of traces of organic carbon compounds makes it unsuitable for use as feedstock in the electrolytic production of chlorine and caustic soda. The dissolved organics appeared to be a very complex mixture of substances of which only three compounds have been identified. These are 3,4-dihydroxybutene-1; 1,4-dihydroxybutene-2; and 1(or 2)-chloro-4-hydroxy-4-vinylcyclohexene-1 and together account for about 35% of the dissolved organic carbon. The removal of the dissolved organics from the effluent would make it a convenient feedstock for electrolysis thereby reducing the disposal problem created by the untreated effluent.

Removal from Water

Treatment of aqueous solutions and water with either ozone or chlorine for the purpose of removing bacteria, odors and organics is well known. It has however been found that treatment of the alkaline effluent from the chloroprene process with ozone converted some of the organics into carbon dioxide and reduced the organic carbon only to about 250 to 350 ppm. It has also been found that treatment of the effluent with chlorine followed by filtration reduced the organic carbon only to 500 to 600 ppm.

Even saturation of the alkaline (pH 12 to 13) effluent with chlorine to a constant pH of about 5 achieved no further reduction in the organic carbon content. It was also found that treatment with either ozone or chlorine removed the same three compounds which have been identified above to account for about 35% of the dissolved organic carbon. Ozone did not appear to react with the effluent which had been acidified prior to the treatment. Thus, the conventional techniques failed to resolve the problem of reducing to a sufficient extent the organic carbon content in the effluent.

In a process developed by *P.J.N. Brown and C.W. Capp; U.S. Patent 3,952,088; April 20, 1976; assigned to Petro-Tex Chemical Corporation,* the aqueous effluent from the dehydrochlorination of 3,4-dichlorobutene-1 is treated with ozone, and is then treated with chlorine until the pH is acidic.

The effect achieved by a double treatment with ozone and with chlorine is surprising. As pointed out above, ozone or chlorine when used separately remove the three specific compounds identified above but very little else and yet, after an ozone treatment, when the three identified compounds have already been removed, chlorination gives a marked further reduction in the organic carbon content of the effluent. The process is further illustrated with reference to the following example in which ppm is part by weight per million parts by weight.

The following is one specific example of the conduct of the process. Ozone in oxygen was passed into filtered aqueous effluent (containing 857 ppm total organic carbon) from the dehydrochlorination stage of the chloroprene process at 40°C until the absorption of ozone was virtually complete. Then, chlorine gas was passed into the solution at room temperature until the pH of the solution fell to 6. The total organic carbon remaining in solution was then 23 ppm.

CHROMIUM

Chromate ions are produced by many industrial processes, particularly chrome plating and chrome dipping processes. In chrome plating processes, chromium ions are electroplated onto an object, while chrome dipping simply involves the dipping of an object into a solution of chromate ions without any external electrical potential being applied.

Because of the widespread use of chromium plating and dipping as a method for improving the appearance and/or corrosion resistance of metal objects, the delivery of chromium-containing anions, often along with other ions produced in the plating operation, to the environment has become a serious problem. This problem arises primarily from the rinsing step employed in the plating procedure, wherein the plated article is rinsed after being removed from the plating bath. Heretofore, the contaminated rinse water was often delivered directly to the environment. Although the concentration of chromium-containing anions in the rinse water might be low, even in low concentrations such ions can have serious adverse environmental effects. In addition, more concentrated chromate solution is produced when it becomes necessary to dump the plating bath itself.

Removal from Water

A process developed by *W.A. Wachsmuth; U.S. Patent 3,989,624; November 2, 1976; assigned to Ecodyne Limited, Canada* is one in which chromium-containing anions are removed from the rinse water by an anion-exchange resin. The resin is periodically regenerated by backwashing it, delivering a dilute alkali metal hydroxide solution to the resin, delivering a more concentrated solution of alkali metal hydroxide to the resin, and by rinsing the resin. Effluent from the backwashing and the initial portion of the regeneration procedure is retained and later delivered to the resin.

Effluent which contains a higher concentration of chromate ions, displaced from the resin, is delivered to a concentrated chromate solution storage tank. As the regeneration progresses, the effluent becomes richer in hydroxide ions, and this effluent is delivered first to the dilute and then to the concentrated anion regenerant tank for reuse. Finally, the resin is rinsed, and the final effluent is delivered to the dilute chromate solution tank. Water which has passed through the resin is used for regenerant solution makeup and for rinsing and backwashing, as well as for rinsing plated articles.

Solid industrial wastes containing minor amounts of water-soluble chromium, such as those mineral wastes resulting from the manufacture of chromic acid and chromates, present a serious disposal problem long recognized by the industry. These wastes are generally discarded by stockpiling outdoors. If they consisted only of inert insoluble residues, they would pose no problem or threat to the ecology, but unfortunately, no matter how efficient the leaching process, some soluble toxic chromium salts remain.

When such stockpiles are exposed to the elements and wetted by rain, these salts are gradually leached from the residue over long periods of time to pollute the groundwater. It is clear that to stockpile these wastes where they would not be wetted, would be difficult and expensive, therefore some means of minimizing or eliminating the tendency of these wastes to pollute their environment with toxic chromium compounds is badly needed.

A process developed by *A.B. Gancy and C.A. Wamser; U.S. Patent 3,981,965; September 21, 1976; assigned to Allied Chemical Corporation* involves treating solid waste material containing minor amounts of water-soluble chromium compounds with a reductant, particularly sulfide ions, to convert the soluble chromium to an insoluble state, and produce a solid waste from which substantially no chromium can be leached by water, as by exposure to rain.

A process developed by *W. Izdebski; U.S. Patent 3,869,386; March 4, 1975; assigned to Schlage Lock Company* involves the treatment of plating effluents, including at least chromium plating effluent and, in plants having it, zinc plating effluent as well. Hexavalent chromium ions are removed directly, without first reducing them to trivalent forms, by adding barium acetate. When other plating effluents are treated, the formed barium compounds are used as a coagulant for their hydroxides, which may be formed by mixing any of these other effluents with the formed barium compound and adjusting the pH to the value enabling precipitation of the hydroxides. Sulfuric acid may be used for this pH adjustment, thereby also precipitating out any remaining barium ions in solution.

An improved process has been developed by *E.J. Feltz and R. Cunningham; U.S. Patent 3,969,246; July 13, 1976* for removing and recovering chromium from wastewater in the form of chromic acid and/or metallic chromate salts by the direct precipitation of chromium using barium carbonate in aqueous solutions acidified with glacial acetic acid at an acidic pH preferably ranging from 4.5 to 4.7 followed by filtering the resultant chromium material with an acid resistant filter media having openings preferably in the 2 to 4 micron size range and acid-resistant within the pH range of 2 to approximately 5.

The weight ratio of barium carbonate to chromium material contained in the waste liquid is preferably 2:1 and the preferable weight ratio of the barium carbonate to the acetic acid is preferably 3:1. The barium carbonate and acetic acid are freshly mixed in aqueous media prior to either addition to the chromium waste liquid or addition of the chromium waste liquid to the freshly prepared treating material in aqueous media.

Figure 21 illustrates the application of this process in treating chromium waste or rinse waters. As a result of a chromium plating process, the plated parts are rinsed in one or more rinse tanks, the last of which is usually a hot rinse tank.

As parts are rinsed in the rinse tank(s) **1**, the solution therein builds up with contaminating chromium ions. In this process, the contaminated rinse water from one or more rinse tanks is removed for treatment through line **2** or by free flow from the rinse tank(s) to treatment tank. The rinse water in the treatment tank is then treated with the barium carbonate and acetic acid constituents in the proper concentration relative to the amount of chrome in the contaminated water as indicated by the arrow thereon.

FIGURE 21: BLOCK FLOW SHEET OF PROCESS FOR CHROMIUM REMOVAL AND RECOVERY FROM WASTEWATER

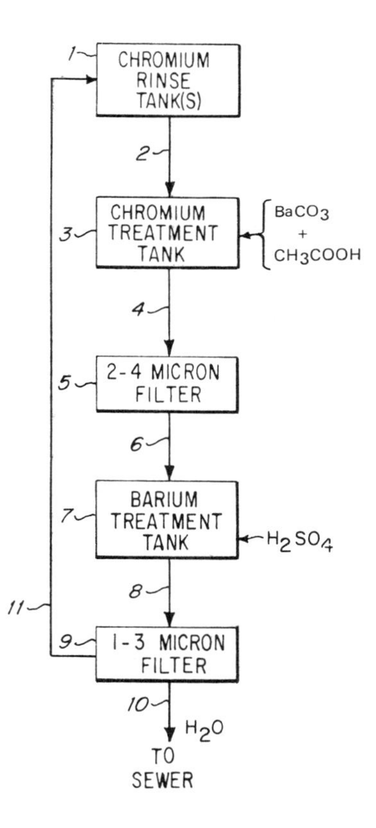

Source: U.S. Patent 3,969,246

The treatment tank **3** is equipped with conventional agitation, preferably air, and the agitated solution in the treatment tank is then fed through line **4** to a 2 to 4 micron filter **5**. If further processing is required, the filtered effluent is then fed through line **6** to an additional holding tank **7** or self-contained filtering system where sulfuric acid or similar material is added to precipitate the slight remaining barium, e.g., as barium sulfate, as indicated by the arrow.

The resulting effluent is fed through line **8** to a 1 to 3 micron filter **9** to capture the barium sulfate. The resulting effluent can be removed directly to the sewer, river or stream through line **10** or, in a continuous operation, back to the original rinse tank(s) through line **11** for reuse.

See Cooling Tower Blowdown Effluents (U.S. Patent 3,810,542)
See Cooling Tower Blowdown Effluents (U.S. Patent 3,901,805)
See Iron Cyanides (U.S. Patent 3,819,051)
See Leather Processing Effluents (U.S. Patent 3,950,131)

COAL GASIFICATION PROCESS EFFLUENTS

In one process the coal is pyrolized to form a char which is then fed to the gasification zone. In other processes raw coal is fed to the gasification zone and pyrolysis also occurs in that zone. Pyrolysis reactions release tars, oils, tar acids and bases, water, hydrogen sulfide, organic sulfides, ammonia and organic nitrogen compounds. During gasification of a char, oxygen, nitrogen and sulfur compounds in the coal can react to form water, ammonia and hydrogen sulfide.

Thus the gaseous streams taken from the gasification zone and from any preliminary pyrolysis zones contain water, including unreacted steam from the gasification zone. Condensation of this water, in the course of purifying the gaseous streams, results in the formation of highly contaminated wastewater containing particulate matter, dissolved carbon dioxide, hydrogen sulfide and ammonia, and, depending on the particular process, dissolved organics (tar acids and bases) and traces of oil. Disposal of this wastewater through ordinary channels can create serious environmental problems.

One of the problems associated with the gasification of coal and similar carbonaceous solids is that of preventing the discharge of toxic trace elements into the environment. Studies have shown that most coals contain small amounts of cadmium, cobalt, lead, zinc, mercury, antimony, arsenic, and other elements which are toxic in low concentrations and could become hazardous pollutants. Some of these elements are retained for the most part as insoluble compounds in the ash formed during gasification and combustion operations but others, such as mercury, are volatile enough to be present in trace quantities in the product and flue gas streams produced during such operations.

These streams are normally cooled for the recovery of heat and the removal of condensed steam and then scrubbed with aqueous solvents and water washed to remove carbon dioxide, hydrogen sulfide, hydrogen cyanide and similar acidic constituents. As a result of these gas cleanup operations, the volatile trace element constituents may be transferred into process water streams where they may tend to accumulate. Because of the low concentrations in which these materials are normally present in the gases and the limited use of coal gasification and related processes in recent years, there has been relatively little attention directed to the elimination of these materials from the aqueous effluents. It can be shown, however, that large gasification plants and similar installations may produce such materials in quantities sufficient to create serious problems if they are not removed from the effluents.

Removal from Air and Water

A process developed by *L.D. Friedman; U.S. Patent 3,966,633; June 29, 1976; assigned to Cogas Development Company* is one in which wastewater formed in the gasification of coal and containing ammonia, H_2S, and phenolic compounds is treated to remove H_2S and a substantial proportion of its ammonia, while leaving sufficient ammonia in the water to maintain a pH of at least 8 (such as up to about 10.5, preferably about 8.5 to 9). The ammoniacal water is then flashed into a stream of superheated steam being fed to the gasification zone. At the high temperatures in the gasification zone the organic impurities are decomposed.

Figure 22 illustrates the operation of the process. In the process, shown in the drawing, vapors resulting from low temperature pyrolysis at **39** may be led to a separation zone **41** in which they are cooled to condense oily liquids and water and the aqueous phase is separated from the oily phase.

This aqueous phase is typically, say, about 4 to 12% by weight of the coal fed to the pyrolysis zone and contains fairly high concentrations of water-miscible organic compounds (such as phenol, cresols, xylenols, resorcinol, methyl dihydroxybenzene), hydrogen sulfide (e.g., in amount in the range of about 0.1 to 1%, such as about 0.3 to 0.5%) and ammonia (e.g., in amount in the range of about 0.1 to 0.5%, such as about 0.2 to 0.4%), together with water-dispersed higher alkylated phenols, such as a broad spectrum of mixed phenols of the type having two or more carbons in one or more substituents (which may be cyclic) and/or three or more methyl substituents, the individual components of this mixture being present in such small proportion as to be dispersed or dissolved in the water.

Thus, such compounds as ethyl phenol, propyl phenol, hydroxyindane, dihydroxyethylmethylindene, dihydroxylnaphthalene, trimethyl phenol, tetramethyl phenol and dimethyl ethyl phenol may be present, among others.

In the process illustrated in the drawing, the oily liquids are then purified at **42** to remove heteroatoms and reduce their viscosity. One method for doing this involves hydrogenation which converts combined nitrogen, oxygen and sulfur to ammonia, water and hydrogen sulfide, respectively, and yields a two phase mixture comprising an aqueous phase and an organic phase, the latter being a combustible light hydrocarbon oil, which may be further refined or treated to produce typical petroleum products such as gasoline, etc.

The synthesis gas stream produced by the gasification (at **43**) of the char contains not only carbon monoxide and hydrogen but unreacted steam, particulate material (such as char fines), CO_2, a little ammonia, hydrogen sulfide (e.g., up to about 1% depending on the sulfur content of the coal) and traces of phenolic materials. In the embodiment illustrated in the drawing this gas stream is subjected at **44** to a purification step after it has been mixed with uncondensed material (gas) from the separation step applied to the volatilized products of the pyrolysis; the latter gas (from the separation step) may contain C_1 to C_4 hydrocarbons, CO, H_2, CO_2, H_2S, NH_3, COS.

The purification at **44** may be effected, for instance, by scrubbing and cooling the gas with plain water (e.g., to reduce the gas temperature to a temperature of about 25° to 200°C, preferably about 40°C, at a pressure of about 25 to 150 psig, preferably about 50 psig); this yields an aqueous waste stream containing dissolved hydrogen sulfide, carbon dioxide, ammonia and particulates and, often, water-soluble or water-dispersed organic compounds (such as phenolic compounds) and traces of water-insoluble oily material. After scrubbing and cooling, the gas still contains such impurities as H_2S and it is preferably given a further treatment, e.g., a solvent extraction (using such solvents as potassium carbonate solution or alkanolamine solutions). The gas may then be subjected to a shift reaction, desirably after reducing the sulfur content of the gas to a very low level as by contact with a suitable material such as zinc oxide.

FIGURE 22: PROCESS FOR TREATING EFFLUENTS FROM COAL GASIFICATION

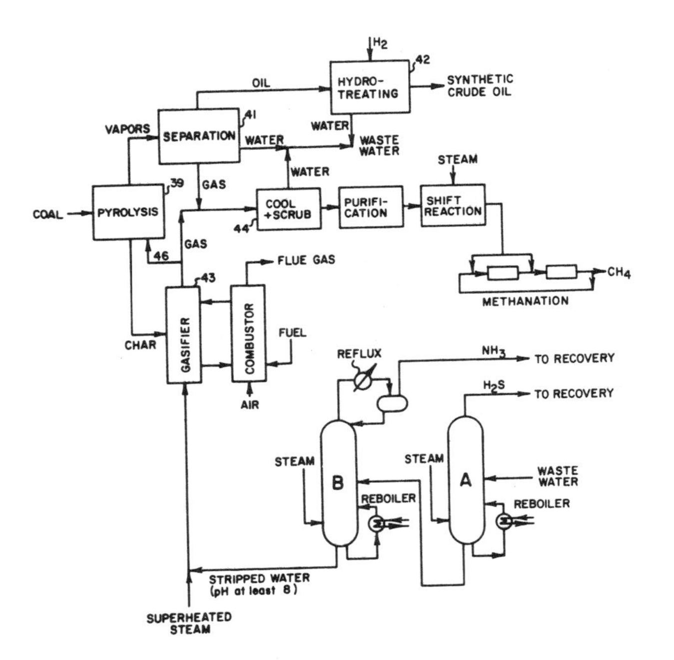

Source: U.S. Patent 3,966,633

As illustrated, in one preferred embodiment a portion **46** of the crude synthesis gas stream (e.g., about 15 to 30%, such as about 25%, thereof) from the gasification zone is fed to one or more of the pyrolysis zones **39** to serve as a fluidizing medium therein and its constituents will thus be incorporated with the pyrolysis products.

Also, instead of adding the relatively impure pyrolysis gas from separation zone **41** to the synthesis gas, the pyrolysis gas may be separately treated for removal of H_2S (and CO_2), e.g., by solvent extraction as described above, and then washed, as with a liquid hydrocarbon, to remove C_2 to C_4 hydrocarbons. The resulting purified pyrolysis gas may then be mixed with the purified synthesis gas and the resulting gas mixture may then be subjected to a shift reaction, desirably after reducing the sulfur content to a very low level as by contacting the gases, individually or in admixture, with a suitable material such as zinc oxide.

The shift reaction is carried out at a temperature of say about 250° to 550°C desirably at a relatively high pressure, such as 500 psig, in the presence of added steam to convert some of the carbon monoxide in the gas to carbon dioxide and hydrogen, e.g., to give a 1:3 $CO:H_2$ mol ratio. The gas may then be cooled to condense out some of the water content to adjust the water content prior to methanation.

The gas may then be subjected to methanation in which the carbon monoxide and hydrogen react in the presence of a suitable catalyst (such as the known nickel catalyst) to form methane and water. The gas is then cooled to condense out the water. Owing to the purity of the feed gas at this stage, the condensed water is relatively pure and suitable for use in a conventional steam boiler to make steam for the process. In addition the methanation reaction is very exothermic and may be used as a source of heat (by conventional heat-exchange) to produce steam for the process.

The methanation reaction is preferably carried out in stages, as is known in the art. Thus the feed gas stream may be divided into several smaller substreams. One substream is diluted with a stream of recycled methane and fed to a first methanation reactor. The hot gaseous product at a temperature of, say, about 500°C is then cooled, by heat-exchange, to a temperature of, say, about 300°C, and the second substream of feed gas is mixed therewith and fed to a second methanation reactor, and so forth.

For each 100 parts by weight of water fed to the gasifier, the amount of the wastewater which is most highly contaminated with organic compounds, i.e., the aqueous pyrolysis liquor generated from the pyrolysis of the coal, generally is in the range of about 8 to 15 parts (e.g., about 11 parts). When a water-containing gas is used for fluidization in the pyrolysis step, (such as the gas stream **46** from the gasification zone) the amount of wastewater from the pyrolysis step, e.g., the aqueous phase from separation zone **41**, may be about doubled, e.g., it now amounts to about 15 to 20 parts per 100 parts of water fed to the gasifier.

The amount of wastewater from the purification of the crude gas (e.g., from purification **44**) may be in the range of, say 20 to 30 parts; the total amount of wastewater from these three steps (pyrolysis, hydrotreating, crude gas purification) is generally below 50 parts such as in the range of about 35 to 45 parts (again per 100 parts of water fed to the gasifier) and the amount of relatively pure water from the methanation step may be relatively large such as about 25 to 35 parts.

A process developed by *M.L. Gorbaty; U.S. Patent 3,975,168; August 17, 1976; assigned to Exxon Research and Engineering Company* is one in which toxic trace element pollutants present in the raw product gas and raw flue gas streams produced during the gasification of coal or similar carbonaceous solids containing sulfur and such trace elements are recovered by separately scrubbing the product gas and flue gas with water, combining the resulting aqueous effluents, and removing the pollutants from the combined aqueous stream as insoluble metal sulfides.

Figure 23 illustrates such a process. In the process shown, a solid carbonaceous feed material such as bituminous coal, subbituminous coal, lignite, coke or the like which has been crushed to a particle size of about 8 mesh or smaller on the Tyler Screen Scale is fed into the system through line **10**.

The feed solids introduced through this line are fed into a closed hopper or similar vessel **11** from which they are discharged through star wheel feeder or equivalent device **12** in line **13** at an elevated pressure sufficient to permit their introduction into the gasifier at the system operating pressure or a somewhat higher pressure.

A carrier gas stream is introduced into the system shown in the drawing through line **14** to permit the entrainment of coal particles or other solid feed materials from line **13** and facilitate introduction of the solids into gasifier **15**.

The feed stream prepared by the entrainment of coal or other solid particles from line **13** in the gas introduced through line **14** is normally fed into the gasifier through one or more fluid-cooled nozzles not shown in the drawing. Cooling fluid will normally be low pressure steam but may also be water or the like. This fluid may be circulated in the nozzles for cooling purposes or injected into the gasifier around the stream of feed gas and entrained solids to control entry of the solids into a fluidized bed in the gasifier. In the system shown, the gas and entrained solids flow into injection manifold **16** and then pass into the gasifier through four injection nozzles **17** spaced about the gasifier periphery.

The gasifier employed in the system depicted in the drawing comprises a refractory lined vessel containing a fluidized bed of char particles introduced into the lower end of the system through bottom inlet line **18**. The inlet line extends upwardly through the bottom of the gasifier to a point above an internal grid or similar distribution device not shown in the drawing. Steam for maintaining the char particles in a fluidized state and reacting with the char to produce a synthesis gas containing hydrogen and carbon monoxide is introduced into the lower portion of the gasifier below the grid or other distribution device through manifold **19** and steam injection lines **20**.

The raw product gas from the fluidized bed in gasifier **15** moves upwardly from the upper surface of the bed, carrying entrained solids with it. This gas is withdrawn from the gasifier through overhead line **21** and passes to a primary cyclone separator or similar device **22** where the larger entrained solids are separated from the gas. In lieu of an external separator as shown in the drawing, the gasifier may contain one or more internal cyclones or similar devices for the removal of entrained solids from the upflowing gas stream. The solids removed from the gas in the separator are conveyed downwardly through dip legs **23** and **24** for reinjection into the system as described hereafter. The overhead gas from the separation unit is passed through line **25** to a secondary cyclone or equivalent separation unit **26** where additional entrained solids are removed from the gas. The fines thus recovered are withdrawn by means of

FIGURE 23: ALTERNATIVE SCHEME FOR TREATING COAL GASIFICATION EFFLUENTS WITH PARTICULAR EMPHASIS ON TOXIC METALS

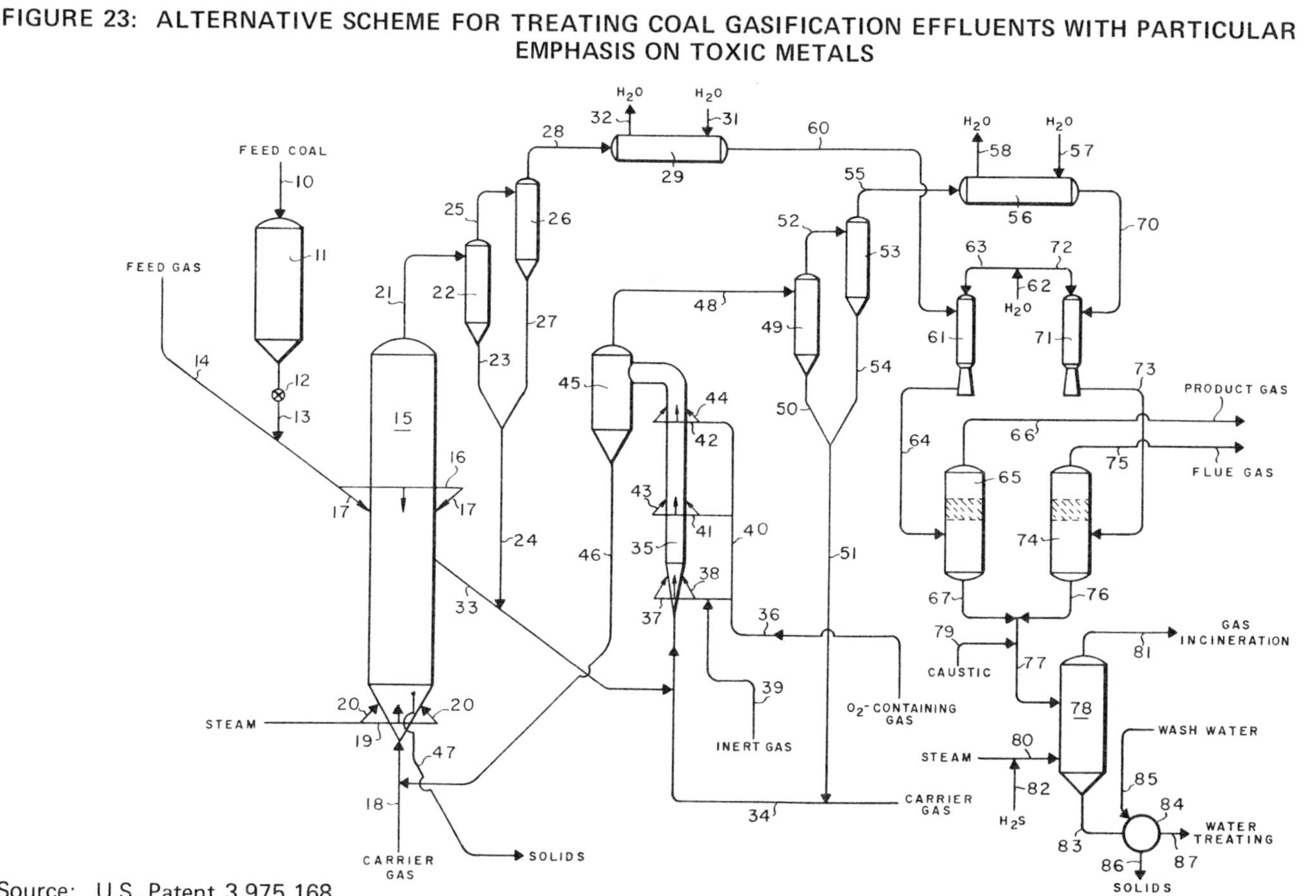

Source: U.S. Patent 3,975,168

dip leg **27** and may be passed with the solids from the first separation unit through dip leg **24** for injection into a transfer line burner as shown in the drawing or for reinjection into the gasifier. The raw product gas taken overhead from unit **26** through line **28** may be passed through heat transfer unit **29** for the recovery of sensible heat in the gas by indirect heat transfer with water or other cooling fluid introduced through line **31** and withdrawn through line **32**. Although only a single heat transfer unit is depicted, it will be understood that a battery of heat exchangers or similar devices may be employed for the recovery of heat from the gas stream if desired.

The heat required for the gasification process shown in the drawing is generated by continuously withdrawing char particles from the fluidized bed in the lower portion of the gasifier by means of line **33**, passing these particles and fines from dip leg **24** into an upflowing stream of carrier gas introduced into the system through line **34**, and injecting this stream containing entrained solids into the lower end of transfer line burner **35**. The carrier gas employed may be recycled flue gas, inert gas or the like. An oxygen-containing gas, normally air, is injected into the system through line **36** and introduced into the lower end of the burner through manifold **37** and peripherally spaced injection lines **38**.

It is generally preferred to dilute the oxygen-containing gas introduced at the bottom of the burner with recycle flue gas or inert gas introduced through line **39** so that the oxygen content of the gas entering the burner at this point is about 15% or less, preferably less than about 6%. Additional oxygen-containing gas, normally air, is introduced into the upper portion of the burner through line **40**, manifolds **41** and **42**, and peripherally spaced injection lines **43** and **44**. The combustion of carbon as the solids move upwardly through the burner in the presence of the oxygen-containing gas results in heating of the solid particles to a temperature in excess of that within gasifier **15**.

It is generally preferred to control the operation of the transfer line burner **35** so that the solid particles leaving the upper end of the burner have a temperature of about 50° to about 300°F above the fluidized bed temperature in gasifier **15**. Solids leaving the burner enter cyclone separator or similar device **45** where the larger particles are removed from the gas stream and conveyed downwardly through line **46** for reintroduction into the gasifier with the carrier gas introduced through line **18**. This circulation of hot solids between the gasifier and the transfer line burner maintains the fluidized bed in the gasifier at the required operating temperature and supplies the heat necessary for the endothermic reactions taking place within the gasifier.

The buildup of ash within the fluidized bed in the gasifier can be avoided by the periodic or continual withdrawal of solids from the gasifier through line **47**. These solids may be conveyed to a fluidized bed vessel not shown in the drawing for cooling and then transferred to a second vessel which is not shown for their removal from the system as a slurry in water. The solids withdrawal rate can be controlled by controlling the pressure within the fluidized bed vessel or by other means.

The raw flue gases from cyclone separator **45** are taken overhead through line **48** and passed to a primary burner cyclone separator or similar device **49** where entrained fine solids are removed and conveyed downwardly through dip legs

50 and **51**. These fine particles may be introduced into a stream of carrier gas such as that in line **34** and reintroduced into the burner with the solid particles from line **33** for combustion in the burner.

The raw gas taken overhead for separation unit **49** through line **52** is passed through a secondary burner cyclone or similar device **53** where additional fines are removed. These fines may be discharged downwardly through line **54** and combined with the solids in dip leg **51** for reintroduction into the burner. The overhead gases from separator **53** are passed through line **55** to heat transfer unit **56** where sensible heat is removed by indirect heat exchange with water or other fluid introduced through line **57** and withdrawn through line **58**. Again, a battery of heat exchangers or the like may be employed in lieu of the single unit shown in the drawing if desired.

The raw product gas, which emerges from the gasifier at a temperature between about 1300° and about 1900°F, depending upon the gasifier operating conditions, is cooled to a temperature between about 450° and about 1000°F in heat transfer unit **29** and then passed through line **60** to a scrubber **61**, preferably a venturi scrubber where the hot gas is contacted with water introduced through lines **62** and **63**. Here the water is entrained in the gas and the resulting fluid is passed through line **64** to separation vessel **65** from which the gas, now generally at a temperature between about 200° and about 450°F, is taken off overhead through line **66** for downstream processing.

Such processing may include contacting of the gas with an alkali metal compound or similar shift conversion catalyst to adjust the hydrogen to carbon monoxide ratio, treatment of the gas stream with a solvent such as monoethanolamine, diethanolamine, hot potassium carbonate, methanol or the like for the removal of acid gas constituents, contact with an absorbent for the recovery of light hydrocarbon liquids remaining in the gas stream, and treatment with zinc oxide or a similar material for the removal of trace quantities of hydrogen sulfide remaining in the gas stream.

Thereafter, the gas can be methanated by conventional means to increase the methane content, compressed and dried, and sent to storage for use as a synthetic natural gas. Alternatively, the methanation step may be omitted and the product gas employed as a low Btu fuel gas or feed stock to a Fischer-Tropsch plant. Other conventional downstream processing such as cryogenic treatment for the recovery of methane, hydrogen and other constituents may also be employed if desired.

The scrubber water from separation vessel **65** is withdrawn through line **67**. This aqueous stream will normally contain trace element constituents removed from the gas, include sulfur and nitrogen compounds absorbed by the water, and have an alkaline pH in the range between about 7.5 and about 9.5.

The flue gas from the transfer line burner cyclones is treated in a manner similar to that described above. The hot gas from the burner, at a temperature between about 1500° and about 2000°F is cooled in heat transfer unit **56** to a temperature on the order of from about 450° to about 750°F and then injected through line **70** into a venturi scrubber or other scrubbing device **71**. Here water injected through lines **62** and **72** is entrained in the gas and the resultant stream is introduced through line **73** into separator **74**. The overhead gas from the separator, withdrawn through line **75**, may be reheated, expanded to a turbine, and then further processed for the removal of gaseous contaminants before it is discharged into the atmosphere or used as a fuel in a carbon monoxide boiler to supply additional heat for the process.

The scrubber water recovered from the flue gas scrubber separation vessel **74** is withdrawn through line **76**. This aqueous stream will contain trace element constituents removed from the gas, will contain sulfur and ammonium compounds in somewhat lower concentrations than the product gas scrubber water, and will usually have an acid pH.

The two scrubber water streams produced as described above are combined in line **77** and fed into steam stripper or similar device **78**. The combined stripper water fed to the stripper as described above is contacted in the stripper with steam or other stripping gas introduced into the system through line **80**. The stripping gas removes hydrogen sulfide, carbon dioxide, hydrogen cyanide, ammonia and other dissolved gases from the aqueous stream and carries them overhead through line **81**, from which the gas may be passed to a gas incineration unit or other downstream facilities designed to permit eventual disposal of the noxious constituents without atmospheric pollution.

The stripping action taking place within vessel **78** will, of course, tend to reduce the hydrogen sulfide content of the water within the vessel and promote a change in pH. To compensate for this in cases where the amount of dissolved hydrogen sulfide introduced with the water through line **77** is relatively low, additional hydrogen sulfide may be introduced into at least one of the scrubber water streams if desired, through line **82** for example in quantities sufficient to effect essentially complete precipitation of the trace elements present in the water.

Water containing precipitated sulfides, as well as any other solids that may have been carried over with the aqueous scrubber effluent, is withdrawn from stripper **78** through line **83** and passed to a rotary filter or similar device **84** where the solids are removed. Wash water may be supplied to the filter as indicated by line **85** and the solids may be disposed of as indicated by line **86** by land fill or other means. The stability and insolubility of the sulfides permits their use in land fill operations with essentially no danger of pollution. The water from which the solids have been removed is withdrawn from the filter through line **87** and may be sent to a water treating plant for the elimination of other undesirable constituents before it is recycled in the system.

Removal from Water

A process developed by *C.W. Matthews; U.S. Patent 3,971,637; July 27, 1976; assigned to Gulf Oil Corporation* is one in which coal is converted to carbon monoxide and hydrogen by a process which exhibits a minimum potential for polluting. Essentially no water effluent is produced. Water makeup for use within the process as steam for gasification or as wash water may include polluted, solids-containing water from other processes. As a result, process requirements for fresh water are greatly reduced, and conventional requirements for purification and discharge of process wastewater are similarly reduced.

Ash, entering as part of the coal feed, is removed from the process in the oxidized form as solidified slag, suitable for landfill or for additional processing to recover valuable minerals. Noncombustible solids introduced in water makeup from other processes or in raw water are also removed as part of the oxidized, solidified slag. Essentially no ash or other solids are rejected to the atmosphere.

Gaseous impurities, having a potential for pollution, which are generated within the process are treated within the process and converted into acceptable forms for sale or disposal, or the impurities are destroyed within the process. For example, sulfur compounds entering the process are converted to hydrogen sulfide directly, or to sulfur dioxide and then to hydrogen sulfide; the hydrogen sulfide is recovered by known processes; and the recovered hydrogen sulfide is converted to elemental sulfur for sale or storage by use of known processes.

Nitrogen compounds entering the process are converted mainly into ammonia, or to nitrogen gas, or to nitrogen oxides and then to ammonia or nitrogen gas; the ammonia is recovered and purified by known processes for sale. Gas streams before venting are first water scrubbed within the process to remove all dust and particulate contaminants.

Any traces of oils and tars which may be formed within the process are treated at high temperature to cause thermal cracking and are thereupon converted to gaseous or solid materials which are further reacted to form the desired gas product. At the same time, the improvements of the process enhance process economy, especially in water usage, in process heat utilization, and in reliability.

A process developed by *P. Wiesner, F. Wöhler and H.-M. Stönner; U.S. Patent 3,972,693; August 3, 1976; assigned to Metallgesellschaft AG, Germany* is one in which dissolved organic compounds in wastewater resulting from coal degassing or gasification processes are removed by extraction with a solvent which is substantially water-immiscible, the solvent residues in the wastewater are removed by inert gas scrubbing, and CO_2, H_2S and NH_3 in the wastewater are removed by heating and steam contacting.

COAL HANDLING EFFLUENTS

In recent years, approximately half of the coal mined in this country has been subjected to some form of wet cleaning treatment. A variety of treating methods are used including jigs, concentrating tables, classifiers, launders and dense media processes. One thing all of these wet cleaning processes have in common is the generation of wastewater streams. These coal washery wastewaters, often called black water, contain in suspension very finely divided coal particles, clays, shale particles and other mineral constituents associated with coal or coal ash. Particle size of the coal and inorganic constituents suspended in the wastewaters is typically very small with a substantial portion being finer than 325 mesh or about 44 microns.

Removal from Water

These wastewaters are difficult to treat and are usually passed to settling ponds or lagoons, often arranged in series, where most of the large particles slowly settle. Wastewaters so treated are usually satisfactory for recycle to the cleaning treatment being utilized but seldom are the waters sufficiently clarified as to allow disposal into streams or other surface waters. Only infrequent attempts have been made to recover the coal content of these waters.

Generally such attempts have involved some type of froth flotation process. One example of a system for disposing of such wastewaters involves passing the waste through a spiral classifier to recover any large particles of coal. The wastewater is then passed to a thickener where a fine sludge fraction is removed and a recycle water stream is recovered. The fine sludge is pumped to a retaining basin which is periodically dredged with the waste being transported to a disposal site.

The solids which settle in the ponds or lagoons are slow to consolidate and dry and present a potential water pollution hazard so long as they remain in place. Most settling ponds or lagoons are formed by building an earthfill type dam often constructed of coal mine refuse. Since most disposal areas are placed in natural drainage ways such as stream valleys, they are vulnerable to storm damage and flooding and so present a semipermanent hazard to downstream areas.

A process developed by *R.H. Shubert; U.S. Patent 3,856,668; December 24, 1974* is a two-step process by which coal washery wastewaters are clarified. The coal particles are first agglomerated and removed from the water. Thereafter, the inorganic mineral constituents contained in the wastewater are separated to leave a clarified water stream suitable for reuse. Agglomeration of the coal particles is accomplished by intense mixing of a heavy hydrocarbon, preferably in the form of a water emulsion, with the washery wastewater.

Figure 24 is a flow diagram of this process. A hydrocarbon stream **10**, such as a heavy fuel oil or waste lube oil, and a water stream **11** are introduced into emulsifying means **12**. This means preferably comprises an inline motionless mixer which is a no-moving-part duct-like mixing device having stationary flow dividing and mixing elements. Other apparatus such as homogenizing valves or mechanical mixers may be used as well. An oil-in-water emulsion **13**, produced in means **12**, is then passed to agitation means **14** along with a wastewater stream **15** containing coal particles and inorganic mineral constituents in suspension. The agitation means preferably comprises an inline mixer of the motionless type of similar design but larger capacity than the mixer. However, means **14** may also comprise a dynamic inline mixer or a mixing device of the turbine or propeller type.

As a result of agitation in means **14**, hydrocarbon droplets contained in the emulsion are dispersed on the surface of coal particles. Inorganic particles are unaffected by the hydrocarbon, probably because of the preferential wetting characteristics of the hydrocarbon toward the coal particles. Coal particles are agglomerated into larger masses and the hydrocarbon treated slurry **16** is then passed to separation means **17** wherein agglomerated coal particles are removed from the other slurry constituents.

The separation means may comprise a quiescent zone such as a settling vessel wherein the agglomerated coal masses float to the surface and the inorganic constituents tend to settle to the bottom. Detention time within such a settling vessel need be but a very short time, on the order of a few minutes, since separation of the coal flocs from the other constituents occurs very rapidly. A coal fraction **18** is removed from separation means **17** and may be further processed by dewatering and drying if desired.

FIGURE 24: METHOD FOR TREATMENT OF COAL WASHERY WATERS

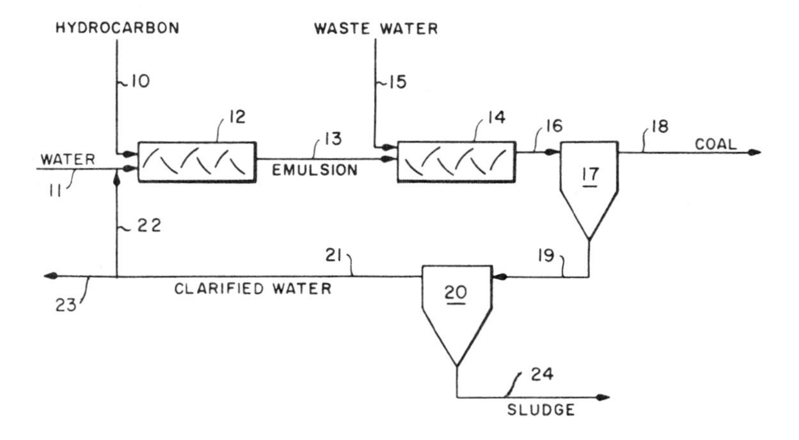

Source: U.S. Patent 3,856,668

Instead of a settling vessel, means **17** may comprise other apparatus for removing agglomerated coal particles from the other slurry constituents. It is possible to perform an effective separation by screening or filtering slurry **16** through a relatively coarse medium such as stainless steel or fabric screening of about 100 to 200 mesh. Essentially all of the coal agglomerates are retained on such a screen while the water containing inorganic constituents in suspension passes readily through.

A water suspension of inorganic constituents **19** is removed from a lower portion of the separation means and is passed to settling means **20**. This means may comprise a settling vessel similar to means **17** or may comprise a settling basin or lagoon. Detention times in means **20** may be relatively short but are greater than those required in means **17**. Settling times of one-half to several hours produces a water stream **21** substantially devoid of suspended materials without the use of flocculating aids. A portion **22** of clarified water stream **21** may be used to form the hydrocarbon-water emulsion while the remainder **23** may be reused in a coal processing plant.

A settled sludge fraction **24** may be removed from a lower portion of means **20**. This sludge fraction contains most of the noncoal mineral constituents of the wastewater stream and comprises a waste material. It is preferred to dispose of this waste by pumping or otherwise transporting it to a section of abandoned mine workings. Alternatively, sludge fraction **24** may be allowed to accumulate and consolidate in a lagoon or settling basin.

If desired, coal fraction **18** may be further treated by reslurrying it in water and subjecting it to a second separation such as that performed in means **17**.

This further reduces the amount of inorganic or ash constituents in the coal fraction and enhances its value as a fuel. The coal fraction dewaters easily by vacuum filtration or other conventional techniques. It may then be thermally dried without the severe dusting problems normally encountered with coal of this particle size.

A process developed by *J.A. Notary and D.E. Metheny; U.S. Patent 3,805,713; April 23, 1974; assigned to Heyl & Patterson, Inc.* is one in which the ash in coal preparation plant tailings is converted into inert water-insoluble pellets by continuously delivering the tailings in the form of a slurry of water and particles of combustible material and ash to a combustion zone in which a fluidized bed of those particles is maintained. Self-sustaining combustion of the combustible material in the bed is established and maintained at a temperature that causes the ash particles to agglomerate to form the pellets, which are removed from the bed at a rate that maintains the desired pressure drop across the bed. Figure 25 is a flow diagram of such a process.

FIGURE 25: DISPOSAL OF ASH FROM COAL PREPARATION PLANT TAILINGS

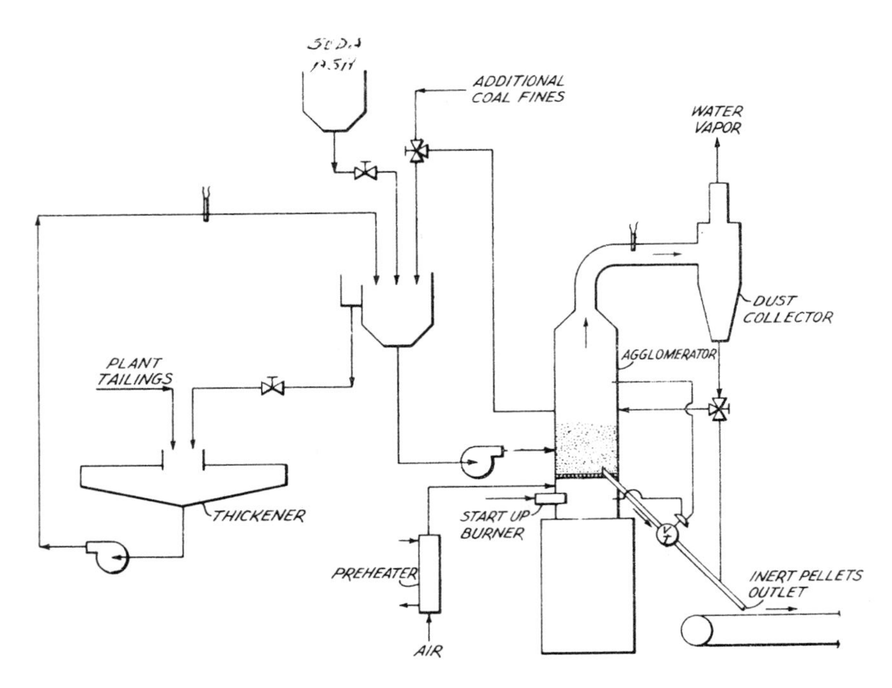

Source: U.S. Patent 3,805,713

It is common practice to supply coal-burning electric generating plants with fuel by fluidized transport of the coal. According to this method, coal is pulverized and formed into an aqueous slurry at the mine. The slurry is pumped through a pipeline to the electric generating plant, the coal is separated from the aqueous medium of the slurry (usually by a cyclone separator), the aqueous underflow from the separator is discharged into a nearby river, lake or arm of the ocean, and the coal is dried and fed into the burners.

A disadvantage of the process is that most solid separators are incapable of removing coal particles of colloidal size. Hence, the effluent normally contains its full complement of colloidal coal which is a significantly large amount. Typically it amounts to 3 to 5% of the weight of the water, and in instances has been as much as 10% of the weight of the water. Recent legislation had made the discharge of this type of fluid (which is jet black in color) into streams and lakes an illegal act of pollution, and a demand has arisen for a process which will remove this colloidal coal leaving a substantially clean aqueous phase. Absent the coal, the liquid phase is essentially water having a mildly alkaline pH (7.5 to 8.5) and is nonpolluting.

A process developed by *M.F. Werneke; U.S. Patent 3,717,574; February 20, 1973; assigned to American Cyanamid Company* is one in which colloidal coal particles in aqueous alkaline medium are flocculated by the combined action of a low molecular weight water-soluble anionic polymer at least 50 mol % composed of acrylic acid linkages and a high molecular weight anionic polymer at least 95 mol % composed of acrylamide linkages, and adjusting the pH of the aqueous medium to a value below 7. A water-soluble anionic starch may be added as supplementary flocculant.

COAL MINING EFFLUENTS

Dust particles floating in air in an underground mine working constitute a nuisance to mining operations. In the instance of combustible dust particles, combustion of the particles resulting in an explosion is a constant hazard, and numerous proposals have been made to reduce or eliminate the hazard.

Removal from Air

One approach authorized for reducing the explosion hazard is to add rock dust in the mine working to inhibit coal dust combustion and absorb heat energy which may be released by the ignition of the combustible dust particles. This addition of rock dust adds to the unpleasantness of the mining operation as well as being a health hazard. The dust can cause silicosis or pneumoconiosis in workers exposed to the dust for long periods of time. An alternative to rock dusting has been known for quite some time. This alternative embodies the coating of exposed surfaces in a mine working with a dust entrapment agent.

Even though the wall coating system has an apparent advantage over the rock dusting technique, it has not been accepted by the mining industry because of side effects of the process and its relative ineffectiveness unless properly utilized. The materials used in preparing the wall coating substance often have a corrosive effect on mining machinery. Experience has also shown the difficulty in balancing the desired properties of the wall coating substrate, and compromises in these properties have lessened the overall effectiveness of the coating.

An improved process developed by *C.H. Jacoby; U.S. Patent 3,896,039; July 22, 1975; assigned to Akzona Incorporated* involves coating exposed surfaces in mine workings with a dust binding agent. Additives to the binding agent enhance entrapment and may be coated on the exposed surfaces after an initial layer of binding agent has been structured in the mine workings.

In the process, a first layer comprising NaCl, clay (attapulgite or bentonite), and water in a paste solution is placed on the walls, ceiling and exposed beams. The mixture is placed in such a manner that a crust is formed on the wall surface. The mixture may be placed on the wall in a number of known ways. Spraying the paste on the wall with conventional gunite machines is the preferred method, however.

Additives to the first layer such as sodium nitrite or sodium dichromate may be used as a corrosion inhibitor. Surface active agents may also be employed. Although prior art teaching indicates nonionic surfactants should be utilized, it has surprisingly been discovered that a more dust receptive crust surface may be formed if the use is made of anionic surfactants in the paste in a percentage of 0.01 to 0.05% with maximum effectiveness occurring when 0.02 to 0.03% anionic surfactant is incorporated in the paste.

The first layer may be allowed to accumulate dust for a period of time before application of the second layer. The use of an anionic surfactant such as dioctyl ester of sodium succinic acid (Aerosol OT-75%) is helpful in forming a crust having better entrapment characteristics than nontreated salt crusts. Of lesser, but still increased effectiveness are cationic surfactants similar to Ethomeen S/25 and Ethomeen T/15, tertiary amine ethylene oxide condensation products of primary fatty amines.

The second layer containing a dendritic crystal forming salt paste may be added to increase the dust entrapment capabilities of the first paste layer. A preferred method of preparing this paste is to add a water-soluble complex iron cyanide selected from the group consisting of alkali metal ferrocyanide salts, alkaline earth metal ferricyanide salts, alkali metal ferrocyanide salts and alkaline earth metal ferricyanide salts to salt paste similar to the first layer paste. The addition of the complex iron cyanide tends to modify the crystal structure growth of NaCl into a dendritic form. This paste containing the complex iron cyanide may be applied in a light coat over the first layer using similar spreading techniques.

As is known in the prior art, the salt crust formed in mine workings must be rewetted at intervals to regenerate its dust entrapment capability. It is thought the rewetting is needed to urge previously trapped dust particles further into the interstices of the salt crust and further to enhance formation of salt crystals on the surface of the crust to cover existing dust particles and capture surfaces for the particles. The process utilizes a brine solution to regenerate the crust formed in the mine workings.

Removal from Water

See Mining Effluents for treatment of acid mine waters.

COFFEE ROASTING EFFLUENTS

The effluent from a coffee roaster is air carrying particulate matter and a variety of gaseous organic compounds, such as esters, aldehydes, ketones, and acids.

Removal from Air

With effective direct flame incineration, these organic pollutants can be oxidized to form carbon dioxide and water, which with the effluent air can then be discharged harmlessly into the atmosphere.

A device developed by *L.C. Griffin; U.S. Patent 3,960,504; June 1, 1976; assigned to Griffin Research & Development, Inc.* involves direct flame incineration of a polluted air effluent, such as the effluent from a coffee roasting oven. The apparatus comprises a cylindrical housing containing three concentric longitudinally aligned shells, each of which defines a respective combustion zone, and with the outermost shell defining with the housing an effluent heat exchange passageway to receive the effluent from an effluent inlet.

A fuel gas or oil nozzle directs a spray of fuel forwardly into the first innermost shell, with primary air also flowing into the first shell to provide initial combustion of the resulting air-fuel mixture. The effluent flowing from the heat exchange passageway flows partly through an annular secondary inlet into the combustion zone of the second intermediate shell, and partly through an annular tertiary passageway into the final combustion zone of the third outermost shell. The combustion products from the final combustion zone pass out an exhaust stack essentially as carbon dioxide and water.

COKE OVEN EMISSIONS

In the art of making coke in a by-product oven, which is one of many ovens situated side-by-side in battery form, the coal is charged by means of a charging car through apertures or holes in the roof and the coal is heated indirectly by means of heated refractory walls which in turn are heated by the burning of fuel. During the baking of the coal or the coking which lasts about 17 hours, by-product gases evolving from the coal leave the oven by means of an ascension pipe which delivers these gases into a collecting main which connects the battery of ovens to the by-product coke plant where these gases which are rich in chemicals, are processed. The battery of ovens has two sides to it, the pusher side and the coke side and each oven has two doors, one on the pusher side and one on the coke side.

The pusher-machine is located on the pusher side and runs on rails; it is equipped with means to level the coal during charging, take off the door of the oven, push the coke through the oven to the coke side after the coking cycle and put the door back on the oven after the push. The coke-guide door-machine which is located on the coke side also runs on rails and is equipped to take off the door of the oven, align a guide means with the opening of the oven to be pushed and put the door back on the oven after the push.

A quenching car receives the incandescent coke from the guide means which car is then propelled by an engine to a tower where the coke is quenched for about 90 seconds with about 8,000 gallons of water to drown the coke and drop the temperature thereof; then the coke is transported to a wharf where the quenching car discharges it for inspection and delivery to the screening and storage area of the blast furnace department. During the charging of the coke, the removal of the doors, the pushing of the incandescent coke and the quenching thereof a very serious problem of pollution is created.

Further there is a pollution problem created by the combustion of the coke oven gas in the flues of the oven and the leakage of gases from the oven chambers into the flues which gases are ejected out of the stack of the battery. The technology is such that there are several disadvantages from the standpoint of health to the environment, difficult operating conditions and detrimental economic considerations. Because of the hot, dirty and simply miserable working conditions, good operations are hampered and good quality of manpower refuses to accept such occupation in many cases.

Removal from Air

Attempts are being made to solve the above-mentioned problems such as the collecting hood installed at Ford Motor Company, Dearborn, Michigan, to collect fumes during the push but at best this can only solve the pollution problem during the push. Such hood is not capable to solve the problem of pushing green coke which is very common in many pushes. The AISI in cooperation with Jones & Laughlin, Pittsburgh, devised a system for smokeless charging by ejecting steam in the standpipes to create a negative draft in the oven and thus aspirate the smoke into the collecting main.

This has not been a successful installation because of the limitations of a single main which most batteries in this country possess, the pulling of coal dust into the main, the fighting of the positive pressure in the main and tar deposits occurring during coking presenting problems to the steam ejection system. The AISI installation is again limited to charging only. An installation at Weirton, West Virginia contemplates the collection of gases during the push and the quench but this installation is very massive and requires great changes to existing facilities including very high investment costs. This system again is limited to the push and the quench.

Other schemes are being proposed such as dry quenching in a complete and separate refractory lined retort used in Russia and some countries in Europe. This system was mainly developed for heat recovery and still has the limitation of smoking during the push of green coke and has no provision to take care of the charging problem. Another is quenching in the car after the push using a trailer with water wherein the excess water is recirculated. Here again no provision is made for the handling of the excessive smoke caused by green coke and no provision is provided to take care of the gases during charging.

It is questionable whether a gas cleaning plant big enough to handle the steam and gases during quenching can be mounted on a moveable trailer. Pipeline charging is being installed by some companies but this system is limited to taking care of gases during charging but not emissions during the push and the quench. Some other schemes are proposed in Europe such as pushing the incandescent coke into a skip-bucket design with a hood and then dumping the contents of this skip-bucket into a rotary drum in which quenching will take place. This system again will only be limited to the push and the quench but not to take care of emissions caused by charging, pushing, quenching and stack emissions from the flues in one single solution.

A process developed by *A. Calderon; U.S. Patent 3,972,780; August 3, 1976* is claimed to solve all the above-mentioned pollution problems in the order in which they occur:

(1) Pollution during charging
(2) Smoke from the green coke during door removal
(3) Smoke during the push
(4) Gas and particle evolution during the quench
(5) The treatment of gases from the flues

To solve these problems, the process provides

(a) A centralized dust collection system which is stationary and of adequate size equipped with a gas cleaning scrubber, a main duct leading to all the ovens of the battery with aperture means for staying connected, and a fan to insure the putting of any particular oven to be evacuated in a negative draft.

(b) A secondary door on each oven door of every oven on the coke side similar to the leveling door on the pusher side through which charging gases are evacuated.

(c) An apparatus having in combination, an evacuator as well as quencher, adapted to travel on wheels on the coke side which apparatus makes connection in a sealed fashion to the main duct in such a manner that the connection to the duct remains fixed despite the traveling motion of the apparatus. This apparatus possesses three main components:

    (1) A fume evacuator to extract fumes during charging from the secondary door disposed to the oven door on the coke side.

    (2) A door extracting mechanism of conventional design.

    (3) A quenching guide enveloped by a chamber adapted to quench the coke during the push.

    These main components are mounted on a carriage for movement from oven to oven along the entire battery.

(d) Along the length of the battery a water distribution system is provided to make possible the availability of water for the apparatus.

(e) Provisions are made on the apparatus to evacuate gases during the door removal on the coke side prior to the push.

(f) Gases from the flues are diverted to the main duct after being fully burned in an afterburner for treatment in the centralized dust collection system instead of ejecting the gases into the atmosphere by the conventional battery stack.

With the provision of the steps and the means labeled above, pollution control of smoke, fumes, gases and steam will be made possible in order to render the production of coke noninjurous to the health of the workers as well as the environment.

A process developed by *E.F. Lowe, Jr.; U.S. Patent 3,862,889; January 28, 1975; assigned to Interlake, Inc.* provides a system for preventing effluents from being emitted from coke ovens upon charging certain of the coke ovens with coal while coking coal in others of the coke ovens. Effluent collectors and a source of subatmospheric pressure are provided for each of the coke ovens and communication is established between a coking oven and the oven to be charged with coal during the charging thereof. Valves are provided to control communication between the charging oven and the collecting means and to control communication between the coking oven and the source of subatmospheric pressure.

A process developed by *F.G. Krikau; U.S. Patent 3,926,740; December 16, 1975; assigned to Interlake, Inc.* involves collecting emissions from coke ovens through the doors thereof during the coking of coal. A chimney is provided for each coke oven to create an upward flowing draft for conveying emissions escaping through the associated coke oven door through the chimney into a manifold and then to a cleaner for treating the emissions to remove pollutants therefrom. The cleaned emissions may then be exhausted to the atmosphere.

At the end of the carbonization period, hot coke is pushed from one side of the oven through the slot-type door at the other side of the oven into an open railcar. Large quantities of atmospheric pollutants, such as smoke and combustion fumes, are released in the general area of contact between the hot coke and air as the hot coke emerges from the oven and bursts into flame.

The consideration of environmental quality has required that as much atmospheric pollutants as feasible should be kept from entering the atmosphere. Since most of the batteries of coke ovens already in existence do not have provision for containing the pollutants produced in the discharge of coke from an oven, an important consideration in the design of pollution control equipment is that it can be added to the coke handling equipment already in use.

A scheme developed by *C.C. Jakimowicz, L.C. Jones and D.D. Raugh; U.S. Patent 3,951,751; April 20, 1976; assigned to National Steel Corporation* is one in which a coke guide carriage is provided with an exhaust hood which encloses the coke guide and extends over the transfer car thereby enshrouding the discharging coke and resulting fumes and smoke during the discharge operation. The exhaust hood is movable along the length of the oven battery.

A stationary exhaust manifold extends along the battery in spaced relation above the oven exhaust hood. The exhaust hood is equipped with connecting means which are adapted to register in coplanar sealing engagement with selected gated openings in the manifold while simultaneously opening the selected gates. Pollutants are exhausted from the discharge site and conducted through the manifold and associated treating apparatus with draft induced by an exhaust means connected to the manifold and treating apparatus.

An improved hot coke transfer car has been developed by *J.D. Sustarsic, R.C. Kinzler and W.D. Edgar; U.S. Patent 3,868,309; February 25, 1975; assigned to Koppers Company, Inc.* It is one in which a finely woven wire mesh screen, suitably supported transversely, is wound on a large reel, and the screen can be pulled down over an opening in the top of a hot coke transfer car to prevent the emission of fumes and particulate matter. A traction car or traction vehicle is coupled to the transport vehicle or transfer car and carries gas scrubbing equipment that exhausts the fumes and particulate matter from the interior of the transfer car; ambient air, being drawn in through the wire mesh screen, displaces the fumes and also cools the wire mesh screen through which it passes.

Crude coke oven gas produced during the carbonization of coal contains components such as hydrogen, ammonia, hydrogen sulfide, naphthalene and other hydrocarbons that are normally removed from the coke oven gas in a by-product recovery, or gas treatment system. The purified coke oven gas is thereafter sent to a gas distribution system and is used as a fuel for the underfiring of coke ovens or as a fuel for various steel plant furnaces or for other industrial purposes. The chemicals recovered from the crude coke oven gas are disposed of by a variety of methods. Ammonia, for example, may be disposed of simply by burning. Hydrogen sulfide, on the other hand, is usually converted to valuable by-products including sulfuric acid and elemental sulfur.

The processes for converting the hydrogen sulfide to sulfur may produce objectionable amounts of sulfur dioxide. One such process, conventionally termed the Claus Process, produces a tail gas containing large amounts of unconverted hydrogen sulfide, sulfur dioxide, and other sulfur-containing compounds that are generally incinerated and discharged through a stack of sufficient height to provide extremely low sulfur dioxide concentrations at grade, or ground level so as to comply with pollution codes. The incinerator and stack required for the incineration of the tail gas are major investment items.

A process developed by *B.F. Tatterson; U.S. Patent 3,798,308; March 19, 1974; assigned to Koppers Company, Inc.* is one in which the sulfur-containing components of the tail gas are removed by first combining the tail gas with crude coke oven gas in the by-product recovery system. Sulfur dioxide and the other gaseous sulfur-containing components of the tail gas are removed from the combined gases by the various gas-treatment phases of the by-product recovery system or pass through the system and are subsequently burned or vented to the atmosphere at less objectionable concentrations.

A process developed by *G. Wunderlich, H. Weber, G. Choulat and D. Laufhutte; U.S. Patent 3,822,337; July 2, 1974* is a multistage process for completely eliminating ammonia and hydrogen sulfide from coke oven gases, their condensates, desorption gases or vapors. This is accomplished by the combustion of ammonia so as to form nitrogen and water, and the combustion of hydrogen sulfide to form elementary sulfur in a process in which the air, necessary for sustaining the combustion, is added to the above gases and vapors in several, preferably two, combustion stages and the gases from the combustion stages are likewise cooled in at least two stages, with at least one cooling stage following the first combustion stage, and at least a second cooling stage following the second combustion stage.

High pressure steam is obtained as a by-product from the process. The decomposition of NH_3 in this process is complete, leading to generation of nitrogen and water, which are worked up separately, while H_2S and SO_2, the latter formed during the process, are used for the recovery of elementary sulfur.

A process developed by *E. Haese; U.S. Patent 3,950,492; April 13, 1976; assigned to Dr. C. Otto & Comp. GmbH, Germany* is one in which coke oven gases are washed with an aqueous metallic salt solution of sulfuric acid or sulfurous acid to absorb ammonia, hydrogen sulfide and hydrogen cyanide. The washing solution is then oxidized to recover elementary sulfur. A portion of the oxidized washing fluid is returned for continued washing of gas and a portion of the washing fluid is diverted to a separator where solid compounds which include metal hydroxide and metal cyanide compounds are removed leaving ammonium sulfate solution.

The ammonium sulfate solution is heated with combustion air and in a heating agent to produce combustion products including nitrogen, hydrogen and an acid anhydride. The solid compounds from the separator are conducted as an aqueous suspension to a reaction vessel wherein, at elevated temperature and pressure, hydrolysis products are formed including free ammonia, metal hydroxide and formate salts. The metal hydroxide is reacted with the acid anhydride in a condensation tower to form reaction products including an aqueous solution of the metal salt which is then conducted together with the formate salt products for further washing of the gas. The ammonia product from the hydrolysis process is returned for washing by the aqueous metallic salt solution and subsequent removal by combustion.

Such a process is shown in Figure 26. With reference to the drawing, line **1** conducts gas, especially coke oven gas, to undergo processing. The gas is conducted by this line to a washing tower **2**. The purified gas after washing is exhausted through line **3**. Line **4** delivers washing fluid to the washing tower. Regenerated washing solution is conveyed by line **5** into line **4** for entry into the washing tower. In the washing tower the gaseous ammonia, hydrogen sulfide and hydrogen cyanide are transformed into ammonia salts, sulfides and cyanide compounds.

Line **6** is used to conduct the washing fluid bearing the absorbed, transformed gases to the lower portion of oxidizing tower **7**. Line **8** is used to charge air into the oxidizing tower wherein as a result thereof sulfur is precipitated by oxidizing the sulfur compounds resulting from the hydrogen sulfide gas. Bivalent iron hydroxide is transformed into trivalent iron oxide which subsequently plays an important part in the absorption of H_2S gas. The elementary sulfur precipitated in the oxidizing tower is floated by the rising air and leaves the head of the oxidizing tower in the form of flowers of sulfur via line **9** for further treatment.

Line **10** is used to discharge waste air from the oxidizer to the outside. The oxidized washing solution is conducted by line **4** and a portion of this solution is returned to the washing cycle. Line **11** is used to draw off a portion of the washing solution and conduct that portion to a filtering apparatus **12**. After separating the solid materials from the oxidized washing solution, the clear fluid that remains is discharged to a combustion installation **14** where, with the addition of a heating agent which is added through line **15**, and the addition of combustion air which is added through line **16**, the clear fluid is burned at 900° to 1200°C. Known forms of apparatus may be employed to carry out this combustion process.

FIGURE 26: PROCESS FOR REMOVAL OF IMPURITIES FROM COKE OVEN GASES

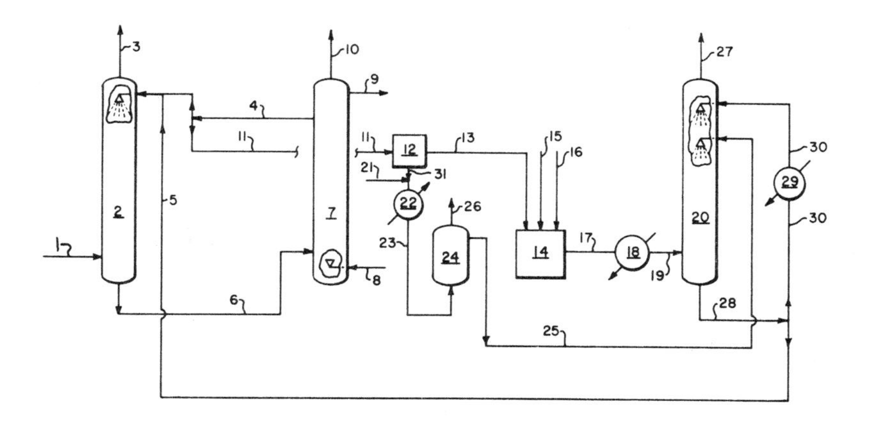

Source: U.S. Patent 3,950,492

It has been shown that when pure ammonium sulfate solution is burned, the specific heat output may be increased to two- or threefold from prior heat output value due to the absence of metal oxides during the combustion process. Furthermore, there is no longer a need to place limitation with regard to the size of the combustion apparatus. The waste gases from the combustion process, which contain water vapor, nitrogen, carbon dioxide and the acid anhydride are conducted by line **17** through a heat recovery system **18**, then by line **19** to a condensation tower **20**.

Line **31** is used to conduct solid substances separated by the filtering device **12**. These solid substances are carried by line **31** as a suspension in condensate or solvent which can be added through line **21**. A heat exchanger **22** is used to heat the suspension to 190° to 280°C after which it is transferred by line **23** to a reactor **24**. In the reactor, a hydrolysis reaction of the cyanide compounds takes place. The hydrolysis reaction forms ammonium formate and bivalent iron hydroxide. Trivalent iron hydroxide is introduced along with the suspension into the reactor but this trivalent iron hydroxide does not play a role in the reaction.

The hydrolysis reaction also frees ammonia in a gas phase which is drawn off from the reactor by line **26**. This gaseous ammonia is introduced into line **1** where it is combined with the coke oven gases to be treated upon entering the washing tower **2**. Line **25** is used to conduct the reaction products from the hydrolysis process to the condensation tower. In the tower these reaction products are brought into immediate contact with the waste gas from the combustion process. The chemical reaction in the tower produces a renewed washing solution. This reaction is between the metal hydroxide and the acid anhydride.

Liquid drawn off at the bottom of the condensation tower is conducted by line

28 from where a portion is recirculated by line **30** after it passes through a cooler **29** into the condensation tower. The cooler is used to extract residual heat from the fluid in line **30**. A portion of the circulating fluid in the condensation tower **20** is drawn off from line **28** by line **5**. The liquid in line **5** is the actual renewed aqueous metallic salt solution which is then added, as previously indicated, to the oxidized washing solution in line **4** for the further washing of gas in the washing tower. Line **27** is used to expel waste gases freed from the acid anhydride to the outside.

Heretofore, severe atmospheric corrosion of structures, buildings, and process equipment has occurred at coke processing plants for many years. The problem has become more critical as a result of the installation of additional chemical processing facilities thus placing a high concentration of complex process equipment in an area exposed to this atmospheric corrosion. A major cause of this problem is the use of aqueous waste streams to quench the coke. These waste streams contain dissolved inorganic salts that become air-borne with the steam produced in the quenching operation, and subsequently upon cooling, rain out of the atmosphere on adjacent buildings and equipment.

Unfortunately, these salts are very corrosive to structures, buildings, and process equipment that are in the path of this salt fallout. This situation has been tolerated because no economically attractive alternative method of disposing these waste streams was available. Disposal of the aqueous waste streams in a deep well has been considered and is being practiced in some coke plants. However, deep well disposal at some coke processing plants requires a disposal well of about 19,000 feet; and the probability of such a well being able to handle the large volume of water has been reported to be fifty percent. Further a danger exists that underground streams will be contaminated by this method of waste disposal.

Removal from Water

A process developed by *W.J. Didycz and D. Glassman; U.S. Patent 3,790,448; February 5, 1974; assigned to United States Steel Corporation* involves purifying waste fluid containing corrosive salts with waste steam at essentially atmospheric pressure from a continuous coke quenching apparatus to produce substantially pure salt-free condensate. The thus purified wastewater may be used in relatively pollution-free coke quenching operations.

As is already well known, it is highly desirable to make use of heat in coke discharged from a coke oven, so that the energy of this heat which otherwise is wasted can be used to advantage. For this purpose the coke which is discharged from the coke oven may be placed in contact with a gas so as to carry out dry-quenching of the coke, and the gas which extracts heat from the coke in this way is then directed through a heat exchanger so that in the heat exchanger heat is taken from the dry-quenching gas which may then be returned to the coke from the coke oven in order to further dry quench coke and extract additional heat therefrom. A second gas which is heated by the heat exchanger may be used for purposes such as drying and preheating coal.

However, arrangements of this type are extremely expensive in that the heat exchanger used to take heat from the dry-quenching gas and deliver the heat to the coal-drying and preheating gas is an extremely expensive unit. In addition such a heat exchanger is complex in its construction, requires a relatively large amount of space, and creates problems with respect to installation and maintenance.

Proposals have already been made for preheating coal, prior to delivery thereof to a coke oven, by placing the coal directly in contact with a dry-quenching gas, but the gas as it issues from a dry-quenching bunker has never been used both for drying wet coal and for preheating the coal. In known installations where preheating of coal takes place by directing a dry-quenching gas through a body of coal, there is an unavoidable generation of undesirable water gas. The gas after preheating the coal is circulated back to the coke to carry out dry-quenching, but the gas is not in a dry condition so that considerable disadvantages result from placing such a moisture laden gas in contact with the hot coke.

A process developed by *R. Kemmetmueller; U.S. Patent 3,888,742; June 10, 1975; assigned to American Waagner-Bird Company* is one in which an inert completely dry gas is directed through a body of hot coke which has just been discharged from a coke oven in order to cool the hot coke, which is thus dry-quenched, while the inert gas becomes heated. The hot inert gas is then placed directly in contact with the wet coal so that the wet coal is in this way dried and preheated with heat extracted from the coke by the inert gas. After the coal is thus dried and preheated, the inert gas is cleaned, dried and then returned to flow again through a body of hot coke, so that in this way the inert gas is continuously circulated along a closed path extracting heat from hot coke and delivering the heat to wet coal.

A process developed by *P. Diemer, G. Preusser and P. Radusch; U.S. Patent 3,840,653; October 8, 1974; assigned to Heinrich Koppers GmbH, Germany* is one in which hydrogen sulfide is removed from coke oven gas by washing the gas with a liquid ammoniacal washing medium. The washing medium containing the hydrogen sulfide is then treated to oxidize the hydrogen sulfide to elemental sulfur. Air in the presence of a catalyst is preferably used as the oxidizing agent. The washing medium, after oxidation of the hydrogen sulfide and separation of the elemental sulfur, is recycled to remove hydrogen sulfide from other coke oven gas. The recycled washing medium contains contaminants such as sulfates, thiosulfates, polythionates and the like. A portion of the recycled washing medium containing these contaminants is withdrawn from the recycle stream and replaced with fresh uncontaminated liquid washing medium.

The withdrawn washing medium containing the contaminants is subjected to an elevated temperature sufficient to volatilize selected volatile compounds. The volatile compounds are separated from the remaining liquid residue and are mixed with unwashed coke oven gas. The remaining liquid residue from the withdrawn portion of the contaminated liquid washing medium is concentrated by evaporating at least a portion of the water and the thickened residue is added to the coal to be coked. A substantial portion of the sulfur compounds in the liquid residue added to the coal are converted during the coking process to hydrogen sulfide and only a relatively small portion of the hydrogen sulfide reacts chemically with the coke formed during the coking process. A substantial portion of the thiocyanates in the residue are, it is believed, converted to nitrogen, ammonia and carbon dioxide with little, if any, of the thiocyanates forming hydrocyanic acid.

A process developed by *H. Grulich, E. Hackler and M. Galow; U.S. Patent 3,915,655; October 28, 1975; assigned to Didier-Kellogg Industrieanlagenbau GmbH, Germany* is a process for removing and burning ammonia vapors and the like which are stripped from a waste liquor such as ammonia water which may be produced during the purification of distillation gases obtained from coke gases.

It utilizes a stripper column means for stripping ammonia vapors from a waste liquor having an inlet for communicating with a waste liquor supply and having an outlet in the top portion for stripped ammonia vapors; and a burner member having a plurality of feed ducts, each of the feed duct members separated from the other feed duct members and being arranged to form a cone of flame having a hollow interior region; a furnace means having the burner member located therein at the upper portion thereof and having an exit for exhaust gases; feed means for supplying the stripped ammonia from the stripper column to the burner member; and a closed system heat transfer means operatively connected between the furnace means and the stripper column for transferring heat for use in the stripper column.

Coke oven gases contain noxious compounds such as H_2S, HCN, HCNS, C_5H_5N (pyridine), NH_3 and the like. The coke oven gases are scrubbed with liquids, such as waste ammonia liquor, to remove a substantial portion of such compounds from the coke oven gases. Many of the compounds which are stripped from the coke oven gases can be stripped from the wastewaters thus formed, rather easily.

However, the wastewaters, such as waste liquor from the ammonia still, benzol wastes and scrubbing liquids, contain relatively large concentrations of various cyanide compounds, both complex and simple, and complex organic compounds which discolor the wastewaters. Obviously, these types of wastewaters cannot be disposed of by conventional means, for example, by discharging the untreated wastewaters into environmental surface or ground waters. It is, therefore, necessary to reduce the cyanide content to a level which is not harmful to plant, animal or human life, and the objectionable color in the wastewaters to an acceptable color prior to disposal thereof.

A process developed by *S.T. Herman and R.J. Horst; U.S. Patent 3,847,807; November 12, 1974; assigned to Bethlehem Steel Corporation* is one in which wastewaters, from a coke oven by-product plant, containing cyanides and objectionable complex organic compounds are treated by a high density sludge process. The wastewaters are mixed with an aqueous high calcium lime slurry and a portion of the sludge formed in the process, which portion is recycled in the process. The aqueous high calcium lime slurry and the portion of recycled sludge are mixed for a time to obtain a uniform mix. A solution containing iron values is added to the uniform mix. Iron cyanide compounds and a portion of the complex organic compounds are precipitated. The precipitate is flocculated and is passed to a settling tank wherein the precipitate settles out to form a high density sludge.

A process developed by *P. Marecaux; U.S. Patent 3,855,076; December 17, 1974; assigned to Societe Pour L'Equipment Des Industries Chimiques Speichim, France* is one in which effluents containing phenols and ammonium salts, e.g., from coke ovens, are purified by adding alkali to release ammonia from the ammonium salts with strong acids, partially evaporating the mixture to remove volatile constituents and incinerating the vapors so produced. The remainder is evaporated to produce vapor which is scrubbed with alkali to remove phenols and a concentrate which is incinerated.

A process developed by *P. Diemer, G. Preusser and P. Radusch; U.S. Patent 3,923,965; December 2, 1975; assigned to Heinrich Koppers GmbH, Germany* is a process for eliminating waste liquors accumulated in the desulfurization of coke oven gas with a washing solution containing an organic oxygen carrier which includes evaporating between 30 and 70% of the water from the waste liquor to provide a concentrated waste liquor solution. The concentrated waste liquor is introduced into a combustion chamber with air and coke oven gas. The concentrated waste liquor is thermally decomposed in the combustion chamber and eliminated sulfur is separated from the gaseous products of combustion. The gaseous products of combustion are mixed with the coke oven gas evolved during the coking process.

COLORED MATERIALS

See Coke Oven Emissions (U.S. Patent 3,847,807)
See Dyestuff Wastes (U.S. Patent 3,979,285)
See Dyestuff Wastes (U.S. Patent 3,803,030), etc.
See Kraft Paper Mill Effluents (U.S. Patent 3,945,917)
See Kraft Paper Mill Effluents (U.S. Patent 3,833,464)
See Kraft Paper Mill Effluents (U.S. Patent 3,833,463)
See Kraft Paper Mill Effluents (U.S. Patent 3,758,405)
See Kraft Paper Mill Effluents (U.S. Patent 3,736,254), etc.
See Nitroanilines (U.S. Patent 3,458,435)

CONCRETE BATCHING EFFLUENTS

In the construction of roadways and buildings of concrete, the batching of the cement and aggregates and their discharge into the mixer is carried out with equipment which is frequently moved from place to place. It is estimated that five pounds of dust emanates from an ordinary plant with each 4,000 cubic yards of mixed concrete produced by the plant. To a nearby resident, this is considerable; to an observer it appears as occasional puffs of smoke and over a long period, some settled fines in the very immediate area of the plant.

Removal from Air

A process developed by *R.W. Strehlow; U.S. Patent 3,812,889; May 28, 1974; assigned to Rexnord Inc.* is one in which a concrete batch plant is provided with a subatmospheric air carried dust collection system which discharges under positive pressure into the body of the aggregates being processed through the plant. This aggregate body functions as a continuously renewed filter bed.

A system developed by *A.A. Mills, Jr. and N.H. Koerner, Sr.; U.S. Patent 3,868,238; February 25, 1975; assigned to The Columbus Bin Company, Inc.*

is a dust control system for collecting dust as a transit mixer is charged from a batch plant. It includes a shroud which is moved into cooperation with the inlet of the mixer and a dust-filtering system connected to the shroud. The filtering system includes a pair of filtering chambers, through which the dust is passed for collection during the loading of the mixer and in which the flow is reversed in alternate chambers during cleaning periods.

CONTAINER INDUSTRY EFFLUENTS

In the container manufacturing industry a class of printing inks has come into wide spread usage. These inks are water based, pigmented compositions, are commonly available commercially, and are typically fast drying.

In a typical printing operation the printing press is washed down after a color change and at the end of a working day. A typical press may thus be washed down anywhere from 1 to 10 times a day using from 20 to 50 gallons of water per washdown. Each such washdown operation produces as waste a diluted ink water which heretofore has commonly been discharged into the nearest sewer at a rate which typically can range from about 200 to 3,000 gallons per day. Such diluted ink water contains, among other contaminating impurities, iron, lead, and copper salts, as well as cyanides, chromates, and the like. Organic matter used in formulating such inks, including phenol and certain other hydrocarbons, may be present. Commonly, such ink water also contains finely divided colloidal particles, such as pigments. Such wastewater has become a significant environmental pollutant.

Also, the container manufacturing industry has come to make extensive use of starch based adhesives, particularly in the manufacture of corrugated paperboard and the like. A starch based adhesive is charged to a pan on a corrugator, applied therefrom to paper, the paper is contacted with other paper to make corrugated board, and the board product is heated to set and dry the adhesive. Typically, about once a day, starch pans are washed which produces a wastewater containing starch. In addition, sometimes a partial batch of a starch based adhesive is dumped for various reasons. Commonly, these operations produce about 300 gallons of wastewater per day.

Further, at a typical manufacturing site, adhesive making equipment is found, usually located in a so-called starch room. The starch room and the equipment therein are commonly washed down about once a week using about 400 to 500 gallons of water. The wash water from such starch adhesive clean-up operations usually contains not only starch and other organic materials employed in starch adhesives (such as phenol, resorcinal, and resinous materials sometimes conventionally used in starch adhesive formulations), but also oily hydrocarbon materials from the machines used to produce corrugated and other containers.

This wastewater has an extremely high content of suspended solids, being on the order of from about 2,000 to 60,000 parts solids per million parts water. The BOD levels in such starch wastewater commonly range from about 2,000 to 14,000 ppm. Measurements herein are generally made according to the procedures given in *Standard Methods for the Analysis of Water and Waste Water,* 13th Edition, U.S. Public Health Service. Such wastewater has thus become a significant environmental pollutant which has been heretofore commonly discharged into the environment through the nearest sewer.

Recently, direct discharge of both such wastewaters into sewers has become a violation of law, as has direct discharge to the ground or surface waters. Normal biological bacterial degradation is not suitable because of the high toxicity and concentration of such wastewaters. Such wastewaters actually dye and discolor river and sewage streams upon discharge thereinto. No acceptable means has been heretofore available for treating such wastewaters to avoid environmental pollution therewith.

There is a strong need in the art for a system which will handle all wastes generated in a container manufacturing facility either singly or in combination. Such system must produce an effluent which can be discharged into substantially any approved sewer treatment system or, alternatively, which can be recycled for the manufacture of starch adhesives or recycled for washdown of water-based or commonly termed flexographic inks.

Recycling is particularly attractive since if a manufacturer recycles he can never be in violation of law because no contamination is leaving his plant(s). Solid wastes recovered can leave the plant(s) for use as sanitary landfill or the like without polluting water provided that this fill contains no leachable substances which can contaminate run-off water. Development of a suitable wastewater disposal system for container plants and the like has proven to be a formidable problem to solve.

Removal from Water

A process developed by *H.R. White and A.J. Doncer; U.S. Patent 3,959,129; May 25, 1976; assigned to Alar Engineering Corporation* employs chemical treatment of such a starting aqueous waste to facilitate removal of colloidal particles and heavy metals followed by physical removal of suspended solids. Optionally, the resulting treated water is further subjected to decolorization.

Figure 27 shows such a process. Wastewater **10** is collected in a holding tank **11**. The size of the holding tank can range widely, a particularly convenient size being in the range from about 200 to 10,000 gallons. The treating chemicals are separately prepared in two different tanks **12** and **13**. Thus, in tank **12**, a water-soluble iron salt is dissolved in water to prepare a solution whose concentration preferably ranges from about 10 wt % up to the saturation point in water of the particular iron salt employed.

Conveniently and preferably a solution of from about 15 to 40 wt % of water-soluble iron salt is prepared. Typical suitable iron salts include ferrous and ferric (preferred) sulfate, ferrous and ferric chloride, and the like. Iron sulfates are generally preferred over iron halides since the halides appear to be more corrosive to equipment than the sulfates. Furthermore, iron chloride appears to be somewhat hygroscopic, making shipment thereof somewhat inconvenient.

In tank **13**, one prepares a slurry of calcium hydroxide. Conveniently the amount of calcium hydroxide in the slurry ranges from about 5 to 25 wt % and preferably from about 10 to 15 wt %. Preferably, the particle size of the calcium hydroxide is initially in the range of from about 50 to 100 μ.

FIGURE 27: PROCESS FOR CLEANUP OF CONTAINER INDUSTRY WASTEWATER CONTAINING STARCH AND INK WASTES

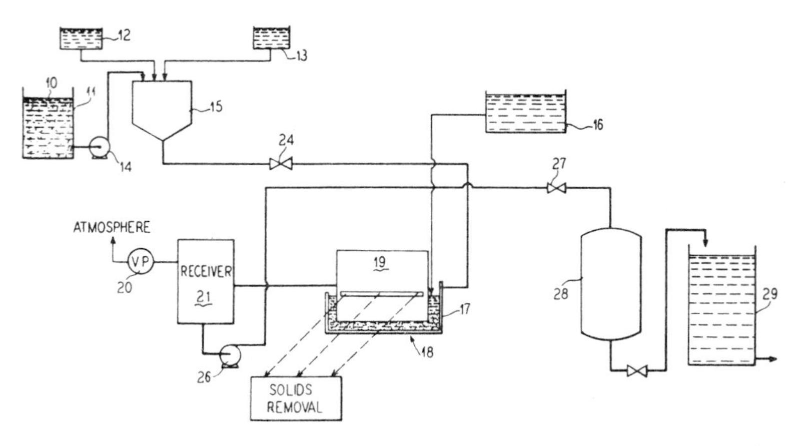

Source: U.S. Patent 3,959,129

The wastewater composition **10** in tank **11** is conveniently pumped, as by a pump **14**, into a treatment tank **15**. Conveniently the amount of waste liquid employed in the treatment tank ranges from about 500 to 4,000 gallons though larger and smaller batches can be employed. Next, one gradually adds from tank **12** the solution of iron salt to the treatment tank. Preferably the addition of such iron salt occurs over a time interval ranging from about 10 to 20 minutes, although longer and shorter times may be employed, as those skilled in the art will appreciate.

After addition of iron salt, it is preferred to allow the combined mixture to undergo a period with continuous agitation in order to achieve a maximum possible effect from the dissolved iron before further processing is carried out. It is believed that the iron tends to break down the colloidal dispersion existing in the starting wastewater. It is preferred that the composition of dissolved iron plus wastewater have a pH in the range of from about 3 to 6 and more preferably about 4 to 5, and most preferably about 4.5. In a next or subsequent step, the lime slurry from tank **13** is mixed with the resulting iron-containing mixture produced as just described.

Preferably, while the addition of chemicals as above-described is progressing, but, as a practical matter, at any convenient preceding time, the filter aid for the rotary vacuum filter is prepared. In general, the filter aid is an inert, particulate, substantially completely water-insoluble material having a particle size below about 250 μ. A particularly convenient such material is an aluminum calcium silicate, such as fuller's earth, bentonite, diatomaceous earth, amosite asbestos, pulped paper, synthetic or natural fibers, or the like. Conveniently, the filter aid is prepared as a slurry of from about 5 to 25 wt % (total slurry weight) of such particulate material in water. The filter aid is conveniently prepared in a tank **16** following which it is discharged into the tank **17** of the rotary vacuum filter **18**.

As the filter aid slurry is discharged into the tank **17**, the drum **19** is vacuumized, as by means of a vacuum pump **20**. Preferably a screen member is interposed over and about a drum before the filter aid is introduced as those skilled in the art will appreciate. A receiver **21** is positioned between the pump and the drum, and the pump is interconnected to the interior of the drum by means of appropriate tubing. In addition, the drum is revolved by a mechanical drive means (not shown). Typical (vacuum) pressures maintained on the exterior surfaces of the drum at this time range from about 4 to 12 psig, and typical drum rpm values range from about 0.5 to 2.0.

The filter aid is deposited as a layer upon the cylindrical working surfaces of the drum portion of the rotary vacuum filter **18** and held to such surfaces by the subatmospheric pressures used. Typical starting thicknesses of the layer of filter aid composition upon cylindrical surface portions of the drum range from about ½" to 6", and preferably is from about 1" to 2" and the amount of filter aid slurry added is chosen so as to be sufficient to produce a layer of this thickness. After the filter aid has thus been deposited upon cylindrical surface portions of the drum, the chemically treated waste liquid in treatment tank **15** is allowed to pass into the tank **17** of the rotary vacuum filter as by opening the valve **24**. A pump (not shown) may be used to transfer the liquid system from the treatment tank into filter tank **17**.

Using the above indicated pressures and drum rpm's, the chemically treated waste liquid undergoes filtration to separate waste solids from waste liquids. The liquids or effluent is drawn off and collected in the receiver **21**. The solids are collected as a deposit upon the cylindrical surface portions of the drum as a filter cake and are continuously removed along a longitudinal position relative to one side of the rotating drum.

While any conventional scraping arrangement can be employed to operate the rotary vacuum filter, it is much preferred to use a scraper blade arrangement which systematically removes the filter cake and a small portion of the filter aid particulate material by cutting action as a layer ranging from about 0.0001" to 0.0010" in thickness measured radially relative to the axis of the drum though thinner and thicker layers may be taken off. Such an arrangement is particularly satisfying for this purpose, since the solid material is characteristically in a slimy form which makes separation and removal thereof difficult from the drum of the rotary vacuum filter without the use of the filter aid layer deposited as above described upon the cylindrical surface portions of the drum.

Solids so removed from the drum are found to be in a nearly dry condition characteristically and may be used directly for sanitary landfill. Preferably the chemically treated wastewater is continuously fed to the rotary vacuum filter until all of a given batch of such treated wastewater in tank **15** has been charged to the filter. Preferably, the interrelationship between the amount of waste composition processed and the thickness of filter aid on the cylindrical surface portions of the drum is such that the filter aid is not consumed before the batch is processed. At the end of the processing, the vacuum pump is turned off, the drum is flushed with clear water to remove any filter aid material remaining thereon, and this wash effluent is conveniently returned to the treatment tank.

It is preferred, as shown, to employ decolorization to remove a characteristic and typical reddish color associated with effluent water collected in the receiver.

For this purpose, the water in receiver **21** is conveniently pumped, as by means of a pump **26**, past a now opened valve **27** into a column **28** filled with activated carbon (not shown). The activated carbon preferably functions to substantially completely decolorize the liquid so that effluent from the column is passed as a clear colorless liquid into a storage tank **29** or the like. Preferably the decontaminated water in the tank is recycled in subsequent cleanup operations in a container manufacturing plant, or is used in starch based adhesive preparation for operations of such plant.

See Flexographic Printing Process Effluents (U.S. Patent 3,970,467)

COOLING TOWER BLOWDOWN EFFLUENTS

Water cooling towers are used in many applications in which it is necessary to remove large quantities of heat at temperatures somewhat above ambient wet bulb temperatures. The use of the water cooling tower conserves the use of water by permitting it to be recirculated for reuse. The recirculated water removes heat from a process stream. The hot water is returned to the water cooling tower at a temperature higher than it leaves the tower.

Heat is removed from the water by evaporation of a portion of it into the air. Approximately 1,000 Btu are removed for each pound of water evaporated. This evaporation of the water must be replaced by makeup water which carries dissolved salts into the cooling water circuit. These salts could build up to the point where they exceed their solubility in water and deposit on the heat exchangers, cooling tower components, etc. It is normal procedure to control the buildup of solids by purging a portion of the recirculating water; this purged stream is termed the blowdown.

The water in a cooling tower system must be treated to reduce scale formation and/or corrosion. A common effective treatment includes the addition of chromates either alone or in combination with other compounds. However, chromium is toxic. It must be removed from the purge stream before the latter can be discharged into a river, or other stream.

Removal from Water

One method of removing chromium is to adjust the pH to 2 or 2.5, to add ferrous sulfate or sulfur dioxide which reduces the hexavalent chromium to trivalent chromium, to add an alkali such as lime to obtain a pH of about 8 which causes chromium hydroxide to precipitate and to remove the chromium hydroxide by settling. At a pH of about 10, phosphorus and zinc as well as most other heavy metals that may be present are removed.

A process developed by *A. Gloster and H.G. Bocckino; U.S. Patent 3,810,542; May 14, 1974; assigned to Texasgulf, Inc.* involves combining the streams from a cooling tower and an acid plant scrubbing tower to reduce the pollution problems of these industrial wastewaters. The cooling tower stream is characterized by the presence of chromium products in which the chromium is in the hexavalent state, such as chromic acid or a chromate salt which are common corrosion inhibitors in cooling waters. The acid plant scrubbing tower is characterized by an acidic component such as sulfur dioxide dissolved in water. By combining these two streams the hexavalent chromium under acidic conditions is converted to trivalent chromium. The combined streams are then neutralized to precipitate chromium hydroxide and other compounds which are undesirable in the effluent. After settling out the chromium hydroxide and other compounds the waste stream may be discharged without presenting a pollution problem.

Figure 28 is a block flow diagram of the process. A cooling tower system of conventional design **101** is fed with plant service water **1** which circulates through the system and passes out by evaporation **2**. Corrosion and scale prevention reagents **3** are added to the system as necessary in accordance with known techniques. In the process, one such reagent is a chromium compound which must eventually be removed from the process stream prior to discharge. As the concentration of salts increases in the cooling tower water a portion of the water must be discharged and replaced by fresh water. This discharge, or blowdown, is identified as the cooling tower purge stream **4**.

FIGURE 28: PROCESS FOR REMOVING CHROMIUM FROM COOLING TOWER BLOWDOWN STREAMS

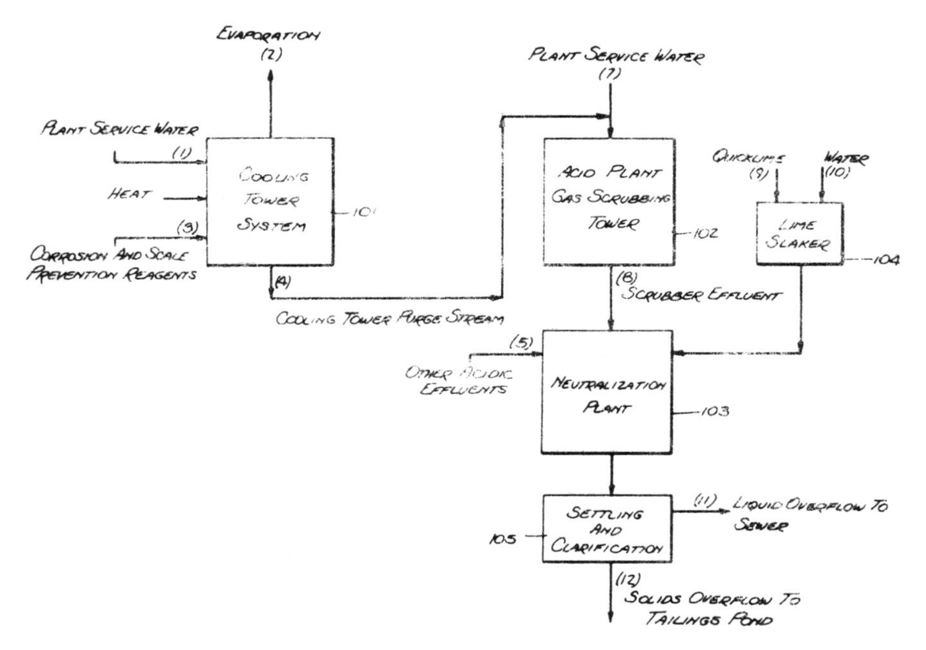

Source: U.S. Patent 3,810,542

The acid plant gas scrubbing tower **102** is used for cleaning gases containing sulfur dioxide. Conventionally this is carried out by contacting the gases with plant service water **7** in accordance with known techniques. In the process the cooling tower purge stream is added to, and at least partially replaces, the plant service water used to contact the acid plant gas. The scrubber effluent **8** from the acid plant gas scrubbing tower is passed into a neutralization plant, **103**. The neutralization plant, which is of conventional design for this purpose, is also the treating means for other acidic effluents **5** which are waste streams from many processes such as roasting operations. Quicklime **9** and water **10** are reacted in a lime slaker **104** and fed to the neutralization plant. The neutralization plant may comprise a series of staged vessels with both acid streams and lime entering a first vessel and the combined streams then passing into one or more vessels, where more lime can be added for precise pH control.

The aforementioned operations may be carried out on a batch, semicontinuous or continuous basis in which the scrubber effluents and other acidic effluents are at a pH below 4, preferably at or below 2 to 2.5, for example, a pH of 1. If additional acidity is required to reach the low pH, the acid effluents may be used to assist in the acidification of the scrubber effluent by causing stream **5** to mix with stream **8** before entering the neutralization plant. The quicklime solution is used to raise the pH above 7, preferably about 8 or above.

The stream from the neutralization plant is then passed through a settling and clarification vessel **105**. The liquid overflow **11** from the settling and clarification tank may be discharged to a sewer since the liquid overflow is virtually free of the chromium pollutants initially added to the cooling tower system as a corrosion and scale prevention agent. These chromium products are passed as a solids underflow **12** from the settling and clarification tank to a tailings pond.

A process developed by *R. Stewart; U.S. Patent 3,901,805; August 26, 1975; assigned to Dow Badische Company* is one in which a fluid reducing agent (e.g., sulfur dioxide) is injected into an essentially unimpeded stream of an acidifed aqueous industrial effluent such as a cooling tower blowdown stream flowing in a simple conduit which is a mere passageway having no baffles or other obstructive or detaining means therein. The stream contains soluble chromium(VI) and has a turbulence characterized by a Reynolds Number of at least about 5,000. The resulting chromium(III) is then precipitated as $Cr(OH)_3$, which is allowed to settle and is subsequently separated from the essentially chromium-free supernatant.

A process developed by *J.M. Dulin, E.C. Rosar, H.S. Rosenberg and J.M. Genco; U.S. Patent 3,876,537; April 8, 1975; assigned to Industrial Resources, Inc.* involves insolubilizing water-soluble sodium sulfur oxide wastes resulting from backwash of process feedwater demineralizers and cooling tower blowdown wastes. The sodium sulfur oxide wastes, typically sodium sulfate and sulfite, are reacted in solution with ferric ions and sulfuric acid to produce an insoluble, basic hydrous or anhydrous sodium hydroxy ferric sulfate or sulfite compound.

The reaction takes place at an acid pH in a temperature ranging from about 50° to 300°F and may occur in single or multistage reactors. Air and/or bacterial activation at a pH of less than about 5.5 may be employed. The end-product basic, sodium hydroxy ferric sulfate and sulfite compounds are substantially water-insoluble, having a solubility of less than the standard calcium sulfate, and may be disposed of by simple landfill without the water pollution hazards inherent with landfilling of wet or dry sodium sulfite or sulfate wastes.

A process developed by *J.R. DeMonbrun, C.R. Schmitt and E.H. Williams; U.S. Patent 3,989,608; November 2, 1976; assigned to U.S. Energy Research and Development Administration* is an electrochemical process for removing hexavalent chromium or other metal-ion contaminants from cooling tower blowdown water. In the conventional process, the contaminant is reduced and precipitated at an iron anode, thus forming a mixed precipitate of iron and chromium hydroxides, while hydrogen being evolved copiously at a cathode is vented from the electrochemical cell. In the conventional process, subsequent separation of the fine precipitate has proved to be difficult and inefficient.

In the improved process, the electrochemical operation is conducted in a manner permitting a much more efficient and less expensive precipitate recovery operation. That is, the electrochemical operation is conducted under an evolved hydrogen partial pressure exceeding atmospheric pressure. As a result, most of the evolved hydrogen is entrained as bubbles in the blowdown in the cell.

The resulting hydrogen rich blowdown is introduced to a vented chamber, where the entrained hydrogen combines with the precipitate to form a froth which can be separated by conventional techniques. In addition to the hydrogen, two materials present in most blowdown act as flotation promoters for the precipitate. These are (1) air, with which the blowdown water becomes saturated in the course of normal cooling tower operation; and (2) surfactants which commonly are added to cooling tower recirculating water systems to inhibit the growth of certain organisms or prevent the deposition of insoluble particulates.

COPPER

In the manufacture of printed circuit or wiring boards, the print-etch process is most generally employed. Such a process starts with a nonconductive substrate, such as of phenolic resin, phenol fiber, ceramic or some other electrically insulating material, having a thin layer of copper either electroplated or applied as an adhesive-backed foil on one or both sides of the substrate, depending on whether single or double-sided circuitry is desired. Copper-clad, phenolic resin substrates have become the most extensively used printed wiring substrates to date, primarily because of physical, electrical and chemical inertness characteristics, as well as cost.

The thin layer(s) of copper is normally selectively coated with a protective etch resist to form a positive pattern which conforms to the desired printed circuit configuration. The copper not covered by the resist (negative pattern) is then chemically removed by immersing the circuit board in a suitable etchant which chemically reduces the exposed copper. Removal of the resist is accomplished by either a vapor degreasing or a solvent cleaning operation. This yields the final desired conductive printed circuit pattern. As a final circuit processing step, a thin layer of gold is normally plated on the terminal ends or fingers of the circuit.

While the print-etch process readily lends itself to the mass production of printed circuits, primarily by taking advantage of reliable and precision photo printing and engraving techniques, there is a serious collateral disadvantage that arises when utilizing this process, namely, the problem involved in disposing of the spent etchant.

The etchant most commonly employed heretofore has been a ferric chloride-hydrochloric acid solution. Such an etchant has a number of disadvantages, the most important of which is that it is not capable of being readily regenerated and, as such, normally necessitates continuous disposal of the spent solution.

While certain solvent extraction and copper salt reducing techniques have been proposed for use in the regeneration of spent ferric chloride solutions and the recovery of dissolved copper therein heretofore, they involve systems which are quite complex, expensive, and entail careful handling of certain of the spent constituents and critical control of a number of sensitive operating parameters. Even then, a substantial portion of the original etchant, separated during the extraction process, must be disposed of, with considerable expense being involved in both rejuvenating the spent etchant with additives and in the disposal of that portion not reusable.

Removal from Water

A process developed by *G.D. Parikh and W.C. Willard; U.S. Patent 3,784,455; January 8, 1974; assigned to Western Electric Company, Incorproated* involves recovering copper metal from an electrolyte and/or an etchant on specially constructed and configured cathodes. The metal deposits advantageously may be subsequently removed from the cathodes with no mechanical or chemical cathodic stripping operation being required.

In a more specific application, the process relates to a continuous cupric chloride-hydrochloric acid etching system for use in recovering etched copper from articles, such as printed circuit boards, wherein the spent etchant is continuously, or intermittently, electrolytically regenerated and etched copper simultaneously recovered as loose, spongy, fine grain deposits on elongated cathodes preferably having arcuate or curvilinear plating surface profiles. The recovered copper deposits advantageously may be readily removed from the cathodes by passing them through a water spray.

A process developed by *E.H. Newton, J.M. Ketteringham, J.L. Sienczyk and C.M. Isaacson; U.S. Patent 3,783,113; January 1, 1974; assigned to The Shipley Company, Inc.* is a process for extracting copper as metal from a used etchant solution containing complexed cupric ions as an oxidant while simultaneously regenerating the etchant for further use. The etchant treated may have a pH from below 0 to 13, but preferably has a pH of at least 3 and more preferably has a pH between about 4 and 10. The process comprises electrowinning a portion of the copper from solution under conditions effective for electrowinning but not etching.

These conditions include a substantial freedom from oxygen in the vicinity of the cathode, high current density, a substantial freedom from solution agitation during the winning operation and preferably for good efficiency, low solution temperature at the interface of the solution and the cathode, typically below 140°F. The process is economical because in the preferred embodiment, a portion only of the copper in solution is removed, the remaining copper being left in solution and available as a source of cupric ion for reuse of the etchant. The process is an important contribution to pollution abatement efforts as it eliminates the need for dumping copper and other wastes resulting from an etching operation.

A process developed by *D.E. Garrett and J.P. McKaveney; U.S. Patent 3,988,221; October 26, 1976; assigned to Occidental Petroleum Corporation* is one in which a cathodic bed of essentially inert, low cost silicon metal alloys in particulate form is used for the removal of ionic impurities such as copper from aqueous solutions by electrolytic deposition. The alloys used are substantially inert to the strong acid used to regenerate the alloy surface and remove the deposited metals.

A process developed by *C.H. Roy; U.S. Patent 3,816,306; June 11, 1974* is one in which the copper content of aqueous effluent containing the rinsings from the etching of copper containing substrates is reduced to less than about 5 ppm by the steps of: (a) adjusting the pH of the effluent to no more than about 4; (b) providing in the effluent, while vigorously agitating, hydrogen peroxide and a water-soluble compound such as calcium chloride; and (c) adding to the reaction mixture as required an alkaline material such as lime or a water-soluble salt of carbonic acid, in an amount effective to adjust the pH to at least about 8.

Upon reaching pH 8 in step (c), the copper begins to precipitate from the reaction mixture, probably in the form of copper carbonates, hydroxides or hydrated oxides, leaving a supernatant from which copper and many other metal compounds have been substantially eliminated. The precipitation can be accelerated by the addition of known flocculating materials.

A process developed by *J. Hulsebos; U.S. Patent 3,900,314; August 19, 1975; assigned to Cities Service Oil Company* involves recovering copper values from acidic solutions containing dissolved copper. The process and apparatus comprises reacting in an intake line and feed pump powdered iron in the size range of up to -40 mesh with the solution to displace the copper therefrom as metallic copper, and separating the copper from the solution by settling the metallic copper in the bottom of a vessel known as a polishing tower. Preferably, the solution is clarified by passing it upwardly through a bed of scrap iron, after which the solution is centrifuged to remove particles therefrom.

Figure 29 shows a suitable form of apparatus for the conduct of the process. The copper is precipitated according to the equation:

$$Fe + Cu^{++} \longrightarrow Cu\downarrow + Fe^{++}$$

Excess acid in the solution also reacts with the iron thus aiding in neutralization of the solution. Powdered iron in the size range of up to -40 mesh on the U.S. Sieve Scale is mixed with the copper bearing solution in amounts up to the stoichiometrically required quantity in an extremely turbulent mixing zone. For instance, as shown in the drawing the powdered iron is introduced into a hopper or feeder **12** from a source, not shown, and metered through pipe **14** to a feed pipe **16** through which the copper bearing solution enters the system.

The feed pipe is connected to the intake of a pump **18** which serves to pump the slurry of solution and powdered iron into a reaction pipe **20** connected to the outlet of the pump. The reaction pipe is any suitably turbulent reaction zone wherein the copper bearing solution and the iron powder are rapidly and intimately mixed so that the reaction between the iron powder and the dissolved copper is substantially complete within the reaction pipe.

FIGURE 29: PROCESS FOR RECOVERING DISSOLVED COPPER FROM SOLUTIONS CONTAINING COPPER

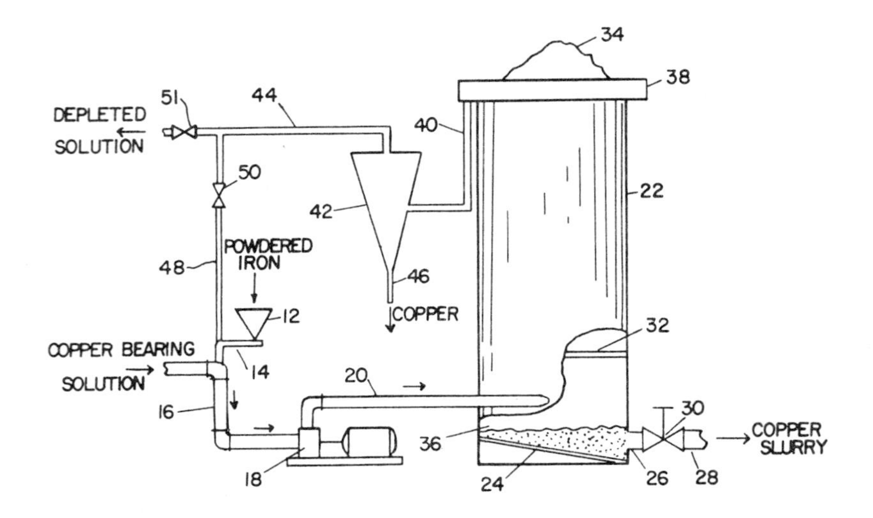

Source: U.S. Patent 3,900,314

The outlet of the reaction pipe is connected tangentially to the bottom of a polishing tower **22**. As shown, the reaction pipe connects to the side of the tower in a tangential position so that the reaction products and fluid enter the tower circumferentially to reduce the flow velocity of the solution containing metallic copper and allow metallic copper to settle into the bottom of the polishing tower. This tower is a large volume cylindrical vessel which will flow toward a discharge port **26** located in the vessel side adjacent the low end of the tower bottom. A discharge pipe **28** is connected to the discharge port and serves to withdraw the settled copper slurry from the tower. A valve **30** is mounted in the discharge pipe to regulate the level of the copper slurry in the tower.

The reaction pipe is attached to the tower at an elevated level several feet above the upper level of the inclined bottom and opens tangentially into the tower. At a spaced distance above the inlet, a support grid **32** is horizontally mounted within the tower and serves to support a packed mass of scrap iron **34**. The mass of scrap iron may be for instance detinned, shredded steel cans. In terms of function, the volumetric space within the tower and below the support grid constitutes a settling zone **36** for copper particles precipitated out of solution in the reaction pipe. The reaction product from the reaction pipe is therefore tangentially introduced into the settling zone, to rapidly dissipate the flow velocity of the fluid and thereby allow a maximum amount of copper particles to settle by both centrifugal force and gravity to the bottom of the zone.

The fluid leaves the settling zone by passing upwardly through the support grid into the volume of scrap iron and is withdrawn from the tower via an overflow trough **38** surrounding the rim of the tower. The scrap iron serves the dual purpose of acting as reactant and as a filter, so that any copper particles entrained in the upwardly flowing solution are trapped or disengaged from the fluid before the fluid is withdrawn as overflow. The iron will react with any dissolved copper still present in the solution to further recover copper and neutralize the solution.

An overflow withdrawal pipe **40** is connected at one point to the overflow trough **38** and at its downstream end to a cyclone separator **42**. The cyclone separator serves to remove substantially all the remaining solid particles entrained in the fluid, which fluid is then discharged overhead through a fluid discharge pipe **44**. The solids separated from the fluids in the cyclone are discharged through pipe **46** and commingled with cemented copper recovered elsewhere, if desired. A recycle pipe **48** connects the discharge pipe **44** to the feed pipe **16** from the feed hopper. A valve **50** is mounted in pipe **48** and a valve **51** is mounted in pipe **44** to monitor and control recycled liquid. The recycle rate of liquid to the feed of copper bearing solution is no more than 1:1; however, it is preferred that there be as little recycle of liquid as is possible.

A process developed by *E.A. Tomic, U.S. Patent 3,196,107; July 20, 1965; assigned to Du Pont* takes note of the fact that as little as 2 to 4 micrograms of copper per liter of aqueous solution in streams flowing through aluminum processing equipment causes extensive corrosion damage. Thus, a process for the removal of copper from nonaqueous and aqueous solutions employing a polymeric agent especially selective for copper has been proposed which employs a chelating polymer having a repeating thiosemicarbazide unit of the structural formula:

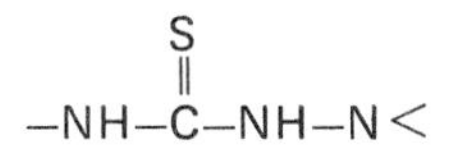

A process developed by *L.E. Lancy; U.S. Patent 3,218,254; November 16, 1965; assigned to Lancy Laboratories, Inc.* for treating ammoniacal copper solutions is nonelectrolytic and simply involves adding an excess of caustic soda to the toxic solution and heating the solution, to convert the ammonium and copper content into ammonia gas and copper oxide, and to drive off the ammonia gas and precipitate out or remove the copper oxide from the solution to form a copper-free and ammonia-free supernatant liquid.

A process developed by *E.B. Saubestre; U.S. Patent 3,666,447; May 30, 1972; assigned to Enthone, Inc.* involves the removal of substantially all copper from a substantially cyanide-free alkaline waste solution containing ionic copper. A reducing agent, for example, formaldehyde, for the ionic copper is added to the solution if not already present therein, and usually a complexing agent for the ionic copper. The process provides for the recovery of marketable zero-valent metallic copper.

The alkaline waste solution is contacted with a material or metal catalytic to the reduction of the ionic copper by the reducing agent to zero-valent metallic copper, and the catalytic contacting continued until the desired amount of copper is precipitated as zero-valent metallic copper. The metallic copper precipitate is then separated, for instance, by filtration, from the solution. Exemplary of the catalytic metals for use in catalyzing the reduction of the ionic copper to zero-valent metallic copper are palladium, platinum, gold and silver.

See Adipic Acid Process Effluents (U.S. Patent 3,673,068)

CORN MILLING EFFLUENTS

Odorous organic compounds are often produced during, or as a consequence of, the processing of organic substances. Where liquids are present, or are employed during processing, the resulting products may be subjected to drying operations in order to drive off undesired fluid residues. This can result in gaseous exhausts that carry odorous organic compounds.

Thus in the wet milling process for the recovery of various products from corn, such as corn oil, gluten and starch, kernels of corn are initially steeped in water. Sulfur dioxide gas is added, and the kernels are subjected to a coarse grinding and pulping operation. This allows the germ of the kernels, from which corn oil is extracted, to be separated from the remainder of the pulped corn. The latter is then subjected to further stages of processing to recover such products as gluten and starch. Since each of the products has a residual wetness, it is generally subjected to a drying operation. This produces a gaseous exhaust which contains a variety of odorous organic contaminants. Typical contaminants associated with the wet milling process are organic acids, aldehydes, mercaptans and possibly amines.

Removal from Air

In order to abate the odors caused by the organic contaminants in a gaseous stream, it has been common practice to use activated carbon, clay or molecular sieves to absorb the odorous substances. This technique has the disadvantage that the absorptive materials become saturated in due course and require replacement. In addition, these materials present disposal problems of their own after they have become saturated with odorous contaminants.

Another technique for odor abatement in the case of organic contaminants has involved the scrubbing of the gaseous stream with water or some other suitable solvent. This technique also poses the problem of the need for replacing the solvent after it becomes laden with organic contaminants. There is also a disposal problem. Another technique for dealing with odorous organic contaminants in gaseous streams has involved incineration. This technique has the disadvantage of requiring large quantities of fuel at the point of odor abatement.

A process developed by *L.M. Adams and H.F. Hamil; U.S. Patent 3,875,034; April 1, 1975; assigned to Corn Refiners Association, Inc.* is one in which the gaseous stream is subjected to a silent electric discharge. This brings about at least the partial decomposition of the contaminants to an odor-abated form. The silent electric discharge is established in a reaction chamber to which the gaseous stream is applied.

Figure 30 shows such an apparatus. The system **10** employs a reaction unit **20** containing a reactor **30** to which an odorous gaseous stream is applied from a source **40**. The source may be a gaseous exhaust at any of various stages in the processing of organic substances, including the various driers used in the recovery of products resulting from the wet milling process. Alternatively the source may be a gaseous stream into which one or more liquid organic contaminants have been injected.

FIGURE 30: APPARATUS FOR ODOR ABATEMENT IN GASES FROM CORN MILLING PROCESSES

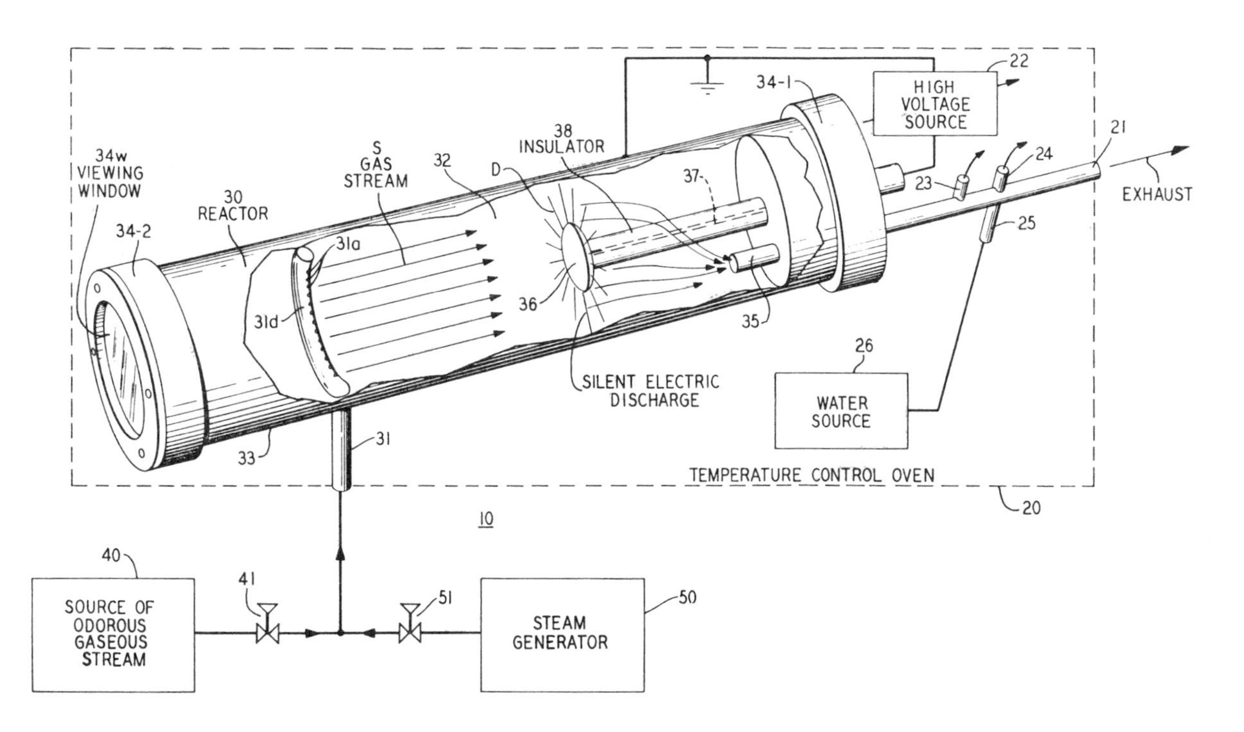

Source: U.S. Patent 3,875,034

In the reactor **30**, the gaseous stream is subjected to a silent electric discharge by which odorous organic compounds of the stream become at least partially decomposed, so that the residual gaseous stream discharged at a vent **21** of the reactor is odor-abated. The term silent electric discharge is used in the sense of any electron discharge between two electrodes short of the point of arcing.

From the source **40**, the odorous stream is applied through a control valve **41** to an inlet **31** of the reactor. After entering a chamber **32** of the reactor, the inlet takes the form of a semicylindrical tube **31d** with a set of apertures **31a**. The latter promote the uniform flow of gaseous substreams **S** into the chamber to the vicinity of an end cap **34-1** from the vicinity of an end cap **34-2**.

The substreams pass axially along the interior of the chamber and converge at an outlet tube **35**. In so doing, they pass through a silent electric discharge **D** that is established between an interior electrode **36** and the shell **33**. The discharge subjects the odorous organic compounds of the substream to electron bombardment and brings about at least their partial decomposition. In the case of gaseous streams from a wet milling process, the odorous organic compounds become highly reactive free radicals or ions which decompose to odor-abated substances.

As a result of the decomposition of the odorous organic contaminants, the gas stream that emerges from the vent is considerably abated in odor. As shown, the interior electrode is in the form of a disk with a relatively sharp outer periphery to facilitate the formation of the silent electric discharge. A conducting rod **37** extends into the chamber and is affixed to the disk electrode **36** in order to energize it with respect to the shell. The conducting rod is covered by an insulator **38** to confine the silent electric discharge to a precisely defined region extending from the periphery of the disk electrode to the outer shell. It is to be understood that the electrode structure is merely illustrative.

The electrodes **36** and **33** are energized from a variable, high voltage source **22**, with the shell taken as a point of reference or ground potential. The voltage from the source **22** may be positive or negative and be directly applied with respect to ground, or the voltage may alternate in polarity.

For the most effective operation of the reactor, i.e., the decomposition of the contaminants carried by the gaseous stream, the voltage of the source **22** is adjusted to a level where the discharge is just below the point of arcing between the electrodes. This provides the greatest density of electron discharge. With other conditions held constant this provides the greatest incidence of activation. During the adjustment of the voltage, it is initially increased from zero to a relatively low magnitude where a dark corona discharge is produced. This discharge is not visible to the eye, but its presence is confirmed by the flow of current in the high voltage source **22**. As the voltage of the source **22** is increased further, a glow corona discharge appears.

This discharge is in the visible range of the electromagnetic spectrum and is viewable through a window **34w** in the end cap **34-2** at the inlet end of the reactor. A still further increase in the voltage of the source **22** results in a brush discharge for which streamers or brushes extend between the center electrode and the shell. A further increase in voltage produces an arc discharge. The latter represents a breakdown over a localized path between the center electrode and the shell. Such a breakdown is undesirable because it prevents the gaseous stream from being subjected to a uniform discharge. The brush discharge, on the other hand, is not only relatively uniform, it produces a relatively high density of discharge electrons for initiating the decomposition of odorous organic compounds.

The viewing window **34w** in the inlet cap **34-2** permits visual monitoring of the adjustment of the variable voltage source **22** as the discharge passes from the glow to the brush condition. Because of the semicircular form of the inlet **31d** within the chamber **32** there is an unobstructed view of the discharge **D** through the window. However, the governing consideration with respect to the form of the inlet is that the substreams of the odorous gas be suitably emitted into the chamber, and it will be understood that numerous other forms of inlet may be employed. The end caps **34-1** and **34-2** of the chamber are removable to permit access to the interior.

In a test embodiment of the reactor **30**, the end caps were fabricated of a relatively nonodor-releasing metal such as aluminum to prevent the reactor from itself augmenting the odor constituents in the gas stream. Alternatively, if the end caps are of a porous, absorptive material, such as Teflon, they can contribute to odor abatement by absorbing some of the contaminants in the gaseous stream.

In the case of the wet milling process, the organic contaminants encountered in the gaseous exhaust from the various dryers are organic acids, aldehydes, mercaptans and possibly amines. Representative acids are butyric and n-caproic. Representative aldehydes include isobutyraldehyde and furfural. Other representative aldehydes are pyruvaldehyde and methional, which is an aldehyde containing sulfur. A representative mercaptan is methyl mercaptan.

The reactor is shown placed within a temperature control oven represented by a dashed outline. While the decomposition of odorous organic compounds using a silent electric discharge may be practiced at ambient temperatures, there can be situations where the decomposition is enhanced by elevating the temperature of the gaseous stream. The temperature control oven serves that purpose. Further, a heater may be applied at the inlet **31** for additional temperature control. The temperature of the gaseous stream within the chamber is indicated by a thermocouple serving as a dry-bulb thermometer in a housing on the vent **21**.

In addition to the gaseous stream from the source **40**, the inlet is able to receive an input from a steam generator **50**, by the way of a valve **51**. The steam generator is used for controlling the humidity of the stream applied to the chamber of the reactor. Where the gaseous stream is air, an increase in humidity reduces the incidental production of ozone, which is an acrid by-product of the discharge. The amount of vapor added to the gas stream from the generator is controlled by a valve **51**. An indication of the humidity of the gas stream is given by a thermocouple serving as a wet-bulb thermometer in a housing **24** of the vent tube **21**. The wet-bulb thermometer is moistened in conventional fashion through housing **25** from a water source **26**.

Other factors that affect the extent of odor abatement and decomposition are: (1) the residence time of the gaseous stream in the discharge and (2) the power (wattage) of the discharge. In general, as residence time increases, there is an increase in the extent of decomposition of the odorous organic substances in the gaseous stream. Test results have further demonstrated that an increase in power in the electric discharge has a corresponding effect on the percentage of decomposition.

CUPOLA FURNACE EMISSIONS

In cupola furnace systems, draft effecting arrangements have been employed to control the varying melting processes within the furnace and to exhaust furnace gases through particle separating and collecting apparatus to clean the exhaust gases prior to emission thereof into the atmosphere. Such prior art systems may include a cap structure disposed atop the furnace flue, and a suction fan or the like disposed downstream from the cap structure and operated to maintain a desired flow of air through the furnace charge door and into the flue and cap and then into the separator apparatus and exhaust stack.

In certain instances, the prior art system includes the provision of a damper controlled opening in the cap to permit the exhaust of furnace gases directly to atmosphere or to permit induced flow of air into the cap structure by the exhaust fan for cooling the exhaust gases as they travel along the path to the separator apparatus and the exhaust stack.

Cupola emission control systems of the above character fail to provide satisfactory furnace process control commensurate with desirable furnace gas emission control. Moreover, the induced cooling air arrangements necessitate the use of expensive high-alloy stainless or refractory-lined takeoff ducting from the cap assembly because of the inability of these systems to otherwise accommodate the extremely high emission gas temperatures realized during certain periods of furnace operation.

Removal from Air

A device developed by *T.D. Barnes, Jr.; U.S. Patent 3,744,215; July 10, 1973; assigned to Don Barnes Ltd., Canada* is a cupola emission control system which includes a cylindrical cap assembly adapted to be mounted on top of a cupola flue so that gases from the cupola enter the assembly through an opening in the bottom thereof. The top of the cap assembly is defined by a pair of doors adapted to be opened to open the cap to atmosphere.

Opposed side wall portions of the cap are provided with openings, one of which is an ambient air opening and the other of which is a mixture outlet opening. Fan means is disposed in a ductway leading to the air inlet opening so that ambient air under positive pressure can be introduced into the cap for dilution and cooling in the cap of gases from the furnace. The outlet opening is connected to a pair of separator units which are serially arranged between the cap housing and an exhaust stack for the system. An exhaust fan is associated with the output side of the second separator leading to the stack in order to induce flow of furnace gases through the system to the stack.

An apparatus developed by *W.R. Allen, Jr.; U.S. Patent 3,773,308; November 20, 1973; assigned to Research-Cottrell, Inc.* is a gas quencher and scrubbing apparatus for use in operative association with a plurality of furnace cupolas or other sources of hot, particle-laden gases. The quencher has a three-fold purpose, namely, (1) to quench or lower the temperature of the very hot discharge gases from a plurality of cupolas; (2) to separate a large portion of the particles from the gas stream; and (3) to provide a water seal damper to either one of two gas sources so that one cupola can operate while another is shut down.

Such an apparatus is shown in Figure 31. Water drained from valve controlled drains **84, 86** located in the bottom of the quencher **32**, is carried by conduit **90** to a settling or separation tank **92** where the heavier particulate matter falls to the bottom, is progressively drained away through drain **94** and disposed of. A float-valve **96** monitors the level of the water in the tank permitting introduction of additional water when required.

FIGURE 31: GAS QUENCHER-SCRUBBER AND WATER SEAL APPARATUS FOR CUPOLA FURNACE EXIT GASES

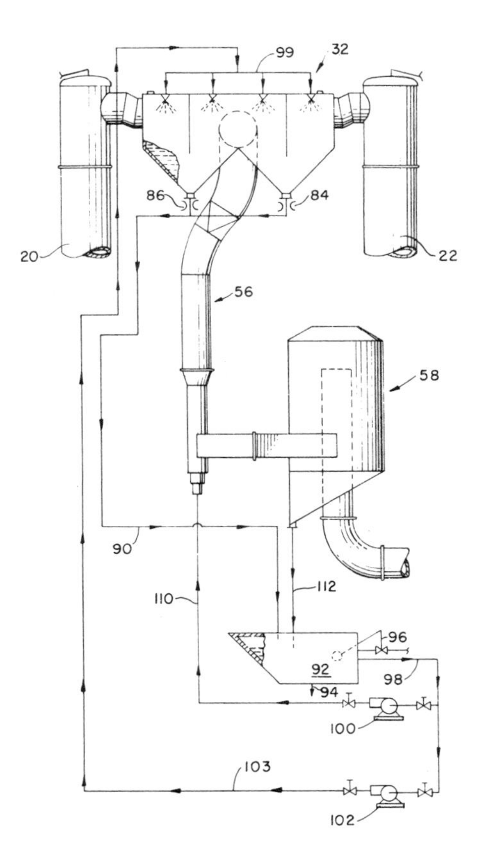

Source: U.S. Patent 3,773,308

Water for recirculation is drawn off through conduit **98** and is returned to the headers **99** on top of the quencher by pump **102** and conduit **103**. Water is also recirculated to the flooded disc scrubber **56** by conduit **110** and pump **100** while the scrubber water is returned to the settling tank **92** through conduit **112** from the mist-eliminator **58**. If necessary, the chemical nature of the water may be continuously monitored and chemicals added to maintain the pH at the desired level.

This apparatus will collect about 50% of the particulate matter moving in the gas stream of about 60 fps and lower the temperature of the gas from 1400° to a range of 200° to 400°F as the gas leaves the outlet opening. The performance of the quencher connected with the flooded disc scrubber system is increased to remove about 95 to 99% of the total particulate matter in the gas stream.

A process developed by *T.C. Sunter; U.S. Patent 3,972,518; August 3, 1976; assigned to Fuller Company* involves conditioning the gases discharged from a cupola. The apparatus consists of three flow connected vessels comprising first the cupola with improved gas conditioning means therein, secondly, a water spray tower and thirdly, a baghouse dust collector. The improved gas conditioning method and apparatus comprises a movably mounted conduit and nozzle for spraying cooling water through the cupola charge door during the burndown period of cupola operation when the bed gas is not used to preheat the new charges in the cupola.

The cooling water is used instead of large quantities of dilution air to temper the discharge gas at this point. Therefore, by the improved method, during burndown the temperature of the discharged gas can be maintained at slightly above the normal cupola operational temperature without requiring discharge ducts and dust collection capacity for the additional dilution air. The elimination of the additional dilution air results in a corresponding decrease in the size and expense of the discharge ducts and the dust collecting system.

CYANIDES

There is a widespread problem of cyanide pollution in industrial waste streams, particularly with chemical, steel and electroplating industries. These streams are aqueous and generally contain large quantities of inorganic solids (1,000 to 10,000 ppm); oxidizable organics other than cyanides; and relatively low levels of cyanide (<2,000 ppm). However, these cyanide levels are still well above all pollution standards which generally allow 1.0 ppm cyanide or less.

Industry is making major and expensive investments to minimize stream pollution in this area, as governmental authorities adopt and enforce stricter water quality standards. No methods are known to be able to reduce generally the cyanide levels of such streams to extremely low levels.

Removal from Water

The classical and most common approach is to oxidatively destroy the cyanide by the use of alkaline chlorination. However, this method has several limitations because of:

(1) Inability to perform this oxidation upon complexed cyanides already existing in water, or being formed upon making alkaline

(in either case, indigenous transition metal ion presence will be the causal factor);

(2) The inherently inefficient use of chemicals because of theoretical considerations running as high as $1/lb of cyanide removed, i.e., 5½ equivalents Cl_2 required for 1 equivalent of CN, and 4 equivalents NaOH required for 1 equivalent of CN;

(3) The general need for even more chemical usage because of the nonselective nature of oxidation (other stream constituents both organic and inorganic must be oxidized also); and

(4) The distinct possibility of creating other toxic chemicals such as cyanates and chlorinated hydrocarbons.

The only other commercially acceptable technique presently used is the complete removal of all ions by ion exchange. However, this technique is not generally applicable because of its low capacity. The large salt background in most streams preclude its use (commercially again) because of the large chemical usage for regeneration purposes.

Furthermore, the overall accomplishment is the concentrations of all solids; the percentage of cyanide and the form of the cyanide is identical to the original stream. With streams of high total solids, concentration factors greater than 10 are impossible. The cyanide regenerated from the ion exchange system will still be identical in nature to that originally in the waste stream, and thus still present a disposal problem.

A process developed by *W. Fries; U.S. Patent 3,788,983; January 29, 1974; assigned to Rohm and Haas Company* involves the detoxification of cyanide-containing effluents by a treatment with a metallic ion, such as ferrous iron, which forms stable anionic complexes of cyanide, and the complete and selective removal of the complexes by an anion exchange resin.

Sodium cyanide is used in the electroplating industry as one of the components in the plating bath. During the plating operation, this material is converted into various cyanide waste products which are extremely toxic. Therefore, the disposal of these toxic waste products creates a problem. It is desirable to convert such cyanide waste products back into sodium cyanide which can be used in the plating operation and thus eliminate this toxic waste disposal problem.

A process developed by *L.F. Scott; U.S. Patent 3,744,977; July 10, 1973; assigned to Franke Plating Works, Inc.* utilizes an apparatus for converting cyanide wastes into sodium cyanide in which a heat exchanger is interconnected between a plurality of reaction tanks and a first tower adapted to hold a charge of sodium hydroxide. The tower is interconnected to one compartment of a chambered tank, and a second compartment in the tank is connected to a second tower adapted to hold a charge of sodium hydroxide. A pumping assembly is provided for withdrawing the sodium cyanide from the first compartment and for recycling the reaction mixture through the towers and the compartmented tank.

Such an apparatus is shown in Figure 32. As shown, the cyanide wastes to be treated **10** are emptied into a holding tank **12**. The wastes are pumped as by a pump **14**, from the tank through a pipe **15** containing a flow meter control **17**

FIGURE 32: PROCESS FOR CONVERTING AQUEOUS CYANIDE WASTES INTO SODIUM CYANIDE SUITABLE FOR REUSE

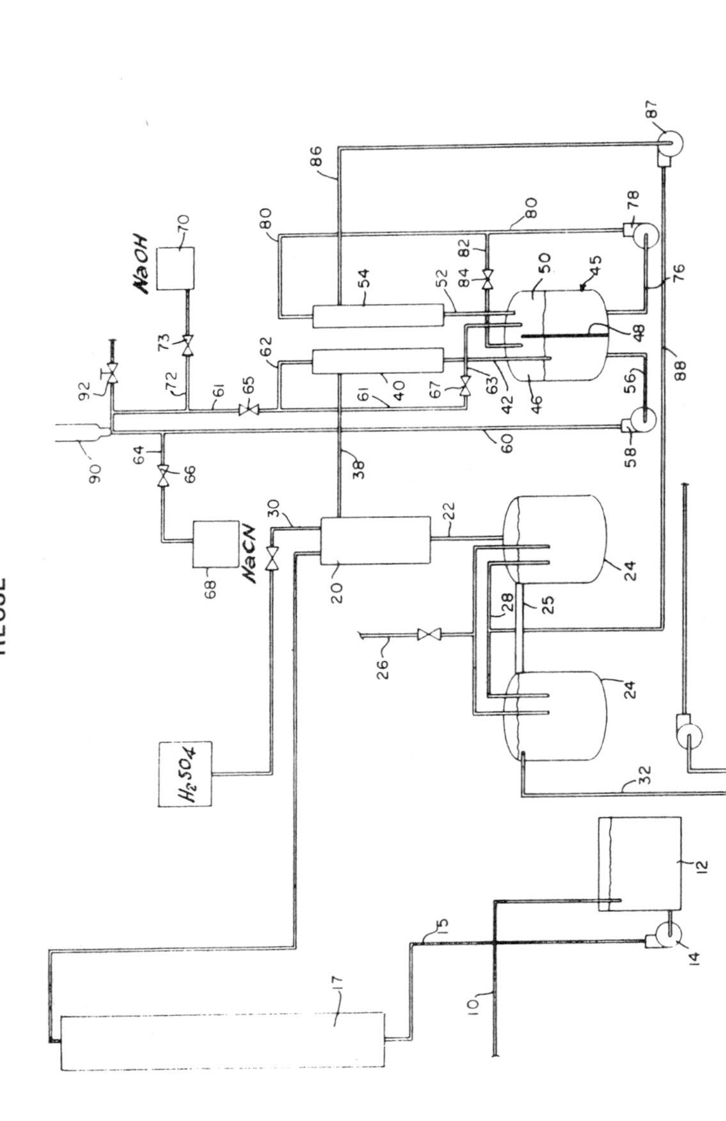

Source: U.S. Patent 3,744,977

into a steam heated heat exchanger **20** where the wastes are preheated to a temperature of about 185°F. The wastes, while still in liquid form, are discharged from the heat exchanger through a pipe **22** into a plurality of reaction tanks **24.**

The tanks are connected in series by a conduit **25** having a diameter and a position of connection with respect to the tanks such that there is a free flow of both gas and liquid between the tanks through the conduit. A steam line **26** is also connected to the tanks for maintaining them at the desired temperature. As will become apparent hereinafter, pipes **28** are also connected to the tanks for supplying gaseous agitation thereto.

The tanks contain sulfuric acid which is supplied thereto through a pipe **30** connected between a source of the acid and the heat exchanger. The cyanide wastes and sulfuric acid produce a reaction mixture in the tanks having a pH of from about 2 to about 4. The reaction is accelerated by the tanks being maintained at a temperature of about 210°F, the tanks being heated by the steam introduced into the tanks through the line **26**. The reaction of the acid converting the wastes into hydrogen cyanide is completed in about 2 hours, and with the elevated temperatures in the tanks, hydrogen cyanide gas is evolved which flows through the conduit **25** and then through the pipe **22** to the heat exchanger. The liquid overflow in the tanks is removed through pipe **32** connected to a sump **35**.

The hydrogen cyanide gas, together with water vapor, exits the tanks through the conduit **22** and the heat exchanger into a pipe **38** connected to a reaction tower **40** containing sodium hydroxide. The tower is connected by a pipe **42** into a chambered tank **45**. This tank has a first compartment **46** containing a solution of sodium hydroxide, the incoming hydrogen cyanide, and sodium cyanide produced by the sodium hydroxide-cyanide reaction. The unreacted hydrogen cyanide gas in this compartment escapes over a baffle **48** in this tank into a second compartment **50** and exits this compartment through a pipe **52** connected to a second tower **54** of sodium hydroxide.

The compartment **46** has an outlet **56** connected to a pump **58**. The outlet **60** of the pump **58** is connected to a return line **61** provided with a first connection **62** to the top of tower **40** and a second connection **63** to the compartment **50**. In order to control the flow through the line **61**, a valve **65** is connected in the line upstream of the connection **62** and a valve **67** is connected in the connection **63**. A discharge line **64** provided with a valve **66** is connected between the outlet **56** and a sodium cyanide collecting tank **68** for withdrawing the sodium cyanide from the tank **45** as will be more fully described hereinafter.

In order to introduce the sodium hydroxide into the system a tank **70** containing a saturated solution of sodium hydroxide is connected to the return line by a pipe **72** having a valve **73**. Thus, when the system is in operation, the valves **66, 73** and **67** are closed and the valve **65** is opened to permit the pump **58** to recirculate the reaction mixture in the compartment **46** as a spray through the tower **40** in the same flow direction as the cyanide vapors entering the tower and compartment.

The compartment **50** also has an outlet **76** connected to a pump **78** having its outlet **80** connected to the top of the tower **54**. A by-pass **82** provided with a valve **84** is connected to the outlet **80** and empties into the compartment **46.** Thus, with the valve **84** closed, the solution in the compartment **50** consisting primarily of the saturated sodium hydroxide solution is pumped as a spray through the pump **78** into the top of the tower **54**. In this manner, the sodium hydroxide flows through this tower in a countercurrent flow to any unreacted cyanide vapors moving up through this tower. The sodium cyanide produced in this tower drops back down through the tower into the compartment **50**.

Any noncondensable gases in the tower exit to this tower through a pipe **86** connected to a vacuum pump **87** and are returned to the reaction tanks **24** through a pipe **88** connected to the pipe **28** to thus agitate the reaction mixture in these tanks. As will be apparent, the vacuum pump is connected to the reaction tanks through the heat exchanger **20**, pipe **38**, tank **45**, and pipe **86** to thus place the tanks **24** and **45** under a reduced pressure of about 5" of mercury and pull the cyanide gas from the tanks **24** through the towers **40** and **54** and tank **45**.

An air cushion **90** and petcock **92** are interposed between the outlet **60** and return line **61**. Samples of the recirculating reaction mixture in the tower **40** and compartment **46** can be withdrawn through the petcock. When it has been determined, as by laboratory analysis of the samples, that the conversion of the hydrogen cyanide into sodium cyanide has been completed, the resulting sodium cyanide is withdrawn from the compartment **46** by closing the valves **65** and **73** and opening the valve **66** to the collecting tank **68**.

When this compartment is emptied, the valve **84** is opened to permit the pump **78** to pump the solution in the compartment **50** into the compartment **46**. The compartment **50** is refilled with the saturated sodium hydroxide solution by deenergizing this pump and opening the valves **73** and **67** to permit the saturated sodium hydroxide solution from the tank **70** to flow into the compartment **50**.

As thus will be apparent, the reaction system consisting of the tower **40** and compartment **46** together with their pump **58** can operate independently of, or in combination with, the reaction system consisting of the tower **54** and compartment **50** together with their pump **78**, and each system can be operated independently of, or in combination with, the reaction tanks. This permits the sodium cyanide to be withdrawn from the system and the compartment to be recharged while the acidification reaction is being carried out in these tanks.

A process developed by *U. Schindewolf; U.S. Patent 3,945,919; March 23, 1976; assigned to Deutsche Gold- und Silber-Scheideanstalt vormals Roessler, Germany* is one in which solid and/or liquid cyanide waste is destroyed under environmentally favorable conditions by adding water to the cyanide and heating the mixture at high temperature and with the use of pressure.

It has long been known to destroy hydrocyanic acid with water to form ammonia and formic acid.

$$KCN + 2H_2O \longrightarrow NH_3 + HCOOK$$

This reaction, however, is extraordinarily slow and at normal pressure in a boiling water solution only takes place with a speed which is not sufficient for the destruction of cyanides on an industrial scale. For example, the complete conversion of 1 gram of KCN in 100 ml of water takes several hours.

The problem is to permit the reaction to take place in an industrially productive manner. This problem has been solved in a surprisingly simple manner by adding water to cyanide, so far as it is not present in aqueous solution and the mixture or solution is heated to high temperatures while using pressure. Thus there can be used 1 to 1,000 parts of water per part of cyanide. The pressure should be from 5 to 100 atm above normal atmospheric pressure.

Perhaps the most common practice for removing cyanide from waste solutions has been the use of an alkali-chlorination treatment as noted above. Another known system employs an electric current to break down the cyanide compounds and the cyanide ion to release free nitrogen and carbon dioxide to the atmosphere. This latter system has not heretofore been economical, however, to reduce the concentration of cyanide to an innocuous level. Rather, electrolysis, as this system is called, is successfully used only to reduce the cyanide concentration and the solution must then be subjected to an alkali-chlorination or other treatment.

A process developed by *J.F. Zievers and C.J. Novotny; U.S. Patent 3,756,932; September 4, 1973; assigned to Industrial Filter & Pump Mfg. Co.* involves concentrating the waste solution to increase the concentration of cyanide to about 2,000 ppm or more, then electrolyzing the solution by passing a direct electric current therethrough to break down the cyanide compounds and cyanide ions and reconcentrating the solution to maintain the concentration of cyanides in the solution through which current is passed at a value of about 2,000 ppm or more.

A process developed by *J. Fischer, H. Knorre and G. Pohl; U.S. Patent 3,970,554; July 20, 1976; assigned to Deutsche Gold- und Silber-Scheideanstalt vormals Roessler, Germany* is one in which wastewater containing cyanides, cyanohydrins and/or organic nitriles is detoxified by adding a peroxide compound in the presence of iodide ion or free iodine, in a given case in the presence of silver ions.

See Acetone Cyanohydrin (U.S. Patent 3,984,314)
See Acrylonitrile Process Effluents (U.S. Patent 3,940,332)
See Coke Oven Emissions (U.S. Patent 3,847,807)
See Hydrogen Cyanide (U.S. Patent 3,935,188), etc.
See Nitriles (U.S. Patent 3,756,947)
See Nitrites (U.S. Patent 3,502,576)

CYANOHYDRINS

See Acetone Cyanohydrin (U.S. Patent 3,984,314)
See Cyanides (U.S. Patent 3,970,554)

CYANURIC ACID

Cyanuric acid (either free or in salt form) is present in the waste liquors from processes which manufacture it, and those which convert it into di- and trichlorinated derivatives, widely used as a source of active chlorine in various household detergents and in swimming pools. The discharge of such wastes is expected to be prohibited by pending governmental regulations, and the question arose as to whether it is possible to remove cyanuric acid from such waste liquors in a reasonable length of time by biological means, or whether it is necessary to go to expensive chemical treatment, or to even more expensive evaporation and incineration.

Removal from Water

A process developed by *J. Saldick; U.S. Patent 3,926,795; December 16, 1975; assigned to FMC Corporation* is one in which cyanuric acid is removed from aqueous chemical plant wastes containing it by treatment of the wastes with active bacteria derived from sewage or soils while maintaining relatively anaerobic conditions, supplying nutrients to the system, while preferably holding the pH fairly close to the neutral point (about 5.0 to 8.5) and the temperature close to ambient.

The scheme of operations is shown in Figure 33. A stream of plant waste **10** is fed into a hydrolysis vessel **12**, to which is fed activated sludge from line **16A** along with bionutrient from line **14**. The vessel may be of any convenient design; pilot work has been in a column packed with Raschig rings or Berl saddles. The waste is maintained in contact with the active biomass long enough to complete the hydrolysis; about 12 to 24 hours seem indicated.

FIGURE 33: SCHEME FOR BIOLOGICAL TREATMENT OF PLANT WASTE-WATER STREAMS TO REMOVE CYANURIC ACID

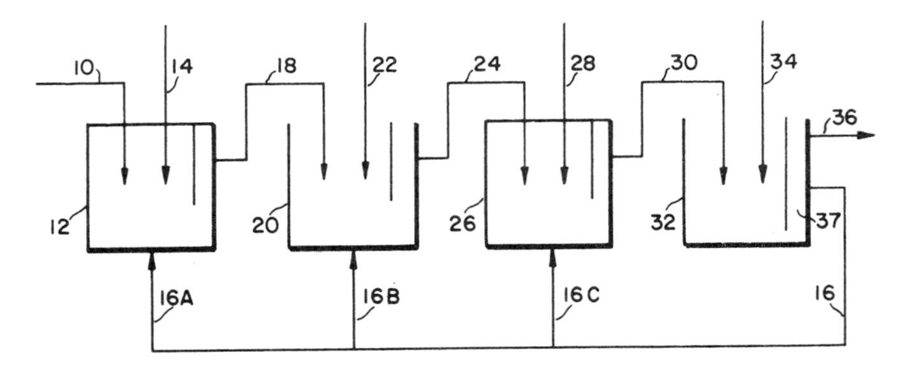

Source: U.S. Patent 3,926,795

The effluent from the vessel goes through line **18** into another vessel **20**, fed with biomass from line **16B** and air from line **22**. This is a standard process for sewage treatment, requiring about 24 hours, to convert the ammonia to nitrates. The effluent from vessel **20** enters vessel **26** through line **24**; biomass from line **16C** and bionutrient from line **28** act on the nitrates to reduce them to nitrogen, again in known manner. This denitrifying reaction takes about 4 hours.

The effluent from vessel **26** passes through line **30** into an aerating unit **32** containing a clarifier **37** where the mixed liquor is activated with air from line **34**, and clarified in section **37**. Clarified purified liquid is discharged to waste through line **36**, and the activated sludge is recycled into line **16** to the feed vessels **12**, **20** and **26**.

CYCLOHEXANE OXIDATION WASTES

In the cyclohexane oxidation to cyclohexanone with oxygen-containing gases, secondary oxidation products form which are present as acid materials in the waste liquor during the separating and refining steps following oxidation. Working off of these by-products and their removal has presented grave problems to the industry.

Removal from Water

It has been attempted in the past to make use of these waste liquors by separating them into their components, esterification, hydrogenation, solvent extraction, conversion of lactones into lactams, etc. However, the results have been most unsatisfactory. Even burning of the wastes is impractical because it is very expensive. The unconverted products cannot be dumped into rivers or lakes because of the ensuing pollution, so that some disposal is necessary.

A process developed by *W. Griehl et al; U.S. Patent 3,523,064; August 4, 1970; assigned to Inventa AG* is a method for the production of cellular protein from the waste liquors obtained in the cyclohexane oxidation. Microorganisms of the family of Pseudomonadaceae convert these waste products to proteins in alkaline medium at ambient temperatures at 24° to 30°C and pH levels of substantially 6.7 to 9.2 within 50 to 90 hours. The process has the twofold advantage of manufacturing a useful product from wastes and preventing water pollution.

DAIRY EFFLUENTS

Dairy effluents consist for the most part of whey or milk serum which is a liquid phase obtained after precipitation from the milk (by the addition of rennet or acid) of casein which is used in the manufacture of cheese. The average composition of milk serum is as follows:

Water	93.5%	Fats	0.3%
Lactose	4.5%	Minerals	0.6%
Nitrogenous materials	0.9%	Lactic acid	0.2%

One of the principal constituents of the nitrogenous materials is the protein lactalbumin.

The logical outlet for dairy effluents is traditionally the nearest watercourse. However, disposal in this way results in changes in both the chemical and microbiological composition of the watercourse and promotes the development of microorganisms releasing acid or toxic products which cause ecologically unacceptable changes including the elimination of fish and modification of flora.

Removal from Water

One method of eliminating milk serum is to use it either directly or after concentration or drying for feeding livestock. The direct use of milk serum necessitates the establishment of a livestock-breeding station (for example a pig farm) in the proximity of the dairy farm.

On the other hand, it is only economical to concentrate on dry milk serum in cases where the quantity of milk serum to be treated exceeds about one million liters per day. In fact, most dairy farms do not produce sufficient quantities of milk serum (30,000 liters per day for an average dairy farm) to make it economical to install treatment installations and, for various reasons, cannot be run adjacent to pig farms.

Another method is to use the milk serum as a manure. In most cases, however, direct disposal is still the most economical solution although it is the most dangerous so far as the environment is concerned. Accordingly, a process for purification before disposal would be extremely desirable, more particularly a microbiological purification process which can be adapted to suit the type of milk serum to be treated.

A process developed by *A. Ullmann and M. Schwartz; U.S. Patent 3,930,028; December 30, 1975; assigned to Agence Nationale de Valorisation de la Recherche, France* involves directly fermenting milk serum in particular with proteolytic bacterial strains of the Enterobacter or Serratia type belonging to the family of Enterobacteriaceae. It is possible by this process to eliminate more than 80% of the dry solids from the milk serum under highly economical conditions.

DEGREASING PROCESS WASTES

There are a number of industrial processes by which halogenated compounds are employed as cleaning or degreasing solvents, or otherwise become admixed with high-boiling-point liquid compounds, such as oils. For example, in the metal processing industry, oil-contaminated metal parts are often degreased through the employment of halogenated compounds, such as trichlorethylene, perchlorethylene, carbon tetrachloride or fluorocarbons, chlorocarbons or chlorinated or fluorinated hydrocarbons. After use, the halogenated compounds become contaminated with waste products and industrial oils from the parts which are degreased and form an oil-halogenated sludge mixture which often contains from 40 to 90% volume of the halocarbon.

In addition, halocarbons are also employed in dry cleaning operations where such halocarbons also can become contaminated with a wide variety of waste products and oils, as well as in other industrial solvent or vapor extraction processes. The oil halocarbon mixtures have, in the past, been discarded as water products; however, the discharge of such material as waste products serves as a pollutant in the waterways of the nation.

It would be desirable to provide a process for the separation of the halocarbons from the contaminating oil and waste products. However, for example, the recovery of trichlorethylene from hydrocarbon oils has not been effective enough to render the discharged oil free of trichlorethylene.

The primary difficulty associated with these methods of recovery and separation, particularly the trichlorethylene, is due, in part, to the heat sensitivity of halocarbons at elevated temperatures. For example, trichlorethylene has a critical decomposition temperature which, when exceeded in the presence of air or moisture, may induce violent explosions or give off toxic gases.

The critical decomposition temperature of trichlorethylene depends on pressure, the amount of oxygen and moisture present, as well as the type and quantity of stabilizers, inhibitors, and other additives present. Typically, the critical decomposition temperature of trichlorethylene is about 265°F at atmospheric pressure. A similar problem concerning heat sensitivity and decomposition temperatures also affects other halocarbons. Due to such decomposition temperatures, any processes for the separation and recovery of the halocarbon from such mixtures must be done employing suitable safety devices and at relatively low temperatures. Such operating restrictions have, in the past, resulted in the employment of such low temperatures that separation of the halocarbon and the oil is not efficient enough, either in single or multiple evaporation systems, to recover or remove the oil or the halocarbon in sufficient quantity and purity to make such processes economical or to provide for the reuse of the halocarbon or oil. Accordingly, there exists the need for a process which would separate and recover simply, economically, and efficiently trichlorethylene hydrocarbon oil waste sludge mixtures.

Removal from Water

A process developed by *J.G. Miserlis and A. Petrou; U.S. Patent 3,803,005; April 9, 1974; assigned to Silresin Chemical Corporation* is a process for removing trichlorethylene solvents from a degreasing plant sludge oil mixture, which process comprises: employing an evaporation two-stage stripping operation to include a directly contacted heat interchange involving recirculation of a liquid oil solvent stream to avoid exceeding the critical decomposition temperature of the trichlorethylene, which process provides for the separation of solvent-free oil containing less than 50 parts per million solvent, the separation and recovery of the trichlorethylene solvent.

Figure 34 shows the arrangement of equipment used in the conduct of the process. Referring to the drawing, the first and second stage distillation columns, respectively, are shown at **10** and **12**. The first stage is divided into upper and lower sections **14** and **16**. The upper section is comprised of a packed heat exchange section and the lower section includes a tray **18** and a liquid holdup section **20**. Feed from storage tank **22** enters a preevaporator **24** under pumped flow control **26**, by pump **28** and leaves the evaporator **24** as a mixed liquid-vapor stream. The steam to the evaporator is controlled by the pressure control valve **30** and the condensate leaves the evaporator through steam trap **32**.

The liquid-vapor stream enters the first stage distillation column **10**, where it is disengaged into a liquid stream and vapor stream. The solvent vapor mixes with steam and trichlorethylene vapor that is coming from the bottom section **16** and which travels upwardly through the packed section **14** of the first distillation column **10**.

The steam and trichlorethylene vapor enter the shell side of a water-cooled shell and tube condenser **34** and the two-phase liquid trichlorethylene-water system enters decanter **36**. The heavier trichlorethylene phase leaves the boot of the decanter **38** as finished 99% plus trichlorethylene product; and the lighter water phase leaves the top of the decanter.

The liquid which disengaged from the vapor-liquid mixture entering section **14** drops onto distillation tray **18** where it blends with a recirculation stream that has drained down from the packed section above it, and by that time is brought up to a suitable temperature. Stripping steam bubbles through the liquid and carries out a substantial quantity of the dissolved trichlorethylene. The liquid then drops through the downcomer into the liquid holdup section **20** of the first distillation column **10**. The liquid in section **20** is then pumped through a shell and tube heat exchanger **40** by pump **42** where a sufficient amount of heat is transferred into it to provide the heat for the first distillation column **10**. The quantity of steam required is controlled by the temperature indicator controller **44**. The bulk of the heated liquid, whose flow is recorded by a flow recorder at **46**, flows through a hand control valve **48** into a nozzle header **50**, where it is sprayed onto the top of the packed section **14** of the first distillation column **10**.

A bleed stream is removed from this column **10** through a liquid level controller **52** and serves as feed for the second stage distillation column **12**. This feed enters the second stage distillation column and drops down through a suitable number of distillation trays **54** or a suitable type of packing. The liquid which has accumulated at the bottom of this column is now stripped of substantially all of the trichlorethylene solvent, and is recirculated by pump **56** through a shell and tube heat exchanger **58** to provide the heat for the second stage distillation column **12**. The quantity of heat transferred is controlled by temperature controller **60**. Oil which has been stripped of trichlorethylene to the

FIGURE 34: METHOD FOR RECOVERY OF TRICHLORETHYLENE FROM OIL WASTE BY PLURAL STAGE DISTILLATION

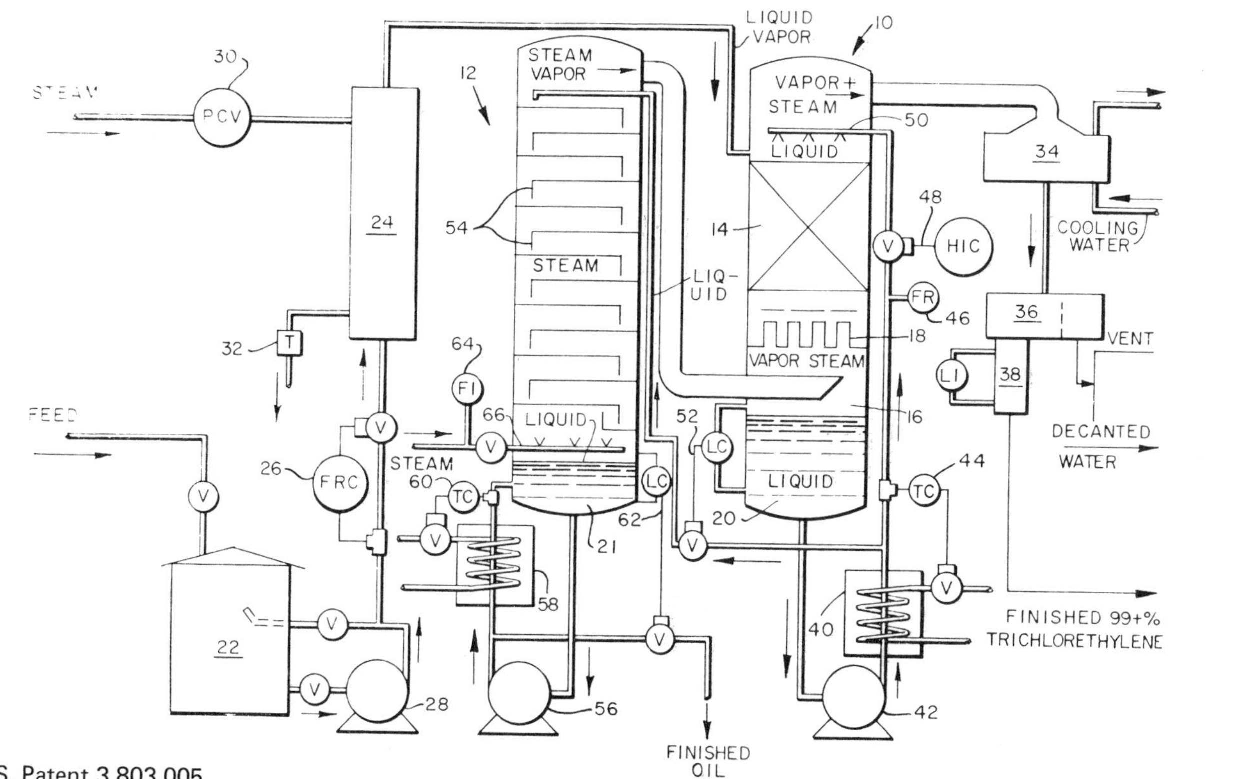

Source: U.S. Patent 3,803,005

required levels is removed from the system through liquid level controller **62**.

Stripping steam flowing through flow indicator **64** enters the bottom of the second stage distillation column **12** through a sparging system **66** and effects the required stripping. The steam and trichlorethylene vapor which is produced travel upwardly countercurrent to the liquid through the second stage distillation column **12**. The steam and trichlorethylene vapor generated leave the top of the second stage distillation column **12** and enter section **16** of the first stage distillation column **10**.

The steam and trichlorethylene vapor bubble through distillation tray **18** of distillation column **10**, reclaiming more trichlorethylene vapor, and meet with the vapors that were generated in evaporator **24** and disengaged from the liquid in the first column. These vapors comprise the stream which leaves the top of this column.

DETERGENTS

The increased use of surface active agents, particularly as detergents and soaps, has created significant pollution difficulties from the point of view of adequate waste disposal. The treatment of fluid drainage containing large amounts of anionic surfactants has been a very serious problem because of the foams which are formed in drainage treating plants and when the drainage is discharged into rivers and because of the toxicity of the drainage to fish.

Removal from Water

One method for removing these surface active agents from drainage and sewage streams has been to inject air into the streams so as to cause foaming whereby they can be removed by absorption. This technique, however, has several serious disadvantages. For one, it requires an undesirably extended period of time for complete removal of the surface active agents from the waste streams, which necessitates the use of large-size treating apparatus and containers. For another, the separated surface active agent solutions usually contain less than 5,000 ppm of the surface active agent so that the solution must be concentrated before it can be reused or further disposed of. This means that additional concentration and storage towers must be supplied.

Still another difficulty with the prior art techniques is the necessity to break the foam in order to accomplish separation. Depending upon the particular surface active agent being recovered, breaking of the foam may be quite a difficult task.

Another method of separating surface active agents has been the removal of alkyl benzene sulfonates by the use of aluminum sulfate and activated carbon. In that process, however, not only large-scale treating apparatus is required, but the cost of recovery of the activated carbon is undesirably costly.

A process developed by *A. Okabe and T. Ishii; U.S. Patent 3,898,159; August 5, 1975; assigned to Lion Fat and Oil Co., Ltd., Japan* is one involving admixing 0.5 to 10 mol equivalents of a polyvalent metal ion per mol of anionic surface active component in the waste stream to convert the surface active component to an insoluble polyvalent salt, and injecting a gas, such as air, into the stream to cause foaming, whereby the insoluble salt is adsorbed.

DIISOPROPYL AMINE

See Polyester Manufacturing Effluents (U.S. Patent 3,867,287)

DIMETHYL SULFATE

Dimethyl sulfate is used widely as a methylating agent. Due to its high toxicity it has to be removed from the indoor air, and in particular from the exhaust air, of plants where it is utilized.

Removal from Air

Dimethyl sulfate may be removed from gases by adsorption on active charcoal. The regeneration of active charcoal charged with dimethyl sulfate by simple desorption at elevated temperatures is not practical. The filters are regenerated, therefore, by rinsing them with bases, preferably with a solution of ammonia, washing with water, drying in a hot gas current and cooling to the adsorption temperature.

It has been found that, surprisingly, drying of the charcoal after washing is unnecessary and that the charcoal can be employed again immediately afterwards for the adsorption of dimethyl sulfate.

A process developed by *R. Grimm, W. Herzog and R. Lakemann; U.S. Patent 3,736,726; June 5, 1973; assigned to Farbwerke Hoechst AG, Germany* is one in which gases are conducted through active charcoal to absorb dimethyl sulfate. The active charcoal is regenerated with a base, then washed with water and reused for the absorption of dimethyl sulfate without any intermediate drying.

DITHIOCARBAMATES

Industrial effluents containing dithiocarbamates have long been discharged and it has commonly been thought that this is a tolerable situation in view of the relatively low proportion of dithiocarbamates even in dithiocarbamate-rich industrial effluents when discharged into watercourses or when mixed with other effluents and in view of the possibility of decomposition.

It has been found however that purification of industrial effluents containing dithiocarbamates is, in fact, very necessary before discharge if the discharge of such effluents is to be satisfactory. Thus, although dithiocarbamates are of quite low toxicity to mammals and birds, water-soluble dithiocarbamates are very toxic to fish; they have a significant and undesirable inhibiting action on the nitrification bacteria whose action is valuable in sewage treatment and they are very toxic to grazing fauna (such as white worms) that play a vital part in controlling fungal growth in bacterial sewage beds.

Removal from Water

Water-soluble dithiocarbamate is not biologically degraded in sewage treatment processes and thus, if present in the matter to be treated by such a process, will also be present in the treated effluent discharged and will therefore be liable to poison any fish in the watercourse into which the treated effluent is discharged.

A process developed by *A. Stevenson and N. Harkness; U.S. Patent 3,966,601; June 29, 1976; assigned to Robinson Brothers Ltd, England* is a process for purifying industrial effluent containing dissolved dithiocarbamate which comprises, before discharging the effluent, mixing the effluent at a pH of 6 to 8 with soluble heavy metal salt and thereby precipitating heavy metal dithiocarbamate, separating the precipitated heavy metal dithiocarbamate from the effluent and discharging the effluent.

The following is a specific example of the conduct of this process. A sample 750 gallon batch of liquid effluent from a plant manufacturing dithiocarbamates was analyzed and found to contain 406 ppm of dissolved dithiocarbamate. The pH of the effluent was found to be 10.0. The batch of effluent was fed to a tank, adjusted to a pH of 7.2 by controlled addition with stirring of 20% w/v sulfuric acid (about 2.5 gallons) and 8 gallons of a 10% w/v solution of ferrous sulfate heptahydrate added. Mixing was aided by means of a stirrer. The mixture was allowed to stand for 2 hours and separation of the precipitate formed achieved by settlement.

Analysis of the treated effluent showed that the dissolved dithiocarbamate content was 12 ppm. Another 750 gallon batch of the same effluent, having the same dissolved dithiocarbamate content and pH was treated in the same manner but using 38 gallons of a 20% w/v solution of cupric sulfate pentahydrate instead of the ferrous sulfate solution. The mixture was allowed to stand for 2 hours. In this case the treated effluent had a dissolved dithiocarbamate content of 1 ppm.

DRY CLEANING PLANT WASTES

Apparatus for the chemical cleaning (dry cleaning) of garments and other textile material or systems for the degreasing of metals using an organic solvent, are generally provided with means for introducing the solvent in a liquid state into the vessel, means for agitating the liquid solvent in contact with the objects to be treated, and means for draining the solvent from the vessel. Subsequently, an air stream is passed through the vessel to remove traces of the organic solvent.

The solvents which are employed are not only expensive but are also more or less toxic. They include primarily hydrocarbons and especially chlorinated and/or fluorinated hydrocarbons. The solvents may be of the relatively high boiling type, e.g., perchloroethylene and trichloroethylene, or of a low boiling type, such as trichlorotrifluoroethylene and trichloromonofluoromethane. It has already been pointed out that a drying operation requires passage of an air stream, especially air, through the treatment vessel to remove traces of the solvent and subsequent recovery of the solvent from the air stream so that such toxic and expensive impurities will not be discharged into the environment.

Removal from Air

A process developed by *H. Führing and J.H. Sieber; U.S. Patent 3,883,325;*

May 13, 1975; assigned to Böwe Böhler & Weber Maschinenfabrik, Germany is one in which air stream from a treatment vessel, e.g., the drum of a dry cleaning machine or a device for the degreasing of metals using an organic solvent, is passed through a cooler to reduce condensable-component level in the air stream before being introduced into an adsorber in which solvent residues are removed.

Between the cooler and the adsorber, the air stream is reheated so that the relative humidity of the gas entering the adsorber is well below 100%. Preferably, the air stream is circulated through the cooler and the heater for a period sufficient to reduce the level of condensable components in the air stream to the point that the adsorber will not be immediately saturated thereby, whereupon the adsorber is brought into play.

Figure 35 is a vertical elevation in diagrammatic form of a dry cleaning apparatus using this process. The dry cleaning machine comprises a drum housing **1** in which the perforated basket or drum carrying the fabric or metal parts to be degreased is rotatable in the usual manner and can serve to bring about intensive contact between the solvent and the objects.

Upon discharge of the liquid solvent, a drying air stream is induced to pass through the drum housing in the direction of the arrow by a blower **2** which draws the air through the intake duct **3** and a lint filter **4**, directing the air stream into the cooler **5**. The cooler **5** includes a water cooler **5a** and a refrigerant cooler **5b**, connected in series, the latter cooler being provided with the compressor **6**.

FIGURE 35: APPARATUS FOR SOLVENT GAS RECOVERY FROM AN AIR STREAM FROM A DRY CLEANING MACHINE

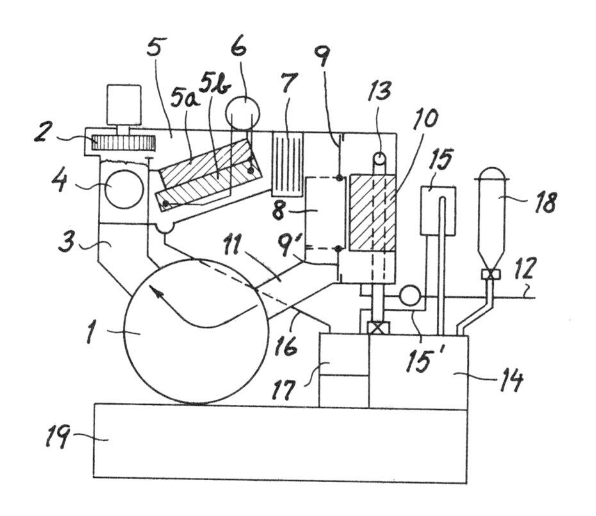

Source: U.S. Patent 3,883,325

The cooler/condenser **5** is provided upstream of the air heater **7** which is traversed by the cooled air emerging from the cooler. With the flaps **9, 9'** in their illustrated positions, the air stream is returned directly to the drum housing **1** and circulated until there has been a substantial drop in the solvent concentration by condensation in cooler **5**. At this point, the valve flaps **9, 9'** are shifted into their broken-line positions and the adsorber **10** is connected in the circuit. The residual solvent is thereupon picked up by the adsorber bed. The air is returned via duct **11** to the drum housing **1**.

For recovery of the solvent from the adsorbent, a valved line **12** induces steam to flow upwardly through the adsorber **10** and the steam, and extracted solvent vapors are collected in the distillation vessel **14**, condensed at the condenser **15** and discharged in the solvent water separator **17** into respective compartments. The line leading to the condenser **15** is represented at **15'** and the distillation vessel **14** is provided with an adsorber **18** communicating with the atmosphere and designed to prevent the development of elevated pressures within the still **14, 15** without permitting solvent vapors to enter the atmosphere.

Once the steam desorption step is completed, blower **2** circulates air through the units **1, 2, 5, 7, 10, 1** so that air is heated at **7**, contacts the adsorber at an elevated temperature and hence expedites discharge of residual mixture and/or solvent. The moisture-laden drying air is cooled at **5** to condense water from the system and reheated at **7**. The drying current flow of air is terminated when no further moisture is collected at the cooler **5**.

DRYING OVEN EFFLUENTS

Dryers for products containing volatile hydrocarbons such as impregnated hard board, electrical insulation having insulating varnish, etc., normally produce during drying gaseous hydrocarbons which, for ecological reasons, cannot be discharged directly into the atmosphere but which must be treated.

Removal from Air

A device developed by *H. Vits; U.S. Patent 3,875,678; April 8, 1975; assigned to Vits-Maschinenbau GmbH, Germany* for effecting such treatment is a two-sectioned dryer in series in which the material being dried, passes from section to section; in the first section the more volatile hydrocarbons are driven off and are after-burned with the heat of combustion being recovered and used to heat the dryer sections; in the second section less volatile hydrocarbons are driven off which are washed or condensed; the hydrocarbons precipitating out during washing being used as a fuel to assist in the after-burning of the more volatile hydrocarbons. The washed exhaust from the second section and exhaust gases from after-burning of the more volatile hydrocarbons are combined after washing and after-burning for dehumidification of the washed exhaust and are discharged directly into the atmosphere.

See Solvents (U.S. Patent 3,486,841)
See Solvents (U.S. Patent 3,868,779)
See Wood Dryer Effluents (U.S. Patent 3,853,505)
See Wood Dryer Effluents (U.S. Patent 3,945,331)

DUPLICATING MACHINE EFFLUENTS

In electrostatic copying machines, a photoconductive surface first is charged by exposure to the action of a corona or the like and then is passed by an imaging station at which the charged surface is subjected to an image of the original to be copied to produce a latent electrostatic image. After production of the image, the surface passes through a developer system in which toner particles are applied to the image so as to be deposited in the regions which retain charge, thus to develop the image.

In some systems, the copy material itself carries photoconductive material. In other systems, the photoconductor is on the surface of a drum or belt or the like and is subjected to the action of a tacky toner which later is transferred to ordinary paper to produce the copy.

Many of the machines now in use employ a liquid developer in which the toner is suspended in a light hydrocarbon carrier liquid. The carrier liquid may be of any suitable type such, for example, as Isopar G. This developer incorporating Isopar G as a carrier liquid may be applied to the photoconductive surface in a number of ways. Conventionally, the photoconductive surface is moved past the developing system. The developing system directs developing liquid upwardly and into contact with the image as the photoconductive surface moves past the developer unit. Some machines remove excess developer from the surface as it leaves the unit.

Following development of the image it moves either to a transfer location at which the tacky image is transferred to a sheet of paper or in the case in which the paper itself has a photoconductive surface, it moves to a drying or image-setting station. In the course of this movement the hydrocarbon carrier liquid volatilizes. At any point at which the machine is open to the atmosphere the air polluted with hydrocarbon liquid is circulated into the room in which the machine is installed. While this consideration is not an important one for a relatively small machine having only periodic use, it assumes significance in a large machine producing a high volume of copies. The problem of pollution from volatilized carrier liquid is aggravated by the production of aerosols by the action of the air knife or the like for removing excess developer.

Removal from Air

A number of systems or devices have been tried in attempts to solve the problem of pollution in a large, high volume electrostatic copying machine. Catalytic devices with and without heat exchangers were tried. These devices per se involved the defect that an excessive amount of heat was rejected into the room in which the machine was used. They did not permit recovery of the hydrocarbon carrier.

Where the devices were provided with heat exchangers they required outside air or water necessitating the use of either an air duct or a water pipe through the wall of the room for effective removal of the large amount of heat generated. They were, consequently, too expensive and too inconvenient to install. As an alternative to the catalytic devices, mechanical filters were tested. However, the filters did not operate sufficiently well to reduce the pollution to an acceptable level. Such filters are extremely expensive in that they require pumps to force vapor through a high pressure-drop filter.

A system developed by *P.I. Brown and A.L. Brown; U.S. Patent 3,767,300; October 23, 1973; assigned to Savin Business Machines Corporation* is a pollution control system for an electrostatic copying machine employing a liquid developer in which polluted air from adjacent the surface of a photoconductive element in a generally closed cabinet is drawn into a cold trap to produce a condensate made up of the liquid hydrocarbon carrier and water, which condensate is carried to a separating system which separates the carrier liquid from the water and returns the carrier liquid to the supply. The cleaned air is circulated back to an air knife or the like directed against the photoconductive surface as it emerges from the developer system.

A process developed by *H. Hasimoto, et al; U.S. Patent 3,679,369; July 25, 1972* utilizes a deodorization device particularly adapted for use with ammonia process diazo copying machines which includes an odor removing conduit having two odor removing sections, through which ammonia containing exhausts are caused to flow. A filler is provided in the first odor removing section, a deodorizing chemical being applied to the filler, while an ammonia adsorbent is provided in the second odor removing section, the ammonia containing exhausts flowing first through the first section and then through the second section.

See Tetrabromomethane (U.S. Patent 3,437,429)

DYESTUFF WASTES

Removal from Water

Various proposals have already been made for removing residues of dyestuffs and auxiliaries from industrial effluents. Thus, for example, it has been provided that the residual liquors, including wash waters, are collected in collecting tanks and the residues of dyestuffs and auxiliaries are precipitated by addition of suitable flocculating agents and separated out by sedimentation and filtration. However, these processes suffer from various disadvantages. The volumes of water to be treated are extremely large and sedimentation is frequently protracted.

A process developed by *H. Wegmüller and J. Haase; U.S. Patent 3,979,285; September 7, 1976; assigned to Ciba-Geigy Corporation* is based on the discovery that a complete or at least very extensive purification, including decolorization, of industrial effluents is achieved if these are brought into contact with adsorbents which consist of cellulose pretreated with precipitants. The process according to this method is above all suitable for the removal of anionic dyestuffs, optical brighteners, dyeing auxiliaries and washing agents, and for the elimination of residues of tanning agent.

A process developed by *R.A. Montanaro and H.B. Moreau; U.S. Patent 3,803,030; April 9, 1974; assigned to Fram Corporation* provides a high degree of removal of color bodies and metals from a dyehouse waste stream, while minimizing consumption of adsorbent and regeneration materials. Figure 36 shows such a process.

Referring to the drawing, columns **10** and **12** contain, between screens **14**, beds **16** and **18** of macroreticular polymer resin adsorbent particles. Macroreticular

FIGURE 36: PROCESS FOR REMOVING COLOR BODIES FROM DYEHOUSE WASTEWATERS

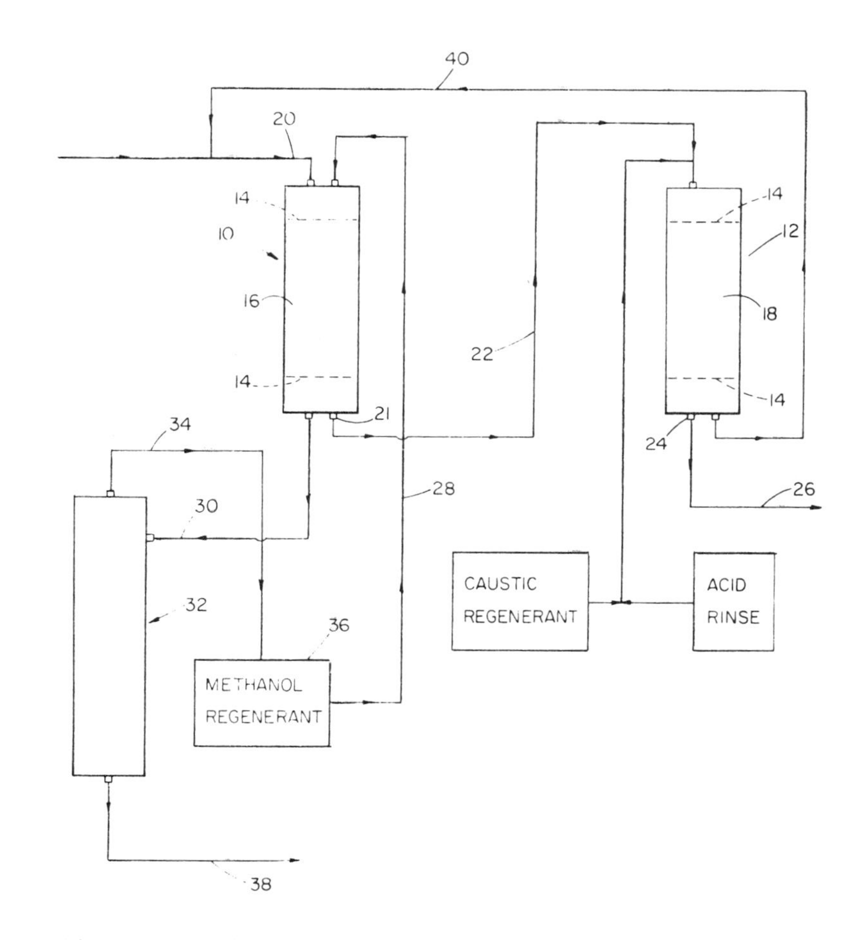

Source: U.S. Patent 3,803,030

polymer resins in general consist of beads composed of microspheres, with large pores (e.g., of the order of 100 A pore diameter) and large surface area (e.g., of the order of 100 m^2/g). Amberlite and Duolite are two commercially available products suitable for this purpose.

In this instance, bed **16** is Amberlite XAD-7 (a solvent regenerable resin) and bed **18** is Amberlite XAD-12 (a base regenerable resin). However, Duolite S-37 may be substituted for the Amberlite XAD-12 and is presently preferred. The beds **16** and **18** preferably contain equal volumes of resin although other ratios are suitable (e.g., for a bed **18** of two units, the range of volumes for bed **16** which will yield good results is at least one unit to four units).

The waste stream is supplied to the top of column **10** through conduit **20** and percolates through bed **16** to provide a partially clarified effluent at outlet **21** which flows through conduit **22** to the top of column **12** for percolation through bed **18**.

The affinity of the adsorbent in bed **16** for the contaminants to be removed is insufficient to provide a high degree of purification at outlet **21**. Thus, in the first typical run referred to above, the effluent at outlet **21** had color density of 20 optical density units, COD (i.e., Chemical Oxygen Demand) of 682 ppm, chromium of 6.5 ppm, cobalt of 15.6 ppm, and TOC (i.e., Total Organic Carbon) of 255 ppm.

However, after passage of that effluent through bed **18**, in which the adsorbent has a higher affinity for the contaminants, highly purified effluent is produced at outlet **24** and is carried away through conduit **26**. In a second typical run, the effluent at outlet **21** had color density of 1.9 optical density units, COD of 2,330 ppm, copper of 210 ppm, TOC of 780 ppm, and total suspended solids of 52 ppm.

The ease of regeneration of beds **16** and **18** is in inverse relation to their affinity for contaminants. Thus, when bed **16** becomes fully loaded, it is regenerated by elution with hot (60°C) methanol supplied to the top of column **10** through conduit **28**. The methanol, entrained water, and eluted contaminants are taken from the bottom of the column through conduit **30** to a still **32**, where the methanol is recovered by distillation and returned through conduit **34** to reservoir **36**. The methanol left entrained on bed **16** is steam stripped from the resin and taken off through the top of the column.

A waste stream, in which the concentration of contaminants is one hundred times as great as in the original stream in conduit **20**, is carried from still **32** through conduit **38** for final disposal. The more tightly held contaminants in bed **18** (containing the polar adsorbent) are removed by passing through caustic solution (0.1 to 4.0% concentration) through column **12** from top to bottom, followed by an acid rinse to restore bed **18** to essentially its original capacity.

The caustic, acid rinse, and desorbed contaminants are carried through conduit **40** to the main waste stream for subsequent passage therewith through columns **10** and **12** in the manner already described. The pH in conduit **20** is kept on the acidic side (e.g., between 3 and 5) to facilitate adsorption.

The described cycles of adsorption and regeneration are repeatedly carried out, so that contaminants initially adsorbed on bed **18** are repeatedly recycled and eventually adsorbed on bed **16** (where the absolute range of pore size overlaps the average pore size associated with bed **18**). The solvent regeneration of bed **16** isolates all removed contaminants in the solvent. Distillation of the solvent provides the final degree of contaminant concentration.

This process achieves the desired high degrees of waste purification and contaminant concentration. It has been determined that, for an acceptable degree of waste purification, neither a typical nonpolar nor a typical polar adsorbent alone would approach the efficiency desired. For 100 gallons of dyestuff wastewater, a volume of five units of nonpolar adsorbent was required to achieve 99+% color removal and contaminant concentration (after adsorbent regeneration) was only

20 to 1. One unit of polar resin achieved the same degree of color removal, but resulted in only an 8 to 1 concentration. Combining the resins in series in a procedure in accordance with the process produced the same degree of color removal in 100 gallons of dyestuff wastewater with only 1.4 units of resin and with a concentration of 100 to 1. Thus, not only does this process achieve better results, but it does so economically (i.e., a relatively small amount of resin is required).

A process developed by *M. Ohkawa and Y. Sawaguri; U.S. Patent 3,966,594; June 29, 1976; assigned to Sumitomo Chemical Company* is one in which organic wastewater containing water-soluble organic anionic substances is treated by

(1) Contacting an acidic wastewater with a water-insoluble organic solvent solution of at least one amine represented by the formula,

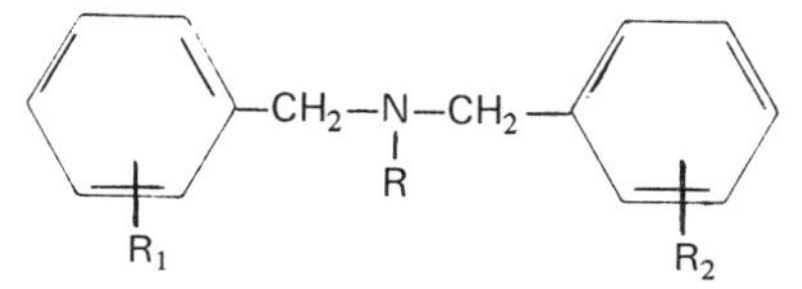

wherein R_1 and R_2 are each a hydrogen or halogen atom or a C_1 to C_4 alkyl group, and R is a C_8 to C_{18} alkyl or alkenyl group,

(2) Separating the aqueous layer from the organic layer, and

(3) Contacting the organic layer with an aqueous alkali solution whereby the substances are transferred to the aqueous alkali layer and the amine is recovered as the organic solution thereof.

Wastewaters which can be treated by this method are specifically those which are mainly discharged from dye works, industrial chemical works, dyeing works, medicine works and the like, for example wastewater containing sulfonated aromatic compounds such as benzene-, naphthalene- and anthraquinone-sulfonic acids, dyestuffs having at least one sulfonic group, nitro compounds, and phenols. By this method, even wastewater having a COD value as high as 15,000 to 20,000 parts per million can be treated efficiently, with a subsidiary effect of largely reducing the color depth of wastewater in most cases.

In the practice of the process, the pH of the wastewater to be treated is first adjusted preferably to less than 2, particularly preferably to 0.5 to 1.5 by the addition of hydroxides of alkali metal and alkaline earth metal or aqueous ammonia, or acids such as sulfuric acid and hydrochloric acid. Therefore, in a multistage extraction process, initially acidity of wastewater to be treated must be adjusted so that the wastewater may have a pH of less than 2 at the end of the process. When a large amount of insoluble matter is produced by adjusting the pH of the wastewater to be treated, it is preferred to remove the matter by filtration prior to the amine-extraction.

A solution of the amine in organic solvent is added to the acidic wastewater and the mixture is stirred or shaken for several minutes or more. The amount of amine added is 1 to 50% by weight, generally 5 to 15% by weight, based on the weight of the wastewater. The addition and stirring (or shaking) are carried out at 10° to 80°C, usually at 20° to 40°C.

The abovementioned cationic surfactants are preferably added to the wastewater just before or during addition of the amine solution. The amount thereof is about 0.001 to 0.1% by weight based on the weight of the wastewater. A larger amount may be added, but with no effect. Furthermore, the amount of surfactant added is so small that an effect of the surfactant on COD value is negligibly small.

Separation into aqueous and oily layers occurs very rapidly, and the oily layer containing organic anionic substances is separated from the aqueous layer and then subjected to subsequent alkali back-extraction using an aqueous alkali solution. The alkalis used for the back-extraction are water-soluble ones such as hydroxides of alkali metal or alkaline earth metal and aqueous ammonia. That is, the organic layer is stirred or shaken for several minutes or more together with the aqueous alkali solution in such amount that pH of a lower aqueous layer is made 7 or more, then it separates into an upper organic layer containing the amine and a lower aqueous layer containing water-soluble organic substances (hereinafter referred to as a high COD-liquor).

The temperature at which the alkali back-extraction is carried out depends upon the concentration of alkali, and it is generally from room temperature (10° to 20°C) to 80°C. The concentration of the aqueous alkali solution is preferably 10 to 50% by weight (converted to caustic soda), and a higher concentration is advantageous because the volume of a high COD-liquor which is finally separated from the amine-containing organic layer becomes smaller. But, the aqueous alkali solution of too high concentration causes an increase in the viscosity of high COD-liquor so that it is necessary to warm the liquor. The high COD-liquor contains most of the organic anionic substances in the initial wastewater and has a small volume relative to the wastewater, e.g., at its smallest, a volume ratio of high COD-liquor to initial wastewater of 1:30. It is treated by, i.e., combustion.

Thus, separation into two layers can be achieved very rapidly with no formation of intermediate layer, and wastewater can be reduced in its COD value with removal of 75 to 90%.

A process developed by *S. Oohara; U.S. Patent 3,829,380; August 13, 1974; assigned to Kanebo Ltd., Japan* is one in which liquid containing diluted anionic dyes or disperse dyes is cleaned by specific polyamide fibers having a high diluted dye absorption coefficient for the dye of at least 0.5 and at least 150 meq/kg of amino group, which are contained in an absorption column through which the liquid is circulated, and the cleaned liquid is returned into the dyeing system for the next process.

A process developed by *S. Oohara and K. Nakashima; U.S. Patent 3,822,205; July 2, 1974; assigned to Kanebo Ltd., Japan* is one in which a waste colored aqueous liquid containing nonanionic coloring substance is cleared by coagulating the nonanionic coloring substance with a coagulating agent. The coagulating agent may consist of condensation products of sulfonated dihydroxydiphenylsulfones or sulfonated dihydroxydiphenyls with lower aliphatic aldehydes or sulfonated condensation products of dihydroxydiphenylsulfones or dihydroxydiphenyls with lower aliphatic aldehydes. A final step involves removing the coagulated dye from the waste aqueous liquid.

A process developed by *M.J. Hurwitz, D.C. Kennedy and C.J. Kollman; U.S. Patent 3,853,758; December 10, 1974; assigned to Rohm and Haas Company* is one in which effluents from dye manufacturing and dyeing operations which contain waste dyestuffs are decolorized and their oxygen demand substantially reduced by passing at least a major part of the effluent through a bed of essentially nonionogenic, macroreticular, water-insoluble, crosslinked polymeric adsorbent resin followed by contacting the partially decolorized effluent with a weak acid and/or aliphatic weak base ion exchange resin.

A process developed by *F.N. Case and E.E. Ketchen; U.S. Patent 3,912,625; October 14, 1975; assigned to the U.S. Energy Research and Development Administration* involves treating waste organic dye material dissolved or dispersed in an aqueous effluent and comprises contacting the effluent with an inert particulate carbonaceous sorbent at an oxygen pressure up to 2,000 psi, irradiating the resultant mixture with high energy radiation until a decolorized liquid is produced, and then separating the decolorized liquid.

A process developed by *J.B. Powers; U.S. Patent 3,947,248; March 30, 1976; assigned to The Dow Chemical Company* involves the treatment of wastewater streams from the dyeing of polyester yarns and fabrics with organic pigments.

The waste streams from this process contain residual amounts of color bodies which are removed by adding a sufficient amount of a water-soluble cationic polymer (e.g., polyethylenimine) to flocculate the color bodies.

The manufacture and use of sulfur dyes results in the formation of a toxic and highly alkaline waste liquor which cannot be disposed of in a conventional manner, since it unduly pollutes and stagnates fresh water rivers and streams. These waste liquors are highly colored and odoriferous. The sodium sulfide or other alkali sulfides contained therein hydrolyze with water and form hydrogen sulfide which is toxic to marine life and has a foul, disagreeable odor. No wholly suitable means for the treatment or disposal for these liquors is known.

Neutralization treatments of these alkaline liquors, such as with sulfuric acid, result in the formation of a highly turbid suspension and the evolution of large quantities of hydrogen sulfide gas. Such acid treatments do not suitably improve the color of these waste liquors.

A process developed by *J.B. Story; U.S. Patent 2,877,177; March 10, 1959* relates to the treatment of waste liquors containing alkali sulfides, and more particularly to the treatment of sulfur dye wastes.

The following is a specific example of the conduct of the process. A volume of sulfur black dye waste liquor (T-1636 Sulfogene carbon HCF grains) containing sodium sulfide (1,000 parts by weight of liquor) having a foul odor and a pH greater than 11 was treated at room temperature with sulfurous acid to coagulate the dye constituents thereof and to eliminate the stagnating and polluting contents of the liquor. The liquor contained about 2 parts of sodium sulfide.

In the treatment, the acid was poured into the waste liquor, and mixed thoroughly within five seconds. A sufficient amount of sulfurous acid (65 parts by weight of 0.5 molar solution) was required to give a 6.8 pH for the treated liquor, immediately following the mixing. Particles immediately appeared in the mixture.

To so-treated liquor was then allowed to settle, leaving a clear, colorless, odorless and essentially neutral supernatant liquid and a black, curdy precipitate. Upon separation in a filter, the supernatant liquid was disposed of in a conventional sewerage system. The precipitate is thereafter redissolved with sodium sulfide solution and reused for further dyeing operations.

A process developed by *G. Hertz, U.S. Patent 3,485,729; December 23, 1969; assigned to Crompton and Knowles Corp.* is one in which waste dye liquor is decolorized by electrolytic treatment of an aqueous solution containing chloride ions. The organic dyestuff is oxidized by hypochlorite formed thereby permitting the treated material to be discharged into sewers and sewage systems.

ELECTRIC ARC MELTING FURNACE EFFLUENTS

In an electric metal melting furnace of the type which comprises a crucible tiltable about one or more axes for tapping and slagging operations and further containing orifices for such operations, the exhausting of fumes generated during charging tapping, slagging and melting is a problem which has caused great concern for environmental and health reasons. Generally, such fumes tilt forwardly for tapping and rearwardly for slagging such that the pouring spout extends forwardly and the slagging opening opens rearwardly. Since the top of such a furnace is generally closed by a roof having perforations therethrough through which a plurality of electrodes protrude for providing electrical power to melt metal with which the furnace is charged, it is necessary to provide means closely spaced above the roof to provide the necessary fumes exhaust for the considerable amount of fumes generated during the melting operation.

Removal from Air

An apparatus developed by *R.C. Overmyer and P. Nijhawan; U.S. Patent 3,979,551; September 7, 1976; assigned to Hawley Manufacturing Corporation* is an integral fumes exhaust system for an electric metal melting furnace which provides fumes exhaust during tapping, slag discharge, charging and melting and oxygen lancing operations.

ELECTROPLATING WASTES

Removal from Air

A process developed by *L.R. Myers; U.S. Patent 3,985,628; October 12, 1976* is one in which air above an electroplating bath is scrubbed with the plating rinse water to effect the transfer of chemical values from the air to the rinse water and to effect transfer of water to the air. The chemically enriched rinse water is added to the plating bath. In the system the waste heat generated in the plating bath is the only heat used in water removal, substantially no water is circulated from the plating bath, water added to the rinse is restricted by using a relatively high pressure-low volume spray rinse, and substantially no auxiliary air is used in addition to that required to sweep over the plating baths.

Removal from Water

See Chromium (U.S. Patent 3,869,386)
See Cyanides (U.S. Patent 3,788,983)
See Cyanides (U.S. Patent 3,744,977)
See Nickel (U.S. Patent 3,630,892)

EMULSIFIERS

See Detergents (U.S. Patent 3,898,159)
See Neoprene Production Process Effluents (U.S. Patent 3,778,367)
See Neoprene Production Process Effluents (U.S. Patent 3,890,227)

EPOXIDE RESIN MANUFACTURING EFFLUENTS

Removal from Water

A process developed by *A. Renner; U.S. Patent 3,716,483; February 13, 1973; assigned to Ciba-Geigy, Switzerland* is a process for removing dissolved, emulsified or suspended organic substances, for example fat, resin, oil or dyestuff, from water. The contaminated water is brought into contact with a highly disperse, solid, water-insoluble organic polymer, for example melamine-formaldehyde resin or polyacrylonitrile of average molecular weight greater than 1,000 and a specific surface area greater than 5 m^2/g, and the polymer charged with the contamination is separated from the water.

In a typical example of the conduct of this process, the effluent from a factory manufacturing epoxide resins contains 1% by weight of emulsified resin which essentially corresponds to the diglycidyl ether of bisphenol A. If 100 parts of this effluent are treated with 0.5 part of a highly disperse urea-formaldehyde polymer, an absolutely resin-free effluent is obtained after filtration, as can be seen from the light absorption measurements at 460, 530 and 650 $m\mu$.

ETHYL CHLORIDE MANUFACTURING EFFLUENTS

Removal from Water

A process developed by *J.M. Collins; U.S. Patent 3,536,617; October 27, 1970; assigned to Ethyl Corp. of Canada, Ltd., Canada* neutralizes and cleans an oil contaminated acid solution, such as results from the hydrolysis of a heavy ends stream from a process for the manufacture of ethyl chloride by reaction of ethylene and hydrogen chloride in the presence of aluminum chloride catalyst, by passing the solution downward through a flooded bed containing a carbonate, such as in limestone or clam shells, reacting the acid with the carbonate to form carbon dioxide, and allowing the carbon dioxide to pass upward through the bed. The oil is stripped from the water and held at the surface of the water as froth for removal. If particulate matter is present in the acid solution, it will be removed with the oil.

ETHYLENEDIAMINETETRAACETIC ACID (EDTA)

Until recently one of the most common class of bleaches comprised alkali metal ferricyanides, especially potassium ferricyanide. Much research has been performed to define substitutes for these materials in view of the potentially relatively high toxicity of the ferrocyanide compounds which form the bulk of the waste discharged after bleaching in view of the reaction products formed by such waste upon exposure of same to sunlight in streams, etc. If waste processing solutions containing hexacyanoferrate ion (the products of ferricyanide bleaching) are discharged to the sewer without treatment, these ions are slowly oxidized in the presence of ultraviolet radiation to cyanide ions, which are probably the most toxic to fish and other aquatic life of any of the chemicals discharged from photographic processes. A valuable substitute for the ferricyanide bleaches which is well known to those skilled in the photographic art is the ferric EDTA complexes, for example ferric ammonium EDTA or tetrasodium EDTA which, while producing excellent bleaching effects, do not yield the potentially toxic ferrocyanides of earlier ferricyanide systems.

Thus, the most commonly used bleaching agent is rapidly becoming ferric ammonium EDTA which avoids the potential toxicity problems of the ferricyanide and demonstrates excellent bleaching properties, but which has been found to result in the discharge of materials requiring high chemical oxygen demand.

Removal from Water

A process developed by *T.W. Bober, T.J. Dagon and I. Slovonsky; U.S. Patent 3,767,572; October 23, 1973* is one in which waste photographic processing solutions which contain ethylenediaminetetraacetic acid (EDTA), such as exhausted ammonium iron EDTA bleaching or bleach-fixing solutions, are chlorinated to destroy EDTA and thereby increase the biodegradability of the solution. Chlorination can be effected by introduction of chlorine gas or by the use of hypochlorite solution. Since EDTA and complexes thereof account for a large portion of the oxygen-consuming material present in photographic processing effluent, a significant source of water pollution is substabtially eliminated by this method.

FATS AND FATTY OILS

The water pollution problem in the food-processing industry (including particularly meat-packing houses, slaughtering houses, meat canning, and the like meat processing operations) is becoming more and more serious as new regulations are being imposed and existing regulations strictly enforced. Regulations limit oils and grease to 100 ppm. Proposed regulations will put limits on oils and grease which are substantially lower than present standards.

The cleanup of equipment used in this industry generates wastewaters which characteristically contain significant amounts of fats and greases (mainly mixed triglycerides) of animal or plant origin, sometimes termed hexane solubles, as well as miscellaneous contaminants, such as cleaning agents, surfactants, alkalies, acidic materials, and the like. Such wastewaters contain both suspended and dissolved solids and tend to be quite stable, and such waters characteristically contain levels and impurities far above those permitted under federal and state pollution standards. Thus, such wastewaters must be treated to clean them up sufficiently for discharge into sewers of conventional sanitation systems.

Removal from Water

The task of clarifying and purifying such wastewaters, at least to an extent sufficient to produce a product water which meets the minimum standards for sewering, constitutes a major problem particularly when such task is to be accomplished in an economical, practical and reliable manner. Heretofore, the art has commonly attempted to effectuate a separation of water from animal derived fatty materials using gravity separation devices which are dependent upon differences in the specific gravity of water relative to animal fats and greases in admixture therewith.

Thus, it is contemporarily common to find in conjunction with food-processing operations, a so-called grease pit into which wastewaters from such operations are discharged. Such a grease pit as often as once a day may be cleaned manually by skimming to remove floating solids. Waters passing through the grease pit are simply routinely sewered. Available evidence indicates, however, that during cleanup periods such a grease pit cleanup water can commonly contain more materials (usually in an emulsified form) than was present in the initial input wastewater, and such emulsified materials may be more stable than the initial fatty wastewater. Simple gravity separation is inadequate to meet conventional effluent requirements.

Perhaps the most prevalent system currently on the market is Dissolved Air Flotation. This system can meet the effluent quality requirements; however, large quantities of sludge are produced which need additional treatment to avoid exorbitant waste disposal costs. These systems also require chemical pretreatment, surge tanks, pumps, controls, and large amounts of space which result in an expensive and complicated operation.

A pollution control system designed for this industry must consider all of the following factors:

(1) The waste stream from the plant will vary in flow rates, temperature, pH and composition.

(2) The effluent from the treatment system should meet the present criteria and proposed federal guidelines.

(3) The sludge from the system should be relatively dry and free of excess water to reduce operating costs, eliminate additional treatment, and facilitate handling.

(4) The unit should be fully automatic to reduce labor and minimize operational errors.

(5) The unit should be compact and suitable for outdoor installation as space is at a premium.

(6) The initial cost, installation cost and operating costs must be reasonable.

A process developed by *A.J. Doncer and H.R. White; U.S. Patent 3,951,795; April 20, 1976; assigned to Alar Engineering Corporation* is one in which the wastewaters which typically contain significant amounts of fat and grease materials of animal and plant origin together with other contaminants are charged to the system at a nonuniform rate as generated and are continuously first chemically treated in two successive zones, then are passed into a quiescent holding zone wherein settling can occur and wherein variations in wastewater input volume or output volume can be smoothed out. Fluid from the bottom regions of such holding zone is continuously charged at a uniform rate to an operating rotary vacuum filter assembly to accomplish separation of solids. The product water is highly purified.

A process developed by *T. Miyazawa; U.S. Patent 3,940,334; February 24, 1976; assigned to Kayaba Industry Co., Ltd., Mitsubishi Industries, Ltd. and E.C. Chemical Industries Co., Ltd.; Japan* is one in which in the treatment of wastewater containing fats and oils, paraffinic hydrocarbons of low specific gravities, as extractive solvents for fats and oils, and inorganic or organic coagulants are added to the wastewater. The mixture is then stirred and allowed to stand; the oil-containing scum separating as an upper layer is then removed so as to separate the oil from water, and furthermore, the solvent previously added is recovered from the oil-containing scum so removed by means of distillation.

Figure 37 shows a suitable form of apparatus for the conduct of this process. Fat- and oil-containing wastewater discharged from a plant is sent to the original wastewater storage pool **2** through the drain pipe **1** and stored. The wastewater stored in the above original wastewater storage pool **2** is pumped by pump **3** into the treatment vessel **4**, and an appropriate amount of paraffinic hydrocarbons of low specific gravity, namely the fats and oils extracting solvent, is added from the solvent tank **6** by manipulating valve **5**. (Although such paraffinic hydrocarbons as n-hexane with a specific gravity of less than 0.8 are most desirable, other hydrocarbons with a specific gravity of less than 0.8, such as cyclohexane, pentanes, or octanes, can also be used effectively.)

As soon as the above paraffinic hydrocarbons are introduced, the stirrer **7** fitted to the treatment vessel **4** is started, and after an appropriate time, a certain amount of coagulant is added with stirring from the coagulant tank **9** through pump **8**. (Such a coagulant may typically be obtained by dissolving the condensation products of benzaldehyde or its derivatives with polyalcohols containing four or more hydroxyl groups, such as dibenzylidene sorbitol, into a polar organic solvent together with a surface active agent, and further dispersing the solution so prepared homogeneously into water. However, other organic coagulants consisting of water-soluble long chain high polymers containing functional groups such as carboxyl groups, amide groups, or amino groups, and other inorganic coagulants such as aluminum sulfate and iron salts, and polyaluminum chloride can also be used.)

FIGURE 37: SCHEME FOR SOLVENT EXTRACTION REMOVAL OF FATS AND FATTY OILS FROM WASTEWATERS

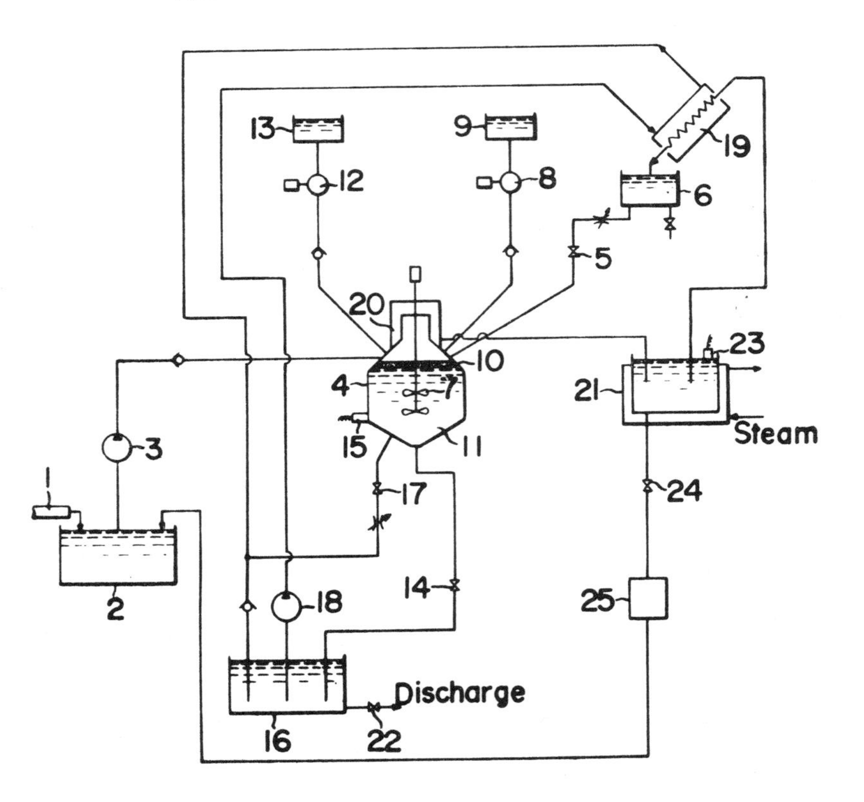

Source: U.S. Patent 3,940,334

After a short while, agitation is stopped and the mixture is allowed to stand for a certain period of time. Through such treatment, the fine paraffinic hydrocarbon particles formed by agitation are adsorbed over the deemulsified fats and oils in the wastewater, and they move upward in the form of bulky flocs under the action of the coagulants to form well-defined layers, namely the oil-containing scum **10** and the treated water **11**. In this case, the floating behavior of the oil-containing scum is greatly affected by the specific gravity of the solvent paraffinic hydrocarbons. When the specific gravity exceeds 0.8, the floating of the oil-containing scum becomes extremely slow and layer formation requires a long time. Therefore, the use of paraffinic hydrocarbons with a specific gravity of less than 0.8 is desirable from the standpoint of operational efficiency.

When the pH in the original wastewater storage pool **2** falls in the acidic range, the pH value must be adjusted after transfer into the treatment vessel **4** by the addition of an appropriate amount of aqueous hydroxide from the neutralizer tank **13** through the pump **12**.

When layer separation is completed in the first run as described, valve **14** is opened in order to withdraw the treated water **11** in the treatment vessel **4** into the treated water pool **16** until the floating oil-containing scum **10** reaches the level of detector **15**, and again a certain volume of the wastewater in the original wastewater pool **2** is charged into treatment vessel **4** through pump **3** to repeat the same treatment described above. This treatment is repeated five to seven times.

When the oil-containing scum **10** in the treatment vessel **4** reaches a certain thickness, the treated water, which is continuously circulating between the treated water pool **16** and the condenser **19** through pump **18**, is introduced into the bottom of the treatment vessel **4** by opening valve **17** so as to raise the floating oil-containing scum **10** until the scum overflows into chamber **20** attached to the top portion of the treatment vessel **4**.

The oil-containing scum is then withdrawn into the distillation vessel **21** and the treated water already pumped into treatment vessel **4** is again withdrawn back into the treated water pool **16** by opening valve **14**. This treated clean water may then be discharged through the valve **22** as desired.

The oil-containing scum transferred into the distillation vessel **21** was heated externally by steam and distilled at about 65°C, and the paraffinic hydrocarbons contained therein were vaporized into the condenser **19** to be cooled by treated water supplied from treated water pool **16** and condensed before returning to solvent tank **6** for recycling.

When the oil-containing scum from which the paraffinic hydrocarbons had been removed reached a certain volume as detected by detector **23**, it was transferred into filtration basket **25** through valve **24** and converted into a cake containing about 80% water for subsequent incineration or discarding. The water obtained by filtration was returned to the original wastewater storage pool **2** for repeated treatment.

FEEDLOT INDUSTRY WASTES

In recent years the raising of animals in confined areas such as feedlots and similar high animal concentration facilities has become commonplace. This trend is a result of several factors, including increased technology in the livestock industry, and increased population and weight gain of confined animals over pasture and yard animals. A principal drawback of animal confinement raising is the intense odor emanated from the animal wastes which accumulate in the relatively small areas occupied by the animals.

Until relatively recently, animal waste odors have not been a problem because the enterprises raising the animals were located at some distance from residential communities, and hence odor dilution by the prevailing winds was adequate. Urban growth, however, has put residential communities much nearer to the animal enterprises and the same odors previously ignored or unnoticed are now offensive and a serious problem.

Removal from Air

Numerous procedures have been developed to physically dispose of the animal wastes. Some of these include solid or slurry spreading on fields, oxidation ditches, incineration, and anaerobic or aerobic lagoons or digestion systems. Except for field spreading, these treatments involve significant capital expense and/or operating expenditures. Furthermore, these procedures do not eliminate the objectionable odors evolved from the animal wastes except in the aerobic treatment process.

Numerous procedures or systems for dealing with animal waste odors have also been developed. Some involve the use of specifically designed water scrubbers located in the ventilating system of a confinement area. Other procedures involve masking the odors with various scents; oxidizing the odors with oxidizing chemicals such as ammonium persulfate; and selectively inhibiting the formation of the odorous compounds with sulfa drugs.

A process developed by *E.T. O'Neill and W.H. Kibbel, Jr.; U.S. Patent 3,966,450; June 29, 1976; assigned to FMC Corporation* is one in which the odor of an animal waste slurry is controlled and the plant nutrient values of the slurry are increased by contacting the animal waste slurry with about 5 to 500 ppm hydrogen peroxide, adjusting the pH of the slurry to between about 4.0 and 8.0 with a mineral acid, mixing the slurry until the odor is no longer objectionable, and recovering the animal waste slurry which contains increased amounts of ammonium salt values.

FERTILIZER MANUFACTURING EFFLUENTS

Removal from Air

Elimination of finely dispersed particulates in exhaust stack gases of fertilizer manufacturing plants is most difficult. However, apparatus and methods for reducing the concentration of particulates in flue stack gases are well known in the fertilizer manufacturing industry and include electrostatic precipitators, fabric filter collectors, scrubbers, and mechanical collectors such as wet and dry cyclone separators. In a process developed by *J.E. Seymour; U.S. Patent 3,885,946; May 27, 1975; assigned to Royster Company* air pollution caused by emission of fine particulates of fertilizers in exhaust stack gases is substantially reduced by spraying aqueous solutions of nitrogen-containing compounds into the feed or intake to the air cooler for the furnace-dried and granulated fertilizer compositions.

FIGURE 38: PROCESS FOR REDUCING THE CONCENTRATION OF FERTILIZER PARTICULATES IN EXHAUST STACK GASES

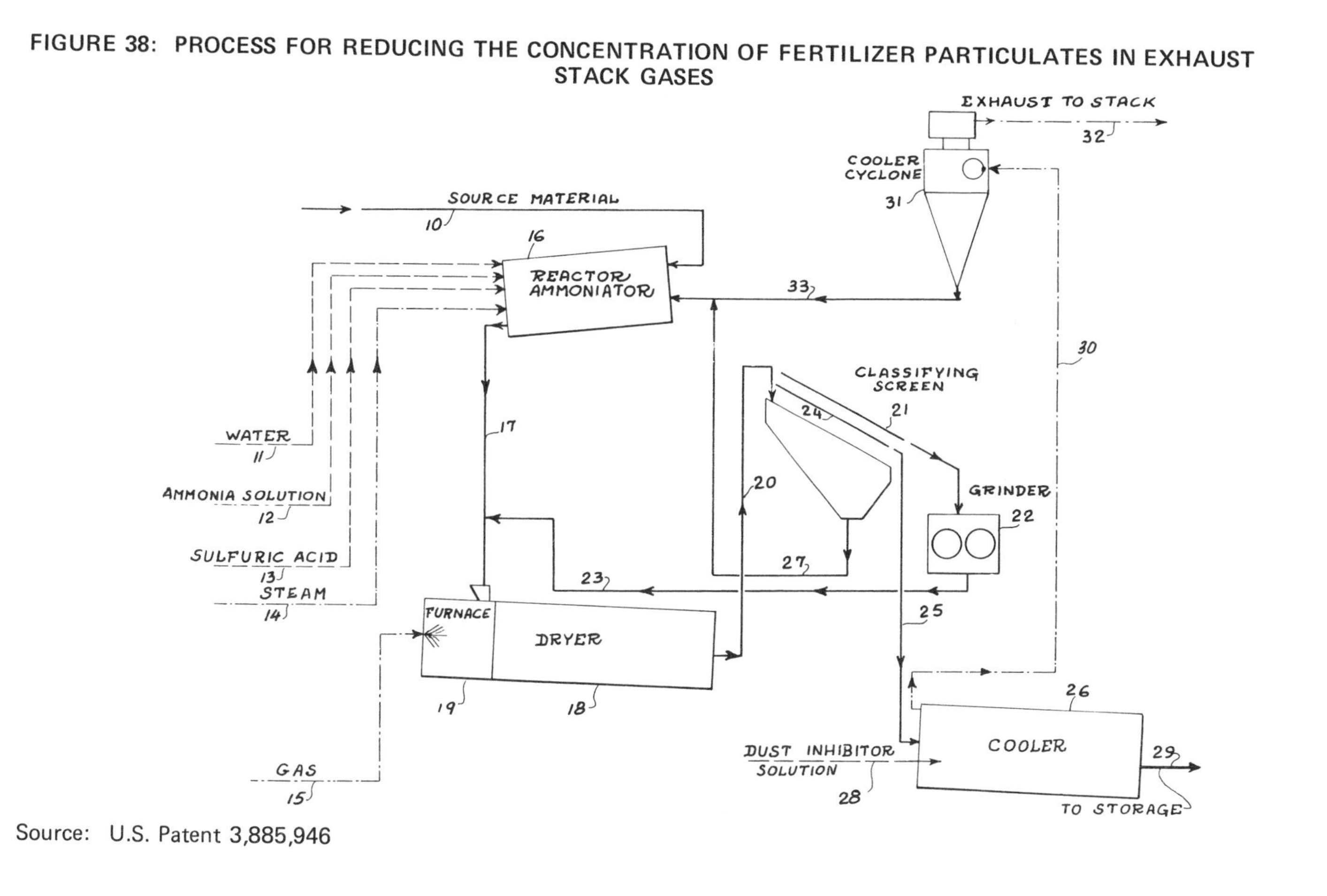

Source: U.S. Patent 3,885,946

Removal from Water

A process developed by *C.C. Legal, Jr.; U.S. Patent 3,725,265; April 3, 1973; assigned to W.R. Grace & Co.* is a multistage liming process which purifies wastewater by controlling the pH of the wastewater, e.g., as from a phosphate fertilizer plant, and thereby permitting selective precipitation of impurities in the wastewater and their recovery as useful by-products of the process. In one embodiment the main product of the first stage is a low silica calcium fluoride, and the main product of the second stage is a low fluorine dicalcium phosphate. Such a process is shown in Figure 39.

FIGURE 39: MULTISTAGE LIME TREATMENT PROCESS FOR PURIFICATION OF FERTILIZER PLANT WASTEWATERS

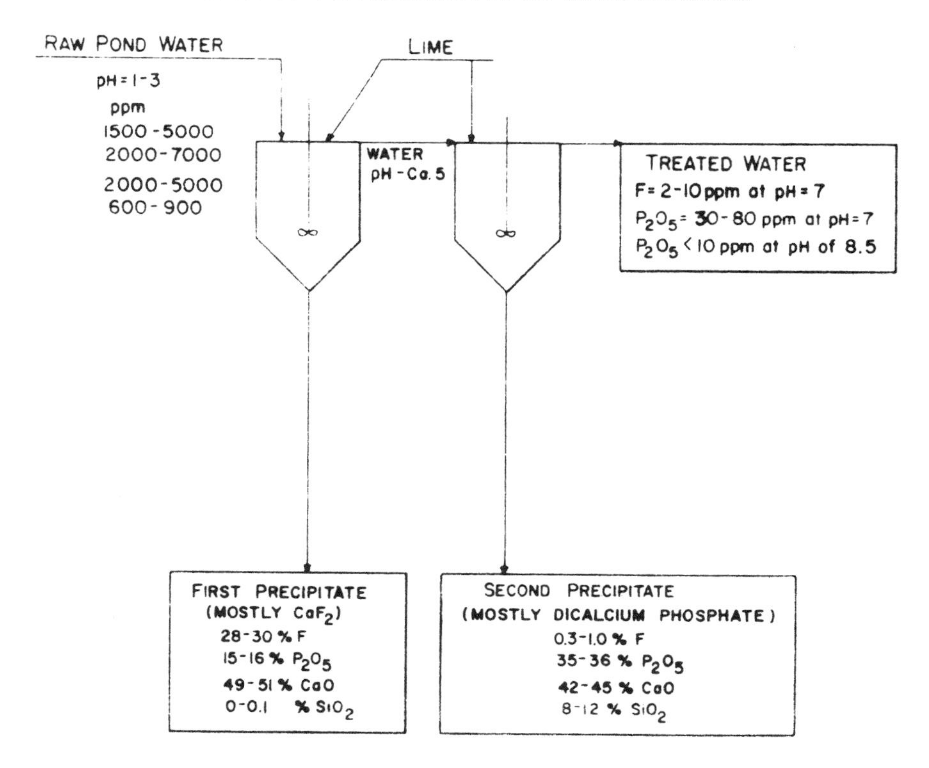

Source: U.S. Patent 3,725,265

FIBERGLASS PRODUCTION EFFLUENTS

In the glass fiber insulation process, the principal pollutants of interest are organic compounds such as p-hydroxybenzyl alcohol, o-hydroxybenzyl alcohol, and subsequent prepolymer resoles. These principally result from the evaporation of incompletely reacted resin binders sprayed on newly formed, semimolten glass fibers during the total process of manufacturing insulation. The resin monomers generally include phenol, formaldehyde, melamine and urea. The emission rate of the monomers, and resoles depends on the degree of polymerization of the binder prior to the spraying operation, the amount of excess monomer in the binder, and the type of insulation that has been produced.

Removal from Air

A process developed by *J.R. Richards; U.S. Patent 3,984,296; October 5, 1976* is a process for controlling air pollution and basically relates to a photochemical method of removing contaminant compounds, such as sulfur dioxide and nitrogen oxides, from polluted effluent gas systems. Such contaminant compounds are first formed into complexes (such as an electron donor-acceptor molecular complex) and clusters. After the contaminant compounds become associated with these complexes and clusters, ultraviolet light is introduced into the system resulting in the complexes or clusters being photooxidized.

The photooxidation of the complexes or clusters generally tends to internally rearrange the complexes or clusters and to form nonvolatile acid products that readily condense on available nuclei. The acid-like products resulting from the photooxidation treatment of the complexes or clusters can be removed from the gas system by conventional particulate removal techniques.

Figure 40 illustrates the application of this process to a fiberglass production plant. Two separate effluent systems of gas are generated within the manufacturing process. These two systems of effluent gas are:

(1) the gases passing from the insulation formation chamber on the left of the diagram; and

(2) the curing chamber on the right of the diagram.

The gases being emitted by these two sources are mixed together and generally assume a temperature within the range of 70° to 200°F. The control process is provided with a water and/or steam injection system adapted to produce saturation temperatures at a minimum gas temperature of approximately 120°F.

After the effluent gas has been cooled as provided for in the above step, a positive corona electrostatic precipitator is used to produce complexes which tend to tie up (or form complexes) with the organic compounds of interest, such as p-hydroxybenzyl alcohol, o-hydroxybenzyl alcohol, and subsequent prepolymer resoles.

Once these complexes have been formed, ultraviolet light of a spectral range of 1500 A to 5000 A is irradiated into the system and exposed to the formed complexes. Particularly, the ultraviolet light acts to photooxidize the formed complexes and to initiate chemical reactions which act to generally internally rearrange the complexes to form tightly bound, high energy complexes which again assume the form of nonvolatile acids.

The nonvolatile product acids or product aerosols are then collected from the system by a conventional particulate technique such as by the use of an electrostatic precipitator.

FIGURE 40: PROCESS FOR AIR POLLUTION CONTROL IN FIBERGLASS MANUFACTURE

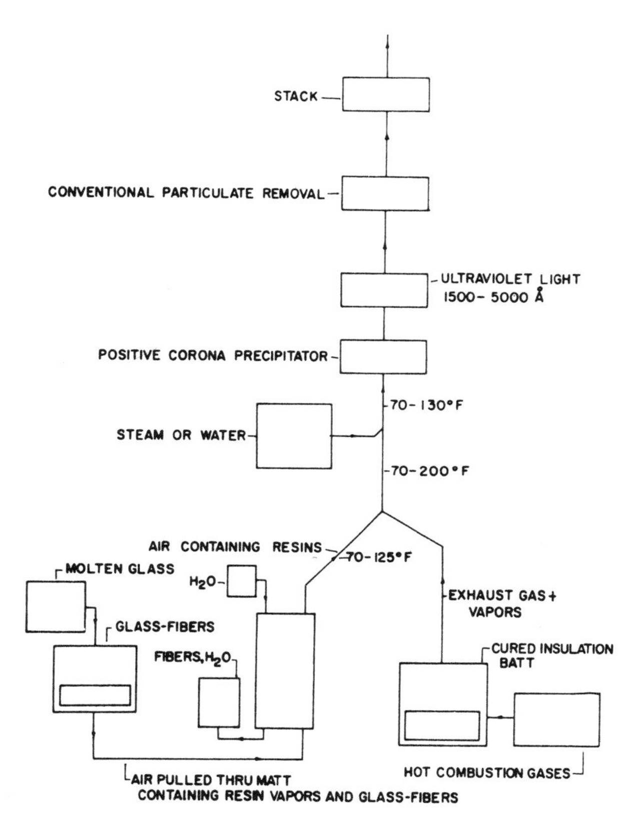

Source: U.S. Patent 3,984,296

A process developed by *R.E. Loeffler; U.S. Patent 3,865,540; February 11, 1975; assigned to Johns Manville Corporation* is one in which a binder-impregnated glass fiber blanket is cured and shaped by passing the blanket through a series of spaced-apart heated platen assemblies. The curing of the blanket produces vapors, fumes, odors and other pollutants which must be prevented from escaping to the surroundings of the platen assemblies. Consequently, these gases are purged from the blanket as the blanket passes between adjacent platen assemblies.

Hot air is introduced from a plenum chamber onto one side of the blanket. The hot air and pollutants from the binder are drawn through and from the blanket by a suction chamber on the opposite side of the blanket. The gases from the suction chamber are then passed through an air filler and discharged to the atmosphere.

A process developed by *H.J. Hoag, Jr. and A.M. Rawlinson; U.S. Patent 3,902,878;*

September 2, 1975; assigned to Owens-Corning Fiberglas Corporation involves forming fibers of glass by engaging centrifuged streams or primary filaments of glass by an annular blast to attenuate the streams or primary filaments to fibers. A heat-absorbing vaporizable liquid is delivered onto the high temperature fibers at the fiberizing zone and evaporating substantially all of the applied liquid by the heat from the fibers and the fiberizing environment to cool the fibers so that binder may be delivered onto the cooled fibers with a minimum of volatilization of the binder constituents thereby reducing the discharge of volatiles and particulates into the atmosphere.

A process developed by *F.E. Warner and A.P. Rice; U.S. Patent 3,528,220; September 15, 1970; assigned to Fibreglass Limited, England* is a process for removing phenolic air pollutants in the production of glass fiber products. The polluted air is passed sequentially through at least two low energy contacting zones in each of which it is contacted with descending scrubbing liquor, each succeeding contacting zone having at its bottom individual liquor collecting means.

There are four major uses for water in a fiberglass manufacturing process. Firstly, cooling water is required for furnaces, compressors and the like. Secondly, water is used to quench the molten glass should the process have to be stopped for any reason. When the molten glass is quenched in this manner it rapidly solidifies and fractures into fragments called cullet. Thirdly, water is used to carry away waste glass fibers and binder from the area in which the fiberglass is formed. Fourthly, water is used in a scrubber system to reduce air pollution. The water used in these four areas makes up the majority of the wastewater of a fiberglass manufacturing process.

The principal contaminants carried by the wastewater used in the above procedures are glass fiber particles and waste binder. The binder is normally a urea-formaldehyde resin or phenol-formaldehyde resin, and gives rise to dissolved and suspended solids in the wastewater. The resin is formed by reacting urea or phenol with formaldehyde, and heating the mixture to give varying degrees of polymerization.

The dissolved solid contamination in the water results from resin solids with a low degree of polymerization which are water-soluble. Such resin solids, if recovered, would be reusable in the process. The suspended solid contamination comprises glass particles and those resin solids which have been polymerized by the heat in the forming process to such an extent that they are water-insoluble.

It is the presence of the organic material in the wastewater which gives rise to the high chemical oxygen demand (COD), and a reduction of the organic material leads to a reduction in COD. The reduction of COD is very important if the treated wastewater is to be discharged into a river, since a high COD would mean that the wastewater consumed the oxygen in the river water which supports the plant and animal life therein.

If the wastewater were not treated, it would be possible to recycle it to a very limited extent until the level of contamination rose to an unacceptable level. However, the solids content rapidly builds up to an unacceptable level, even as high as 7 to 8%, and it is then necessary to shut down the plant and clean out the water system. If this is not done the efficiency of the process suffers badly, and in particular the spray system, by which water is sprayed onto the fiberglass to remove waste materials, is reduced in efficiency, since a high concentration of resin or fiberglass particles blocks the spray nozzles.

Accordingly, it is highly desirable to treat the wastewater in order to maintain the contamination at an acceptable level for as long as possible, so as to maximize the time between stoppages, or eliminate stoppages completely. In the past a number of chemical treatments have been utilized for such a purpose, although none have been entirely satisfactory in treating the wastewater from a fiberglass manufacturing process.

Removal from Water

One treatment that has been employed to a significant extent uses alum as a coagulating agent to promote precipitation of solid contamination. However, the use of this compound did not give entirely satisfactory results, and also suffered from the additional problem that it introduced aluminum ion into the water which is not desirable because aluminum shows some toxicity to plants.

In another process for the recycling of wastewater from a fiberglass manufacturing plant, barium hydroxide is added to the wastewater to improve its detergent properties, and a part of the water treated with barium hydroxide is drawn off and used in the formation of binder for the process. This draw-off portion of the wastewater is acidified with a weak acid such as ammonium sulfate to reduce the alkalinity of the water to about pH 7 before it is employed in binder makeup. However, this process allows the dissolved ions in the bulk of the wastewater to rise to a high level, and it should be pointed out that barium ion is also an undesirable constituent of the treated wastewater because of its high toxicity.

A process developed by *T.N. Crowley and D.M. Urbanski; U.S. Patent 3,966,600; June 29, 1976; assigned to Amchem Products, Inc.* for the treatment and recycling of wastewater from a fiberglass manufacturing process contaminated with both water-soluble and insoluble resins together with nonresin particulate material comprises:

(a) acidifying the wastewater at a temperature from about 70° to 200°F by the addition of an inorganic acid selected from the group consisting of sulfuric, carbonic and phosphoric acid in an amount sufficient to give acidified water having a pH of from about 3.5 to 4.5;

(b) neutralizing the acidified wastewater by the addition of a base selected from the group consisting of calcium oxide, calcium hydroxide and magnesium oxide, in an amount sufficient to give neutralized water having a pH of from about 7.5 to 8.5, the acid of step (a) and the base being selected so as to form a water-insoluble inorganic salt;

(c) adding an anionic polyelectrolyte flocculating agent to promote separation of solid material obtained from step (c);

(d) separating the solid material from the neutralized water, and adding a part of the solid material to the acidified water between steps (a) and (b); and

(e) returning the neutralized water separated from the solid material to the fiberglass manufacturing process.

Figure 41 is a block flow diagram of such a process.

FIGURE 41: PROCESS FOR THE TREATMENT OF WASTEWATER FROM A FIBERGLASS MANUFACTURING PROCESS

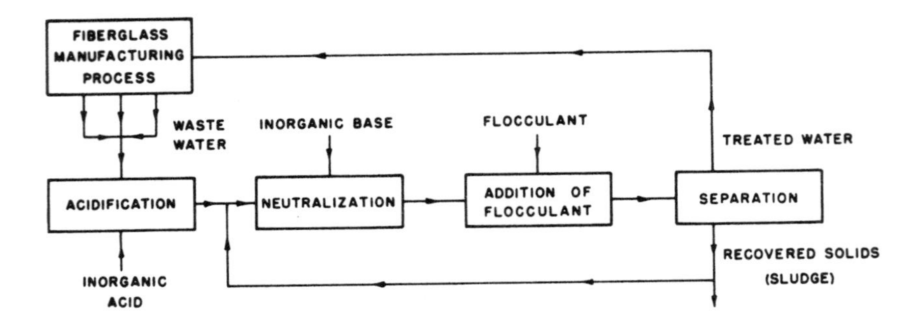

Source: U.S. Patent 3,966,600

A process developed by *J.E. Etzel, C.H. Helbing and C.A. Justus; U.S. Patent 3,791,807; February 12, 1974; assigned to Certain-teed Products Corporation* is one in which the wastewater is treated both mechanically and chemically to remove solid materials contained in the wastewater such as fiber glass, dust and other like materials. The waters are also chemically treated with high molecular weight cationic polymers to reduce substantially the phenolic resin content of the waters.

Recirculation of various water streams for utilization in the fiber glass manufacturing process including utilization of treated waters in the manufacture of binder solutions is included in the process. The process is such that wastewaters used in the manufacturing process can be totally reused thus requiring no disposal of waters to the environment.

Figure 42 is a flow diagram of such a process. In that figure, glass fiber threads **1** are shown being drawn into the face of a forced gas flame emanating from a burner **2**. The fibers attenuated in the face of the flame are blown in short staple lengths onto a forming chain **3**. Prior to their deposition on the surface of the forming chain **3** a phenolic type resin binder is applied from spray head **4** to the fibers.

The mat **5** formed on the forming chain **3** is passed over a roller **6** to a suitable oven. The forming chain bends around the roller **6**, passes over a series of rollers **7** and **8** and is admitted to tank **10** containing soft water. The chain **3** during its passage over the rollers to the tank **10** is sprayed with water from a spray **11**. The chain passes through tank **10** over rollers **9** and **12** and then passes over rollers **13, 14, 15** and **16** and is subjected to high pressure water spray from spray device **18**. The chain then passes over rollers **17, 18'** and **19** and is returned to the mat forming area for the collection of further mat **5**. All water introduced into the system onto the forming chain **3** via nip roll spray **11**, soft water bath

FIGURE 42: PROCESS FOR WASTEWATER RECLAMATION IN FIBERGLASS MANUFACTURE

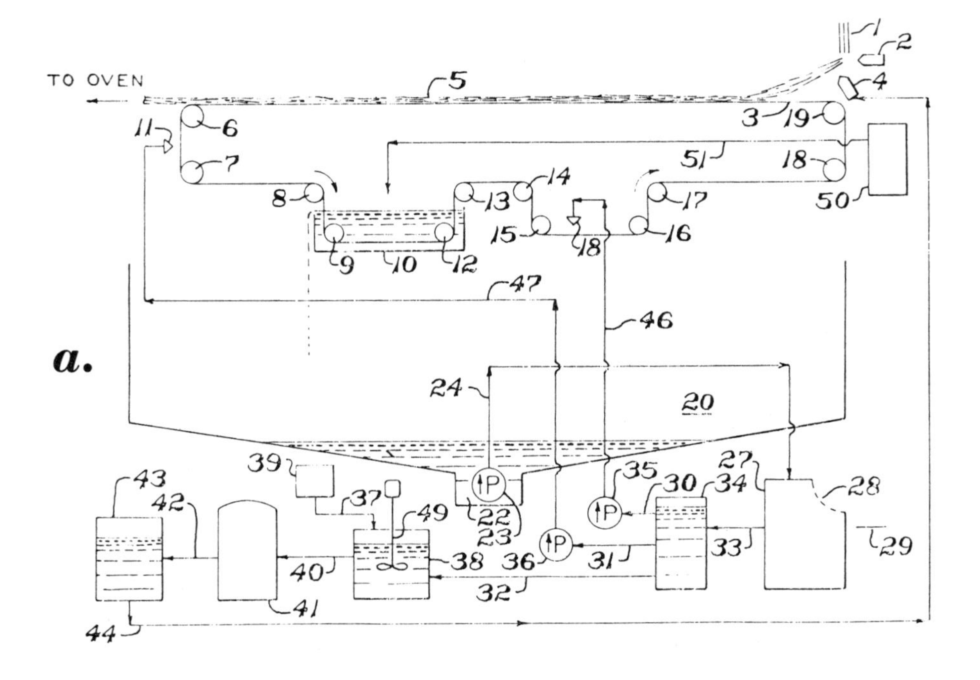

Wastewater Reclamation System

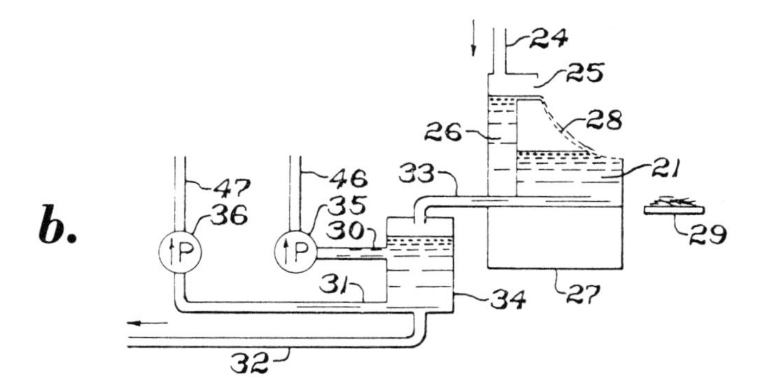

Enlarged View of the Solids Dewatering System

Source: U.S. Patent 3,791,807

10 and the high pressure spray **18** is collected in a large floor reservoir or sump generally indicated as **20.** The reservoir **20** is provided in the central portion thereof with a ditch or sump **22.** Located within sump **22** is pump **23.** All water collected in the ditch or sump area **22** is pumped into a solids separation device **27** via line **24.**

Solids are removed at the face **28** of separator **27** and flow onto a conveyor **29** for removal. Liquid from the solids-liquid separator is removed via line **33** and is passed to an effluent tank **34.** The liquid material from tank **34** is removed in the preferred embodiment in three lines **30, 31** and **32.** Lines **30** and **31** are connected to pumps **35** and **36**, respectively. Line **32** feeds liquid from tank **34** to the liquid treatment tank **38.** The tank **38** is provided with a line **37** running from a tank **39** to supply chemicals to tank **38** to treat the water contained therein.

The treated water from tank **38** is fed to filter **41** via line **40** and the filtrate is removed via line **42** and is fed to a binder mixing tank **43.** Binder solutions from tank **43** are fed to spray **4** via line **44.** Pump **35** feeds the water from line **30** via line **46** to the high pressure spray system **18** and pump **36** feeds water from line **31** via line **47** to the spray system **11.** Makeup water to the system is supplied in the preferred embodiment from water softener **50** via line **51** to the tank **10.**

Turning now to Figure 42b which is an enlarged view of the preferred solids separator system, the material pumped from the concentrated ditch or sump pool **22** by pump **23** is passed via line **24** to the solids separator **27.** In the preferred embodiment the separator **27** is a sieve device having positioned on a generally vertical face thereof a screened element **28.** Water is fed into chamber **26** of separator **27** and overflows at lip **25.** The solids and liquid pass down the face of screen **28** and solids are removed from the screen and pass onto a dewatered solids collector **29.** The liquid passes through the screen into reservoir **21** and is passed via pipe **33** to screened effluent tank **34.**

An effluent pipe **30** is connected to the high pressure spray system **18** via pump **35** and line **46** where it is used as wash water for the forming chain **3.** Line **31** feeds water via pump **36** and line **47** to spray system **11** for use in cleaning the forming chain **3.** A blowdown line **32** is also provided in the screened effluent tank **34** which feeds the remainder of water to a coagulation tank **38** provided with a suitable mixer **49.**

FIREFLOOD OPERATION EFFLUENTS

The production of hydrocarbons from subterranean hydrocarbon-bearing formations by means of a fireflood is well known to those skilled in the art. Concisely, a fireflood operation is effected by igniting hydrocarbons in a subterranean formation, injecting air through an injection well to sustain the combustion, and producing hydrocarbons freed by the heat of combustion from a production well. Combustion product gases are also produced from a production well. When the subterranean hydrocarbon-bearing formation also contains sulfur compounds, H_2S is sometimes formed and must be removed prior to exhausting into the atmosphere.

Removal from Air

A combustion product gas from a fireflood operation is comprised largely of nitrogen and carbon dioxide. However, the combustion product gas contains up to 10,000 parts per million H_2S and 150,000 to 180,000 parts per million carbon dioxide.

A process developed by *V.W. Rhoades; U.S. Patent 3,845,196; October 29, 1974; assigned to Cities Service Oil Company* is one in which an oxygen-containing gas is heated by a combustion engine employed to drive a compressor employed to inject air into a fireflood operation. The heated oxygen-containing gas is contacted with H_2S containing combustion product gas from the underground combustion of the fireflood operation. The oxygen and the H_2S react to form elemental sulfur, which can be recovered. Pollution of the atmosphere with H_2S is mitigated.

Figure 43 shows a suitable arrangement of apparatus for the conduct of the process. In the figure, an injection well **105** is cased and completed through cement completion **104** from the earth's surface **101** through overburden **102** into subterranean reservoir **103** which contains a viscous crude oil. Perforations **106**, for example, are provided in order to admit air **131** injected from compressor **130** connected through well head tubing **129** into reservoir **103**. The injected air forms an air bank **107** within the formation supplying oxygen for sustaining the combustion front **108** burning the residual oil **109** remaining after primary and/or secondary oil recovery.

Residual oil **109** and noxious production gases resulting from the in situ combustion fireflooding of the reservoir **103** are produced through production well **110**, cased and completed by cement **111**, from wellhead **112** and are passed to and separated in a gas-liquid separator **113** to yield recovered liquid **115** from the lower section of the gas-liquid separator **113** and H_2S containing combustion product gases from the gas outlet **114** located in the upper portion of the gas-liquid separator.

The H_2S-containing combustion product gases are introduced into a catalytic reactor **116** through inlet **150**. The catalytic reactor contains a catalyst to promote the reaction of H_2S with oxygen to produce sulfur and water vapor. The catalyst can also be effective to promote the conversion of CO and oxygen to form CO_2. A stack gas of water vapor, nitrogen, and carbon dioxide **118** is produced from the exhaust stack **117** of the catalytic reactor **116**. The catalyst of the catalytic reactor is continuously recirculated and regenerated for the removal of sulfur.

This is accomplished by removing the catalyst from the lower portion of the catalytic reactor **116** through exit **119**, passing the catalytic material through the pump **120** and introducing it into a sulfur recovery unit **121** into which steam **122** is introduced. Water and elemental sulfur **124** are produced through exit line **125** controlled by valve **123**. Regenerated catalyst **132** is then reintroduced into the upper portion of the catalytic reactor **116** through entry **154**. Also charged to the catalytic reactor through entry **155** via line **126** entirely insulated by insulating covering **156** is an oxygen-containing gas.

FIGURE 43: PROCESS FOR RECOVERING H_2S FROM FIREFLOOD OPERATION VENT GASES

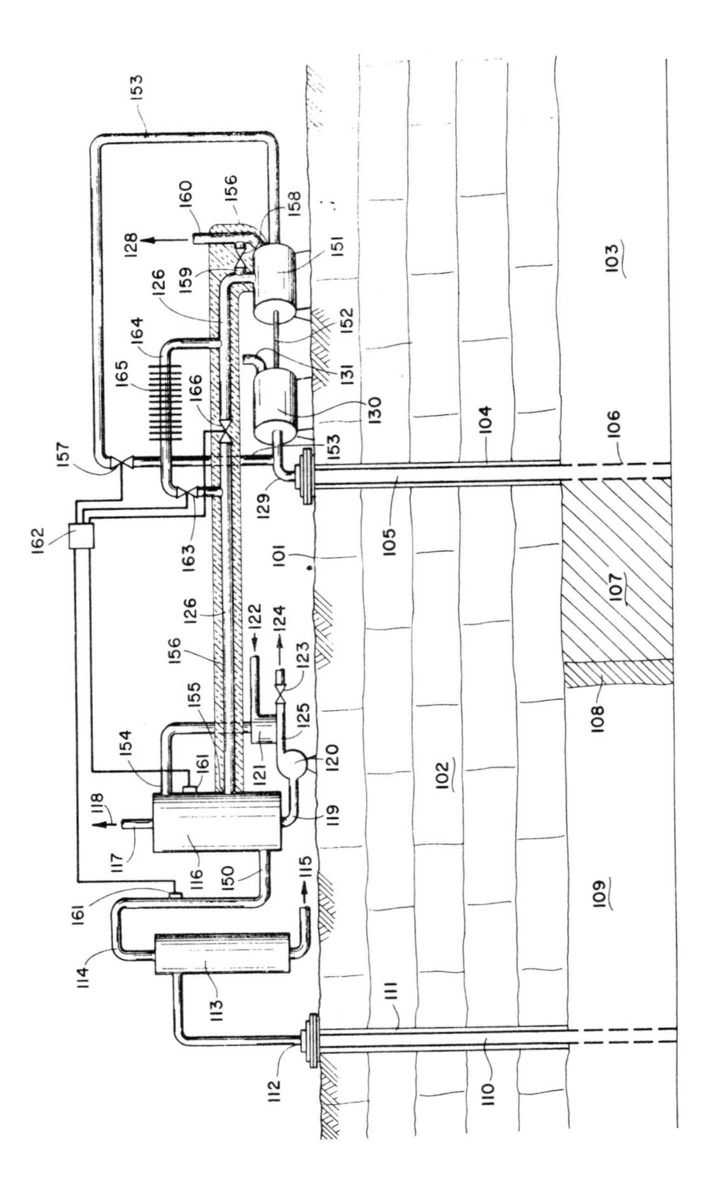

Source: U.S. Patent 3,845,196

The oxygen-containing gas which is charged to catalytic reactor **116** through entry **155** is taken from high-pressure air line **129**. The air from high-pressure air line **129** is passed via line **153** through control valve **157** in heat exchange relationship with combustion engine **151** via line **126** to catalytic reactor **118**.

Combustion engine **151** drives compressor **130** by means of driveshaft **152**. Exhaust gases **128** are exhausted from combustion engine **151** by way of exhaust pipe **160**. Valve **158** can be opened if desired to pass exhaust gases through line **159** into line **126** and thence to reactor **116**. The system including line **160** and **126** is enclosed by insulative jacket **156** to conserve heat.

A control system provides for optimum heat control in the reactor **116** as well as optimum oxygen content for most efficient reaction of oxygen with H_2S and CO in the H_2S-containing combustion product gases from the production well. Thus, temperature and H_2S plus CO content sensors **161** are connected to controller **162** which automatically activates valve **157** controlling the amount of air fed to the reactor, and valves **163** and **166** which control the temperature of air passed to the reactor **116** through entry **155**.

Valves **163** and **166** control the amount of oxygen-containing gas passing through shunt **164** and heat radiator **165** thus controlling the temperature of oxygen-containing gas passing to the reactor **116** through entry **155**, and thus the reaction temperature in the catalytic reactor **116**.

FLEXOGRAPHIC PRINTING PROCESS EFFLUENTS

Flexographic printing processes are used extensively in the corrugated paperboard industry and such use is increasing. Accompanying the extensive use of the flexographic printing process is a problem in the disposal of the waste wash-up water from cleaning the presses. The presses periodically are cleaned by flushing with large quantities of water. The used wash water contains highly visible contaminants and presents a very difficult disposal situation in many areas.

The product obtained from flushing flexographic presses with water to remove unused or unwanted ink is known as flexo waste wash-up water. This water contains quantities of ink, is generally extremely dark, is usually of high tinctorial strength, and presents a difficult disposal problem in many areas. An average sample of wash water has a pH of about 8.5.

Usually some extraneous materials are present, mostly paper fibers, but this accounts for only a small percentage of the total solids weight. The solid material is composed mainly of finely divided particles, most of which are less than a few microns in diameter. The pH, because of the high ink dilution, usually approaches that of the water used for wash-up.

Removal from Water

A process developed by *J.E. Voight, E.M. Bovier and C.J. Liebman; U.S. Patent 3,970,467; July 20, 1976; assigned to Anheuser-Busch, Incorporated* involves treating the wash water in flexographic printing processes to remove ammonium ions and subsequently reusing the water in preparing the adhesives. The preferred process includes the steps of separating the ink solids from the wash water, recovering a liquor, and removing ammonium ions from the liquor to condition the liquor for subsequent use in preparing corrugating adhesives. Such a process is shown schematically in Figure 44.

FIGURE 44: PROCESS FOR WASTEWATER PURIFICATION AND REUSE IN PAPER BOX MANUFACTURE

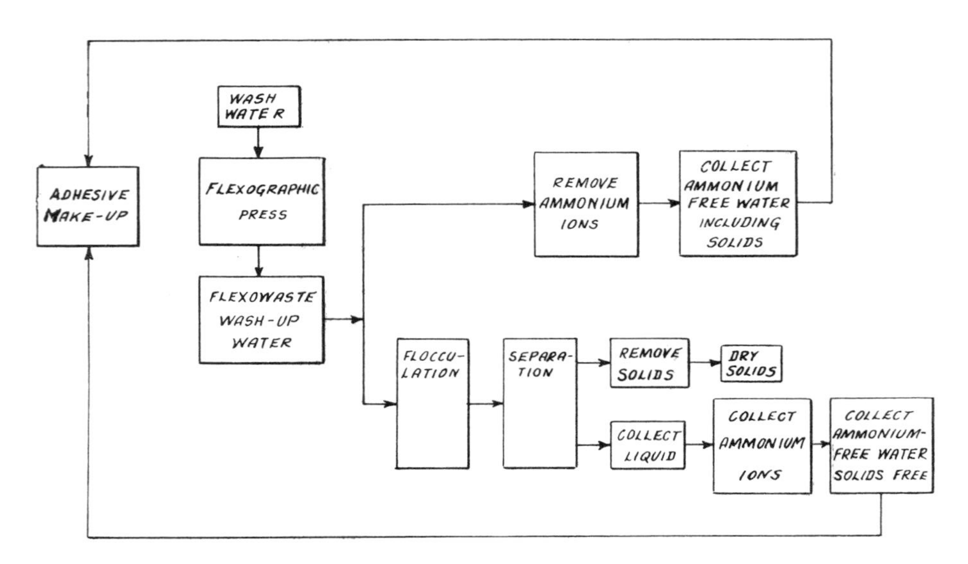

Source: U.S. Patent 3,970,467

See Paper Box Plant Effluents (U.S. Patent 3,868,320)

FLUOBORATES

Lead and tin fluoborates and fluoboric acid are widely used in lead and tin plating processes and appear as contaminants in the water used to rinse the plated parts. The problem of finding a satisfactory method for removing these contaminants is a difficult one because the appreciable solubility of fluoborates in aqueous media makes it impossible to precipitate them on anything approaching a quantitative basis. For example, the least soluble of the fluoborates, potassium fluoborate, is soluble in water to the extent of 4.4 grams per liter. Sheer volume contributes to the magnitude of the problem. A plating operation can generate 50,000 to 100,000 gallons or more of fluoborate contaminated wastewater daily.

Removal from Water

A process developed by *J. Singh; U.S. Patent 3,959,132; May 25, 1976; assigned to Galson Technical Services, Inc.* is based on the discovery that fluoborates can be rapidly and efficiently removed from aqueous media by first hydrolyzing the fluoborates to fluorides with aluminum or an aluminum salt and then reacting the resulting fluorides with a calcium salt. This converts the fluorides to calcium fluoride, which is sparingly soluble and accordingly precipitates and can be readily separated from the aqueous liquid. Such a process is shown in Figure 45.

FIGURE 45: METHOD FOR REMOVING FLUOBORATES FROM AQUEOUS MEDIA

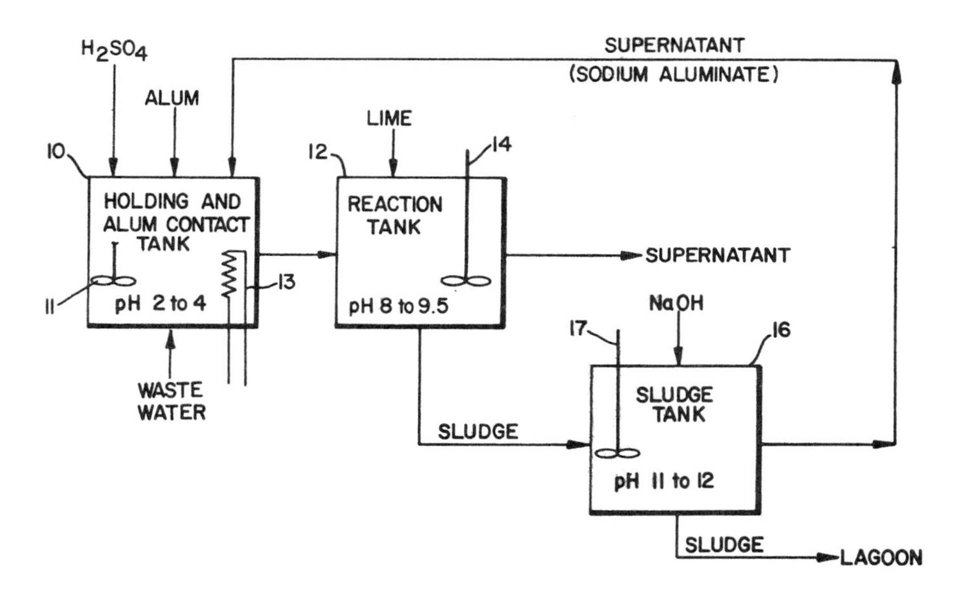

Source: U.S. Patent 3,959,132

The wastewater to be treated is introduced into a holding tank **10** equipped with an agitator **11** along with alum with whatever agitation may be necessary to produce uniform dispersion of the alum. Sulfuric acid or other acidic material is also introduced into holding tank **10**, if necessary, to adjust the pH of the solution in the holding tank to between 2 and 4.

The solution is held in tank **10** for a period of 4 to 8 hours. At the end of this period the hydrolysis or decomposition reactions will be essentially complete (the holding time can be reduced by employing an electrical resistance heater **13** or other heating device to elevate the temperature of the holding tank contents and speed the decomposition reactions).

The contents of holding tank **10** are then transferred to a reaction tank **12**. Lime is also added to reaction tank **12** until the pH of its contents has been raised to the selected level in the 8.0 to 9.5 range. Reaction tank **12** is preferably equipped with a two-speed agitator **14**. Initially, the agitator is operated at high speed to thoroughly mix the contents of tank **12**. Thereafter, the mixture is slowly stirred. This promotes floc formation and drives the precipitation reactions toward completion.

In tank **12** the fluorides produced by hydrolysis in holding tank **10** react with the calcium in the lime and precipitate as calcium fluoride. The aluminum precipitates as aluminum hydroxide; and tin, lead, copper, and any other heavy metals present in the mixture also precipitate, forming a sludge. The sludge is allowed to settle and the clear supernatant is discharged. The fluoride content of the supernatant will typically be found to be at least 95% lower than that of the liquid supplied to holding tank **10**.

The sludge can be lagooned, dewatered, or otherwise processed in a conventional manner. Alternatively, if the initial fluoborate content was high and large amounts of aluminum are therefore present in the sludge, the sludge can be processed to recover the aluminum. This may be accomplished by transferring the sludge to a sludge tank **16** equipped with an agitator **17**. Here, a dilute solution of sodium hydroxide is added with agitation until the pH in sludge tank **16** reaches a level of 11 to 12.

At this pH, the calcium fluoride remains insoluble but the aluminum hydroxide dissolves, forming a solution of sodium aluminate supernatant. This supernatant can be recycled to holding tank **10** to provide aluminum for hydrolyzing the fluoborates.

An acidic material such as sulfuric acid will typically have to be added in this event to decrease the pH in holding tank **10** to the desired level. The sludge remaining in sludge tank **16** after the removal of the sodium aluminate supernatant is treated as suggested above.

FLUORINE COMPOUNDS

Removal from Air

A process developed by *V.S. Kalach and L.I. Burlakova; U.S. Patent 3,966,877; June 29, 1976* involves the processing of waste gases by absorption of hydrogen fluoride and silicon tetrafluoride or hydrogen fluoride, silicon tetrafluoride and sulfur dioxide from the waste gases by water solutions containing ammonium compounds such as ammonium carbonate, ammonium bicarbonate and ammonium fluoride. In addition to these ammonium compounds the absorption solutions contain sodium fluoride and ammonia. The absorption process produces a water solution containing ammonium fluoride and the precipitate of sodium fluosilicate. This solution is treated with sodium carbonate after which the precipitate of sodium fluoride is separated from the solution and the latter is delivered for reuse in the absorption process or for other use.

This method can be utilized in processing the waste gases liberated in the production of defluorinated fodder phosphates obtained by hydrothermal decomposition of phosphate ores. In addition to hydrogen fluoride these waste gases contain small proportions of silicon tetrafluoride and phosphate ore dust which cannot be trapped by special dust collectors.

This method can also be utilized in processing waste gases liberated in the production of double superphosphate and extraction of phosphoric acid. In addition to hydrogen fluoride the waste gases contain silicon tetrafluoride and phosphoric acid mist.

In addition, the method may be utilized in processing waste gases in the aluminum industry, the gases containing equal parts of hydrogen fluoride and sulfur-bearing gas (sulfur dioxide) with a small admixture of silicon tetrafluoride, and dust particles of cryolite, sodium fluoride and alumina. The method can also be used, in particular, for processing of waste gases in the production of concentrated hydrofluoric acid, which contain hydrogen fluoride and silicon tetrafluoride. Figure 46 is a flow diagram of the process.

FIGURE 46: SCRUBBING FLUORIDES FROM WASTE GASES

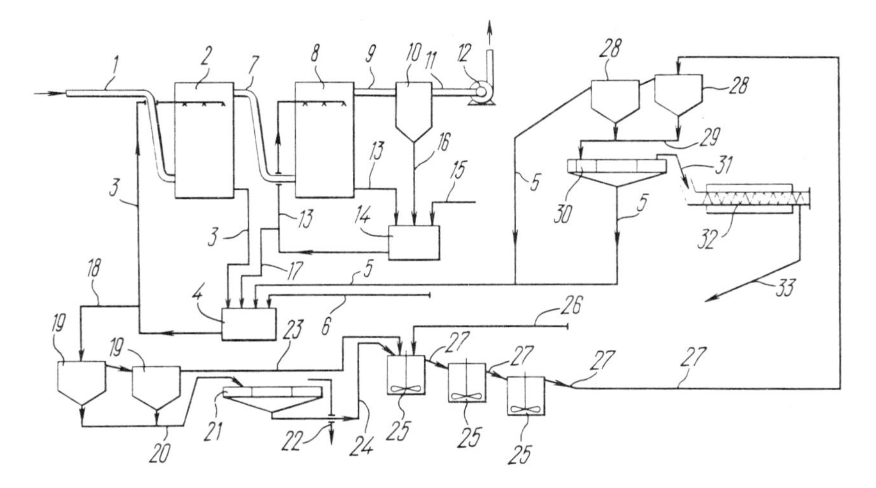

Source: U.S. Patent 3,966,877

According to the drawing, the waste gas containing hydrogen fluoride and silicon tetrafluoride as well as such unwanted admixtures as phosphate ore dust, cryolite or alumina dust, sulfur-bearing gas (sulfur dioxide) or phosphoric acid mist flows through gas pipe **1** into the first stage absorption apparatus **2** which is sprinkled with a water solution circulating through pipeline **3** and delivered from the first stage circulation tank. The water solution fed into the absorption apparatus contains sodium fluoride within its solubility limits, ammonium compounds (ammonium carbonate, ammonium bicarbonate), ammonia and ammonium fluoride.

In the course of absorption, ammonium carbonate, ammonium bicarbonate and ammonia interact with gaseous hydrogen fluoride and form ammonium fluoride which readily dissolves in water; this proceeds according to the following reactions:

$$2HF + (NH_4)_2CO_3 \longrightarrow 2NH_4F + CO_2 + H_2O$$

$$HF + NH_4HCO_3 \longrightarrow NH_4F + CO_2 + H_2O$$

$$HF + NH_3 \longrightarrow NH_4F$$

If the waste gases to be processed contain sulfur dioxide, in addition to hydrogen fluoride and silicon tetrafluoride, the process of absorption produces monosubstituted ammonium sulfite according to the reaction:

$$SO_2 + (NH_4)_2CO_3 \longrightarrow NH_4HSO_3 + CO_2 + NH_3$$

Simultaneously, the sodium fluoride contained in the absorption solution interacts with silicon tetrafluoride which is delivered with the gas flow, and forms the precipitate of sodium fluosilicate according to the reaction:

$$SiF_4 + 2NaF \longrightarrow Na_2SiF_6$$

A small excess of ammonium compounds in the absorption solution produces a neutral or alkaline medium wherein the phosphate ore dust and other solid particles entrained by the gas flow are not dissolved and precipitate together with sodium fluosilicate. The pH value of the circulating absorption solution may vary from 5.0 (slightly acid) to 9.0 (slightly alkaline).

However, the most acceptable pH is in the neutral range of from 6.5 to 7.5. These predetermined parameters are maintained by continuous or intermittent supply into the first stage circulation tank **4** of a water solution containing ammonium compounds and sodium fluoride through a pipeline **5**; sometimes it is sufficient to supply a water solution of ammonia delivered through a pipeline **6**.

After the major proportion of fluoride compounds has been absorbed in the first stage absorption apparatus **2** the gas moves through a gas pipe **7** for secondary cleaning (to meet sanitary requirements) in the second stage absorption apparatus **8** from which it moves through a gas pipe **9** into a spray separator **10**. From the spray separator **10** the gas is discharged into the atmosphere through a gas pipe **11** and a fan **12**. The second stage absorption apparatus **8** is also sprinkled with absorption solution which circulates through a pipeline **13** and is delivered from the second stage circulation tank **14**.

The process of absorption of hydrogen fluoride and silicon tetrafluoride from waste gases is accompanied by partial evaporation of water and by discharge of vapors into the atmosphere together with the purified gases through the fan **12**; therefore, the second stage circulation tank **14** is supplied with clean water through a pipe **15** to make up for the water losses at all the stages of the technological process. Simultaneously, the same tank **14** is supplied through a pipe **16** with liquid from the spray separator **10**.

A part of the solution is pumped from the tank **14** through a pipe **17** into the first stage circulation tank **4**. A part of the circulating absorption solution is, in turn, taken from the first circulation tank **4** and delivered through a pipe **18** into tandem-connected settlers **19** for settling the suspended particles of sodium fluosilicate, phosphate dust and other insoluble admixtures.

The settling rate of these particles ranges from 0.1 to 0.3 m/hr. The sludge with a solid-to-liquid phase ratio of 1:10 is delivered through a pipe **20** to a vacuum filter **21** from which the filtered sludge is discharged from the process through a line **22**. The sodium fluosilicate removed from the process can be used as a commercial product.

The clarified solution containing ammonium fluoride is delivered from the settlers **19** through a pipe **23** and the filtrate is delivered through a pipe **24** into the first one of the three successively located reaction vessels **25** with agitators. Simultaneously, sodium carbonate (soda ash) is delivered through a line **26** into the first reaction vessel **25**.

The other two reaction vessels **25** are designed to adjust the predetermined relation of source components (ammonium fluoride and sodium carbonate) and to stir the reaction material. The amount of delivered sodium carbonate is smaller than that of ammonium fluoride, constituting 80 ± 10% of the amount determined by calculations.

Interaction of soda ash with the clarified solution containing ammonium fluoride and the unreacted part of sodium fluoride, ammonium carbonate and bicarbonate, and ammonia produced sodium fluoride and ammonium carbonate according to the reaction:

$$2NH_4F + Na_2CO_3 \longrightarrow 2NaF + (NH_4)_2CO_3$$

It must be noted that the produced ammonium carbonate is hydrolyzed to a considerable extent and produces ammonium bicarbonate and ammonia according to equations:

$$(NH_4)_2CO_3 + H_2O \rightleftharpoons NH_4HCO_3 + NH_4OH$$

$$NH_4HCO_3 \rightleftharpoons NH_4OH + CO_2$$

$$NH_4OH \rightleftharpoons NH_3 + H_2O$$

Nevertheless, to simplify calculations, the materials required for the technological process are expressed in terms of ammonium carbonate only. It can be seen from the last three equations above that considerable concentrations of ammonium carbonate will cause losses of ammonia into the atmosphere, therefore concentration of ammonium carbonate in absorption solutions is limited to 3 to 5%. It has been mentioned above that ammonia losses are made up for by delivering an aqueous solution of ammonia through the pipe **6** into the first stage circulation tank **4**.

If the waste gas contains sulfur dioxide, the mono-substituted ammonium sulfite is transformed into sodium sulfite. Solubility of sodium fluoride in water is not over 4.2 wt % while in the industrial solutions it does not exceed 3.0 to 3.5 wt % so that the major part of sodium fluoride produced according to the reaction of NH_4F and Na_2CO_3 precipitates. If the waste gases contain sulfur dioxide in addition to hydrogen fluoride and silicon tetrafluoride, the main proportion of sodium sulfite contained in the absorption solution precipitates together with sodium fluoride.

The suspension of sodium fluoride is fed from the reaction vessels **25** through a pipe **27** into the tandem-arranged settlers **28** to thicken the precipitate until the solid-to-liquid phase ratio reaches 1:3. The thickened mass is delivered through a pipe **29** to a vacuum filter **30**.

The clarified solution and the filtrate saturated with sodium fluoride and containing ammonium carbonate, ammonium bicarbonate and ammonia are delivered, respectively, from the settlers **28** and vacuum filter **30** through the pipe **5** into the first stage circulation tank **4** from which they can be used again for absorption of hydrogen fluoride and silicon tetrafluoride from waste gases. The filtered precipitate of sodium fluoride is delivered through a line **31** into a drying oven **32** from which it is discharged through a line **33**, packed and delivered to the consumers as a commercial product.

See Alkylation Process Effluents (U.S. Patent 3,972,956)
See Aluminum Cell Effluents (U.S. Patent 3,876,394)
See Aluminum Cell Effluents (U.S. Patent 3,827,955)
See Aluminum Cell Effluents (U.S. Patent 3,907,971)
See Phosphoric Acid Process Effluents (U.S. Patent 3,811,246)

FLY ASH

Removal from Air

A process developed by *J.B. Dunson, Jr. and R.L. Lucas; U.S. Patent 3,969,094; July 13, 1976; assigned to E.I. Du Pont de Nemours and Company* is one in which baffle tray columns are designed and operated so as to function as highly efficient scrubbers for the cleaning of flue gases subject to wide variations in flow. The columns are designed and operated such that areas within the scrubbing area operate with a liquid phase continuous froth as the contacting medium. The columns can also contain an integral nonclogging separator for entrainment control. Figure 47 shows two suitable types of baffle tray column design.

The view at the left of the figure is a cross-section of a baffle tray column of the disc-and-donut type. The baffle tray scrubber is a vertical column **1** essentially divided into three areas, a deentrainment area **2**, a scrubbing area **3**, and a bottom area **4**. The scrubbing area **3** is further divided into a top portion **5** and a bottom portion **6**.

The integral entrainment separator **8** illustrated is of the cap type. This type of deentrainment means is preferred as it can accommodate a wet/dry interface without creating a plugging problem. Another type entrainment separator could be employed, such as a cyclone, or a zigzag vane type, or a mesh pad, the latter being preferred when there are no suspended solids in the gas stream. If desired, a cap entrainment separator containing a vaned annulus could be employed.

The liquid scrubbing stream **7** is introduced on the outside wall above the top of the cap type entrainment separator **8**. The supply means is not critical and an overflow weir or a group of spray nozzles can be employed. The stream flows down the wall of the column past the cap as a falling film, thereby washing the outside wall so as to prevent the formation of a mud ring or a ring of crystal growth adjacent to the lower lip of the cap **9**.

To avoid possible plugging problems, a smaller spray **10** of the scrubbing liquid can be used to wash away the solids that may collect on the riser **11** and the inside of the cap. The liquid scrubbing stream drains via **13** from the entrainment separator down through a seal loop **23** to feed weir **14** on the top tray, from which it cascades down through the scrubbing area. A portion **12** may be drawn off to allow use of more liquid for wall washing than is needed for column operation at very high gas throughputs.

The scrubbing area **3** has the minimum free open area for flow at the top **5**, with lower trays having progressively more free open area **6**. Note, however, the progressive change in free open area need not be uniform.

FIGURE 47: ALTERNATIVE BAFFLE TRAY COLUMN DESIGNS FOR FLUE GAS SCRUBBING

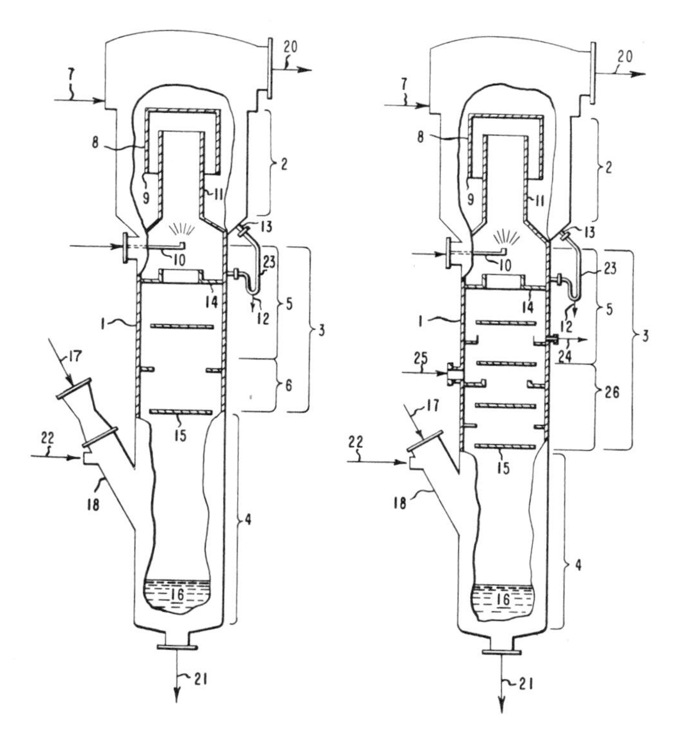

Source: U.S. Patent 3,969,094

This particular design allows smooth operation with whatever number of trays desired from the top of the column down in froth operation, while those below remain in spray operation. This concept, which provides control of normally unstable froth condition, allows higher liquid holdup and consequently much better gas scrubbing than is practical with other kinds of scrubbers suitable for hot slurry service.

It is not necessary that the froth condition be maintained in the top of the scrubbing area; it could be maintained at various zones throughout the area. In fact the entire scrubbing area could be operated under froth condition provided the gas flow can be controlled and will not increase unexpectedly. Preferably for stability, at least some portion of the scrubbing area is operating under spray conditions. The important feature is that the froth condition be maintained in at least part of the scrubbing area, and this area be from the point of liquid addition downward.

In the version at the right of the figure, a mid-column feed **25** is also employed and the froth condition is maintained both at the top of the column in area **5** and also in area **26**. The use of a mid-column feed in conjunction with a top feed is a preferred embodiment. It is not necessary that the two liquids be the same, in which case a mid-column drawoff of liquid **24** may be used. This embodiment would be particularly useful to water scrub fly ash separately and prior to scrubbing SO_2 or the like with a regenerable absorbent such as Na_2SO_3, $MgSO_3$, citric acid, or FeS.

If desired, more than two liquid feed points could be employed. Thus, the liquid could be introduced into the scrubbing area at four or five different locations. Near each of these points of liquid introduction the froth condition would be maintained. The use of multiple liquid feeds has the advantage that the system can compensate for maldistributions of gas and/or liquid.

The bottom tray **15** is a disc, which puts the liquid against the wall as a falling film as it drains to the bottom sump **16**. If a side-by-side column is employed, the last side-by-side tray should be placed vertically opposite the gas inlet, with a small gap between it and the wall. This feature provides a wetted wall thus preventing crystal growth. Dirty gas **17** enters from an inclined nozzle **18** below the bottom tray of the column. This nozzle can impart a swirling motion to the gas stream; such a flow of gas and liquid will reduce plugging and scaling problems at the entrance.

Heavy solids drop out into the liquid of the bottom sump **16** as the gas turns to go up through the column. The gas stream flows up through the column, moving radially in and out to pass the baffles. Cleaned gas is discharged above the top of the cap separator through the clean gas exit **20**. The liquid collected in the bottom sump can be removed through the liquid exit **21**. This liquid can then be reconstituted with fresh limestone and recycled to the column.

If desired, a portion of the liquid scrubbing stream **22**, or merely an aqueous stream, can be introduced into the gas entrance nozzle **18** as a swirling film along the sides. This film of liquid will eliminate a buildup of solids at the wet/dry interface where the entrance enters the column, thus eliminating a plugging problem.

In operation the gas stream, e.g., a hot flue gas containing SO_2, is fed into the entrance **18**. Into this same entrance is fed a portion of the liquid scrubbing stream, i.e., a slurry containing about 10% by weight of suspended solids. The gas stream flows into the column in a downward direction and the liquid stream flows along the wall of the entrance as a swirling film.

Upon entering the column the gas stream abruptly changes direction and flows upward through the column. This abrupt change in direction causes some of the particulates to fall into the liquid sump **16** at the bottom of the column. Generally, the particulates that will be removed in this area are those having a particle size above 100 microns.

The fractional collection efficiency for particulates is illustrated by the following example. 1160 acfm (exit conditions) of flue gas is cleaned by the type of column shown at the right hand side of the figure. 32 gpm of a scrubbing slurry containing limestone is fed into a Venturi quencher, and 110 gpm to three feed points on the column. Venturi pressure drop was 0.7 inch H_2O; column pressure drop was 6.2 inches H_2O. This column is operating under conditions of the process. SO_2 removal efficiency was about 95%. Overall particle collection efficiency was about 93% of an inlet grain loading of about 0.36 grain/standard cubic foot.

See Sodium Sulfur Oxide Wastes (U.S. Patent 3,962,080)
See Steam-Electric Industry Effluents (U.S. Patent 3,726,239)
See Steam-Electric Industry Effluents (U.S. Patent 3,890,207)

FOREST INDUSTRY WASTES

After an area has been logged the area is burned so as to reduce the possibility of a later noncontrollable fire. In order to realize maximum productive area for growing the next crop of trees, it is advisable to remove the stumps from the ground and to bunch together the logs, stumps, branches and limbs and to burn as much as possible. In a highly forested region, there will be many bunched logs, stumps, trees and branches. For example, there may be a crane at a landing site for bunching the logs and stumps and there may be many of these bunches.

This necessitates a crane and a crane operator. Further, there is a fire tender who oversees the burning of these bunches and who adds fuel such as gasoline, kerosene and diesel oil and even old rubber tires and other burnable material. In addition, there are two or three assistants who are continually scouring the countryside to extinguish small fires which are set near these bunches by sparks flying from the bunches. In all, there may be five or six or seven men working to burn these bunches of logs, stumps, limbs and branches in addition to the heavy machinery involved.

As an inherent result of burning these bunches or trash accumulation, there is produced a thick blue smoke due to the incomplete combustion of the wood and diesel oil. The wood comprises cellulose, lignin, resins and the like. In the burning of the wood and the diesel oil and other fuel, there are produced solid particulates in the smoke, water, carbon dioxide, carbon monoxide, and pyrolysis products of the burning of the wood.

Many of these pyrolysis products are carbon products. Except for the carbon dioxide, and the water, resulting from the burning of the wood, these other products are pollutants in the air. In the burning of these bunches and trash accumulation the time required is a minimum of approximately two days with observers being present all of the time to reduce or lessen the possibility of starting a forest fire. In the Pacific Northwest, the average airborne pollutant resulting from the burning of trash accumulation on cutover land is approximately 200 tons per acre of cutover land.

Removal from Air

A process developed by *A.S. McCorkle and C.L. Hoar; U.S. Patent 3,964,716; June 22, 1976* reduces such air pollution by making it possible to economically take the cull logs, broken logs, stumps, limbs and branches and remove these from the land so as to make it possible to have more available land for growing the next regenerative cycle of trees. The apparatus and method make it possible to take a large cull log or a large stump and in 15 minutes, reduce the size of this large cull log or large stump to a size which can be readily handled.

This will make it possible to readily dispose of the reduced log and stump over the ground to act as a mulch and a fertilizer or, preferably, to take the reduced log and stump and transport this reduced log and stump to a central processing plant and further process it into a more useful product. The apparatus may be a mobile apparatus for travelling to the cutover land and processing the logs and stumps on the cutover land or it may be a stationary apparatus so that the stumps and logs are brought to it and then processed at a central location.

FORMALDEHYDE

Removal from Air

See Aldehydes (U.S. Patent 3,909,408)
See Phenolic Resin Process Emissions (U.S. Patent 3,741,392)
See Phenolic Resin Process Emissions (U.S. Patent 3,911,046)

Removal from Water

A process developed by *W. Riemenschneider and O. Probst; U.S. Patent 3,328,265; June 27, 1967; assigned to Farbwerke Hoechst AG, Germany* permits obtaining a sewage that is practically free from formaldehyde when concentrating aqueous formaldehyde solutions by pressure distillation. The process comprises introducing boiling water into the sump of the distilling zone, continuously feeding in the aqueous formaldehyde solution to be concentrated at a height of one-fourth of the total height of the distilling zone and discharging a product that is practically free from formaldehyde from the sump of the distilling zone. The discharged sump product can be drained off without additional intermediate treatment as sewage which is unobjectionable from a biological point of view.

See Aldehydes (U.S. Patent 3,929,636)

FOUNDRY CASTING OPERATION EFFLUENTS

Many metal-melting furnaces are mounted to tilt about a substantially horizontal axis to discharge the molten contents. Whenever the furnace mouth is open, and particularly when material to be melted is charged into the furnace or when molten material is being poured, heavy, noxious fumes arise from the furnace into the circumambient atmosphere. The resultant pollution of the working environment in foundries has long been a serious problem. For many years, huge exhaust systems have customarily been used in an effort to maintain tolerable conditions for foundry workers by frequently changing the entire atmosphere within the foundry. Such systems, though they require the rapid movement of tremendous volumes of air, have been found to be quite ineffective even to maintain an average breathable atmosphere in a foundry and, of course, are utterly incapable of protecting the workers against the sudden intense pollution which occurs in the immediate vicinity of a furnace when it is thus fuming.

Removal from Air

A device developed by *R.C. Overmyer and J.R. Scheel; U.S. Patent 3,756,582; September 4, 1973; assigned to Hawley Manufacturing Corporation* is one in which an exhaust system is provided to maintain a subatmospheric pressure within a hood, and when the furnace is of the tilting variety with a pouring spout, the exhaust system will include swivel connections and telescoping ducts permitting the hood to shift on the furance and to tilt with the furnace in its shifted position. The hood may include a section movable about a vertical axis to provide greater access to the mouth of the furnace.

In general practice, sand cores and molds are made by mixing sand with from 1 to 5% liquid binder to obtain a free-flowing mix which is formed around a pattern in a flask or in a core box to the desired shape. Binders, depending upon the type used, may be dried or cured to harden the sand form by heat or by the use of reagent gases. In all instances, the hardened sand forms are held together by the cured or dried binder to provide operative working surfaces which, in the case of cores, form cavities in metal castings or, in the case of molds, form the outer or finished surfaces of metal castings.

It is important in founding metals that the cores which are used have certain desired characteristics which enhance the economics of the operation. Among these characteristics is: the necessity that the curing process must be very rapid and such as to minimize the cost of pattern equipment. The core which is produced must retain its form and strength until the metal stabilizes or freezes and then the core should disintegrate as rapidly and completely as possible to minimize the cost of removing the residual sand component of the core from the internal voids and cavities in the casting.

Other properties of a core which are deemed necessary are that it not cause hot tears in the castings, pinhole porosity or other surface defects which would subtract from dimensional tolerances or the physical strength of the casting, and that the core retain its properties in highly humid atmosphere. Cores formed by the use of resin binders excel in these properties, as compared with cores formed with conventional nonresin binders.

Many processes are used for the production of cores, but resulting advantages increasingly favor cores which are cured in the core box or pattern before being removed therefrom. These are called precision cores and are distinguished from cores which are transferred from boxes into dryers for drying in ovens, which result frequently in warping or abrading in handling. The economy of production and improved quality of cured-in-the-box precision cores greatly exceeds conventional cores and great emphasis is being put on methods and processes which can improve the production speed and minimize the pattern cost to produce such quality cores.

Various resinous binders have been discovered to be useful which can be cured very rapidly with basic or alkaline reagent gases to produce cores that have good physical properties and have excellent collapsibility after the molten metal has been stabilized or frozen. These two properties are in many situations superior to those of cores and molds produced by sand which includes sodium silicate reacted with CO_2 reagent gas as a binder.

The various resin compositions of the binders currently used require the use of NH_3 or, preferably, one of the amine gases or vapors such as trimethylamine or triethylamine as curing agents. These gases are basic and do not injure the internal part of vacuum pumps, automatic valves, patterns or core boxes employed in the process and apparatus, and therefore are desirable from this standpoint. However, these basic gases or vapors are toxic and noxious, and it is therefore necessary that concentration thereof in the work area be maintained below 25 parts per million of air so as to be reasonably tolerable for comfort and safety of the operators.

It has been extremely difficult to develop pattern equipment or core boxes with seals that can dependably contain these gases while they are pressurized to permeate completely through the sand-resin mix. In addition to this difficulty, the use of a mixture of air and basic triethylamine, according to usual pressurizing methods, leaves a residue in the cured core or mold of triethylamine condensate which slowly emanates from the core as a noxious vapor after the same has been cured and removed from the box.

As a result, cores cured by such reagents require very expensive pattern equipment and continue to release concentrations of triethylamine into the work area to such extent that the 25 parts per million concentration established as a minimum health standard cannot be maintained economically. The toxicity hazard and odor characteristics of this process therefore have handicapped its acceptance by the industry.

While 25 parts per million of trimethylamine or triethylamine are deemed adequate as a concentration to meet the minimum health standards, the realities are that such amines are the same gases as are given off by decaying fish or meat and are so noxious that much lower concentrations must be achieved if the processes using these reagents are to gain acceptance.

As an example, a person walking through an area with a concentration of 25 parts per million of such reagent gases can absorb enough of the odor in his clothing that it can only be removed by aerating the clothing for two or three days in a well ventilated space. While some other amines are less noxious than trimethylamine and triethylamine, the difference is only a matter of degree, since all amines have this characteristic of objectionable odor.

A process developed by *L.R. Zifferer and L.F. Stump, Jr.; U.S. Patent 3,795,726; March 5, 1974; assigned to Alphaco, Inc.* for controlling such undesirable polluting effects involves a control system which operates positively and in a foolproof manner by means of sensing the pressures in the chamber and thus minimizes the total machine cycle time irrespective of the size of core or mold being cured or of the size of the chamber employed and produces a predictable concentration level of noxious gases in the void spaces between the sand grains of the cured core, the desired steps of the process occurring strictly and automatically in an uninterruptible manner until the cycle is completed to provide mold and core products which present no health hazard and are technically satisfactory.

In the above-described operation, it is preferred that when the pump evacuates the chamber and the molds or cores contained therein, the noxious and toxic reagent gases withdrawn therefrom may be rendered harmless and unobjectionable by passing the pump discharge into a bath. The bath may be of a neutralized nature, such as a solution of phosphoric acid, and the neutralized product then may be discharged to the atmosphere. Otherwise, if desired, the exhaust from the pump may be burned, such as by combining it with acetylene.

A technique developed by *H. Riester; U.S. Patent 3,941,868; March 2, 1976; assigned to Georg Fischer AG, Switzerland* is one in which in casting operations using molds made up plastic-bonded sands, the exhaust gases generated from the plastic binder are collected within an enclosure to which the mold passes from a casting location. The gases are mixed with combustion air supplied into the enclosure and an ignition source is arranged within the enclosure to ignite the combined exhaust gases and combustion air.

The enclosure is insulated to retain the heat generated in the combustion operation. An outlet is provided from the enclosure so that any unburnt gases remaining from the exhaust gases can be collected and harmful substances removed to avoid any pollution of the atmosphere.

A process developed by *L.E. Flora and B.G. DuBois; U.S. Patent 3,937,272; February 10, 1976; assigned to Sutter Products Company* is one which involves supplying during an accurately adjustable controlled cycle time an accurately adjustable controlled volume of atomized cold set catalyst into a high pressure dry air, nitrogen and/or CO_2 gas stream passing through the core box to cure the resin followed by a continuing air, nitrogen or CO_2 gas purging of the core box.

The fumes from the core box are drawn through a natural gas or propane fired fume incinerator wherein an excess air-fuel mixture causes oxidation of the fumes during a controlled temperature and residence time. The fume incinerator operates continuously at a constant temperature under control by an automatic temperature sensor regardless of fume input and includes suction means to prevent back flow of gases from the incinerator to the core box.

FRUIT PROCESSING WASTES

The value of citrus (mainly orange and grapefruit) cannery residue as animal feed has been recognized by various state experiment stations, by dairymen and cattlemen located in the eastern half of the United States and in California and Texas. Because of the high water content and perishable nature of the residue, however, it cannot be transported economically for feeding purposes. The fresh material is difficult to handle, ferments rapidly, and sours, resulting in the attraction of insects and the emanation of objectionable odors. Its preservation by pressing and ensiling has been reported, but the practice has not been widely adopted.

Processing citrus cannery residue to form a dried citrus pulp is now a widely accepted means of preparing the valuable animal feed. Dried citrus pulp is one of the most digestible ingredients in feed available for cows. When this dried citrus pulp has been fed to dairy herds, milk production has been maintained and animals have been kept in a thrifty condition.

Dried citrus pulp is a citrus cannery residue to which nutrients and other additives can be added, and from which a major amount of liquid has been removed. The citrus residue contains pectin, which binds water in the residue. In the preparation of the dried citrus pulp, the citrus residue is generally limed to destroy the hydrophilic nature of the pectin. The moisture content of the citrus residue is then reduced by draining in bins and/or pressing with auxiliary equipment. The resulting solids or press cake are converted into dried citrus pulp by heating in a rotary kiln (drier) using a direct, hot air or steam tube device.

The liquids from the bins and/or presses are called peel liquids, which are comprised of water, carbohydrates, essential oils and d-limonene. It is customary in the citrus by-products industry to recover d-limonene from peel liquids and distillation operations. However, d-limonene in the solids (press cake) is generally carried into the atmosphere with waste gas vapors from the rotary kilns.

d-Limonene is a liquid terpene, $C_{10}H_{16}$, which occurs in citrus fruit. d-Limonene finds use as a solvent, as a wetting and dispersing agent and in the manufacture of resins. Thus, it is desirable to recover this material rather than venting it to the atmosphere. More importantly, it is characterized by an objectionable odor, and is now considered to be an air pollutant. In fact, it is believed that the city of Los Angeles, California, is actively seeking to reduce d-limonene discharges into the atmosphere to tolerable levels.

Removal from Air

A process developed by *E.L. Cocke and F.W. Muncie; U.S. Patent 3,966,984; June 29, 1976* involves the reduction of air pollution by recovering d-limonene from the process in which citrus residues are treated to yield animal feed. Figure 48 is a flow diagram of the process.

Citrus residue **28** comprising peel, rag and seeds passes under a lime feeder **29**, then it is conveyed to a size reduction unit in a manner similar to that employed in the prior art. The citrus residue is discharged onto a conveyor **31** and fed into a pug mill **32**. The limed citrus residue **33** is then fed into an enclosed, heated, agitating screw, ribbon or cut flight conveyor **34** (with or without hollow flights).

It has been found that the combination of heat in the enclosed agitation conveyor **34** and lime in the limed citrus residue **33** disintegrates the cell walls of such firm pieces of material as the flavedo, bleeding out the cell contents and bringing about a marked increase in absorptive capacity of the material. It has also been found that subsequent drying of the concentrated solids component **35** discharged from heated conveyor **34** is greatly accelerated in drier **36**. In contrast, prior art processes produce unlimed and only superficially limed chunks of flavedo which present a drying problem.

The vapors from the heated agitating screw conveyor **34** pass into condenser **41** to collect the d-limonene. Vapors usually pass through cyclones, scrubbers or both such devices before being exhausted into the atmosphere.

FIGURE 48: METHOD OF REDUCING AIR POLLUTION BY RECOVERING d-LIMONENE FROM CITRUS PULP PROCESSING OPERATION

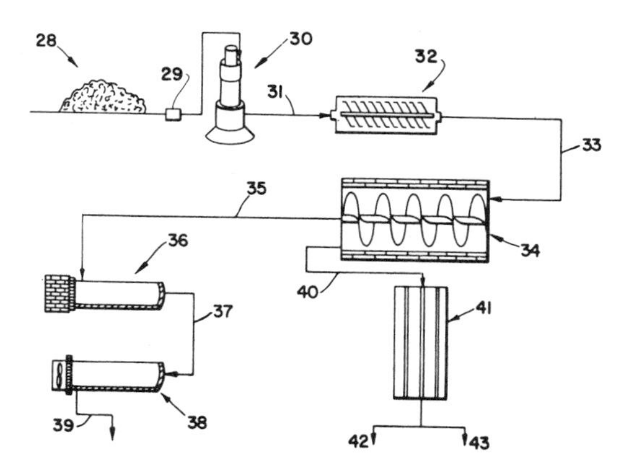

Source: U.S. Patent 3,966,984

While removal of d-limonene and the precipitation of calcium citrate might be accomplished by steam distillation of limed residue **33**, it has been discovered that bound water released by heating the limed residue in conveyor **34** can be removed readily by heating the citrus residue without the introduction of live steam. Thus, it is essential that the bound aqueous material be bled out of the cells of the citrus residue during heating of the limed residue to provide sufficient moisture to support distillation and evaporation.

The process also enables the removal of a major portion of the d-limonene present in the citrus residue. The d-limonene is contained in a vapor component **40**, which also comprises citrus oil and water. By collecting and condensing the vapor component **40** in a vapor condenser **41**, the escape of d-limonene into the atmosphere is prevented. By selectively condensing the substances in vapor component **40**, one can obtain stripper oil **42** and water **43**. For reasons of safety and economics, condenser **41** is preferably operated at about normal barometric or atmospheric pressure.

The process not only permits the removal of a major portion of the d-limonene from the limed citrus residue, but the moisture content of the limed citrus residue is so reduced in conveyor **34** that the concentrated solids portion **35** can be introduced directly into a drier, such as **36**. Then the dried citrus residue (pulp or meal) **37** passes through cooler **38** and is ready for storage or sale **39**.

FURFURAL

See Pulp Mill Digester Effluents (U.S. Patent 3,819,812)

GALVANIZING PROCESS EFFLUENTS

In galvanizing plants, a vessel is charged with an acid having a pH of about one. Typically sulfuric acid is used, though hydrochloric acid can be used. The vessel or container is heated to raise the temperature to a point just short of boiling. Metal sheets and plates are dipped therein to clean dirt and scale therefrom. The plates may be dipped again after improper galvanizing to clean the surface of the zinc coated plates.

After a period of time, the vessel contains a substantial amount of dirt, contamination, and various quantities of metal ions. The metal ions typically include iron and traces of metal elements such as manganese, copper, chromium, lead, tin, molybdenum, titanium, nickel, strontium and zinc. Other trace elements may also be found in the acid bath after the cleaning of metals.

The acid bath is quite effective when the acid is first placed in the vessel. However, after a period of time and use, the acid bath loses some of its effectiveness. The rate at which the acid bath becomes unusable is dependent on many variables, including the relative size of the vessel, rate at which it is used, degree of contamination on the plates dipped in the bath, and many other factors. Eventually the acid bath becomes quite contaminated, and galvanizing suffers in quality. When the acid is contaminated, the effectiveness of the bath is reduced and it may fail in its intended purpose.

When the liquor in a vessel has been contaminated to an extent that it can no longer be used, it is preferably replaced. There is a great deal of difficulty in disposing of a substantial amount of strong acid.

Removal from Water

A process developed by *W.L. Eddleman; U.S. Patent 3,801,481; April 2, 1974* provides a means of treating a galvanizing plant pickle liquor whereby the liquor can be purified, permitting the acid to be reused, and as a consequence, the cost of galvanizing is reduced and the pollution of sewage and surface water is eliminated.

The method involves placing an anode and a cathode in the pickle liquor. A DC current is passed through the terminals and the bath. Elemental metals can be recovered in the vicinity of the cathode by the use of an electromagnet at or above the surface of the pickle liquor.

These materials attach to the cathode momentarily, but do not plate thereon as that term is ordinarily understood in the art. The current flow through the pickle liquor changes the valence of the metal ions at the anode enabling the formation of insoluble salts or oxides which can be recovered from the pickle liquor by filtration.

GAS TURBINE ENGINE EFFLUENT

The present day emphasis on the elimination of air pollution has resulted in a great deal of work and effort by aircraft gas turbine engine manufacturers who have succeeded in significantly reducing most forms of polluting emissions. Nitrogen oxides, however, are one form of pollutant emitted from a gas turbine engine which have not been satisfactorily reduced. Although it is not fully understood how oxides of nitrogen are formed, it is believed that such oxides are produced by the direct combination of atmospheric nitrogen and oxygen at the high temperatures occurring in primary combustion zones. The rates with which nitrogen oxides form depend upon the flame temperature and consequently a small reduction in flame temperature will result in a large reduction in the nitrogen oxides.

Removal from Air

One proposed solution for reducing the emission of nitrogen oxides involves the introduction of more air to the critical primary combustion zone during peak periods of nitrogen oxide formation. The introduction of more air operates to reduce the fuel air ratio and corresponding flame temperature. The formation of oxides of nitrogen is generally most severe at the high power settings of the engine such as during takeoff. However, if the engine were designed to provide an excess of air during the peak power period at takeoff, then the fuel to air ratio at low power settings or at lightoff would likely be too lean to sustain or initiate combustion.

Therefore, it becomes necessary to provide variable geometry apparatus to modulate the flow of air to the combustor in a manner which provides an excess of air during high power settings of the engine and reduces the airflow at low power settings to prevent the combustor flame from blowing out. Variable geometry, however, generally adds weight and complexity to a gas turbine engine and consequently is not an entirely satisfactory solution to the problem of eliminating nitrogen oxide emissions.

Another proposed solution for reducing the emission of nitrogen oxides relates to the effect of vitiating the combustion air with inert products such as recirculated cooled exhaust products or steam, wherein the flame temperature is reduced mainly by dilution and by the increased specific heat of the mixture. One difficulty with recirculating cooled exhaust products or steam, however, is that the combustor is generally at a higher pressure than the exhaust necessitating the addition of a pump or blower to introduce the exhaust products directly into the combustor. The addition of a pump or blower again adds weight and complexity to the gas turbine engine with an attendant reduction in engine efficiency.

The addition of a pump or blower could be eliminated by reintroducing the exhaust products directly into the compressor inlet; however, this method also incurs certain disadvantages. Foremost is the increased risk that air contaminated by exhaust products will be circulated through the aircraft from malfunctioning air conditioners which utilize compressor bleed air. Also, introducing exhaust products into the compressor inlet would accelerate the overall corrosion within the compressor.

Still another proposed solution for reducing the emission of nitrogen oxides relates to the introduction of a spray of water into the primary combustion zone in order to lower the flame temperature. This, however, necessitates that water storage tanks and pumps be added to the aircraft which could significantly increase the weight of the aircraft.

A technique developed by *F.F. Ehrich; U.S. Patent 3,842,597; October 22, 1974; assigned to General Electric Company* for reducing the formation and emission of nitrogen oxides in a gas turbine engine involves bleeding and cooling a portion of the airflow pressurized by the compressor. The cooled compressor bleed airflow is then introduced into the primary combustion zone of the combustor in order to reduce the flame temperature effecting a reduction in the rate of formation of oxides of nitrogen.

A process developed by *S.M. De Corso, C.E. Hussey, Jr. and M.J. Ambrose; U.S. Patent 3,826,080; July 30, 1974; assigned to Westinghouse Electric Corporation* is one in which water is supplied to a fuel injection nozzle via the atomizing air passages disposed therein providing a coolant fluid directly to the primary combustion zone to reduce the content of nitrogen oxides in the exhaust of a gas turbine engine.

GASOLINE SERVICE STATION EFFLUENTS

When a motorist stops to refuel, he may take, typically, about 7½ U.S. gallons, which means that 7½ gallons of gasoline vapor and air are displaced during the refueling. Additionally, there is at least an equal evaporation loss due to agitation of the fuel, which may be warmer than the ambient temperature. This means that 15 gallons of vapor or two cubic feet of vaporized gasoline and air are vented to the atmosphere.

Recent California laws limit vapor pressure of gasoline seasonally, so that "summer grade," by area and date, may not exceed 9 pounds Reid vapor pressure at 100°F. This lower vapor pressure and the evaporation control systems on late-model automobiles have reduced losses when filling automobile tanks, and the lower vapor pressure has helped to reduce losses when filling tanks for storage in stations. The late-model cars commonly use a charcoal canister to absorb and store gasoline vapors from the fuel-tank and carburetor until their use as fuel is permissible. Thus, loss in filling automobile tanks has been substantially reduced to about 0.5 gallon or less per 1,000 gallons transferred, but this is additive to the other losses in the filling station—the other main loss occurring when filling main storage tanks, where loss is in the range of 1 to 3 gallons per 1,000 gallons transferred. Thus, typical overall loss is at least 2 or more gallons per 1,000, amounting to 5,500 grams or more per 1,000 gallons transferred.

It is apparent that controls will have to be placed on the handling of gasoline at service stations and at bulk distribution centers to meet the stringent standards set by the 1970 Clean Air Act.

Removal from Air

A system developed by *C.K. Viland; U.S. Patent 3,815,327; June 11, 1974* is a self-contained vapor recovery system for gasoline service stations and for similar applications. Displaced hydrocarbon gases are collected at the point of entry when a vehicle's fuel tank is being filled, or when the service station's main storage tanks are receiving a fresh loading of gasoline. These vapors are collected under controlled pressure conditions, dehydrated, and passed through a refrigerated condensation or absorption zone, and the recovered liquid is returned to the service station's storage tank, preferably below the liquid level there. The essentially hydrocarbon-free gas, now mainly air, is discharged into the atmosphere.

A scheme developed by *P. Casteline, Jr.; U.S. Patent 3,911,973; October 14, 1975; assigned to Cities Service Oil Company* is shown in Figure 49.

FIGURE 49: CAR FUELING SETUP WITH VAPOR RECOVERY

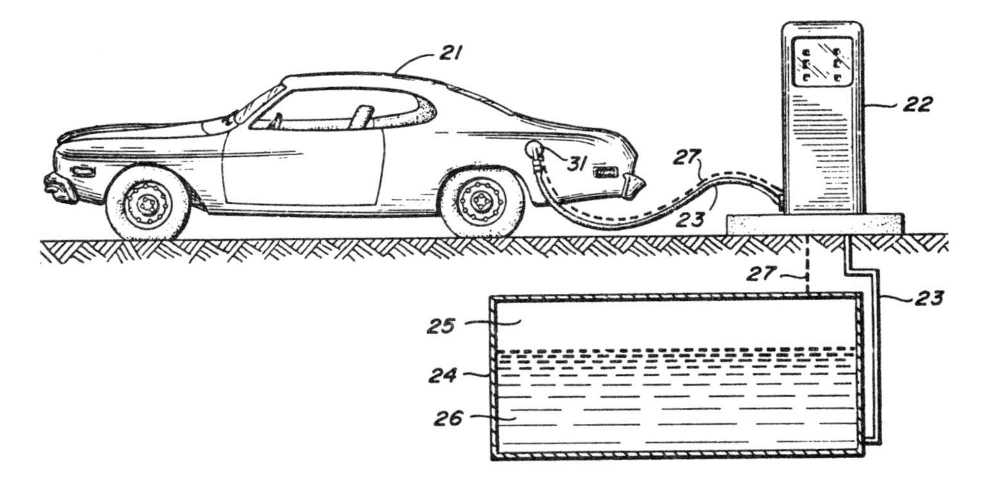

Source: U.S. Patent 3,911,973

As illustrated, the process of refueling a motor vehicle with gasoline requires a pump **22** to meter out the liquid fuel **26** from storage reservoir or vessel **24**. The fuel traverses through tubular member **23** entering the fuel tank of motor vehicle **21**. Heretofore, hydrocarbon vapors in the head space of such a tank would be displaced by the inflow of liquid fuel and discharged into the atmosphere. However, the present scheme creates a positive vapor seal at junction **31** of the fuel tank fill pipe and fuel dispensing apparatus, collecting virtually all displaced vapor which is channeled via vapor return conduit **27** to the headspace **25** of the storage reservoir. The vapor is retained without adversely affecting the surrounding atmosphere.

A system developed by *M.L. Stary, E.L. Brown and E.L. Pridonoff; U.S. Patent 3,926,230; December 16, 1975* is a system for preventing escape of vapors into the atmosphere when filling gasoline or the like into a tank, and includes means for collecting vapors from the vicinity of the filling nozzle, or nozzles, adsorbing the vapors onto an adsorbent substance such as activated charcoal, and then utilizing the vapors in admixture with air as the sole fuel for driving a combustion engine. The engine may be started intermittently, preferably in response to each actuation of the dispensing nozzle.

A system developed by *Y.C. Lee, J.M. Kline and M.L. Stary; U.S. Patent 3,918,932; November 11, 1975; assigned to Environics, Inc.* is designed to prevent the pollution of air caused by hydrocarbon fuel vapors escaping during fueling of vehicles and other fuel transfers. The noxious fumes are collected and converted into the nonpollutants of water vapor and carbon dioxide which may be later discharged into the surrounding atmosphere. Vapor collection conduits having a partial vacuum formed therein serve to draw off the otherwise escaping fuel vapors at the filler opening of the vehicle, or at the vent piping of an underground storage tank, whereupon the collected vapors are passed into canisters containing carbon beds for temporarily storing the peak vapor emissions by an adsorption process. Thereafter, the vapors are retrieved from the storage canisters by pumping air therethrough to form a suitable air and vapor fuel mixture which is passed into a catalytic reactor. The fuel mixture is substantially completely oxidized in the reactor to form the water vapor and carbon dioxide end products.

Figure 50 shows the overall system (at the top of the figure) and a detail of the vapor collection device at the vehicle fueling point (at the bottom of the figure).

A typical service or gasoline station is illustrated including first and second gasoline pumps or dispensers **11** and **12** for pumping fuel from an underground bulk storage tank **13**. Dispensers **11** and **12** are each provided with a manually operated valve and nozzle assembly **16** and **17** which have been specially modified to capture or collect fuel vapors emitted during fueling operations. Also, a vent piping or conduit **18** provided for storage tank **13** is connected to the vapor collection and disposal apparatus illustrated to capture excess fuel vapors otherwise emitted into the surrounding atmosphere during and after bulk transfers of fuel to tank **13** through a filler opening **19**.

Both the refueling operations provided by nozzle assembly **16** and **17** of dispensers **11** and **12** and the bulk transfers or bulk fuel drops to underground storage tank **13** represent in general the transfer of fuel from one container to another. The present system in general is capable of preventing undesirable hydrocarbon fuel vapor emissions normally occurring during such fuel transfers.

For this purpose, vapors otherwise escaping during the use of dispensers **11** and **12** or during a fuel drop to storage tank **13** are captured and passed on downstream of the system via vapor collection lines or conduits **21** and **22**. The collected vapors are thereupon fed into a temporary storage chamber or zone which includes a substance for adsorbing the gasoline or fuel vapors. In the embodiments described herein, one or more canisters **23** containing carbon granules are employed to provide the temporary vapor storage zone.

The beds of carbon granules serve to store peak fuel vapor emissions for later retrieval and combustion at a low fuel to air ratio. Following the temporary storage in the carbon bed canisters **23**, the fuel vapors are forced over a connecting conduit **24** by an air pump **26** to a combustion chamber for complete oxidation. Preferably, the oxidizing chamber is provided by a catalytic reactor **27**

FIGURE 50: APPARATUS FOR RECOVERY AND INCINERATION OF VAPORS FROM VEHICLE FILL POINT AND UNDERGROUND STORAGE TANKS

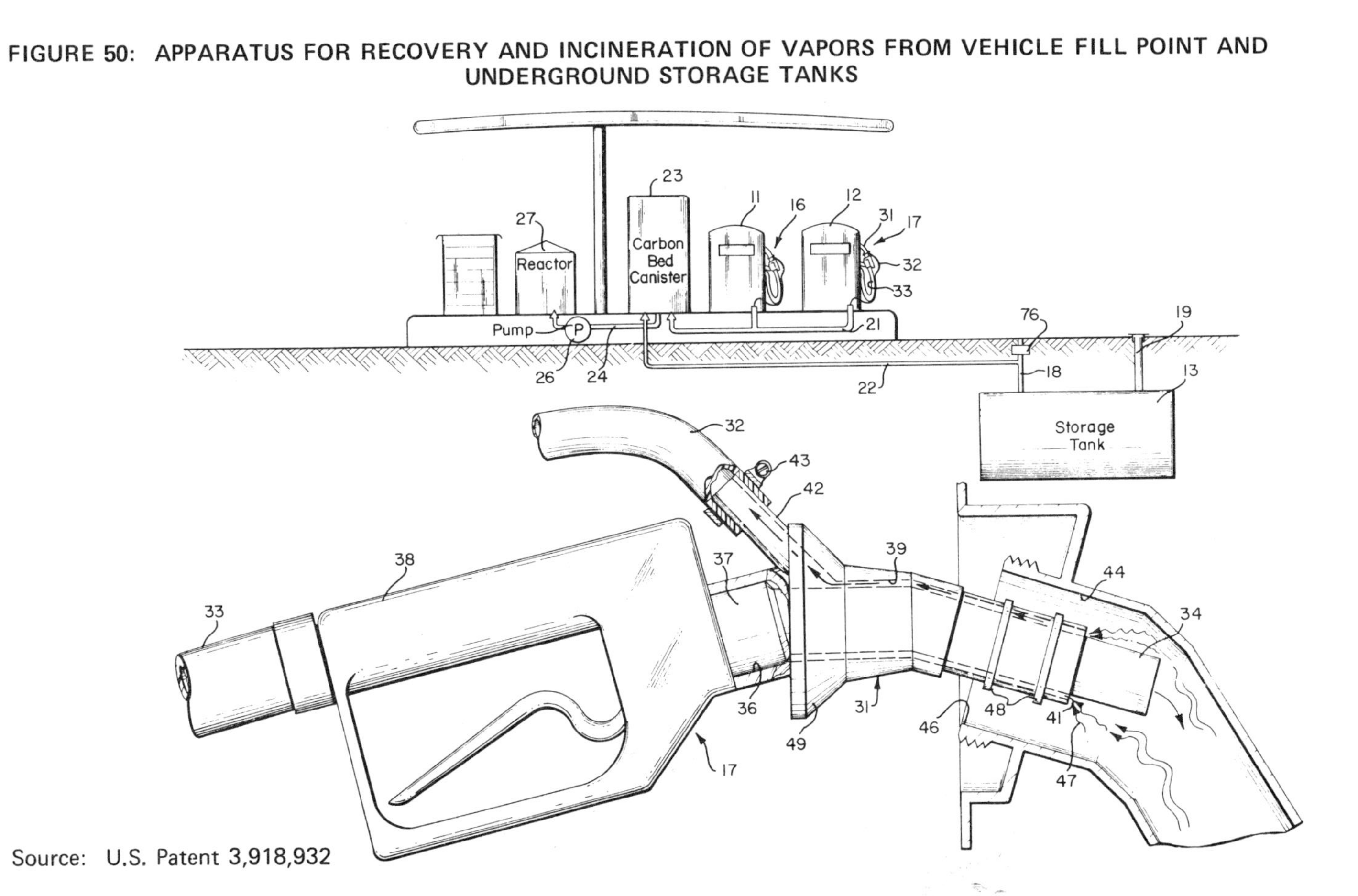

Source: U.S. Patent 3,918,932

which is operated at a low fuel to air ratio for substantially complete oxidation of the hydrocarbon vapors into water vapor and carbon dioxide.

To capture substantially all of the hydrocarbon vapors emitted during refueling operations, dispenser nozzle assemblies **16** and **17** are specially modified as best shown in the detail at the lower part of the figure to include a nozzle attachment **31**. This attachment operates with a source of low pressure or partial vacuum available via a flexible hose **32** which is extended to the vapor collection conduit **21** for dispensers **11** and **12**. In this instance, attachment **31** and hose **32** are illustrated in combination with nozzle assembly **17** of dispenser **12**. Hose **32** may be dressed alongside the fuel hose **33** as best shown in the overall view of the system at the top of the figure.

Attachment **31** has a generally elongated hollow cylindrical form, slightly bent at one or more places along the axis thereof so as to fit coaxially about the elongated nozzle **34** of the usual configuration. The standard nozzle assembly may include a wire spring-like structure adjacent the assembly handle and such structure may be removed prior to the mounting of attachment **31**.

An interior cylindrical wall portion **36** of attachment **31** is adapted to be affixed to an exterior cylindrical portion **37** of nozzle **34** adjacent handle **38**. By proper dimensioning of attachment **31**, the remaining interior wall portion **39** thereof is selectively spaced away from the exterior cylindrical surface of nozzle **34** to form an annular passage along the body of the nozzle. More particularly, by providing a length of attachment **31** which is less than that of nozzle **34**, a forward annular inlet **41** is formed between wall portion **39** of the attachment and the adjacent exterior wall of nozzle **34** for drawing or sucking in substantially all of the fuel vapors present in the zone surrounding this annular inlet.

A partial vacuum or pressure differential is formed at the annular inlet relative to the surrounding or ambient pressure by connecting hose **32** to a tubular connector **42**, here integrally formed with attachment **31**, and having an interior passage communicating with the annular passage associated with inlet **41**. A suitable clamp **43** may be provided to secure hose **32** to connector **42**.

With the entire nozzle assembly including the forward portion of attachment **31** inserted into the neck **44** of a vehicle fuel tank filler opening **46**, all the hydrocarbon vapors **47** drifting upwardly toward opening **46** are sucked into the annular inlet **41** by the partial vacuum supplied over hose **32**. Attachment **31** may be provided with one or more exterior annular ridges **48** provided in lieu of the aforementioned nozzle spring structure to prevent the nozzle assembly from slipping out of the filler opening if left unattended. Additionally, the attachment **31** may include a rearwardly flared annular flange portion **49** positioned forwardly of the rearwardly inclined connector **42**. Flange **49** serves as a shield protecting the rearwardly extending connector **42** and also as a stop for engaging filler opening **46**. As an advantage of this collection means, no seal is required between the attachment **31** and the filler opening **46** thus simplifying its use and enhancing its reliability.

Usually, each dispenser nozzle assembly will be provided with a vapor collection attachment **31** as shown for assembly **17**. The associated collection hoses, such as hose **32** for assembly **17** are jointly fed to the vapor collection conduit **21** for subsequent disposal of the vapors. The number of nozzle assemblies and dispensers will, of course, vary from installation to installation, and in general the apparatus may be adapted to serve any number of dispensers and dispenser nozzles.

An apparatus developed by *J.F. Straitz, III; U.S. Patent 3,914,095; October 21, 1975; assigned to Combustion Unlimited Incorporated* is suitable for use at automobile and truck fuel tank filling stations, at tank truck loading stations and other storage and dispensing facilities. It utilizes the pressure attendant upon the handling of combustible vapors such as those of gasoline, fuel oil, jet fuel, and other volatile combustible liquids, separating the vapor from the combustible liquid, and smokelessly burning the combustible vapor. The structure for burning is adapted for disposal of vapors over a wide fluctuation of firing rates and compositions, and has provisions to avoid flash back, to protect the system in the event of excessive operating temperatures and provisions for noise reduction.

A system developed by *A. Kattan and J.E. Gwyn; U.S. Patent 3,979,175; September 7, 1976; assigned to Shell Oil Company* is a system for preventing gasoline vapors contained in air vented from gasoline storage tanks from entering the atmosphere. It includes a conduit to pass gasoline vapor-laden air to a bed of adsorbent for gasoline and, when the absorbent approaches saturation, subjecting it to back-flushing, with or without heat, to an extent and for a time adequate to remove enough gasoline from the adsorbent to restore it to a regenerated condition. The gasoline removed from the adsorbent bed is incinerated.

GLASS PRODUCTION EFFLUENTS

Glass manufacture results in two principal pollutants from the glass furnaces: (1) fine particulates of glass batching components blown through the furnace by the furnace draft or flame pressure, and (2) SO_2 and SO_3 (SO_X) evolution from the Na_2SO_4 fining agent used in the glass batch.

Removal from Air

The particulate fines are difficult to collect in electrostatic precipitators for two major reasons. First, the fines blown through the furnace tend to be the smaller submicron sized particulates since larger, heavier particles cannot easily be carried by the flue gases and fall out in the furnace. These fines are in the size range of least collection efficiency of electrostatic precipitators. Second, the resistivity of the fines is high, rendering collection by electrostatic methods inherently inefficient. The very fine particulates that pass through such collectors are proving to be significant in their adverse health effects.

Other particulates collection devices, such as cyclone collectors tend to be even less efficient, especially for such small fines, and have a very high energy requirement, particularly in the large pressure drop.

Wet scrubbers using lime, limestone, sodium hydroxide or sodium carbonate, have been proposed and are under development for SO_X control in power plants, where, typically, the amounts of SO_X in the flue gases are far greater than in glass manufacture. While sodium alkalis are better for internal scrubber operation since they do not contribute to caking or scale formation, the end product sodium sulfur oxide salt, e.g., sodium sulfate or sulfite is a water pollutant. The result is to go from the frying pan into the fire, trading a water pollution problem for an air pollution problem when employing a sodium alkali.

To overcome the sodium scrubber wastes water pollution problem, lime or limestone has been used in scrubbers. However, the insoluble calcium carbonate or calcium oxide causes high abrasion of scrubber, piping, nozzle and pump parts, and is a source of scale or sludge that plugs the scrubber or the downstream demisters. Further, reduced reactivity, as compared to sodium alkalis, and the need to keep the amount of undissolved calcium alkali particles low, results in high liquid to gas ratios, on the order of 50 to 100 gmcf.

A hybrid approach is the so-called double alkali process. A soluble sodium alkali is used in the scrubber, and the water pollution problem is solved by reacting the sodium sulfate or sulfite (or bisulfite or bisulfate) with lime or limestone external to the scrubbers. The process chemistry has proven difficult to put into operation on a continuous basis, requires two reactants, complex plumbing and process control sensors, controllers and other devices, and has an increased capital investment as compared to once-through throw-away processes.

Wet scrubbers also are energy hungry, having a high pressure drop, high pumping requirements since liquid rather than air must be circulated, are sensitive to chlorides in the water, and require reheat of the cleaned flue gases in order to get efficient stack ejection or avoid a plume.

Alkali sprays have been tried or proposed for use in coal or oil-fired power plant flue gas SO_x control, alone or in conjunction with wet scrubbers as a mode of flue gas-alkali solution contact. Typical use in scrubbers is for demister vane washing to prevent or reduce scaling or plugging. Sprays alone followed by two-stage collection, a mechanical cyclone as stage 1 and an electrostatic precipitator as stage 2, have been proposed and pilot tested. This latter assembly is expensive, requiring two collection devices in series, and has a high overall pressure drop, principally through the cyclone. Further, it does not appear to have been tried for glass making, possibly because of inefficiency of electrostatic precipitators to collect high resistivity, very fine glass batching ingredients.

Still other alkali recycle processes such as the Wellman-Lord sodium sulfite-bisulfite process, the MgO process, the molten carbonate process and the like are under development. These also have difficulties of process operation. In addition, the capital investment required for recycle is usually more than double that for throw-away once-through processes.

Finally, baghouses are useful for particulates control but are not generally operable for sustained periods much above the 450° to 550°F range without very special bag materials. In contrast, glass furnace flue gas is generally in the range of 500° to 1500°F. The bags can be "burned up" when an upset condition of overheated flue gas contacts the bags, and may be prone to bag blinding when the temperature is dropped below the dew point. Further, calcium alkali sorbents, useful for SO_x emissions control in a wet scrubber, are essentially unreactive in a dry state at the low temperatures of baghouse operation. Synthetic, commercial sodium bicarbonate is a more reactive dry sorbent than sodium carbonate, lime, or limestone, but is expensive and most efficient at the upper limits of permissible baghouse temperature operation, 450° to 550°F.

Also, sodium carbonate or bicarbonate results in a serious disposal problem in that resultant sodium sulfite or sulfate, being water soluble, are water-pollutants. Sodium hydroxide apparently is not used in a baghouse because of expense, handling difficulties, e.g., water vapor adsorption, and causticity.

There is thus an unsolved need to control both particulates and SO_x emissions present in glass manufacture flue gases by a process that is efficient, simple, inexpensive, does not result in a water pollutant, does not require extensive modifications to glass processing operations, can be easily retrofit onto existing glass plant equipment and available space, and which pollution control process can permit glass manufacturers to meet particulates and SO_x emissions codes.

A process developed by *J.M. Dulin and E.C. Rosar; U.S. Patent 3,880,629; April 29, 1975; assigned to Industrial Resources, Inc.* is a process for manufacture of glass using a wet or dry sodium alkali source, preferably of the carbonate type such as crude Nahcolite ore, either raw or calcined, sodium carbonate or the like for SO_x and furnace particulates emissions control and for Na_2O values. The SO_x and particulates are collected in a baghouse in which dry powdered Nahcolite ore is used in the dual function of a particulates filter aid and as an adsorbent which reacts withtthe SO_x to form $Na_wH_yS_zO_x$, principally Na_2SO_3/SO_4 or the corresponding hydrosulfite, sulfide, hydro-, tetra-, and pentasulfides, bisulfite, bisulfate, peroxydisulfate, pyrosulfate, or pyrosulfite. In a second embodiment an aqueous sodium alkali, such as an $NaHCO_3$ or Na_2CO_3 solution, preferably raw or calcined crude Nahcolite ore dissolved in water, is sprayed at a controlled rate into the hot (200° to 1500°F) flue gases coming off the glass furnaces.

The SO_x reacts with the sodium alkali solution and the heat of the flue gas dries out the resultant sodium sulfur oxide salts which, along with particulates are collected in the baghouse. The baghouse filter cake is recycled to the glass furnace so that the process is in material balance without producing potentially water pollutable wastes. Figure 51 illustrates such a process.

Nahcolite (Na_2CO_3) ore is shipped in by rail **1** or truck **2** and the ore as received is placed in surge bin **3**. Turning first to the use of the Nahcolite ore as the air pollution control material, the ore as received, typically $-1\frac{1}{2}$ inch size, is passed via material handling means (conveyor) **4** to crusher **5** which crushes the ore, typically to $-\frac{1}{8}$ inch size. The crushed ore is passed via material handling means **6** to a grinder **7** which reduces the ore to the desired size range, typically 90% passing 200 mesh. Other typical mesh sizes could be a mixture of sizes, or a single size on the order of 325 mesh. The crushed material is split into two streams **8** and **9** and passed to surge bins **10** and **11**, respectively. Typically 20% of the crushed Nahcolite feed is stored in surge bin **10** and 80% would be stored in surge bin **11**.

Ground ore from surge bin **10** is used to precoat the bags in the baghouse **13** via fan **12** with approximately 20% of the total Nahcolite ore feed. This precoat is a thin layer of ground Nahcolite ore coated on the bags so that the hot flue gas in line **26**, when it first contacts the bags, will have some Nahcolite ore with which the SO_x will react to form $Na_wH_yS_zO_x$, typically Na_2SO_3/SO_4. The precoating is done with fresh air drawn in through forced draft fan **12**.

FIGURE 51: GLASS PRODUCTION PLANT SHOWING RECOVERY AND RECYCLING OF PARTICULATES

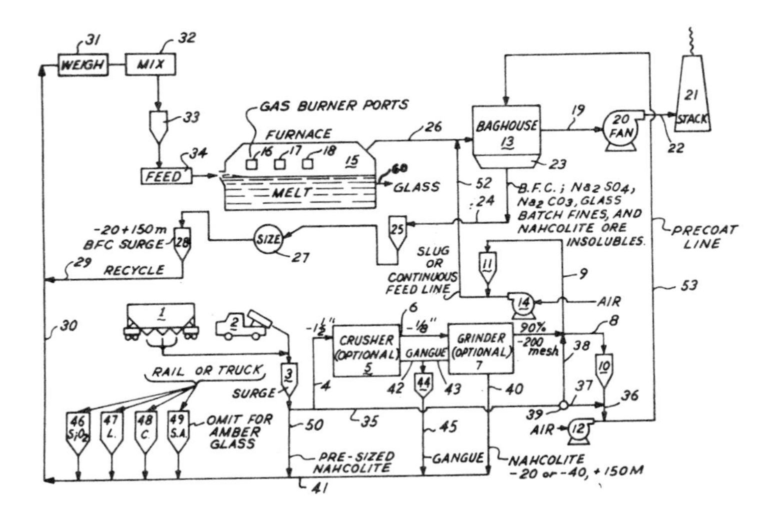

Source: U.S. Patent 3,880,629

After the bags have been precoated, the remaining 80% of the Nahcolite ore is slug fed or continuously fed into the flue gas stream by means of fan **14.** The flue gas stream carries the powdered Nahcolite ore onto the bags where it reacts with the SO_2 contained in the flue gas stream. By "slug fed" is meant feeding the powdered ore into the flue gas as fast as the flue gas can carry it without substantial drop-out of the ore prior to reaching the bags. Typical grain loadings in either feeding mode would be below about 20 grains/scf (dry basis).

The Nahcolite ore cake on the bags also acts as a filtering aid for trapping the particulate material which has been blown through the glass furnace **15** by the high pressure developed by the gas burner ports **16, 17** and **18.** Normally, the heat source in a glass furnace is gas, so that the flue gas in duct **26** contains only sulfur dioxide and glass batching fines as the particulate material. This is in contrast to a coal fired boiler wherein the particulates are predominately ash from the coal. In the case of oil firing, the content of particulates other than from glass batching fines, may be sufficiently low to be acceptable for some green or amber glasses.

Thus, the baghouse **13** with the Nahcolite ore layer on the bag serves the dual function of collecting the glass batch particulates and also removing the SO_2. The cleaned flue gas, from which up to about 99.5% or greater particulates have been removed, is withdrawn by ID fan **20** and passed to the stack **21** by duct **22.** The baghouse filter cake is periodically removed from the baghouse by mechanical shaking, reverse air, pulse jet, or combination of those procedures (not shown here in detail since it is conventional). The baghouse filter cake falls down into the hopper area **23** of the baghouse **13** and may then be removed by appropriate conveying means **24** to a storage hopper **25** (optional).

A top or bottom entry baghouse may be employed, but bottom entry with a combination pulse jet and mechanical cleaning procedure is preferred. Bags are typically 11 to 11½ inches in diameter and 30 to 40 feet long. The superficial velocity, expressed as an air to cloth ratio, may be in the range of 0.1 to 20 ft/sec, with the preferred range being 0.5 to 5.0, it being understood that the lower the A/C ratio, the more efficient the particulates removal. The flue gas temperature in the baghouse, and at the point of injection of the dry powdered Nahcolite into flue gas duct **26** are kept below about 500°F when using treated glass bag fabric.

The flue gas duct may employ dilution air dampers or fresh water sprays (not shown) to reduce or control flue gas temperatures above that level. The baghouse may be insulated or uninsulated, but insulation is preferred for more consistent operation, particularly through seasonal weather variations. The bags may be treated with one or more layers of silicone, graphite, Teflon, Kel-F or other conventional coatings. Other fabrics such as polyester, Nomex, nylon, treated cotton, metal mesh and the like, may be used, depending on the temperature range of operation. Cycle times may range from 20 minutes to about 2½ hours between cleanings, with 30 to 60 minutes being preferred.

When precoating is used, the precoating portion of the cycle takes from 1 to 30 minutes, with 5 to 20 minutes being typical. It is preferred to adjust the water vapor content of the flue gas in the range of 5 to 15% with greater than 2% being a critical threshold at lower temperatures (200° to 300°F), and the preferred range being 5 to 10%. Higher moisture content does not adversely affect the reactivity, but bag weeping should be avoided. Very high particulate grain loadings, on the order of 16 grains/scf (dry), have not proven to interfere with the SO_x-Nahcolite ore reaction.

The baghouse filter cake (BFC) contains the $Na_wH_yS_zO_x$, typically sodium sulfite of sulfate, produced by the reaction of sodium bicarbonate with the SO_2 in the flue gas coming into the baghouse in duct **26**. In addition, some of the unreacted sodium bicarbonate will be roasted to produce sodium carbonate. The remaining material in the baghouse filter cake will include glass batch fines and Nahcolite ore insoluble materials which include dolomites, silica, potassium feldspar, calcite and dawsonite.

The flue gas in duct **26** is adjusted to be in the area of about 500°F, which is the optimum operating temperature for the baghouse in terms of efficiency of removal of SO_2. If the incoming flue gas is of higher temperature than that, a water spray can be used to cool the flue gas down to the bag operating temperature. The efficiency of the Nahcolite reaction in the dry state drops off with temperature and is not recommended below about 200°F. However, this should be well below the operating flue gas temperature of most glass plants. Even where operating glass plant flue gas temperature is in the 200° to 300°F range, loss of efficiency can be tolerated since the by-product sodium carbonate produced by roasting of the Nahcolite ore can be used directly as a feed in the glass manufacturing process. The flue gas heat will evolve both water and CO_2 from the sodium bicarbonate content of the Nahcolite Ore converting it to sodium carbonate.

As can be seen from a typical assay, the Nahcolite ore contains about 0.8 to 2% carbon as organic material. The organics need not be removed, and can be passed into the glass melt wherein they will provide a percentage of the carbon normally used as a fining agent or colorant for the glass manufacture. The FeS_2 in the Nahcolite ore can also act as a colorant.

Since the baghouse filter cake collected in surge bin **25** will generally have component particles in a size range normally too small for glass batching operations, it should be sent through a sizing operation **27**, which typically may be a rotary pan agglomerator or a pelletizer. Thereafter, the appropriately sized baghouse filter cake composite material, in the range of -20+150 mesh would be retained in SO_x control by-product surge bin **28.** If desired, the intermediate surge bin **25** may be omitted. The pelletized baghouse filter cake material is then conveyed in materials handling means **29** to be blended with other glass batching materials in **30,** and eventually passed through weighing means **31,** mixing means **32,** surge bin **33,** glass furnace feeding means **34,** and finally fed into the glass furnace **15.**

At this point, it will be appreciated that all of the by-product material from the pollution control operations, in this example the baghouse **13**, will be recycled as to the glass melt. This is made possible by the unique nature of Nahcolite ore, and the fact that the baghouse permits dry collection of fine particulate glass batching materials that are blown through the furnace. SO_2 evolved from sodium sulfate, a normal component in glass operations, currently represents the main, if not the only, source of SO_2 in glass manufacturing stack flue gases. This SO_2 is reconverted to sodium sulfate or sulfite in the baghouse filter cake, which is recycled to the glass furnace; thus the loop has been closed. Perhaps more severe is the particulates emissions problem, but the baghouse permits collection of both particulates and removal of SO_2, while at the same time converting unreacted sodium bicarbonate in the Nahcolite to sodium carbonate, known to be useful as glass batch component. There is thus no inefficiency in the use of Nahcolite ore material.

If Nahcolite ore is desired to be delivered to the glass plant already ground, as a 200 or less mesh sized material, it would then pass by conveying means **35** from surge bin **3** directly to the precoating duct **36** via duct **37**, and to the slug or continuous feed duct **9** via duct **38.** The same 80/20 proportioning, or other desired proportioning, may be obtained by adjusting proportioning feed valve **39.**

The Nahcolite also offers the potential of being a substitute, as such, for soda ash as a glass batching material, and thus may be passed directly to the glass batching operations after passing through the grinder **7** via line **40.** The oversize from the pollution control grinding operation may be passed as a -40+150 mesh, or -20+150 mesh material to the feed line **41** via the oversize overflow from grinder **7** in line **40.**

Alternately, Nahcolite ore can be delivered to bin **3** in the size range -20+150 mesh and a portion of this to be used in the baghouse can be further ground at the plant to -200 or smaller mesh for SO_x emissions control usage.

Since the Nahcolite ore gangue material, dolomite, silica, potassium feldspar, calcite and dawsonite, are acceptable in many glasses, e.g., green and amber glass, the oversize gangue material from the crushing operation **5** and the grinding operation **7** may be passed via lines **42** and **43**, respectively, to a surge bin **44** and then metered via line **45** into the glass batching operation feed conveyor **41**. Where the Fe and S content of the ore or gangue is low, they may be used in flint and plate glass.

Normally, in glass operations, silica, lime, cullet, and soda ash are brought in by rail or truck and stored in surge bins **46, 47, 48** and **49**, respectively. Part or all of the soda ash **49** may be omitted, when Nahcolite ore is passed to the conveyor feed **41** via line **50** from surge bin **3**, or when the grinder oversize **40**, or when crusher and grinder gangue **45** is used. In addition, it will be recognized that use of the oversize **40** and/or gangue **45** permits omitting some of the silica, lime and cullet called for in bins **46** to **48**.

The amount of dry powdered Nahcolite ore fed into the baghouse, by both precoat line **53** and feed line **52**, depends principally on the flue gas temperature, the amount of SO_x removal desired, and the A/C ratio. For example, where the A/C ratio is around 2.0 and the flue gas temperature is in the 450° to 550°F range, an amount of Nahcolite ore to provide 1.0 times the stoichiometric amount of sodium bicarbonate according to the equation:

$$(1)\ 2NaHCO_3 + SO_x \longrightarrow Na_2SO_x + H_2O + 2CO_2$$

is used, where SO_x is SO_2 and/or SO_3 forming Na_2SO_3 and/or Na_2SO_4 as the case may be, and removes 90% or more of the SO_x at an efficiency of 80 to 95% in the dry state. At a lower temperature, in the range of 350° to 450°F, the same SO_x removal may be obtained by using 1.5 to 2.0 times the stoichiometric amount of Nahcolite ore. Alternatively, the same 1.0 SR (stoichiometric ratio) at lower temperature will result in 70 to 80% removal of the SO_x emissions.

An important option is use of calcined Nahcolite ore, shipped either finely ground or in larger size on the order of one-half inch to surge bin **3**. This calcined Nahcolite ore is then fed via lines **50** and **40** to the glass batching operation feed line **41**. The calcined Nahcolite would consist primarily of a sodium carbonate material including the Nahcolite insolubles, dolomite, silica, potassium feldspar, calcite and dawsonite, as well as the organic carbon.

It is not preferred to use the calcined Nahcolite ore in the baghouse, since tests show that sodium carbonate is less reactive with SO_2 than the sodium bicarbonate values of raw Nahcolite ore. However, the calcined Nahcolite ore may be used since no penalty will be paid for inefficiency in the baghouse reaction; the unreacted sodium carbonate in the baghouse cake can be used in the glass via lines **24, 29, 30,** etc.

A process has been developed by *G.A. Bowman; U.S. Patent 3,728,094; April 17, 1973; assigned to Bowman and Associates, Inc.* in which by a combination of integrated steps, waste chemicals in the form of vapor or particulates lost up the glass furnace stack to atmosphere are recovered and recycled into either agglomerated or unagglomerated glass batch. Also chemicals normally discharged to the sewer in a waste water stream are recovered and recycled into either agglomerated or unagglomerated glass batch. The process also enables recovery and recycling into the glass batch of a substantial percentage of the Btu content of the fuel supplied to melt the glass batch in a standard furnace. The process enables compliance with most regulations regarding air and stream pollution. Figure 52 illustrates such a process.

FIGURE 52: ALTERNATIVE PROCESS FOR RECOVERY AND RECYCLE OF PARTICULATES AND GLASS MANUFACTURE

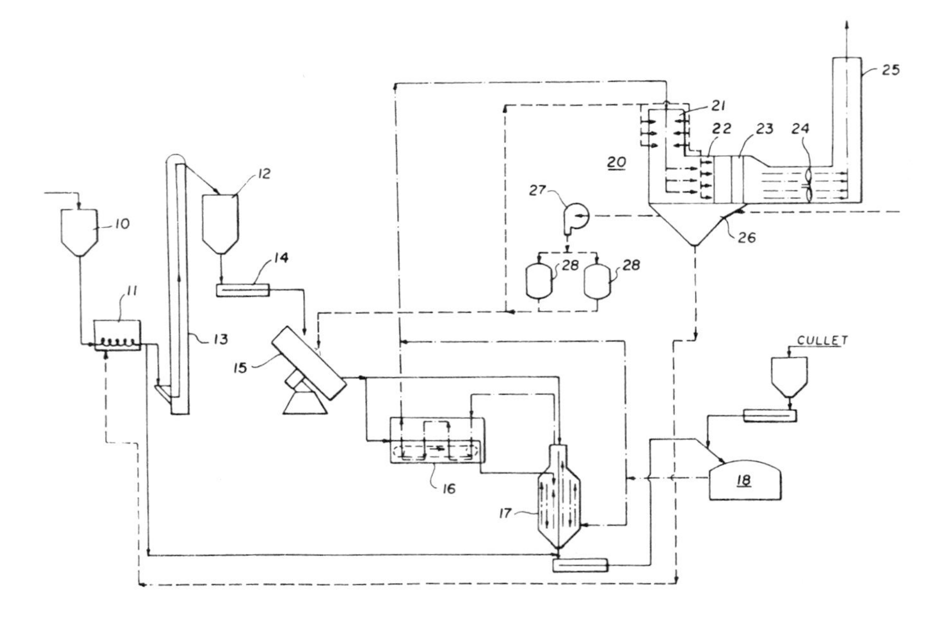

Source: U.S. Patent 3,728,094

Referring to the drawing a batching bin **10** delivers powdered raw materials to a mixer **11** which thoroughly mixes the powdered raw materials and delivers them to a holding bin **12** by means of elevator **13**. The mix is then fed using an accurate weigh feeder **14** to an agglomerating unit such as the disc pelletizer **15** where water is added. The pelletized particles may be of any size or shape that fits the batch mixer and equipment used.

The agglomerated particles are fed to a gas dryer **16** which may be any conventional type, preferably compartmentalized, normally waste gas heated unit, wherein the agglomerated particle is placed on moving belts to a depth of at least 4 inches. The actual bed depth will depend on the residence time in the dryer, temperature of input hot gas, the temperature desired for the exhaust gas and the pressure drop that can be tolerated across the sectionalized moving beds. Gas flow is preferably directed up and down through the bed as it enters different compartments. The gas flow is counter-flow to the movement of the agglomerated particles on the belt.

The hot gases normally will be the waste flow from glass furnaces which have been passed through another higher temperature waste heat exchanger stack furnace **17**. The stack furnace **17** receives the dried agglomerates from the dryer **16** where they come in contact with the hotter gases from the glass melting furnace. Here more heat is absorbed in the agglomerated particle, increasing the temperature of the particle and cooling the waste gas. The transfer of heat depends on the Btu content and temperature of the incoming gas, residence time of gas and particles in contact with each other, specific heat and temperature of the incoming particle. The flow of the particles is vertically downward, counter current to the upward flow of waste gas.

Under some conditions of operations, the compartmentalized dryer may be bypassed or eliminated. In this event, the top portion in the stack of the stack furnace is used as the dryer for incoming agglomerated particles. Care must be taken to make certain that moist particles are not dried too rapidly, which would cause steam to form inside the particle. Cracking would follow as the pressure relieves itself. After severe cracking, particles may tend to disintegrate before melting as they move through the remainder of the process. In addition cracked smaller chips from originally larger particles will fill voids between agglomerates thereby increasing the pressure drop in the gas flow.

From the stack furnace **17** the heated pellets are weighed or measured and conveyed to the glass melting furnace **18** at a selected rate. The cullet is also carefully weighed and fed at a controlled rate to the glass furnace at this point in the process. The temperature of the heated particle of agglomerated glass batch as it enters the glass furnace should be from 100° to 200°F below the temperature of the hot gas exiting from the glass melting furnace. The preheated particles of glass batch feed are now heated to reaction temperature and vitrification takes place forming glass.

The waste gases normally leave the checkers of the glass melting furnace at about 1000° to 1200°F. As they proceed upward through the stack furnace **17** heat is transferred to the particles coming down the stack. The gas leaves the stack furnace and enters the dryer **16** if one is used. In the dryer, the gas passes through the bed of particles on the compartmentalized belt and heat is absorbed by the particles from the gas. The gas normally leaves the dryer at a temperature of between 200° to 300°F.

The gas leaving dryer **16** is conducted to a wet scrubbing system **20** consisting of a wet quenching elbow **21**, scrubber **22**, moisture eliminator or separator **23**, fan **24** and motor, chimney stack **25**, thickener **26**, pumps **27** and water filters **28**. In the wet elbow **21**, the hot gases are quenched to adiabatic saturation. The scrubber will be selected to remove vapors and all particulate matter to meet the codes of regulatory authorities at the plant site. The water flow through the scrubber is collected in a thickener **26** located under the scrubber. The moisture separator **23** removes condensation as droplets, also collecting them in the thickener **26**. The fan **24** provides suction for transfer of exhaust gases from the glass melting furnace through the stack furnace, insulated ducts, dryer, wet elbow, scrubber, moisture separator and out the chimney stack to the atmosphere. The pressure drop across the fan will vary between 15 to 50 inches water gauge depending on the system designed and especially the type of scrubber selected.

Sodium sulfate is one of the most prevalent chemicals in the waste gas stream. This chemical leaves the glass furnace as a vapor. When the gas stream temperature cools to about 400°F, the sodium sulfate begins to condense out of the gas stream as very fine solid particles. Some of these fine chemical particles adhere to the agglomerated briquette or pellet, thereby causing the chemicals to be removed from the gas stream and remain in the bed of agglomerated particles. The remainder of the chemicals are carried into the scrubber **20** where they are wetted by the sprays, nozzles or other wetting procedures by which the selected scrubber operates. Here the chemicals are dissolved by the scrubbing solution. As the solution is recycled through the scrubber, it becomes more and more concentrated as the soluble chemicals dissolve.

Sodium sulfate, one of the most prevalent chemicals is very soluble, and readily dissolves into the water solution. The scrubber also removes nondissolving chemicals as solids at this point. Thus, substantially all of the polluting chemicals are removed from the waste gas stream going to the atmosphere.

A thickener **26** is placed under the scrubber. It acts as a gravity receiver of the chemical solution and nondissolved suspended solids coming out of the scrubber as a slurry. This latter class of materials consists of such things as silica, lime cores, insoluble or slow dissolving sulfates. By properly sizing the thickener with regards to flow and temperature, clarified water rises to a weir at the top and solids settle to the bottom. The solids on the bottom are removed as blowdown. The blowdown solids slurry will be piped back to the batch mixer where water is desirable for improved mixing procedures. In this manner, the chemicals in the blowdown from the thickener are restored to the batch and are recycled through the glass making process. There is no loss or pollution to the environment.

The overflow of clarified water, from the thickener **26**, containing dissolved chemicals is piped through filters **28** to sprays at the pelletizer or briquetting machine. Here the water and chemicals are absorbed in the batch. In this manner, the chemicals dissolved in the water solution are restored to the batch and are recycled through the glass making process, without loss or pollution to the environment.

The overflow from the thickener **26** consisting of settled water is filtered to remove any particles remaining in this clarified water flow. While any of several types of filters may be used, the preferred type is one containing deep sand beds which is insensitive to thickener upsets caused by variation in water temperature. Therefore, filters are used containing deep sand beds. They filter in vertical cross section of the filter bed, and use air to loosen and scour the solids from the carefully sized relatively large sand grain media. The air scour takes place simultaneously with an upward backwash rinse flow which carries the solids up and out of the filter tank.

The loosened solids, having been agglomerated in the filter is piped to the thickener inlet where it now settles to the thickener bottom and is removed with the other solids in blowdown. It is recycled to the mixer, combining here with the glass batch. Here, too, there is no leakage or loss to the environment. Filtrate from the filters **28** is used on the pelletizer disc sprays or on the nozzles in certain types of scrubbers. Clean filtrate will help give trouble free nozzle spray operation.

A process developed by *W.P. Mahoney; U.S. Patent 3,789,628; February 5, 1974; assigned to Ball Corporation* involves the scrubbing of furnace emissions, and particularly glass furnace emissions, from furnace exhausts, by spraying an aqueous solution of sodium silicate into the hot exhaust gases as the exhaust gases are being exhausted through the glass furnace exhaust stack, collecting the precipitated prill and recycling the precipitate back into the glass batch materials.

A process developed by *R. Hirota, H. Kakuta, S. Matoba, K. Shimizu, R. Kanno and M. Narita; U.S. Patent 3,944,650; March 16, 1976; assigned to Asahi Glass Company, Ltd., and Mitsubishi Kakoki Kaisha, Ltd., Japan* involves the removal of oxides of sulfur, dust and mist from a combustion waste gas such as the waste gas evolved from a glass melting furnace. The waste gas is contacted with an absorbing solution containing an alkaline absorbent such as NaOH, Na_2CO_3 and Na_2SO_3, whereby the oxides of sulfur are absorbed and the temperature of the waste gas is decreased and the humidity of the waste gas is increased.

The treated waste gas is then passed through a glass fiber filter which is maintained under moist conditions in order to efficiently remove dust and mist particles from the waste gas. Crystalline sodium sulfate hydrate is recovered from the absorbing solution contacted with the waste gas.

A process developed by *H. Kakuta, S. Matoba, K. Shimizu and S. Yamashita; U.S. Patent 3,966,889; June 29, 1976; assigned to Asahi Glass Co., Ltd., Japan* is one in which selenium is recovered from combustion waste gas, especially the combustion waste gas evolved from glass melting furnaces by a process in which the waste gas is contacted with an absorbing solution containing an alkali metal sulfite or bisulfite, whereby metallic selenium and selenium compounds are absorbed, the temperature of the waste gas is decreased and the humidity of the waste gas is increased. The treated waste gas is then passed through a moist glass fiber filter which collects the remaining amounts of selenium from the waste gas.

The absorbing solution is combined with the solution used to wash the glass fiber filter and the combined solutions are treated with an acid which reduces all of the selenium present in the solution to metallic selenium which precipitates from solution.

Selenium has been widely used as a coloring agent in the manufacture of colored glass products such as plate or sheet glass having a bronze or neutral gray color. The combustion waste gas which is discharged from the furnaces wherein such colored glass or glass containing selenium is melted or from a pyrite calcination plant, contains selenium either in the form of elemental or metallic selenium or as a compound such as selenium dioxide. Because selenium containing materials are toxic, it is necessary to separate and recover these materials from combustion waste gases in order to prevent air pollution. A suitable form of apparatus for the conduct of the present process is shown in Figure 53.

A waste gas **2** containing selenium materials which is discharged from glass melting furnace **1** is passed through a waste heat boiler **3** and is cooled to about 250° to 300°C and then is fed to a lower portion of the absorption tower **4**. The absorption tower **4** consists of a lower zone **5**, an upper zone **6**, and a glass fiber demister **7** consisting of a glass fiber bed in the upper portion of the tower. An absorbing solution is fed from tank **8** to spray **9** situated above the lower zone **5**, and is sprayed into the tower to contact the upflowing waste gas and is returned to the tank **8**. A portion of the absorbing solution flows down from the upper zone **6** to the lower zone **5** and is contacted with the upflowing waste gas and returned to the tank **8**. The absorbing solution is fed from tank **10** to spray **11** in the upper zone **6**. The absorbing solution is contacted with the waste gas and then is returned to the tank **10** and recycled. Fresh absorbing solution **12** is continuously fed to the absorbing solution in tank **10** and the combined solutions are recycled to the upper zone **6**. As stated previously, the fresh absorbing solution **12** is preferably a solution of sodium hydroxide or sodium carbonate although it can be a sodium sulfite or sodium bisulfite solution.

FIGURE 53: SCRUBBER FOR REMOVING AND RECOVERING SELENIUM FROM GLASS FURNACE OFF-GASES

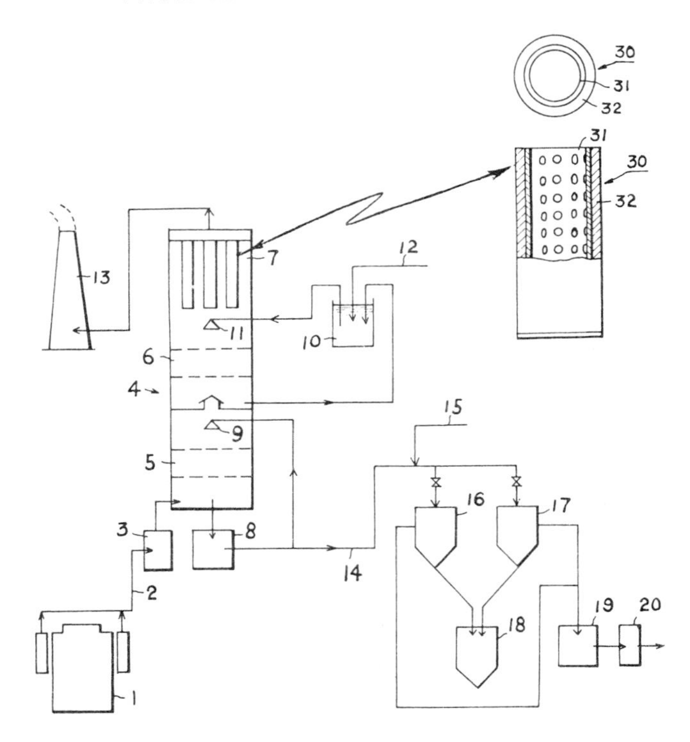

Source: U.S. Patent 3,966,889

The combustion waste gas discharged from the glass melting furnace contains large amounts of CO_2 so that it is important to prevent formation of Na_2CO_3 when the waste gas is contacted with the absorbing solution. In order to prevent the formation of Na_2CO_3, the pH of the absorbing solution recycled in the lower zone is preferably adjusted to from 5 to 7, especially to about 5.5. On the other hand, the pH of the absorbing solution discharged from the upper zone is preferably adjusted to from 6 to 9, especially 7 to 8. The selenium materials, especially SeO_2, in the waste gas react with sulfite or bisulfite ion in the absorbing solution and dissolve to form selenosulfate.

In the practice of the first step, a simple spray tower can be used wherein a spray is provided in the upper regions of both the upper and lower zones. A blank tray is situated in the tower which separates both zones, and the tray is provided with an opening which allows free passage of the waste gas up into the upper zone and free passage of the flowing absorbing solution down into the lower zones when the level of the solution reaches a certain depth.

However, in actual practice, a sieve tray tower consisting of an upper zone and a lower zone divided by a blank tray of the above-mentioned type is preferably employed. In the sieve tray tower each zone is provided with at least one sieve tray, preferably two or four sieve trays. The sieve trays are provided with a plurality of apertures ranging in diameter from 5 to 20 mm distributed uniformly across the surfaces of the trays.

The waste gas which has passed through the absorbing solution, is fed to the glass fiber demister **7**. In one embodiment of the structure of the demister, cylindrical glass fiber filters **30** are suspended vertically in the upper regions of the absorption tower **4**. The detailed structure of the glass fiber filters **30** is shown in the detail at the upper right of the figure. Each demister consists of a perforated chemical resistant metal or plastic cylinder **31** having a bottom portion and a glass fiber layer **32** wrapped around the cylinder. Waste gas that contacts the glass filters diffuses through the glass fiber layer and passes through the perforations in the plastic or metal cylinder into the interior regions of the cylinder. Thereafter, the filtered gas rises through the core of the cylinder.

The waste gas from the glass fiber demister **7** is discharged from stack **13** to the atmosphere. The absorbing solution in the tank **8**, contains the selenium materials which have been scrubbed from the waste gas as selenosulfate. A portion of the absorbing solution **14** is fed to a zone wherein the absorbed selenium materials are recovered. The discharged absorbing solution **14** is neutralized by addition of a solution of NaOH or KOH **15**, and then is fed alternately by either tank **16** or **17**. In the tanks an acid such as H_2SO_4 is added to the absorbing solution to adjust the pH of the solution to from 1 to 3, preferably about 2. Consequently, the metallic selenium precipitates from solution.

A portion of the selenium materials is present in the absorbing solution as selenite ion ($SeO_3^{=}$). The selenite ion can be reduced to metallic selenium by SO_2. When the acid is added, the absorbing solution is allowed to stand at 50° to 90°C for 3 to 5 hours. During this time, most of the dissolved selenium materials in the absorbing solution precipitate as metallic selenium. If the absorbing solution is heated or boiled before the addition of acid, the reduction reaction is promoted. The optimum results are obtained when the solution is boiled for more than 10 minutes or when it is heated to 50° to 90°C for more than 1 hour.

The reduction reaction is prevented when a large amount of sulfite ion is present in the absorbing solution, and accordingly oxygen or air is preferably injected into the absorbing solution before adding acid in order to oxidize sulfite ion to sulfate ion. After the reduction reaction, the supernatant liquid is fed to tank **18** and is passed through tank **19** and bag filter **20** which filters the fine suspended metallic selenium particles. The solution is neutralized and then discharged. Alternatively, the precipitated metallic selenium slurry is fed to the tank **18**, washed with water and dried to recover the precipitated selenium.

GLASS TREATMENT EFFLUENTS

When filling glass bottles with beverages, or the like, at high rates of speed, the bottles may be handled very roughly and may be subjected to excessive impact forces as they are bounced around, thus causing some of the bottles to break or crack. As a result, most of previously known bottle filling operations could only proceed at a rate of about 150 to 200 bottles per minute without encountering any surmountable breakage difficulties. However, in order for bottles to effectively compete with other types of containers such as cans, it is desirable that bottles be filled at a rate of about 1,000 bottles per minute. At that high rate of filling the bottles are handled very rapidly, are subjected to rough treatment and frequently are incapable of withstanding the impact of the blows imposed upon them. Consequently, the number of bottles broken or cracked during filling is unacceptably high. Furthermore, broken bottles cause work stoppages which are costly.

To solve this problem it has been proposed to coat the glass bottles with an external metallic coating which reduces the scratching and chipping of the bottles. Since most of the strength of thin wall bottles can be lost by a scratch, the coating imparts overall strength and stability to the bottles and enables them to withstand the high impact blows which may occur at the high rate of filling. The coating step is performed during the bottle manufacturing operation as the bottles pass from a bottle forming station to an annealing lehr. At the coating step stage of the operation, the formed bottles are at a temperatures of about 800° to 900°F.

As they pass through a hood located at the coating station, a metallic chloride spray, preferably anhydrous stannic chloride ($SnCl_4$), a liquid of low vapor pressure, is sprayed on the bottles. The anhydrous stannic chloride ($SnCl_4$) will be in vapor and/or droplet form, with the droplets being 0.4 to 0.6 micron in size. Because of the high temperature of the bottles, the $SnCl_4$ vapor and/or droplets contacting the bottles are burned to provide a stannic oxide (SnO_2) coating on the external surface of the bottles. This SnO_2 coating substantially increases the impact strength of the bottles and consequently enables a subsequent bottle filling operation to proceed at a high rate, competitive with that of other containers such as cans.

While the coating step increases the strength of the bottles, unfortunately, it also creates undesirable dangerous pollution problems in the overall bottle manufacturing operation. A substantial percent of the $SnCl_4$ spray does not adhere to the bottle but passes through the coating hood and is subsequently carried away by an air exhaust stream. In addition, the $SnCl_4$ successfully coated on the hot bottles as SnO_2 also generates hydrogen chloride gas (HCl gas) as a reaction product which is removed in the exhaust stream. Thus, exhaust air from the coating station cannot simply be emitted into the atmosphere surrounding the bottle manufacturing facility because of the large amount of anhydrous stannic chloride ($SnCl_4$) vapor and the HCl gas entrained therein, the latter of which reacts with water vapor or droplets to form hydrochloric acid.

In addition, once the anhydrous stannic chloride vapor is in the atmosphere, the stannic chloride forms the hydrate, settles, and is slowly hydrolyzed to hydrated stannic oxides and hydrochloric acid which is corrosive and reacts with the surface of the material onto which it settles or adheres. The stannic oxide hydrates formed also act as suitable surfaces for absorbing water vapor, thus retaining the hydrochloric acid formed in a concentrated form, similar to the manner in which dust forms the core in the formula of rain drops. This problem is greatly intensified on days when the relative humidity in the atmosphere is high.

To alleviate the aforedescribed pollution problems, it is necessary to treat the polluted air exhausted from the coating hood to remove the solid and gaseous contaminants therefrom before the air is emitted into the atmosphere.

A system developed by *R.S. Lyon and R.L. Lyon; U.S. Patent 3,885,929; May 27, 1975; assigned to United McGill Corporation* is a system for cleaning air polluted, for example, with metallic chlorides and hydrogen chloride gas. The air is heated to convert the metallic chlorides to metallic oxides and hydrogen chloride gas and then conveyed to a gas scrubber. Before entering the scrubber, fresh water is sprayed into the air to rapidly cool the air and the water reacts with the hydrogen chloride gas to form hydrochloric acid. The scrubber employs water to wash and scrub the air and remove the pollutants from the air. Such a system is shown in Figure 54.

FIGURE 54: SCRUBBER FOR TREATING EXHAUST AIR FROM GLASS BOTTLE TREATING OPERATION

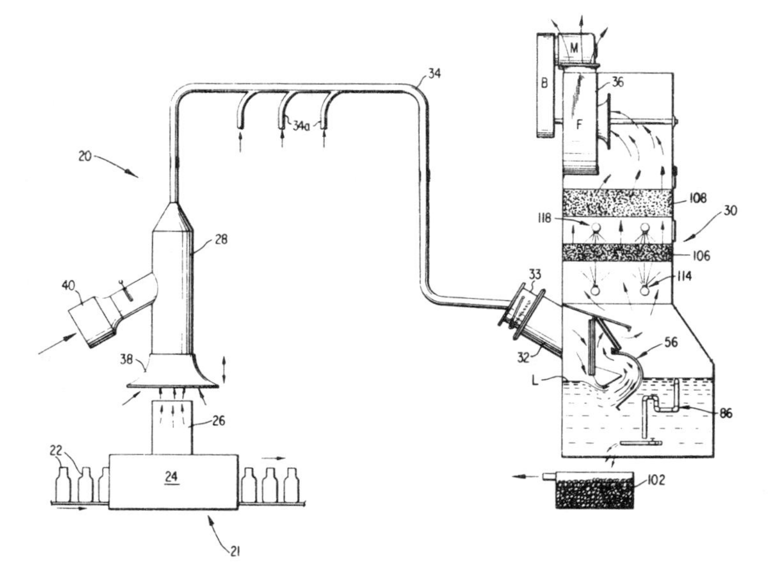

Source: U.S. Patent 3,885,929

The exhaust scrubbing system **20** is shown in use with a hot end coating station **21** positioned along one of the production lines of a glass bottle manufacturing facility. As the bottles **22** are conveyed from a forming station to an annealing lehr, they pass through a coating hood **24** wherein a metallic chloride, preferably anhydrous $SnCl_4$ mist of 0.4 to 0.6 micron in diameter, is sprayed against the external surfaces of the bottles. Because the bottles are at the temperature of about 800° to 900°F, the $SnCl_4$ reacts at the surface of the bottles to form a SnO_2 coating on the bottles and at the same time release hydrogen chloride gas (HCl gas) which is carried away in an air exhaust stream.

Scrubbing system **20** generally comprises a gas collection assembly including hood **24,** its vertical standpipe **26** and an enlarged gas collection conduit **28** mounted above hood **24.** A swirl orifice type scrubber **30** has a gas inlet duct **32,** the outer end of which has a fresh water spray duct **33** flange connected thereto, and conduit **28** is connected to spray duct **33** by a heat insulated duct **34.** Conduit **28** and ducts **32, 33,** and **34** are shown schematically. In practice, conduit **28** may be 8 inches in diameter, ducts **34, 33,** and **32** may be about 6 inches, 11 inches and 12 inches in diameter, respectively.

Scrubber **30** also includes a fan **36** mounted at its upper end. The fan creates the exhaust stream of air which flows in the open ends of hood **24** and out through standpipe **26** into conduit **28,** and from conduit **28** through ducts **34, 33** and **32** into scrubber **30** wherein the air is cleaned before being exhausted via the housing of fan **36.** It should be particularly noted that the fan **36** and its drive assembly are the only mechanically moving parts in scrubber system **20.** Consequently, the system requires a minimum of maintenance over long periods of usage.

The volume rate of flow of air through hood **24** is variable by vertically adjusting a bell-mouth fitting **38,** mounted on the lower end of conduit **28,** relative to the hood standpipe **26.** This adjustment also varies the amount of induced air flow created between the lower end of fitting **38** and the upper end of standpipe **26.** In a prototype of the process constructed for experimental purposes, wherein scrubber **30** had a capacity of about 3,000 scfm (standard cubic feet per minute), fitting **38** was positioned to produce an exhaust air stream of 250 to 300 scfm through hood **24** and a total air flow within the lower portion of conduit **28** of about 500 scfm.

It has been found that the anhydrous $SnCl_4$ vapors or droplets entrained in the exhaust air stream collected from hood **24** are very difficult to separate or retain in scrubber **30,** primarily because of the low vapor pressure of anhydrous $SnCl_4$, i.e., lack of affinity of the molecules for one another, and the brief residence time in the scrubber which probably does not permit the less volatile hydrates to be formed. Consequently, the vapors and droplets do not tend to agglomerate into larger aggregates which could be more easily removed by scrubber **30.**

To alleviate this problem a gas burner **40** is mounted on the side of conduit **28** and injects a raw gas flame directly into the conduit. The exhaust air is thereby heated to a temperature high enough to initiate the following chemical reaction:

$$SnCl_4 + (x+2)H_2O \overset{\Delta}{\rightarrow} SnO_2\downarrow + xH_2O\uparrow + 4HCl\uparrow$$

The $SnCl_4$ will be in vapor or droplet form, the SnO_2 will be in solid particle form, and the HCl will be in gaseous form, and the $SnCl_4$ must be heated to a temperature above 350°F.

In the prototype, it was found that the air preferably should be heated to a temperature of at least 500°F to produce the above chemical reaction and to ensure that the hydrogen chloride is maintained in its gaseous state as the air passes through duct **34** to scrubber **30**. The heat insulating duct **34** preferably includes an inner stainless steel pipe surrounded by an outer galvanized pipe, with a layer of heat insulating material positioned between both pipes. Such a pipe construction ensures that the temperature of the air passing through the duct is kept high so that HCl acid is not formed in the duct, thereby substantially eliminating any corrosion problems in the duct. As the air leaves duct **34** and enters the spray duct **33**, it preferably should be at a temperature of about 350° to 400°F so that the HCl is still in a gaseous state. In passing through spray duct **33**, the air is rapidly cooled to a temperature of about 140° to 160°F before entering scrubber **30.**

The heating of the air stream by burner **40** is beneficial in another respect. Without the heating step, the anhydrous $SnCl_4$ in the air stream passing through duct **34** may form the stannic chloride hydrates, which will settle in the duct and hydrolyze to hydrated stannic oxides and hydrochloric acid. Consequently, solids and semisolids would tend to collect on the walls of the duct and the hydrochloric acid would corrode the duct. The heat from burner **40** avoids these problems.

Duct **34** may be connected by branch conduits **34a** to additional coating stations **21** so that a single scrubber **30** may be used to treat the exhaust air from a plural number of coating stations located along the several production lines which generally are included in a bottle manufacturing facility.

A scroll baffle assembly **56** is fixed within lower scrubber section to cause the air stream entering the scrubber from duct **32** to flow along a tortuous scroll type path as generally illustrated by the flow arrows.

The level of bath **L** must be closely controlled to ensure optimum efficiency of the scrubber under operating conditions. To accomplish this, a level control PVC piping assembly **86,** acts as an overflow opening for bath **L,** thereby establishing and maintaining the level of the bath during operation of the scrubber.

As shown, the acidic liquor drained from bath **L** may be passed through a limestone bed **102** which neutralizes the liquor, thereby enabling the neutralized effluent to be further treated in conventional water treatment equipment that may be available in the manufacturing facility.

Returning now to the air flow through scrubber **30**, as the air leaves the baffle assembly **56** it is drawn upwardly by fan **36** and passes through a final scrubber pad **106** and a demister pad **108** before it exits scrubber **30** by way of the fan housing.

A lower fresh water spray assembly **114** is mounted beneath pad **106** and includes suitable piping and a plurality of atomizing nozzles which direct a continuous pressurized spray of fresh water upwardly against the bottom of pad **106.**

Similarly, an upper fresh water spray assembly **118** is mounted above pad **106** and includes suitable piping and a plurality of atomizing nozzles which direct a continuous pressurized spray of water downwardly against the top of pad **106.**

As a result of spray assemblies **114** and **118,** the pellets in pad **106** are highly wetted and the pad acts as a final transfer or separation stage for any contaminants remaining in the air. The water from assemblies **114** and **118** continuously washes pad **106,** keeping it clean and effective as a final filtering and separating stage. Demister pad **108** removes the water from the air stream before it enters the housing of fan **36** and is exhausted to the atmosphere.

A process developed by *A.B. Scholes and B.F. Semans; U.S. Patent 3,919,391; November 11, 1975; assigned to Ball Corporation* permits the removal of tin or titanium compounds from hot exhaust gases associated with glass making operations. The apparatus comprises exhausting the treatment gases through a plurality of directed fluid treating patterns whereby the hydroscopic materials are hydrated forming aggregates and the suspended particles are wetted passing the exhausted gases over a supply of a treating fluid whereby some of the aggregates and wetted particles are removed, exposing thereafter the exhausted gases to ions in an electrostatic field so that the remaining aggregates and wetted particles are charged and travel under the influence of the field, and removing the charged aggregates and particles by a descending fluid film. Figure 55 shows such an apparatus.

FIGURE 55: ELECTROSTATIC DEVICE FOR REMOVING TIN AND TITANIUM COMPOUNDS FROM EFFLUENT GASES FROM GLASS TREATMENT

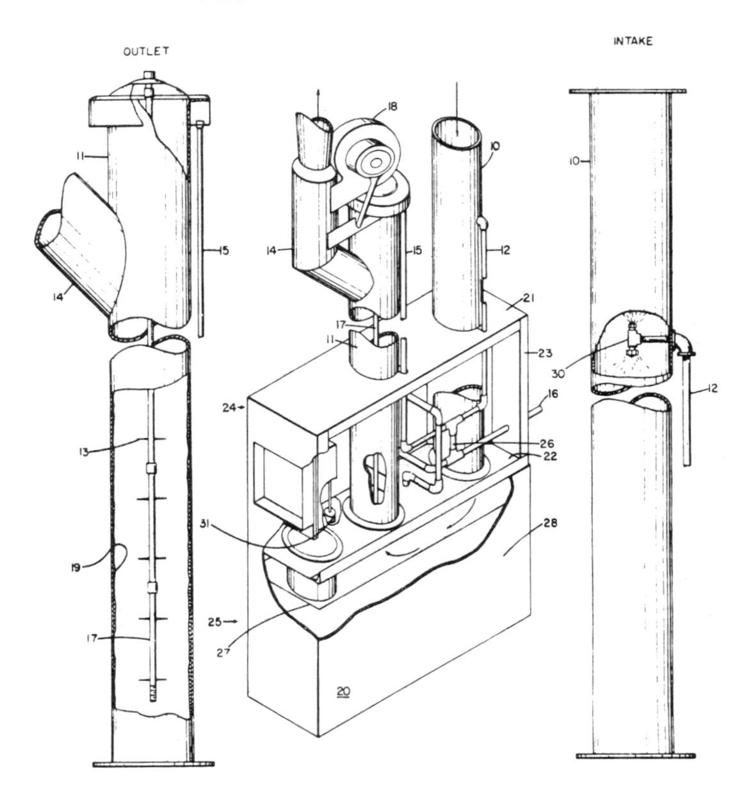

Source: U.S. Patent 3,919,391

The device comprises a major housing **20** which may be divided into two sections, an upper section **24** and a lower section **25**. The lower section **25** is a rectangular body used to confine a collecting fluid **27** employed in the apparatus. The upper section **24** is provided with two lengths of cylindrical columns **10** and **11** which are supported and attached to a table plate **21** and a supporting table **22.**

Provided within housing **20** is outlet tubing **16** which is supported by an upright wall **23** and is fitted via a tee-shaped tube **26** to the upper portions of column **10** on the one side and column **11** on the other by tubing **12** and **15**, respectively. At the terminal portion of tubing **12** there is provided within the internal surface of column **10** a flare nozzle **30.** Although only one pair of nozzles is shown, it is understood that several may be used.

As regards cylindrical column **11** there is provided centrally an electrode **17** having a plurality of pointed members **13** attached to and spaced apart from electrode **17.** The electrode **17** is placed well above the level of fluid in reservoir **28.** Column **11** is provided with a suction fan **18** which communicates with and is attached to an angled extension of column **11**, an exhaust stack **14.**

Means are associated with the device for maintaining a constant fluid level within the reservoir. A number of such means are conventional and well known to those skilled in the art. A regular valve and float assembly **31** has been successfully employed with the subject device.

In operation, hot gaseous materials which may be generally anywhere between a temperature of about 100° to about 600°F, are drawn into cylindrical member **10** by means of the suction fan **18.** The hot gaseous material is impinged by a basic fluid, i.e., a fluid having pH greater than 7, issuing from nozzle **30.** It is advantageous that the hot gases drawn into member **10** be adequately cooled below about 160°F. In cylindrical member **10**, the gaseous material is not only cooled but the volume being drawn becomes saturated with water vapors. Thus, the gaseous material is conditioned in that it is cooled and humidified. It follows that the organic and inorganic halides within the gaseous material are hydrolyzed producing products which are solids.

It will also be appreciated that any hydrochloric acid will be taken up by the saturated water atmosphere within member **10.** Essentially, the gaseous material comprising the organic and inorganic halides undergoes a phase change from that of a gaseous state to that of a solid state. During the cooling process the gases are drawn downwardly through the full extent of member **10** into the upper portion of section **25.** Although the organic and inorganic materials are at once hydrated and hydrolyzed in the scrubbing operation resulting in the formation of solid particles there is also agglomeration of these particles. These particles and agglomerates thereof are generally larger than 1 micron and are mostly entrained and captured in the sweep of droplets containing the particles.

Larger heavy particles which have not been entrained fall by gravity into the collecting fluid **27** forming a reservoir **28.** The remaining gases are drawn across reservoir **28** and upwardly into the column **11** where they are at once subjected to the influence of an electrostatic field created by electrode **17** and column **11.** The particles are at once charged and drawn to the walls of column **11** where they are swept by a collecting fluid **19** which presents a thin film circumferentially within column **11.** Thus, charged particles are drawn and washed downwardly into the reservoir **28** and are precipitated therein. It will be appreciated that column **11** also functions as a condensation column in that the saturated water vapor in passing upwardly through column **11** becomes supersaturated. Thus, the column **11** functions as a condensation tower in that the highly humidified stream of vapors reaches its dew point and there occurs a raining effect within column **11**. The formation of this rain or fine droplets within the column **11** is advantageous in that it removes acidic materials therefrom.

It has been found that generally about 40% of the acidic materials is removed by the scrubbing operation which occurs in cylindrical column **10** and thereafter about 50% of the remaining acidic material is removed by the condensation of the highly humidified stream in column **11**. It should be recalled that acidic gases which are produced in the apparatus and which are captured and drawn thereto pass through column **11** and are absorbed into the fluid whereby the acidic materials are neutralized by basic ingredients dissolved therein. Accordingly, it has been found that the apparatus described herein is capable of removing about 90% of all the acidic fumes which are channeled therethrough.

Although the exact dimensions of the device are not critical and will depend upon sundry parameters it has been found advantageous for most industrial applications that the columns **10** and **11** be about 10 feet in length and about 8 inches in diameter. Generally, the first column **10** is provided with a bank of spray nozzles which continuously flush and scrub the downwardly drawn treatment gases. It has been found that approximately 600 gallons per hour of aqueous solution made basic with dissolved ingredients such as sodium carbonate, sodium bicarbonate, sodium hydroxide and the like, will serve several hot-end glass forming machines. Generally, it has been found advantageous when the pH of the solution is approximately 8 or higher whereby proper neutralization of acidic material can be accomplished.

GLYCOLIC ACID

Removal from Water

A process developed by *C.B. Murchison, R.E. Bailey and R.W. Diesen; U.S. Patent 3,819,516; June 25, 1974; assigned to The Dow Chemical Company* is one in which an aqueous liquor containing oxidizable organic materials such as glycolic acid, salicylaldehyde production wastes or brewery wastes is treated to oxidize the organics. There is provided in solution, in the organically polluted aqueous liquor, a catalytic quantity of iron ions. The pH of the aqueous liquor is adjusted to 4 or below and then treated with oxygen and light energy having a wave length of 5800 A or less. The principle products of the oxidation process are CO_2, water and lower molecular weight organics.

GRAIN DRYER EFFLUENTS

In agriculture, for example, it is the practice to harvest forage crops, such as alfalfa or the like, and to dry the leaves and stems (in chopped or unchopped form) by means of dehydration apparatus which removes all but about 8% of the moisture and leaves a concentrated, uniform feed product. Some prior art dehydration apparatus comprises an oil or gas-fired furnace which exhausts hot air (typically up to about 2000°F) into a dryer which comprises a rotating sealed drum through which the matter to be dried is passed. The material in the drum continually gives off moisture and consequently, is normally maintained at a temperature of about 140° to 150°F.

A dryer fan is connected to the discharge end of the drum, removes hot, moist gases and also advances the matter through the drum and discharges it into a first cyclone separator. The stems, which are moister and heavier than the leaves and require more drying time, automatically move more slowly through the drum. A second fan transports the dried matter from the first cyclone separator to a second cyclone separator, cooling and further drying it in the process. The dried cooled matter is removed from the second separator and is then ground or pelleted, bagged and stored. A foreign material separator is provided at the outlet end of the first cyclone separator to remove debris such as stones and tramp metal. Vents at the top of the first and second cyclone separators allow hot, moist gases, fine particulate matter and other waste products extracted by the dryer fan and the cyclone or cooler fan, respectively, to be discharged to atmosphere.

While such prior art dehydration apparatus is generally satisfactory for its intended purpose, the exhaust to atmosphere from the cyclone separators is a source of objectionable odors, fine particulate matter, and sometimes smoke and other gases produced in the dryer which may violate pollution codes or otherwise be offensive. Formerly, these odors and fines could only be eliminated by means of inefficient after-burners or high-cost heat exchangers. It is desirable, therefore, to provide improved dehydration apparatus and methods which eliminate the aforesaid problems and have other advantages.

Removal from Air

A process developed by *W.A. Arnold; U.S. Patent 3,738,796; June 12, 1973* effects two-stage dehydration of wet particulate matter (such as green forage crops) and disposition of waste products produced during the second stage of dehydration, such as gases, smoke and charred particles. Figure 56 shows a suitable form of apparatus for the conduct of such a process.

There is shown a dehydration apparatus which is used to dry wet particulate matter, such as freshly-harvested chopped alfalfa (hereinafter referred to as "chops"). The apparatus comprises furnace means including a first furnace **10** having a burner assembly **12** and a combustion chamber **14**. The burner assembly **12** is understood to operate on either gas, coal, oil or other suitable fuel which is burned in combustion chamber **14** to supply a quantity of heated air in the temperature range of about 1000° to 2000°F to a first dryer **16**.

Burner assembly **12** is controlled by a thermostat **18** which senses air temperature at the discharge outlet of first dryer **16** and regulates burner **12** accordingly, as regards fuel and air, to maintain a constant desired temperature at the discharge outlet of the first dryer. Burner assembly **12** and combustion chamber **14** are designed so that exhaust products therefrom are kept at a minimum and they burn clean.

FIGURE 56: APPARATUS FOR CROP DEHYDRATION AND USE OF WASTE PRODUCTS AS DEHYDRATOR FUEL

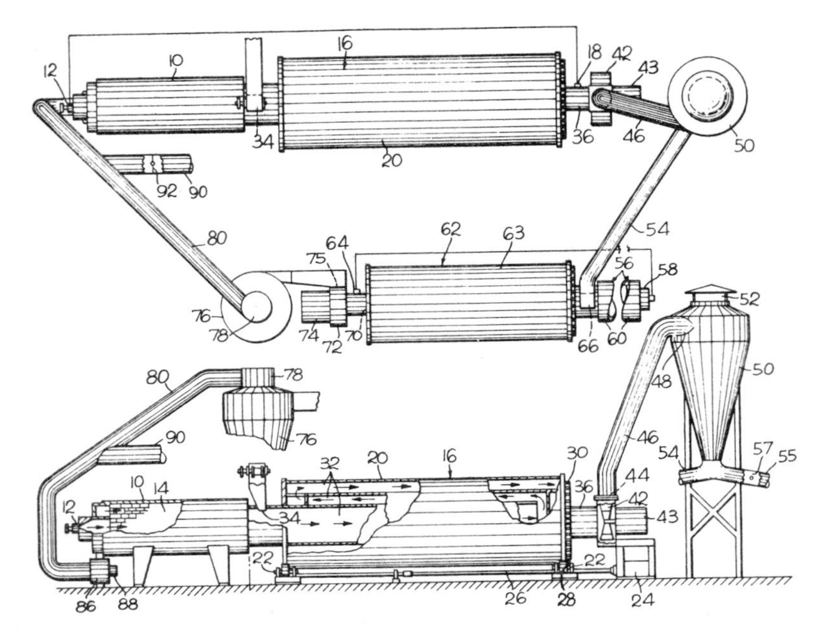

Source: U.S. Patent 3,738,796

First dryer **16** is shown in the drawings as a rotatable drum type dryer but it is to be understood that dryer **16** could be any type of enclosed dryer such as, for example, a flash-type dryer or a conveyor-type dryer wherein a conveyor is used to move material through a fixed enclosure.

First dryer **16** may be of the single pass or multiple pass type. However, in the figure, first dryer **16** takes the form of a multiple pass type. First dryer **16** comprises a rotatable drum **20**, suitable bearing means **22** for supporting the drum for rotation, and an electric motor **24**, a power shaft **26**, a chain drive **28**, and a rotation gear **30** for rotating the drum. Drum **20** is provided on its interior with a series of interconnected passages **32** through which the chops pass in traveling from a wet feed inlet **34** of the drum to a discharge outlet **36**. In the embodiment shown, it is understood that the chops pass through a distance approximately three times the overall length of the drum. Chops are supplied to inlet **34** of first dryer **16**.

Outlet **36** of first dryer **16** is connected to a first dryer fan **42** which creates an air flow and serves as a means to draw the chops through the passageways **32** in drum **20**, as well as to draw hot, moist air from the dryer. First dryer fan **42** is driven, for example, by an electric motor **43**. A discharge outlet **44** of first dryer fan **42** is connected through a conduit **46** to an upper inlet **48** of a first cyclone separator **50**. Chops and hot, moist air drawn from dryer **16** by first dryer fan **42** are expelled into cyclone separator **50**. The hot, moist air is discharged to atmosphere through a vent **52** at the top end of cyclone **50** and the chops fall to the bottom of cyclone separator **50** from whence they are discharged by gravity through a conduit **54.** It is to be understood that the chops are only partially dried in first dryer **16** and consequently, the only waste product produced is hot, moist air which is discharged to atmosphere as hereinbefore explained. In practice, it is contemplated that about 80 to 90% of the moisture be removed from the chops in passing through first dryer **16** and the temperature of the chops in the first dryer averages, for example, about 150°F in the case of alfalfa and, consequently, no charring or burning or smoke is produced.

As the figure further shows, the furnace means comprises a second furnace **56** having a burner assembly **58** and a combustion chamber **60.** Structurally, second furnace **56** is similar to first furnace **10;** however, second furnace **56** is smaller and adapted to provide hot air heated only to a range of about 500° to 1000°F to a second dryer **62.** Second furnace **56** is controlled by a thermostat **64** similar to thermostat **18** hereinbefore described. In practice, both thermostats and both furnaces are adjusted so that the system is balanced to provide a desired percentage of drying in each of the dryers. It is to be understood that second furnace **56** also burns clean and provides no substantial waste products to the atmosphere.

Second furnace **56** is adapted to supply hot air to second dryer **62** which is similar in construction and operation to first dryer **16,** except as hereinafter explained. Dryer **62,** which like dryer **16** could be any suitable type of enclosed dryer, could also be of the single pass or multiple-pass type, but is shown in the drawings to comprise a rotatable drum **63** having an air-handling capacity which is substantially smaller than the air-handling capacity of drum **20** of dryer **16.**

For example, drum **63** of dryer **62** may be considered to have an air-handling capacity of about only one-third to one-half of that of drum **20** of first dryer **16.** Drum **63** of second dryer **62** has a feed inlet **66** to which is connected the discharge end of conduit **54** from cyclone separator **50** and partially-dried chops are supplied to dryer **62** by gravity feed. Partially-dried chops enter dryer **62** at substantially the same temperature at which they left first dryer **16,** i.e., about 150°F in the case of alfalfa.

In the course of passage through second dryer **62,** for example, all but about 2 to 15% of the moisture is removed from the chops. In addition, charring of some of the chops occurs because the chops are no longer as wet as when passing through first dryer **16** and, consequently, waste products such as smoke, gases and charred particles may be produced in second dryer **62.**

Second dryer **62** is provided with a discharge outlet **70** to which a second dryer fan **72** is connected. Second dryer fan **72,** which is driven by an electric motor **72,** may be similar in construction and operation to first dryer fan **42,** hereinbefore described, and serves as a means to move the chops through second dryer **62** and also to remove the waste products produced in second dryer **62.** Second dryer fan **72** has a discharge outlet **75** connected to a second cyclone separator **76** and discharges the finally-dried chops, as well as all waste products, thereinto.

The second cyclone separator **76** is provided near its top with a discharge vent **78** which is connected through an enclosed conduit or pipe **80** to combustion chamber **14** of first furnace **10.** Second cyclone separator **76** is also provided with a discharge outlet at its lower end. The finally-dried chops settle to the bottom of second cyclone separator **76** and pass through the discharge outlet by gravity into a hammer mill or other device for further processing or final disposition. However, the waste products produced in second dryer **62** are discharged through discharge vent **78** in second cyclone separator **76** and through conduit **80** to combustion chamber **14** of first furnace **10** wherein they are consumed.

Conduit **80** has a booster fan **86**, driven by an electric motor **88**, connected to it just upstream of the point where it is connected to combustion chamber **14** of furnace **10**. Booster fan **86** causes dispersion of waste products entering combustion chamber **14**, especially solid particles present in the waste products, and thereby assures complete combustion.

Conduit **80** is also provided with an auxiliary waste material inlet duct **90**, possibly having a control damper **92** therein through which other waste products may be introduced. For example, auxiliary inlet duct **90** may be connected to an outlet to receive waste material, such as fines and dust, which are produced in the hammer mill. These waste materials are also then consumed by furnace **10** and do not contaminate the atmosphere. It is to be understood that burner **12** of furnace **10** is adjusted so as to allow for entry of such makeup air as is necessary to ensure complete combustion of all products entering combustion chamber **14** through conduit **80**, either from cyclone separator **76** or auxiliary inlet duct **90** or both.

A device developed by *E.E. Alms; U.S. Patent 3,747,225; July 24, 1973* provides a simple and inexpensive antipollution screen for dryers of particulate matter such as grain. It makes use of existing dryer frame members for supporting the screen and takes advantage of the heat produced by the drying operation to prevent sticking of the unwanted particles to the screen. The normal vibration produced in the frame by the fan and burner associated with the dryer shakes down any unwanted particles lodged in the screen. Readily releasable means are provided for removing a portion or all of the screen to gain access to the accumulated unwanted particles at the bottom of the dryer.

An apparatus developed by *S.P. Thompson, W.T. Thompson, D.L. Boyert and T.B. Swearingen; U.S. Patent 3,883,327; May 13, 1975; assigned to Thompson Dehydrating Company* operates by combining solid particulate matter entrained in the stack gas of a dryer utilizing a combustible fuel with water to remove the solid matter from the gas stream. Present regulations of the Environmental Protection Agency applicable to alfalfa dehydrating plants require that solid atmosphere pollutants be held to no greater than 25 pounds per ton of alfalfa being dehydrated. Using the described apparatus in the patent a reduction of 79% below the permissible level has been achieved in actual test results.

GREASE

Removal from Air

A process by *N.F. Costarella and A.A. Giuffre; U.S. Patent 3,628,311; December 21, 1971; assigned to Nino's, Inc.* provides a ventilating hood and duct structure for stoves wherein a baffle and water spray arrangement within the hood forces the smoke to pass through successive water curtains. Grease and other impurities are drained from the hood structure through a rectangular trough which overlies the fire area.

See Restaurant Effluents (U.S. Patent 3,802,158)

Removal from Water

See Fats and Fatty Oils (U.S. Patent 3,951,795)
See Fats and Fatty Oils (U.S. Patent 3,940,334)
See Sewage Treatment Effluents (U.S. Patent 3,894,833)

HEAT TREATING FURNACE EFFLUENTS

Various means have been employed to control and maintain pressure of the flow of exhaust gases from treatment retorts. The exhaust gases themselves may be deleterious to the articles treated in the retort, and thus must be withdrawn continually. Certain waste gases contain matter which is objectionable when discharged into the atmosphere and these gases require scrubbing before such discharge. In the process of chromizing, wherein steel strip is chromized in an annealing retort, a positive pressure should be maintained at all times within the retort. Positive pressure is generally maintained by means of a liquid seal. Because of the high temperature of the effluent gases, and the chemical constituents which these waste gases contain, conventional liquid seals have not proved entirely satisfactory.

Removal from Air

A system developed by *H.K. Young and H. Wald; U.S. Patent 3,721,429; March 20, 1973; assigned to Bethlehem Steel Corporation* involves exhausting waste gas from a metal strip heat treating retort into a jacketed vessel where the gas is cooled somewhat while circulating in the vessel, and is then exhausted from the vessel to a liquid sealpot. The gas enters the sealpot below the liquid level, and is well distributed through the liquid by means of a gas diffuser. The gas is scrubbed in the liquid and leaves the sealpot by way of a tube, or pipe, having its entry end located above the liquid level. The liquid level in the sealpot is maintained by a liquid communication therewith. Liquid is withdrawn slowly and uniformly from the sealpot to remove impurities therefrom, and liquid is introduced into the liquid level tank at a slightly higher flow rate.

The system provides for a constant pressure in the heat treating retort and simultaneous removal of impurities from the foul waste gas, the soluble impurities being discharged with the scrubbing medium. The gas, which is ultimately exhausted to the atmosphere, is relatively free of objectionable components. The sealing and scrubbing device requires very little maintenance.

HYDRAZINE

Removal from Air

A process developed by *W.M. Gardner and W.H. Revoir; U.S. Patent 3,489,507; January 13, 1970; assigned to American Optical Corporation* utilizes a filter for removing contaminants of hydrazine and its organic derivatives from fluids. The filter operates by having at least two distinctly different porous materials which are known for their sorbing properties arranged in layers or in a mixed relation. In the layer method the upstream layer is impregnated with a strong oxidizing agent while the downstream layer remains unimpregnated. The vapors of hydrazine or organic derivatives of hydrazine on first entering the filter react with the oxidizing agent to form volatile and nonvolatile substances. The nonvolatile substances are sorbed by the impregnated material and the volatile substances that pass through the impregnated material are then sorbed by the nonimpregnated material.

The material in the upstream layer is selected from the group consisting of activated charcoal, silica gel, alumina and molecular sieves, and the material in the downstream layer is selected from the remainder of the group, and the oxidizing agent is selected from the group consisting of iodine (I_2), potassium permanganate ($KMnO_4$), potassium dichromate ($K_2Cr_2O_7$), potassium bromate ($KBrO_3$), potassium iodate (KIO_3), sodium permanganate ($NaMnO_4$), sodium dichromate ($Na_2Cr_2O_7$), sodium bromate ($NaBrO_3$), and sodium iodate ($NaIO_3$).

HYDROCARBONS

Air pollution regulations require that when hydrocarbons vented in an emergency are burned, there be no emission of smoke as the hydrocarbon burns. Since, of the known hydrocarbons, methane alone burns in open air at flares without smoke production, the problem of smoke suppression in flare operation is demanding, since hydrocarbons other than methane must be vented.

Smoke can be suppressed by increased turbulence in the burning zone, by air injection to the burning zone, and by high velocity injection of steam to the burning zone, and to combine air injection with increased turbulence by other means known to those versed in the art. However, the effectiveness of such smoke prevention measures is hindered if through wind action the temperature of the flame is decreased.

Removal from Air

A device developed by *R.E. Schwartz and R.K. Noble; U.S. Patent 3,982,881; September 18, 1976; assigned to John Zink Company* is an improved low pollution invisible flare burner which comprises a tall stack lined with ceramic. The stack is supported above the ground level and has a wind screen surrounding the open portion below the stack floor. Primary air is introduced under pressure in a tube below and coaxial with the stack. The top of the tube contains a burner for the vented hydrocarbon gases. The top of the tube and the burner are at the level of the floor of the stack. Secondary air is introduced into the stack in the annular space between the primary air conduit and an opening in the floor of the stack. Turbulent mixing of the primary air and the vent gas, plus the availability of sufficient atmospheric air for complete combustion, plus the effect of heated ceramic in the vicinity of the flame, provide means for complete combustion of the vented gases with low emission of smoke and light.

A process developed by *H.W. Haines, Jr.; U.S. Patent 3,907,524; September 23, 1975; assigned to Emission Abatement, Inc.* is a vapor recovery method wherein gases contaminated with vapors from volatile organic liquids are recovered by contacting the vapor-containing gas in an absorbing tower with a sponge oil which absorbs the vapors. The sponge oil rich in absorbed vapors is conveyed to a flash tank wherein the absorbed vapors are removed and recovered. The clean gas is vented to the atmosphere and the sponge oil can be successfully reused.

See Drying Oven Effluents (U.S. Patent 3,875,678)
See Duplicating Machine Effluents (U.S. Patent 3,767,300)
See Scrap Melting Furnace Effluents (U.S. Patent 3,869,112)
See Petroleum Storage Effluents (U.S. Patent 3,778,968)
See Petroleum Storage Effluents (U.S. Patent 3,979,175)
See Petroleum Storage Effluents (U.S. Patent 3,714,790)

Removal from Water

Contamination of ground water by petroleum products such as crude oil and gasoline has been reported on many occasions. The amount of gasoline contamination in water need not be great to be detected by taste, as little as 0.005 mg/l being detected by sensitive people. Since the solubility of modern gasolines in water is about 20 to 80 mg/l, as little as one-hundredth the average solubility of gasoline in water will cause a taste or odor problem. Also, biodegradation occurring from natural causes cannot be depended upon to remove petroleum contamination in a reasonable time.

A process developed by *R.L. Raymond; U.S. Patent 3,846,290; November 5, 1974; assigned to Sun Research and Development Company* is based on the discovery that hydrocarbon contaminants of underground water sources can be quickly disposed by providing nutrients and oxygen for hydrocarbon-consuming microorganisms normally present in the underground waters, the nutrients and oxygen being introduced through wells within or adjacent to the contaminated area and removing water from the contaminated area until the contaminating hydrocarbons are no longer present or are reduced to an acceptable level. Such a process is shown diagramatically in Figure 57.

As shown in the figure, a contaminated producing well **11** will yield hydrocarbon contaminated water due to the presence of hydrocarbons (such as crude oil, fuel oil, gasoline, or other hydrocarbon) in the water. A second well **12** which, if necessary, may be specifically drilled for this purpose, is within the contaminated area and adjacent the producing well **11** and will be employed as an injection well. At ground level, the injection well will be equipped with a mixing tank **13** for the nutrients and an injection tube **14** extending from the tank to the water level for introducing the nutrients to the subsurface water. An air pump **15** pumps air through a conduit **16** to below the water level, the conduit being preferably fitted with a sparger device **17** to effect smooth and uniform distribution of the air throughout the water area.

A pump **18** at the producing well **11** removes water from the well through a submerged conduit **19** thus causing a flow of subsurface water from the injection well area to the producing well area. The normal microorganism flora between the injection well and the producing well will thus have sufficient nutrients and oxygen to effectively feed on the hydrocarbon substrate present in the area between the wells and decomposition of the unwanted hydrocarbon will proceed in good order.

The distance between the injection well and the producing well will depend upon the area of contamination and the porosity of the formation. The injection well will preferably be on the outside perimeter of the contaminated area so that nutrients move through all of the contaminated area to the producing well. It will be understood that the injection well need not necessarily be within the contaminated area, but may be outside it and merely in the vicinity of the area. The water flow created by the output of the producing well will bring the nutrients and oxygen into the contaminated area where the biodegradation of hydrocarbon will occur.

FIGURE 57: SCHEME FOR RECLAMATION OF HYDROCARBON CONTAMINATED GROUND WATERS

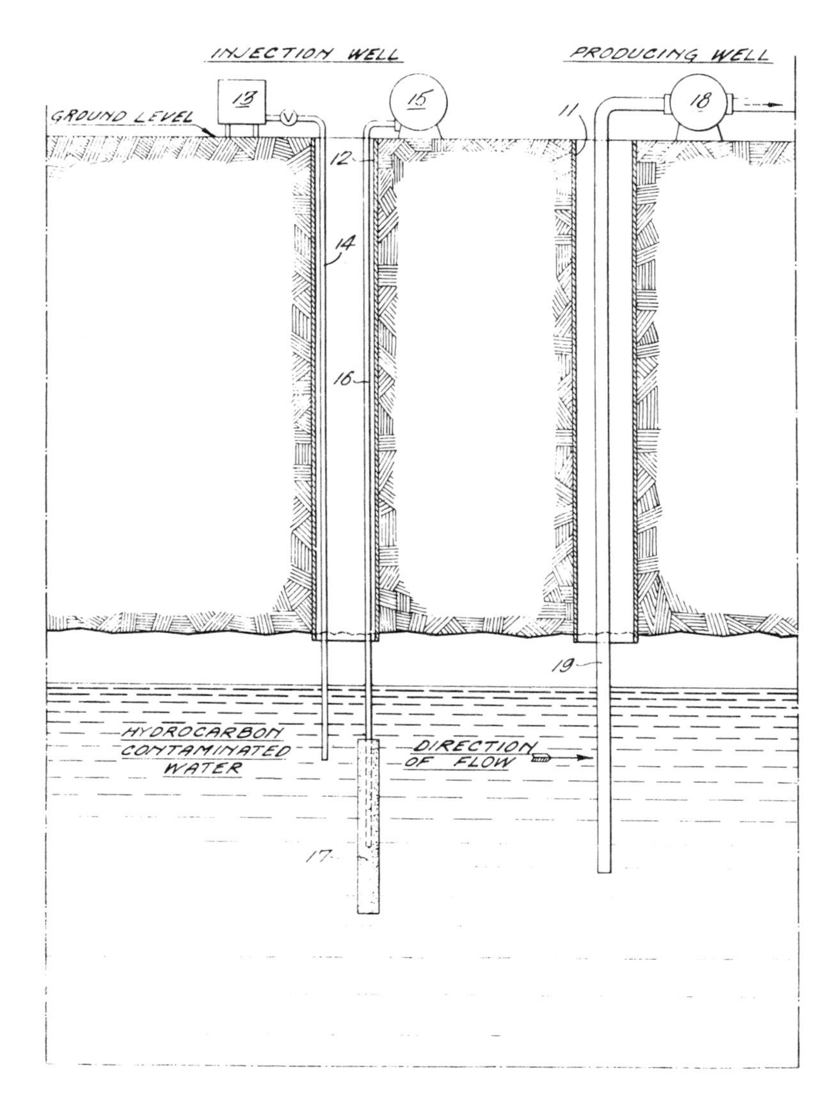

Source: U.S. Patent 3,846,290

Where the formation involved is dense, the distance between the two wells may be closer than where a porous formation is involved. In a typically porous formation, a single producing well will effect significant water movement for a radius of greater than 600 feet. On the other hand, if the producing well and injection well are too close, significant losses of nutrient may occur due to its being pumped out of the producing well. To avoid such nutrient losses, a minimum distance between the two wells of about 100 feet should be maintained. The pumping rate should also be carefully balanced with the regeneration of the water table. If the pumping rate is too high, there is the possibility of having the hydrocarbon trapped in the formation out of contact with nutrients. The following is one specific example of the conduct of this process.

An essentially circular area of land (average radius of about 350 feet) having a number of water wells throughout (about 20 million gallons of water, was contaminated with about 54,500 gallons (344,000 pounds) of high octane gasoline from a broken pipe line which carries gasoline through the area to a remote point. Ten wells on the periphery of the contaminated area were prepared as injection wells by erecting a mixture tank at each well and providing an injection tube and an air pump, conduit and sparger. Two wells in the central portion of the contaminated area were selected as producing wells.

A solution of 80 parts by weight of $(NH_4)_2SO_4$, 48 parts of Na_2HPO_4 and 32 parts of KH_2PO_4 in 834 parts of water is injected into each injection well at a rate of 5 gallons per hour. Air is simultaneously pumped into each injection well through a porous sparger at a rate of 3.5 cubic feet per minute in order to introduce oxygen into the formation water. The pumps of the two producing wells remove from the formation a total of about 1.5 million gallons of water per day. In this way water movement from the injection wells to the producing wells is achieved and the approximately 1.25 tons of bacterial cells produced each day effectively remove the gasoline from the water and surrounding formation in about 100 days.

In many instances it is desirable that an aqueous stream be generally oil- or organic-free, and that when such a stream becomes contaminated with an organic or oil-like liquid that it be shut off as soon as possible in order to prevent spread of organic contamination. Such a valving situation is particularly desirable on water discharge lines which go to natural streams and on water input lines to processing facilities wherein organic contamination is observable. It would be desirable if there were available an improved valve for aqueous lines which would be responsive to organic liquid contamination and which would close in the presence of organic liquids.

Such a valve has been developed by *R.H. Hall, D.H. Haigh, R.L. Derby and W.E. Jennings; U.S. Patent 3,750,688; August 7, 1973; assigned to The Dow Chemical Company* and it may be installed in a line carrying an aqueous stream. The valve is in the form of a permeable bed of particulate swellable polymer particles which imbibe organic materials and on contact therewith will swell to provide a positive shutoff. A sectional view of such a valve design is shown in Figure 58.

In the figure valve **12a** comprises a housing **16**. The housing **16** defines an internal cavity **17**. A first conduit or inlet **18** is in operative communication with the cavity **17** and a second conduit or outlet **19** is also in communication with the cavity **17** and is generally remotely disposed from the conduit **18**. A first retaining means **21** is disposed adjacent the inlet **18** and a second support member **22** is disposed adjacent the outlet **19**. Beneficially, the retaining means are of conventional structure such as screen, particulate material such as sand, pebbles and the like. Disposed between the support members **21** and **22** is a body **24** of a polymer which is capable of swelling on contact with organic liquids. The support members **21** and **22** prevent any significant axial movement of the body **24** within the conduit.

FIGURE 58: VALVE FOR AUTOMATICALLY SHUTTING OFF AQUEOUS STREAM WHEN HYDROCARBONS ARE PRESENT

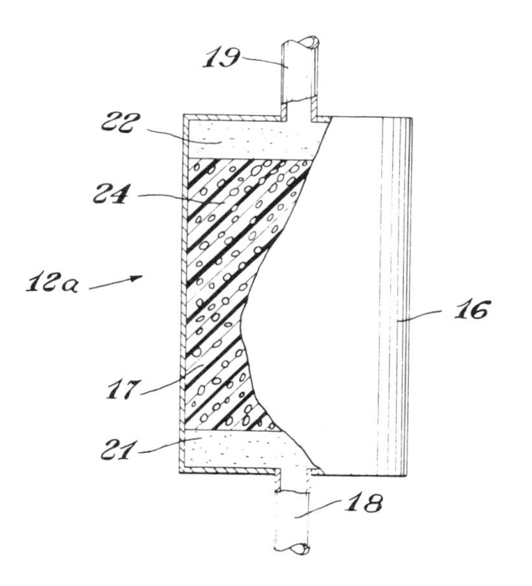

Source: U.S. Patent 3,750,688

When organic contamination occurs in the stream entering the valve, the organic material, or at least a substantial portion thereof, is imbibed by the body **24** causing the body to swell. As the body is confined and restrained within the cavity **17** the spaces and passageways within the body are decreased in size and finally closed completely, thereby providing an effective shutoff valve. The addition of instrumentation or controls (which are responsive to pressure increase as the body swells) for any particular situation is well within the means of anyone skilled in the art.

Polymers useful in the construction of this unusual valving means are any polymers which are water-insoluble and which swell on contact with organic liquids. Useful polymers may swell on contact with water. However, additional swelling must occur when contacted with an organic liquid. Selection of a polymer for use with any organic liquid is readily accomplished by determining a swelling index for the polymer particles. Beneficially, such a swelling index is readily determined by immersing a particulate polymer to be evaluated in water until the polymer has reached equilibrium swelling and subsequently adding the desired organic liquid and determining the volume per unit weight of polymer after a period of 30 minutes with water and organic liquid and the volume per unit weight of the polymer when in equilibrium with water. The ratio of the volume per unit weight with organic liquid and water to volume per unit weight of the polymer with water provides the swelling index. If the polymer is soluble the swelling index is infinite. If the swelling index is greater than about 1.2, the polymer particles are useful in the construction of this device.

Beneficially for most applications a swelling index of at least 1.5 and preferably greater than about 3 is desirable. It is not critical to employ a crosslinked polymer which swells but does not dissolve. If the polymer swells in the presence of the organic liquid and water it is suitable. However, for most applications it is desirable to employ a polymer which is crosslinked to a sufficient degree that it exhibits a swelling index between about 1.5 and 50 and preferably between about 3 and 50. By utilizing the crosslinked polymer the hazard of dissolution of the polymer over extended periods of time is eliminated.

However, for many applications, particularly those wherein instrumentation is employed to detect the pressure drop and an organic liquid contamination in water stream will appear in relatively large quantities, uncrosslinked polymer is eminently satisfactory. A wide variety of polymeric materials are employed with benefit. Such polymers include polymers of styrenes and substituted styrenes.

A process developed by *A. Tribellini; U.S. Patent 3,905,901; September 16, 1975; assigned to Creusot-Loire, France* for purification of hydrocarbon-contaminated wastewaters involves intimately mixing with the effluent a hydrocarbon solvent insoluble in water, and passing the mixture obtained through a bed of a solid complex consisting essentially of a solid substrate bearing free hydroxyl groups activated with 0.01 to 10 parts by weight of an acid halide per one part of the substrate, and to which is covalently bound from 0.01 to 10 parts by weight of an amine selected from the primary aliphatic amines having from 6 to 20 carbons and the primary aromatic amines having, attached to a phenyl ring, a linear hydrocarbon chain of 1 to 10 carbons bearing the amine group so as to separate the hydrocarbon solvent, containing the hydrocarbons and other organic materials in the dissolved state, from the purified effluent. Such a process is shown in Figure 59.

The effluent to be treated, for example, coming from a decantation tank, is supplied by a pipe **1** to the top of two prepurifying columns $\mathbf{C^1}$ and $\mathbf{C^2}$ which are provided with a bed **2** of a granular solid complex and operate in an alternating manner. The supply of the effluent to the columns is regulated by valves **3** and **4** respectively. Periodically, each column is alternately cleared by sending liquid to the lower part of the latter by way of a conduit **5** provided with valves **6** and **7**. When one column is being cleared it is the other column which purifies and vice versa. In order to clear the column either the effluent to be purified itself may be employed or any water from which materials in suspension have been removed. The water which has been used for clearing the column leaves the column by way of a pipe **8** in which are included valves **9** and **10**.

The prepurified effluent leaves the columns $\mathbf{C^1}$ and $\mathbf{C^2}$ by way of a pipe **11**, the flow being controlled by valves **12** and **13**. It arrives in a mixer-homogenizer unit **14** in which it is intimately mixed with hydrocarbon solvent which is immiscible with water arriving by way of a pipe **15** having a metering pump **16**. The effluent-solvent mixture is sent by way of a pipe **17** to a finishing column **C**^{3} filled with a bed **18** of solid granular complex. The pipe **17** opens out at a certain distance (for example 10 to 50 cm) below the surface of the bed **18**. It is in this column that the separation between the solvent and the effluent occurs. The solvent leaves the column by way of a pipe **19** located at the top of the latter and the purified effluent leaves at the lower part by way of a pipe **20**.

The following is one specific example of the conduct of this process in the treatment of the effluent from an oil cracking and reforming unit. After a first decantation, the aqueous effluent still contains 60 to 70 ppm of total hydrocarbons and a similar proportion of materials in suspension. The effluent passes through a first column (**C**^{1}-**C**^{2}) containing 70 cm of a crushed porcelain-n-tetradecylamine complex having a particle size of 1 to 3 mm. The flow is 1.5 l/hr/cm^2 and every 3 hours the column is cleared for 10 minutes with a flow of 4.5 l/hr/cm^2 with the same effluent which is thereafter returned to the decantation tank.

At the outlet of the first column, the effluent contains only 13 ppm of total hydrocarbons. It contains practically no more materials in suspension and no phenol. There are then intimately mixed with the effluent issuing from the first column 2,000 ppm of an aliphatic paraffinic hydrocarbon cut of C_{12} to C_{18}. The mixture is homogenized by mechanical agitation so that there is obtained an emulsion which is stable for 12 hours. The mixture obtained is then supplied to a second column **C**^{3} containing a layer of 50 cm of the same complex but having a particle size of 0.6 to 1.2 mm. An effluent is obtained at the base of the column which contains only 3.2 to 3.4 ppm of total hydrocarbons.

FIGURE 59: SOLVENT EXTRACTION PROCESS FOR REMOVING HYDROCARBONS AND OTHER ORGANIC PRODUCTS FROM AQUEOUS FLUIDS

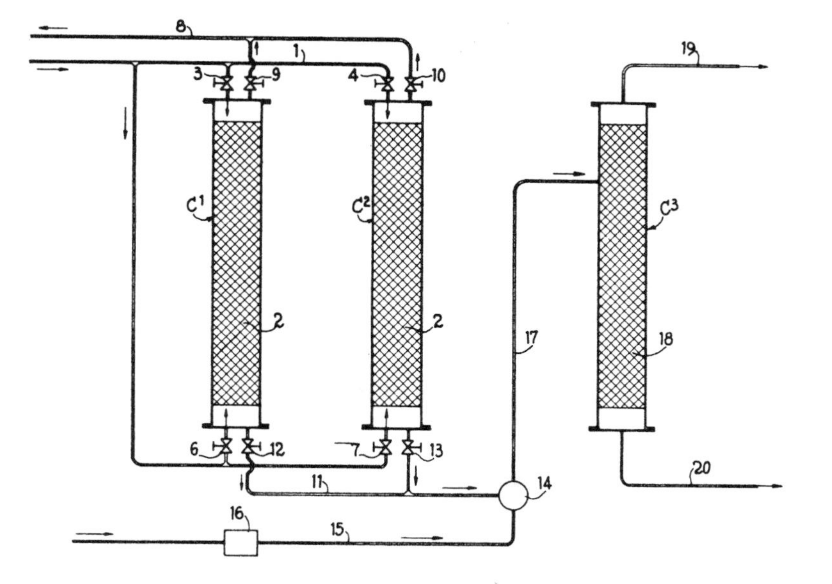

Source: U.S. Patent 3,905,901

HYDROGEN CHLORIDE

The use of halogen-containing plastics such as polychloroprene, polyvinyl chloride and polyvinylidene chloride has increased in recent years and is expected to increase at an accelerated rate in the future. One of the major advantages of these plastics is that they may be discarded and incinerated after use. The Battelle Memorial Institute has estimated that one percent of the municipal wastes in the United States is halogen-containing plastics. Most municipal wastes are burned in incinerators, with the combustion of the halogen-containing plastics resulting in the halogens being released, usually in the form of the hydrogen halide. Such combustion products are undesirable due to the air pollution and corrosion which they cause. Hydrogen chloride may of course occur in the vent gases from other industrial operations as well as indicated in the cross-references at the end of this section.

Removal from Air

A process developed by *W.R. Fuller, B.H. Bieler and D.C. Morgan; U.S. Patent 3,556,024; January 19, 1971; assigned to The Dow Chemical Company* involves reducing the amount of halogen halide emitted during the incineration of halogen-containing plastics. The method involves applying an alkali to the plastic before it is burned. The method can reduce the emission of halogen halide by greater than 75% when properly employed.

The high solubility of hydrogen chloride in water and the low vapor pressures of even 20% hydrochloric acid solutions make the collection of hydrogen chloride from gases in water an effective and inexpensive method of control.

A process developed by *J.E. Seebold; U.S. Patent 2,545,314; March 13, 1951; assigned to Hercules Powder Company* relates to the manufacture of hydrochloric acid by absorption from waste gases containing hydrogen chloride. In this process, gases rich in hydrogen chloride may be passed through an absorption tower without external cooling means, the tower being of such size, velocity and temperature that 20° Baumé hydrochloric acid is produced.

A process developed by *H.C. Wohlers, et al; U.S. Patent 2,730,194; January 10, 1956; assigned to Michigan Chemical Corporation* relates to the separation and recovery of hydrogen chloride, hydrogen bromide and other gaseous substances from waste gases discharged such as from brominators or from chlorinators engaged in the preparation of chloral and related products by alcohol chlorination, paraldehyde or acetaldehyde chlorination. Recovery of hydrogen chloride from the vent gases of a chlorinator, as in the manufacture of chloral, poses many of the problems characteristic of the recovery of hydrogen chloride from other gas systems but, in addition, a number of particularly difficult problems are imposed because of the presence also of large amounts of free chlorine and relatively large amounts of chloral hydrate, chloral and other impurities such as monochloroacetaldehyde, dichloroacetaldehyde resulting from the chlorination of acetaldehyde or paraldehyde or ethyl chloride resulting from chlorination of ethyl alcohol.

In order to achieve efficient and effective removal of the organic impurities soluble in hydrochloric acid at an initial stage of the process, it is desirable to carry out the scrubbing action in a scrubber such as a packed column having hydrochloric acid passing downwardly therethrough by gravitational force from an acid inlet at the top to a discharge opening at the bottom while the vent gases discharged from the reactor or chlorinator are fed through a gas inlet near the bottom of the column and rise upwardly through the column to a gas outlet at the top.

A process developed by *R. Krumböck and W. Kühn; U.S. Patent 3,980,758; September 14, 1976; assigned to Hoechst AG, Germany* is a process for the combustion of chlorine-containing residues and wastes with simultaneous recovery of the hydrogen chlorine thereby obtained, which comprises burning the chlorine-containing residues with an excess of air and simultaneous addition of azeotropically boiling hydrochloric acid formed in the work-up of the combustion gases. Figure 60 is a flow diagram of such a process.

FIGURE 60: PROCESS FOR COMBUSTION OF CHLORINE-CONTAINING WASTES WITH HCl RECOVERY

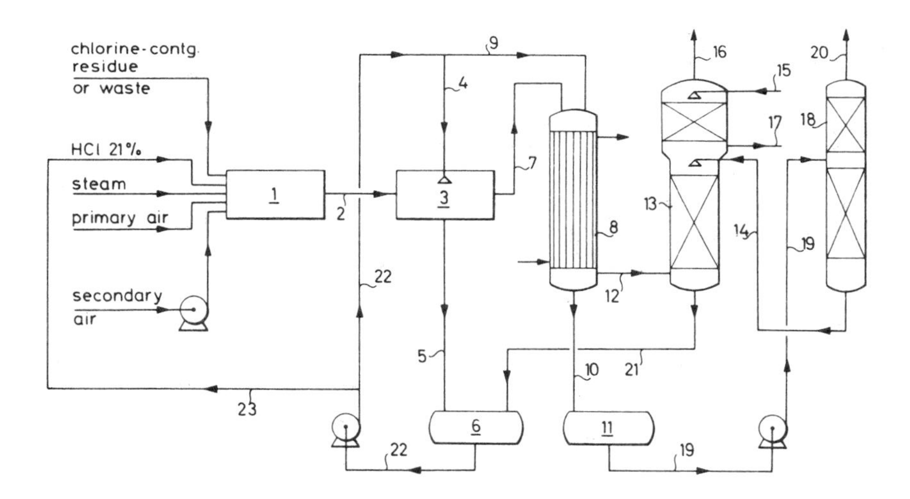

Source: U.S. Patent 3,980,758

The hot combustion gases leaving the combustion chamber **1** are forwarded, via duct **2**, to a quenching chamber **3**, where they are cooled to 90°C by means of azeotropic HCl fed in via ducts **22**, **4** and let off via duct **5** into a reservoir **6**. Subsequently, the gas is forwarded, via duct **7**, to an absorber **8** and washed there with azeotropic acid fed in via ducts **22, 9**. In this operation, the greater part of the hydrogen chloride contained in the combustion gases is absorbed by the hydrochloric acid (its hydrogen chloride content thus being concentrated to about 30%), which is let off via duct **10** into a reservoir **11**. The gases are forwarded from absorber **8** via duct **12** into a countercurrent scrubber **13**, where they are again washed with azeotropic hydrochloric acid, duct **14**, and then with pure water, duct **15**. The gases leave the scrubber column as waste gas via duct **16**, and the washing water used in scrubber **13** leaves it via duct **17**. The hydrogen chloride contained in the 30% hydrochloric acid obtained in absorber **8** and stocked in reservoir **11** is recovered in desorber **18**; the hydrochloric acid being fed in via duct **19**. At the top of the desorber, hydrogen chloride is taken off via duct **20**. In the sump of the desorber, azeotropic hydrochloric acid is collected, which is then forwarded via duct **14** to scrubber column **13**, from where it is discharged via duct **21** into reservoir **6**. Since, according to the state of the art, water was continuously injected into the combustion chamber **1**, of course large amounts of azeotropic hydrochloric acid were formed which were collected in the sump of desorber **18** and stocked in reservoir **6**. Generally, there was no practical application for these amounts which, apart from the portion forwarded via duct **22** to quenching chamber **3** and absorber **8** were discharged into the sewage.

Thus, large amounts of hydrogen chloride are lost without any practical use, and additional chemicals are even required to make them innocuous for the sewage; It has been found that it is possible to avoid the continuous formation of azeotropic hydrochloric acid by introducing the azeotropic hydrochloric acid collected in the sump of absorber **18** and stocked in reservoir **6**, instead of water, into combustion chamber **1** via duct **23**, thus recovering nearly all hydrogen chloride obtained in the combustion and making possible its profitable reuse, while simultaneously contributing to a decontamination of the sewage from chemicals.

A process developed by *W. Kunzer et al; U.S. Patent 3,036,418; May 29, 1962; assigned to Badische Anilin- & Soda-Fabrik AG, Germany* involves working up waste gases containing hydrogen chloride and impurities and originating from organic chlorination reactions. When working in this way there is the disadvantage that the waste gas passing to the further working up is in general saturated with the vapors of impurities contained therein and originating from the organic chlorination reaction and that there is a risk of these vapors being condensed upon the entry of the waste gas into the absorption zone of the tower.

The disadvantage can be avoided by heating the contaminated gas containing hydrogen chloride, prior to the absorption, to a temperature which is above the dew point of the impurities. This heating above the dew point of the impurities may advantageously be effected by leading the gas coming from the chlorination apparatus in heat exchange relation to the hot acid flowing from the absorption zone of the tower. For this purpose it is thus only necessary to install in the absorption zone a suitable device for heating up the gases to be worked up, through which the acid produced flows downwardly.

See Glass Treatment Effluents (U.S. Patent 3,885,929)

Removal from Water

See Pickle Liquors (U.S. Patent 3,468,797)
See Iron and Steel Pickle Liquors (U.S. Patent 3,896,828)
See Iron and Steel Pickle Liquors (U.S. Patent 3,745,207)

HYDROGEN CYANIDE

It is common for a wide variety of gases to contain acidic gaseous impurities. For example, gaseous fuels, e.g., coke oven gas, gases from coal gasification processes, gases from distillation operations, gases from partial oxidation operations and even natural gas contain, in addition to hydrocarbons, hydrogen sulfide, carbon dioxide, carbonyl sulfide (carbon oxysulfide), and frequently other components which are particularly troublesome during further processing, such as hydrogen cyanide in particular.

Removal from Air

To remove these acidic components, it is known that the gas can be subjected to a solvent scrubbing step with a scrubbing agent which dissolves such components as hydrogen sulfide from the gas and is subsequently recycled into the process after regeneration. A special problem in this procedure is caused by hydrogen cyanide, on the one hand because it is readily soluble in numerous solvents and correspondingly strongly retained in the solvents and, on the other hand, because of its corrosive action and extraordinary toxicity. Therefore, the cited process provides, prior to the actual main scrubbing step which removes the hydrogen sulfide, a preliminary scrubbing process with alkaline-adjusted water in a separate preliminary scrubbing tower which serves to remove the hydrogen cyanide before all other gas components are scrubbed out.

This is done, however, at a considerable expense, since the preliminary scrubbing column must be designed to accommodate the entire quantity of gas irrespective of the relatively minor amount of hydrogen cyanide contained in the gas. Furthermore, a scrubbing step with alkaline-adjusted water is ineffective with exhaust gases having high CO_2 and H_2S partial pressures, since carbonic acid displaces hydrocyanic acid from KCN solutions. For this reason, preliminary scrubbing steps with untreated water are more often utilized in practice. However, in these processes, the hydrocyanic acid is obtained as a highly dilute aqueous stream giving rise to either a serious water pollution problem or, when the scrubbing water is subsequently stripped with air or the like, a serious air pollution problem.

An improved process developed by *H. Karwat; U.S. Patent 3,935,188; January 27, 1976; assigned to Linde AG, Germany* embodies the improvement wherein the hydrogen cyanide is scrubbed out simultaneously with the other acidic components. The resultant loaded scrubbing agent is mixed with an aqueous alkali metal or alkaline earth hydroxide solution, and the thus-formed cyanide salt solution is thermally converted into ammonia and formate.

When sulfurous gases are converted to elemental sulfur as in a Claus sulfur recovery unit, for example, the presence of hydrogen cyanide, which is incompletely burned in the limited oxygen atmosphere of the burner, results in extensive corrosion of the subsequent apparatus and the production of dark, contaminated sulfur. When sulfuric acid is the desired product, the presence of hydrogen cyanide results in the formation of nitrogen oxides, which, together with unburned hydrogen cyanide, results in deactivation of the vanadium oxide catalyst of the contact process. The nitrogen oxides also cause extensive corrosion of equipment due to formation of nitroso compounds. Furthermore, the nitrogen oxides cause the continued formation of sulfuric acid in the tail gas stack of the chamber process. This results in heavy corrosion of this equipment and emission of sulfuric acid to the environment.

A process developed by *O.A. Homberg, C.W. Sheldrake and J.B. Lynn; U.S. Patent 3,923,957; December 2, 1975; assigned to Bethlehem Steel Corporation* provides a method of eliminating hydrogen cyanide, particularly hydrogen cyanide present in foul gas streams by reacting hydrogen cyanide with hydrogen sulfide and oxygen to produce ammonia and carbon sulfides. A preferred embodiment destroys the hydrogen cyanide present in foul acid gas streams recovered from industrial gas desulfurizers, and the process is particularly useful when employed prior to a Claus or similar sulfur recovery unit. The process can also be used to produce commercial quantities of carbon disulfide. Figure 61 is a schematic representation of this process, when used in conjunction with a Claus sulfur recovery unit.

FIGURE 61: PROCESS FOR CONVERSION OF HYDROGEN CYANIDE IN FOUL GAS STREAMS TO CARBON DISULFIDE

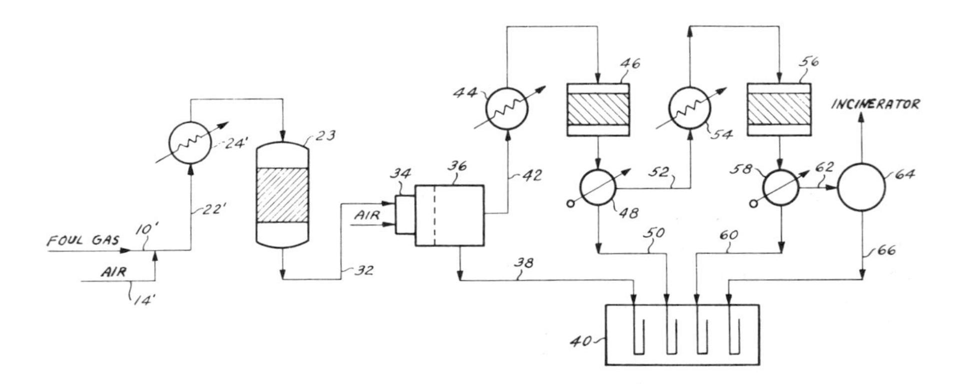

Source: U.S. Patent 3,923,957

A foul gas derived from a coke oven gas desulfurizer (not shown) at about 9.0 pounds per square inch gauge, in line **10** is combined with air from line **14'** and the combined gases pass through heat exchanger **24'** in line **22'** where they are heated to about 100°C. Proceeding from the heat exchanger, the gases enter catalytic reactor **23**. Catalytic reactor temperatures range up to 250°C and, after passing through the reactor the exit gases emerge in line **32**. From line **32** the exhaust gases are directed to hydrogen sulfide burner **34** and thermal reactor **36**. Beginning with burner **34** the succeeding units of the process are often collectively referred to in the prior art as the Claus process. In accordance with the usual Claus process, burner **34** generates SO_2 from H_2S present in the gas feed and the SO_2 reacts in thermal reactor **36** with the H_2S according to:

$$2H_2S + SO_2 \longrightarrow 2H_2O + 3S$$

Sulfur is produced in thermal reactor **36** and leaves the reactor in line **38** to sulfur storage unit **40**. Gases exit from thermal reactor **36** in line **42** at about 160°C, are heated to about 235°C in heater **44** and enter catalytic reactor **46**. Emerging from catalytic reactor **46**, where sulfur is produced according to the above equation, the reactor gases are cooled to about 160°C, in cooler **48**, causing more sulfur to precipitate. This sulfur enters sulfur storage unit by line **50**. In a manner similar to the one just described, the exit gases from reactor **46** leave cooler **48** in line **52**, are reheated in heater **54**, and enter catalytic reactor

56. The product of reactor **56** is cooled in cooler **58** to about 160°C, causing sulfur again to precipitate. The sulfur travels to sulfur storage unit **40** in line **60** and the tail gases leave cooler **58** by line **62** and enter tail gas separator **64**. In tail gas separator **64**, residual sulfur precipitates, is sent to sulfur storage unit **40** in line **66** and the tail gases emerging from unit **64** proceed to a stack gas incinerator (not shown) for removal of trace H_2S prior to discharge of the purified tail gas. An analysis of the gases during the course of the Claus reaction indicates that the ammonia produced by this method is disassociated in burner **34** to hydrogen and nitrogen, the hydrogen then burning to water. The carbon sulfides (either the CS_2 or the COS) are also largely destroyed in the burner. The Claus process apparatus shows none of the corrosion apparent before the implementation of this HCN destruction system.

A process developed by *I. Ooka, N. Tomihisa, Y. Nogami and K. Katagiri; U.S. Patent 3,953,577; April 27, 1976; assigned to Osaka Gas Company, Ltd., Japan* is a process in which gases containing hydrogen cyanide, ammonia and hydrogen sulfide are purified by first washing the gas with a suspension of solid sulfur to fix the hydrogen cyanide as ammonium thiocyanate and ammonium thiosulfate. The resulting suspension is then subjected to a wet-oxidation wherein the thiosulfate and a part of the thiocyanate are converted into sulfuric acid and ammonium sulfate after which the resulting liquid containing unreacted thiocyanate is decomposed into an ammonium salt and is recovered from the oxidation and decomposition steps. A flow diagram of such a process is shown in Figure 62.

FIGURE 62: PROCESS FOR PURIFYING GASES CONTAINING HCN AND H_2S BY CONVERSION TO AMMONIUM SULFATE

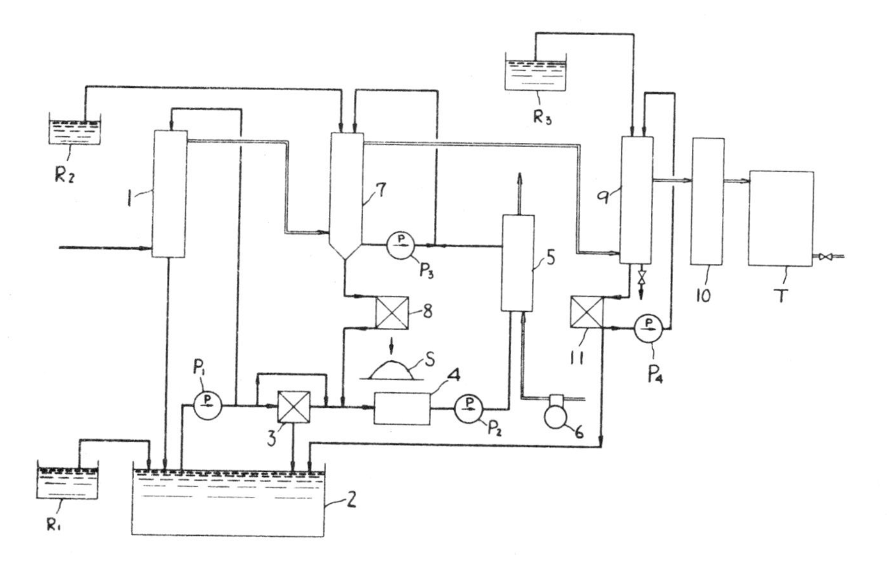

Source: U.S. Patent 3,953,577

Gas containing HCN, NH_3 and H_2S is supplied to an absorption tower **1** of the gas-liquid countercurrent type, where the gas is washed with a recycling suspension of sulfur particles from a tank **2** to fix HCN in the form of NH_4SCN. Since the ability of the suspension to fix HCN is reduced corresponding to the increase of the concentration of NH_4SCN and $(NH_4)_2S_2O_3$ in the suspension, part of the suspension may preferably be discharged from the tank **2**, with a fresh sulfur suspension supplied from a tank R_1. The discharged suspension is sent by way of a pump $\mathbf{P_1}$ to a filter **3**, where a great part of the sulfur is separated off. The liquid is then sent to a tank **4**. The separated sulfur may be sent back to the tank **2** for reuse.

The suspension discharged may be caused to bypass the filter **3** to convert the unrecovered sulfur directly to sulfuric acid in the subsequent oxidation decomposition step so that the acid can be favorably utilized in the ammonia removing step. By means of a pump $\mathbf{P_2}$, the suspension containing NH_4SCN, $(NH_4)_2S_2O_3$ and solid sulfur is sent from the tank **4** to oxidation decomposition means **5**, where it is brought into contact with air supplied from a compressor **6** with heating. In the oxidation decomposition means **5** almost all amounts of $(NH_4)_2S_2O_3$ and solid sulfur and about 50 to 70% of NH_4SCN are oxidized with the air into $(NH_4)_2SO_4$ and H_2SO_4. After decomposition, the reaction mixture is led to a reactor **7** to come into contact with a solution of inorganic acid from a tank $\mathbf{R_2}$, whereby unreacted substances principally comprising NH_4SCN are decomposed. In the case where H_2SO_4 is used as the inorganic acid, NH_4SCN will be decomposed as represented by the following equations:

$$2NH_4SCN + H_2SO_4 + 2H_2O \longrightarrow (NH_4)_2SO_4 + 2H_2S + 2HCNO$$

$$2HCNO + H_2SO_4 + 2H_2O \longrightarrow (NH_4)_2SO_4 + 2CO_2$$

As an advantageous device to be used for the reactor **7**, an ammonia saturator may be used which is employed for removing hydrogen sulfide from gases. If, however, it is difficult to effect complete decomposition of NH_4SCN only by the ammonia saturator, one or more reactors may be additionally used. Part of the inorganic acid-containing liquid discharged from the reactor **7** is returned to the reactor **7** again by means of a pump $\mathbf{P_3}$ so as to be reused for the decomposition of NH_4SCN. The liquid containing high contents of $(NH_4)_2SO_4$ produced in the oxidation decomposition means **5** and ammonium salt of inorganic acid produced in the reactor **7** is discharged from the bottom of the reactor **7** and fed for example to a centrifugal separator **8** to collect $(NH_4)_2SO_4$ produced in oxidation decomposition means **5** and ammonium salt of inorganic acid produced in the reactor **7**.

Ammonium salts thus separated are shown at **S**. The resulting liquid is sent back to the tank **4** if unreacted NH_4SCN, $(NH_4)_2SO_4$ and ammonium salt of inorganic acid still remain in the liquid. The gas discharged from the absorption tower **1** where HCN has been removed therefrom is introduced into the ammonia saturator by an inorganic acid as in a conventional manner, for which the reactor **7** is advantageously used and where NH_3 is removed from the gas in the form of an ammonium salt of inorganic acid. The gas is then subjected to a desulfurization step with an aqueous solution of catalyst such as picric acid, meta-nitrophenol, hydroquinone, etc. from a tank $\mathbf{R_3}$ in a conventional wet-type desulfurizer **9** and, where required, it is further led to a dry-type desulfurizer **10** where nearly complete desulfurization is achieved. The gas is then stored in a purified gas tank **T**.

Solid sulfur recovered by the wet-type desulfurizer **9** is separated from the catalytic solution, for example, by a centrifugal separator **11** and then sent back to the washing liquid tank **2** for reuse, while the catalytic solution is returned to the desulfurizer **9** through a pump P_4.

A process developed by *M. Matumoto and T. Aono; U.S. Patent 3,855,390; December 17, 1974; assigned to Nippon Steel Corp., Japan* is a process for purifying a gas containing hydrogen cyanide wherein the gas is brought into contact with an alkaline absorbing solution containing polysulfides to remove hydrogen cyanide in the form of thiocyanates, comprising withdrawing a side stream of the solution in order to maintain the pH of the absorbing solution above 7.7, supplying the side stream with a gas containing more than 2 mols of free oxygen per mol of thiocyanate in the stream, thermally treating the stream at a temperature above 150°C and under such a pressure as to keep the reaction mixture in the liquid phase and thereby hydrothermally oxidizing the thiocyanates to ammonium sulfate and CO_2.

A process developed by *S. Kumata, Y. Shimoi, T. Hirabayashi and Y. Hiwatashi; U.S. Patent 3,887,682; June 3, 1975; assigned to Nittetu Chemical Engineering Limited, Japan* is a wet-type desulfurization process for purifying gases containing hydrogen sulfide and hydrogen cyanide contaminants. The gas is washed with an alkaline aqueous solution to absorb the contaminants. The wash solution containing the dissolved contaminants is oxidized to form free sulfur from any hydrosulfide salt that might be present. The sulfur is removed as a precipitate. The oxidized wash solution is then roasted to convert the thiocyanates, sulfates, sulfites, and thiosulfates to carbonates, sulfides and hydrosulfides. These salts are then recovered in an aqueous solution, which solution is recycled for use in the absorption or oxidation steps.

A process developed by *D.K. Beavon; U.S. Patent 3,878,289; April 15, 1975; assigned to The Ralph M. Parsons Company* is one in which hydrogen cyanide contained in gas streams such as coke oven gas streams, is eliminated by catalytically hydrolyzing the hydrogen cyanide in the presence of water to ammonia and using as the catalyst one or more of the alkali metal hydroxides supported on alumina, alumina-silica, or silica. The formed ammonia is readily separated from the gas stream by any conventional technique. When unsaturated hydrocarbons and oxygen are present in the treated gas stream, the gas stream may be hydrogenated to eliminate oxygen and at least partially saturate the unsaturated hydrocarbons to prevent catalyst degradation.

A process developed by *T. Nicklin and P.S. Clough; U.S. Patent 3,859,415; January 7, 1975; assigned to North Western Gas Board, England* is one in which hydrogen cyanide is removed from gas mixtures containing hydrogen cyanide and hydrogen sulfide by treating the gas mixture with a catalyst comprising the elements nickel, uranium and thorium disposed as their oxides on a gamma-alumina support. Preferably the gas mixture is saturated with water vapor before being subjected to the catalytic treatment.

HYDROGEN FLUORIDE

See Alkylation Process Effluents (U.S. Patent 3,972,956)

See Aluminum Cell Exit Gases (U.S. Patent 3,907,971)
See Aluminum Cell Exit Gases (U.S. Patent 3,827,955)
See Aluminum Cell Exit Gases (U.S. Patent 3,919,392)
See Aluminum Cell Exit Gases (U.S. Patent 3,876,394)

HYDROGEN SULFIDE

The modified Claus process to convert hydrogen sulfide obtained from sour natural gas typically emits 3,000 ppm SO_2, a level which is substantially higher than the 250 to 300 ppm SO_2 maximum standard adopted in various parts of the country. This pollution control problem also applies to numerous industrial operations where hydrogen sulfide is a by-product.

Removal from Air

A process developed by *J.A. Connell; U.S. Patent 3,864,460; February 4, 1975; assigned to NRG Incorporated* is one in which hydrogen sulfide is removed from gas streams by liquid absorption, the absorbed hydrogen sulfide is partially converted to sulfur as in a Claus plant and the residual hydrogen sulfide in the tail gas is concentrated by molecular sieve adsorption and recycled into the absorption stage at higher concentration such that all the hydrogen sulfide is removed from the gas stream as sulfur, thus eliminating atmospheric pollution which normally results from release of hydrogen sulfide containing tail gas.

Manufactured synthesis gases and fuel gases must be desulfurized as a rule before they are used. This is required to prevent a poisoning of catalysts, which in most cases are susceptible to sulfur, and to minimize the sulfur dioxide content of the flue gases of combustion. Such gases may be desulfurized by being scrubbed with liquid absorbents (aqueous solutions of alkali salts of weak acids or of strong organic bases). Carbon dioxide is often scrubbed from the gas with the sulfur compounds.

A process developed by *K. Bratzler, A. Doerges and J. Schlauer; U.S. Patent 3,896,215; July 22, 1975; assigned to Metallgesellschaft AG, Germany* provides for the removal of hydrogen sulfide from a gas stream or for the conversion of hydrogen sulfide to elemental sulfur according to the Claus process. It comprises passing the gas stream through a Claus process reactor having at least one contact stage thereby transforming much of the hydrogen sulfide into elemental sulfur. The effluent gas, containing residual sulfur and hydrogen sulfide is afterburned in the presence of a coke layer with a quantity of oxygen sufficient to react stoichiometrically with the sulfur and hydrogen sulfide and the resulting sulfur dioxide is scrubbed from the gas by an absorber. The sulfur dioxide is desorbed with steam and recycled.

Figure 63 shows a plant for the conduct of such a process. The plant consists essentially of a Claus reactor **1**, a sulfur condenser **2**, a sulfur separator **3**, an afterburning chamber **4**, a coke filter **5**, an absorption tower **6** and a regenerator **7**. From the sulfur separator **3**, the residual gas from the Claus process is passed in conduit **8** to the afterburning chamber **4** and is burned in the chamber, which is fed through conduit **9** with a preferably gaseous fuel and through conduit **10** with a surplus of air. By this combustion, noncondensed sulfur vapor and nonreacted hydrogen sulfide are converted into sulfur dioxide and small quantities of SO_3 are formed. The afterburned exhaust gas is passed through conduit **11** into the coke filter **3** after being subjected to interstage cooling. The coke filter **5** is operated at temperatures of 200° to 500°C to reduce the SO_3 to SO_2 and to consume the surplus oxygen by reaction with coke to CO_2.

FIGURE 63: PROCESS FOR THE REMOVAL OF HYDROGEN SULFIDE FROM A GAS STREAM

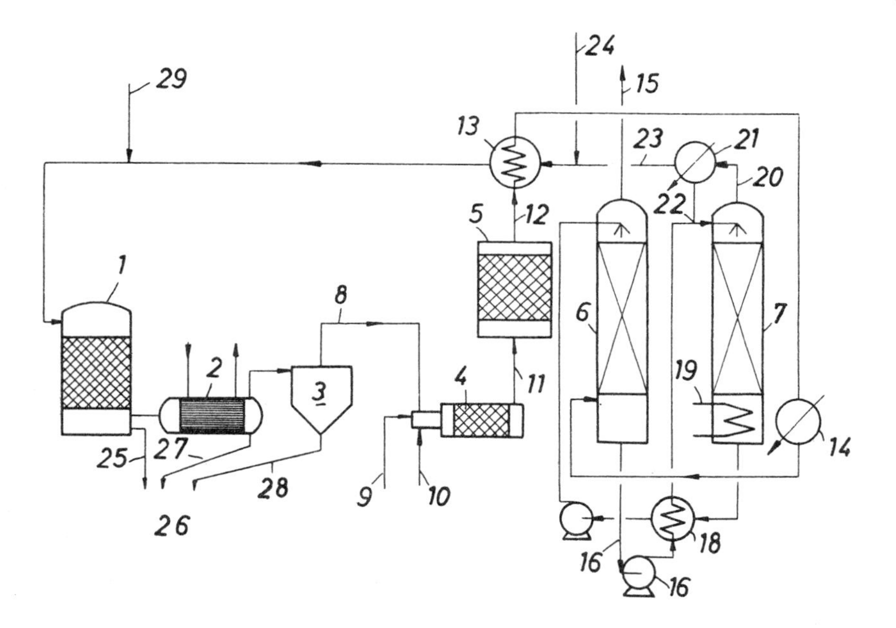

Source: U.S. Patent 3,896,215

At the outlet of the coke filter the gas is at a higher temperature than at the inlet thereof and contains neither SO_3 nor free oxygen. Via a conduit **12**, the gas is passed through a heat exchanger **13** and a cooler **14** and approximately at the ambient temperature is supplied into the absorption tower **6** above the sump thereof. In the tower **6**, the gas is scrubbed with regenerated absorbent solution. An exhaust gas which contains less than 200 ppm sulfur dioxide leaves the absorption tower **6** at the top thereof and may be discharged into the atmosphere through the chimney **15**.

The absorption solution which is laden with SO_2 is withdrawn from the sump of the absorption tower and via a conduit **16** is fed by a pump **17** through a heat exchanger **18** to the top of the regeneration tower **7**, in which the solution flows down over packing to the sump. In the latter, the solution is heated to the boil by a heater **19** so that the absorbed SO_2 is entirely expelled and by the rising steam is stripped from the solution which is trickling down. From the top of the regeneration tower **7**, the exhaust gas consisting of sulfur dioxide and water vapor is conducted through a conduit **20** to a condenser **21**, from which the collected condensate is recycled in conduit **22** to the top of the regeneration tower. The residual concentrated sulfur dioxide is conducted in a conduit **23** through the heat exchanger **13** to the Claus process reactor **1**. The hydrogen sulfide-containing gas to be processed is added through a conduit **24** to that SO_2 stream before it enters the heat exchanger **13**. The Claus process reactor **1** and associated sulfur condenser **2** and sulfur separator **3** may consist of a plurality of stages in known manner. In the Claus process reactor, H_2S and SO_2 are reacted in known manner in contact with a granular catalyst which consists preferably of bauxite. The elementary sulfur formed by the reaction is drained in a liquid state from the reactor **1** and is conducted via a conduit **25** to the collecting container **26**, which through conduits **27** and **28** receives also the liquid sulfur which is collected in the sulfur condenser **2** and the sulfur separator **3**.

In the processing of hydrogen sulfide-containing gases having a moderate H_2S content up to about 10% by volume, the SO_2 may be recirculated at the rate which is required for the reaction. A surplus of SO_2 is desirable because it increases the conversion to elementary sulfur. In the processing of gases rich in H_2S it may be desirable to burn part of the gases with air and to supply the combustion gas, e.g., through a conduit **29** into the conduit **23** before the Claus process reactor.

A process developed by *H. Tsuruta, Y. Hiwatashi, T. Hirabayashi and S. Kumata; U.S. Patent 3,941,875; March 2, 1976; assigned to Nittetu Chemical Engineering Ltd., Japan* is a process for treating a hydrogen sulfide-containing gas in a closed loop system wherein the gas is passed through and absorbed by an alkaline aqueous absorbent containing an alkali carbonate and an oxidation catalyst. The solution containing the dissolved hydrogen sulfide is oxidized with an oxygen-containing gas to convert the absorbed hydrogen sulfide into elementary sulfur and sulfur salt compounds.

After separation of the elementary sulfur from the solution, the solution is recirculated for use as alkaline absorbent. A part of the recirculated solution is diverted and subjected to mixed-combustion with an auxiliary fuel in a combustion furnace at an air ratio lower than 0.9 and at a temperature of 700° to 1100°C to thermally decompose the sulfur compounds into hydrogen sulfide and an alkali carbonate. The gaseous products of decomposition are brought into direct contact with boiling water to collect the alkali carbonate in the form of an aqueous solution and leave hydrogen sulfide in the gas.

Figure 64 shows such a process. In the following, the process is described as applied to a coke oven gas (which will be hereinafter referred to as COG for brevity) containing hydrogen sulfide, but it should be noted that this process is not limited to the treatment of COG. COG is fed into a lower section of an absorption tower **1** and an absorbent liquid containing sodium carbonate is introduced into an upper section of tower **1** through line **30**. In this manner, the COG is scrubbed with the absorbent liquid in a countercurrent flow relationship. In general, 1 m^3 of COG contains 5 to 6 grams of H_2S and about 2 grams of HCN and a significant amount of CO_2 as an acidic gas.

It is possible to raise the absorption rate of the hydrogen sulfide and hydrogen cyanide to substantially 100% and to minimize the absorption rate of CO_2 by properly selecting: the ratio of sodium carbonate to sodium bicarbonate in the absorbent liquid and the concentration of these two compounds; the ratio for the amount of COG gas feed to the amount of Na_2CO_3 containing absorbent liquid; and the ratio of the amount of oxidation catalyst to the amount of oxygen in the liquid. The liquid containing the absorbed gas passes from the lower section of absorption tower **1** through line **34** to an oxidation tower **2**, wherein it is oxidized by bubbled air which is injected into the liquid by means of an air compressor **3**. The absorbed hydrogen sulfide is converted into elementary sulfur in the oxidation tower **2**. A portion of liquid which contains the precipitated sulfur is removed from the top portion of the tower **2** and is introduced into a filter **4** to separate the sulfur from the liquid. The filtrate is returned to line **30** for reuse as an absorbent and the separated sulfur is collected. The liquid which has been treated in the oxidation tower **2** is returned through line **36** and circulating line **30** to the top of the absorption tower **2**.

FIGURE 64: ALTERNATIVE PROCESS FOR THE REMOVAL OF HYDROGEN SULFIDE FROM A GAS STREAM

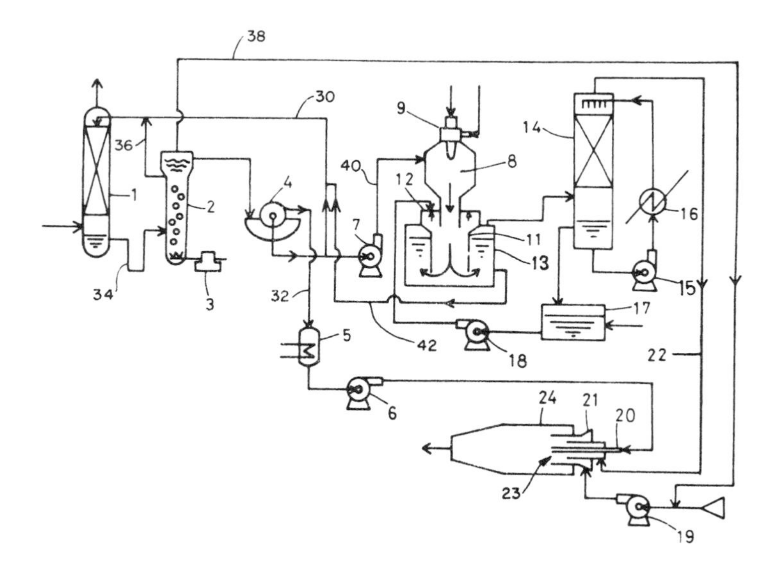

Source: U.S. Patent 3,941,875

The liquid exiting through line **36** has a reduced concentration of sodium hydrosulfide and a high concentration of sodium carbonate; thus the absorptivity of the liquid is restored. The gas exits from the oxidation tower **2** through line **38**. Reactions occur simultaneously to produce sulfur oxide-containing salts and thiocyanates. Since the concentrations of these salts would increase in the absorbent liquor with time, a waste liquor delivery pump **7** is provided to divert a portion of the absorbent liquor and to maintain the concentration of such by-product salts below a predetermined value. The portion of the circulating absorbent liquid which is removed from the system via line **40** by means of the pump **7** is sprayed into a combustion furnace **8** as is, or after being condensed by a suitable condenser (not shown). Since the water content of the waste liquor is as much as 50 to 80% by weight and the solid components of the waste liquor are not self-combustible, an auxiliary fuel burner **9** is provided in the furnace. The burner **9** burns COG or other fuels in a reduction atmosphere having an air ratio less than 0.9. During combustion, water in the waste liquor is evaporated and solid components are thermally decomposed. Thus, the decomposition by mixed combustion is conducted, on the whole, in a reducing atmosphere. The mixed-combustion decomposition is carried out on the whole at an air ratio below 0.9 and at a temperature of from 700° to 1100°C. Preferred results can be obtained by using a residence time for the gases within the furnace **8** of longer than 0.5 second, a temperature of 800° to 850°C and an air ratio ranging from 0.6 to 0.7. Under these conditions, most of the organic substances are gasified and the greater part of the sulfur-containing salts, cyanide and thiocyanate, are thermally decomposed to sodium carbonate and hydrogen sulfide.

The decomposition products mixed with the combustion gas pass through the bottom of the furnace **8** to a quenching chamber **12** where they are rapidly cooled by contact with water supplied by pump **18**. Then, the combustion products pass through a downcomer tube **11** in a mixed phase flow into the liquid which fills a soda recovery tank **13**. The mixed phase flow reaches almost complete thermal and physical equilibrium while passing through the liquid in tank **13**. The boiling water in recovery tank **13** dissolves the sodium carbonate which has been produced in the preceding thermal decomposition within furnace **8**. The liquid in tank **13** also dissolves a small amount of carbon dioxide, thereby causing the sodium carbonate to be converted into sodium bicarbonate. At this stage, hydrogen sulfide present in the mixed gas is only slightly soluble in the liquid within tank **13** and, therefore, passes through, remaining in the gaseous state. The liquid containing sodium carbonate and sodium bicarbonate is routed via line **42** for use as an absorbent in tower **1**.

The temperature of the free gas exiting tank **13** is substantially the same as that of the liquid in the recovery tank **13**, the gas being saturated with steam vapor. In order to remove steam from the gas, the gas is cooled by contact with cool water in a condensing tower **14** which is provided with a circulation pump **15** and a condenser **16**, thereby condensing the steam to a degree that a partial vapor pressure for water corresponding to the scrubbing water temperature is attained. The condensed water is then transferred to a circulation liquid collection tank **17** for supplementing the feedwater to quenching chamber **12** by means of a pump **18**.

The waste gas, after removal of water vapor in tower **14**, is mixed, if necessary, with an auxiliary fuel to make the gas self-combustible and fed through line **22** to nozzle **20** of combustion burner **23**. The waste gas is mixed with air supplied by a blower **19** and burned within chamber **24** to oxidize the hydrogen sulfide to sulfur oxides.

The elementary sulfur which is recovered in the sulfur filter **4** is heated and melted in a sulfur melting tank **5**. The molten sulfur is transferred from tank **5** to the mixed combustion burner **23** by means of a pump **6**. The mixed combustion burner **23** is hermetically connected to one end of the waste gas combustion furnace **24**. Air exhausted from the top of the oxidation tower **2** via **38** is used to burn molten sulfur and waste gas within the combustion furnace **24**. The exhaust air from tower **2** is supplemented with fresh air supplied from a secondary blower **19** and fed into the mixed combustion burner **23** through a window box **21**. The amount of secondary air is controlled so that the ratio of air to the total combustibles in the waste gas, the auxiliary gas fuel and melted sulfur is maintained at a predetermined constant value. In this connection, it should be noted that where an excess of oxygen is supplied, the sulfur is oxidized to sulfuric acid gas, and where a stoichiometric amount of oxygen is used, sulfur dioxide is obtained. In addition, it is possible to control the burning system in such a manner as to obtain a gaseous mixture of hydrogen sulfide and sulfur dioxide suitable for use in producing free sulfur.

See Blast Furnace Slag Quenching Effluents (U.S. Patent 3,900,304)
See Blast Furnace Slag Quenching Effluents (U.S. Patent 3,941,585)
See Fireflood Operation Effluents (U.S. Patent 3,845,196)
See Petroleum Production Effluents (U.S. Patent 3,547,190)
See Pulp Mill Digester Effluents (U.S. Patent 3,842,160)
See Pulp Mill Digester Effluents (U.S. Patent 3,701,824)
See Coke Oven Emissions (U.S. Patent 3,840,653)
See Coke Oven Emissions (U.S. Patent 3,798,308)

Removal from Water

A process developed by *K. Tsutsumishita, Y. Yokoyama and T. Egawa; U.S. Patent 3,903,250; September 2, 1975; assigned to Daikyo Oil Company Ltd., Japan* is one in which a wastewater effluent which contains hydrogen sulfide, mercaptans and other sulfur compounds, is allowed to come in contact with a hydrocarbon oil having a molecular weight preferably of 100 to 300 such as kerosene in a preferable mixing ratio of the hydrocarbon oil to the wastewater effluent of 10 to 1:1 by volume thereby to extract and transfer the sulfur compounds into the hydrocarbon oil.

The mixture is separated in a settler. The separated water is used in contact with another hydrocarbon oil or with crude oil or the like in a desalting means or in a gas scrubbing means. The separated oil is fed to a hydrogenation-desulfurizing means to recover the absorbed sulfur compounds as hydrogen sulfide.

A process developed by *D.V. Gosden, A.R. Marshall and D.G. Robson; U.S. Patent 3,904,734; September 9, 1975; assigned to Woodall-Duckham Limited, England* provides a method of decontaminating the circulating reagent in a gas purification process in which the H_2S and/or HCN contaminant in a foul gas stream is absorbed by a liquid reagent, H_2S thereby being converted to sulfur and HCN being converted to a thiocyanate, the reagent being subsequently regenerated and recycled for absorption of further contaminant. The process involves bleeding off a sidestream of the liquid reagent and subjecting the same to reduction in a fluidized bed, recycling the gaseous products of such reduction to the foul gas stream being subjected to purification by the process, and returning any solids residue of the reduction to the reagent.

Figure 65 illustrates such a process. The plant illustrated in the drawing comprises a fluidized bed reactor vessel **1** within which a bed of, e.g., synthetic rutile having a mean particle size of about 155 μ (about 90% in the range 89 to 211 μ) is heated by combustion in the lower part of the bed of a mixture of air and coke oven gas in proportions such that the combustion products are reducing in character. For this purpose, a compressor **2** feeds air via line **3** to a premixer **4** which also receives coke-oven gas at suitable pressure via a line **5**.

The reactor vessel **1** is in circuit with a quench/leach vessel **6** through which the bed material is continually recycled by means of a slurry pump **7**. Thus bed material is withdrawn from the upper levels of the bed in vessel **1** through a line

8 to the vessel 6 where it is quenched and leached with the liquid reagent used in an H_2S removal state (not shown) for a coke oven foul gas stream. For the purposes of illustration the liquid reagent may be considered to be that typically used in the Stretford Process, namely an alkaline solution of one or more anthraquinone disulfonic acid (ADA) salts (e.g., sodium salts) and containing vanadium with carbonate (e.g., Na_2CO_3) as alkali, and contaminants comprising thiosulfate, sulfate and thiocyanate.

FIGURE 65: PROCESS FOR DECONTAMINATING SCRUBBER LIQUID USED IN H_2S REMOVAL FROM GASES

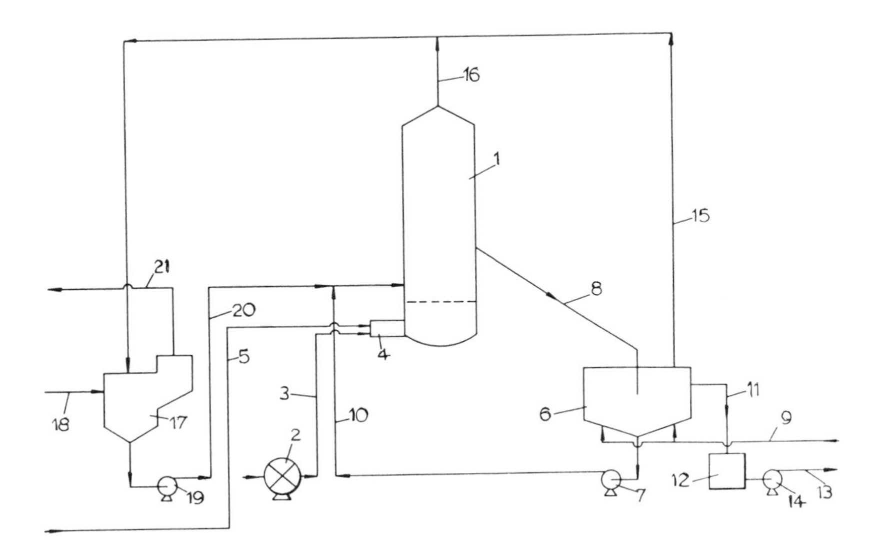

Source: U.S. Patent 3,904,734

This reagent enters the vessel **6** from a line **9** in a manner to mix intimately with the bed material. The reagent may, for instance, be pumped into the vessel **6** through one or more sparge pipes and/or the vessel may be equipped with stirring arrangements. The pump **7** draws a slurry of bed material and liquid reagent from the vessel **6** and feeds this slurry, via a line **10**, to the lower levels of the bed in the vessel **1**. Some of the slurry may be returned to the vessel **6** to accomplish stirring of the contents of the latter by repeated mixing of slurry with incoming bed material from line **8** and reagent from line **9**. Conditions in the vessel **6** are such that the solid bed material settles to enable liquid which has quenched and leached the incoming bed material to be decanted from the upper part of the vessel and to flow via a line **11** to a tank **12** from which it is pumped back to the H_2S removal stage via a line **13** by means of a pump **14**. Some of the liquid reagent in tank **12** may be recycled to the vessel **6**.

In the reactor vessel **1**, various constituents of the liquid reagent entering the bed with the slurry through the line **10** will be decomposed. The ADA will decompose to steam, CO, CO_2 and some carbon, the vanadium, entering as $NaVO_3$, will either be unchanged or reduced to a lower state of oxidation, and the carbonate will be unchanged. The thiosulfate will be reduced, usually to the sulfide but sometimes further and the sulfate will be similarly reduced to sulfide and as in the case of the sulfide, sometimes further. Thiocyanate is reduced to carbonate and/or sulfide and hydrogen, nitrogen, CO, CO_2 and H_2S. The solid products, other than carbon, are mainly salts, e.g., sodium salts, which are soluble in the liquid reagent entering the vessel **6**. Therefore in the vessel **6** these salts are mainly leached out of the incoming bed material and contained in the outflowing decanted liquid in the line **11**.

Some of the gases, both fluidizing and product, in the vessel **1** will escape to the vessel **6** via the line **8**. These gases are vented from the vessel **6** via a line **15** and are conveyed, with the bulk of the gases from the vessel **1** escaping via a line **16**, to a preconcentrator vessel **17** where they give up heat to incoming wash liquor bled off from an HCN removal stage in which the foul gas stream is washed with an alkaline polysulfide wash liquor.

In the vessel **17**, the wash liquor sidestream entering through a line **18** is heated and partially evaporated to raise its solute concentration. For the purpose of illustration, the wash liquor may be regarded as a solution of ammonium polysulfides contaminated by ammonium thiocyanate. After concentration, the wash liquor is pumped from the vessel **17** by a pump **19** and fed via a line **20** to join the slurry entering the bed of the vessel **1** from the line **10**. In the vessel **1**, the constituents of this liquor are decomposed, the polysulfides being reduced to hydrogen, nitrogen and H_2S, and the thiocyanate to hydrogen, nitrogen, CO, and H_2S.

The gases leaving the preconcentrator vessel **17** through a line **21** are conveniently returned to the foul gas stream upstream of at least the H_2S removal stage so that the H_2S content of the gases may be extracted. Preferably line **21** would convey the gases to the foul gas stream at the primary cooler for the latter so that the returned gases should not add unwanted heat and water vapors to the foul gas stream entering the purification process.

See Sulfides (U.S. Patent 3,806,435)
See Sour Water (U.S. Patent 3,754,376)
See Sour Water (U.S. Patent 3,984,316)
See Sour Water (U.S. Patent 3,804,757)

IODINE

Iodine 131 is a gaseous nuclear fission by-product found in effluent gases associated with the atmosphere in a reactor containment system, particularly from fuel rupture or fuel melt down incidents. It is also found in the effluent gases associated with the processing of spent nuclear fuels.

The iodine generated under such conditions takes on two principal forms, as molecular iodine, hydrogen iodide, and a small but radiologically significant fraction consisting of low molecular weight alkyl iodides, principally methyl iodide. It is known that radioactive elemental or molecular iodine, when taken into the body, concentrates in the thyroid gland. Subsequent studies have also shown that low molecular weight iodides, particularly methyl iodide, behave in the same way as elemental iodine in the human body. It is therefore imperative from a radiological health standpoint to control the concentration of iodine, in any form, present in gaseous effluents of the type described.

Removal from Air

A process developed by *F.N. Case; U.S. Patent 3,429,655; February 25, 1969; assigned to U.S. Atomic Energy Commission* is one whereby iodine removal from an iodine-containing off-gas stream is provided by contacting the stream with a metallic fatty acid solution wherein the metal is selected from silver, copper, or mercury. Also provided is an iodine removal filter medium which comprises a polyurethane backing impregnated with copper oleate.

A process developed by *B.A. Soldano et al; U.S. Patent 3,630,942; Dec. 28, 1971; assigned to U.S. Atomic Energy Commission* utilizes a solution which can effectively remove molecular iodine as well as organic iodine compounds from gaseous atmospheres. Such a solution is an aqueous solution adjusted to a pH in the range 9 to 10 with an alkali metal hydroxide such as potassium hydroxide or sodium hydroxide, the solution containing (a) up to 0.3 wt % boron as borate; (b) a reducing agent for molecular or atomic iodine selected from the group consisting of sodium thiosulfate and formaldehyde and (c) from 0 to an effective amount of a free radical getter, or a material which reacts with hydrated electrons.

A process developed by *W.J. Maeck; U.S. Patent 3,658,467; April 25, 1972; assigned to U.S. Atomic Energy Commission* involves absorbing and retaining airborne inorganic iodine and organic iodine species by passing a gaseous stream containing these iodines through a filter bed of synthetic zeolite in a metal ion exchanged form, which metal is reactive with iodine.

A process developed by *J. Wilhelm, H. Schüttelkopf, L. Dorn and G. Heinze; U.S. Patent 3,838,554; October 1, 1974; assigned to Bayer AG and Gesellschaft Fur Kernforschung mbH, Germany* is a process for the substantial removal of iodine and/or an organic iodine compound which has a low number of carbons from a gas and/or vapor, wherein a gas or vapor containing the iodine and/or iodine compound is passed through a layer of porous particles of a sorption agent which particles comprise amorphous silicic acid and are impregnated with a metal salt, and which have only a low water adsorption and are resistant to hot steam and acid vapors.

See Nuclear Industry Effluents (U.S. Patent 3,964,887)

IRON AND STEEL PICKLE LIQUORS

Removal from Water

As is known, when metallurgical semifinished products, such as sheets, bands, wires or profiles, are descaled with sulfuric acid, iron sulfate enters into solution, which can be separated as ferrous sulfate heptahydrate ($FeSO_4 \cdot 7H_2O$) upon cooling or by subjecting the waste liquor to vacuum treatment. Since the decrease in sulfuric acid content of the pickling solution also decreases the rate of pickling, the waste liquors, even those obtained after exhaustive pickling, always contain unconsumed acid.

The acid content of the waste liquor is at least 20 to 30 grams of sulfuric acid per liter of solution, but, depending on the method of pickling, the residual concentration of the acid may reach or even exceed several hundred grams per liter. In order to increase the rate of pickling it is generally desirable to increase the acid concentration, and to avoid the exhaustion of the solution by almost total consumption of sulfuric acid. The productivity of pickling plants increases with the increase in the pickling rate.

The efforts to accelerate pickling are limited, however, by the fact that with increasing acidity of the waste liquor the amount of unconsumed acid leaving the system in the waste liquor is increased proportionally. This increases the operational costs and causes problems of environmental pollution, since disposal of acidic sewages is strictly prohibited by regulations.

The difficulties and costs of neutralizing the waste liquors, usually performed with lime, also increase in parallel with the acid content of the waste liquor. It is a further disadvantage of the lime treatment that neutralization proceeds in a heterogeneous reaction, and complete neutralization cannot be achieved even with lime in excess. Moreover, the handling, transportation and disposal of lime sludge are difficult and expensive operations, providing no refunds (e.g., recovery of chemicals, etc.) at all.

An industrially acceptable method of processing sulfuric acid-containing waste liquors is the separation of ferrous sulfate heptahydrate by vacuum crystallization. The acidity of the waste liquor increases as a consequence of the withdrawal of water of crystallization and steam, thus the resulting solution can be reintroduced into the pickling procedure. Because the demand for ferrous sulfate heptahydrate is rather limited, the major part of the thus obtained solid is discarded, so that the problems of environmental pollution cannot be overcome by this method either. Therefore the large-scale use of this method is very much restricted.

Thus, the attention of research workers has turned toward the development of a method by which the acid content of the waste liquors formed in pickling can be recovered, and iron sulfate can be reconverted simultaneously into sulfuric acid.

A process developed by *J. Kerti, A. Mándoki and M. Széky; U.S. Patent 3,969,207; July 13, 1976; assigned to Licencia Találmányokat Értékesitó Vállalat, Hungary* is a process for the cyclic electrochemical processing of pickle waste liquors containing a maximum 100 grams of sulfuric acid and a minimum of 25 grams of iron ions per liter, produced by the pickling of iron.

According to the process the ammonium, magnesium or alkali-metal sulfate concentration of the waste liquor is adjusted to 0.5 to 1.0 mol/l by introducing an appropriate salt, and the acid content of the waste liquor is reduced to a value not exceeding 100 grams of sulfuric acid/liter, then the waste liquor is fed into and passed through the cathode chambers of an electrolysis apparatus consisting of a series of dual electrolytic cells with diaphragms, whereupon the iron content of the waste liquor is reduced to a value of 7 to 15 grams of Fe^{2+}/l, thereafter the waste liquor is fed into and passed through the anode chambers of the electrolysis equipment in parallel with the cathode flow, and a current density of 15 to 22 amp/dm^2 of diaphragm surface and an electrolysis temperature of 70° to 90°C is maintained.

During the production of metal strapping surface impurities and the like must be removed from the metal strapping prior to shipment thereof. The strapping is passed through a hydrochloric acid pickling bath which must be rinsed from the metal strapping to prevent the metal strapping from rusting. Rinsing the metal strapping usually involves a continuous process wherein large amounts of water are used to rinse the acid from the strapping with the wastewater being discharged to disposal systems such as city sewers and the like.

A process developed by *J.J. Foster, R.J. Sittema and R.E. Nelson; U.S. Patent 3,896,828; July 29, 1975; assigned to Interlake, Inc.* is a pollution control process involving treatment of the waste rinse water with sodium hydroxide which results in the use of lower temperatures, shorter residence time and enlarges the acceptable pH range for the reaction.

A suitable arrangement of apparatus for the conduct of this process is shown in Figure 66. Referring to the drawing, there is disclosed a system **100** for rinsing metal strapping **105** transported through the system by means of powered rollers **106**. The metal strapping passes from an acid tank **110** through a first rinse tank **120** into and out of a second rinse tank **130**. More particularly, there are provided squeegee rollers **111** and **112** and squeegee rollers **113** and **114** adjacent the exit of the metal strapping from the acid tank to minimize the amount of acid drag-over from this tank to the bath **120**.

The metal strapping passes from the acid tank and the hydrochloric liquid **115** therein to the tank **120** wherein it passes underneath an immersion roller **121** so that the strapping is completely submerged within the liquid **125** in this tank. Squeegee rollers **122** and **123** are provided at the exit end of this tank to minimize the amount of liquid drag-over from this tank to tank **130**. To this end, a tray **124** is positioned so that liquid removed from the strapping by the squeegee rollers **122** and **123** falls back into the tank **120**.

FIGURE 66: PROCESS FOR THE PURIFICATION OF WASTEWATERS FROM HYDROCHLORIC ACID PICKLING

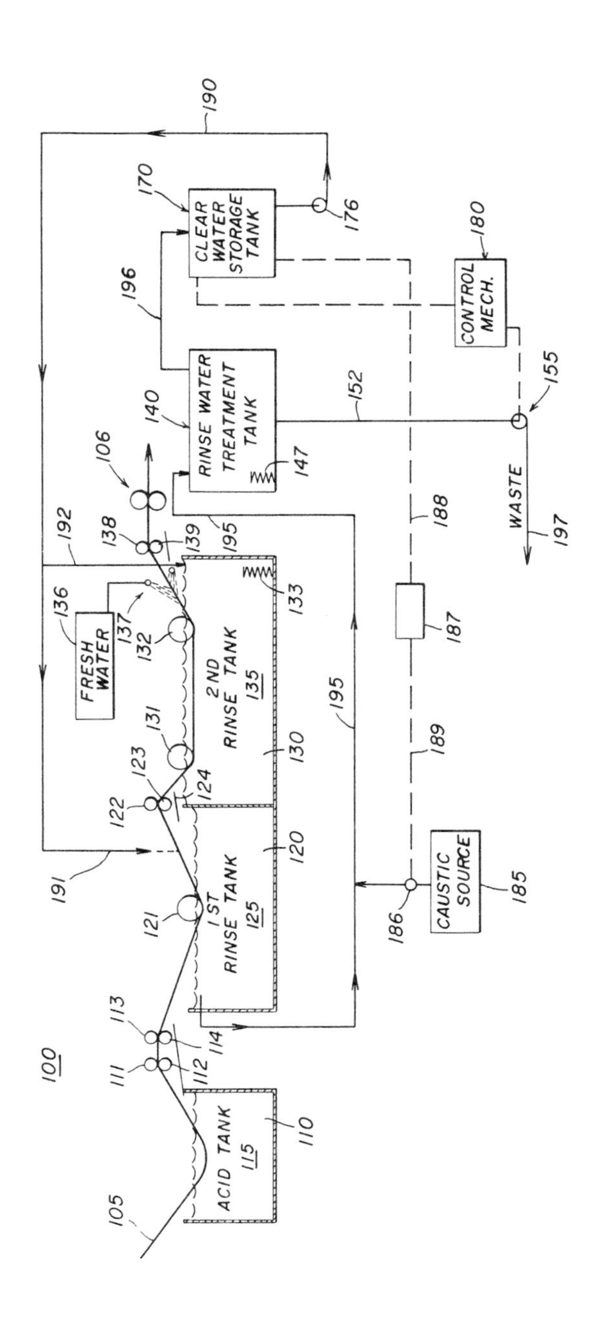

Source: U.S. Patent 3,896,828

The metal strapping **105** passes from the tank **120** into the tank **130** and more particularly underneath a pair of immersion rollers **131** and **132** positioned so that the strapping is completely submerged in the liquid **135** present in the tank **130**. A heater such as a steam sparge **133** is positioned within the second tank **130** to provide the necessary operating temperatures for the liquid in the tank. A fresh water source **136** is positioned above the exit end of the tank **130** to contact and rinse the metal strapping as it emerges from the liquid in the tank **130**.

The fresh water is applied to the strapping in the form of spaced apart sprays **137** positioned to spray the water on the strapping in a direction as shown, that is toward the first tank **120**. Spraying the strapping with fresh water in this manner provides for significantly improved rinsing of the strapping as compared to other configurations wherein rusting may occur at slow strap travel speeds or when the line is shut down. Spaced apart squeegee rollers **138** and **139** are positioned adjacent to the exit end of the tank **130** to minimize the amount of liquid carried out of the tank **130** by the strapping. The strapping after it leaves the second rinse tank may be coiled or otherwise treated.

A rinse water treatment tank **140** is provided and has a total capacity of about 8,600 gallons. Two receiving containers **141** and **142** are positioned near the entrance end of the tank **140** with each of the receiving containers being cylindrical in shape and being open at the top thereof. A scum baffle **143** is positioned adjacent to the exit end or weir **144** of this tank and extends entirely across the width thereof and below the normal operating liquid level in the tank, the liquid **145** being maintained at a level as hereinafter set forth. Due to a chemical reaction which takes place in the tank **140**, magnetite, that is Fe_3O_4, precipitates from the liquid in the tank and forms a solid phase at the bottom of the tank. A heater **147** such as a steam sparge is provided in this tank.

This tank has outlet pipes in the form of apertured conditions, the apertures being of substantial dimension to accommodate the passage of the solids therethrough. The outlet pipes are located at the bottom of this tank and are connected by a pipe **152** to a pump **155** which serves to suck the sludge or solids out of the bottom of this tank. It is noted that two outlet pipes are employed to remove the solids and to prevent solids accumulation in this tank. Similarly, spaced apart apertured pipes are provided on the other side of this tank, the apertures being of the same dimension as the apertures on the other side. These outlet pipes are connected to a line in communication with a suction pump which removes the solids in the form of a sludge from the bottom of this tank as hereinbefore set forth.

A clarified water storage tank **170** having a capacity of 1,400 gallons is positioned adjacent to the tank **140** and in fact is fed from that tank by overflow of the liquid in that tank over the weir and into the tank **170**. The clarified water in this tank is pumped out of the tank by two pumps **176**. These pumps operate continuously so long as metal strapping is passing through the system **100**.

A control mechanism **180** in the clarified water storage tank includes a level sensing mechanism. The control device is operatively connected to the pumps **155** which function to remove the solids in the bottom of the treatment tank. A caustic soda source **185** is operatively connected to a line **195** which leads from the outlet of the first rinse tank **120** to the inlet of the rinse water treatment tank **140**. A valve **186** is intermediate the caustic soda source **185** and the line **195** and is operatively connected to a control mechanism **187**. The control mechanism is connected to a pH meter (not shown) situated in the clarified water storage tank **170**. Suitable electrical leads connect the pH meter in the clarified water tank with the control mechanism.

The system **100** and particularly the individual components thereof are connected as follows: A pipe **190** is connected to the pump **176** in the clarified water storage tank and extends from the clarified water storage tank to the first rinse tank where a tap **191** is positioned above the metal strapping **105** as it exits or emerges from the liquid **125** in the first rinse tank. Clarified water flowing from the tap rinses the metal strapping and thereafter falls into the first rinse tank, the tray **124** and the squeegee rollers **122** and **123** serving to prevent dragover from the first tank to the second tank **130**.

The pipe **190** has another tap **192** which serves to introduce clarified water as make-up to the second rinse tank, the clarified water being introduced directly to the rinse tank **130** without contacting the metal strapping emerging therefrom. The line **195** previously discussed leads from the exit of the first rinse tank to the rinse water treatment tank. A line **196** represents a flow-over from the rinse water treatment tank to the clarified storage tank.

A process developed by *Y. Morimoto; U.S. Patent 3,743,484; July 3, 1973; assigned to Daido Chemical Engineering Corporation, Japan* is one in which sulfuric acid pickling waste is regenerated by mixing it with circulating sulfuric acid. The resultant liquid mixture is heated prior to concentration in a vacuum evaporator to obtain sulfuric acid of a concentration of 40 to 55% by weight and ferrous sulfate monohydrate crystals dispersed therein in a slurry concentration of 10 to 30% by weight. Finally the sulfuric acid is separated from the ferrous sulfate monohydrate crystals for recovery, and part of the recovered sulfuric acid is circulated.

Suitable apparatus for the conduct of such a process is shown in Figure 67. A vacuum evaporator **1** of external heating type is provided with an evaporation chamber **2** whose lower end is connected to the lower end of a heater **3** by a liquid circulating pipe **4** and whose top is communicated to a condenser **34** connected to a vacuum pump (not shown). An intermediate portion of the pipe is connected to a line **5** for supplying a liquid to be fed into the evaporator. The pipe is further connected, also at its intermediate portion, to a line **6** for discharging a concentrated liquid from the evaporator. The extreme end of the line **5** is connected, by way of a pump **7**, to a mixture tank **8**.

The extreme end of the line **6** is connected to a settling tank **10** by way of a pump **9**. The settling tank is provided with a baffle plate **11** for preventing turbulence of the liquid at the surface thereof and an overflow barrier **12** disposed at the upper end of the side wall, a liquid chamber **13** defined by the overflow barrier and the inner wall of the tank communicating with the mixture tank by means of a discharge pipe **14**. The mixture tank is connected to a sulfuric acid pickling waste reservoir **15** by a line **17** having a pump **16**. By way of a line **18**, the bottom end of the settling tank is connected to a tank **20** for a rotary vacuum filter **19**. Crystals separated by the filter are recovered through a hopper **21** into a storage tank (not shown).

FIGURE 67: PROCESS FOR REGENERATION OF SULFURIC ACID PICKLING WASTE

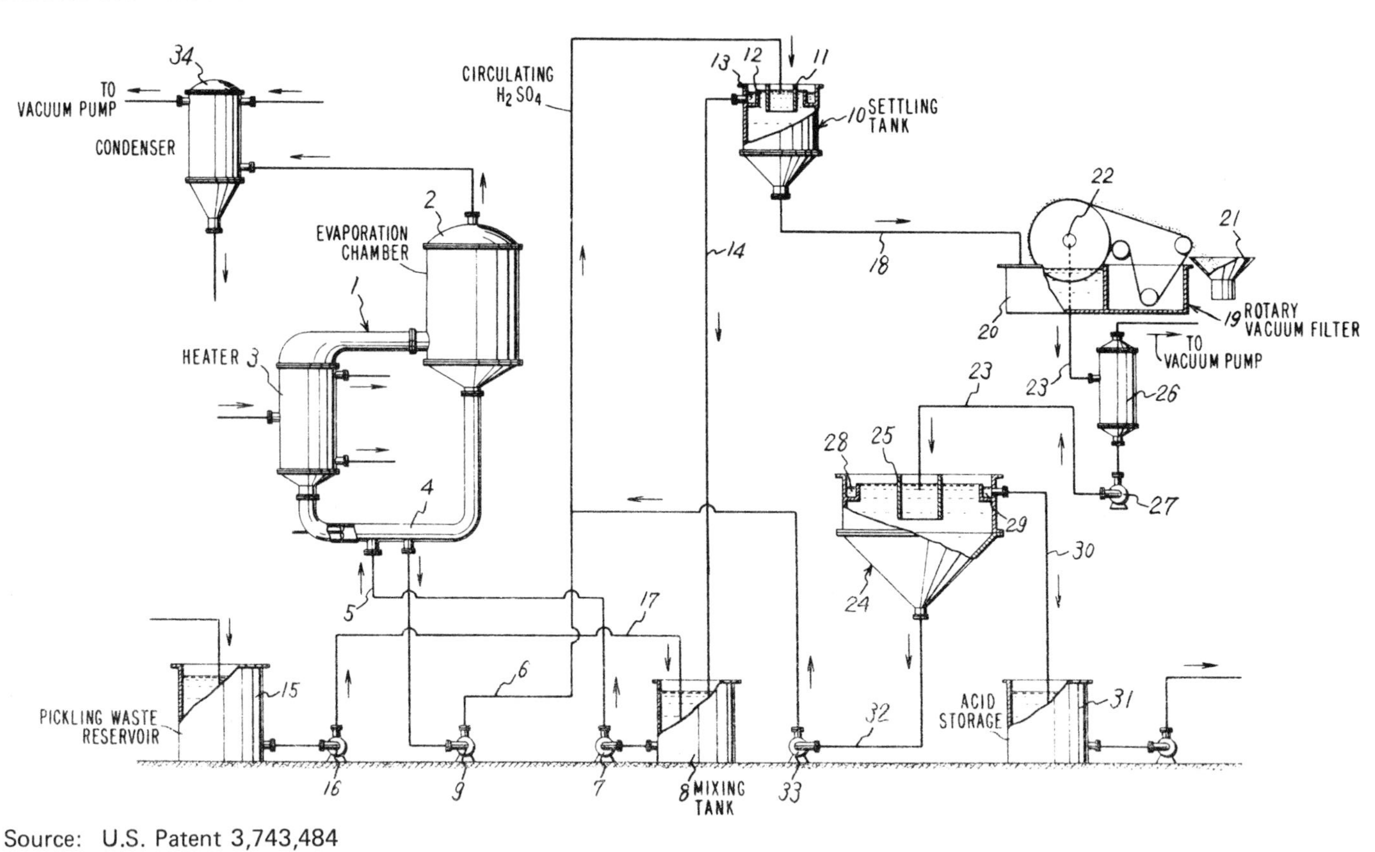

Source: U.S. Patent 3,743,484

On the other hand, sulfuric acid is recovered at a liquid chamber **20** of the filter **19**. Since almost all crystals have been separated from the sulfuric acid, the sulfuric acid may be used satisfactorily for pickling operation as it is, but where it is desired to obtain an acid of a higher purity, the sulfuric acid thus recovered may further be introduced into a crystal-growing tank to effect growth of a small amount of fine crystals contained therein, whereupon the crystals are allowed to settle and separated to obtain pure sulfuric acid. The embodiment shown is an example of the apparatus for carrying out such operation, wherein the sulfuric acid separated by the filter is sent from the liquid chamber of the filter into a crystal-growing tank **24** through a line **23**.

The line **23** connecting the chamber to the tank is provided with a filtrate tank **26** subjected to a reduced pressure by a vacuum pump (not shown) and a pump **27**. The crystal-growing tank has a baffle plate **25** for eliminating turbulence at the liquid surface and an overflow barrier **28** disposed at the upper end thereof, a liquid chamber **29** defined by the overflow barrier and the inner wall of the tank **24** communicating with a tank **31** for storing recovered acid through a line **30**.

The crystal-growing tank is provided, at its bottom, with a line **32** for taking out grown crystals whose distal end is connected, by way of a pump **33** to an intermediate portion of the line **6** so as to filter out again the grown crystals which get settled at the bottom of the crystal-growing tank. In the case where the crystal-growing tank is not employed, the line **23** may be connected to the tank **31** for the recovered acid.

The operation of the process will now be discussed with reference to the figure. By way of the line **17** with the pump **16**, a sulfuric acid pickling waste containing 10% by weight of H_2SO_4 and 15% by weight $FeSO_4$ is supplied from the reservoir **15** into the tank **8** at a rate of 1,000 kg/hr, while circulating sulfuric acid containing 50% by weight of H_2SO_4 and 1% by weight of $FeSO_4$ is sent out from the settling tank **10** into the tank **8** at a rate of 1,500 kg/hr, wherein two liquids are mixed. Even if the circulating sulfuric acid contains a small amount of fine ferrous sulfate monohydrate crystals they will be dissolved in the sulfuric acid pickling waste, whereby the adverse effect of such fine crystals on the growth of ferrous sulfate monohydrate crystals newly formed is eliminated.

The mixture in the tank **8** is then introduced into the circulation pipe **4** of the vacuum evaporator **1** by means of the line **5** with the pump **7** and is passed through the heater **3**, evaporation chamber **2** and the liquid circulation pipe **4** at about 700 mm Hg below atmospheric pressure while thereby being subjected to heat concentration. During this step, water is evaporated off at a rate of 635 kg/hr and ferrous sulfate monohydrate crystals are precipitated to obtain a slurry containing 10% by weight of ferrous sulfate monohydrate crystals dispersed in 50% by weight sulfuric acid. Through the line **6** with the pump **9**, this slurry is fed into the settling tank at a rate of 1,900 kg/hr.

While allowing the crystals to settle, the supernatant liquid is caused to flow over the overflow barrier **12**, through the discharge pipe **14** and into the tank **8** at a rate of 1,500 kg/hr for mixture with a sulfuric acid pickling waste. At a rate of about 400 kg/hr, the concentrated slurry of ferrous sulfate crystals is taken out from the tank and is supplied through the line **18** into the liquid chamber **20** of the vacuum filter, where the crystals are separated from the sulfuric acid. As a result, ferrous sulfate monohydrate crystals are recovered from the hopper **21** at a rate of about 200 kg/hr, and the sulfuric acid is recovered from the line **23**, at a rate of about 165 kg/hr.

The recovered sulfuric acid contains only about 0.5 to 1.0% by weight of ferrous sulfate monohydrate crystals, which insures effective reuse for pickling operation. In order to further improve the purity, the sulfuric acid thus recovered is introduced through the line **23** into the crystal growing tank **24**, where it is left to stand for 10 to 30 hours, for example, for 20 hours. The supernatant liquid is then led through the line **30** into the tank **31** at a rate of 165 kg/hr to recover high purity sulfuric acid which contains only about 0.1 to 0.3% by weight of ferrous sulfate monohydrate crystals. On the other hand, the grown crystals settled in the tank **24** are returned through the line **32** to the line **6** together with a portion of the sulfuric acid for further treatment in the settling tank **10**.

A process developed by *L.J. Hansen; U.S. Patent 3,745,207; July 10, 1973; assigned to Environmental Technology, Inc.* is one in which waste acid pickle liquor is converted into iron oxide and acid by depositing it on a compact moving bed and sweeping the bed with hot oxidizing gases.

The contact material is itself preferably a form of iron oxide. Mill scale, which is substantially Fe_3O_4, may be used as a start up material; subsequently the oxide product may be recycled in part, as contact material. The term compact bed as used herein has the meaning usual in the process industries, i.e., a bed of discrete solid elements supported by one another, rather than by a percolating fluid, as in a fluidized bed.

Figure 68 shows the application of such a process to hydrochloric acid waste liquor. The pickle liquor, which has preferably been concentrated in a manner to be described, is delivered through a line **10** to a reactor **11**. A nozzle **9** may be used to spray the liquor into the reactor. The reactor is preferably an elongated chamber of rectangular cross section. It is mounted at an angle of 20° to 40° to the horizontal and means (not shown) may be provided for enabling this angle to be adjusted. On the lower surface of the reactor is mounted a refractory brick platform **12**.

This platform rests on supports **13** which are constructed of a high temperature wear-resistant material such as tungsten carbide, and which give the platform a capability for movement, axially of the reactor, the movement being limited by a stop **13a**. A vibrator **8** is suspended from the platform by a shaft **7** and serves to vibrate the bed to facilitate movement of the material down the bed, and also to churn up the bed so that all particles are exposed on the surface to the same degree. The shaft may be internally water-cooled and hinged as at **6** to accommodate changes in the inclination of the reactor.

At the bottom of the reactor is located a solids offtake duct **5** which is water-cooled by a sleeve **4**. The duct, in turn, empties through airlock **3**; located immediately below the airlock is a vibratory inclined screen **2**, activated by vibrator **40**.

A hopper **14** for contact material is provided above the reactor. A vibrator **15** is situated on the floor of the hopper to cause the hopper contents to move

FIGURE 68: PROCESS FOR RECOVERY OF HCl PICKLE LIQUOR

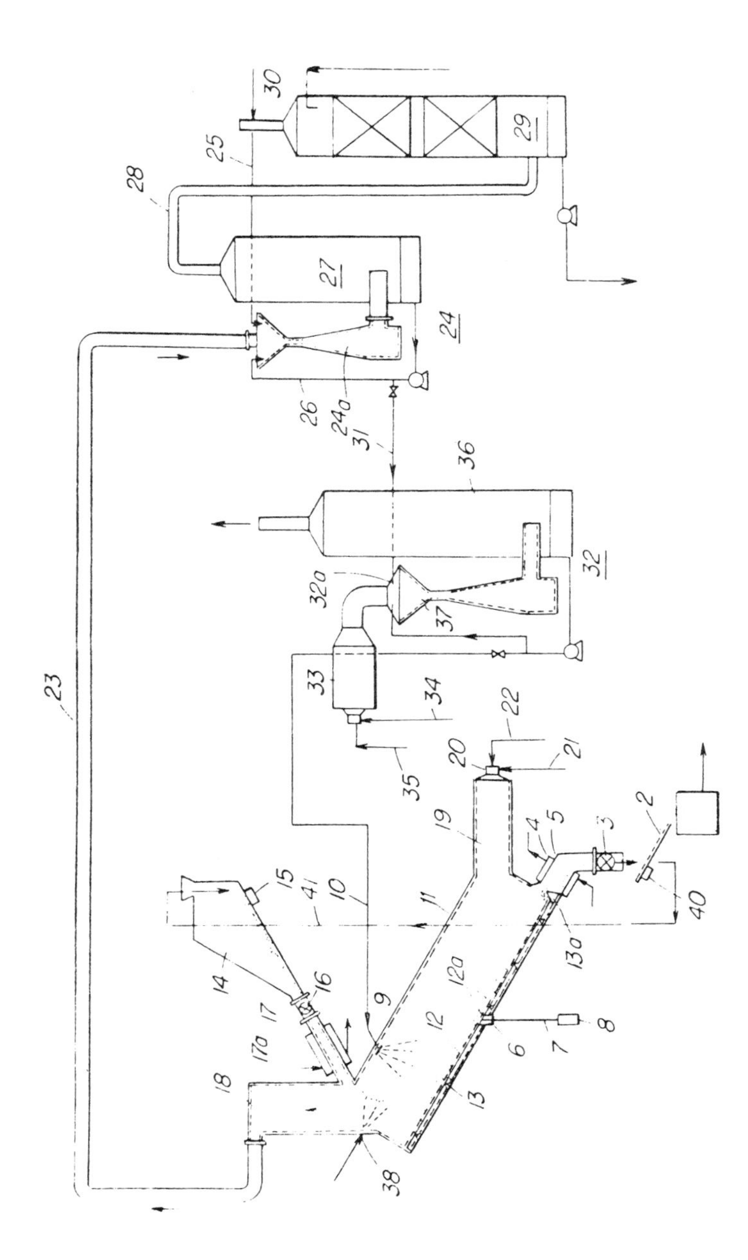

Source: U.S. Patent 3,745,207

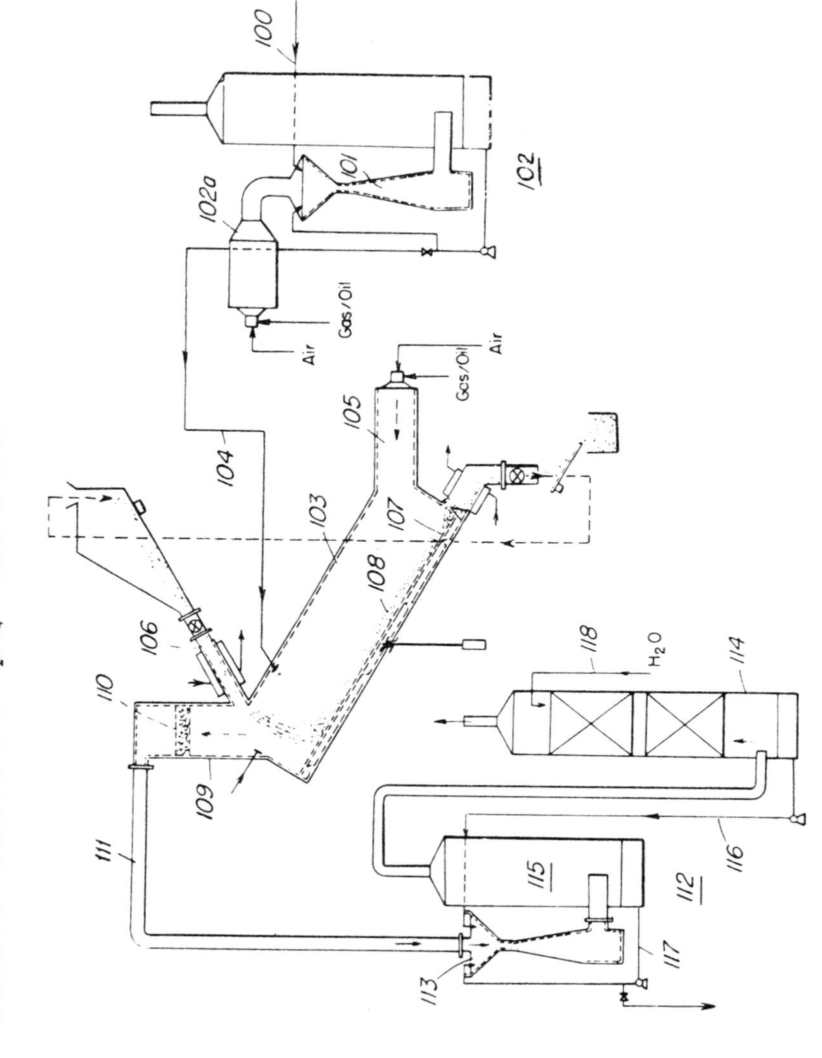

FIGURE 69: PROCESS FOR RECOVERY OF H_2SO_4 PICKLE LIQUOR

Source: U.S. Patent 3,745,207

down through an airlock **16** into a conduit **17**, water cooled by sleeve **17a**, which empties into the upper end of reactor **11**.

At the lower end of the reactor and emptying into the reactor is a burner tunnel **19**. A burner **20** is located at the end of the tunnel. Fuel, such as gas or oil, is fed to the burner through line **21** and air through line **22**.

The upper end of the reactor is provided with a gas offtake duct **18**. This offtake duct empties into line **23** which, in turn, delivers into the venturi section **24a** of a venturi economizer **24**. To the venturi is also delivered fresh feed liquor via line **25** and economizer recycle liquor through line **26**. The venturi discharges into a separation tank **27**. The overhead gases from this tank are carried via line **28** to an absorber tower (normally a packed tower) **29**. A line **30** is provided for furnishing water to the top of the tower.

A line **31** is provided for transporting a portion of the economizer recycle liquor to the venturi section **32a** of a venturi evaporator **32**. Hot gases for this evaporator are generated in a burner **33** which may be fed with fuel through line **34** and air through line **35**. This venturi discharges into a separation tank **36**. Line **10** connects this tank with the reactor. A line **37** is provided for recycling a portion of the bottoms from this tank to the venturi **32a**.

In operation, spent acid pickle liquor enters the system through line **25**. In economizer venturi **24a** it is immediately contacted with hot combustion gases drawn from the reactor. By this means it is heated and concentrated, for example from, say, 15% $FeCl_2$ to, say, 17 to 18% $FeCl_2$, and substantially all volatile free acid is eliminated. At the same time, the hot gases from the reactor are reduced in temperature so that refractory-lined conduits and the like are no longer required.

The concentrated liquor is then delivered through line **31** to evaporator venturi **32**. Here it meets hot combustion gases at, say, 2000°F, and is concentrated to, say, 30 to 45% $FeCl_2$. It is then delivered through line **10** to the reactor via nozzle **9**.

If the unit is just being started up, hopper **14** will be filled with a contact material from some external source such as mill scale, ground to an average particle size of, say, 1/16" to 1/8". The contact material **14a** is caused to move through airlock **16** into duct **17** by the action of vibrator **15**. This duct, cooled by water sleeve **17a**, is funnel-shaped so that the contact material leaving the duct falls as a curtain across the width of the reactor and builds up as a relatively shallow bed, say 1" to 3" thick, on the refractory brick platform **12**.

As it falls across the reactor, the contact material is contacted by liquor sprayed through the nozzle and by hot combustion gases entering the reactor through the burner tunnel. In the arrangement illustrated, the liquor is first contacted with hot gases so that a further concentration occurs and, at the time the liquor contacts the solid particles, it is in the nature of a thick syrup. The degree of concentration obtained at this point can be adjusted to some extent by varying the point at which the liquor is injected into the reactor.

For example, an alternate position is indicated at **38**. In any case, the liquor is picked up or absorbed by the solid particles of contact material and is carried by them as they fall to the platform **12** to form the bed **12a**. Continued contact with the hot gases emanating from burner **20**, which are at a temperature of, say, 800° to 1400°C, causes the further evaporation of water and free acid gases with the deposition of salt, typically $FeCl_2$ in and on the particles of contact material. Further contact with the oxygen-containing gases, as the solid contact material moves down the inclined reactor under the action of vibrator **8**, converts the salts to oxides.

The stop **13a** extends above the level of platform **12** and this causes the solids to dwell at this point and insures total conversion. Ultimately, the solids move over the stop and fall through offtake duct **5** which is kept cool by water sleeve **4**, through airlock **3** and onto screen **2**. The screen is inclined and fitted with a vibrator **40**. The coarser material moves off the screen into a storage bin **41**, where it may be delivered to a steel furnace or disposed of as desired. Smaller size material (1/16" to 1/8" and less) is recycled back to hopper **14** by a conveyor **41** where it serves as make-up contact material.

The overhead gases from the reactor in the case of hydrochloric waste liquor, pass directly through duct **18**, and line **23** to the economizer. These gases will consist chiefly of water, oxides of carbon, HCl, fixed gases and some oxygen. They will be at, say 600° to 700°C. In the economizer, their temperature is reduced to, say, 95° to 105°C. They are then carried through line **28** to absorber **29**. Here they are contacted with water and substantially all the hydrochloric acid is absorbed, though additional scrubbing may be carried out if desired. The bottoms stream from the absorber contains on the order of 20% HCl and may be used directly for pickling.

A slightly modified version of this system as shown in Figure 69 may be used for processing sulfuric acid waste liquor. The raw liquor is first fed through a line **100** to the venturi section **101** of an evaporator **102** which is identical to the venturi evaporator **32** of the system already described. Here it is met with hot combustion gases from burner **102a** and heated and concentrated. The concentrated liquor containing, say, 12 to 20% $FeSO_4$ and 8 to 10% free acid is then sent directly to reactor **103** via line **104**.

This reactor is identical to reactor **11** in the preceding figure, and again the liquor is discharged into the reactor, is contacted with hot combustion gases issuing from burner tunnel **105** and, after evaporation of water and acid gases, is picked up by particulate solid contact material falling in a curtain from duct **106**, and is absorbed thereby. The contact material falls to platform **107** and forms a moving compact bed **108** thereon. Further evaporation occurs in the bed with precipitation of iron sulfates and subsequent decomposition of the sulfates to pure iron oxide and SO_2 and SO_3. The latter gases pass through outlet duct **109**. In this duct is inserted a packed bed **110**.

This comprises packing material such as berl saddles upon which is deposited an oxidizing agent such as V_2O_5 capable of converting SO_2 in the gases to SO_3. The hot gases are then conveyed through line **111** to the venturi economizer **112**. In the venturi section **113** of this economizer, the hot gases are contacted with sulfuric acid from a downstream absorber tower **114** and recycle liquor from the economizer tank **115** delivered through lines **116** and **117**, respectively.

By this means the hot gases are cooled and a certain portion of the SO_3 present therein is absorbed. The venturi section **113** discharges into a separator tank **115**. Overhead gases from this tank are sent to the absorber tower **114** where they are contacted with water supplied through line **118** for final absorption of SO_3. Product H_2SO_4 is drawn as a bottoms stream of economizer separation tank **115**.

A process developed by *H. Silby, M. Silby and J.H. Krause; U.S. Patent 3,712,940; January 23, 1973; assigned to Wire Sales Company* is one in which pollution of rivers, lakes and streams caused by the dumping of waste sulfuric acid pickle liquor, can be completely eliminated by a process including the steps of crystallization, to recover and recycle most of the spent acid, contacting the filtrate with NaOCl, separation of precipitate followed by drying to obtain ferric oxide which is a commercially useful product. The filtrate obtained after treatment with NaOCl contains the corresponding sodium salt, depending upon the composition of the waste pickle liquor, which can be recovered as a further useful product.

A process developed by *S. Bastacky; U.S. Patent 3,813,321; May 28, 1974* is a process for treating industrial wastes containing sulfuric acid and metal constituents comprising subjecting the wastes to a cell equipped with a plurality of alternately positioned positive and negative lead-antimony alloy electrodes and charging the electrodes. The liquid portion of the waste is dissipated and a residue including the metallic constituents forms.

A secondary process is also disclosed comprising the steps of placing additional pickle liquor in a secondary processing tank, contacting the pickle liquor in the secondary processing tank with the exterior wall of the cell, and conducting thermal energy through the cell wall to the pickle liquor in the secondary tank wherein the liquid portion of the liquor in the secondary processing tank is dissipated and a residue including the metallic constituents forms within the secondary tank. The metallic constituents may be reclaimed by a drying step or by an ignition step.

See Galvanizing Process Effluents (U.S. Patent 3,801,481)
See Rolling Mill Effluents (U.S. Patent 3,986,953)

IRON CYANIDES

There are a goodly number of industrial applications in which dissolved ferricyanide is used to great advantage including photography and metal finishing. This complex iron cyanide is not a toxic material, but under certain conditions it has a tendency to decompose to cyanide which is very toxic. Such conditions can exist in sewage plants, streams, lakes and other bodies of water into which liquid waste effluents containing the aforesaid iron cyanide in dissolved form are discharged or find their way.

The presence of extremely small amounts of cyanide, for example a few parts per million, can be fatal to microorganisms, fish and other aquatic life that live in such environments. Despite the harm that can emanate from discharging liquid effluents containing ferricyanide into sewage systems and natural bodies of water, it is, nevertheless, common practice to do so. A reason for this is that there has not heretofore been available a practical and economical method for removing this complex iron cyanide from the liquid effluent.

An example of an industrial process in which both ferricyanide and hexavalent chromium are used is in the application to aluminum surfaces of coatings which are corrosion resistant and which have characteristics such that paint adheres readily and strongly to the coatings. Hexavalent chromium and ferricyanide are ingredients of a popularly used acidic aqueous coating solution which forms such coatings on aluminum surfaces. Such an acidic coating solution also contains fluoride and may contain other ingredients such as sodium fluoborate, potassium fluozirconate and nitric acid.

Removal from Water

A process developed by *T.D. Henley and R.F. Reeves; U.S. Patent 3,819,051; June 25, 1974; assigned to Amchem Products, Inc.* is one in which dissolved hexavalent chromium and dissolved complex iron cyanide are removed from a liquid waste effluent by adding thereto a reducing agent having an anion capable of reducing the hexavalent chromium to trivalent chromium and a cation which forms a solid or precipitate with the complex iron cyanide. The preferred reducing agent is zinc hydrosulfite.

The trivalent chromium can be removed from the liquid effluent by precipitating it in the form of chromium hydroxide. An alkaline material, preferably lime, is added to the effluent to accomplish this. Trivalent chromium is also a toxic material, but is not considered to be quite as toxic as hexavalent chromium. Some governmental regulations permit higher concentrations of trivalent chromium in a discharged liquid effluent, for example, 1 ppm to a few ppm, than hexavalent chromium, the concentration of which should be less than 1 ppm, for example, 0.05 ppm or less. The chromium hydroxide precipitate and the solid iron cyanide, and any other solids that are present in the liquid effluent can be separated therefrom by any suitable means, such as filtration, centrifugation, etc.

A composition for use in treating the aforementioned type of waste effluents comprises about 20 to about 80 wt % of the reducing agent and about 20 to about 80 wt % of an alkaline material such as lime. A preferred composition comprises: (A) about 20 to about 75 wt % zinc hydrosulfite; (B) about 20 to about 75 wt % lime; and (C) about 5 to about 50 wt % of a nontoxic salt of a strong mineral acid, preferably sodium chloride, calcium chloride or calcium sulfate dihydrate.

A process developed by *I. Shimamura and H. Iwano; U.S. Patent 3,931,004; January 6, 1976; assigned to Fuji Photo Film Co., Ltd., Japan* is one in which ferricyanide ions and/or ferrocyanide ions are removed effectively from a photographic reducer or waste liquid containing them by contacting the waste liquid with a weakly basic anion exchange resin in the presence of ammonium ions and thiosulfate ions.

Another process developed by *I. Shimamura and H. Iwano; U.S. Patent 3,869,383; March 4, 1975; assigned to Fuji Photo Film Co., Ltd., Japan* is one in which ferricyanide and/or ferrocyanide ions are effectively removed from waste photographic treating solutions by contacting the solutions with weakly basic anion exchange resins. Improved results are obtained with the free base form of the weakly basic anion exchange resin when contacting takes place in the presence of a compound having a buffering action at a pH of 7 to 9, e.g., in the presence of boric acid, metaboric acid or a water-soluble borate. Upon regeneration of the anion exchange resin the eluted ions can be reused as a photographic processing solution, e.g., a bleaching solution, with the addition of make-up components.

A process developed by *A. Abe and Y. Usui; U.S. Patent 3,909,403; Sept. 30, 1975; assigned to Fuji Photo Film Co., Ltd., Japan* involves removing ferricyanide ions and/or ferrocyanide ions from a waste solution or water containing these ions formed in photographic processing by bringing the waste solution in contact with a weakly-basic anion-exchange resin to absorb the ferricyanide ions and/or ferrocyanide ions on the anion-exchange resin.

An alkaline concentrate of the ferricyanide ions and/or ferrocyanide ions is formed by immersing the weakly-basic anion-exchange resin having absorbed thereon the ions and after adding to the concentrate a strong alkali, a hypochlorite, and bromide ions to increase the pH of the concentrate to above 12, the mixture is heated to temperatures over 50°C at normal pressure, whereby the ferricyanide ion and/or ferrocyanide ions are decomposed.

A process developed by *H.C. Baden; U.S. Patent 3,772,194; November 13, 1973; assigned to Eastman Kodak Company* is a process for conditioning waste containing complex cyanide compounds and oxidizing at least the complex iron cyanides, e.g., hexacyanoferrates, comprising chlorinating the waste in the presence of a silver ion catalyst (present as a soluble silver salt such as $AgNO_3$, $AgCl_2$, etc.) to convert the cyanide to a nontoxic compound. Chlorination is preferably carried out at a temperature of between about 25° and 68°C in an alkaline medium, using an excess of free chlorine or hypochlorite as the chlorinating agent.

A process developed by *M. Ichiki and M. Ishii; U.S. Patent 3,816,275; June 11, 1974; assigned to Mitsui Mining & Smelting Co., Ltd., Japan* involves treating a waste liquor containing a difficultly decomposable cyano-complex such as ferrocyanides and ferricyanides and/or cyano ions, by electrolyzing the waste liquor by employing iron as an anode to thereby form water-insoluble colloid, floating and concentrating the colloid by an action of bubbles formed during the electrolysis to thereby convert it to a scum, and removing the scum from the waste liquor in an electrolytic cell to thereby obtain a purified liquor free of cyano-component.

A process developed by *R.L. Garrison, H.W. Prengle, Jr. and C.E. Mauk; U.S. Patent 3,920,547; November 18, 1975; assigned to Houston Research, Inc.* is one in which cyanides in an aqueous cyanide solution, particularly an aqueous solution of cyanides complexed with iron, are destroyed by contacting an aqueous solution thereof with an ozone-containing gas, while simultaneously irradiating the aqueous cyanide solution with ultraviolet light. The method is preferably carried out while maintaining the pH of the aqueous cyanide solution within the range of pH 5 to 9. Increased reactivity can also be achieved by heating the aqueous cyanide solution.

The method is preferably carried out by contacting the aqueous cyanide solution and an ozone-containing gas in a plurality of separate contact zones, countercurrently or with parallel flow, with the irradiation of the aqueous cyanide solution with ultraviolet light being carried out in at least one of the separate contact zones, preferably at least the final contact zone.

A suitable form of apparatus for use in the conduct of the process is shown in Figure 70. As illustrated, a supply of aqueous cyanide solution, specifically, a cyanide complexed with iron, is pumped from feed tank **10** through line **12** by means of metering pump **14** to the top of reaction tower **20**. Typically, an aqueous cyanide solution obtained as an industrial effluent from such processes as photo processing and metal finishing operations will contain a concentration of cyanide ions of from about 1 to about 4,000 mg/l although the method is applicable to lower and much higher cyanide ion concentrations also.

FIGURE 70: PROCESS FOR DESTROYING IRON CYANIDES IN WASTEWATERS USING OZONE IN THE PRESENCE OF ULTRAVIOLET LIGHT

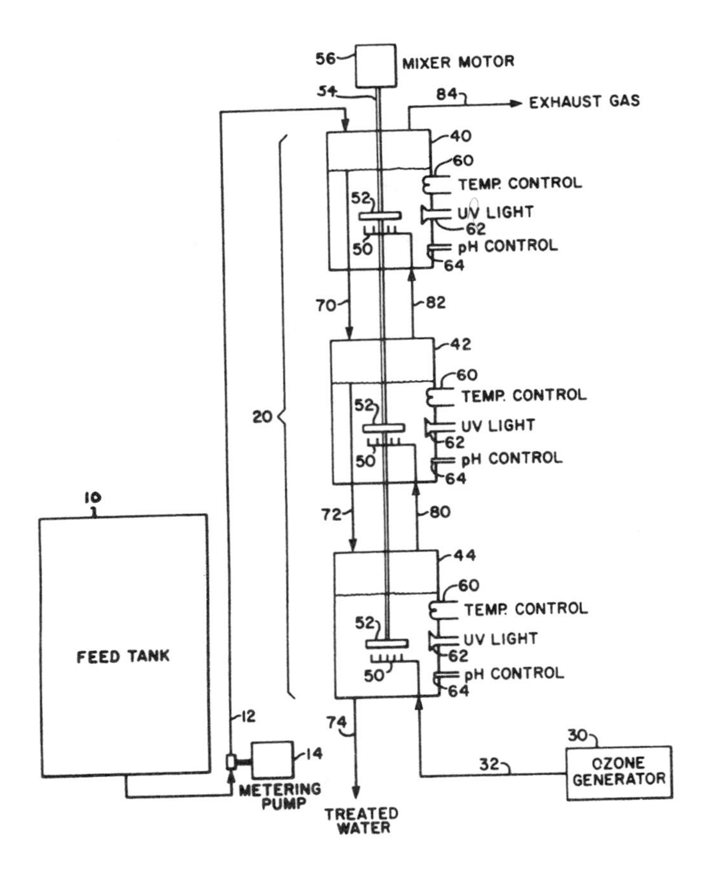

Source: U.S. Patent 3,920,547

The rate of flow of the aqueous cyanide solution from feed tank **10** through line **12** into reaction tower **20** can vary over extremely wide limits, depending upon many factors, including the ability of the equipment to handle the necessary flow rate, the number of contact zones employed for contact between the aqueous cyanide solution and ozone-containing gas and the concentration of cyanide ions within the aqueous cyanide solution.

As illustrated, an ozone generator **30** is provided to generate an ozone-containing gas, which is introduced to the bottom of the reaction tower through line **32**. Any of the conventionally employed ozone-generating devices can be applicably employed. One typical ozone-generating device generates a gas which comprises from about 1 to about 8% by weight ozone, from about 20 to about 99% by weight oxygen, and up to about 80% by weight nitrogen with possibly small quantities of other gases which are normally present in air, such as carbon dioxide, argon, and the like.

As was the case with regard to the feed of the aqueous cyanide solution, the flow of ozone into the bottom of the reaction tower through line **32** can be varied within extremely wide limits with any amount effective to destroy the cyanides being applicable. The amount of ozone, however, is generally controlled, based upon the concentration of cyanide ions in the aqueous cyanide solution feed, and based upon the desired final concentration of cyanide ions in the treated water.

While any particular flow of ozone will depend upon additional factors including the number of contact zones employed, satisfactory results are achieved when the ozone-containing gas is supplied to the reaction tower in an amount, in terms of the ozone contained therein, within the range of 1.8 to 4.7 mg per 1 mg cyanide ion at 4,000 mg/l of cyanide ion concentration. Where the initial cyanide concentration is greater than 4,000 mg/l, a greater supply of ozone may be necessary, and, also, the process may be very satisfactorily carried out using less ozone under certain circumstances. Operation within the foregoing limits provides efficient destruction of cyanide ions and the production of a treated water meeting all pollution standards.

As illustrated, reaction tower **20** is separated into three contact zones **40**, **42** and **44**, with the feed of aqueous cyanide solution through line **12** being introduced into the top of contact zone **40** and the feed of ozone-containing gas being introducted into the bottom of contact zone **44** through line **32**. This arrangement, as illustrated, provides for countercurrent contact of the aqueous cyanide solution and ozone-containing gas in each of the contact zones in accordance with the preferred embodiment.

While the arrangement illustrated shows each of the contact zones as part of the same reaction tower, the different contact zones employed can be physically separated. In addition, the method can be carried out utilizing parallel contact of the ozone-containing gas and cyanide solution where desired. All that is required is that efficient utilization of the ozone be achieved through the required intimate contact between the ozone-containing gas and aqueous cyanide solution.

As illustrated, contact zones are identically constructed. These contact zones can take varying forms, although it is preferred that the individual contact zones be constructed of material which is resistant to oxidation and ozonation. Moreover, within the same tower, the contact zones can be of varying sizes and dimensions, as well as include different elements based upon different concentrations of cyanide ions within the separate contact zones.

Each of reaction zones **40, 42** and **44** can be a simple reaction vessel in which the ozone-containing gas is bubbled through the aqueous cyanide solution or alternatively can be in the form of a packed contact zone in which each of the zones contains a packing material.

Each of the contact zones in the embodiment illustrated is provided with a suitable means **50** to disperse the ozone-containing gas into the aqueous cyanide solution. Such means can be a porous stone diffuser, ejector, or any other suitable means of providing a large number of small bubbles to obtain the desired good mass transfer from the gas phase into the liquid phase. Optionally, each contact zone can be provided with a blade mixer **52** on shaft **54**, driven by motor **56**. Alternatively, packed contact zones can be used to achieve such intimate contact.

Each of the contact zones is similarly provided with a temperature control element **60**, a source of ultraviolet light **62**, and a suitable means to control the pH of the aqueous cyanide solution, indicated as element **64**.

The ozone-containing gas which enters the bottom of contact zone **44** and is dispersed in the aqueous cyanide solution by means of a suitable dispersing means **50**, exits this contact zone through line **80**, whereupon it is introduced into contact zone **42** by means of a similar dispersing element. The ozone-containing gas exits this contact zone through line **82** and is introduced into contact zone **40** by means of a similar dispersing element **50**, the exhaust gas being exited to the atmosphere by means of exit line **84**. The concentration of ozone in the system will decrease from the maximum concentration delivered by ozone generator **30** to the minimum concentration in exit gas exiting via line **84** due to consumption of the ozone in its reaction with the cyanide ions in the various contact zones.

This method has advantages over conventional processes in that the cyanide ions can be substantially completely destroyed. This is achieved without the necessity of adding chemicals to the aqueous cyanide solution and the substantial total destruction avoids the discharge of cyanides with the treated water. In addition, no toxic sludge remains.

A process developed by *T. Yamada, S. Kondo, S. Yoshihara, T. Sugahara, S. Murakami, M. Ito and K. Hirata; U.S. Patent 3,843,516; October 22, 1974; assigned to Toyo Gas Chemical Industry, Inc., Japan* is one in which waste solutions containing cyano-heavy metal-complex compounds are treated by adding thereto phosphoric acid, sulfuric acid and an oxidizing agent and then effecting a decomposition reaction of the cyano-heavy metal-complex compound or compounds in the waste solution to separate the cyanide moiety and at the same time to convert the heavy metal ion into a soluble heavy metal salt.

According to the process, the cyanide moiety is separated and at the same time iron ions are converted into soluble iron salts by adding to the solution (1) a mixed acid comprising phosphoric acid and sulfuric acid and further (2) an oxidizing agent, to convert a salt of bivalent iron into a salt of trivalent iron and thereafter heating and agitating the solution, with air or other suitable gas such as nitrogen.

IRON OXIDES

Removal from Air

A process developed by *H. Werner; U.S. Patent 3,169,054; February 9, 1965; assigned to Metallgesellschaft AG* is one in which the flue dust contained in the waste gases of metallurgical furnaces, preferably converters for the manufacture of steel, is precipitated in an electric filter, and the precipitated dust, in accordance with its hydrated component parts, is pelletized and the dust in the form of fresh pellets immediately recycled into the furnace.

A process developed by *N.F. Tisdale; U.S. Patent 3,365,340; January 23, 1968* is a method for the economical recovery of iron oxide particles suspended in fumes produced in the melting, refining and production of steel, sintering of ores, or blast furnace operations. It also relates to the recovery of iron from iron oxide scale, produced in the heating operation of iron and steel production. The oxygen type furnace, which is being widely adopted, is a producer of copious iron oxide fumes. The open hearth furnace, with or without the oxygen lance, also produces these iron oxide fumes. Several other melting and refining processes all produce similar fumes.

Communities which have grown up around such melting plants have enacted laws to force the reduction of these fumes and their solids from going into the air because of their adverse effect on health. Since there is approximately 1½% of the charge volatized into the air as iron oxide during the melting operation, this becomes a substantial economic loss unless recovered and a large amount of iron oxide is put into the atmosphere. Methods to recapture the particles from these fumes require high initial expense, substantial operating cost and an elaborate cooling system to cool the fumes. There are several methods which are used singly or in combination, in an attempt to do an efficient job.

The scheme developed by Tisdale is to take the fumes and have them impinge in a heated liquid bath of silica slag which has been altered just a small amount by materials to make it more fluid. The furnace is so constructed that there may be one or more baffles along its length with the result of forcing impingement of the gas on the surface of the slag. This can be augmented by pressure or by a pump to force this material to hit the slag.

A process developed by *J.E. Cooper; U.S. Patent 2,810,633; October 22, 1957* involves joint treatment of the extremely fine dusts produced by the blast furnace and the pickle liquor produced by the sulfuric acid pickling of steel.

The dust blown out of a blast furnace during operation has substantially the same chemical composition as the charge, and varies in size from sizable chunks down to particles of submicron size. By weight, by far the greater portion of the blast furnace dust is recovered in the initial dry dust catcher. However, to clean the blast furnace gas sufficiently to permit its efficient utilization as a furnace fuel, resort is had to three further cleaning apparatuses.

Arranged in series, these are first, an ordinary wet gas scrubber; second, a so-called disintegrator; and third, an electrostatic Cottrell precipitator. In each of these pieces of apparatus, the dust is finally removed as a suspension in water. In a typical blast furnace installation producing 3,000 tpd of pig iron, the water flow through these three final gas cleaners amounts to about 8,250 gal/min and is burdened with approximately 229 grains per gallon of solid matter. Over a 24 hour period these wash waters remove from the three final cleaners about 194 tpd of dry solids.

In conventional blast furnace practice, these wash waters are combined and processed through a settling chamber and the effluent from the settling chamber is discharged into the sewer or other adjacent water course. The average dry solids content of this effluent which is discharged into the sewer is about 8 grains per gallon or, in the course of 24 hours, about 7 tons of dry solids are so lost.

The iron value of these dry solids is not particularly serious economically even though its recovery would, of course, be desirable. The unfortunate feature of this process is that the material escaping in the effluent from the settling chamber represents the very finest dust produced in the blast furnace, and hence the dust which is the most objectionable from a standpoint of disposal in natural water courses, since these dusts are the most objectionable from a chromogenic standpoint. These dusts impart a persistent iron red color to the water course which is highly objectionable to the general public, despite the fact that quantitatively their effect upon the water course may be insignificant.

In the process developed by Cooper, there is produced a substantially clear filtrate and a sludge having a water content between 50 and 80%, which is sufficiently dense to enable its transportation by conveyor to a sintering plant. Here it may be sintered along with the dry pulverulent solids recovered from the blast furnace gas by the dust catcher.

See Steel Converter Effluents (U.S. Patent 3,908,969)
See Steel Converter Effluents (U.S. Patent 3,976,454)
See Steel Converter Effluents (U.S. Patent 3,788,619)

Removal from Water

A process developed by *M. Steinberg et al; U.S. Patent 3,537,966; November 3, 1970; assigned to U.S. Atomic Energy Commission* provides a method for removing dissolved iron oxides from acidic aqueous solutions, such as mine wastewaters, which comprises exposing the aqueous solution to gamma irradiation while aerating and contacting the solution with calcium carbonate to induce precipitation of the contained iron oxides from the solution.

The process is especially useful in treating acidic mine wastewater which has soluble ferrous iron compounds contained therein. It will remove substantially all ferrous iron compounds from the aqueous solution. Exemplary of ferrous iron compounds which can readily be removed from aqueous solutions are ferrous sulfate, ferrous bicarbonate, ferrous chloride, etc. When the process is employed, the ferrous iron compound concentration can efficiently and economically be lowered from the saturation point to under six parts per million. Conventional nuclear and chemical engineering techniques and equipment can be employed to carry out the practice of this process.

A process developed by *F.J. Sines; U.S. Patent 3,351,551; assigned to United States Steel Corporation* relates to the removal of small scale particles from the flushing water of a steel mill after the recovery of larger scale particles for use in the blast furnace or open hearth.

See Steel Mill Effluents (U.S. Patent 3,844,943)

IRRIGATION PUMPING ENGINE EFFLUENTS

Irrigation wells and pumping facilities have come to be an accepted part of general crop planting. In areas where rainfall is limited to a level below that required by the crops, water has to be supplemented from other sources. Typically, water is supplied from distant sources by irrigation canals or brought up from underground springs. Thus in arid regions such as the southwest and midcentral states, there are numerous water wells and pumping stations. In any event, such supplemental bodies of water are a recognized part of the grower's armamentarium where rainfall is characteristically limited or erratic with respect to predictability.

At each water well, the underground water is brought up by pumping. These irrigation pumps are commonly driven by engines. Some of these engines are designed specifically for driving pumps but many are converted from automobile engines. Some of them are fueled with gasoline whereas others may be fueled with diesel fuel, propane or natural gas. In all cases, these irrigation engines are sources of air pollution as many known pollutants such as nitric oxide, hydrocarbons and carbon monoxide are produced in appreciable amount from the engine exhaust.

Removal from Air

A scheme developed by *G.G. Yie, S.J. Cummingham and R.E. Rosenberg; U.S. Patent 3,888,652; June 10, 1975; assigned to Institute of Gas Technology and Southern California Gas Company* involves fertilizing crop areas by dispensing aqueous ammonia solutions from irrigation engines. The ammonia is formed in place by catalytically converting nitrogen oxide emissions present in the engine exhaust. The ammonia is thereafter dissolved in the body of water which moves through the irrigation pump driven by the engine.

Such a scheme is shown in Figure 71. The schematic drawing illustrates in general fashion an irrigation internal combustion engine **2**, which operates pump **3** to deliver water from source or well **4** along line **5** where it is delivered from discharge line **6** to the field or crop area **7**.

Exhaust from engine manifold **8** is delivered along exhaust line **9** to a catalytic conversion zone **10** where under conditions of exhaust temperature the nitrogen oxide pollutants are treated to obtain ammonia therefrom. The catalytic conversion zone may simply be an enclosed chamber to which exhaust gases are routed, such chamber containing a catalyst bed which is made of selected catalyst. The formed ammonia may be contacted with the body of irrigation water

FIGURE 71: IRRIGATION ENGINE EXHAUST CONVERSION TO OBTAIN FERTILIZING SOLUTION

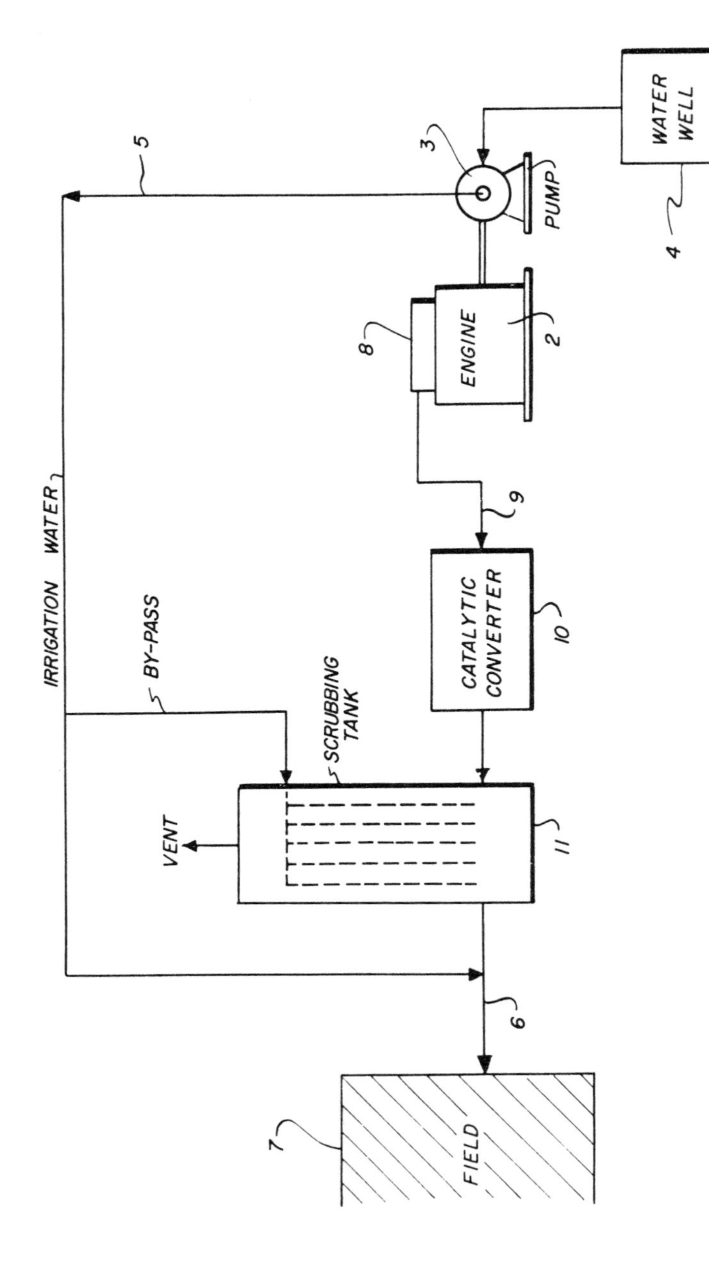

Source: U.S. Patent 3,888,652

in the irrigation engine directly by gas-liquid contact at the discharge line **6**, or a portion of such body of irrigation water may be routed to a scrubbing zone or tank **11** for more efficiently dissolving the ammonia, whereupon such scrubbing zone water may be returned to the irrigation water or dispensed directly onto the crop area.

KETONES

See Acrolein Process Effluents (U.S. Patent 3,923,648)

KRAFT PAPER MILL EFFLUENTS

Kraft pulping, also referred to as sulfate pulping, and sulfite pulping, although somewhat different in process, possess what may be called a common problem-causing disadvantage. As is well known, both pulping processes give rise to waste by-products which must be dealt with effectively. Current legislation on the municipal, state and federal levels has prohibited or at least qualified the type effluent which may be discharged to natural waterways.

Since this is the case, the paper industry is faced with the problem of providing for the effective and legal discharge of these waste products. This problem is complex technically because these wastes are extremely resistant to biodegradation and thus are not amenable to conventional biological treatment methods. In addition, they are highly colored so adsorption or precipitation techniques of removal must be designed to be exceptionally efficient to accomplish adequate treatment.

High treatment levels and correspondingly high losses during adsorbent or precipitant recovery contribute to the high cost of these methods. As is apparent, if there is a cost incurred in providing for the disposal of wastes from the pulping process, then the cost of the pulp derived must necessarily reflect this cost. As can be appreciated from the knowledge of the amount of pulp produced, the amount of waste lignin produced is extremely large. For example, in 1967 the average kraft mill output was 400 tons of paper per day with the concomitant generation of 192 tons of waste lignin per day.

At that time it was estimated that by 1977 these figures would be doubled. Nationally, this would mean that in 1977 there would be an estimated production level of over 29×10^6 tons of paper with a corresponding discharge of 14×10^6 tons of waste lignin into various bodies of water serving as waste diluents. To merely treat that volume of wastewater with any chemical prior to its disposal obviously would require a huge expenditure.

Removal from Water

A process developed by *A.M. Stern and L.L. Gasner; U.S. Patent 3,737,374; June 5, 1973; assigned to Betz Laboratories, Inc.* involves improving the color characteristics of these wastes and utilizing the product or material obtained thereby to produce valuable and utilizable materials or chemicals. Basically, the method entails contacting an oxidatively treated waste and, more specifically, an oxidatively treated waste lignin with a microbial population for a time and at conditions necessary to convert the oxidatively treated lignin to the utilizable end products. Examples of utilizable products which can be obtained in this manner are organic acids, biopolymers, proteins, antibiotics, steroids, vitamins, fertilizers, etc.

A process developed by *G.G. Copeland; U.S. Patent 3,862,909; January 28, 1975; assigned to Copeland Systems Incorporated* affords an aqueous waste treatment method, especially suitable for use in connection with pulp and paper mill operations. Waste effluents containing carbonaceous substances are subjected to a controlled oxidation process in a fluid bed reactor to effect autogenous combustion and pyrolysis of carbonaceous substances to activated carbon products suitable for such standard uses as purification and heat generation.

Prior to treatment in a fluidized bed reactor, the aqueous wastes may be subjected to standard solids concentration processes. Recovery of activated carbon products may be by flotation. The process is especially suitable for use in treatment of wastes from pulp and paper mill operations, whereby bark and other carbonaceous mill wastes may be converted to activated carbon products useful in purification of effluents from associated bleach plant operations. The process is illustrated in Figure 72.

FIGURE 72: FLUIDIZED BED AUTOGENOUS COMBUSTION AND PYROLYSIS OF AQUEOUS KRAFT MILL EFFLUENTS TO PREPARE ACTIVATED CARBON

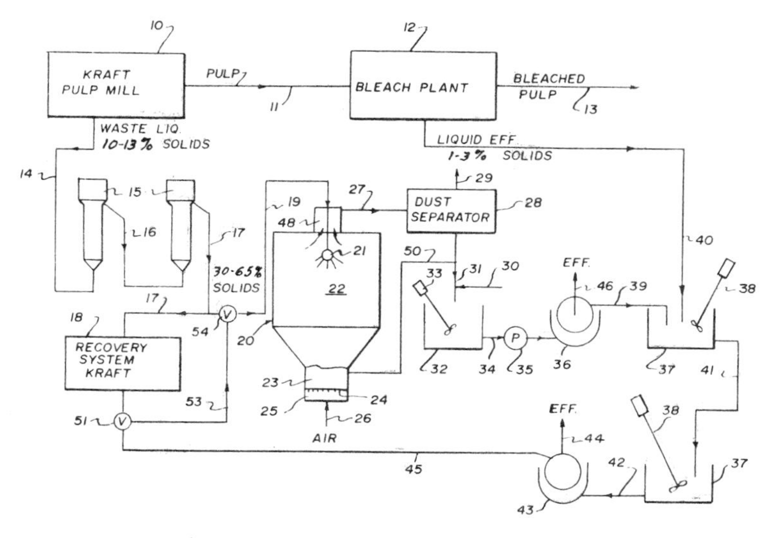

Source: U.S. Patent 3,862,909

In the kraft pulp system illustrated, there is provided a kraft pulp mill **10** of customary construction. Pulp produced is transferred as indicated by the line **11** to a bleach plant **12** in which the pulp is bleached and from which the kraft bleach pulp is removed as indicated by the line **13**.

The pulp mill **10** produces an aqueous waste liquor comprising carbonaceous waste materials, such as lignin, and inorganic salts either derived from the wood or employed during the pulping process. Ordinarily the aqueous liquor contains from about 10 to about 13% solids and in the customary system is transferred through line **14** to a series of evaporators **15.**

As the waste liquor passes through the evaporators **15** as indicated by the line **16,** the liquor is concentrated to a solids content of about 30 to 65 weight percent, for example, with the remaining 35 to 70 weight percent being essentially water. This concentrated waste liquor is customarily transferred, as by a line **17,** to a kraft recovery system **18** which normally uses a combustion process to oxidize the kraft waste and recover the pulping chemicals.

According to the present process, at least a portion of the concentrated (about 30 to 40 weight percent solids) waste liquor from the line **17** is diverted as indicated by the branch line **19** and directed into a fluidized bed reactor **20** of the type described in U.S. Patent 3,309,262.

In this reactor **20** the portion of the concentrated liquor from line **19** is forced through a spray head **21** located in the reactor freeboard space **22** onto the fluid bed **23** of solid material which is stable at the temperature of reaction. The fluid bed **23** is supported on a gas pervious plate **24** which separates the wind box or plenum chamber **25** from the fluid bed **23.**

The particles in the bed **23** are maintained in a fluidized condition by means of a gas that contains oxygen, such as air as illustrated, being directed up through the bed. In the illustrated embodiment fluidization is accomplished by pumping air from a line **26** into the space **25** for flow upwardly through the supporting plate **24,** through the bed **23** and into the freeboard space **22.** This air may be preheated before it is passed through the bed.

Due to the concentrated condition of the waste liquor sprayed onto the particles in the bed **23** for contact with the oxygen of the air, combustion of carbonaceous substances in the bed is autogenous, and no fuel need be introduced except, perhaps, for a brief period on start-up of a cold bed.

When liquors such as kraft (sulfate) or sodium base sulfite liquors are treated according to the method, the particles in the fluid bed **23** are preferably a mixture of sodium carbonate and sodium sulfate, which salts are the end product of oxidation of the inorganic materials contained in those liquors.

The water content of the feed is preferably such that autogenous combustion of carbonaceous materials is possible while temperatures in the fluid bed zone **23** remain below the melting or fusion point of the inorganic salts which may make up the particles of the bed. As an example, kraft liquors injected into the reactor **20** at 32% solids will burn autogenously given a temperature of about 1300°F in the fluid bed **23.** In the course of injection of liquor at about 32% solids through the freeboard space **22,** considerable heat exchange is experienced between the exiting gases and the incoming feed material, resulting in a lowering of exit gas temperature to 1150°F.

The rate of feed to reactor **20** is controlled to stoichiometric requirements of the air passing through the system so that there is substantially no oxygen present in the freeboard zone **22**. As a practical matter, oxygen in the freeboard space **22** is ordinarily less than 2% by volume.

Evaporation of water from the feed in the freeboard space provides the atmosphere of water vapor, which at 1150°F in a deficiency of oxygen, promotes pyrolysis of particles of incompletely burned carbonaceous substances to activated carbon products.

Particles so pyrolyzed are fine in size, thus producing maximum surface for subsequent activated carbon use. The gaseous combustion products with entrained activated carbon products and inorganic salts are carried from the top of the reactor through a conduit **48.**

The mixed gases, inorganic salts, and activated carbon particles are directed by a line **27** into a dust separator **28** which may be of the cyclone type. The gases, principally nitrogen, carbon dioxide,and oxygen, are vented to the atmosphere or carried to scrubbing equipment and/or heat recovery devices (not shown) through a vent line **29.**

A portion of the activated carbon particles and inorganic salts may be directed through a line **50** to the reactor bed **23** if desired, and the balance mixed with water from the line **30.** The resulting aqueous slurry of activated carbon products is directed through a line **31** into a mixer tank **32** where it is thoroughly mixed by a motor driven mixer **33.** In this mixer tank **32,** the inorganic salts are dissolved in water, the carbon remaining in an insoluble state. From tank **32,** the mixture is transferred through a line **34** by a pump **35** into a vacuum filter **36.** In this filter, which is of customary construction, the carbon solids are separated and the inorganic materials are discharged in the filtrate. The activated carbon products are thereafter directed to a series of mixer tanks **37, 37** for use in purification of effluents from bleach plant **12.** Each mixer **37** is provided with a motor driven agitator **38** similar to the mixer agitator **33.**

The activity of the carbon products recovered is a function of the controlled conditions under which they are produced as described above. Products derived from treatment of sulfate and sulfite liquor mill wastes in the above manner generally have an activity of about 30 to 40% of that of commercially available activated carbon. Activity of the products can be increased to that of commercial carbon by activation in a separate step consisting of treatment in a fluid bed reactor where there is exposure to temperatures of 1200° to 1600°F in an oxygen deficient atmosphere. An alternative procedure for deriving additional activity would be treatment with activating salts such as zinc chloride, caustic soda, and the like, according to known processes.

It should be noted that high activity is not necessary where activated carbon products are employed to purify a bleach plant effluent and recycled to existing combustion processes. Low activity can be offset by using more activated carbon per gallon of liquid or alternatively increasing the contact time in mixers **37.**

Where concentrated aqueous waste materials containing bark and wood and little or no inorganic salts are treated, the activated carbon products derived may have an activity 80 to 95% of that of commercial carbon.

In the series of mixer tanks **37**, the carbon solids conveyed from the filter **36** as indicated by the line **39** are mixed with liquid effluent from the bleach plant **12** flowing through a line **40** into the first mixer tank **37** and from there into successive mixers **37**. In this thorough mixing of the activated carbon solids with the liquid effluent from the bleach plant, the activated carbon or "char" absorbs oxygen-demanding pollutants and removes BOD (biological oxygen demand) and COD (chemical oxygen demand) as well as color.

From the last mixer tank **37** the mixed liquid effluent and carbon is conveyed as by a line **42** to another filter **43** which is also of the customary vacuum type. Here the clarified filtrate is directed to a place of disposal as through a line **44** and the filtered solids which now may contain 90% or more of each of the BOD and COD and up to 100% of the color from the effluent may be conveyed as indicated by the line **45** to the recovery system **18** for use as fuel. If desired, this loaded carbon could be diverted by valve **52** into line **53** by way of valve **54** for burning in the fluid bed reactor **20**.

A process developed by *O.E. Rolfe; U.S. Patent 3,833,464; September 3, 1974; assigned to Owens-Illinois, Inc.* involves mixing with a colored kraft paper mill effluent an excess of calcium carbonate, followed by combination at a pressure above atmospheric with a fluid rich in carbon dioxide, serving to convert the calcium carbonate to soluble calcium bicarbonate which is in turn discharged to atmospheric pressure. The calcium bicarbonate reconverts to calcium carbonate precipitate, the precipitate attracting and sweeping the color-imparting color bodies; the precipitate and color bodies being readily separable from the color-free effluent which may be returned to the source, stream, river or lake or reintroduced as fresh water into the pulping process as needed. Such a process is shown in Figure 73.

FIGURE 73: PROCESS FOR DECOLORIZING KRAFT MILL EFFLUENT

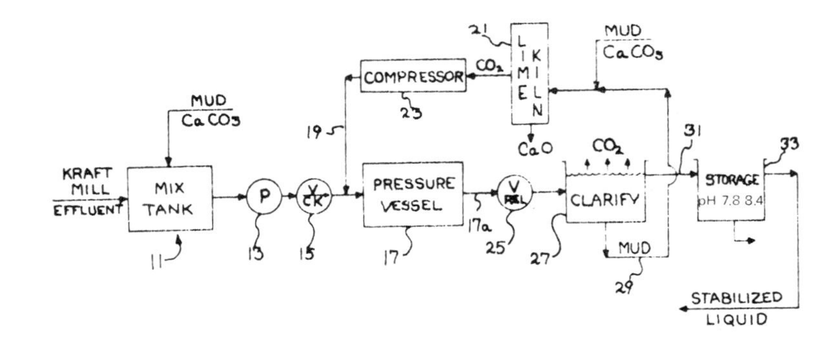

Source: U.S. Patent 3,833,464

As can be seen, starting at the left, the discolored mill effluent is directed together with a slurry of calcium carbonate mud to mixing tank **11**. The resulting mixture is pumped by pump **13** through pressure-maintaining check valve **15** to pressure vessel **17** into which is essentially simultaneously introduced, through line **19**, pressurized carbon dioxide present in by-product stack gases emanating from lime kiln **21**; the stack gases containing the carbon dioxide being pressurized by compressor **23**. The pressure vessel is designed in terms of capacity or volume to provide a residence time, of the mixture of kraft effluent and calcium carbonate in admixture with pressurized carbon dioxide, of from about 2 to 5 minutes, whereby essentially all of the calcium carbonate is converted at the higher pressure to water soluble calcium bicarbonate.

Pressure is maintained within the pressure vessel by a pressure relief valve **25** in the outlet line **17a** of vessel **17**; the valve being predeterminedly set to maintain a desired pressure desirably within the range of about 20 to 100 pounds per square inch. It will be appreciated that the quantity of effluent to be treated together with the amount of calcium carbonate slurry and quantity of carbon dioxide, bearing in mind the concentration of the carbon dioxide in the lime kiln exhaust gases, will determine the size and capacity of the pressure vessel **17**.

The pressure relief valve **25**, in addition to maintaining the desired pressure within the vessel, serves to release the liquid, e.g., aqueous calcium bicarbonate, to clarifier tank **27** maintained at atmospheric pressure, the release in pressure initiating the formation of insoluble calcium carbonate. The calcium carbonate precipitate increases in amount and propagates, gradually falling to the bottom of the tank, sweeping with it as it grows in size the color bodies. The efficiency of color removal according to this method is believed in part due to the fact that the initial precipitate is quite small and serves to attract in admixture therewith the extremely small, in fact colloidally sized, color bodies.

The precipitate continues to form and grow, carrying with it differing sized color body entities. In accordance with a preferred embodiment, the removal of the color bodies is very efficiently accomplished by an excess of calcium carbonate which is first solubilized in the pressure vessel and then is reprecipitated in the clarifier tank, providing thereby a vast multitude of formed insoluble molecules of insoluble calcium carbonate which propagate and grow in the manner above described, resulting in a mass sweeping action upon the body of liquid in the clarifier tank. The precipitate falls, as indicated, to the bottom of the tank, carrying with it the color bodies; all of which are removed through line **29** as mud and redirected to calcium carbonate mud storage and/or to the lime kiln for recovery of calcium oxide and reproduction of carbon dioxide. Simultaneously, with the initiation of precipitation of the calcium carbonate, there is initiated a release of carbon dioxide as a gas which bubbles, as it were, to the surface and passes harmlessly to the surrounding atmosphere.

Liquid constituting the "tops" in the clarifier tank **27** is passed via line **31** to a secondary treatment and storage tank **33**, providing additional time for stabilization of the liquid to a final pH ranging from about 7.8 to about 8.4. Here again, any residual precipitate which may and frequently does form is withdrawn at the bottom and directed to the calcium carbonate while effluent water is drawn off as stabilized, essentially color-free water adapted for reintroduction into the stream, river or lake from which originally obtained or available for reintroduction to the mill operation at whatever point fresh color-free water is desired or needed. The stabilization of the clarified liquid in tank **27** may be hastened by utilizing a violent mixing operation.

A process developed by *P.D. Foster; U.S. Patent 3,945,917; March 23, 1976; assigned to Westvaco Corporation* is one in which considerable reduction in color of a kraft paper mill effluent has been found with the addition of barium ions (Ba^{++}) to the effluent. Unlike the familiar lime decolorization process for kraft paper mill effluents, decolorization using barium is substantially independent of the pH of the effluent and the color removing mechanism is in the form of an occlusion of the color bodies from the effluent into the barium sulfate precipitate ($BaSO_4$) which forms from the sulfate ($SO_4^{=}$) already present in or added to the effluent.

The amount of decolorization produced by the process depends upon the amount of sulfate ions in the effluent, the amount of barium-ion-containing material added to the effluent and the relationship of the amount of barium to the amount of sulfate in the effluent. A slight excess of sulfate over the stoichiometric equivalent of barium is desired, and the total amount of barium-ion-containing material and sulfate that must be added to the effluent depends on the initial color of the effluent and the degree of decolorization desired.

A process developed by *H.A. Fremont; U.S. Patent 3,758,405; September 11, 1973; assigned to U.S. Plywood-Champion Papers, Inc.* involves removing and disposing of color bodies in aqueous effluents from kraft pulp manufacturing operations containing such bodies by adjusting the pH of the effluents to from about neutrality to about 9, subjecting the effluent to ultrafiltration to form an aqueous permeate and a retentate containing substantially all the color bodies in the effluent in a solids concentration of at least about 15%, and thermally oxidizing the retentate at a temperature and for a time sufficient to oxidize the color bodies to colorless inorganic salts and colorless inorganic gases which can be safely disposed of.

Figure 74 shows the overall process scheme (at the left) as well as a detail of a stirred ultrafiltration cell used in the process (at the right).

While the process is applicable to all effluent streams from kraft pulping operations, it will be described in connection with first stage caustic extraction filtrate from a kraft process plant. Such filtrate generally comprises only a minor portion of a kraft mill's total wastewater (approximately 3 to 6%), but it contains more than 50% of the color bodies (color units) in kraft mill wastewater. Such color bodies have been found to be organic in nature.

The process comprises adjusting the first stage caustic extraction effluent by addition of acid. This pH adjustment can be accomplished using any acid, such as sulfuric, and by adding the acid and effluent to a conventional in-line mixing device or a pipe of sufficient length to provide complete mixing. The rate of acid addition is controlled by conventional in-line pH probes located after such mixing means. Depending upon the temperature of the effluent, it may be necessary to cool it to a temperature below 125°F prior to processing the ultrafiltration units in order to avoid damage to the membranes in the cells.

After the pH adjustment (and temperature adjustment if necessary), the effluent is preferably filtered through a nonmolecular filter which will remove the suspended solids. This can be any conventional filter of the rotary vacuum drum type, gravity type, or sand bed type. In addition, polishing filters can be provided to insure the removal of the very small suspended solids in the order of microns. It is desirable to use one or both of these preliminary filtering operations since this helps to minimize any possible fouling of the ultrafiltration membranes by such solids. The thus filtered effluent is then pumped at the required pressure through ultrafiltration cells, more particularly described below, having membranes of a pore size to retain the color bodies in the effluent while permitting lower molecular weight materials, such as inorganic salts, to pass therethrough. In order to have a steady source of material for the ultrafiltration cells it is preferred to place the effluent either before or after coarse filtration into a surge tank or feed reservoir.

FIGURE 74: FLOW DIAGRAM OF KRAFT MILL EFFLUENT DECOLORIZATION AND DETAIL OF ULTRAFILTRATION CELL

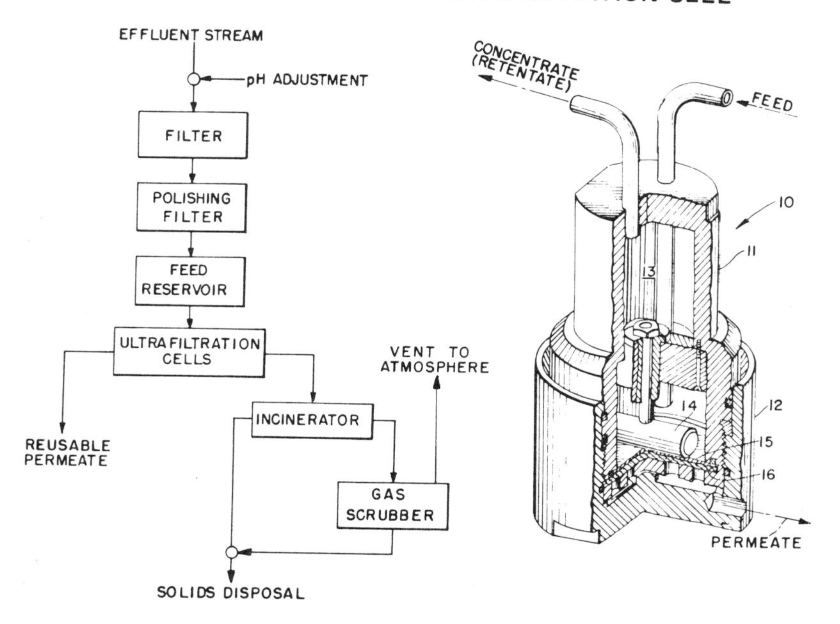

Source: U.S. Patent 3,758,405

The permeate from the ultrafiltration cells can be recycled for further use and the concentrate or retentate is burned in any incinerator adapted to burn high-solids liquids. The residual ash, volatile acids, such as hydrochloric acid, and other volatiles are removed from the off gases by scrubbing prior to venting. This will act to eliminate any possible or potential air pollution. The solid wastes removed from the gas scrubber are combined with the solids formed by incineration and can be disposed of in any conventional manner. Thermal oxidation is important in order to render the color bodies colorless by the oxidation that occurs during combustion. The colorless inorganic solids formed by this oxidation also require very little, if any, biological oxygen.

Ultrafiltration is a membrane process for the concentration of dissolved materials in aqueous solutions. A semipermeable membrane is used as the separating agent and pressure as the driving force. In an ultrafiltration process a feed solution is introduced into a membrane unit or cell where water and certain solutes pass through the membrane having a predetermined pore size under an applied hydrostatic pressure. Solutes whose sizes are greater than the pore size of the membrane are retained and concentrated. The pore structure of the membrane thus acts as a molecular filter, passing some of the smaller size solutes and retaining the larger size solutes.

The pore structure of this molecular filter is such that it does not become plugged because the solutes are rejected at the surface and do not penetrate the membrane. Furthermore, there is no continuous buildup of a filter cake which has to be removed periodically to restore flux (rate of solution transport through the membrane) since concentrated solutes are removed in solution.

It has been found that regulation of the pore size to retain the high molecular weight organic color bodies and permit the lower molecular weight organics and inorganics to be passed through the membrane into the permeate has many advantages. For the color bodies ordinarily found in kraft mill effluent, a membrane with a pore size of from about 0.01 to 0.05 micron is used. This pore size can be varied dependent upon the size of the color bodies in the particular stream being treated. Such regulation of pore size permits high efficiency in retention of the organic color bodies and allows concentration to very high volumetric reduction ratios since the inorganic salts which comprise the bulk of the dissolved solids are not retained in the concentrate. This avoids both a buildup of a high osmotic pressure in the concentrate which would inhibit concentration to high solids levels, and fouling resulting from increased concentrations of both organics and inorganics (e.g., residual calcium salts).

Furthermore, a concentrate is obtained in the ultrafiltration process which contains primarily organic materials and disposal of this concentrate by incineration is substantially facilitated by the removal of inorganics, especially chlorides. It is possible to consider use of this purified organic concentrate as a suitable base for the production of by-product chemicals. The ultrafiltration cell used can be any commercially available with a wide variety of membranes available for use. A conventional stirred cell with a diaphragm-type membrane is depicted.

As is known, the membranes are usually modified cellulose acetate films approximately 100 microns in thickness with an active (dense) surface layer of approximately 0.2 micron. These membranes are manufactured by dissolving cellulose acetate in a mixture of solvents and then casting the mixture as a thin film. The solvents are evaporated and the film gelled. The surface of the film from which the evaporation first took place becomes the dense, active layer. The gelled film (or membrane) is then treated with hot water and is thereby caused to shrink to its final state.

In the view at the right of Figure 74, there is shown a stirred ultrafiltration cell **10** having an outer jacket **11** and cap **12** defining an interior passage **13**, with a magnetically controlled stirrer **14**, attached to the interior of jacket **11** in the passage **13**. A feed inlet and concentrate outlet are provided at the top of the cell and a permeate outlet at the bottom of the cell. Spaced below the stirrer **14** and disposed between the feed inlet and permeate outlet in the passage **13** is membrane **15** supported on a sintered disc **16**.

The operation of the cell will be largely evident from the description given. The effluent is pumped through the feed inlet into the passage **13.** The bulk of the water and smaller molecular weight material will pass through the membrane **15** and sintered disc **16** and be discharged through the permeate outlet. The color bodies and minor amount of water are discharged through the retentate outlet. The stirrer is operated by magnetic means (not shown) at the bottom of the cap **12.**

A process developed by *W.G. Timpe and E.W. Lang; U.S. Patent 3,763,040; October 2, 1973; assigned to the U.S. Administrator of the Environmental Protection Agency* is one in which contaminated water is subjected to a continuous countercurrent multistage treatment with particulate activated carbon in a plurality of tanks in each of which a high slurry density, for example, between 50 and 200 grams per liter of the slurry, is maintained. Particles of activated carbon, intermediate in size between those used heretofore in such processes for the reduction of the organic-carbon content of such wastewaters, are advantageously used. A suitable form of apparatus for use in the conduct of the present process is shown in Figure 75.

FIGURE 75: PROCESS FOR REDUCING THE ORGANIC-CARBON CONTENT OF WATER CONTAMINATED WITH ORGANIC COMPOUNDS BY CONTINUOUS COUNTERCURRENT MULTISTAGE TREATMENT WITH ACTIVATED CARBON

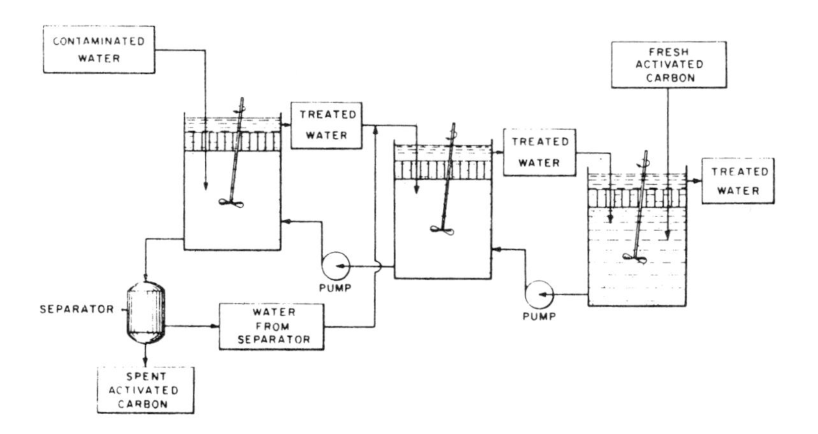

Source: U.S. Patent 3,763,040

In an example of the conduct of the present process, a kraft mill effluent water having a total organic carbon content of 220 milligrams per liter and a color equivalent to 1,100 APHA color units per liter was treated with activated carbon, the particles of which passed through a No. 40 standard sieve and were retained on a No. 100 standard sieve, namely, particles having sizes between 149 and 420 microns.

The kraft mill effluent water was charged at the rate of 2,880 milliliters per hour, corresponding to a retention time of 102 minutes, and 6.75 grams of fresh activated carbon particles were charged at each 30 minute interval to the third beaker in the series, corresponding to a retention time of 35 hours in the system. The characteristics of the water in each of the three beakers in the series, after equilibrium had been established in each, were as follows:

	Total Organic Carbon, milligrams per liter	APHA color units per liter
Untreated water	220	1,100
Beaker 1	120	483
Beaker 2	77	213
Beaker 3	45	90

A process developed by *M. Ichiki and M. Ishii; U.S. Patent 3,793,174; February 19, 1974; assigned to Mitsui Mining & Smelting Co., Ltd, Japan* for treating wastewater containing ligninsulfonate, wherein wastewater contains sodium salt, calcium salt, potassium salt, etc. of ligninsulfonic acid, comprises the steps of: conducting electrolysis with a ferrous anode plate utilizing the wastewater as an electrolyte which has been made to have pH within the range from 5 to 9 prior to the start of electrolysis, while air is being blown therein by a suitable means, whereby the ligninsulfonate is rendered insoluble; and subsequently separating the insoluble substances from the mother liquid by floating them.

See Particulates (U.S. Patent 3,957,464)
See Pulp Mill Effluents (U.S. Patent 3,796,628)

LACQUER SPRAY EFFLUENTS

See Surface Coating Effluents (U.S. Patent 3,861,887)

LATEX

When latex is used in a rug finishing process for attaching burlap or foam rubber backing to rugs, for example, the latex mixing tanks, dip troughs, pipe line, etc., must be frequently cleaned and washed out with water resulting in large amounts of wastewater with varying quantities of latex and latex foam therein which heretofore have been dumped into the sewer or river prior to or after some batch type of settling techniques used in the industry.

Removal from Water

A system developed by *W.H. Scott, Jr.; U.S. Patent 3,726,794; April 10, 1973; assigned to Clarkson Industries, Inc.* senses the flow of latex laden wastewater above a predetermined minimum, mixes and treats same with chemicals in a compartment prior to storage of a batch of latex laden wastewater and chemicals in a primary settling tank. The wastewater is pumped from the primary tank upon filling of same above a predetermined level and chemicals are fed thereinto prior to being supplied to a polishing tank for final clarification and thereafter discharged into the sewer.

LAUNDRY WASTES

Removal from Water

A process developed by *W.H. Hoffman; U.S. Patent 3,350,301; October 31, 1967* is concerned with an effective method of breaking emulsions during the treatment of waste laundry water without the use of either expensive additives or expensive equipment.

This process for the purification of wastewater containing fat, oil, greases, or the like in emulsion includes the step of treating the water in a flotation vessel in the presence of lime and air bubbles and the step of recycling partially purified water to the flotation vessel. One improvement comprises the step of diluting the wastewater entering the flotation vessel with a sufficient volume of the recycled water to break the emulsion. More particularly, the improvement comprises the steps of diluting each volume of wastewater before entering the flotation vessel with a first dilution stream which consists essentially of about an equal volume of water containing a substantial proportion of the soil-lime agglomerates which form in the flotation vessel; diluting each volume of wastewater entering the flotation vessel with a second dilution stream which consists essentially of at least one volume of recycled water for each one thousand parts per million impurities in the wastewater, the second dilution stream being substantially purer water than the first dilution stream; and adding the air and lime to the second dilution stream.

See Carwash Effluents (U.S. Patent 3,911,938)
See Carwash Effluents (U.S. Patent 3,774,625)
See Carwash Effluents (U.S. Patent 3,904,115)

LEAD

Lead is used as an industrial raw material for battery manufacture, printing, paint and dyeing processes, photographic materials and matches and explosives manufacturing.

Removal from Water

Despite its wide use by industry, and the industry associated lead-bearing wastes which must result from its use, there is extremely little information in the technical literature on lead concentrations in wastewaters, or on methods of treating lead-bearing wastes.

Precipitation by lime or dolomite is an indicated route, however. The extreme insolubilities of both lead hydroxide and lead carbonate, the two most common precipitation products, would indicate that good conversion of dissolved lead to insoluble lead should be achieved. These precipitated lead compounds can then be removed by settling, with or without a coagulant aid or removed by filtration. Capital and operating costs should be equivalent to those reported for lime precipitation of other heavy metals (e.g., copper, zinc, nickel).

LEAD ALKYLS

See Tetraalkyllead Manufacturing Effluents (U.S. Patent 3,308,061)
See Tetraalkyllead Manufacturing Effluents (U.S. Patent 3,770,423)
See Tetraalkyllead Manufacturing Effluents (U.S. Patent 3,403,495)

LEATHER PROCESSING EFFLUENTS

The loss of soluble chromium compounds in tannery effluents has gradually increased over a period of years because it was cheaper to buy fresh supplies of chrome salts than to incur the expense of chrome recovery. Another factor contributing to this discharge of chromium salts was the desire to speed up the tanning process by forcing penetration of chromium salts by using higher concentrations of chromium salts. This results in an increased tanning rate with commensurate economics in equipment cycling and labor per unit of production. At the same time, however, it increases the amount of chromium discharged in the tannery effluent as well as the percentage of chromium lost from the leather making process. From a number of surveys made for various tanneries, it has been found that the most efficient tanneries use about two-thirds of the total chromium purchased and discard one-third in their plant effluent. Less efficiently managed tanneries may discard up to one-half of the chromium purchased.

All tannery waste generated by washing, soaking, dehairing, bating, pickling, chrome tanning, dyeing, fat-liquoring, and finishing combines to form an effluent that is predominantly alkaline and contains the waste chromium in its precipitated hydroxide or metal chromite form. This can be settled by conventional catch basin systems resulting in the production of sludge containing not only chromium compounds but grit, manure, grease, and other suspended or settleable solids. However, the collection of sludge for disposal entails a great deal of expense. If the chromium is diverted from the total plant effluent, then the volume of sludge to be disposed of is significantly reduced.

Removal from Water

A process developed by *H.H. Young; U.S. Patent 3,950,131; April 13, 1976; assigned to Hoffmann-Stafford Tanning Company* provides a continuous method for reclaiming chromium compounds which heretofore were discharged as industrial waste from chrome liquor tanneries. The method includes continuous agitation of chrome sludge along with continuous filtration and movement through successive filtering zones. This method is both economically attractive and vital from an ecological standpoint. The reclaimed chromium values are in condition for reconstitution into fresh chrome-tanning liquors. Figure 76 is a flow diagram of such a process.

FIGURE 76: CONTINUOUS RECLAMATION AND REUSE OF CHROMIUM HYDROXIDE FROM SPENT TANNING LIQUORS

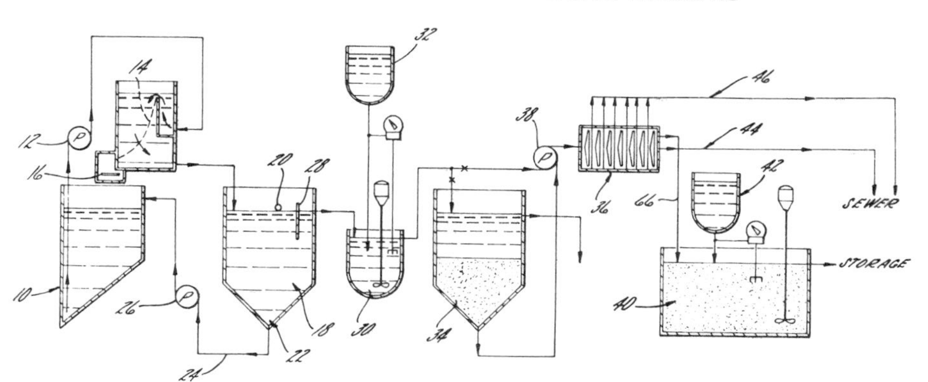

Source: U.S. Patent 3,950,131

In the drawing, **10** is the sump or well into which all spent chrome liquor is conveyed as collected in various parts of the plant. It is pumped by pump **12** over the Hydrasieve **14** (a cascade sieve screen), screenings (solids) being caught on an endless belt conveyor **16** for discharge into a receptacle (not shown). The screened liquor flows into a separator **18** where fat and grease accumulate at the surface, eventually overflowing through port **20**. Any sediment sinks to the cone bottom **22** and is drawn off through line **24** by pump **26** and returned to sump **10** for rescreening. The intermediate layer of clear spent chrome liquor in tank **18** exits under the baffle **28** into a mixing tank **30** where liquid caustic soda or other suitable alkali solution is introduced from feeder **32** to adjust the pH in tank **30** to the range of 8.0 to 9.5. If the concentration of precipitated chromium hydroxide is high enough to preclude settling in the clarifier **34**, it by-passes directly to the continuous pressure filter **36**. But, if the suspension of chromium hydroxide is sufficiently dilute to permit settling, it is detained in the clarifier **34** so as to permit the precipitate to collect in the cone bottom.

Simultaneously, clear supernatant salt solution is decanted from the clarifier at predetermined levels by a simple decanting line (not shown), thereby reducing to a minimum the volume of chromium hydroxide suspension which must be filtered. In either case, the suspension is pumped under pressure by pump **38** into the continuous filter **36** where the filter cake is maintained at a constant thickness and the chromium hydroxide is concentrated to a predetermined concentration. At this desired concentration (14 to 15%), the concentrated paste of chromium hydroxide is ejected into holding tank **40** where a proportionate quantity of sulfuric acid from feeder **42** is mixed in until the desired basicity is achieved and the precipitate has been redissolved. The clear filtrate is conveyed through lines **44** and **46** to the plant sewers, having less than two parts per million of total chromium equivalent.

A process developed by *E.R. Ramirez; U.S. Patent 3,969,245; July 13, 1976; assigned to Swift & Company* is one in which a wastewater flow passes through an electrocoagulation cell having rod-shaped electrodes longitudinally disposed therein. The electrodes are oriented in one or more circles or portions of circles. The wastewater flow remains within the cell for about 1/10 to 2 minutes, during which time pollutants within the wastewater are turbulently mixed and combined with bubbles produced at the electrodes, thereby forming an embryo floc. The embryo floc flows out of the cell with the treated wastewater from the top portion of the downstream end of the cell, after which it may be treated with a flocculant to form a full floc. The full floc and thus clarified wastewater flow into a flotation basin where the full floc undergoes a laminar flow to the top of the clarified wastewater and is skimmed off. Figure 77 shows such an apparatus in schematic form.

Influx conduit **11** directs the flow of raw wastewater into the electrocoagulation cell, indicated generally by reference numeral **12**. The cell **12** includes a gas vent **13** along the top thereof and is in communication with a transfer conduit **14**. The other end of transfer conduit **14** is in communication with large tank or flotation basin **15**, the downstream end thereof having a clarified wastewater efflux conduit **16**. Positioned along the top of flotation basin **15** is a skimming means **17** for conveying the full floc from the surface of the basin **15** for disposal, storage, or further treatment. Skimming means **17** preferably is directed such that the full floc leaves basin **15** from its end opposite that in communication with efflux conduit **16**.

The preferred, but not essential, flocculant injector **21** is shown in communication with transfer conduit **14**. Optional coagulant injector **22** and injector **23** for adding a compound to adjust the pH of the wastewater are shown in communication with influx conduit **11**.

FIGURE 77: ELECTROCOAGULATION SYSTEM FOR REMOVING POLLUTANTS FROM LEATHER PROCESSING WASTEWATER

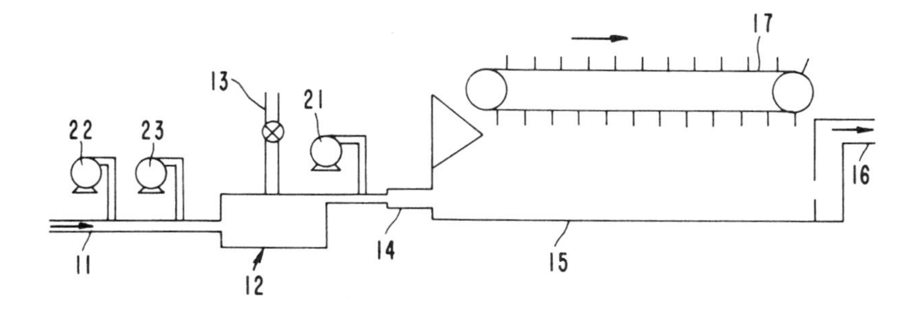

Source: U.S. Patent 3,969,245

In a pilot arrangement of the apparatus, four electrodes made of Duriron were grouped in a single circular fashion. Each electrode was 5 feet long and 1½ inches in diameter. The tank was 6 feet 6 inches long and was designed for a flow rate of only 10 gallons per minute but exhibited the ability to easily handle up to 18 gallons per minute. A total of 42 amperes of current was generated by a 9 volt power source. A ferrifloc coagulant was added in the amount of 1,100 parts per million, along with 850 parts per million of calcium hydroxide to a wastewater flow from a leather treatment plant. Such water flowed through the electrocoagulation cell, after which 12 ppm of an anionic polymer flocculant such as those described herein was then added, after which the treated wastewater entered a flotation basin.

The raw wastewater contained 1,100 ppm BOD (biochemical oxygen demand), 220 ppm fats and oils, and 680 ppm suspended solids. The treated wastewater was found to contain 260 ppm BOD, 18 ppm fats and oils, and 130 ppm suspended solids. Thus, the BOD content was reduced by 77%, the fats and oils content by 92%, and the suspended solids content by 80%. Also, close inspection of the apparatus after several days of operation showed substantially no settled floc at the bottom thereof, indicating the superior lifting power of the electrocoagulation cell. The cell itself required little or no attention throughout this period.

A process developed by *B. Hoke; U.S. Patent 3,817,862; June 18, 1974; assigned to Fried. Krupp G.m.b.H., Germany* provides a method for treating wastewaters containing emulsified and dissolved, oxidizable, organic substances which are free of cyanide, nitrite, hydrazine, dimethyl hydrazine, and nitrogen tetroxide. Wastewater is added to carbon or ion exchangers serving as a catalyst and the interfacial area between the air and the resulting mixture of wastewater and catalyst is increased. The following is a specific example of the application of this process to the wastewater from a plant processing and coloring animal skins.

This water had a dark red appearance when diluted. Undiluted it was black and could not be seen through. In its undiluted state, its COD was 1,200 mg/l. The undiluted water was mixed with powdered activated carbon in the ratio of 300 grams of carbon per cubic meter of water. The resulting mixture was then allowed to fall through a cooling tower utilizing a countercurrent air draft. The fill of the tower was designed to prevent its collecting dirt. After passing through the tower, the water was taken from the collecting basin and pumped back to the top of the tower. A portion of this recirculated water equal to the amount of water coming into the mixing basin was directed off to a filter, with the filtrate being passed to a drainage ditch. The carbon-powder sludge collecting in the filter was recirculated, with less than 5% loss, back into the mixing basin before the top of the tower. A stirrer provided for good mixing of wastewater and carbon in the mixing basin. The wastewater, originally strongly loaded with wetting agents and dyes as well as other organic contaminants, left the filter clear and colorless; it did not foam, and its COD lay at 300 mg/l. Due to the quantity of water recirculated by the pump from the collecting basin back to the top of the tower, a theoretical aeration time of 4 hours was calculated.

A process developed by *J.M. O'Donnell; U.S. Patent 3,976,465; August 24, 1976; assigned to Orgonics, Inc.* involves prereacting the organic waste material with a water-soluble methylol compound subject to condensation, such as a methylolurea compound, under alkaline pH conditions; and, thereafter, condensing the methylol compound by establishing an acid pH condition to form a solid waste product comprising a condensation polymer containing methylene bridges and a sterile solid waste material. The following is a specific example of the application of this technique.

A sample of tannery waste sludge containing about 10% solids and composed of such materials as fleshings, hair, entrails, and general hide scrappings (in addition to these organic constituents there was a sufficient amount of sulfide contamination to cause considerable odor problems) was subjected to treatment by adding 166 grams of urea-formaldehyde solution to 1,000 grams of the tannery waste sludge. The resulting slurry was maintained at a pH of 8.0 and a temperature of 30°C for a period of 30 minutes.

The pH was then reduced to 3.0 by the addition of a dilute solution of sulfuric acid and the temperature was raised to 60°C. Sufficient agitation was supplied to to maintain a state of equilibrium between solid and liquid phases. Methyleneization was allowed to continue to a point where no free formaldehyde was detected. The resulting mixture was then neutralized with a sufficient quantity of a dilute sodium hydroxide solution (1% NaOH) to raise the pH level to 7.5, dried by subjecting the mixture to a continuous stream of hot air (110°C) while tumbling in a rotating cylinder for a time sufficient to reduce the moisture content to about 5% and then ground to a uniform particle size (-10 to 20 mesh). The product is a useful organic fertilizer.

LIGNIN SULFONIC ACIDS

See Kraft Paper Mill Effluents (U.S. Patent 3,793,174)
See Kraft Paper Mill Effluents (U.S. Patent 3,737,374)

LIME KILN EFFLUENTS

Removal from Air

An apparatus developed by *J.P. Tomany and G.H. Cash; U.S. Patent 3,793,809; February 26, 1974; assigned to Universal Oil Products Company* is a multiple-stage venturi-type scrubber unit. The constructional details of such a unit are shown in Figure 78.

FIGURE 78: VENTURI-SPHERE HIGH ENERGY SCRUBBER DESIGN APPLICABLE TO LIME KILN EFFLUENTS

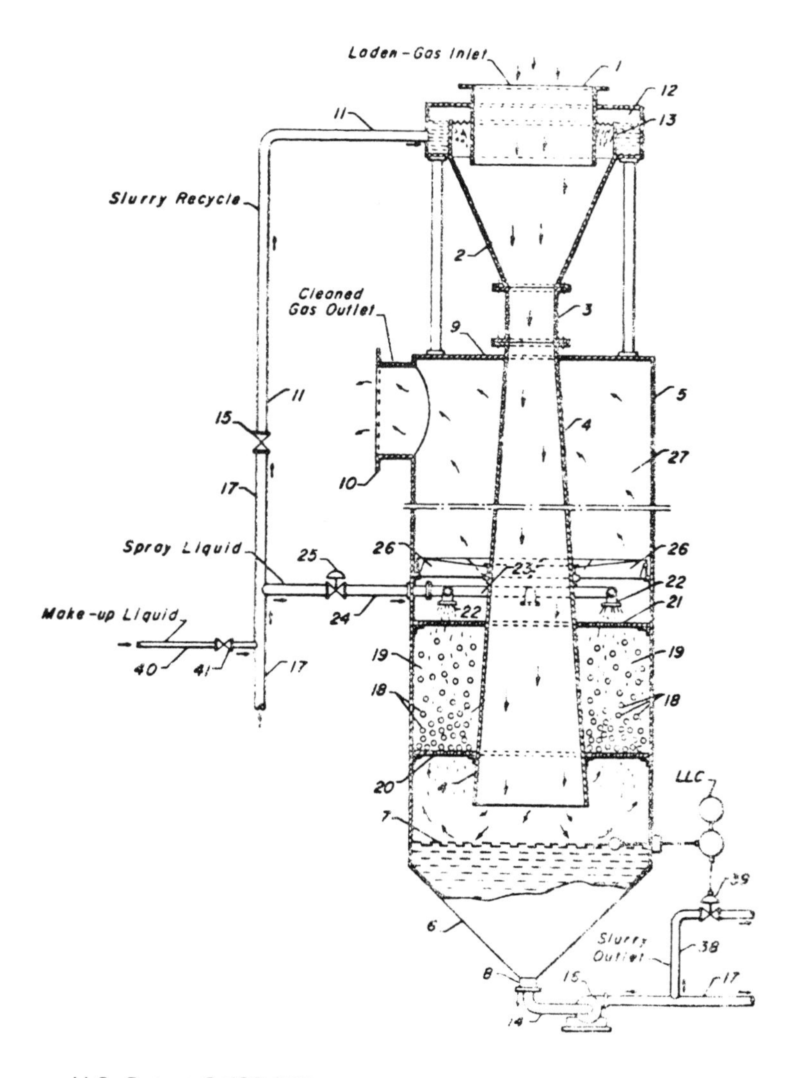

Source: U.S. Patent 3,793,809

There is indicated a vertically oriented combined unit having a flanged inlet means **1** for a particle laden stream and adapted to feed downwardly into a funnel shaped venturi inlet section **2** which in turn feeds into venturi throat section **3** and then into diffuser section **4**. The latter is shown to be positioned substantially axially within a vertically elongated housing section **5** which has a conically shaped lower end portion **6** to provide a liquid or slurry collection section **7**. The lower end of the housing cone section **6** is provided with a slurry outlet means **8**, while the upper portion of the housing **5** is provided with a top closure plate **9** which encompasses the diffuser section **4**. Also, there is indicated an upper side outlet means **10** from housing **5** for the removal of the cleaned gas stream from the unit.

Various means or systems have been provided in combination with venturi-scrubber units to effect the introduction of a liquid stream into the funnel inlet section of the venturi or directly into the throat section thereof; however, a preferred arrangement provides for the uniform introduction of a liquid slurry stream over the entire venturi funnel section **2** so as to preclude the adherence or sticking of any particulates within the inlet section of the venturi. In the embodiment shown, there is shown a slurry recycle line **11** feeding into a liquid reservoir section **12** such that there may be a 360° flow of liquid over a notched weir member **13** which is indicated as an upper extension of funnel section **2**. Thus, in the operation unit, there will be a continuous flow of liquid down over the inside wall of venturi inlet section **2** and into the restricted area of venturi throat section **3**.

At this restricted area entrance to the venturi throat, the liquid or slurry and the laden gaseous stream will meet and proceed through the throat section under high pressure conditions. As a result, liquid droplets are formed and are caused to move rapidly with high velocity action from the gas stream such that there is instant saturation thereof. Also, there is a reduction in the temperature of such stream where it is being introduced as a hot gas. In any case, the entrained finely divided particles are in effect forced into contact with water particles or droplets such that all the entrained particulates are directly wetted and caused to subsequently agglomerate.

Once the laden gas stream and saturated particles leave the venturi throat section and carry down through the diffuser tube there is some regain of energy, which may be of the order of 20 to 30%, or more, of the pressure drop which was sustained across the venturi throat section **3**. The saturated gas stream and the wetted particulates are continuously discharged from the lower end of the diffuser tube **4** at a high velocity such that most of the particulates and excess water are caused to impact with the liquid interface in the collection zone **7**. For example, the stream leaving the lower end of diffuser tube **4** may be of the order of 4,000 feet per minute and the mean velocity reaching the gas-liquid interface at the top of the sump section **7** may be of the order of 2,500 feet per minute.

In order to preclude the liquid level in sump section **7** from being excessive, there will be a constant withdrawal of slurry from the outlet means **8** and, where desired, a suitable liquid level controlling means associated with the lower end of the housing **5**. A slurry withdrawal line **14** connects with suitable pump means **16**, which in turn discharges into line **17**, with valve **15**, for effecting a recycle of slurry to the upper end of the unit and into the venturi funnel section **2**. In this instance, slurry discharge from the unit is indicated by way of line **38** and valve **39** with the latter being controlled responsive to a liquid level control system **LLC**. Also, liquid make-up for the system is indicated as being made by way of line **40** and valve **41** into the slurry recycle line **17**. In the system there is effected a 180° reversal in direction for the gaseous stream leaving the lower end of the diffuser tube **4** such that the gas stream necessarily passes upwardly through a multiplicity of mobile contact elements **18** maintained in a contact zone **19**. In this instance, there is shown a lower perforate plate **20** and an upper screen or perforate retainer plate **21**. In operation, the upflowing gaseous stream with liquid droplets and some remaining entrained particles will cause the turbulent random motion of the light weight fluidizable fluid contact elements **18** in the presence of a descending liquid flow maintained from spray nozzles or other distributor means **22**. The latter are spaced around and above annular contact zone **19** so as to provide a uniform distribution of contact liquid through the entire zone and in turn insure the desired countercurrent scrubbing effect over the surfaces of all of the multiplicity of elements **18**.

The distributors **22** are indicated diagrammatically as depending from an internal distributor or header means **23** which in turn connects with a suitable liquid inlet line **24** having control valve means **25**. The gaseous stream reaching the upper end of the annular contact zone **19** preferably passes through or over a flow deflecting means, such as a plurality of spaced vanes **26** which are arranged in a radial pattern to provide a spiralling or spinning flow for the upwardly moving gas stream whereby there is a resulting spin-out of any entrained water droplets to the inside wall surface of zone **27** prior to the discharge of the cleaned gas stream by way of outlet means **10**. In other words, the vanes or spinner means **26** will provide a further deentrainment of water in the elongated upper spin-out section **27** and such water that is collected along the inner wall of the housing portion of housing **5** will drain down into the lower portion of the unit through the mobile packing stage **19**.

The spheres or other contact elements which are maintained in the second stage contact section **19** of the multiple stage scrubber system may be of a size generally from about ½ to 3 inches in diameter, with the optimum size being selected with regard to the size of the chamber or the size of the contact section being used in the system. There may be a variety of shapes and sizes and various materials for the elements; however, conventionally they will consist of plastic hollow spheres or hollow balls being formed from a thin polypropylene wall or skin.

As a specific example of the operation of the two-stage high energy venturi-scrubber system, there may be the scrubbing of a hot gas stream in the 400° to 600°F temperature range carrying entrained lime particles from a kiln. The laden gas stream will carry down through the unit to enter the venturi funnel at a velocity of the order of 3,300 feet per minute. Recirculated slurry is introduced into the top end of the funnel, at a rate of approximately 10 gallons per 1,000 cubic feet of saturated gases. The liquid slurry is obtained from the lower end of the unit, and may contain some 20 to 30% of solids in the recirculated slurry stream carrying down into the funnel section.

Preferably, the pressure drop through the venturi section is of the order of 15 inches of water for this particular service which will in turn insure an approximate 99% collection efficiency. By utilizing an elongated diffuser tube from the venturi throat section there may be an energy regain of the order of 20 to 30% so that there is only an overall approximate 10 to 12 inches of water pressure drop for the stream leaving the venturi diffuser section. The velocity of the mixed phase stream contacting the liquid level of the slurry collection zone at the lower end of the unit is of the order of 2,500 feet per minute so that a majority of the wetted solid particles are deposited and entrapped in this lower sump section. The reverse upward flow of the gas stream through the enlarged cross-sectional area provided by the mobile contact elements in the second stage of scrubbing is somewhat reduced and generally of the order of 1,000 feet per minute but may be in the range of about 500 to 1,000 feet per minute.

The mobile packing in the turbulent scrubbing section may, for example, comprise lightweight polyethylene spheres of approximately 1½ inches in diameter such that they must be in a static depth of approximately 12 inches in a zone providing 3 to 4 feet between upper and lower retaining screens. The liquid introduced to the top of the countercurrent mobile contact element section will be of the order of 5 to 10 gallons per minute for each 1,000 cubic feet per minute of inlet gas. As noted hereinbefore, a particular advantage of the countercurrent scrubbing in the presence of the mobile contact elements further results in a constant scrubbing together of such elements and the relatively large surface area that they provide for contact with the upwardly moving gas stream and the continuously descending liquid stream will carry along to the lower end of the scrubber section any entrained and wetted particulates that would otherwise carry upwardly to the outlet section of the unit.

As a result of the two-stage of scrubbing as provided by this compact unitary system there is a highly efficient 99% overall removal of both the plus-micron and sub-micron particles from the gas stream.

LIMONENE

See Fruit Processing Wastes (U.S. Patent 3,966,984)

LINOLEUM MANUFACTURING EFFLUENTS

Removal from Air

A process developed by *N.J. Handman; U.S. Patent 3,618,301; November 9, 1971; assigned to Clermont Engineering Co., Inc.* involves preventing hot noxious vapors, comprising aerosols of organic materials emanating from an oven in a manufacturing process, from polluting the atmosphere through oven-stack emissions. The apparatus used comprises a spray chamber, finned cooling coils disposed within the spray chamber, and a mist eliminator communicating with the spray chamber. Hot gases emanating from the oven are cooled by the finned coils to provide a two-phase liquid condensate in the spray chamber. The finned coils are sprayed by the two-phase condensate, at high pressure, to scrub and clean the finned coils of accumulated tars and to agglomerate the aerosols in the spray chamber. Any aerosols emanating from the spray chamber are filtered by the mist eliminator. Figure 79 shows such an apparatus. Referring to the figure, there is shown an oven **10** of the type used in the manufacture of linoleum **12** communicating with the input of a spray chamber **14** through a conduit **16** and a valve **18**. The output of the spray chamber **14** is connected to the input of a mist eliminator **20** through a conduit **22**, and the output of the mist eliminator **20** communicates with the atmosphere through a conduit **24**, such as a stack vented to the atmosphere.

FIGURE 79: APPARATUS FOR REMOVING AEROSOLS FROM LINOLEUM CURING OVEN EFFLUENT GASES

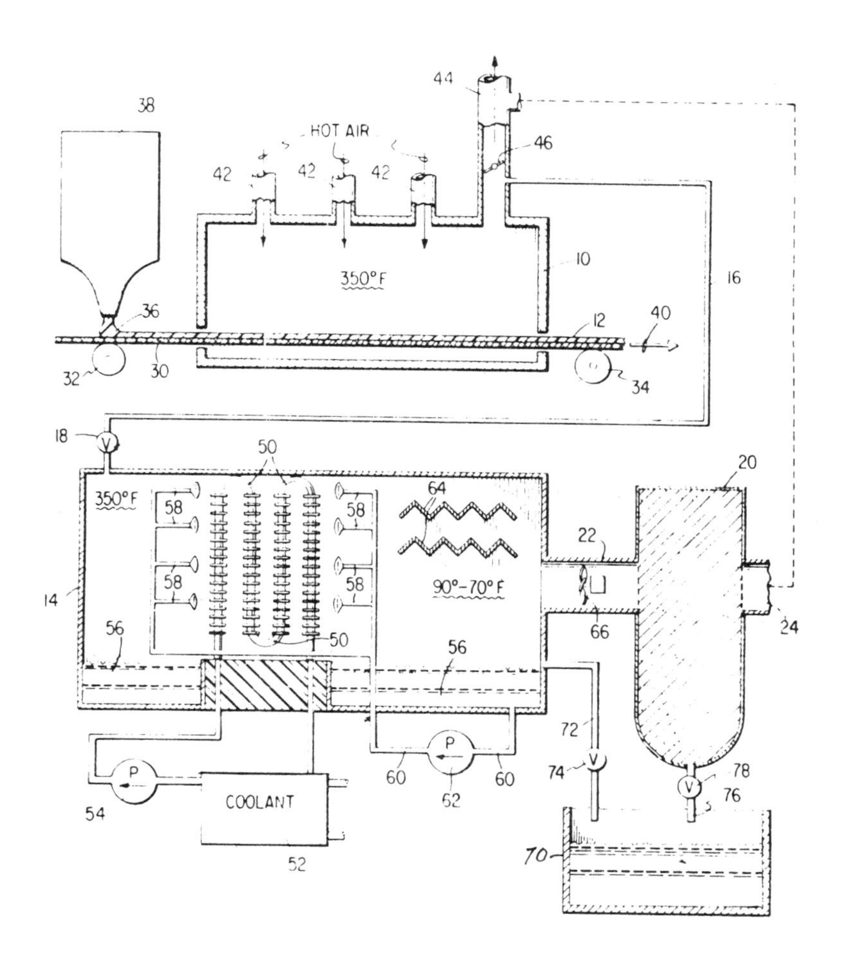

Source: U.S. Patent 3,618,301

In the manufacture of the linoleum **12**, a cloth **30**, such as hessian fabric, for example, is passed through the oven **10** by any suitable conveying means such as rollers **32** and **34**, for example. Linoleum cement **36** from a tank **38** is deposited on the cloth **30** as the cloth **30** enters the oven **10**, the cloth **30** being moved in the direction of the arrow **40**. The linoleum cement **36** may comprise oxidized linseed oil mixed with resins, such as kauri gum and fillers which must be cured. Plasticizers, such as dioctyl phthalate and mineral spirits, for example, are added to the linoleum cement for curing purposes. Hot air at a temperature of about 350°F is introduced into the oven **10** through ports **42** to cure the liquid cement **36**. During the curing process, aerosols, consisting of particles of the organic materials of the plasticizers, mineral spirits and resins from the coating **36**, are formed within the oven **10**. If these aerosols were permitted to pass directly into the atmosphere through a stack **44** communicating with the oven **10**, the atmosphere would become dangerously polluted and present a serious health hazard to humans and animals that would be exposed to it. The aerosols of the organic particles are dispersed in the hot air and other gases, such as carbon dioxide, for example, formed in the oven **10** during the manufacturing process.

Means are provided to recover the pollutants, that is, the noxious vapors (aerosols) from the oven-stack emissions so as to render harmless the ultimate emission to the atmosphere. To this end, the hot noxious aerosols of organics are prevented from being emitted from the stack **44** by a closed damper **46**, and the hot aerosols are passed into the inlet of the spray chamber **14** through the conduit **16** and the open valve **18**. The hot aerosols are passed over a plurality of finned cooling coils **50**. The finned cooling coils **50** are connected to a source **52** of a coolant at a temperature of about 55°F through a closed system including a pump **54**. The coolant may be any suitable liquid or gas, such as water or ammonia, for example, which may be reused continuously since it does not come in contact with the aerosols and is in a closed system.

Hot aerosols entering the spray chamber **14** at about 350°F are cooled to between 90° and 70°F after passing over the finned cooling coils **50**. The hot oven gases passing over the finned cooling coils **50** cause the plasticizer fumes, solvents, and stream from the moisture in the air to condense and form a two-phase condensate **56**, comprising a layer of organics floating on water at the bottom of the spray chamber **14**. The two phases of the condensate **56** may form an emulsion.

The condensate **56** is pumped at high pressure to spray nozzles **58** through a conduit **60** and a high pressure pump **62**. The nozzles **58** are directed onto opposite sides of the finned cooling coils **50** to increase the heat transfer efficiency between the aerosols and the finned cooling coils **50**, to scrub and clean the finned cooling coils **50** of any accumulated tars and resins deposited from the hot vapors, to clean the vapors of any dirt and dust carried over by the oven gases, and to agglomerate the aerosols into much larger droplets so that they can be trapped easily by eliminators **64** disposed within spray chamber adjacent the outlet thereof. To accomplish these functions the pressure of pumping the condensate **56** should be between about 80 and 100 pounds per square inch. This pressure is about four or five times that ordinarily used for air conditioning purposes. Also, the spraying of the condensate **56** should be from opposite sides of the finned cooling coils **50**.

About 80% of the pollutants, mainly the aerosols, have been removed from the oven gases by the time they reach the outlet of the spray chamber **14**. The oven gases are blown to the mist eliminator **20** by a blower **66** disposed within the conduit **22** that communicates with the outlet of the spray chamber **14** and the mist eliminator **20**. The mist eliminator **20** is preferably one of the type generally known as the Brinks demister type for filtering aerosols of about one micron or less in diameter. The gases emanating from the mist eliminator **20** and passing through the stack **24** have less than 10% of the volatilized organics originally present in the gases in the oven **10**. The two-phase condensate **56** can be collected in a recovery tank **70** through a drain **72** and an open valve **74** communicating with the spray chamber **14** and leading into the recovery tank **70**. A drain **76** and an open valve **78** also communicate with the bottom of the mist eliminator **20** for draining the condensate from the mist eliminator **20** into the recovery tank **70**. Thus, at least 90% of the volatile organics that would ordinarily pollute the atmosphere, if not recovered, are recovered by the improved method and apparatus. The recovered organics may be reused in the manufacturing process, thereby providing an economic use of materials.

LITHOGRAPHIC PRINTING PROCESS EFFLUENTS

Lithographed metal sheets are dried and baked in the process of their preparation. The sheets are carried on spaced vertical wickets throughout an elongated heated chamber which form the drying and baking oven. Heated air contacts each sheet to dry the coatings thereon of solvents and other volatile materials in a first part of the oven and to bake the coatings onto the sheets in a second part of the oven. The coated sheets are used to make cans for food products and the like.

The volume of sheets through such an oven is very large. In larger plants, there may be a number of such ovens operating simultaneously. The amount of solvents removed from such ovens can reach as much as 15,000 gallons per day from, for example, 10 such ovens. This solvent is in vapor form and combustible. For safety purposes it must be mixed with an excess of air in order to prevent explosion. The oxidation temperature of this mixture is about 1000° to 1400°F, and the temperature of the mixture is about 400°F, as it is exhausted from the oven.

Removal from Air

Initially, the solvent containing exhausts were merely dumped into the atmosphere. When pollution control standards were adopted in various localities, it was proposed to incinerate the exhausts by feeding them to an outside refractory lined incinerator in which the mixture would be heated to about 1400°F to oxidize the organic solvents in the mixture. An example of such a refractory lined incinerator is disclosed in Hardison et al, U.S. Patent 3,484,189. The expense of such an incinerator is substantial compared to the cost of the ovens. In addition, the cost of the fuel to heat the mixture substantially increases the cost of drying and baking the coatings on the sheets. As a consequence, plants have been slow to install these pollution control devices and, in many localities, the volatile pollutants are still being dumped into the atmosphere.

Solvent drying ovens are subject to shutdown for short periods of time as well as for extended periods of time. Fuel could be conserved if the solvent incinerator could be shut down while the drying oven is shut down or idling. However, conventional refractory lined incinerators depend on the heat of the refractory to maintain the incineration temperatures and therefore have a relatively long lag time which requires such incinerators to be run almost continuously. When the drying oven is shut down for long periods of time, the refractory incinerators can be shut down, but it takes considerable time and fuel to bring the refractory incinerators up to temperature when the ovens are started again. For these reasons, refractory lined incinerators are expensive to operate.

A device developed by *C.B. Gentry; U.S. Patent 3,706,445; December 19, 1972; assigned to Granco Equipment, Inc.* for application to lithographic printing process effluents is an incinerator for a drying oven and the like to remove combustible fumes such as solvent vapors from the exhaust from such ovens. The incinerator is formed from a narrow, elongated, heat-conducting, metal incinerator conduit having at least one U or reverse bend to increase turbulence within the conduit and to increase heat exchange of hot gases with the exterior of the conduit. The fume-laden gases to be incinerated are channelled along the exterior surface of the incinerator conduit for heat exchange with the incinerator. A fan draws the solvent-laden gases through the larger preheat conduit and forces the preheated gases into the incinerator conduit at a higher pressure than that of the preheat conduit.

In one embodiment, a portion of the combusted gases from the incinerator conduit is admixed with the solvent and air mixture in the preheat conduit to assist in preheating the mixture. The incinerator can be built into the oven with which it is used, or can be a separate entity mounted on top of the oven or even separated from the oven by ducting.

A system developed by *M. Farber, A.F. Bush and F.J. Wiens; U.S. Patent 3,768,232; October 30, 1973; assigned to Republic Corporation* is a solvent recovery system comprising an adsorbent bed for stripping the solvent from an air stream and a vacuum distillation and condensation unit for desorbing the solvent from the bed and recovering the solvent for reuse and regenerating the bed for the next adsorption cycle.

Figure 80 shows such a continuous solvent recovery system for a heat set lithographic plant. The system is designed to remove at least 90% of the C_{14} to C_{16} alkane oil solvent from each of four presses. The ovens evaporate about 20 to 25 pounds per hour of the solvent oil. Only the first press **100** is shown for purposes of illustration.

Each press **100** has a printing section **102** and a drier section **104**. The printing section includes an ink reservoir **106** containing an ink in which the C_{14} to C_{16} solvent oil is an ingredient. The reservoir **106** applies the ink to an inking roller **108** which is in turn transferred to the printing roller **110** containing the master and then to an offset roller **112**. A sheet of paper **114** from roll **116** is imprinted between offset roll **112** and an impression roll **118**. The sheet continues its travel through the drier section **104** onto the take up roll **120**.

In the drier section **104** hot air entering through inlet **122** dries the sheet and carries the oil and other vapors and heated gases into the emission stacks **124** of each press **100**. Each stack **124** may be provided with a particulate filter **127**. Each insulated duct **126** may be provided with a fire extinguishing unit comprising a series of CO_2 sparger jets **128** connected to a CO_2 tank **130** by means of a temperature reactive valve **132** suitably set to be activated at a temperature of 500°F. The ducts join a header duct **134** which feeds the combined emissions into duct **136**. A temperature reactive valve **138** connected to a water source may also be provided immediately above each isolation valve **140** on the top of each adsorber tank **142**. The valve **138** may also be set to be activated at 500°F

FIGURE 80: SOLVENT RECOVERY SYSTEM FOR LITHOGRAPHIC PRINTING PROCESS PLANT

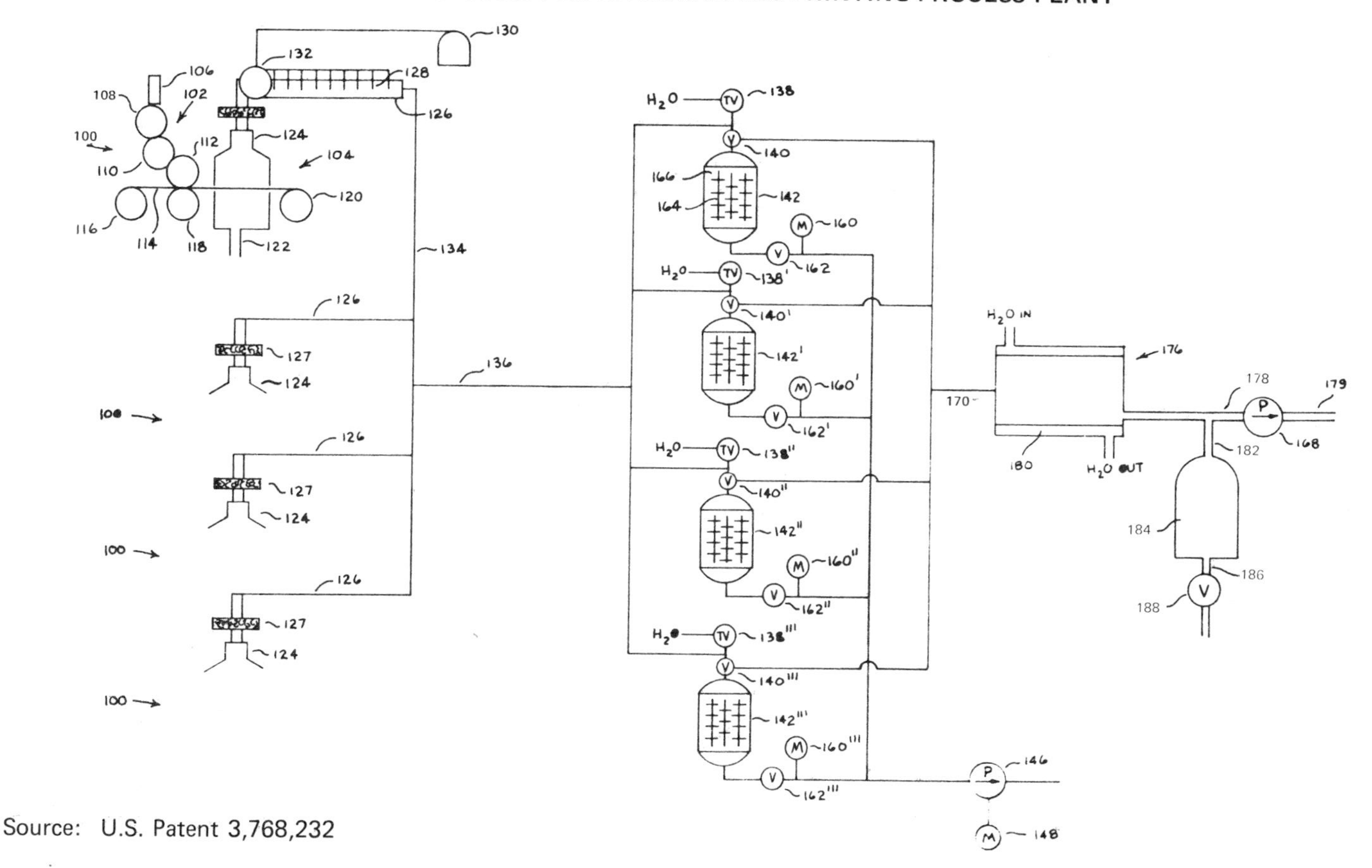

Source: U.S. Patent 3,768,232

so as to flood the tank **142** in case of fire. The hot vapor mixture typically leaves the stack at 300°F and all duct work should be insulated up to isolation valve **140**. The vaporous emission is induced through the tanks by a blower **146** driven by a motor **148**. The blower **146** is connected to the tank isolation valve **162** by duct **172**.

Each tank **142** may contain 6,000 pounds of activated charcoal and three tanks in use simultaneously would be sufficient to remove all hydrocarbon solvent emitted from all four presses **100**. The breakthrough of the tank should be staggered so that each tank may be subjected to a regeneration recovery cycle separately. Each tank may have an adsorption cycle of from 5 to 50 hours, typically about 24 hours. The desorption cycle should be less than the adsorption cycle.

The tanks may be controlled by means of a signal from meter **160** sensing solvent breakthrough or by means of a time-controlled cycle. For example, tank **142** may be set at 0 time or 100% adsorption, when tank **142'** is at ⅓ cycle, tank **142"** is at ⅔ cycle and tank **142"** is at 3/3 cycle and has been switched to desorption. Each ⅓ cycle the central controller switches the next tank onto desorption cycle and returns the regenerated tank to an adsorption cycle.

The tanks **142** contain an isolation valve **162** in the air outlet and finned heating elements **164** embedded within the bed of adsorbent **166**. The inlet side of the condenser **180** is connected to the two-way isolation valves **140** by means of ducts **170**. The output side of the condenser **180** is connected to the storage tank **184** by means of a duct **182**, and to the inlet side of the vacuum pump **168** by a duct **178**.

Cold water is circulated through the annular shell **180** of the condenser and the condensed solvent flows through drain pipe **182** into storage tank **184**. The storage tank should have a 500 gallon capacity sufficient for more than 24 hour continuous operation. The collected solvent may be drained through outlet drain **186** containing valve **188** for reuse when the vacuum pump **186** is off and vacuum on the sytem is released.

MACHINE TOOL OPERATION EFFLUENTS

Removal from Air

One source of industrial air pollution is the cutting oil mist used as a lubricant in machine tool operations. A process developed by *H.W. Weisgerber; U.S. Patent 3,822,532; July 9, 1974; assigned to The Kirk & Blum Manufacturing Co.* utilizes a special apparatus for effecting separation of oil mist from a carrier air stream that is generated in or near factory equipment and to the cleaning and return of the carrier air stream to the factory area.

The apparatus includes a horizontally operable centrifuge chamber disposed above one end portion of an oil collecting pan, a primary oil mist filter located above the opposed end portion of the pan and a secondary oil mist filter positioned above the primary oil mist filter. The relative dispositions of the filters and the centrifuge chamber provide a number of sharp changes in direction of the oil mist laden air stream through the apparatus to effect optimum oil mist-from-air stream separation.

Removal from Water

Manufacturing industries create vast quantities of metal waste in their day-to-day operations, such waste being in the form of chips, turnings, borings and the like. Not only is the metal constituent of this so-called waste of recoverable value but also the lubricating oils coating the metal particles or pieces consequent from the manner in which the metal waste is produced are of value to be reclaimed. In addition to the desired components of the metal and oil, impurities of the type normally attendant operation of industrial facilities are mixed with the metal waste.

In handling industrial waste of this nature, it is desirable to be able to compress the metal constituent of the waste into briquettes or directly feed the metal to a melting furnace. In order to accomplish this, it has been found essential that a high percentage of the oil content of the waste be removed preferably to leave the metal constituent at a level of less than approximately 2% by weight with respect to the residual oil and moisture content on the recovered metal constituent of the waste.

Prior art proposals have not provided the capability of recovering both the metal and the oil from metal waste on an economically profitable basis where the metal constituent in the resulting product has sufficiently low moisture and oil content to enable direct feeding to melting furnaces or compression into briquettes which will possess the desired compressive strength for handling and subsequent use in metallurgical reclaiming processes. Many prior art reclaiming operations for metal waste are directed solely to reclaiming either the metal constituent or the oil constituent and do not provide a combined operation from which overall profitable operation is possible. Other prior art attempts to remove moisture, oil and other impurities from metal chips and the like have been unsuccessful in producing an end product with sufficiently low moisture and oil content that the waste can be formed into briquettes of the high compressive strength of around 10,000 pounds per square inch deemed necessary for practical handling and use in metallurgical reclaiming processes. Other attempts in the prior art to provide commercially usable end products from metal waste have been so complicated as to substantially increase both the required capital investment and the operating and maintenance expenses in carrying out the process.

Figure 81, from *J.R. Keogh, Jr.; U.S. Patent 3,734,776; May 22, 1973; assigned to FMC Corporation* shows the overall layout, in diagrammatic form, of a metal waste cleaning system showing the flow and treatment of the constituents handled in the process of cleaning metal waste.

The system shown commences with the provision of a feeder **10** having a hopper **12** in which the metal waste which is to be cleaned and have the oil and metal constituents separated for reuse is retained. The feeder may take any one of a variety of conventional forms with a screw type feeder having a feed screw **14** driven by a motor **16** being diagrammatically illustrated to control feed of the metal waste from hopper **12** and discharge it from the feeder outlet **18** into the metal waste washer **20**.

The metal waste washer specifically illustrated is made up of two similar wash sections connected in series. However, it is to be understood that depending upon the conditions of the metal waste, detergent solution, desired retention time in the detergent solution for the metal waste and other variables involved in carrying out the waste washing operation, the washer may be constructed of a single section or even constructed of a plurality of sections greater than two.

As heretofore mentioned, the sections of the metal waste washer are similar and accordingly, description of only one section should suffice for understanding of the contemplated construction for the washer. The washer has a longitudinally extending and upwardly inclined tank **22** with a spiral ribbon conveyor **24** operatively mounted longitudinally within the tank to be driven by motor **26** in propelling the metal waste through the flow of detergent solution and agitating such waste so that maximum exposure of the surfaces of the metal waste particles to the cleaning by the detergent solution will occur.

Referring to the sectional detail at the lower left of the figure, it will be seen that the tank is generally U-shaped in cross section having a semicylindrical bottom conforming generally to the exterior of the spiral ribbon conveyor. The tank retains a wash bath of the detergent solution into which the metal waste is discharged from outlet **18** of feeder **10**. The lower rear end of the tank provides a weir over which the spent detergent solution flows into a trough **28** from which such solution is led to the rehabilitation circuit wherein the cleaning capability of the solution is restored as will be described.

Where a multiple section washer **20** such as illustrated is to be used, the sections are mounted in series. As so disposed, the spiral ribbon conveyor in the first section is driven by motor **26** to propel and agitate the metal waste within the wash bath of detergent solution to be moved upwardly along the inclined bottom of the tank and finally discharged at the upper end of the tank through outlet **30**. This outlet is positioned to overlie the lower end of the next succeeding washer section so that the metal waste is discharged into such next section and there exposed to a second portion of the tank bath.

FIGURE 81: APPARATUS FOR METAL WASTE CLEANING SYSTEM

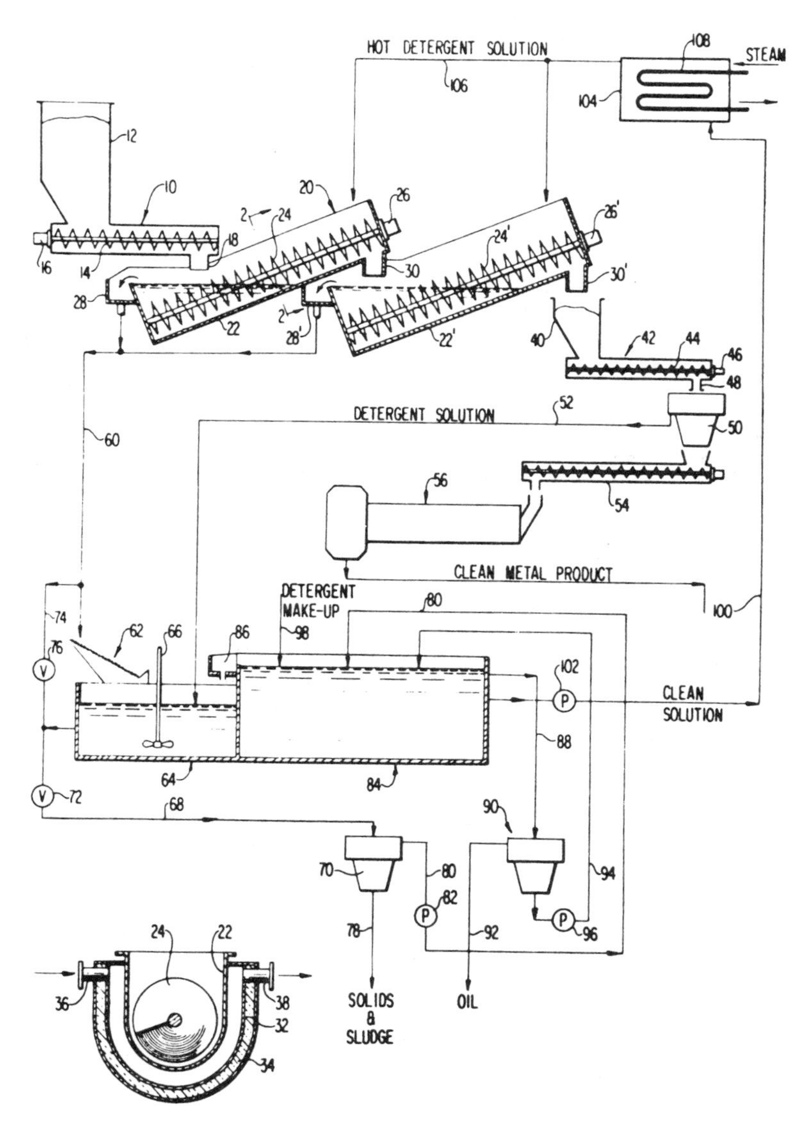

Source: U.S. Patent 3,734,776

In this next section, a similarly formed tank **22′** provided with a spiral ribbon conveyor **24′** driven by a motor **26′** acts to propel the metal waste and agitate it within the second portion of the wash bath moving it up along the inclined wall of tank **22'** out of the wash bath to be finally discharged through outlet **30'** of the second washer section. The lower end of the tank also provides a weir over which the spent detergent solution flows into a trough **28'** to be led to the rehabilitation circuit.

It should be understood that washer **20** may well be constructed so that the detergent solution from tank **22'** will be discharged directly into the upper end of the first section tank **22** of washer **20** and used therein to be finally discharged over the weir of tank **22** into trough **28** rather than being led separately from tank **22'** to the rehabilitation circuit. Also, although the washer sections are illustrated as being similar in length, in some cases it may be desirable to have the sections of different lengths for best cleaning action on the waste in the screw washer.

It will be seen that the generally U-shaped tank **22** is preferably intended to be enclosed along its length by suitable heating means to insure retention of the detergent solution at the desired temperature level for most effective cleaning of the metal waste. In the embodiment illustrated, the exterior of tank **22** has a casing **32** spaced from the exterior of tank **22**. This casing is provided with suitable insulation **34** and an inlet **36** and an outlet **38** for flow of steam through the casing so that heating of the washer **20** may be carried out. As a preferred embodiment, the space between the casing and the tank may contain a plate type internal tube heat exchanger, the tube being connected with inlet **36** and outlet **38**.

When the cleaned metal waste is propelled by spiral ribbon conveyor **24'** upwardly within the second section of the metal waste washer, it is discharged through outlet **30'** from the washer into the surge hopper **40** of a feeder **42**. The feeder may be a conventional screw feeder similar to feeder **10** wherein a feed screw **44** driven by motor **46** withdraws cleaned metal waste from surge hopper **40** and feeds it through outlet **48** to a spinner extractor **50**.

A quantity of cleaned metal waste is discharged from feeder **42** into the bowl of the extractor. As characteristic of such devices, the bowl is rotated at a high speed whereupon the centrifugal forces exerted upon the cleaned metal waste tend to throw off a substantial quantity of the moisture remaining on the cleaned waste after its exposure to the washing action in washer **20**. This separated moisture is detergent solution which is conducted away from the extractor through line **52** and led to the rehabilitation circuit for the detergent solution as will be described in more detail hereafter.

The now substantially dried metal waste in clean form is discharged from the extractor into a screw feeder **54** which transports the cleaned waste by rotation of the motor driven screw of such feeder to the intake of a drier **56**. In the drier, the already partially dried and cleaned waste coming from the extractor is subjected to drying conditions to drive off further of the remaining moisture contained within the cleaned waste.

As used herein, it will be understood that moisture contemplates primarily the detergent solution which is separated from the cleaned waste first in the extractor and subsequently in the drier. However, it is to be recognized that this moisture in both cases would contemplate separation of remaining oil that may be a part of the detergent solution entrapped within the metal waste in the detergent solution as the metal waste is discharged from outlet **30'** of washer **20**.

Whereas feeder **42** provided with surge hopper **40** is advantageous where a batch type extractor **50** is employed, it should be understood that the feeder and surge hopper may be omitted where the extractor is one of the so-called continuous discharge type extractors. However, such a continuous discharge extractor may not always be justified from a cost and operational standpoint and thus, a batch type extractor where the cake of cleaned waste is removed from the extractor bowl at intervals and passed to feeder **54** may be most economically practical. In such event, when the extractor is shut down and the bowl of the extractor is being cleaned, the feeder will be inoperative and the cleaned waste coming from outlet **30'** of the washer **20** can conveniently build up in surge hopper **40**.

Also, although the combination of extractor **50** and dryer **56** in series is shown in the illustrated embodiment, with certain types of metal waste and with certain contemplated further uses for the clean dry metal end product, the provision of the dryer may not be necessary. Of course, the saving in the initial cost of the dryer and the saving in the energy required for its operation would be beneficial if sufficiently dry metal product can be obtained by solely using the extractor. Likewise, it will be recognized that moisture removal may be achieved solely by use of a dryer without use of an extractor.

The clean dry metal product leaving the dryer, or if the dryer is omitted, leaving the extractor, is in a condition of low moisture and low oil content with the sludge or other impurities that originally existed in the metal waste before cleaning removed therefrom. This product is in condition to be fed to a metal melting furnace or to be briquetted for use in other metallurgical processes.

The rehabilitation circuit to restore the detergent solution to its cleaning capability will now be described. The detergent solution which has been used in cleaning the metal waste in washer **20** overflows into troughs **28** and **28'** of the two sections of the washer and is led away from such troughs through line **60**. This solution, of course, carries with it the oil that has been dissolved from the surfaces of the particles of the metal waste together with sludge or other impurities which may have been present in the metal waste prior to its being cleaned.

The contaminated detergent solution is normally discharged from line **60** onto a coarse screen **62**. Of course, the liquid and smaller particles consisting of the detergent solution, oil and sludge will pass through the screen and be accumulated in surge tank **64**. The coarse screen is only needed where the type of the metal waste being cleaned and consequently the flow of liquid from the washer into line **60** is such that coarse or other undesirably large particles flow along with the liquid. Removal of such coarse or undesirably large particles before further rehabilitation of the detergent solution is desirable in connection with obtaining good action in the application of the centrifugal force used in rehabilitating the detergent solution. Where such coarse or large particles are not present in the flow through line **60** from washer **20**, then the coarse screen may be omitted and the liquid from line **60** sent directly to surge tank **64**.

To maintain the detergent solution carrying oil and sludge in suspension, the surge tank **24** may be provided with means to agitate the liquid therein. Such agitation may be obtained by a propeller immersed in the liquid detergent driven through shaft **66** by a suitable motor means (not shown).

The detergent solution carrying sludge and oil is conducted from surge tank **64** through line **68** to a centrifuge **70**. The control of flow to the centrifuge through line **68** is obtained by manipulation of valve **72** in the line. Also it will be noted that a by-pass line **74** containing a control valve **76** is provided communicating with line **60** so that liquid flow therethrough may be directed either to surge tank **64** or through line **74** when valve **76** is open to feed the liquid material directly through line **68** to centrifuge **70**.

The operation of the centrifuge is conventional in that high rotative speed imparted to the liquid material fed to the centrifuge imparts centrifugal forces to such material such that the liquid portion is separated from the solids and sludge. The solids and sludge are then removed through line **78** from the centrifuge while the remaining solution, constituted by detergent and oil, is drawn from the centrifuge through line **80** by means of pump **82**.

The centrifuge may be of the type wherein continuous discharge of the solids and sludge is carried out during centrifuge operation. However, the economics of initial cost and reliability of operation may make the use of a more conventional batch type centrifuge advisable. In a batch type centrifuge, a cake of the solids and sludge is accumulated during operation of the centrifuge and the centrifuge must be stopped and cleaned at intervals varying in time depending upon the rate of accumulation of the solids and sludge in the centrifuge bowl. Where the batch type centrifuge is used, the feed to the centrifuge through line **68** will, of course, be terminated by closing valve **72** when the centrifuge is being cleaned and at such time, the surge tank **64** will serve to accumulate the contaminated detergent solution leaving washer **20** and flowing through line **60**.

The pump **82** draws the liquid portion through line **80** and feeds it to a settling basin **84**. Preferably, the settling basin is associated with the surge tank **64** such that overflow from the basin through trough **86** may be returned to the surge tank. The importance of this relationship arises where a batch type centrifuge is employed in the system. When the liquid is withdrawn from the surge tank to centrifuge **70**, the level in tank **64** will be lowered while the level in settling basin **84** may increase to overflow through trough **86** to maintain a minimum supply in tank **64**. However, when the centrifuge is shut down for cleaning, the liquid from line **60** will tend to accumulate in surge tank **64** and restored clean detergent solution will continuously be drawn from the settling basin requiring adequate volume in such basin.

The liquid portion consisting of detergent solution and oil both with the sludge removed enters basin **84** through line **80**. The oil tends to rise to the surface in this basin with the detergent solution tending to settle beneath the upper strata.

Line **88** is connected to withdraw liquid from basin **84** adjacent the normal surface level of the liquid in such basin. This upper strata will be mostly oil, although a portion of the detergent solution will be present. The liquid flowing in line **88** is led to a centrifuge **90**. This centrifuge operates to apply centrifugal force to the liquid entering from line **88** to separate the oil from the detergent solution. The separated oil is drawn from the centrifuge through line **92**, whereas the detergent solution leaves the centrifuge through line **94** and is pumped back to settling basin **84** by pump **96**.

The operation of centrifuge **90** and pump **96** may be carried out on a timed period. Since only a portion of the overall detergent washing solution is passed through the centrifuge and this portion is drawn from the more concentrated oil layer within setting basin **84**, most effective separation of the oil can be achieved by the operation of the centrifuge.

The restored and clean detergent solution is present in basin **84**. Concentrated make-up detergent solution will be added to the basin through line **98**. The restored detergent solution, having its desired cleaning capabilities, is withdrawn from the basin through line **100** by the action of pump **102** disposed in such line. This clean detergent solution of proper concentration for the desired cleaning effectiveness enters a heat exchanger **104** wherein the temperature of the detergent solution is raised to the desired level to effect best cleaning of the waste material. The heat exchanger may take a conventional form with the detergent solution entering the bottom of the heat exchanger chamber and the desired hot solution withdrawn from the top through line **106**. The heat exchanger casing may appropriately contain a steam coil **108** through which steam is passed to achieve the rise in temperature of the detergent solution.

From line **106**, the hot detergent solution is discharged into the upper ends of the sections of washer **20** in the two-section washer embodiment illustrated. This solution flows downwardly in a direction opposite to the movement of the waste material caused by the spiral ribbon conveyors **24** and **24'**. Thus, most effective counterflow cleaning exposure of the surfaces of metal particles in the waste material to the hot detergent solution is achieved. As noted hereinabove, the temperature of the washer is itself maintained by the heating jacket encasing it so that the detergent solution is maintained at the desired temperature for cleaning.

Although two separate centrifuges **70** and **90**, one for sludge removal and the other for oil removal, have been illustrated in the preferred embodiment, it is to be understood that a single centrifuge type separator may be employed to receive the contaminated detergent solution and separate, in a single operation, the sludge and oil as two constituents then passing on the cleaned detergent solution. It is even possible that such a centrifuge type separator wherein centrifugal forces are applied to the contaminated detergent solution to separate sludge and oil as two constituents apart from the detergent solution could have continuous removal of sludge wherein it is not necessary to shut the centrifuge down to clean the bowl of accumulated sludge.

As a further alternative, rather than utilize a centrifuge **70** of the type wherein intermittent cleaning of the bowl is necessary, duplicate solids removing centrifuges could be used in parallel wherein the contaminated detergent solution would be supplied to one centrifuge while it was in operation while the other centrifuge was being cleaned and vice versa. However, it generally would be more economical to provide the simple surge tank **64** rather than involve the added expense of providing and maintaining two separate solids removing centrifuges.

Also, the oil centrifuge **90** can be sufficiently large to take the full flow from the solids removing centrifuge **70**. In such alternative, centrifuge **90** would be disposed to receive the liquid flowing in line **80**. However, the use of a smaller centrifuge handling partial flow from and to the basin **84** is preferred. If desired, a float-level control for pump **96**, sensitive to the level of liquid in basin **84**, may be provided to start and stop operation of pump **96**. This level control could respond within limits to the contemplated depth of the upper oil strata in basin **84**. Thus, in most cases, the liquid going to the oil separator **90** would be largely oil and also only a small part of the flow which is handled through the sludge removing centrifuge **70**.

MALEIC ANHYDRIDE PROCESS EFFLUENTS

Removal from Air

A process developed by *P. Ackermann; U.S. Patent 3,793,808; February 26, 1974; assigned to Polycarbona Chemie GmbH, Germany* involves washing industrial waste gases which contain foam-forming substances and comprises (1) atomizing an aqueous washing liquid with a centrifugal atomizer into 50 to 150 μ diameter droplets for slowly descending in a tower containing the gas to be washed at a rate requiring greater than two seconds, (2) slowing the vertical descent of the gas relative to the liquid droplets for allowing the droplets to drop into a sump in the bottom of the tower, and (3) recirculating the aqueous washing liquid from the sump up to the atomizer in the top of the tower. The following example illustrates the application of this process.

Example: Steam was injected into the waste gas obtained in the production of phthalic anhydride at a rate of 450 kg/hr. An aqueous liquid obtained as a cleaning liquid was used in a refining plant for the production of maleic anhydride by a gas phase catalytic oxidation of benzene. It was introduced directly into the sump of the wash tower at a rate of 250 liters per hour.

The liquid, which was maintained at a temperature of 80°C or at a higher temperature, contained 4.2% by weight of fumaric acid, 1.6% by weight of maleic acid and 5.2% by weight of an acidic substance which had properties similar to humic acid and which was readily soluble in water. This acidic substance is hereinafter referred to as acid tar. This acid tar caused a particularly persistent foaming and only when used in the method according to the process did it not cause trouble when it formed a constituent of the washing liquid.

A washing solution was sprayed into the wash tower by a distributor or atomizer at a rate of 1,100 liters per hour. The temperature in the washing tower adjusted itself to 36°C. The sediments which were continuously forming in the sump of the tower were continuously being removed therefrom by the conveyor. The dry mass of sediments formed at a rate of about 14 kg/hr and contained about 60% by weight of fumaric acid, 15% by weight of naphthoquinone and 20% by weight of condensation products of naphthoquinone as well as small amounts of phthalic acid and maleic acid.

Water at a rate of about 80 liters per hour was sprayed into the waste gas at a position upstream of the waste gas cyclone or centrifugal separator. The efficiency of the washing of the gaseous phthalic anhydride in the stationary (equilibrium) states was such that the following amounts were washed out: 94.5% of the maleic anhydride, 78% of the phthalic anhydride and 88% of the naphthoquinone.

Washing solution was continuously withdrawn at the rate of 160 liters per hour, the solution containing the following quantities in the dissolved states at 38°C:

	Kg/Hr
Fumaric acid	2.3
Maleic acid	40.8
Phthalic acid	3.9
Naphthoquinone	0.7
H_2SO_4	0.3
Acid tar	13.7
Water	About 110

The wet residue collected on the screen or sieve contained the following amounts of solids:

	Kg/Hr
Fumaric acid	9
Maleic acid	1
Phthalic acid	1
Naphthoquinone	1.3
Condensation products	1.4
Acid tar	0.7

MALIC ACID

In various manufacturing processes, relatively large amounts of organic acids are discharged in the effluent. While in the past such discharge has been economically tolerable, when the added expense of depolluting such effluents is added to the manufacturing costs, concentration and recycling of such acids become an important economic factor.

For example, in the manufacture of malic acid, large quantities of residual malic acid are discharged in the effluent to the extent that a moderately-sized plant has an effluent having an average volume of about 10,000 gallons a day and containing on the order of an average percentage of about 1.3 to 6.0% malic acid and about 0.1 to 0.3% maleic acid and 0.1 to 0.5% fumaric acid, together with lesser amounts, e.g., 10 to 20 ppm chlorine as sodium chloride and iron in amounts of 8 to 19 ppm. Of course, acid streams with greater acid concentrations may also be treated, e.g., streams having an average concentration of malic acid and maleic acid each from about 1.0% to about 10.0%. Fumaric acid may be present in amounts of maximum solubility at room temperature, i.e., about 0.5%.

An effluent of this kind is a substantial pollutant to the extent that its biological oxygen demand by federal regulation must be lowered by 85% in order to comply with acceptable standards. Apart from such regulations, however, by concentrating the acid content of the stream in a plant of the above size to about 30%, a recovery of nearly one million pounds of malic acid per year would result in the overall process.

Removal from Water

While such effluents may be processed in a conventional manner such as by use of lime and ponding in an effort to avoid stream pollution, the cost of such treatment is high, i.e., it is estimated to be more than $100,000 a year for a 10,000 gal/day effluent size plant, and no acid recovery would result from such disposition. Other treatments such as ion-exchange methods, reverse osmosis and evaporative concentration have been suggested for treatment of such effluents; however, none of these methods have appeared suitable because of the excessive costs involved and/or because the conditions surrounding the disposition of the effluent were unsatisfactory.

A process developed by *F.P. Chlanda, H.P. Gregor and K.-J. Liu; U.S. Patent 3,752,749; August 14, 1973; assigned to Allied Chemical Corporation* involves the electrodialytic concentration and removal of acids from aqueous effluents, e.g., malic acid effluent containing less than about 10% malic acid, to produce two streams, a relatively concentrated aqueous concentrate stream of about 30% acid and a relatively dilute stream of less than about 0.3% acid. The process, continuous, semicontinuous or batch, offers a practical solution from the standpoint of meeting pollution control standards and for separating or recovering acids which may be returned to the process by direct recycling.

MEAT PROCESSING WASTES

Removal from Air

A process developed by *A.J. Teller; U.S. Patent 3,183,645; May 18, 1965; assigned to Mass Transfer, Inc.* relates to deodorizing gases from rendering vessels or cookers. In this process, the odoriferous gas is passed into a spray chamber and thereafter is passed through a demisting zone before passage through a bed of adsorbent material such as activated carbon, the noncondensible gases being pulled through the bed by a vacuum applied on the opposite side of the adsorbent bed, preferably by a steam ejector.

See Rendering Plant Effluents (U.S. Patent 3,803,290)
See Smokehouse Effluents (U.S. Patent 3,805,686)

Removal from Water

In the meat industry, the waste from the abattoirs contain animal fats, proteins and other organic materials. Generally speaking, free fat and oil, i.e., not emulsified fat and oil, present no serious problems in regard to separation from water as they generally float to the surface and can be skimmed off. Emulsified fats, on the other hand, stay in solution, causing severe pollution problems.

In this connection, it has been the usual practice in the past to run the wastewater from a packing house to a settling tank or basin having baffles wherein the water would set for an hour or so and the free fat would rise to the top and be skimmed off. The emulsified fat would of course remain in the water and would accompany it to the sewers. Various means, such as aeration and complex apparatus, have been employed in attempts to deemulsify the wastewaters. Usually, however, unless the emulsified oil was very valuable, no effort was made to recover it from the water that was eventually passed to the sewers and hence to the streams and rivers.

In processes where water is reused, the oil can be removed from the system by coagulation with aluminum sulfate and alkali, followed by filtration. The oil is caught in the floc and filtered out of the system. However, periodic backwashes of the filter with hot caustic soda are required. It should be noted, however, that the processes used to completely remove the oil from the water are clearly uneconomical for use in cleaning up wastewater from packing houses.

A process developed by *E.R. Ramirez; U.S. Patent 3,969,203; July 13, 1976; assigned to Swift & Company* permits treating industrial wastewater containing suspended particles such as fats and proteins by exposing the water to a plurality of positive and negative electrodes whereby high current densities are used to create large volumes of microbubbles which float the particles to the surface and periodically reversing the current so as to keep the cathodes clean.

In a specific example, a tank 25 feet long, 6 feet deep and 8 feet wide was divided into 4 sections. Each section was 8 feet by 4 feet leaving 9 feet at the influent end to be used as the floc chamber and for baffles. The first section (nearest the influent end) had nine rod-shaped Duriron electrodes 2⅜ inches in diameter and 7 feet long spaced equally over 4 feet and transverse the flow of water. The second section employed seven electrodes, section 3 had five electrodes while section 4 used four electrodes.

The electrodes in each of these sections were equally spaced apart and were alternately anodes and cathodes connected in parallel. Current drawn in each of the four sections was approximately 150 amperes, 75 amperes, 40 amperes and 20 amperes. Ten volts was employed and common to all four sections. Wastewater from a meat packing plant, high in protein and fats, was treated. The polarity of the anodes and cathodes was reversed every 24 hours. Properties of entering and leaving water from the tank were as follows:

	Influent Wastewater	Effluent Water
Hexane extractables	5,230 ppm	30 ppm
Suspended solids	4,300 ppm	100 ppm
pH	7-12	6.5-7.5

A process developed by *E.R. Ramirez and D.L. Johnson; U.S. Patent 3,959,131; May 25, 1976; assigned to Swift & Company* is one in which a flow of polluted raw wastewater may first be treated with a coagulant, then it is rapidly mixed with very fine bubbles supplied beneath the wastewater flow. It may then be treated with a flocculant, after which the pollutants are separated from the water by skimming off the surface of the wastewater.

Figure 82 shows a suitable form of apparatus for use in the practice of this process. The dimensions of the conduit **11** are such as to accommodate a flow of approximately N liters per minute of raw wastewater. Along the conduit may be provided optional coagulant injector **12** and/or optional further injector **13** for adding a compound to adjust the pH of the wastewater. If desired, either or both of the injectors can be omitted. Such injectors, when included, contain fluid moving devices, such as metering pumps, capable of introducing a predetermined quantity of fluid into the present apparatus and also including a conduit member that connects the injector to the desired location within the apparatus.

FIGURE 82: SCHEME FOR REMOVING IMPURITIES FROM SLAUGHTER-HOUSE WASTEWATER BY FROTH FLOTATION

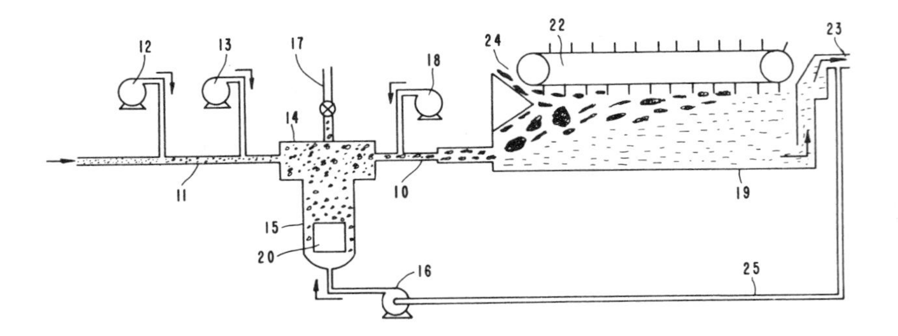

Source: U.S. Patent 3,959,131

Downstream of any such introducer **12** or **13** is a tank **14**. The tank may be of any size and shape, for example, having circular, square or rectangular cross sections. It has a volume on the order of $\frac{1}{10}$ N to 2 N liters, preferably about N liters, to assist in achieving the desired rapid aggregate formation.

Located immediately below the tank member is a cell **15**. More than one such cell may be provided, if desired. The top surface of the cell is in communication with at least a portion of the bottom surface of the tank. Located within the cell is a bubble introduction means **20**.

Bubble introduction means **20** is provided for supplying hydrogen, oxygen or other gas bubbles within cell **15**. The bubble introduction means preferably includes electrodes which are constructed of a conductive material such as a metal and are connected to a power source. When current flows between the electrodes, water within the cell is decomposed into hydrogen and oxygen gas bubbles. Although such an electrode arrangement is preferred, the bubble introduction means may instead supply air bubbles or other gas bubbles, which bubbles are either dissolved in water by a means for pumping gas under pressure into water or are dispersed in water by a blender or a mixer.

An aqueous fluid injector **16** is not an essential feature of the apparatus. When provided, it is in direct communication with the cell **15**. The injector can be utilized to introduce aqueous fluids into the cell in the form of any one or more of tap water containing ionic species, or an aqueous system including a surfactant. The aqueous fluid injector, when provided, has a structure along the lines of injector **12** or **13**, as hereinbefore described. If injector **16** is omitted, the aqueous fluid otherwise provided thereby is supplied from the wastewater itself, entering from the top of cell **15** through tank **14**.

In communication with the top surface of the tank **14** is a gas escape tube **17**, which may be included to vent gases or to relieve excessive gas pressure within the tank by providing a conduit through which gases which have not been combined with particles in the tank may escape.

Flocculant injector **18**, when provided, is downstream of the tank **14** and in communication with a transfer conduit **10**, which conduit permits treated wastewater to flow between the tank and flotation basin **19**, downstream therefrom. In the preferred embodiment, flotation basin **19** is rectangular in cross section and includes a skimming means **22** well above and substantially parallel with the bottom surface of the basin. The flotation basin includes an efflux conduit **23** through which the water clarified by the present apparatus flows. Also provided is skimmings efflux conduit **24** through which the pollutants are moved for storage, safe disposal or further treatment. The following is one specific example of the application of this process.

Example: An industrial installation at a beef slaughtering plant processed wastewater at a rate of 2,270 liters per minute. First added was 350 mg/liter of ferric sulfate as a coagulant. Thereafter, 80 mg/liter of calcium hydroxide was added to adjust the pH. Then, the wastewater was allowed to enter a tank in the apparatus as described.

In this particular apparatus, the cell contained electrodes made from a ferrosilicon alloy (Duriron), there being 70 rod electrodes having a circular cross section of either 1½ or 2 inches in diameter and a length of about 5 feet. The electrodes were suspended within a cell having a square cross section by means of wooden hangers positioned near the top of the electrodes. The conductivity generated in this example was on the order of 700 micromhos per centimeter and 500 amperes of current flowed when applied by a 12 volt DC power source. Satisfactory results were obtained.

It was observed that even better results were obtained and the current flow was substantially increased to about 1,500 amperes, while still utilizing a 12 volt source, by introducing approximately 1.9 liters per minute of 20 weight percent sulfuric acid as an ionic species introduced below the electrodes. Characteristics of the raw wastewater were 1,215 mg/liter BOD (biochemical oxygen demand), 850 mg/liter total suspended solids and 610 mg/liter hexane extractables, while those of the treated wastewater were 95 mg/liter BOD, 88 mg/liter total suspended solids and 15 mg/liter hexane extractables.

See Packing House Effluents (U.S. Patent 3,816,274)
See Proteins (U.S. Patent 3,862,901)

MECHANICAL INSPECTION PENETRANT PROCESS EFFLUENTS

The water-washable inspection penetrant process is used extensively for the nondestructive testing and inspection of critical aircraft parts, such as jet-engine turbine blades, for the presence of potential failure flaws in the nature of cracks, pinholes, forging laps, intergranular corrosion defects and other flaws which are open to the surface. The process, as normally used, includes several steps, as follows: (1) application of the penetrant to test parts; (2) wash-removal of surface penetrant; (3) continuation of washing to deplete background entrapments; (4) drying the parts; (5) development of indications; and (6) inspection for the presence of defect indications.

Most water-washable penetrants which are currently in use throughout industry are the self-emulsifiable type, consisting of an oil vehicle containing a dissolved indicator dye and a combination of detergents and solvent couplers which act to form oil-in-water emulsions upon contact with water. The indicator dye may be visible-color, fluorescent, or both, in accordance with known practices.

In the conventional water-washable inspection penetrant process, a water-soluble or self-emulsifiable penetrant is applied to a test surface by dipping, brushing, spraying or by other convenient means. After a brief dwell time, during which the penetrant liquid is allowed to penetrate into any cracks which are present, the test surface is washed with water, preferably by a spray of water. This spray-wash procedure is usually carried out as a single washing operation, but in reality the washing takes place in two distinct stages. First, the excess surface layer of penetrant is flushed away, and second, depletion of entrapments in cracks or surface porosities takes place upon continued washing.

The test parts are then dried and are allowed to stand so that entrapments of penetrant may exude from cracks by self-development, or alternatively a dry, wet, nonaqueous or plastic-film developer may be used in accordance with known practices. Finally, the test parts are inspected for indications of crack entrapments, under white light or black light, depending on the nature of the indicator dye which is used.

In accordance with the process, the surface penetrant which is removed from test parts becomes mixed with the wash water and is discarded into the nearest sewage or water disposal system. In some cases, efforts are made to extract the penetrant from the wash water so as to minimize pollution in wastewater effluents, but this is quite difficult in the case of readily soluble or easily emulsified penetrants.

Removal from Water

A process developed by *J.R. Alburger; U.S. Patent 3,948,092; April 6, 1976* provides a means for recovery of used penetrant in which penetrant-coated test parts are spray-washed in a prewash stripper step with water saturated with dissolved penetrant liquid. The saturated-water spray removes excess penetrant by the scrubbing action of spray droplets, but it cannot dissolve and deplete penetrant from crack entrapments. The thus-removed excess penetrant may be separated from the prewash water by flotation, and may be recovered for reuse. Following the prewash stripper-recovery step, the normal process of washing, drying and inspection is resumed.

A composition developed by *B.C. Graham and I.Z. Dukats; U.S. Patent 3,716,492; February 13, 1973; assigned to Magnaflux Corporation* is a water-washable, nongelling colored liquid penetrant consisting essentially of a methyl ester of a fatty acid, a nonionic surfactant and a fluorescent dye, the methyl ester and the nonionic surfactant used being readily biodegradable and thereby rendering the penetrant as a whole nonpolluting when subjected, without prior treatment, to the consuming action of microorganisms.

A process developed by *H.N. Skoglund et al; U.S. Patent 3,528,284; September 15, 1970; assigned to Magnaflux Corp.* involves the treatment of dyed penetrant wastes used in metal flow detection. The penetrant waste liquor, containing oily penetrant materials, an emulsifier and wash water, is introduced into an agitation zone where the waste liquor is combined with an electrolyte emulsion breaker and a clay, the mixture is thoroughly agitated, and then stratified to produce a plurality of layers, one of which is essentially an oil-free layer of water which can be reused in the washing zone or disposed of without causing pollution.

MELAMINE PROCESS EFFLUENTS

Removal from Air

A process developed by *R. Mohr; U.S. Patent 3,555,784; January 19, 1971; assigned to Badische Anilin- & Soda-Fabrik AG, Germany* involves the separation of ammonia from off-gas obtained in the synthesis of melamine from urea. The off-gas, after melamine has been separated, is treated with a melt which contains ammonium nitrate and/or ammonium thiocyanate and/or urea. Temperatures which are between the boiling point of ammonia and the decomposition temperature of ammonium carbamate are maintained in the treatment.

It is known that in order to separate the excess ammonia from this off-gas the latter can be scrubbed with water in a column so that almost all the carbon dioxide and some of the ammonia are absorbed. Ammonia containing water vapor is thus obtained and this has to be dried before being reused for example as fluidizing gas in the melamine reactor. This may be done, for example, by cooling the gas mixture in the condenser to temperatures as low as -6°C. For this reason the method is very expensive as regards energy consumption.

It is further known from U.S. Patent 2,950,173 that ammonia can be separated from melamine synthesis off-gas by treatment with an anhydrous solvent, for example dimethyl formamide, ethylene glycol or diethylene glycol, so that at the same time a suspension of ammonium carbamate in the solvent is obtained. The temperature of the solvent is kept at from 0° to 20°C. The ammonium carbamate is separated from the solvent and split, by heating at about 100°C, into ammonia and carbon dioxide from which solvent which has been discharged together with the moist carbamate must be removed by condensation. Moreover, cooling brine has to be used for cooling the liquid.

MERCAPTANS

The removal of hydrogen sulfide and alkyl mercaptans from liquid and gaseous streams, such as the waste gases liberated in the course of various industrial chemical processes, for example, in the pulping of wood, and in petroleum refining, has become increasingly important in combating atmospheric pollution. Such waste gases not only have an offensive odor, but they may also cause damage to vegetation, painted surfaces and wild life, besides constituting a health hazard to humans. The authorities have increasingly imposed lower and lower tolerances on the content of such gases vented to the atmosphere, and it is now imperative in many localities to remove virtually all of the hydrogen sulfide and alkyl mercaptans, under the penalty of an absolute ban on continuing operation of the plant.

The quantities of hydrogen sulfide and mercaptans in waste gases are often not very high. The stack gases obtained in the concentration of black liquor, the waste pulping liquor of the kraft pulping process, contain from 500 to 2,000 parts per million of hydrogen sulfide. However, hydrogen sulfide can be detected by humans at a concentration of approximately 0.01 part per million. The result is that an extremely efficient process for the removal of hydrogen sulfide and alkyl mercaptans is required for effective capture of small amounts of these materials.

Removal from Air

The mercaptans may be oxidized to disulfides and the H_2S oxidized to elemental sulfur by a variety of techniques. One possible technique uses metal chelate catalysts to expedite these reactions.

A process developed by *J.P. Sibeud and C.D. Ruff; U.S. Patent 3,897,219; July 29, 1975; assigned to Rhodia, Inc.* is based on the discovery that the oxidation and catalyst regeneration reactions in the oxidation-reduction metal chelate reaction system can each be made to proceed extremely rapidly if the state of subdivision of the free-oxygen-containing gas is sufficiently fine, although short of foam formation. A suitable apparatus for the conduct of this process is shown in Figure 83.

FIGURE 83: APPARATUS FOR THE REMOVAL OF HYDROGEN SULFIDE AND MERCAPTANS FROM LIQUID AND GASEOUS STREAMS

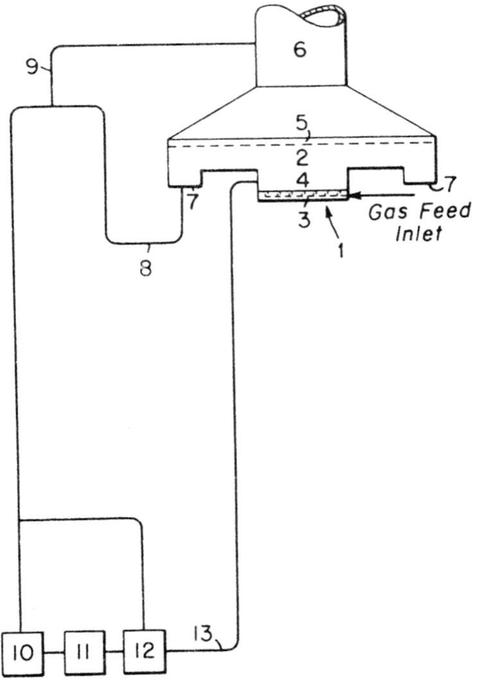

Source: U.S. Patent 3,897,219

The apparatus shown comprises a tubular stainless steel contactor, represented here as defining a relatively short reaction chamber or zone **1**, 10 feet in diameter and 3 feet long, and a cylindrical coaxial stainless steel separator, defining a separation chamber **2**, 30 feet in diameter and 3 feet high, directly above and connected to the reaction chamber. An appropriate gas distribution system **3** is provided at the bottom of the contactor, such as several gas turbomixers with spargers to adequately disperse the gaseous mixture into the liquid catalyst solution. One or more sieve trays **4** are disposed in the contactor, if necessary, in order to ensure proper distribution of flow of the dispersion through the contactor. A gas feed inlet is provided, leading into the gas distribution system **3**, and a liquid catalyst line **13** leads to the reaction chamber above the gas distribution system.

The diameter of the separator chamber **2** can, for example, be from about 1.2 to about 10 times, preferably about 1.2 to about 3 times, greater than the diameter of the reaction chamber **1**, and the height is from about 1 foot to about 10 feet, preferably from about 2 to about 5 feet.

The separator chamber **2** is provided at its upper part with a screen or baffle **5** in order to capture any liquid entrained with the exiting gases, which escape through the outlet **6** above. The annular recess **7** extending circumferentially about the separation chamber is connected to line **8** by means of which liquid catalyst solution containing suspended sulfur and dialkyl disulfides is withdrawn from the chamber. The vent **9** serves to discharge gases which might be entrained by the liquid.

Line **8** leads to an appropriate conventional system for separating the solid suspended sulfur. That shown in the drawing by way of example includes centrifugal separator **10** and a liquid-liquid phase separator **11** for the separation of the dialkyl disulfide liquid from the aqueous catalyst phase. Storage tanks **12** and other equipment are also provided but these can be omitted. The line **13** leads thence to the reaction chamber.

In operation, the gas containing oxygen (air), hydrogen sulfide and alkyl mercaptans is continuously fed via the gas feed inlet line to the bottom of the contactor into the gas distribution system **3**, while the liquid catalyst solution is fed in via line **13** and mixed therewith to form an intimate dispersion short of foam formation in the reaction chamber **1**. A continuing feed of gases and liquid into the gas distribution system **3** and of dispersion thence into the reaction chamber leads to an upward flow of reaction solution through the reaction zone to the separator chamber **2**, the transit time through the reaction zone to the separator chamber constituting the reaction time. This may be very brief, of the order of 0.1 second to several seconds or more, as desired or necessary for the gases being treated.

The gas-catalyst dispersion enters the separation chamber **2** where, due to the greatly increased diameter, and relatively slow flow rate, or relatively quiescent condition, the dispersion breaks and the gas separates, and is vented through the outlet **6**. Such gas is substantially free from hydrogen sulfide and mercaptans. The catalyst solution, containing suspended sulfur and dialkyl disulfides, overflows into the annular recess **7** of the separator chamber, and is withdrawn through the outlet line **8**. The sulfur is separated in the centrifuge **10**, and the dialkyl disulfides in the phase separator **11**. The catalyst solution, which does not require regeneration, is then recirculated to the reaction chamber **1** via line **13**.

A process developed by *H. Friess; U.S. Patent 3,391,988; July 9, 1968; assigned to Gelsenberg Benzin AG, Germany* involves removing mercaptans from gases containing oxygen. The process comprises contacting the gases with an adsorbent which has been impregnated with an alkaline liquid containing a thickening agent such as starch to convert the mercaptans to disulfides and thereafter removing the disulfides from the gases by adsorption.

A process involving water scrubbing followed by aeration and then chlorination of the scrubber water product has been described by *K.G. Trobeck et al; U.S. Patent 3,028,295; April 3, 1962; assigned to ABBT Metodar and Fabricas de Papel Loreto y Pena Sobre.*

See Pulp Mill Digester Effluents (U.S. Patent 3,701,824)

Removal from Water

See Hydrogen Sulfide (U.S. Patent 3,903,250)
See Methionine Process Effluents (U.S. Patent 3,867,509)

MERCURY

In recent years, with the rapid increase in demand for hydrogen gas in the industrial field, high purity has come to be required of the gas. The by-product hydrogen gas generated from mercury electrolytic cells of brine is so much higher in purity than hydrogen gases obtained by other manufacturing processes, that it is being used very preferably in the industrial field. However, this by-product hydrogen gas contains mercury vapor fairly corresponding to the saturated vapor pressure at its temperature.

Accordingly, this by-product hydrogen gas is unsuitable for a semiconductor industry or any catalytic reaction. Moreover, when this gas is used in a food industry such as the manufacture of hardened oil available as the raw material for man-made butter, there is a danger of the mercury contained in the gas mixing with the food and causing harm to a human body. Thus, it is necessary to eliminate the mercury in the by-product hydrogen gas beforehand by some means or other as completely as possible.

Removal from Air

Mercury may be removed from gases (hydrogen in this case) or from gases such as air by the following techniques.

Low Temperature Processing: This process comprises cooling the gas to such a low temperature, as from -30° to -40°C, under a normal pressure and liquifying the mercury vapor contained in it, thereby separating it from the gas. However, as a gigantic scaled refrigerating equipment and an apparatus for separating a liquid phase from a gas phase are required in this process, it is economically difficult to carry it out. Further, it is not easy to bring the mercury content in the gas after the purification according to this process to less than 0.01 mg/m^3.

Adsorptional Processing: This process comprises adsorbing the mercury in the gas to a molecular sieve or an adsorbent such as active charcoal usually under a high pressure. However, this process has the following defects. First, the adsorbing capacity of the adsorbent used for the mercury is so small that a large amount of adsorbent is required. Second, the desorption of the adsorbed mercury is not easy. Third, the concentration of the mercury in the gas after the treatment is more than 0.1 mg/m^3, and it is very far from the standard mark. Finally, the necessity of a high pressure equipment causes this process to be uneconomical.

Washing with Chlorine Water: This process comprises washing the mercury vapor containing gas with chlorine gas dissolved in water to change the mercury vapor into mercury chloride which is, in turn, dissolved in water. However, as some free chlorine gas may remain in the treated gas, another absorption tower is required to remove this free chlorine gas, the equipment and its operation would become very complicated. Moreover, the amount of the residual mercury in the gas treated by this process is from 0.03 to 0.08 mg/m^3, and it is far from the value of less than 0.01 mg/m^3, the standard mark.

Washing with Acidic Permanganate Solution: This process comprises washing the mercury vapor containing gas with a liquor which is obtained by dispersing in a dilute sulfuric acid about 30% in concentration the brown precipitate of a mercury and manganese compound produced by reacting mercury vapor in an aqueous solution containing a compound of manganese of valency greater than 3, and recovering mercury as mercury sulfate in the liquor (cf. Japanese Patent 532,910). However, as the semiconductor industry has a dislike for the mixing of heavy metal element in hydrogen gas to be used, it is not unavoidable that there is left some room for fear regarding this process which has the probabilities of manganese component mixing with hydrogen gas.

A process developed by *B. Kawase, I. Kojima and K. Otani; U.S. Patent 3,725,530; April 3, 1973; assigned to Showa Denko KK, Japan* is one in which mercury vapor contained in various gases, e.g., by-product hydrogen gas generated by a mercury electrolytic cell producing caustic soda, is removed by a simple and effective method which comprises washing the mercury vapor contaminated gas with a dilute acid solution containing persulfate ions.

By this technique, it is possible to reduce the residual content of mercury in the treated gas to less than 0.01 mg/m^3. Furthermore, it is also possible to recover the mercury metal almost completely from the waste acid solution after washing by introducing therein sulfur dioxide gas.

It has been discovered during recent years that the presence of mercury in industrial processes leads to mixed environmental nuisance risks through discharged gases and other waste products and to the contamination of the manufactured product with mercury. These problems are particularly serious in the case of such products as fertilizers and foodstuffs. Since sulfuric acid is being used in ever increasing quantities within the chemical industry, it has become increasingly important that the mercury content of the sulfuric acid produced be also progressively reduced, which means that it must be possible to purify roaster gases containing sulfur dioxide, which is a normal starting material for the manufacture of sulfuric acid, of accompanying mercury and mercury compounds.

A process developed by *K.-A. Melkersson and B.G.V. Hedenas; U.S. Patent 3,786,619; January 22, 1974; assigned to Boliden Aktiebolag, Sweden* is one in which the mercury compounds present in a roaster gas can be removed therefrom with a very high degree of efficiency and in the absence of the disadvantages associated with the use of activated carbon. The gas is passed through a mass of gas purifying material presenting a very wide specific surface, the mass of purifying material comprising either an inert carrier of, for example, silicon dioxide, aluminum oxide, iron oxide or mixtures thereof, impregnated with selenium, selenium sulfide or other selenium compounds or mixtures thereof, or solely of the aforementioned selenium substances, which by granulation, compaction or other suitable agglomerating methods have been converted to the desired particle shape and size.

The gas passed through the mass of purifying material should suitably have a temperature between 20° and 380°C, preferably between 20° and 300°C, and may advantageously contain 3 to 16% by volume SO_2, preferably 5 to 13% by volume SO_2. The gas may be dry or moist, although direct condensation of water on to the mass should be avoided, which means that the purifying mass should work at a temperature higher than the dew point of the gas. The active purifying mass is placed in a stationary bed through which the mercury containing gas is passed. For the purpose of avoiding, as far as possible, a pressure drop in the gas, the cross-sectional area of gas purifying bed presented to the gas flow should be as large as possible.

One embodiment found to afford particularly favorable results is one in which the bed has the form of a hollow cylinder formed by a net structure enclosing the activated mass, the gas being passed into the bed in a direction towards the center of the cylinder and discharged through a tube placed centrally therein.

Removal from Water

The recovery of mercury from chemical process waters, such as brine solutions from mercury-cell chlorine-caustic production, or the mercury-catalyzed sulfonation of anthraquinone, has both economic and ecological importance. Discussions of the environmental aspects of mercury often do not consider the importance of solubility in the discharge of mercury in process wastes. The solubility of mercury in air-free water is only 0.06 mg/l at 25°C, increasing regularly with temperature to 0.3 mg/l at 85°C. More importantly, the solubility of mercury involves oxidation, and the presence of air will increase the solubility of mercury in water to a level approaching the saturation concentration of mercury oxide, about 40 mg/l at 25°C, if chlorides are also present, solubilities as high as 650 mg/l are possible.

The solubility of mercury through its rather easy oxidation means that simple mechanical separation of mercury from process wastes will not guarantee the absence of mercury in effluent, and the discharge of mercury in clear, filtered or settled wastes can reach significant levels. However, if mercury can be kept in its reduced, elemental state, in a reducing environment, its solubility will be depressed to acceptable levels.

Recovery methods usually involve either extraction, insolubilization or reduction.

Extraction of dissolved mercury by adsorption on activated carbon is not economical because of the limited capacity of the carbon and the difficult removal of the mercury from the carbon. Insolubilization as the sulfide is slow, pH dependent, and a soluble complex, $HgS_2^{=}$, is formed with an excess of sulfide. Displacement of mercury by reduction with zinc, copper, iron, etc. is slow and uncertain, and requires further treatment of the amalgam. Formaldehyde and hypophosphorous acid are effective reducing agents, but only with prolonged reaction times and usually some heat.

None of these methods can be applied to solutions of organic mercury compounds: the organomercurials are not adsorbed; the organomercury sulfides are unstable, decomposing to give volatile dialkyl or diphenyl mercury; the R_2Hg compounds are also formed when organomercurials are reduced.

A process developed by *E.L. Cadmus; U.S. Patent 3,764,528; October 9, 1973; assigned to Ventron Corporation* involves the removal of soluble inorganic mercury compounds from waste streams by mixing with the stream an amount of sodium borohydride about 100 to 150% in excess of the stoichiometric amount required to reduce its inorganic mercury content to metal as a precipitate. If organic mercurials are present in the stream, they are converted to soluble inorganic mercury salts by chlorination of the stream before mixing the stream with borohydride.

A process developed by *J.A. Neal; U.S. Patent 3,989,623; November 2, 1976; assigned to Georgia-Pacific Corporation* is a process for recovery of dissolved mercury salts from aqueous solution which involves precipitating the mercury as mercury sulfide and contacting the solution with zinc sulfide particles.

Apparently, the fine colloidal particles of mercury sulfide are adsorbed to the surface of zinc sulfide particles or in some other manner become attached to the particle to be thus easily removed from the effluent. Since zinc sulfide is insoluble, practically no contamination of the effluent with zinc ion is obtained.

In a process developed by *R.P. Williams; U.S. Patent 3,769,205; October 30, 1973; assigned to Phillips Petroleum Company* it was found that by contacting an aqueous mixture contaminated by dissolved mercury compounds with a water-insoluble organic sulfide or disulfide, preferably a hydrocarbyl sulfide or disulfide, all but minor amounts of the mercury contaminant can be removed from the water. The organic layer can be separated from the water layer. If desired, mixtures of one or more sulfides or disulfides or a mixture of a sulfide and disulfide may be employed as the extraction agent.

A process developed by *J.P. Guptill and G.W. Foley; U.S. Patent 3,764,495; October 9, 1973; assigned to Allied Chemical Corporation* is a process for removing mercury from a mercury cell caustic soda or caustic potash solution wherein a vapor or gas is passed through the solution in the presence of a small amount of a reducing agent. Caustic solutions are obtained containing as low a concentration of mercury as 0.01 part per million. Special means are provided for recovery of the mercury. Figure 84 is a flow diagram of such a process.

FIGURE 84: PROCESS FOR MERCURY REMOVAL FROM CAUSTIC SOLUTIONS BY STEAM STRIPPING

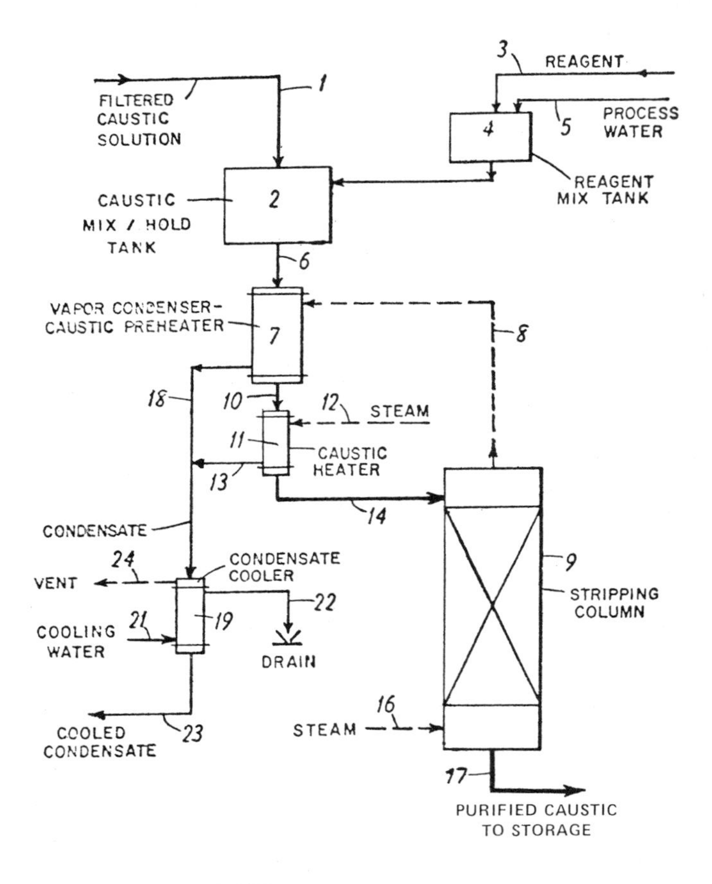

Source: U.S. Patent 3,764,495

Filtered caustic soda or potash solution generally having a concentration of from 35 to 55% obtained from the electrolytic decomposition of brine, and containing mercury as an impurity (generally ranging from about 0.3 to 4.0 ppm), enters through line **1** into caustic mix/hold tank designated by numeral **2**. The reducing agent is introduced through line **3** into reagent mix tank **4** which is also supplied with process water through line **5**. The water and reducing agent are mixed to form an aqueous solution, the concentration of which may vary within a wide range of the order of 5 to 40%. The concentration is not critical since only small quantities of reagent relative to the amount of caustic are employed.

After the caustic solution and reagent are mixed in tank **2**, the mixture is discharged through line **6** on through vapor condenser/caustic preheater **7** wherein it passes in indirect heat exchange with vapors entering through line **8** from stripping column **9**. A preheated caustic solution at a temperature generally about 35 to 90°C dependent on the rate of flow and the temperature of the vapor, then flows down through line **10** and through caustic heater **11**. The caustic is further heated in the caustic heater by indirect heat exchange with superheated steam entering through line **12** and discharging through line **13**, thereby heating the caustic solution to a stripping temperature corresponding to about its boiling point. The thus heated aqueous caustic solution at a temperature close to its boiling point (140°C for 50% NaOH) is introduced through line **14** to the top of stripping column **9** which may be of any conventional packed column type containing Raschig rings, brick work, Berl saddles or the like.

Steam is introduced through line **16** at about 165 psig into the bottom of stripping column **9** and passes upwardly in direct contact with the downflowing aqueous caustic solution, thereby effecting the stripping of mercury from the caustic solution. The purified caustic solution containing as low as 0.0% mercury is discharged to storage through line **17**.

Vapors released from the top of stripping column **9** at about 3 psig and containing mercury, generally in vapor phase, stripped from the caustic solution, pass through line **8** in indirect heat exchange with the incoming caustic solution entering at ambient temperature and passing through vapor condenser caustic preheater **7**. The partially condensed vapor and/or condensate discharges through line **18** into condensate cooler **19** wherein it is cooled by indirect heat exchange with cold water entering through line **21** and discharging through line **22**. The cooled condensate from condensate cooler **19** is discharged through line **23**. Noncondensible gases, if any, are released through line **24**.

In the preferred method of operation, the cooled condensate discharged through line **23**, which contains the mercury removed from the caustic solution, is transferred to the sodium amalgam decomposer, which is a conventional unit employed in the electrolytic processing of brine, employing mercury cathodes, to produce caustic solutions and chlorine. In this way, the mercury remains within the operating cycle.

Another method applicable to the removal of the mercury contaminant in caustic solutions comprises scrubbing the exit gas or vapor with chlorinated spent brine which is then returned to the conventional brine circuit of an electrolytic mercury cell process.

Although reference has been made primarily to the treatment of relatively high concentrations of caustic solutions such as those obtained directly from the operation of electrolytic mercury cells, the method is also applicable to the removal of mercury contained in weak caustic solutions and effluents.

A process developed by *P. De Angelis, A.R. Morris and A.L. MacMillan; U.S. Patent 3,736,253; May 29, 1973; assigned to Sobin Chlor-Alkali, Inc.* is a process for removing mercury from water contaminated with mercury comprising contacting water containing metallic mercury with finely divided anthracite coal. The mercury laden water can be first treated with sodium borohydride to reduce dissolved mercury to the metallic form. The water treated with reducing agent is filtered through a pressure leaf filter coated with a filter aid, contacted with anthracite coal and contacted with chelating resin. Figure 85 illustrates such a process.

FIGURE 85: PROCESS FOR MERCURY REMOVAL FROM WASTEWATER BY BOROHYDRIDE REDUCTION FOLLOWED BY FILTRATION THROUGH COAL AND THEN CHELATING RESIN TREATMENT

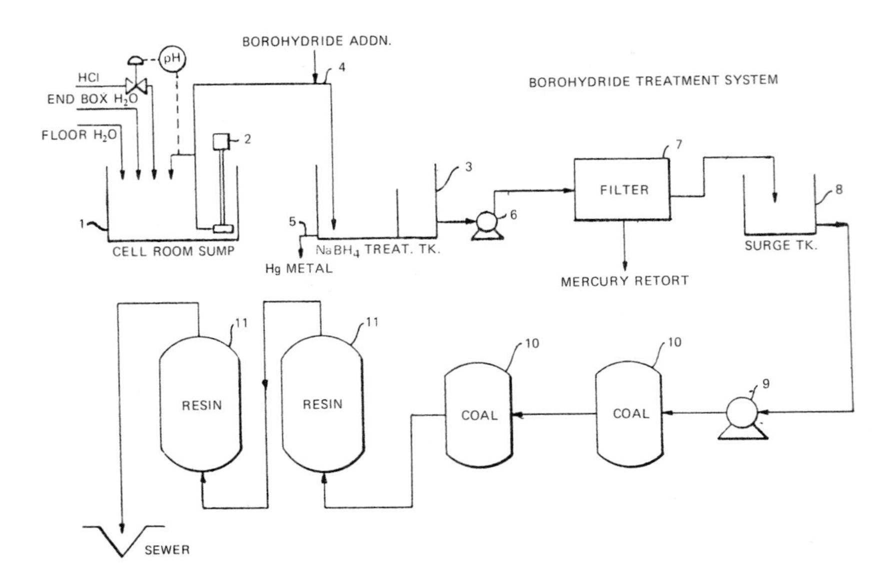

Source: U.S. Patent 3,736,253

Waste process water from the chlor-alkali process is collected from the end boxes of the cells and the floor sumps in cell room sump **1**. Besides containing mercury in solution and suspension, other impurities that might be present in varying amounts are sodium chloride, sodium hydroxide, chlorine, sodium hypochlorite and dirt from floor washing.

The water is adjusted with hydrochloric acid to a pH preferably in the range of 8 to 11. The aqueous stream is pumped through sump pump **2** to tank **3**. While being pumped, alkali metal borohydride is added through mixing T **4** to reduce the mercury to the metallic state. Metallic metal is precipitated and settles to the bottom of tank **3** and is withdrawn through pipe **5** to be reclaimed using standard reclamation procedure. The supernatant liquid that remains is pumped through pump **6** into a filter **7**. This filter is preferably a pressure leaf filter precoated with a filter aid such as diatomaceous earth or cellulose fiber.

The filtrate is then pumped into surge tank **8** through pump **9** into beds **10** of finely divided anthracite coal which adsorb the major amount of mercury remaining in the filtrate. The filtrate is then pumped to beds **11** of chelating resin to remove any residual traces of mercury.

See Paint Manufacturing Effluents (U.S. Patent 3,836,459)

METAL ARC- AND FLAME-CUTTING EFFLUENTS

Conventional flame cutting machines have multiple torches for cutting steel plates into desired shapes. The torches are mounted to traverse the steel plates and the cutting operation results in a considerable amount of cutting waste in the form of relatively small iron oxide particles and solidified molten metal. The small oxide particles are in the form of sparks, some of which tend to become air-borne, causing air pollution which is irritating to the operators and other persons working in the area. The molten metal drops to the floor where it cools and forms slag with the oxides, which when hardened, frequently must be broken up with air hammers or chisels in order to effect its removal.

Also, a plurality of smaller scrap pieces are usually cut from the steel plates and fall to the floor to become part of the slag accumulation and cannot be reclaimed economically for subsequent use. Reclamation of the slag reduces its value to about one-half the value of the salvageable scrap pieces cut from the steel plates due to the presence of oxides therein.

Removal from Air

Various methods have been employed for reducing the air pollution created by such flame-cutting machines. For example, one such method provides a continuous stream of water over an inclined trough disposed beneath the cutting torches. A pump is employed to recirculate the water which is contaminated by entrapped oxide particles. The abrasive nature of the oxide particles causes undue wear to the pump which necessitates its frequent replacement, however.

An apparatus developed by *M.M. Alleman and R.L. Marx; U.S. Patent 3,743,260; July 3, 1973; assigned to Caterpillar Tractor Co.* is an improved waste collector for a burn table having a tank containing a liquid which entraps oxide particles and molten metal waste to reduce the air pollution caused by the oxide particles and to reduce the maintenance cost incurred from removing the slag formed by the molten metal waste.

Plasma-arc cutting is utilized for the removal of selected portions of metal from the metal workpiece and depending upon the particular requirements, such removal may be accomplished by processes such as cutting, gouging, piercing, scarfing, severing and the like. In the plasma-arc process, a gas is introduced into the electric-arc to combine with the arc and form a plasma which is then restricted to a confined area to produce an effluent characterized by stability, high inertia, high energy per unit area and very high temperatures. The effluent generated by the plasma-arc may be the plasma-arc itself as in a transferred arc, or an effluent which is disassociated in space from the actual plasma-arc such as in nontransferred arcs.

In cutting metal, the plasma-arc effluent is brought to bear on the desired area of the workpiece and rapidly heats and melts the metal in such area thus effecting removal of the metal. Although the development of such a high temperature, high energy intensity process for removal of metal has substantially accelerated the rate of metal removal, the plasma-arc operations are noisy, produce large quantities of fumes and cause an extremely bright light which includes ultraviolet emissions which damage the human eye.

Fume suppression has been effected with some success by the introduction of water directly into a fume suppression zone below the cutting zone and this is a standard well-known procedure.

Various methods have been proposed to reduce the noise level of the plasma-arc cutting process which includes the introduction of water curtains around the cutting effluent. However, no apparatus nor method is known which sufficiently reduces the noise level of the plasma-arc cutting operation to a substantially lower level (under 90 decibels) whereby the sound produced by the operation is not likely to injure the hearing of workers in the area over a period of time.

A process developed by *S.L. Miller; U.S. Patent 3,851,864; December 3, 1974; assigned to Lukens Steel Company* is one in which noise, fumes and potentially dangerous light emissions generated by a plasma-arc cutting operation are suppressed by carrying out the operation under a layer of water. To localize the layer of water immediately under the nozzle discharge of the plasma-arc torch, a jacket is provided around the lower end of the torch which continually is receiving water and discharging it within a skirt guidance member depending from the jacket, thus providing a contained volume of moving water around the area involved which immerses the end of the nozzle.

A water layer is also provided under the workpiece and water leaking around the skirt member and through the kerf created by the torch under its nozzle is received in such water layer from whence it is circulated through a separator to remove substances introduced by the cutting operation. The skirt guidance member is preferably composed of a pliant material such as asbestos cloth, rubber or brush bristles. However, more rigid material with a partial bottom may be utilized.

METAL CARBONYLS

Removal from Air

A process developed by *E.R. Breining et al; U.S. Patent 2,985,509; May 23, 1961; assigned to Union Carbide Corp.* relates to the recovery and purification of metal carbonyl compounds and particularly to the recovery of such compounds from the exhaust gases of metallizing operations. It involves water scrubbing of carbonyl-containing gases.

A process developed by *A. Schmechenbecher; U.S. Patent 3,086,340; April 23, 1963; assigned to General Aniline and Film Corp.* is one in which nickel carbonyl vapor can be removed from a gas stream, or a stream containing a mixture of gases, very effectively by passing the gas through any conventional air or gas scrubber containing a solvent which is selective for nickel carbonyl. Such specific and selective solvents are benzene, toluene, tetrahydrofuran, carbon tetrachloride and glacial acetic acid.

If exposure to light is avoided, a substantial part of the nickel carbonyl vapor is recovered from the selective solvent as such. Since these selective solvents do not react with the nickel carbonyl, the latter is not decomposed during the time required to pass the gas through the solvent in any conventional air or gas scrubber washer. In other words, the solution of the nickel carbonyl in the solvent is stable provided undue exposure to light is avoided for a sufficient period of time to permit recovery of the nickel carbonyl by the conventional methods, preferably by fractional distillation.

METAL FINISHING EFFLUENTS

See Chromium (U.S. Patent 3,989,624)
See Chromium (U.S. Patent 3,869,386)
See Electroplating Wastes (U.S. Patent 3,985,628)
See Iron Cyanides (U.S. Patent 3,819,051)
See Paint Spray Booth Effluents (U.S. Patent 3,395,972)
See Paint Wastes (U.S. Patent 3,528,901)
See Surface Coating Effluents (U.S. Patent 3,556,970)
See Surface Coating Effluents (U.S. Patent 3,750,622)
See Surface Coating Effluents (U.S. Patent 3,861,887)

METHANOL

Removal from Water

A process developed by *I.J. Belasco; U.S. Patent 3,928,191; December 23, 1975; assigned to E.I. Du Pont de Nemours & Co.* is one in which biodegradation of the organic content of methanolic wastewater is accelerated by the presence of dispersed particulate attapulgite or montmorillonite.

METHIONINE PROCESS EFFLUENTS

Removal from Air and Water

A process developed by *F. Geiger, T. Lussling and W. Igert; U.S. Patent 3,867,509; February 18, 1975; assigned to Deutsche Gold-und Silber-Scheideanstalt, Germany* is one in which sulfur and nitrogen containing wastewaters and waste gases from such processes as methionine manufacture and including mercaptans are purified by treating with alkali or alkaline earth chlorites in acidic medium at a pH up to 6.

Figure 86a shows a process for treating wastewaters from such a process. The alkaline wastewater is introduced into stirring vessel **1** through conduit **2**. The vessel is equipped with a stirrer **30**. The acid for acidification is introduced through line **3** equipped with control valve **32** and the aqueous chlorite solution is added through line **4** equipped with control valve **34**. The mixture is withdrawn from vessel **1** by line **5** and returned by line **6**. Measuring cells **36** and **38** are provided in line **6** to measure the pH and an amount of chlorite respectively. As soon as the desired potential is reached, the purified wastewater is drawn off by way of line **40** through valve **41** and without further purification can be emptied into the canal, liquid system, sewer, river or the like.

The acid and chlorite introduction lines **3** and **4** are connected by way of valves **32** and **34** with the measuring cells **36** (pH-electrodes) and **38** (redox-measuring cell) with recirculating lines **5** and **6** by way of lines **42** and **44** respectively. Valves **32** and **34** are regulated according to the adjustment of the regulating instruments. In spite of operational variations in the composition of the wastewater, it can be purified continuously.

The following is one specific example of such a process. 500 ml of a bad-smelling, alkaline wastewater from a plant for the production of methionine were brought to a pH of 3 by addition of dilute hydrochloric acid in a three-necked flask provided with a stirrer and dropping funnel. 54.0 ml of aqueous sodium chlorite solution (300 grams of sodium chlorite per liter) were added. The reaction was exothermic. After 5 minutes the solution was odorless and had a potassium permanganate number of 82 mg/l. By iodometric titration it was determined that the sodium chlorite content in the treated wastewater sample was 45 mg of sodium chlorite per liter.

Figure 86b shows a form of process equipment suitable for treating waste gases from such a process. The waste gas is introduced into absorption column **16** by way of line **17**. In the column it is in countercurrent flow to acidified sodium chlorite solution circulating through lines **20** and **21** back to the column **1**. The treated gas leaves the column at the top via line **68**. The introduction of fresh mineral acid takes place via line **18** and with the help of regulating valve **70** and measuring cell **72** is so regulated that the recirculating solution always has a pH value of below 6. The regulating valve **70** is connected by line **74** to the measuring cell **72** provided in line **21**.

FIGURE 86: METHIONINE PROCESS WASTE TREATMENT

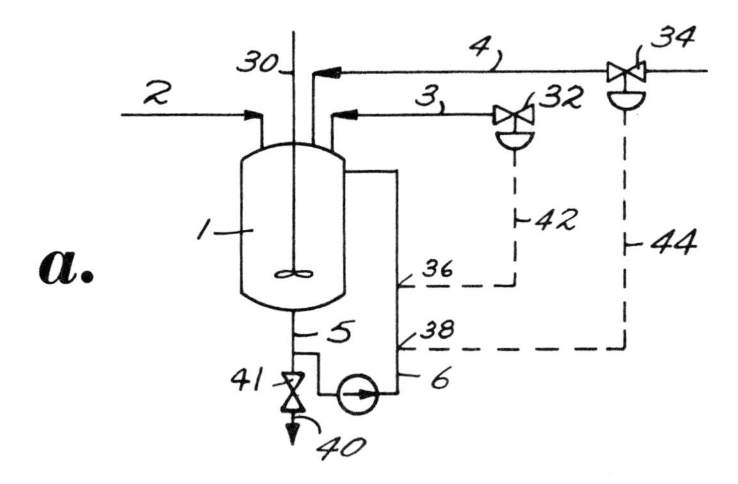

Wastewater Treatment

(continued)

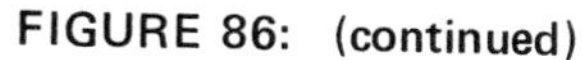
FIGURE 86: (continued)

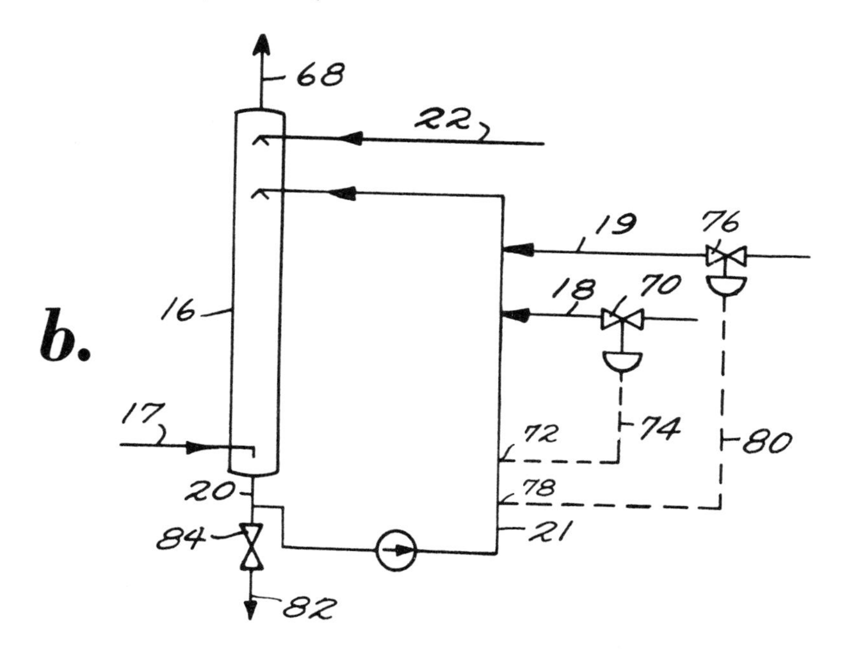

Waste Gas Treatment

Source: U.S. Patent 3,867,509

The addition of fresh sodium chlorite solution takes place via line **19** and with the help of regulating valve **76** and redox measuring cell **78** is so regulated that there is always present a just adequate excess of oxidizing agent. Regulating valve **76** is connected by line **80** with the measuring cell **78** provided in line **21**. At the top of the column there can be introduced by line **22** sufficient fresh water that no disproportionation products of sodium chlorite such as chlorine dioxide occurs in the waste gases that leave via line **68**. The spent recirculating solution can be withdrawn to the wastewater disposal system, e.g., a sewer, by line **82**. A valve **84** can be provided to regulate the removal of spent solution.

The following is one specific example of the conduct of such a process. In the apparatus described, during several weeks there were continuously treated hourly about 60,000 cubic meters of unpleasant-smelling waste gas from a plant for the production of methionine with such an amount of 10% hydrochloric acid and aqueous sodium chlorite solution (300 grams of sodium chlorite per liter) that there was present in the circulating solution in line **20** a pH value of 1 to 4 and 40 to 100 mg/l of sodium chlorite in the wastewater leaving by line **82**. The departing waste gas was odorless. The departing wastewater had a potassium permanganate number of 35 to 42 mg/l.

MINERAL WOOL INDUSTRY EFFLUENTS

The term mineral wool is used generically to denote various types of mineral fibers which are used extensively as insulation against both sound and fluctuations in temperature. The various types of mineral wool include glass wool, slag wool, rock wool and the like. These mineral fibers can be made by several methods from the raw materials such as glass, slag, rock and other fusible raw materials.

In the manufacture of glass wool, for example, individual fibers are made from a single source of liquid material, the fibers or filament being drawn from individual orifices. Another method of making mineral wool involves the disintegration of a molten stream of the raw material simultaneously into several thousand fibers. Conventionally, the molten stream is subjected to the action of a steam blast which shreds the stream into minute droplets which are drawn out into fiber form by the force of the blast. In still another method, the stream of molten material is projected onto the peripheral surface of one or more rotors rotating at high speed, the rotors serving to break up the stream into droplets and to draw them into fibrous form.

It is well known in the art that in the manufacture of mineral wool a great amount of dust is formed which can have harmful effects. The first and most obvious of these harmful effects is the physiological harm caused to the workers in the plant. The workers are highly susceptible to the development of respiratory disorders and, in particular, silicosis. Notwithstanding the seriousness of the harm caused to the workers in the mineral wool manufacturing plants, extensive and irreparable harm is caused to the area surrounding the plant. Enormous amounts of mineral dust are emitted from the mineral wool factories, thereby polluting the surrounding atmosphere in a large radius from the plant. The residents of the area are subjected to respiratory disorders and to various dermatological problems caused by the irritating effect of the dust.

Furthermore, the dust is pervasive and difficult to remove by ordinary sweeping or dusting methods. But, even worse, the dust has an extremely harmful effect on the other plant and animal life within a wide radius of the factory. Thus, a great ecological imbalance can result. Contributing to the ecological imbalance is the fact that the dust settles into, and pollutes, natural waterways and acts as a screen for useful solar rays. In effect, the dust prevents the useful interchange of heat and other solar radiation thereby changing, still more, the ecological conditions.

In an attempt to cure the problem of dusting caused by the manufacture of mineral wool, mineral oil has been extensively used. Mineral wool has long been treated with mineral oil for the purpose of settling the dust. Mineral oil has proven itself to be satisfactory for dust settling, but when a sufficient amount of the oil (say from 4 to 15% by weight) is used, the highly combustible character of the mineral oil renders the product itself combustible in the sense that the oil burns on the mineral wool fibers without, of course, actually burning the fibers.

Due to the highly combustible characteristic of mineral oil it has been responsible for numerous costly and dangerous fires in the production and in the shipment of the treated wool. This fire hazard is chiefly due to the fact that lumps of incandescent coke and lava are often projected from the furnace or cupola into the settling chamber wherein the wool is deposited and these hot lumps vaporize the oil and often ignite it, sometimes causing dangerous explosions.

More important, however, is the fact that while mineral oil settles the dust, the oil itself causes vast amounts of smoke emissions from the plant. When the oil contacts the hot mineral, a portion of the same decomposes to sulfur and nitrogen-containing compounds which vaporize and are emitted from the stacks with unburned hydrocarbons. The resulting smoke pollutes the air, water and vegetation for miles around the plant.

Removal from Air

An improved process developed by *J.H. Tarazi; U.S. Patent 3,861,895; January 21, 1975; assigned to Arthur C. Withrow Company* is one in which mineral wool fibers still hot from being formed by high speed centrifugation are coated with a composition comprising 1 to 30% of a nonionic surfactant and 70 to 99% of a polyalkylene glycol having a viscosity in the range of 35 to 180,000 cs at 100°F. As a result, smoke and dust is reduced, the fibers are annealed and a wetting action is imparted. The preferred surfactants are alkyl phenoxy polyethoxy ethanols.

MINING EFFLUENTS

Removal from Air

A device developed by *T.F. Gundlach and A.L. Hawthorne; U.S. Patent 3,792,568; February 19, 1974; assigned to J.M.J. Industries, Inc.* is an air scrubber that can be used in combination with a continuous mining machine for the removal of pollutants including gases from adjacent the mine face.

The air scrubber includes a duct located adjacent to the mine face for removing the polluted air at the face, a pollutant-removal mechanism operatively connected to the duct for separating the pollutants from the air moving through the duct, and a second duct operatively connected to the pollutant-removing mechanism for directing the resultant clean air back toward the mine face.

The air scrubber has a revolving rotor mounted in a rotor housing, the rotor including a bottom plate, a top plate spaced from the bottom plate to provide a radial passage therebetween, an entrance communicating with the rotor passage, a plurality of substantially radially extending fan blades, a plurality of baffles extending into the rotor passage against which the pollutant-laden liquid impinges, the rotor passage being open peripherally of the rotor for the discharge of clean air and means for permitting discharge of the pollutant-laden liquid through the bottom rotor plate. A spray means sprays the polluted air upon entry into the rotor or at any point in the rotor passage.

Removal from Water

Acid mine water wastes create serious pollution problems in many areas of the country. Abandoned coal mines, as well as active mining operations, continuously leach large quantities of iron and sulfur containing pollutants into adjoining streams, lakes and rivers. Sulfur bacteria feeding on the mine effluents, and/or other oxidizing chemical reactions, create strongly acidic water (as low as pH 2) which kills most aquatic life and renders the water useless for many human needs.

Several methods for correcting this pollution problem have been devised. The most common technique requires the use of lime or limestone as a neutralizing media. This technique creates a serious sludge problem in that calcium sulfate, as well as iron hydroxide, is precipitated. Other techniques of purifying acid mine water such as reverse osmosis, electrodialysis, flash distillation and ion exchange have been tried with varying degrees of success, but better methods of treating acid mine water remain needed.

A process developed by *R.W. Treharne and D.E. Wright; U.S. Patent 3,823,081; July 9, 1974; assigned to Kettering Scientific Research, Inc.* involves electrolytically converting acid mine water to potable drinking water having a neutral pH and a negligible iron content. The electrolytic cell used in the process has a cathode compartment wherein the pH of the acid mine water is driven basic and an iron hydroxide precipitate is formed. A sand barrier separates the cathode compartment from the anode compartment.

In the anode compartment the pH is driven more acidic as sulfuric acid is concentrated. Electrolytic hydrogen evolved in the cathode compartment and electrolytic oxygen evolved in the anode compartment are possible by-products, as are the sulfuric acid produced and iron from the iron precipitate produced. Alternatively, the evolved hydrogen and oxygen may be used within the electrolytic system to increase the efficiency thereof. Figure 87 is a block flow diagram showing the essential features of this process.

FIGURE 87: ACID MINE WATER TREATMENT PROCESS

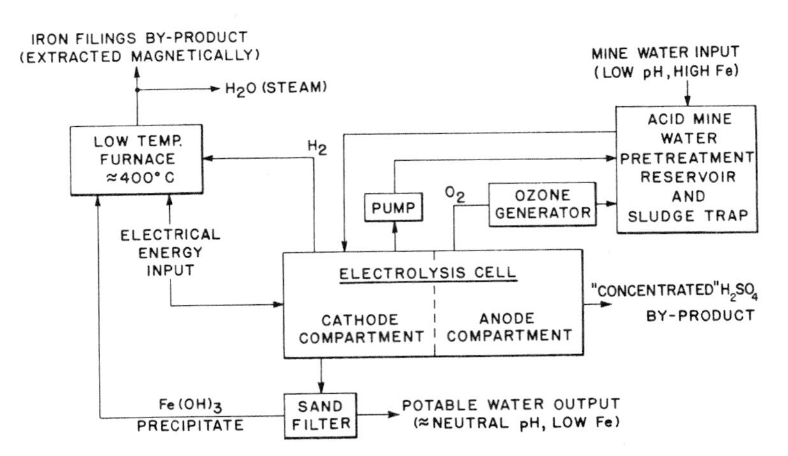

Source: U.S. Patent 3,823,081

A process developed by *R.D. Hill, R.C. Wilmoth and R.B. Scott; U.S. Patent 3,795,609; March 5, 1974; assigned to U.S. Environmental Protection Agency* involves the treating of mineral matter contaminated drain waters, particularly acid mine drainage, and comprises an integrated system. The contaminated water is optionally pretreated to make it more suitable for reverse osmosis separation, then separated by reverse osmosis into a purified water stream containing at least 90% of the feed water, the balance being a brine stream heavily concentrated in the mineral contaminants. Thereafter the brine stream is chemically treated as, for example, by neutralization to produce a sludge product and a recycle brine stream, the recycle brine stream being returned to the reverse osmosis treatment. Desirably the ultimate products of the process are purified water and a relatively small volume of inert sludge, which may be landfilled.

A process developed by *E.A. Pelczerski and J.A. Karnavas; U.S. Patent 3,717,703; February 20, 1973; assigned to Black, Sivalls & Bryson, Inc.* is a process for the purification of mine water whereby potable water is produced and elemental sulfur and iron are recovered as salable products. Solids dissolved and entrained in the mine water are separated therefrom and purified water is withdrawn. The solids are dried and introduced to a molten iron bath containing carbon wherein sulfates contained in the solids are reduced to sulfides, and iron contained in the solids is accumulated. The resultant solids containing sulfides are withdrawn from the iron bath and the sulfides are converted to elemental sulfur.

Fuel for providing heat to the molten iron bath may be high sulfur content coal normally unsuitable as fuel because of the air polluting properties of resultant products of combustion. Waste from the process consists of slag which when deposited in a refuse pile will set up into a hard stable product.

See Coal Mining Effluents (U.S. Patent 3,896,039)
See Iron Oxides (U.S. Patent 3,537,966)

MOLYBDENUM

See Propylene Oxide Process Effluents (U.S. Patent 3,887,361)

NAPHTHOQUINONE

Removal from Air

The organic and entrained substances present in the waste gas from phthalic anhydride manufacture are mainly small quantities of phthalic anhydride, maleic anhydride, benzoic acid, and 1,4-naphthoquinone. Because of the strong corrosive properties of these substances, it is not possible to release into the atmosphere the waste gases of the naphthalene oxidation without previously carefully purifying the same.

According to the prior art, various processes have been recommended for purifying the waste gases. For example, the waste gases have been burned catalytically or thermally directly after they leave the apparatus used for condensing the phthalic anhydride. During such an oxidation process, all organic substances present in the waste gases are eliminated. Such methods are uneconomical and require a high energy expenditure for the necessary heating up of the waste gases.

Depending upon the type of heating means employed in the direct oxidation process, a considerable additional charging of the atmosphere with sulfur dioxide often cannot be avoided. A further disadvantage of burning up the waste gases is that all the potentially valuable reaction products still present in the waste gas are destroyed. This makes it impossible to recover part of the considerable expenditure for this purification process by recovering these reaction products.

A water washing step is often employed for the waste gas purification. However, it is obvious that the water washing method only prevents the contamination of the atmosphere, while the drain water of the washing step is loaded with impurities. As a result, an indispensable second step of this method is the purification of the drain water.

A particular and inherent disadvantage of purifying the waste gases of the naphthalene oxidation by water washing resides in the poor water-solubility of the 1,4-naphthoquinone present in the waste gases in varying amounts. When the water is sprayed into the waste gas, the 1,4-naphthoquinone is eliminated therefrom in the form of a solid, and not in dissolved form.

As a result, not only extensive clogging of pipelines and apparatus occurs, but an imbalance is created in the apparatus and there is danger of destroying or damaging washing devices having rapidly moving parts incorporated therein. Substances which under normal circumstances are quite soluble in water, such as for example, the dicarboxylic acids, are partially prevented from being dissolved by the presence of solid 1,4-naphthoquinone.

A process developed by *G. Hoffman et al; U.S. Patent 3,370,400; February 27, 1968; assigned to Chemische Werke Huls AG, Germany* involves the purification of waste gases resulting from the manufacture of phthalic anhydride by the gas phase oxidation of naphthalene and mixtures of naphthalene and o-xylene. The process is based on the discovery that the waste gases of the gas phase oxidation of naphthalene and mixtures of naphthalene with o-xylene can be treated with concomitant purification by washing the waste gases with an aqueous maleic acid solution, extracting the used washing solution with an organic solvent which is water-immiscible, and subsequently concentrating the residual washing solution by evaporation.

By means of the above described operations, the 1,4-naphthoquinone is easily removed from the wastewater to a degree below the analytical determination of about 5 mg/kg wastewater. From the wastewater freed from the naphthoquinone by the extraction step, all of the acids are recovered, for example by evaporation.

See Maleic Anhydride Process Effluents (U.S. Patent 3,793,808)

NATURAL GAS PROCESSING EFFLUENTS

See Hydrogen Sulfide (U.S. Patent 3,864,460)

NEOPRENE PRODUCTION PROCESS EFFLUENTS

Neoprene, or polychloroprene, is conventionally polymerized in emulsion polymerization utilizing a combination of emulsifying agents. Conventionally, emulsifying agents are such as rosin acid salts and various secondary emulsifiers. Some of the emulsifying agents may remain in the polymer but certain water-soluble emulsifiers are substantially removed before final isolation and processing of the polymer.

These emulsifiers that are removed are sometimes called secondary emulsifiers. The polymer may be recovered by coagulation of the latexes and thereafter the water-soluble emulsifiers removed before final milling. The water-soluble emulsifiers are removed for example, by washing the polymer with warm water on a wash belt.

This washing may be assisted by extraction by use of solvents. Normally, the washing requires the use of large volumes of water and if the emulsifying agents are biodegradable this water may be processed by biooxidative degradation; however, if the wash water contains nonbiodegradable emulsifying agents this presents a major problem because of the restrictions on the releasing of significant amounts of organic compounds into streams and rivers.

Removal from Water

A process developed by *R.D. Pruessner; U.S. Patent 3,778,367; December 11, 1973; assigned to Petro-Tex Chemical Corp.* involves separation of emulsifiers which are salts of the condensation product of naphthalene sulfonic acid and formaldehyde from process water by contacting the process water with an amine having at least two amine groups separated by an aliphatic radical having no more than 5 carbon atoms such as polyethyleneamine with the separation being made at a pH of less than 7.0 and typically 4.0.

A process developed by *P. Merchant, Jr.; U.S. Patent 3,890,227; June 17, 1975; assigned to Petro-Tex Chemical Corp.* is a similar one in which the process water and/or the amine is reduced to a pH of 7.0 or less prior to contacting.

See Chloroprene Manufacturing Effluents (U.S. Patent 3,952,088)

NICKEL

Removal from Water

A process developed by *G. Hirs et al; U.S. Patent 3,630,892; December 28, 1971; assigned to Hydromation Filter Company* involves removing nickel from plating rinse water or the like by upwardly adjusting the pH of an aqueous nickel solution to precipitate the nickel as a colloidal suspension, passing the precipitate suspension through a deepbed filter having a granular, synthetic organic medium therein to deposit the nickel precipitate on the filter medium, backwashing the filter medium to remove the deposited nickel precipitate therefrom and then filtering the backwash liquid through a conventional filtration mechanism.

NICKEL SMELTER EFFLUENTS

A large proportion of the world's nickel resources are contained in oxidic and siliceous ores such as limonite, garnierite and serpentine. The following table illustrates the normal range of composition of each of these types of ores.

	Serpentine (garnierite) Ore percent by weight	Limonite Ore percent by weight
Ni	1.0 - 4.0	0.10 - 3.0
Co	0.05 - 0.08	0.05 - 0.25
Fe	8 - 18	35.0 - 60.0
Cr	0.8 - 2	1 - 3
MgO	20 - 38	0.2 - 6.0
Al_2O_3	1 - 6	0.4 - 10
SiO_2	40 - 35	1.3 - 6
CaO	0.1 - 2	0.06 - 0.1
MnO	0.1 - 1	0.3 - 2.5

The recovery of nickel and cobalt from these materials presents serious problems to the metallurgical industry. The problems result primarily from the fact that ores of this type contain relatively small amounts of nickel and cobalt. It is necessary to treat large quantities of ore for the recovery of the relatively small amounts of contained values. Conventional, relatively inexpensive ore beneficiation methods are not suitable for concentration of the nickel and cobalt values.

One process commercially employed for the recovery of nickel and cobalt values from lateritic and garnieritic ores involves comminuting the ore to substantially 100% minus 65 mesh standard Tyler screen then roasting the ore in a multiple hearth furnace to effect substantially complete reduction of contained nickel and cobalt values to a metallic form with a minimum reduction of iron to a metallic state.

Complete reduction of the ore requires the presence of a large volume of hot reducing gases within the furnace. These gases have a large carrying capacity for ore particles and as a result, many of the particles fed into the top of the furnace, instead of travelling downwardly, are picked up immediately by the rising stream of gases and exit from the furnace therewith.

The recovery of nickel from the ore fed to the furnace will, of course, be adversely affected by large dust losses. Moreover, the reduction roasting operation is greatly complicated by the presence of substantial amounts of dust particles in the gases exiting from the furnace. To minimize air pollution, and to permit reuse of the unspent reducing gases discharged from the furnace, dust collection apparatus must be provided to separate the dust from the gases.

Removal from Air

Various studies have been made of methods of dealing with dust separated from the furnace exit gases. The dust may be simply discarded. If the dust is discarded, not only will nickel and cobalt values in the dust particles be lost but dust collection and disposal facilities must be provided.

Such facilities contribute significantly to the overall cost of the reduction roasting operation. Alternatively, the dust may be returned to the furnace. According to one known operation, the dust is reinjected into the uppermost hearth of the furnace. This practice tends to set up a recirculating load of dust thereby increasing the quantity of dust in the exit gases and decreasing the capacity of the roaster.

A process developed by *J.W. Gulyas and P.T. O'Kane; U.S. Patent 3,775,095; November 27, 1973; assigned to Sherritt Gordon Mines, Ltd., Canada* involves collecting dust particles exhausted with the reducing gases from the furnace and reinjecting these particles into the furnace whereby the amount of reduced ore discharged from the furnace is substantially increased and hence the amount of nickel and cobalt recovered from the feed ore is substantially increased. Figure 88 shows a suitable form of apparatus for the conduct of such a process.

The ore is first dried by a rotary dryer **10** to reduce the moisture content of the ore to below about 5% by weight. The drying step is important in order to enable control of the water vapor content of the furnace atmosphere in the subsequent reduction step. The amount of water vapor present during the reduction step is of importance and must be controlled in order to ensure minimum reduction of iron values in the ore to soluble form.

The dry ore is comminuted to substantially 85% -200 mesh and 98% -65 mesh standard Tyler screen in ball mill **12**. The ground ore may also be pelletized by conventional pelletizing procedures to increase the bulk density and to decrease the amount of dust which must be treated.

The ore is charged into furnace **14**. The illustrated furnace is a Herrshoff multiple hearth furnace which consists of a cylindrical shell **16** within which 16 circular hearths **20–1** to **20–16** are disposed in vertically spaced decks. A vertical central rotating shaft **22** running through the center of the brick arches carries rabble arms **24** having a plurality of downwardly extending teeth. Four rabble arms are disposed about each hearth and upon rotation of the central shaft, the teeth serve to rake the ore across the hearth.

FIGURE 88: REMOVAL OF DUST PARTICLES FROM A DUST LADEN STREAM OF GASES EXITING FROM A MULTIPLE HEARTH FURNACE AND REINJECTION OF DUST PARTICLES INTO THE FURNACE

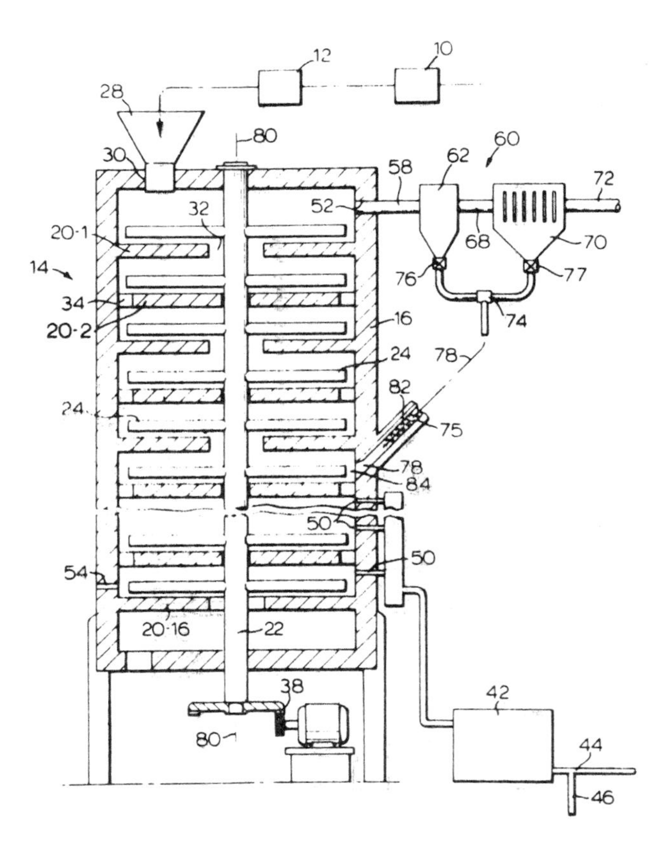

Source: U.S. Patent 3,775,095

A hopper **28** is secured to the furnace at the top. Ore is fed to the hopper and passes downwardly through drop hole **30** to the periphery of hearth **20–1**. The charge is raked inwardly along the hearth to drop hole **32** where it falls downwardly to hearth **20–2**. The ore is raked outwardly along the hearth and falls through drop hole **34** and so on down.

Shaft **22** is rotated by a bevel gear **38** at the bottom, the rate of rotation of the shaft is determined largely by the required capacity of the furnace or the depth of ore on the hearths.

Combustion chamber **42** is fired by fuel which passes through conduit **44** and is vaporized and mixed with air which passes through conduit **46**. Hot gases evolved from the combustion of the fuel are conducted to the furnace through a plurality of conduits **50** vertically spaced about the lower eight hearths of the furnace. Upon entering the furnace, the gases travel upwardly and heat the downwardly moving ore. The gas exits through port **52**.

Fuel may be natural gas or fuel oil such as Bunker C oil. The amount of air mixed with the fuel is restricted so that the fuel burns incompletely and gaseous reductants including hydrogen and carbon monoxide are evolved. The content of reducing gases in the combustion gases may be controlled by adjustment of the amount of oxygen supplied to the combustion chamber.

The quantity of reducing gases introduced into the furnace may be augmented by an additional source of reducing gases produced by steam reforming of hydrocarbons with removal of carbon dioxide. The additional reducing gases may be introduced through conduits **50** or they may be injected through a port **54** directly into the furnace.

Gas which passes upwardly through the furnace exits through conduit **58** which extends laterally from the upper zone of the furnace. The gas contains combustion gases, water vapor and unspent reducing gases. Dust particles carried upwardly with the reducing gases also exit through conduit **58**.

The ore particles in the gas stream are recovered in dust collection apparatus generally designated by the numeral **60**. The apparatus includes a cyclone separator **62** into which the dust laden gases pass directly from conduit **58**. The coarse particles of generally +325 mesh standard Tyler screen are removed from the gas stream and the gases from which the coarse particles have been removed pass through conduit **68** into an electrostatic precipitator **70** in which fine particles of -325 mesh to about 5 microns are removed. The gas from precipitator **70** exits through line **72** to apparatus for recovering unspent reducing gases (not illustrated) then to an exhaust stack.

Particles removed from the gas stream by the cyclone separator and electrostatic precipitator fall through T-joint **74** to a chute **75**. Valves **76** and **77** beneath separator **62** and precipitator **70** respectively are provided so that the quantity of material passing through the chute may be controlled.

The longitudinal axis **78–78** of the chute is inclined at an angle of approximately 45° to the longitudinal axis **80–80** of the shaft **22**. A screw conveyor **82** mounted within the chute moves dust through the chute at a predetermined rate and injects the dust into the furnace through inlet **84**. The ore particles discharged into the furnace from the screw conveyor are in the form of a continuous plug.

NITRATION PROCESS EFFLUENTS

Removal from Air

A process developed by *M.M. Fooladi; U.S. Patent 3,972,824; August 3, 1976; assigned to Vicksbury Chemical Company* is one in which nitrogen oxides mixed with air or other gases are cleaned from such air or other gases by treatment thereof with a mixture composed essentially of chloroform, methyl alcohol, ethyl alcohol, and n-hexane. The composition of the mixture may include isopropyl alcohol in place of methyl alcohol, or it may include in addition to the chloroform and the n-hexane, a mixture of acetone, isopropyl alcohol, and methyl alcohol. The following is one specific example of the conduct of this process.

A stream of nitrogen oxides emitted from the nitration of ortho-sec butyl phenol was introduced into a glass vessel equipped with a cold condenser, an agitator and a thermometer, and containing 30% chloroform, 30% methyl alcohol, 30% ethyl alcohol, and 10% n-hexane (all parts being by volume), the stream being continued to be introduced for a period of 5 hours.

The temperature of the scrubber was maintained at 20° to 30°C during the introduction of the nitrogen oxide gas. A sample from the vent system at the top of the scrubber was taken for analysis every half hour. Infrared analysis, colorimetric analysis, and Bohm procedures all failed to show the presence of nitrogen oxides.

NITRIC ACID

See Adipic Acid Process Effluents (U.S. Patent 3,673,068)

NITRIC ACID PLANT TAIL GASES

It is known that the tail gas from nitric acid plants contains both N_2 and approximately 2.5 to 3% O_2 and, in addition, 1,000 to 2,500 ppm nitric oxides in terms of NO, the latter figures depending on the type of plant and on the mode of operation. Referring to big plants, the emission in terms of HNO_3 may be as high as several 100 kg/hr. This emission is a serious problem from the standpoint of air quality standards. Moreover, it entails naturally a certain reduction of the rated nitric acid output.

Removal from Air

A number of processes have been developed for reducing the nitric oxide concentrations of tail gases. Among these processes, alkaline absorption and catalytic incineration on platinum catalysts may be cited by way of examples.

A process developed by *G. von Semel and E. Schibilla; U.S. Patent 3,809,744; May 7, 1974; assigned to Friedrich Uhde GmbH, Germany* is a process for the removal of nitric oxides from tail gases such as are contained in large quantities in nitric acid plants. The tail gases are scrubbed with a nitric vanadium (V) solution, and the effluent solution is regenerated by boiling point and using air as stripping fluid. The regenerated vanadium solution is cooled and then used as scrubbing fluid.

The particular advantages of this process are that the nitric oxide concentration of the tail gases is substantially reduced. The process does not yield any reaction products that might pollute the environment, such as air and wastewater, or which must be reprocessed in elaborate process steps. The nitric oxides absorbed by the vanadium solution are recovered during the regeneration step in the form of concentrated NO_2. They may be returning to a nitric acid plant where they contribute to raising the HNO_3 yield or may be utilized for other purposes.

A process developed by *D.K. Fleming; U.S. Patent 3,897,539; July 29, 1975; assigned to Institute of Gas Technology* involves abating nitrogen oxide in tail gas from nitric acid production. The system includes a rotary wheel reactor which has a preheat zone to raise the temperature of the inlet tail gas stream to ignition temperature, and a reaction zone wherein the nitrogen oxides are reduced, and where heat of reaction is removed as it is generated for preheating the tail gas stream, and to also prevent heat buildup which would be damaging to the catalysts in the reactor and to ancillary material equipment.

Such a system is shown in Figure 89 as applied to a nitric acid plant where air moves through line **10** to an air heater **12**. Such a heater or preheater previously was used to also preheat the tail gas stream, but now such heat may be advantageously used in its totality for other purposes.

The heated air moves into a mixer **14** which also receives ammonia through line **16**. The mixture is then moved into an ammonia burner **18**. The stream moves to a separator **20**, and then into an absorption tower **22**. Nitric acid leaves the absorption tower through line **24**. Boiler feed water is introduced to the heater through line **28** and steam export occurs through line **30**. Secondary air moves to the absorption tower through line **32** with the help of air compressor **34**. Process water enters the absorption tower through line **36**.

The rotary reactor assembly is provided in such a representative plant as indicated generally at **40** in the schematic. The tail gas stream leaves tower **22** through line **26**, the stream having from 2 to 4% oxygen. The stream enters one face of refractory wheel **42**, hereinafter referred to as the obverse face **41** of the refractory wheel and passes through a preheat zone.

The tail gas stream is preheated in this preheat zone to ignition temperatures from about 800° to about 1000°F and then leaves the opposite or the reverse face **43** of the wheel into a collecting line **44** which empties into mixing chamber **45**. The reducing gas stream enters the mixer chamber through line **46** from a source not shown.

The preheated combined gas stream then leaves the mixer chamber by way of line **47**, and enters the reverse face **43** of the refractory wheel and passes through a reaction zone thereof. This combined gas stream could as well be passed through the reaction zone from the obverse face of the wheel **42**, however, from the description below it will be seen that it is far more convenient to form the ducts to route this stream to enter the reverse face of the wheel. It is understood, however, that both countercurrent (obverse face) and cocurrent (reverse face) combined gas streams may be used.

The catalytic combustion in the reaction zone abates the nitric oxide content at reaction temperatures sufficient to effect desired levels of abatement. The high exothermic heat of reaction is, however, constantly removed as it is

FIGURE 89: TAIL GAS NITROGEN OXIDE ABATEMENT PROCESS

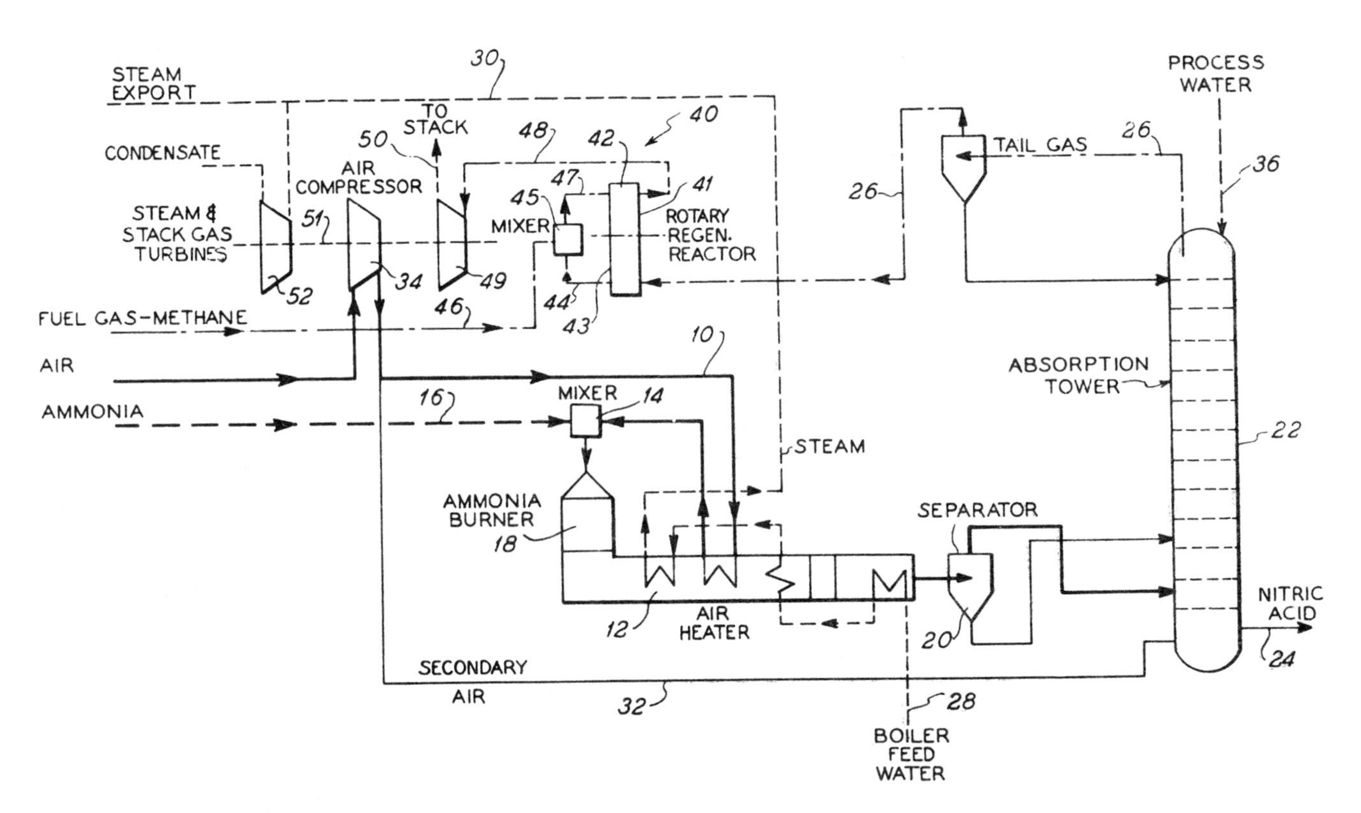

Source: U.S. Patent 3,897,539

generated by the rotating wheel **42**. The waste exhaust stream leaves the obverse face **41** of the wheel through the outlet line **48** which then exhausts such stream in any desired manner, including an expander **49** connected to a stack exhaust line **50**. A turbine **52**, as well as expander **49** and drive compressor **34**, are mounted on common shaft **51**.

A process developed by *M.F. Collins and R. Michalek; U.S. Patent 3,808,323; April 30, 1974; assigned to Engelhard Minerals & Chemicals Corp.* is one in which waste gases containing nitrogen oxides and oxygen are purified in a catalytic reactor. A portion of the effluent from the catalytic reactor is recycled back to the inlet by means of an ejector to effect controlled combustion in the reactor.

NITRIDING PROCESS EFFLUENTS

Removal from Water

A process developed by *E. Mohr; U.S. Patent 3,966,508; June 29, 1976* relates to the treatment of effluent liquid from hardening baths, and more particularly to the treatment of used baths from nitriding processes in which hardened workpieces have been rinsed.

The effluent liquid is introduced in an evaporating boiler for such a time that the watery liquid, or brine, changes, without visible transition, from brine to a salt melt. The salt melt is then heated to a temperature of between about 140° to 350°C. The minor proportion of cyanide compounds are oxidized in this process in the nitrate-nitrite mixture. The temperature range causes separation of low-melting and high-melting salts, and the low-melting nitrite-nitrate mixture can be drawn off and reused.

The evaporation-melting process above has the advantage that it is no longer necessary to introduce detoxified effluent from hardening installations into water treatment plants, or into public waterways; these effluents still contain salts. Additionally, the danger is avoided that breakdown in the detoxification systems might occur so that effluent liquid is actually introduced into water treatment plants, or into public waters, since the boiling off and recovery process no longer results in water being directly discharged after the treatment process.

An additional advantage is obtained in that the salts in the effluent waters can be largely recovered. The heat necessary for evaporation can usually be derived from the exhaust gases of the treatment furnaces for the workpieces themselves. The efficiency of the known process is enhanced since the salts used in the hardening processes are highly water-soluble, so that the rinse waters being used, and discharged, usually already have a high concentration of salts. Since no detoxification chemicals need be used, substantial cost savings can result. A suitable form of apparatus for use in the process is shown in Figure 90.

Rinse liquid, or other effluent to be discarded from hardening processes, generally illustrated at **2**, is contained in a vessel **1**, and conducted by means of piping **3** to an evaporating boiler vessel **4**. The effluent **2**, pumped through pipe **3**, reaches the evaporation boiler vessel **4** in which a melt of nitrite-nitrate compounds is contained. The melt **5** in vessel **4** is maintained at a temperature of about over 400°C, preferably about 500°C. The boiler **4** is located in an evaporation furnace **6**.

FIGURE 90: EFFLUENT TREATMENT INSTALLATION

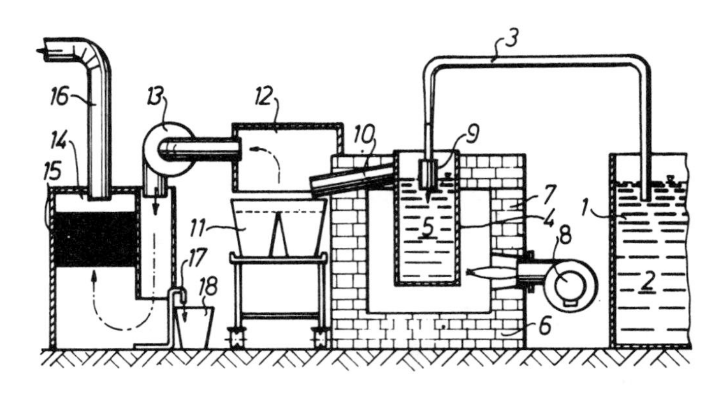

Source: U.S. Patent 3,966,508

An opening **7** in the furnace permits introduction of heat energy, for example by means of a burner **8**, which may be an oil burner, a gas burner, or other heat source. The effluent liquid **2**, conducted through pipe **3**, is finely dispersed by a terminal pipe **9** dipped into the melt **5**.

Pipe **9** is filled with balled, preferably loosely balled steel wool, in a cartridge, or bulb end connected to the pipe **3**, and formed with a sieve, or mesh bottom. The aqueous content of the liquid conducted through pipe **3** immediately evaporates upon being introduced into the melt. The salt components within the effluent liquid, however, merge into, or are absorbed in the melt.

The minor proportions of poisonous cyanides contained in the effluent liquid are oxidized. The liquid level of the salt melt **5** rises as additional effluent, containing salts, is introduced. Eventually, the liquid level will reach an overflow level, or edge of an overflow duct **10**, so that overflowing liquid can be conducted into a catch basin, or trough **11**, in which the salt solidifies. The salts can then be reused, and recycled.

A vapor catch hood **12** is located above the overflow **10**. A ventilator **13** removes steam and vapor from the overflow **10** by means of suction. The vapor, or steam removed by the ventilator is introduced, under overpressure, into a filter housing **14**, to pass through a filter **15**, from which the vapor, or steam can be conducted to ambient air through a duct **16**. The only substance being discharged into ambient air, then, is water vapor.

Any remaining salt particles are trapped in the filter **15**, so that the steam and vapor, being passed through the filter **15** are completely cleansed thereby. Remaining salt remnants, caught in the filter **15**, may drip off the filter together with condensate, to be caught in a catch pan, and conducted by means of a draw-off tube **17** into a condensate vessel **18**.

Necessary pumps, or gravity differentials have been omitted from the showing for simplicity. The contents of the vessel **18** may be reintroduced into pipe **3**, and then into melt **5**, by means of a separate introduction inlet, not shown, or otherwise treated.

If the effluent liquid were directly applied to the surface of the salt melt **5** which, as noted, is heated preferably to about 500°C, then immediately evaporation of the effluent liquid would result under substantial bubble and spray conditions. A substantial portion of the salts within the effluent water would be dispersed into a fine spray thereby, together with water vapor.

This undesirable result is avoided by conducting the effluent liquid directly into the melt. The effluent liquid is dispersed upon introduction into the melt by the special terminal bulb **9**, located at least in part within the melt and containing balled steel wool, so that the liquid from pipe **3** is finely dispersed.

NITRILES

Industrial wastewaters may contain organic cyanide impurities, especially nitriles. Such effluents are found in the discharge from chemical synthesis plants, such as the factories where certain plastic materials or synthetic fibers are manufactured, which discharge waters containing organic nitriles. Among these nitriles, the main pollution agent is acrylonitrile.

Removal from Water

The conventional processes for the purification of these effluents operate particularly by a hydrolysis of the nitrile into amide and into acid by heating at a temperature close to boiling point. In numerous cases, the hydrolysis is accompanied by a more or less intense volatilization of the acrylonitrile, which is discharged into the atmosphere.

Such a process has disadvantages. A complete detoxication necessitates a long treatment time and consequently a considerable consumption of thermal units for heating all the wastewaters to a temperature close to boiling point. Furthermore, the known process causes the secondary pollution of the atmosphere.

A process developed by *J.-P. Zumbrunn; U.S. Patent 3,715,309; February 6, 1973; assigned to L'Air Liquide, Societe Anonyme pour l'Etude et l'Exploitation des Procedes Georges Claude, France* involves rapid degradation of the nitriles into amides and then into salts of organic acids by treatment with hydrogen peroxide in alkali medium.

This process, which has the advantage of saturating the effluents with oxygen, permits the total purification to a residual content lower than 0.5 ppm of any effluent coming from chemical synthesis plants, such as the factories for the production of plastic materials or synthetic fibers, and containing organic nitriles.

A process developed by *Y. Fuji and T. Oshimi; U.S. Patent 3,756,947; September 4, 1973; assigned to Sumitomo Shipbuilding & Machinery Co., Ltd., Japan* is one in which a wastewater effluent containing nitriles and cyanides is treated by passing through an acclimated, activated sludge containing at least one of the microorganisms capable of degrading nitriles and cyanides selected from the genera Alcaligenes and Achromobacter, for example, *Alcaligenes viscolactis* ATCC 21698 and *Achromobacter nitriloclastes* ATCC 21697 thereby to purify the wastewater effluent. The wastewater effluent containing 10 to 50 ppm of cyanide and 1,000 to 2,500 ppm of COD (potassium dichromate method) can be purified with a high efficiency.

See Terephthalic Acid Process Effluent (U.S. Patent 3,836,461)
See Cyanides (U.S. Patent 3,970,554)

NITRITES

Removal from Water

A process developed by *J. Muller et al; U.S. Patent 3,502,576; March 24, 1970; assigned to Deutsche Gold- und Silber-Scheideanstalt* is one in which the detoxification of cyanide and nitrite containing aqueous solutions is accomplished by mixing the cyanide and nitrite solution and reacting the cyanide with the nitrite at a pH of not over 5 in the presence of a contact catalyst.

The cyanides and nitrites react with each other with the result that the nitrites oxidize the cyanides to cyanates or other materials and at the same time they themselves are reduced to nonpoisonous substances, especially nitrogen.

NITRO COMPOUNDS

See TNT Explosive Wastes (U.S. Patent 3,916,805)
See TNT Process Wastes (U.S. Patent 3,954,381).

NITROANILINES

A number of highly colored nitrophenols and nitroanilines are becoming increasingly important commercial products. Of special interest are those products which are commercially available as herbicides. In the manufacture of these nitrophenols and nitroanilines waste streams containing significant quantities of these materials are generated.

Because of their toxicity to plant and aquatic life, it is essential that these nitrophenol and nitroaniline compounds be degraded before the waste streams are released to nature. The intense color of these compounds emphasizes the desirability of their degradation and also acts as a built-in indicator for their presence.

Removal from Water

These nitrophenol and nitroaniline compounds are not satisfactorily degraded by conventional disposal systems. Thus, if waste streams containing such compounds are treated in the conventional manner, the effluent from such treatment contains unacceptably high concentrations of the nitro compound.

A process developed by *R.H.L. Howe; U.S. Patent 3,458,435; July 29, 1969; assigned to Eli Lilly & Company* is based on the discovery that a waste stream containing color-producing nitrophenols or nitroanilines can be disposed of in conventional disposal systems if such stream is first subjected to a pretreatment comprising acidifying to a pH of less than about three, adding an adsorbent material to take up colored components, adding a metallic oxide or hydroxide to adjust the pH to more than about five and also to form a precipitate, and separating the mixture into an effluent and a sludge or foam (scum).

The effluent and sludge (or foam) can then be separately treated by conventional disposal means without interference from the original nitrophenol and nitroaniline compounds. Such a process results in treated wastes free of harmful amounts of nitrophenols and nitroanilines.

NITROCELLULOSE PROCESS WASTES

Cellulose nitrate, hereinafter referred to as "nitrocellulose" is produced in large quantities for use in the manufacture of munitions, synthetic finishes and a wide variety of other products, by the reaction of pure cellulose with an excess of nitric acid and a dehydrating agent such as sulfuric acid.

In the manufacture of nitrocellulose, the freshly nitrated cellulose material is thoroughly washed with substantial amounts of water to remove excess acid present and this wash water carries, in addition to unreacted acid, a suspension of small particles or insoluble nitrocellulose referred to as nitrocellulose fines.

While the wastewater is normally neutralized and treated before flowing into the receiving water, there has, heretofore, been no practical way to eliminate the nitrocellulose fines therefrom, which fines constitute a serious waterway pollutant. Nitrocellulose fines remain in suspension in moving waste streams and receiving waters and this particulate suspension creates a milky appearance in the water, a discoloration that is objectionable from an aesthetic standpoint and one which prevents the use of such waters for domestic consumption or for many industrial uses.

Another problem arising from the discharge of nitrocellulose into receiving waters is the likelihood of concentration of these nitrocellulose fines in pools located along stream and river beds which concentrated deposits could suddenly explode should they become desiccated or come into contact with strong oxidizing agents.

If filtration were to be employed downstream of such a waste discharge for any purpose, the fines would collect on the filter device to an extent sufficient to constitute a serious hazard.

Removal from Water

Attempts to quantitatively separate the nitrocellulose fines from the waste stream output of a producing plant by mechanical means, i.e., sedimentation, filtration and centrifugation have not proved to be fully adequate and even had such efforts been totally successful, would pose a further problem with respect to handling and disposing of the concentrated fines in an environmentally acceptable and safe manner.

Nitrocellulose, being a substituted cellulose is not subject to direct microbiological attack and is not broken down by microbes in receiving water systems. There exists a need for a method for removing and disposing of these insoluble and nonbiodegradable nitrated cellulose products from the waste streams of nitrocellulose producing plants without further reducing the quality of the environment.

A process developed by *T.M. Wendt et al; U.S. Patent 3,939,068; February 17, 1976; assigned to the Secretary of the Army* is one in which total removal of cellulose nitrate particles from a wastewater stream is accomplished by a combination of chemical and biological treatments which result in an effluent having a greatly reduced concentration of nitrogen compounds and having an acceptable BOD. The first step in the process requires that the insoluble, nonbiodegradable cellulose nitrate particles be chemically digested with alkali to produce soluble products.

Following digestion, the waste stream is supplemented with domestic raw sewage and a microbially utilizable carbon source and the supplemented waste stream is subjected first to an anaerobic microbial denitrification treatment to eliminate the oxidized forms of nitrogen and then to an aerobic microbial treatment to reduce BOD and to oxidize reduced nitrogen compounds.

The effluent therefrom is again supplemented with a microbially utilizable carbon source and subjected to a final anaerobic microbial denitrification to remove nitrates present in the wastewater resulting from nitrification processes occuring during the aerobic treatment step.

NITROGEN OXIDES

Nitrogen oxides in stack offgases may be converted in the atmosphere to noxious substances by irradiation with sunlight. These noxious substances are considered to be one of the causes of "photochemical smog" and its mechanism is now being gradually elucidated. The removal of the nitrogen oxides in the stack gases is regarded as extremely important from a standpoint of environmental pollution prevention.

Removal from Air

Heretofore, a variety of methods have been proposed for this purpose and may be grouped into three classes: (1) the oxidation of nitrogen oxides; (2) the catalytic reduction of nitrogen oxides; and (3) the catalytic degradation of nitrogen oxides. For example, concerning class (1) the nitrogen oxides are first catalytically oxidized to form nitric acid which in turn is removed by adsorption on a carrier or absorption in an alkaline solution, and concerning class (2), the stack gases are first desulfurized to be followed by a catalytic reduction with carbon monoxide, methane, hydrogen or the like.

However, catalysts to be used for these methods are likely to be deteriorated by a component, particularly sulfur dioxide, present in the tail gases and the reducing agents have a tendency to react preferentially with oxygen in the tail gases. Accordingly, these methods are often unsatisfactory. On the other hand, the method of class (3) involves the degradation of the nitrogen oxides into innocuous substances or nitrogen and oxygen, so that it is favorable in this respect, but it is disadvantageous because of a long period of time generally required for the degradation of the nitrogen oxides.

A process developed by *M Saito, S. Tani, T. Ito and S. Kasaoka; U.S. Patent 3,981,971; September 21, 1976; assigned to Kurashiki Boseki KK, Japan* is one in which the reduction of nitrogen oxides is effected by passing stack offgases containing nitrogen oxides over a heavy metal sulfide catalyst in an atmosphere where ammonia is present.

Ammonia serves as a reducing agent and the use thereof is advantageous because it is not affected by the oxygen present in the tail gases. The amount of ammonia is from about 0.67 to 4 times, preferably about 1 to 3 times, the stoichiometric amount thereof.

A heavy metal sulfide serves as a catalyst and is desired because it is not easily deteriorated by sulfur dioxide, steam, oxygen or the like, in particular sulfur dioxide. Illustrative of the catalysts are, for example, copper, manganese, nickel, iron and cobalt sulfides. These catalysts may be employed alone or in combination and with a promoter or a carrier, such as, for example, Al_2O_3.

A process developed by *S. Machi, K. Kawamura, W. Kawakami, S. Aoki, S. Hashimoto, K. Yotumoto and H. Sunaga; U.S. Patent 3,869,362; March 4, 1975; assigned to Ebara Manufacturing Company, Ltd. and Japan Atomic Energy Research Institute, Japan* involves treating industrial effluent gases with an ionizing radiation or ultraviolet light so that noxious gas pollutants, especially nitrogen oxides and sulfur dioxide may be changed into particle form or mist, thereby enabling collection of the particles or mist by conventional collecting means such as electrostatic precipitators, filters and the like.

As shown in Figure 91 the effluent gas to be treated is generated by combustion of heavy oil in the heavy oil combustion apparatus **1**. The gas is then delivered to the preliminary dust collector **3** by a blower **2**. The solid pollutants in the gas are removed from the gas in the dust collector **3**, and then the gas moves to a heat exchanger **4**, where the temperature is controlled before the gas enters a reaction chamber **5**. The process is most efficient when the NO_x/SO_2 ratio is 0.5 - 1.5. In the reaction chamber **5**, the gas is irradiated with electron beams from an accelerator **10**. As a result, the gaseous pollutants, SO_2 and NO_x, contained in the gas are converted to mist and solid particles.

FIGURE 91: PROCESS FOR REMOVING NITROGEN OXIDES FROM AIR BY IRRADIATION TREATMENT

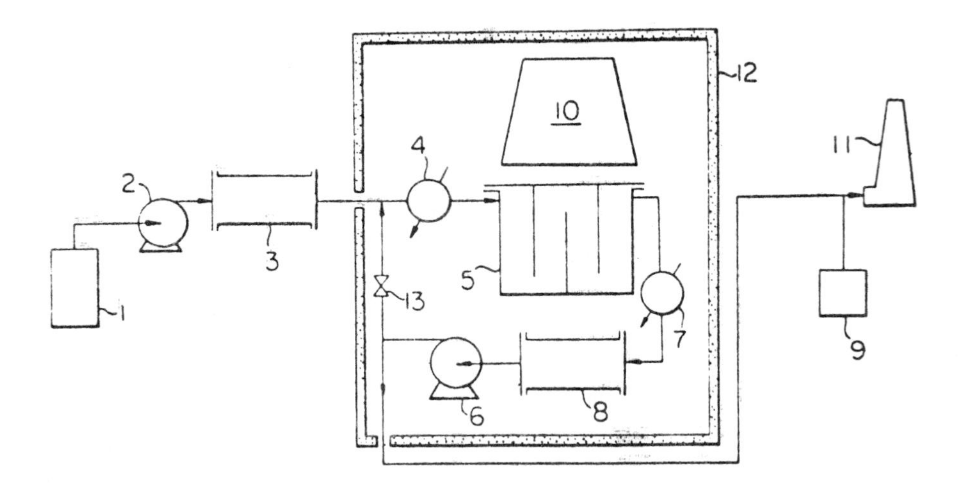

Source: U.S. Patent 3,869,362

The effluent gas containing these reaction products is then delivered by a circulation blower **7** to a collecting means **8** after passing through a heat exchanger **7** to control the temperature of the gas. In the collector **8**, the mist and solid products are removed from the gas.

Then, part of the purified gas is sent to a stack **11** for discard into the atmosphere after passing through a gas analyzer **9** for checking the SO_2 and NO_x content. The balance of the purified gas is returned to the reaction chamber **5** to be treated again by irradiation. The shielding wall to limit the radiation therein is shown as **12**.

A process developed by *Y. Kajitani, H. Ito and S. Mitsuda; U.S. Patent 3,838,193; September 24, 1974; assigned to Kawasaki Jukogyo KK, Japan* involves treating substances which normally generate nitrogen oxide upon combustion, such as ammonia, hydrogen cyanide, other gases containing nitrogen or a mixture of gases containing these materials.

The method comprises combusting these substances by using a combustible gas containing hydrogen, carbon monoxide or gaseous hydrocarbons such as city gas so as not to generate NO and NO_2 or so as to only slightly generate them. More particularly ammonia, ammonia with steam, ammonia with other nitrogen containing gases or another nitrogen containing gas are burned with secondary air and they are introduced into high temperature reducing flame produced by the combustion of a fuel gas with an amount of air below the theoretical amount of air required for complete combustion so as to reduce the amount of nitrogen oxide gas which would be generated by the combustion of ammonia or other nitrogen containing gases.

In this case, the amount of secondary air is preferably 90 to 95% of the amount of air theoretically required for the total combustion of gases after the amount of primary air supplied is subtracted. A further feature of the process is that if it has unburned components, third air is supplied to the relatively high temperature portion out of the flame to burn these components.

A process developed by *W.J. Gilbert, Jr.; U.S. Patent 3,949,057; April 6, 1976; assigned to Croll-Reynolds Company, Inc.* for removing nitrogen oxides from a gaseous mixture comprises contacting the mixture with an aqueous liquid in the presence of a knitted wire mesh packing material of stainless steel containing at least 8% Ni and having a diameter of from 0.003 to 0.015 inch.

More specifically, it is a process for removing nitrogen oxides from a waste gaseous mixture containing oxygen and up to 1% by volume of nitrogen dioxide and/or nitrogen oxide comprising adjusting the oxygen concentration of the mixture at least 10 times greater than the nitrogen oxides concentration of the mixture.

It then involves contacting the mixture with a liquid selected from the group consisting of water and dilute aqueous alkaline solutions, the liquid having a flow rate of from 10 to 60 gallons per minute per 1,000 cubic feet per minute of the gaseous mixture, in the presence of a knitted wire mesh packing material of stainless steel containing at least 8% Ni and having a diameter of from 0.003 to 0.015 inch, the packing material having from 180 to 800 square feet of surface area per cubic foot of volume and maintaining a residence time of from 3.5 to 15 seconds for the gaseous mixture to be in the presence of the packing material.

A process developed by *R.D. Reed, E.C. McGill and C.G. McConnell; U.S. Patent 3,873,671; March 25, 1975; assigned to John Zink Company* permits converting gases containing oxides of nitrogen into gases which may be safely vented to the atmosphere. The gases containing the oxides of nitrogen (NO_x) are mixed with excess combustible products obtained by burning a hydrocarbon fuel with less than its stoichiometric requirements of oxygen, the mixture thus obtained is cooled to avoid temperatures substantially above 2000°F but not below about 1200°F, the combustible material remaining after substantially all of the oxides of nitrogen have been reduced to nitrogen are oxidized, so that the resulting gas, substantially free of NO_x and carbon monoxide (CO), may be vented to the surrounding atmosphere without contamination of the environment and without smoke or other particulate matter.

See Automotive Exhaust Gases (U.S. Patent 3,718,733)
See Automotive Exhaust Gases (U.S. Patent 3,886,260)
See Carbon Black Process Effluents (U.S. Patent 3,835,622)
See Gas Turbine Engine Effluents (U.S. Patent 3,826,080)
See Gas Turbine Engine Effluents (U.S. Patent 3,842,597)
See Irrigation Pumping Engine Effluents (U.S. Patent 3,888,652)
See Nitric Acid Plant Tail Gases (U.S. Patent 3,809,744)
See Nitric Acid Plant Tail Gases (U.S. Patent 3,897,539)
See Nitric Acid Plant Tail Gases (U.S. Patent 3,808,323)
See Nitration Process Vent Gases (U.S. Patent 3,972,824)

NOISE

See Aluminum Cell Effluent (U.S. Patent 3,732,948)
See Metal Arc- & Flame-Cutting Effluents (U.S. Patent 3,851,864)

NONHALOGENATED SOLVENT WASTES

Removal from Water

Until now, the destruction of nonhalogenated organic industrial waste products occurred for the greater part in more or less conventional incinerators in which the solvent to be incinerated is mixed with air, whereinafter the mixture is ignited. A main inconvenience of this method is that solvents with a low heat of combustion (or heat value) or solvents having a relatively high water content may not be burned or incinerated satisfactorily, so that combustion becomes incomplete, causing the liberation of smokes and fumes which may be noxious even at low concentrations.

Moreover, the continued and fluent running of the incineration process itself is only moderately guaranteed. It has been tried to remedy the above-mentioned drawback by providing a kind of after-burner at the exit end of the installation whereby the flue gases are once again submitted to an oxidation process so that the chance of conversion into stable oxides is raised considerably.

Although this method has the advantage of giving more complete combustion as compared with conventional incinerators, the problem of burning liquids having a low heat of combustion still remains. Moreover, since the after-burning installation is located in downstream direction of the incinerator, the temperature of the flue gases may decrease considerably, so that more supplementary caloric energy than necessary must be supplied. Also the necessity to install the after-burner in the exhaust compartment of the installation, for example in the stack, results in greater dimensions of the latter, so that the costs of installation may rise considerably.

An apparatus developed by *P.G. Dierckx; U.S. Patent 3,980,417; September 14, 1976; assigned to Agfa-Gevaert NV, Belgium* is an installation for burning waste solvents which is characterized by the use of a concentrically constructed nozzle by means of which the waste solvents and a mixture of compressed air and fuel may be burned simultaneously.

The compressed air/fuel mixture may either serve to warm up the incinerator prior to the introduction therein of the waste solvents or may be used to support the incineration of waste solvents when the latter have a low heat value. Figure 92 shows such an apparatus.

The incineration installation **10** depicted comprises the following main parts upstream of the burner assembly: a collecting tank **11** communicating with a first sieve tank **12**, a second sieve tank **13** equipped with means for controlling the flow of solvents towards the burner assembly, a source of compressed air **14** and a storage tank **15** in which fuel having a high heat of combustion for example gas oil is stocked.

FIGURE 92: APPARATUS FOR INCINERATING NONHALOGENATED WASTE LIQUIDS

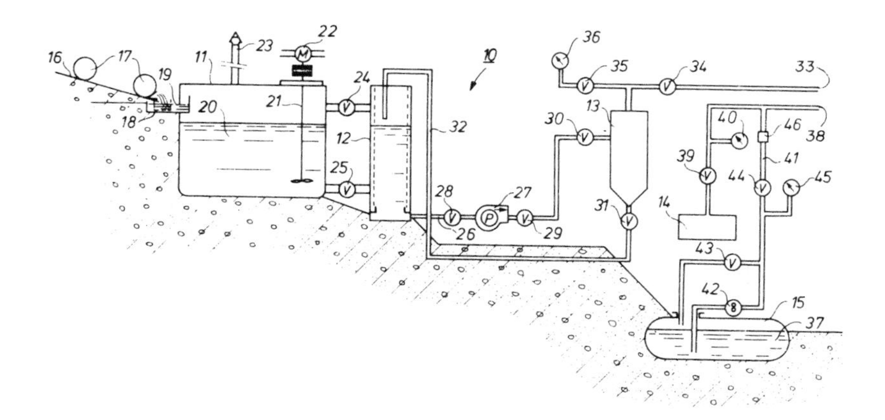

Source: U.S. Patent 3,980,417

At the entry side of the installation is provided an inclined plate **16** which serves to guide the containers **17**, containing the waste solvents, towards the entry of collecting tank **11**. The latter is provided with a coarse sieve **18** in order to retain the very coarse particles, larger than 5 mm diameter such as parts of broken bottles, pieces of cork, etc., which may be present in the waste solvents.

The sieve **18** itself is mounted in a water-sealed casing **19**, masking the opening of collecting tank **11**. The waste solvents are continuously stirred in order to form a mixture **20**. Stirring is done, for example, by means of a stirrer **21**, driven by a motor **22**. Occasionally, a flame protecting device **23** may be provided to prevent an accidental burning of the contents of collecting tank **12** from the outside.

At the top and the bottom of collecting tank **11** a pair of valves **24** and **25** are provided through which, by means of appropriate conduits, the collecting tank **11** is connected with a first sieve tank **12**. This provision permits pouring a given amount of solvents in the tank **12**, whereinafter the latter may be disconnected from the collecting tank **11**. In this way the large collecting tank need not necessarily be permanently connected to the other parts of the installation. In the meantime, solid parts which were not withheld by the sieve **18** at the entry of the collecting tank **11** may be collected.

The first sieve tank **12** is connected by the conduit **26** to a second sieve tank **13** which is equipped with means for flow regulation of the waste solvents. To this end is provided a centrifugal pump **27** which has at its inlet and outlet openings respectively the valves **28** and **29** for the purpose of easy removal of the pump when overhauling of the latter is necessary.

A valve **30** is provided at the inlet of the second sieve tank **13**. At the bottom of the second sieve tank are provided valve **31** and conduit **32**, the latter ending at the top of first sieve tank **12**. In so doing a closed circuit is built up and the returning flow through conduit **32** is regulated by the gradual opening or closing of valve **31**. The net flow towards the burner assembly itself goes through the top of the second sieve tank **13** by means of conduit **33**, which is equally provided with a valve **34**. The pressure by which the waste solvents are fed to the burner may be read on manometer **36**. Therefore, valve **35** is periodically opened.

An auxiliary circuit is provided by means of which a mixture of compressed air and fuel having a high heat of combustion may be directed to the burner assembly. For this purpose a source of compressed air **14** feeds air towards the burner through conduit **38** when opening valve **39**. The pressure itself is indicated by manometer **40**. A second conduit **41** is joined to conduit **38** through which a combined stream of compressed air and fuel from tank **37** is fed.

A high pressure pump, for example a gear pump **42** is provided, and by means of valves **43** and **44** the flow of the fuel (preferably gas oil) may be varied within a wide range. A manometer **45** permits the reading of the pressure at which the fuel is distributed.

NUCLEAR INDUSTRY EFFLUENTS

During the operation of nuclear systems, particularly nuclear power plants, radioactive waste gases develop which contain, inter alia, krypton and xenon nuclides. These radioactive waste gases must not be discharged into the atmosphere at will since the regulating authorities require the observance of maximum permissible values of radioactivity. The discharge of these gases into the atmosphere must thus be delayed for a period of time sufficient for the radioactivity to drop to a value below the maximum permissible values.

Removal from Air

It is known that these requirements can be met if the waste gas which is contaminated with radioactive noble gases is conducted over activated carbon delay paths where the noble gases remain substantially longer than the inactive carrier gas, for example, air, because of the adsorptive forces of the activated carbon.

Such delay systems generally operate at temperatures around 20°C and at pressures of about or slightly less than 1 atmosphere absolute. It has been found, however, that the long-lived radionuclides, particularly ^{85}Kr with a half-life of 10.7 years, are decomposed only insignificantly by such a delay. The danger thus exists that with an increase in the number of nuclear systems the concentration of ^{85}Kr in the atmosphere will continuously increase.

For this reason processes have been developed in which the radioactive noble gases krypton and xenon are separated from the exhaust or waste gas mixture and are stored in a highly concentrated form in a radiation-proof manner. A number of such separating or enrichment processes are known.

All of the prior art separating or enrichment processes have not been able to gain major significance in practice and, in particular, the safe storage of concentrated radioactive gases raises great difficulties. Additionally, the regulating authorities have thus far not required that the long-lived ^{85}Kr be quantitatively separated from the waste gases of the nuclear power plants. Thus activated carbon delay paths or columns are usually employed in practice.

The known activated carbon delay paths must be designed so that delay times of about 2 or 3 days for krypton and of about 30 to 50 days for xenon are achieved. The quantities of activated carbon required for this purpose usually amount to more than 100 m^3 for the usual gas quantities at normal operating pressures and temperatures.

Since the discharge conditions for radioactive substances into the environment will be more strict in the future, this would involve an even further increase in the capacities of the activated carbon systems which should be avoided for reasons of safety and costs, since the costs for enclosed space in nuclear power plants are very high due to the radiation protection measures involved.

An improved scheme developed by *W. Stumpf, H. Queiser, H. Jüntgen, H.-J. Schröter and K. Knoblauch; U.S. Patent 3,963,460; June 15, 1976; assigned to Licentia Paten-Verwaltungs GmbH and Bergwerksverband GmbH, Germany* permits the economical treatment of waste gases containing radioactive contamination, particularly krypton and xenon nuclides. The waste gas stream to be decontaminated is initially conducted through an enrichment system wherein the waste gas stream is divided into at least two partial streams one of which is substantially free from the radioactive impurities and constitutes the poor gas fraction and another of which constitutes the rich gas fraction. The poor gas fraction is discharged, at least in part into the atmosphere while the rich gas fraction is fed into an activated carbon delay path whose output is discharged, either directly or indirectly into the atmosphere.

A device developed by *T.N. Hickey and I.S. Spulgis; U.S. Patent 3,964,887; June 22, 1976; assigned to CVI Corporation* provides a standby gas treatment system and high efficiency charcoal filter for removal of radioactive release within a nuclear containment structure not only during normal purge operations but also in the event of a nuclear failure. More particularly, this device is a rechargeable gasketless charcoal filter which can adsorb radioactive constituents especially iodine and methyl iodide, at an efficiency of better than 99% even should a design basis or loss of coolant accident occur.

The enrichment of uranium customarily takes place through use of the compound uranium hexafluoride so that a process is required for converting uranium hexafluoride into uranium dioxide in a form which can be readily fabricated to shaped bodies having a low fluoride content.

One practice for converting uranium hexafluoride to uranium dioxide employs hydrolysis of uranium hexafluoride to form an aqueous solution of uranyl fluoride. A dilute aqueous solution of ammonium ion is added to the aqueous solution of uranyl fluoride to precipitate ammonium diuranate. After filtration the ammonium diuranate, which has a high fluoride ion content, is dissolved in nitric acid and fluoride decontamination of the resulting uranyl nitrate solution is accomplished by solvent extraction.

From the resulting purified uranyl nitrate solution, ammonium diuranate is reprecipitated and then calcined to give UO_3 which in turn is reduced with hydrogen to give uranium dioxide. In this process scrap fuel is recovered using nitric acid so that the fuel in nitrate form can be reused at the appropriate stage in the ammonium diuranate process.

However, the scrap recovery using nitric acid results in a nitrate waste solution containing, among other materials, ammonium nitrate. Nitric acid is also used in this ammonium diuranate process to flush lines to prevent the accumulation of solid deposits in the line and this gives an ammonium nitrate-containing solution normally discarded as waste.

Further anion exchange resins are employed in the ammonium diuranate process in the clarification step which gives additional nitrate-containing solution normally discarded as waste. In addition, the cladding used to contain the uranium dioxide pellets is typically a zirconium-containing alloy which is etched with nitric acid prior to loading of the uranium dioxide fuel material. After a period of time, there is a build-up of a metallic nitrate concentration in the nitric acid solution.

Removal from Water

A process developed by *J.M. Dotson and T.E. Peters; U.S. Patent 3,862,296; January 21, 1975; assigned to General Electric Company* is one in which liquid plutonium nitrate wastes are converted to releasable gaseous materials, largely gaseous nitrogen and water vapor, and the waste liquid condensate streams have plutonium contamination removed to a concentration less than 10^{-5} grams of plutonium per liter.

This enables release after scrubbing of both the gaseous product and the liquid product from this process without atmospheric contamination. This process involves a fluidized bed calcination of the plutonium-bearing nitrate wastes in the presence of a reducing agent containing the ammonium ion which converts the nitrates substantially to molecular nitrogen. In this process residual plutonium and other trace metallic ions are deposited as oxide calcines in the fluidized bed for recovery by standard recovery processes or for disposal as a solid waste.

Figure 93 is a schematic presentation of the fluidized bed apparatus along with associated equipment adapted for practicing the calcination process for liquid waste plutonium nitrate compounds in which release of plutonium to the environment is avoided.

It shows an apparatus generally designated **40** for conducting a fluidized bed process having a smaller diameter portion **41** containing a fluidized bed and a large diameter portion **42** which is substantially free of the fluidized bed due to its expanded diameter. Portion **41** is heated by external heating means **43** which can be electrical resistance wiring placed adjacent portion **41** or wrapped around portion **41**.

A fluidizing medium such as a gas is introduced into the portion **41** through throat **44** from line **45** which is connected to a source **46** of the fluidizing medium. Flow meter **47**, pressure regulator **48** and valve **49** are provided to enable regulation of the flow of fluidizing medium to the fluidized bed **40**.

FIGURE 93: FLUIDIZED BED PROCESS FOR MAKING AQUEOUS PLUTONIUM NITRATE WASTES HARMLESS

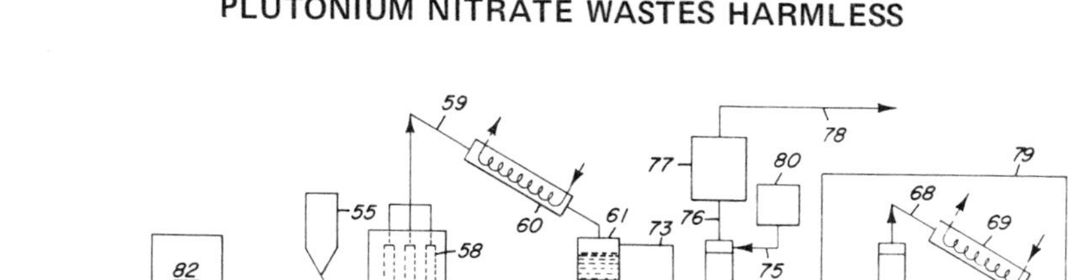

Source: U.S. Patent 3,862,296

The reducing agent from source **81** and waste nitrogen-containing material from source **82** are introduced to tank **50** and then to portion **41** of the fluidized bed **40**. The tank **50** has stirring means **51** for mixing the incoming materials and line **52** leading to entry port **53** near the bottom of portion **41**. Valve **54** is provided in line **52** for purposes of stopping the flow in line **52** when desired, such as for shutdown.

In this process there is the deposition of the cations of the waste nitrogen-containing compound on the particulate medium in the fluidized bed as a result of the decomposition of the nitrogen-containing compound. Intermittent or continuous withdrawal of the larger particles which settle to the bottom of portion **41** is conducted through outlet **57**. New particle medium is introduced to the upper part of portion **41** from reservoir **55** through line **56**. This introduction can be on a continuous basis at a controlled rate or on an intermittent basis.

The expanded portion **42** of fluidized bed **40** is arranged for gas removal with gas filters **58** serving to retain any fine solid particles being carried with the gaseous medium exiting portion **42** in line **59** with representative gas filters **58**.

Condenser **60** is provided in line **59** for condensing water vapor from the effluent gas in line **59** which is collected in reservoir **61** and fed to neutralization tank **62** through line **63**. Appropriate portions of base are added to tank **62** from source **79** to neutralize any acid in the condensate. Line **64** feeds the solution from tank **62** to distillation column **65** which withdraws a plutonium solution of very low concentration for collection in tank **66** which is partially recycled to tank **50** while the distillate in line **68** is condensed in condenser **69** and collected in tank **70** which is connected in turn to quarantine tank **72** by line **71**. At appropriate times the liquid in tank **72** (which contains water free of any plutonium contamination) is discharged to the ground or other desired location.

At tank **61** uncondensed gas from line **59** and tank **61** is fed in line **73** to the bottom of scrubber **74** to which fresh scrub water is fed in line **75** from source **80**. The gas passes out the top of scrubber **74** in line **76** to absolute filter **77** and is then released to the atmosphere from line **78**.

Venting vapors from tank **70** are fed to the middle portion of scrubber **74** in line **79**. This process insures no release of the metallic cation from the nitrate waste material either in the gaseous discharge or the aqueous discharge.

See Alkyl Iodides (U.S. Patent 3,852,407)
See Chlorinated Phenols (U.S. Patent 3,971,717)
See Iodine (U.S. Patent 3,838,554)
See Radioactive Materials (U.S. Patent 3,871,841)
See Radioactive Materials (U.S. Patent 3,983,050)
See Radon (U.S. Patent 3,853,501)
See Tellurium Hexafluoride (U.S. Patent 3,491,513)

ODOROUS COMPOUNDS

Removal from Air

Malodorous particulate matter entrained in gas and/or fume emissions can be removed by combustion, by one of two methods, namely, direct flame combustion and catalytic combustion.

By the direct flame combustion method the odorous effluent gases and/or fumes are raised to temperatures in excess of 760°C, preferably to approximately 900°C. But as the calorific value of the effluent gases and fumes is extremely low, practically all the requisite heat to effect combustion must be supplied by fuel, usually oil. Consequently, this method is fairly costly. It is true that the costs of incineration can be reduced by utilizing waste heat through heat exchangers, but there are not always practical possibilities for so doing, and even if there are the capital costs of the requisite waste heat recovery equipment can be considerable.

By the catalytic combustion method the odorous waste gases and/or fumes are burned at temperatures from 350° to 450°C. The reduction in oxidation temperature compared with direct flame combustion is achieved by the use of catalysts. The savings in fuel permitted by the lower combustion temperatures are substantial but the costs of this method can be considerable, depending on the life of the catalyst.

Many effluent gases or fumes from industrial or chemical processes contain catalyst poisons, sulfur and phosphorus in particular, that greatly reduce catalytic activity. As the precise composition of the effluent gases is frequently not known or cannot be predicted, it is impossible to estimate the life of the catalysts to be employed. As very small amounts of gaseous poisons can have a drastic effect on the activity of the catalyst, the cost of employing this method can be fairly considerable. Thus an absorption process not exhibiting these disadvantages would be desirable.

A process developed by *W. Kötting; U.S. Patent 3,905,774; Sept. 16, 1975; assigned to Steuler Industriewerke, GmbH, Germany* utilizes an absorption tower incorporating a plurality of capillary washing plates arranged one above the other in stages, with streams of washing fluid circulating thereupon at a variable level. The effluent gases and/or fumes with the entrained malodorous particulate matter are fed in from below under positive pressure and impinge upon the underside of the washing plates and penetrate the washing fluid.

In the first stage the washing fluid is an approximately 2 to 6% solution of sodium hydroxide or potassium hydroxide. A quantity of chlorine, determined by the chloroxidizable portion of the entrained contaminants is added to the effluent gases and/or fumes before they enter the first washing stage, so that in the second washing stage the fluid consists of an approximately 3 to 8% solution of sodium hydroxide or potassium hydroxide and that in the third washing stage the washing fluid consists of an approximately 0.5 to 8% sulfamic acid (NH_2SO_3H).

Finally, the effluent gases and/or fumes are discharged into the atmosphere through a separator that removes entrained droplets.

A process developed by *H. Kurmeier; U.S. Patent 3,936,281; Feb. 3, 1976; assigned to Sudoldenburger Tierfrischmehl-Anlagengesellschaft, Germany* involves deodorizing waste gases containing foul-smelling constituents that are chemically decomposable on being washed with aqueous acid formed by dissolving an acid-forming gas, such as chlorine or sulfur, in water.

The waste gas is washed with the aqueous acid containing a suspension of one or more alkaline earth metal carbonates, such as magnesium carbonate or calcium carbonate. In a vessel, the carbonate suspension may be sprayed downward at the top, and the waste gases directed upwardly at the bottom. Both the suspension and the gases pass through loose layers of limestone intermediate the top and bottom of the vessel.

A process has been developed by *P.B. Lonnes, C.M. Peterson, D.A. Lundgren and L.W. Rees; U.S. Patent 3,969,479; July 13, 1976; assigned to Environmental Research Corporation* for oxidizing odorous constituents of contaminated gas with minimum consumption of the oxidizing agent in a countercurrent packed scrubbing tower. The method is primarily useful in controlling odorous emissions from rendering plants, fish processing plants, asphalt plants, and other plants in which aldehydes, fatty acids, ketones, mercaptans, amines, hydrogen sulfide, sulfur dioxide, nitric oxide, phenols or other pollutants are emitted.

Chlorine in the form of sodium hypochlorite in a concentration of 5 to 50 ppm is the preferred oxidizing agent. Consumption of the oxidizing agent is minimized by continuously purging 1 to 5% of the scrubbing liquid to thereby promptly remove precipitates and other solids. A flow diagram of such a process is shown in Figure 94.

Contaminated gas enters scrubber **10** at gas inlet **11** and passes through packed bed **12** of packing material where the odorous constituents are oxidized. The decontaminated gas is discharged from scrubber **10** at gas outlet **13**. An aqueous solution of chemical oxidizing agent is prepared in supply tank **14** and circulated by means of pump **15** through scrubbing liquid line **16** and is discharged in the form of a spray at nozzles **17** above packed bed **12**. The aqueous solution of chemical oxidizing agent flows downwardly over packed bed **12** as the contaminated gas flows upwardly, thus causing a contact in the conventional manner between gaseous fluid and the aqueous solution to bring about oxidation of the odorous constituents.

Spent scrubbing liquid, as well as chemical reaction precipitates and other solids, are collected in scrubbing liquid holding tank **18**. A portion of the spent scrubbing liquid is continuously drained from the scrubbing liquid holding tank **18** and line **19**. The amount of scrubbing liquid drained is preferably within the range of 1 to 5% by weight of the total quantity of circulating solution. Expressed in terms of quantity of scrubbing liquid purged with respect to the flow

FIGURE 94: SCRUBBER SYSTEM FOR ODOR CONTROL

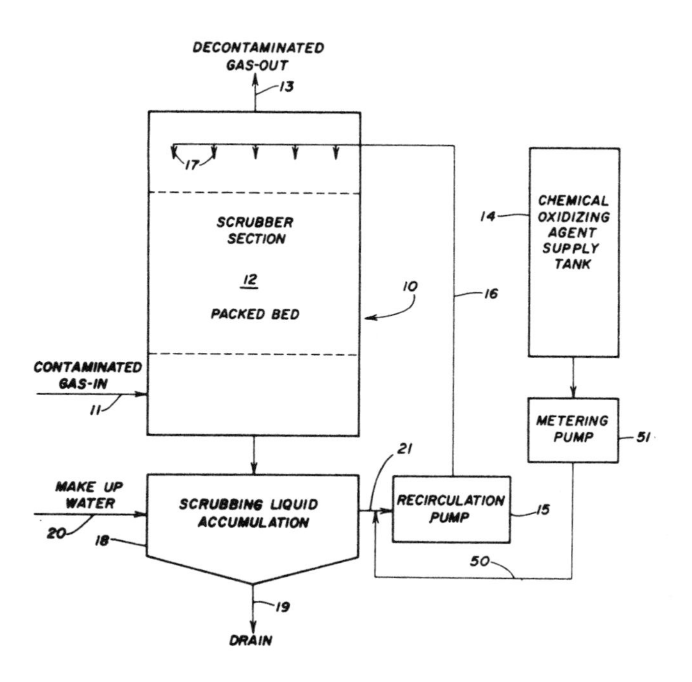

Source: U.S. Patent 3,969,479

rate of the contaminated gas, the purge rate is 1 to 2 gallons per 10,000 cfm of gas. Makeup water is added to scrubbing liquid holding tank **18** at **20**. Spent scrubbing liquid is recirculated from scrubbing liquid holding tank **18** through line **21**.

While chlorine in the form of sodium hypochlorite is the preferred oxidizing agent, other oxidizing agents including sodium and potassium permanganate, ozone, hydrogen peroxide, and other metal hypochlorites may be used. In certain installations for treatment of formaldehyde, hydroxide ions alone may serve as the oxidizing agent, and the typical source is sodium hydroxide.

With potassium permanganate as the oxidizing agent, the concentration of the scrubbing liquid should be within the approximate range of 0.001 to 0.01%. The concentration of potassium permanganate in the oxidizing agent supply tank should preferably be in the range of 0.5 to 4%. With sodium hypochlorite as the oxidizing agent, the concentration of the scrubbing liquid should be within the approximate range of 5 to 50 ppm, and preferably in the range of 5 to 25 ppm. The concentration of sodium hypochlorite in the oxidizing agent supply tank and the feed rate should be in the range of ¼ to 1 gallon of a 15% solution of sodium hypochlorite per 10,000 cfm of contaminated gas at a feed rate of 100 gpm of scrubbing liquid.

The scrubbing liquid should be alkaline, i.e., the pH should be in the range of

7.5 to 10.0. With sodium hypochlorite as the oxidizing agent, no buffer is necessary. The temperature of the scrubbing liquid should lie within the range of 60° to 160°F. The packing material comprising packed bed **12** may range widely in substance, size and density. A suitable substance (Flexirings) may be obtained from Koch Engineering Company.

See Corn Milling Effluents (U.S. Patent 3,875,034)
See Feedlot Industry Wastes (U.S. Patent 3,966,450)
See Methionine Process Effluents (U.S. Patent 3,867,509)
See Pesticide Manufacturing Effluents (U.S. Patent 3,798,877)
See Pharmaceutical Industry Effluents (U.S. Patent 3,923,955)
See Pulp Mill Effluents (U.S. Patent 3,969,184)
See Pulp Mill Effluents (U.S. Patent 3,701,824)
See Pulp Mill Effluents (U.S. Patent 3,794,711)
See Pulp Mill Effluents (U.S. Patent 3,796,628)
See Pulp Mill Effluents (U.S. Patent 3,745,063)
See Rendering Plant Effluents (U.S. Patent 3,803,290)
See Sewage Treatment Effluents (U.S. Patent 3,828,525)
See Tall Oil Processing Effluents (U.S. Patent 3,709,793)

Removal from Water

See Hydrogen Sulfide (U.S. Patent 3,903,250)
See Methionine Process Effluents (U.S. Patent 3,867,509)
See Olive Processing Effluents (U.S. Patent 3,732,911)
See Tall Oil Processing Effluents (U.S. Patent 3,709,793)

OIL

There are many industrial sources of oily wastes. The tabulation below presents the major types of industry producing oil and grease-laden waste streams, and lists characteristic types and sources of oily wastes associated with each industry. By far the three major industrial producers of oily waste are petroleum refineries, metals manufacturers and food processors.

Industry	Waste Character
Petroleum	Light and heavy oils resulting from producing, refining, storage, transporting and retailing of petroleum and petroleum products.
Metals	Grinding, lubricating and cutting oils employed in metal-working operations, and rinsed from metal parts in clean-up processes.
Food processing	Natural fats and oils resulting from animal and plant processing, including slaughtering, cleaning and by-product processing.
Textiles	Oils and grease resulting from scouring of natural fibers (e.g., wool, cotton).
Cooling and heating	Dilute oil-containing cooling water, oil having leaked from pumps, condensers, heat exchangers, etc.

Removal from Water

The use of synthetic polymeric fibers to absorb oil is well known. Much effort has been directed to the clean-up of oil spills from the surface of water, and the use of synthetic polymeric materials for this purpose has been illustrated. Disposal of the oil-bearing synthetic polymeric material receives little attention, although some patents do suggest that oil may be removed by squeezing the polymeric material. Efficiency of oil pick-up by the synthetic material decreases after an initial use, and it has often been found expedient to discard the material rather than attempting to reuse it.

Synthetic fibers have also been used as filters to remove small amounts of oil from oil-water mixtures. In these cases too, the filter is usually discarded rather than cleaned and reused.

Disposal of the oil-bearing synthetic materials has been a continuing problem, particularly in the case of large oil spills as, for example, from an oil tanker. Ecological considerations require that the oil-polymeric material mixture be disposed of in some way which is not harmful to the environment. Thus far, no satisfactory way has been developed. It is the purpose of this process to describe a method for the disposal of oil-bearing synthetic polymeric material which converts such material to a liquid suitable for use as a fuel.

A process developed by *B.P. Martinez and M.D. Zeisberg; U.S. Patent 3,923,472; December 2, 1975; assigned to E.I. Du Pont de Nemours and Company* involves filtering oil from a liquid containing oil by passing the liquid through a filter containing melt-spun thermoplastic synthetic fibers for absorbing the oil. When the fibers become saturated with oil, they are heated at an elevated temperature until they become liquid. The resultant liquid is then drawn from the filter for use as a fuel. Such a scheme is shown in Figure 95.

A mixture of oil and water from a settling basin **10** is directed through valves **2, 4** and **6** by pump **3** into two filter boxes **12, 14**. The filter boxes are arranged so that one is on line while the other is on a stand-by basis. For example, the oil-water mixture passes through the filter box **12** and the oil is absorbed by synthetic fibers (less than 100 denier per filament) in the filter box and retained, while essentially oil-free water is returned to the stream **16** through pipe **13** attached to one outlet of filter **12**.

When the first filter **12** becomes saturated with oil, the oil-water stream is directed into the second filter box **14** by means of the valves **4, 6,** i.e., by closing valve **4** and opening valve **6**. After removal of the oil-saturated filter **12** from service, the entire filter is heated with steam by means of steam coil **18** in filter box **12** to convert the fiber-oil mixture into a liquid.

Advantageously, the temperature may be maintained in the liquid to insure that it will remain pumpable. This liquid is then pumped by pump **20** through pipe **15** from the other outlet of filter **12** to a boiler where it is used to generate steam. The procedure is repeated using steam coil **18** to heat filter **14** when it becomes saturated with oil.

Synthetic fibers will absorb many times their own weight of oil, depending on

FIGURE 95: SCHEME FOR REMOVING OIL FROM WATER USING THERMOPLASTIC FIBERS

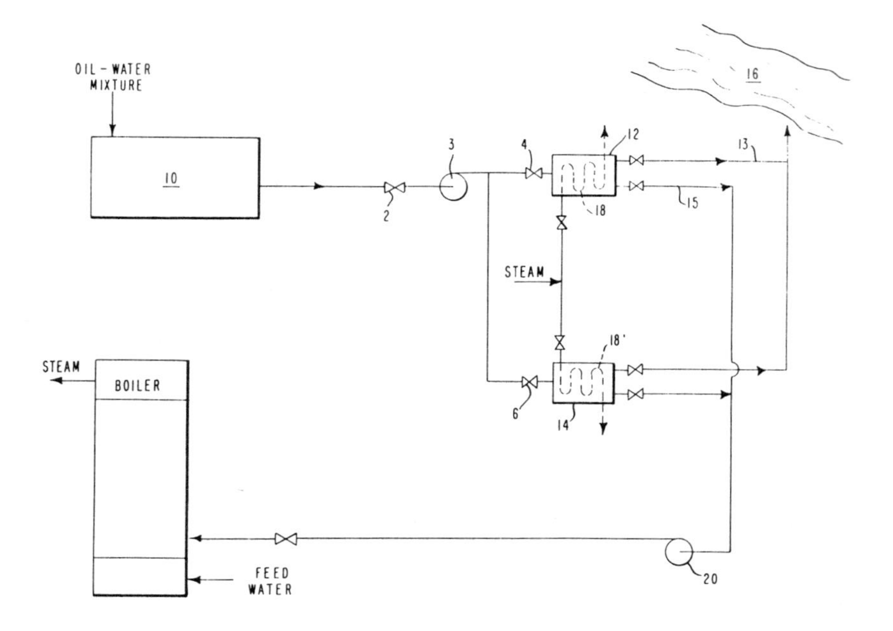

Source: U.S. Patent 3,923,472

type and composition. Typical numbers are 30 to 35 for polyester and 10 to 170 for polyolefins. The oil polymeric material mixture may be converted to a liquid by heating at atmospheric pressure or above, using temperatures in the range from 110° to 300°C. The process may be used with a wide variety of oils, including crude oil, fuel oil, used motor oil and light oils such as textile finishes. The process also may be useful in the disposal of waste synthetic fibers.

In tankers running under ballast, the ballast water becomes contaminated with oil. Also, the engine room bilge water is oily, and the liquid cleaning of the cargo tanks results in the production of slop, which is oily. These various oily waters also commonly contain particulate matter such as silt, rust, waxes, asphalts, etc. Before this ballast, bilge water, or slop is discharged overboard from the tanker, it is essential, or at least highly desirable, that substantially all of the oil be removed therefrom, as well as the particulate matter. The main reason for this is to avoid pollution of the rivers and oceans. By way of example, it may be desired to reduce the oil content of the water to be discharged overboard to less than about six parts per million.

A process developed by *D.C. Garber; U.S. Patent 3,965,004; June 22, 1976; assigned to Sun Shipbuilding & Drydock Company* involves a system for removing oil from oily water which employs an oil droplet coalescer. A ceramic dewaxer is inserted before the coalescer to remove waxes, asphalts, and similar materials which would otherwise quickly clog the fine holes of the coalescer. The "dirty" ceramic elements of the dewaxer may be easily and efficiently regenerated.

OIL SPILLS

Spills of liquid hydrocarbon compounds such as petroleum, crude oil, fuel oil, and the like present a serious water pollution problem and various means have been proposed to quickly remove such spills before contamination of the sea bottom and adjacent shoreline occurs. Such spills occur both in protected waters such as harbors, and also occur in offshore or unprotected waters. Spills in harbors result principally from oil transfer operations, industrial waste discharges, pipeline breaks, collisions, and the like, and generally amount to volumes in the order of 100 barrels.

Spills in offshore waters normally result from collisions or grounding of tankers and merchant vessels, or as a result of malfunctions of an offshore oil rig. Although less frequent, such offshore spills are often very large. The offshore type of oil spill is particularly difficult to deal with because of the wave action which interferes both with efficient consolidation or confinement of the oil slick to a small area, and also with the mechanical separation of the oil from the water surface.

Removal from Water

Various means have been employed in an effort to clean up spilled oil, including the use of chemicals to cause sinking or dispersion of the oil, and also including the distribution of absorbent material on the surface of the slick. In addition, burning of the slick has been attempted, as well as skimming of the surface oil by means of rotating cylinders, suction devices, and the like.

Each of these methods of the prior art has serious limitations. Chemically caused sinking or dispersion of the oil pollutes the water and the harbor or sea bottom. Burning or incineration is objectionable because of atmospheric pollution, and because of the difficulty of maintaining the oil slick at a temperature high enough to sustain combustion. The skimming process, which is capable of removing relatively large quantities of oil at a comparatively high rate, undesirably requires large and expensive settling or centrifuging devices which depend for efficient operation upon a relatively precise orientation of the weir or other skimming device relative to the thickness of the surface film of oil.

Wave action in unprotected waters makes this type of operation very impractical. This is also true of that class of devices which utilize moving drums, belts, or discs disposed at the water surface and moved through the oil film for continuous coating or impregnation with the oil. Such devices must be precisely aligned with the film of oil in order to absorb or carry away a high proportion of oil rather than water.

In summary, the systems of the prior art lack one or more of the following desirable characteristics: high oil recovery rate, minimum inclusion of water with the removed oil, efficient oil removal in the presence of water motion or wave action, relatively inexpensive, and easy to deploy and maintain.

A process developed by *R.E. Hunter; U.S. Patent 3,581,899; June 1, 1971; assigned to Ocean Design Engineering Corporation* involves separating oil from a water surface by distributing many small buoyant bodies of oil absorbent material upon such surface, continuously lifting such bodies from the surface, treating the bodies to remove the absorbed oil, and again distributing the bodies upon the surface for reuse. The apparatus preferably includes booms for gathering the distributed bodies of absorbent material toward a conveyor which lifts the bodies upwardly. The booms are articulated and include floats so that the booms rise and fall with any wave action of the water, such as would exist in the unprotected waters of the open sea. The process is shown schematically in Figure 96.

The oil absorbing material is in the form of a great plurality of relatively small cubes, chunks, bits, or bodies **72** which are open cell or porous, being characterized by a multiplicity of relatively fine capillary passages. The bodies **72** have a density which is less than water, both before and after absorbing any oil. Consequently, the bodies always float upon the water surface **14**, which is extremely desirable in order that absorption be confined to absorption of oil, rather than water.

That is, any pronounced submersion of the bodies beneath the surface of the water would subject the bodies to a hydrostatic head which would tend to force water into the pores of the bodies along with, and perhaps in place of oil.

FIGURE 96: SCHEME FOR OIL SPILL RECOVERY FROM WATER SURFACE USING RECYCLED PLASTIC FOAM PARTICLES

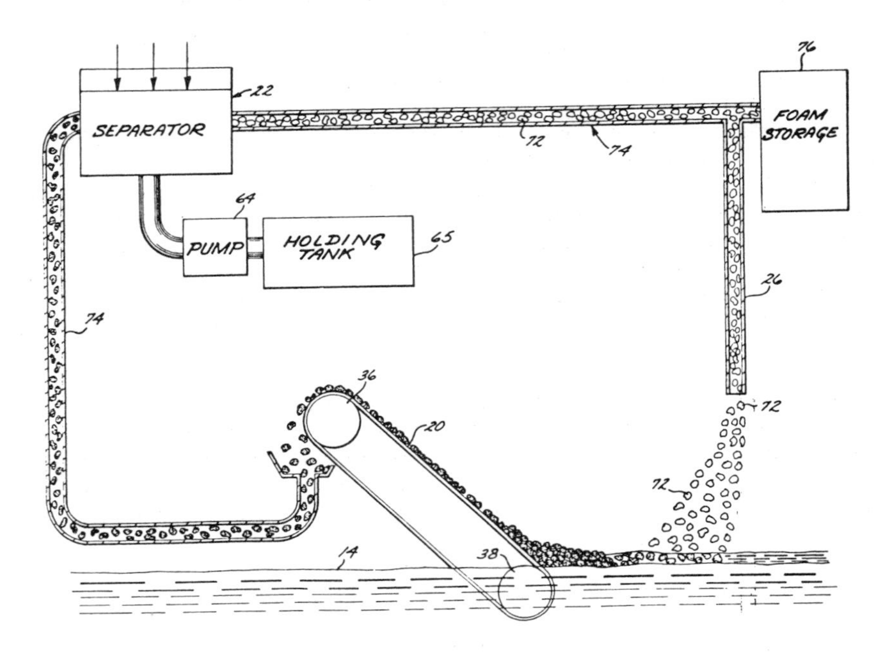

Source: U.S. Patent 3,581,899

Thus, the buoyancy of the bodies results in no appreciable relative movement between the bodies and the water surface **14**, even during conditions of extreme wave action.

As will be seen, the bodies are necessarily resiliently flexible so that any absorbed oil can be extracted from the bodies by compression of the bodies by the compressors **22**.

The bodies are made of a material which is generally hydrophobic and oleophilic so that it will have a tendency to reject water while exhibiting an affinity for oil. A suitable material for this purpose has been found to be flexible urethane or polyester foam having a density of approximately 2 lb/ft^3. The multiplicity of cells in this material is designated as a cell count of so many pores or cells per linear inch, and in the form of urethane foam material successfully utilized, the cell count was 27½ cells per linear inch.

In practice this type of material has been found to preferentially separate normally liquid hydrocarbon compounds from a liquid of higher specific gravity such as water and, since the foam material floats upon the surface of the water the normal affinity of the material for oleaginous materials is not disturbed by the existence of any hydrostatic head, which would exist if, for example, the foam material were to be plunged below the water surface **14** by artificial skimming devices or the like.

In fact, the bodies of this foam material, when spread upon an oil-water mixture and thereafter removed and compressed, have yielded absorbed material which is 95% fuel oil versus 5% or less of the water. Apparently the hydrophobic character of the foam material allows oil to readily displace any water which may be in close proximity to the surface of the foam bodies. However, even when the foam material is completely saturated with oil, it will still float upon the water surface, which is important to this process.

The apparatus basically requires only a conveyor system, generally designated **74** which is effective to lift the oil saturated foam bodies **72** from the water surface **14**, such as by means of the conveyors **20**, deliver the bodies to a foam-oil separator, such as the compressors **22** and, finally redistribute the relatively oil-free bodies back upon the water surface as by means of the distributors **26**.

Preferably some form of reservoir or storage container **76** is provided to compensate for any foam bodies which are not recovered by the conveyors. In addition, suitable pumping equipment **64** would be provided for transferring the oil collected in the holding tank or bunkers **65**.

The circuit of the moving oil-absorbent particles includes recycle tubes **74** and rollers **36** and **38** which support and move the conveyor **20**.

There are three major ways known to recover oil floating on the surface of a body of water. The first is a weir-type skimmer supported on the body of water that permits the uppermost surface of the water to flow into a sump from which the accumulated water is pumped to a separating tank located on a floating vessel or permanently located on shore. This type is basically ineffective during high seas or sea having waves above 2 feet. The second type is a floating suction skimmer. This type sucks the upper surface of the water into a separating tank. The major difficulty with a suction skimmer is the fact that it cannot operate very efficiently in water currents greater than 1 foot per second. The third type of skimmer is an absorbent surface skimmer. The most basic embodiment of this type is that of an endless absorbent belt rotating around two spaced pulleys.

One pulley is located just below the oil slick; the other is near a collection pan above the water surface. The belt in the vicinity of the slick absorbs the oil and then moves toward the elevated pulley to a roller or wiper which squeezes or removes the absorbed oil into a collection pan and then continues on repeating the cycle until the oil slick is removed. The adsorbent surface skimmer, because of the many moving parts, requires surveillance and maintenance.

An apparatus developed by *D.E. Wilson; U.S. Patent 3,983,034; Sept. 28, 1976; assigned to Chevron Research Company* is one for skimming oil spills from water, comprising three pontoons, one located in each of the corners of a triangle made up of interconnecting structural members floating on a body of water.

The corner pontoons are adjustably buoyant permitting the skimmer to move vertically up or down as a unit in the water so as to adapt to the wave height of the body of water. At the base of the triangle is the skimmer mouth sloping upward towards a sump. The two pontoons on each side of the sloped mouth have a clamp easily attachable to an oil boom which guides an oil slick into the skimmer.

The function of the mouth is to skim off approximately two inches of the water surface. The sloped mouth terminates at an impregnable deflector centrally located between two screens that serve as a wave quieting assembly as well as a separator of debris floating on the water.

The skimmed water is then directed through the debris screens into a quieting area. Once in this quieting area, the oil slick flows over a self-adjustable weir into a sump. The weir is made adjustable by a float that modifies the weir elevation with changes of liquid level in the sump.

The skimmed liquid is then pumped to a storage tank from which it is recycled by letting the water at the tank bottom flow under gravity back into the oil boom. Such an apparatus is shown in Figure 97.

In its usual application, the skimmer **50** is towed abreast of a floating vessel **63** equipped with storage tanks **65**, a pump **64**, interconnecting pipes or hoses **81** and a davit **71** for keeping the flexible hose from the aft end of the vessel **63**. An outrigger **68** extending above the water surface **56** and outward from the side of the vessel **63** supports the leading edge of an oil boom **66**.

The oil slick **69** enters the skimmer mouth and flows across an upwardly sloped entrance plate through debris screens (not shown).

Mixed water and collected oil are pumped through a flexible hose **62**, the pump **64** and a rigid pipe **81** to an oil/water separation tank **65** on board the floating vessel **63**. The water which settles to the bottom in the tank is bled through a hose or hoses **70** from the tank **65** back into the area enclosed by the oil booms **66**.

FIGURE 97: VESSEL WITH PROVISION FOR CONTAINING OIL SPILLS AND THEN COLLECTING THE OIL ON BOARD

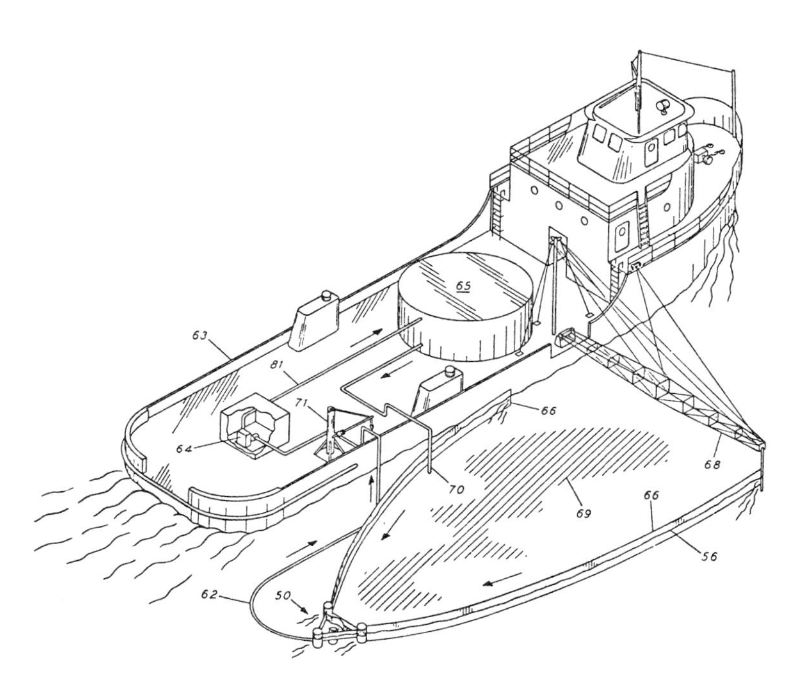

Source: U.S. Patent 3,983,034

A device developed by *R.R. Ayers; U.S. Patent 3,966,614; June 29, 1976; assigned to Shell Oil Company* is a skimmer for removing oil from the surface of a body of water which is articulated from front to rear to be wave comformable and/or has a quiescent collection zone formed by bottom and/or forward baffles.

A flexible skimming head for use with or independently of the above skimmer is composed of a foraminous sheet having an integral chevron flow pattern on its upper surface or, optionally, is a floating skimming head with a central, axially vertically movable cone or other configuration forming a suction mouth. A boom for use with or independently of the above skimmer is composed of converging double booms. Storage capacity for oil collected by the above skimmer is composed of onboard and/or offboard membranes.

A composition developed by *A. Omori, I. Okamura, T. Imoto and T. Katoh; U.S. Patent 3,966,597; June 29, 1976; assigned to Teijin Limited, Japan* is an oil or organic solvent-absorbent sheet prepared by extruding a molten thermoplastic resinous polymer blend of polystyrene and polyethylene containing a foaming agent through a die having a slit aperture of 0.1 to 1.0 mm width, quenching the extrudate at the die exit to a temperature below the glass transition point of the resinous blend, drafting the extrudate at a draft ratio from the maximum draft ratio possible under the operating conditions to one-third the maximum draft ratio, laminating at least two sheets of the resulting unopened, sheet-like reticulated structure having numerous noncontinuous cracks along one direction so that the direction of the cracks of each such sheet is the same, pulling the laminate in a direction perpendicular to the direction of the cracks to separate the constituent fibers from each other, and crimping the opened, sheet-like laminate either alone or together with at least one other sheet-like reinforcing material.

Examples of the reinforcing layers to be employed are papers such as pulp paper and synthetic paper; woven or knitted goods; nonwoven fabrics such as glass mat and nonwoven fabrics of natural, semisynthetic and synthetic fibers; films such as synthetic resin film and regenerated cellulose films; and foamed sheets. The knitted fabrics are preferable.

The absorbent is useful for absorbing and removing oils which have flowed out onto the surfaces of sea or rivers, preventing the intrusion of oils to oyster culture farms, for example, by absorbing and removing waste oils from industrial wastes. The oils to be removed may, for example, be crude oils, heavy oils, light oils, machine oils, kerosene, and vegetable oils.

This absorbing material can also be used to absorb and remove organic solvents, such as aromatic hydrocarbons, for example, toluene, xylene or benzene, ethers, ketones, phenols, halogenated hydrocarbons, or aliphatic hydrocarbons, which are afloat or suspended in air or in wastewater.

A technique developed by *R. Bartha and R.M. Atlas; U.S. Patent 3,959,127; May 25, 1976; assigned to U.S. Secretary of the Navy* is one whereby free-floating oil slicks on bodies of sea and fresh water are disposed of by microbial degradation at a greatly enhanced rate by applying the essential microbial nutrients, nitrogen and phosphorus, to the oil slick in a form that dissolves in or adheres to the oil and thus selectively stimulates the activity of oil-metabolizing microorganisms.

A process developed by *E.C. Peterson; U.S. Patent 3,980,566; Sept. 14, 1976; assigned to Electrolysis Pollution Control Inc.* for removing immiscible fluids such as oil spills from the surface of bodies of water comprises placing an adsorbant compound on the water surface in contact with the immiscible fluid and maintaining contact between the surface of the immiscible fluid and the adsorbant material until the immiscible fluid is adsorbed.

The adsorbant compound consists of an admixture comprising from between about 30 and 70% by weight of lead slag mineral wool, with the balance being a finely divided natural stone substance containing substantial quantities of iron, aluminum, and magnesium oxides, including such natural stones as trap rock, basalt and gabbro. The lead slag mineral wool is treated with a hydrophobic-oil-soluble hydrocarbon chain substance, such as oleic acid to wet the surface of the mineral wool prior to mixing with stone flour.

The lead slag mineral wool is preferably fragmented into nodules having a diameter of, for example, from ½ to 1 inch. The composition may also be utilized for removing oil spills from lake beds or soil surfaces.

OLIVE PROCESSING EFFLUENTS

Ripe olives in their natural (or fresh) state are extremely bitter and thus must be processed for human consumption. To remove the bitter principle olives are soaked in aqueous sodium hydroxide (lye) for extended periods of time. After soaking, the olives are removed from the lye bath, washed with water, neutralized with acetic acid, and then soaked in aqueous sodium chloride (brine) for flavoring.

The lye (or alkaline) solution employed in the initial soaking of the fresh olives becomes contaminated with organic impurities. These impurities impart a deep brown, almost black, color and an extremely unpleasant odor to the lye solution. Because the biological oxygen demand (BOD) of this waste is extremely high, it cannot be released into municipal sewage treatment systems without first being diluted with enormous quantities of water.

Removal from Water

Currently, disposal of the olive-processing fluid is accomplished by forming "ponds" into which the liquid is placed. The water is allowed to evaporate from these ponds, leaving a sludge of lye and the organic contaminants. Because of its unpleasant aroma the used fluid must be piped or trucked many miles to uninhabited areas where pollution consequences hopefully are minimized.

The lye, of course, poisons the soil in and around the ponds. This disposal operation involves considerable expense both for handling the waste and for land needed to house the ponds. Also, the cost of lye has recently increased tremendously. Olive processors fear that expenses will become prohibitive and that they will be forced to discontinue part of their operations.

In addition to salt (about 7 to 9%), spent olive processing brines contain organic substances such as lactic and other acids, proteinaceous materials, pigments, etc. The olive industry in California alone generates about 7 to 8.5 million gallons of used processing brine.

A process developed by *R. Teranishi and D.J. Stern; U.S. Patent 3,975,270; Aug. 17, 1976; assigned to U.S. Secretary of Agriculture* is one in which useful olive-processing liquor is recovered from olive-processing waste solution by a process wherein lime, charcoal, and calcium carbonate are successively added to the olive-processing waste solution to form a mixture and the mixture is allowed to settle. The treated liquor is separated from the settled contaminants and is then immediately recyclable to process fresh olives.

A process developed by *E. Lowe and E.L. Durkee; U.S. Patent 3,732,911; May 15, 1973; assigned to U.S. Secretary of Agriculture* is one in which spent brines, for example, those derived from olive processing plants, are reconditioned by applying a series of steps: Concentration of the brine using submerged combustion, incineration of the concentrate, dissolution of the residue in water, and filtration of the resulting solution. Such a process is shown in Figure 98.

Numeral **1** designates a concentrator wherein the spent brine is evaporated by submerged combustion. The spent brine enters concentrator **1** by pipe **2**; the concentrated brine leaves via pipe **3** equipped with valve **4**.

FIGURE 98: PROCESS FOR RECONDITIONING SPENT OLIVE-PROCESSING BRINES

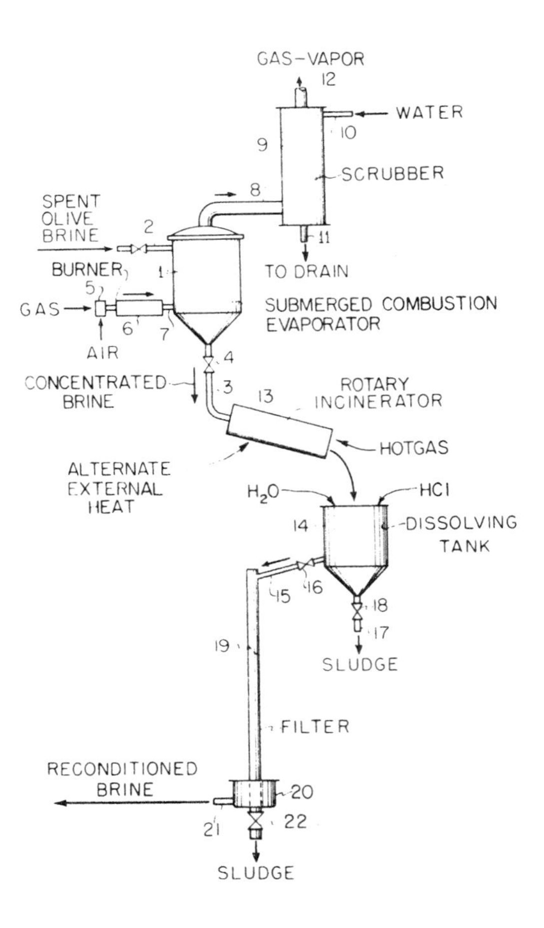

Source: U.S. Patent 3,732,911

To yield the desired concentration by submerged combustion, fuel gas and air are combined in mixer **5** and directed to burner **6** where combustion takes place. The hot gases flow into concentrator **1** via pipe **7** and bubble up through the brine contained therein. The resulting direct contact of the brine with the hot gases results in efficient evaporation without causing any significant scaling effect. It may be noted that the spent brine because of its acid pH, its content of salt, and its content of proteinous materials cannot be effectively concentrated by conventional procedures, those involving contact with a hot metallic surface such as steam coils or jackets. If such procedure is attempted, the surfaces become corroded and fouled with hard scaly deposits so that heat transfer is impeded and the equipment impaired.

Another advantage of this method of concentration is that the evaporation of water is achieved at a temperature below the normal boiling point of the brine. This situation is explained as follows: As the hot products of combustion flow through the brine, heat is transferred from the rising gas bubbles to the liquid through the bubble interface. The partial pressure of the water vapor in the rising bubble is less than the atmosphere so that boiling takes place at a temperature considerably below 212°F. For a saturated NaCl solution, the observed boiling point is about 200°F (in contrast to a normal boiling point of 227.6°F), corresponding to a depressed vapor pressure of about 443 mm Hg, and a heat of evaporation of approximately 978 Btu/lb.

In general, the concentration is continued until there is produced a slurry having a solids content of about 50 to 60%. At this stage much of the salt will be in the form of crystals suspended in the thickened liquid. The slurry, however, retains its liquidity so that it can be readily removed from concentrator **1** and piped to subsequent operations.

The vapors produced during the concentration may be vented to the atmosphere. Usually, however, these vapors have a very disagreeable odor, due to the presence of lactic acid and other volatile organic substances, and it is preferred to scrub the vapors before discharging them. To this end, the vapors are directed by conduit **8** to scrubber **9** where they are contacted with water entering the scrubber via pipe **10**. Water containing absorbed volatiles leaves the system via drain **11**, while residual vapors are released through vent **12**.

The concentrate, a slurry containing salt crystals in suspension, which is formed in the concentrator is directed by pipe **3** to incinerator **13**. Therein, the concentrate is heated at a temperature of about 1000° to 1400°F in the presence of air, whereby organic components are burned or at least carbonized. In conducting the incineration, it is preferred that the temperature be maintained below 1472°F (the melting point of NaCl) to avoid fusing the salt. Incinerator **13** may take various forms. One may use, for example, a rotary kiln which is heated externally, or one which is heated internally by the combustion of natural gas or other gaseous or liquid fuel.

The product of the incinerator is largely (about 95%) salt with a small amount of impurities. It can be stored in that form for future use.

When fresh brine is needed, the appropriate amount of the crude salt (the product of the incineration) is dropped into tank **14** where it is dissolved in water. Since the crude salt contains some alkaline impurities, it is usually necessary to neutralize the solution (bring it to pH 7) by adding a small proportion of an acid such as hydrochloric. Impurities which remain undissolved may be drained out of the bottom of tank **14** through pipe **17** equipped with valve **18**.

Any remaining insoluble material is then removed from the brine by filtration. For this purpose, it is convenient to use the uniflow filter disclosed in U.S. Patent 3,523,077, which apparatus is diagrammatically shown in the figure. The brine is directed by pipe **15** into filter chamber **19**, a length of foraminous hose suspended vertically. The clear brine passes through the interstices of the hose and flows down the outside thereof into receptacle **20** and out of the system via pipe **21**. The insoluble material remains as a thickened sludge within chamber **19** and is removed from time to time by opening valve **22**.

Hereinabove, it has been explained that after the product of the incineration is dissolved in water, the resulting solution is neutralized and filtered. Although this treatment is generally employed, alternative procedures are suitable. For example, the solution may be neutralized after filtration. Also, instead of applying filtration, the solution may be clarified simply by allowing it to stand in a tank or other vessel whereby the undissolved material will settle out. The neutralization may be applied before or after settling. Another plan is to use the neutralized solution directly, i.e., without any filtration or equivalent step, for the treatment of fresh batches of olives. The small content of undissolved material does not interfere with the processing of the fruit.

ORGANIC CHEMICAL WASTES

Removal from Water

A number of methods have been developed for reducing the chemical oxygen demand (COD) of effluent industrial wastes such as biological sludges from chemical and pharmaceutical plants. Among such methods are the wet air oxidation and the activated sludge digestion processes. In the activated sludge processes, organic sludge is subjected to anaerobic and aerobic bacterial action or both. In the wet air oxidation processes, an aqueous slurry of sludge is subjected to oxidation with air at elevated temperature and pressure.

In a newer process, raw primary activated sludge having a chemical oxidation demand of the order of 40 grams of oxygen per liter is oxidized batchwise with the aid of steam injection under relatively quiescent conditions in heavy, thick-walled reactors at around 525°F and 1,750 psig. These reactors are charged once every 24 hours and reduce the chemical oxidation demand of the effluent liquid to around 10 g/l or by about 75%.

All the above outlined processes basically only concentrate the sludge so that it can be disposed of more rapidly, are relatively costly and additionally, the conditions under which they are carried out are not severe enough to dispose of very stable and resistant contaminants such as maleic acid, fumaric acid, phthalic acid, terephthalic acid and the like. Currently, these are handled by bacterial oxidation in ponds but this requires considerable land to hold the waste stream and is slow.

A process developed by *E.L. Cole and H.V. Hess; U.S. Patent 3,772,181; Nov. 13, 1973; assigned to Texaco Inc.* involves heating such a wastewater stream under turbulent flow conditions to temperatures of 400° to 700°F and pressures of 300 to 3,100 psi in the presence of air or of oxygen thereby splitting off carbon dioxide. The oxidized waste stream, which has a considerably reduced chemical oxygen demand, is continuously fed to a hot contacting zone and is in heat-exchange relationship with the incoming effluent stream.

A device developed by *P.S. Sharpe; U.S. Patent 3,892,190; July 1, 1975; assigned to Brule, C.E. & E., Inc.* involves the thermal oxidation of gaseous, liquid or solid organic chemical wastes in a cylindrical oxidation chamber by passing fuel, air and waste through the oxidation chamber in an intimate mixing, spiral, rotating motion at low-pressure to effect high-efficiency oxidation of the waste in the central portion of the oxidation chamber, the vessel walls of the oxidation chamber being cooled by excess air moving spirally in the peripheral portion of the oxidation chamber; and passing the products of oxidation and excess air from the oxidation chamber through an open exhaust end. The oxidation chamber in combination with a water scrubber and fail-safe exhausting means provides a chemical waste disposal apparatus resulting in minimal pollution. Such a device is shown in Figure 99.

The thermal oxidizer shown as **10** comprises cylindrical oxidation vessel wall **11**, having closed feed end **12** and open exhaust end **13** defining oxidation chamber **14**. Closed end **12** has air ports **15** tangential to the axis of chamber **14** and toward the periphery from the center of feed end **12**. The number of air ports toward the periphery of feed end **12** is not critical, but must be adequate to furnish sufficient air for oxidation and combustion with an excess of air for cooling the oxidation chamber walls. Liquid port **16** is located in the central portion of the feed end and has nozzle **17** which provides for injection of liquid fuels and wastes and waste solid containing slurries in a manner which provides conical distribution as the waste progresses away from the nozzle through combustion chamber **14**. Nozzle **17** represents a single nozzle, but multiple nozzles are also suitable.

FIGURE 99: APPARATUS FOR THERMAL OXIDATION OF LIQUID ORGANIC CHEMICAL WASTES

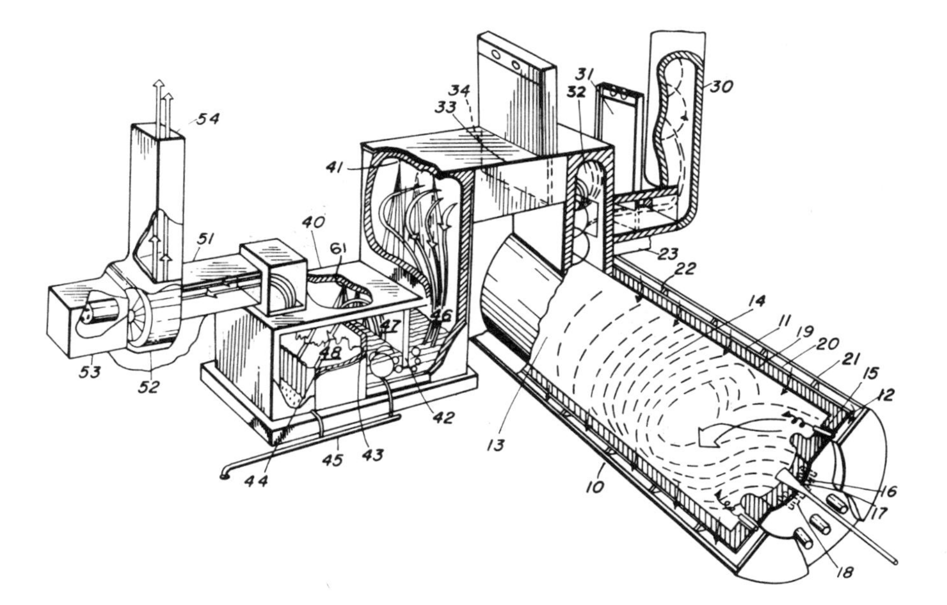

Source: U.S. Patent 3,892,190

Surrounding central liquid port **16** and spaced about the central portion of feed end **12** are a series of gas ports **18** providing injection of gaseous fuel or waste which may be in the same general tangential direction as air from tangential ports **15** or may be parallel to the longitudinal axis of the oxidation chamber. The number of gas ports **18** is not critical but must be adequate to furnish sufficient gaseous fuel for efficient combustion and oxidation when gaseous fuel is utilized.

Cylindrical oxidation vessel wall **11** is shown comprised of fire-brick lining **19** and outside the fire-brick may be a circumferential support of noninsulating material **20**. Jacket **21** surrounds the cylindrical oxidation vessel and is spaced therefrom. Baffles **22** are positioned in the space between jacket **21** and vessel wall **11** so that when air is forced into the space between the jacket and vessel walls at entrance **23** the air is spirally directed along the exterior of the oxidation vessel toward feed end **12**, as shown by the arrows, preheating the air and cooling the exterior of the oxidation vessel.

Air is supplied to entrance **23** by a supply blower, not shown. Air is supplied to entrance **23** at suitable pressures which may be in the range of about 2 to 6 inches of water. The air having a rotating motion imparted to it by the spiral path through the space between the vessel wall and jacket passes through air ports **15** with a rotating motion in addition to the tangential direction imparted by the tangential arrangement of ports **15** with respect to the longitudinal axis of the oxidation chamber. When liquid fuel is used supplemental air may be provided through gas ports **18** to provide sufficient air for combustion.

Thus, the air may enter the oxidation chamber at pressures in the range of about 1 to 4 inches of water and in tangential streams with rotation of the air within each stream. This motion provides intimate mixing and energy exchange between the air, waste and fuel, thereby affecting efficient oxidation of the waste.

Fuel is provided in the gaseous form to gas ports **18** or in the liquid form to liquid ports **16** (from a fuel source not shown) in sufficient quantity to insure complete combustion within chamber **10**.

The open end of cylindrical oxidation chamber **13** is in communication with exhaust means generally shown as **32** for emission of the effluent of the oxidation chamber to the atmosphere. It is desirable to have an expansion chamber located between open end **13** and the final stack open to the atmosphere. It is desirable to pass the products of combustion from the exhaust of the combustion chamber into the expansion volume reducing the velocity of flow of the gases and causing particulate matter to settle to the bottom of the expansion chamber prior to passage into the final stack to the atmosphere.

It is apparent that any pollution control device for removing undesired materials from the stack effluent which operates under low pressure conditions with high volume may be positioned between the exhaust from the combustion chamber and the stack to the atmosphere.

Briefly, the exhaust gas from the oxidation chamber passes through conduit **33** into liquid scrubber **40**. Conduit **33** may contain any suitable energy recovery system to utilize the heat in the exhaust gases. The gas containing particulate matter and/or noxious gases enters the liquid scrubber in section **46** and the spray nozzles **41** remove the larger particulate matter to the bottom of section **46**. The gas moves downwardly in section **46** as the spray from nozzles **41** increases its velocity and strikes the fluid in **46** at a relatively higher velocity. The greater portion of the remaining particulate matter is removed in the liquid by change of direction and relatively high velocity in passing beneath partition **42** into section **47**. The gas rises through the liquid in section **47** at a relatively slower rate permitting solution or desired reaction of the noxious gases in the liquid and the treated gas stream then passes out of the liquid and upward countercurrent to the direction of liquid spray from nozzles **61** in section **47**.

The treated gas passes through the opening above the partition **43** into section **48** wherein the gas is passed through any desired demister and/or a packed column indicated as **44** through conduit **51** and forced by blower **52** driven by motor **53** up the clean effluent stack **54** to the atmosphere.

In order to provide an apparatus of the greatest versatility, that is which is applicable to the oxidation of a wide variety of chemical wastes, effluent stack **30** may be provided to the open atmosphere having shutter **31** open and shutter **34** closed for passage of effluent gases directly from the oxidation chamber to the atmosphere. As pointed out above, the oxidation of some chemicals results directly in gases which may be passed to the atmosphere without further treatment.

For use in such applications, the gases are advantageously directed through stack **30**. For use in oxidations resulting in gases and particulate matter which require further treatment such as the liquid scrubber **40**, shutter **31** is closed and shutter **34** is opened directing the gaseous exhaust from the oxidation chamber through liquid scrubber **40**. Stack **30** is also useful as a fail-safe system permitting uninterrupted operation of the oxidation chamber even in the event of a power failure or over-temperature conditions. The fail-safe system then diverts the flow of gases from the scrubbing systems directly to stack **30** by closing shutter **34** and opening shutter **31** to permit quick venting. The following is a specific example of the operation of this apparatus.

The thermal oxidation apparatus as shown without any scrubber and passing the exhaust from the oxidation chamber directly to the atmosphere was operated with liquid chemical waste from a chemical plant having the following analysis:

Chemical	Volume Percent
Ethyl acetate	76
Ethyl alcohol	16
Methylene chloride	2.1
Acetonitrile	1
Sodium chloride	1
Water	3.9

The liquid waste was delivered to the oxidation chamber at 350 gallons per hour and kerosene fuel was delivered at 150 gallons per hour. The total heat input rate was 50×10^6 Btu per hour. The undesired liquid wastes were completely oxidized in the oxidation chamber.

The stack gas was measured and analyzed generally following Federal EPA test procedures showing the following properties.

Temperature, Average	1030°F
Velocity, Average	25.85 fps
Volume (dry basis)	5.49 x 10^5 scfh
Moisture	6.314%
Particle Concentration (dry basis)	0.03774 gr/scf*
Isokinetic Ratio	95.91%
CO_2	3.4 volume percent
O_2	16.6 volume percent
SO_2	2.94 lb/hr
SO_3	0.90 lb/hr
NO_X	8.19 lb/hr
HCl	14.33 lb/hr
Excess Air Ratio	3.69

*Corrected to 12% CO_2, 0.1332.

ORGANIC VAPORS

Removal from Air

A system developed by *D.C. Kennedy; U.S. Patent 3,798,876; March 26, 1974; assigned to Rohm and Haas Company* is an adsorption system for vaporous organic compounds which system can be efficiently and economically regenerated with either steam, air, or inert gases for multiple cycles of use.

Examples of industrial uses which can be enhanced from an ecological viewpoint by application of this system are: application of paints, lacquers and polymeric films, storage tanks, polymerization and other chemical reactions, dry-cleaning operations, and petroleum refining operations.

The preferred materials of choice for this system application are high-surface area, acrylic, polymeric adsorbents. They appear to offer the greatest range of applicability combining high capacity with superior regeneration properties, and rapid kinetics. They are therefore preferred over the nonpolar aromatic adsorbents in most cases. However, high surface area, nonpolar adsorbents still possess advantages over activated carbon. Indeed, they might be preferred over the acrylic adsorbents for certain applications.

ORGANOTIN COMPOUNDS

See Antifouling Ship Paint Residues (U.S. Patent 3,981,252)

OXO PROCESS EFFLUENTS

Removal from Water

A process developed by *S. Speth, G. Küppenbender and W. Jansen; U.S. Patent 3,953,332; April 27, 1976; assigned to Ruhrchemie AG, Germany* permits the purification of wastewaters used in the Oxo process whereby the waters are purified to such an extent that they can be discharged into waterways without adversely affecting the biological life therein. Figure 100 shows a suitable form of apparatus for the conduct of such a process.

The process is conducted suitably by pumping wastewater into a tilted gravity separator, the water being sucked from a wastewater channel by means of a pump **1**. Initially, it is delivered to the introduction zone or bay **2** of the gravity separator. At such time there is delivered to such introduction zone warm water from a steam distillation process, such warm water entering the introduction zone at a point above the point wherein the wastewater is introduced to the zone.

The warm water from the steam distillation process enters the introduction zone through line **16**. In the introduction zone **2** there is an initial rough separation of an organic product swimming sludge phase which is continuously removed by the aid of a paddle **3** over a downward-sloping discharge **4**. Heavy sludge is separated from organic product and water in a tilted plate pack **5**. Organic product ascends and is continuously drained off over the downward sloping discharge **4**. Heavy sludge deposits in a collecting vessel **6** and is intermittently drawn off by means of pump **10**.

Clarified water ascends in a water compartment **7**. It is raised to overflow a weir and to enter into outlet passage **8**, the latter serving as an additional settling tank. By means of siphon **9**, a definite water level is maintained in outlet passage **8**.

FIGURE 100: APPARATUS FOR TREATMENT OF OXO SYNTHESIS PROCESS WASTEWATERS

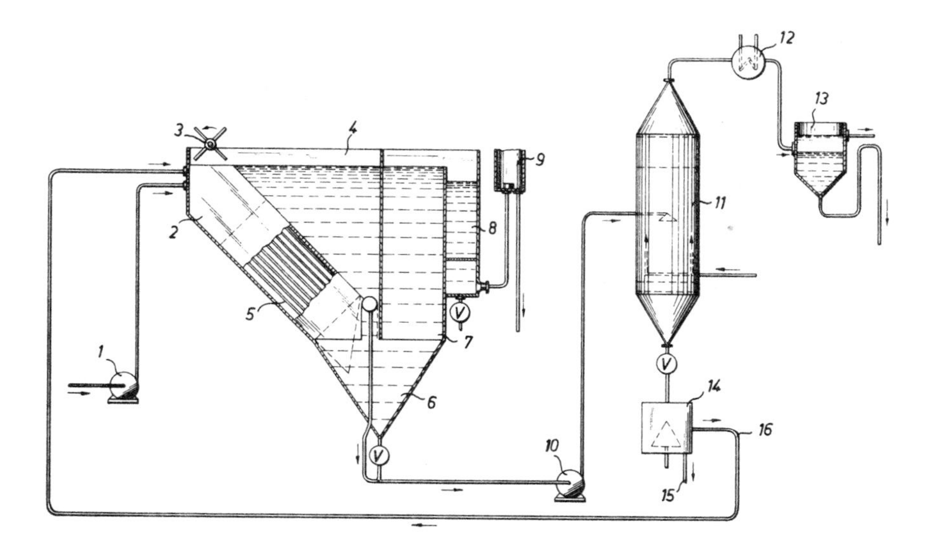

Source: U.S. Patent 3,953,332

With the aid of pump **10**, organic product, swimming sludge and heavy sludge are introduced into steam distillation column **11**. By introduction of steam at 1.6 bars a vapor temperature of 98°C is maintained at the head of the column with the aid of a Samson type regulator. This temperature is sufficient for the evaporation of organic constituents of the product. The level in the steam distillation column **11** is maintained by discontinuous addition of water and heavy sludge from collecting vessel **6** and controlled by a conventional level regulator. The distillation product is cooled in cooler **12**, enters phase separator **13** and is separated into an organic phase, which is drained off for further treatment. The organic phase is separated from water therein.

Water and sludge from steam distillation column **11** are removed as bottoms and introduced into decanter **14**. A sludge phase **15**, containing solids, is drawn off and separately worked up. A decanted water phase **16** is recycled to the introduction zone of the tilted plate separator to heat up the entering unpurified wastewater entering such zone through the line connected to the pump **1**. The following is a specific example of the conduct of such a process.

Wastewater from an Oxo synthesis process is adjusted by addition of 2.5% aqueous sodium hydroxide solution to a pH value of about 11.5 whereby dissolved metallic impurities are precipitated. This initial precipitation serves to insure that all metallic impurities are in a definite form and are in a controllable and uniform state. Thereafter, they are introduced by means of a conventional slow running immersion pump disposed about 20 cm below the water surface into a tilted plate separator of the type shown in the accompanying drawing.

The entering wastewater is uniformly distributed therein by use of a slotted tube. It enters the introduction zone **2** of the separator at a temperature of about 30° to 35°C. At a point above the point at which the entering unpurified wastewater is introduced into the introduction zone there is introduced into such introduction zone water of higher temperature in an amount to maintain the temperature of the wastewater at 33° to 38°C. When treating 70 cubic meters per hour of wastewater there is introduced to the introduction zone 300 liters per hour of water at 70° to 80°C, the water coming from a steam distillation process. The separator employed has 10 elements.

In this separator, water is separated from swimming sludge comprising organic substances flowing upward to the plate packed inlet, and from small amounts of metal oxides descending and being collected as sedimentary sludge below the plate pack. Its content of substances which are soluble in petroleum ether ranges within the limits of the current tentative standard so that it can be introduced into waterways.

Swimming sludge and organic compounds are continuously drawn off by means of a plunger pump having means for controlling the revolutions thereof. The swimming sludge is united with sedimentary sludge withdrawn from the lower end of the separator and united with the swimming sludge. A plunger pump with means for controlling the revolutions thereof can facilitate this operation. The united sludges are introduced into a stripping column at a rate of 500 kg per hour. The stripping column is maintained at a pressure of 1.6 bars by-product steam. The organic constituents which are distilled off in the steam distillation process are condensed in a cooler and are separated from the water. The resulting hot water (300 liters per hour) having a temperature of about 70° to

80°C is recycled to the separator as hereinbefore mentioned in order to regulate the temperature thereof.

The water containing sludge from the sump of the steam distillation column is introduced into a decanter from which hot filtrate is also used for controlling the temperature in the introduction zone of the separator. The hot water introduced into the separator is also uniformly distributed over the entire area of the receiving bay by means of a slotted tube.

By use of the herein described process of separating components of wastewater and purifying such wastewater in a tilted-plate separator the content of petroleum ether soluble constituents is below about 35 parts per million. If, however, the process is conducted without the addition of hot water, i.e., without the regulation of the temperature of the components in the introduction zone the resulting wastewater contains about 120 parts per million petroleum ether soluble constituents.

OXYDEHYDROGENATION PROCESS EFFLUENTS

In an oxydehydrogenation process for diolefin manufacture, as much as 4 mol percent of the olefin feed may be converted to oxygenated hydrocarbons such as carboxylic acids, aldehydes, ketones, etc., especially acetic and propionic acids and acetaldehyde, the nature and quantity of these compounds depending on the conditions under which dehydrogenation is effected. Under normal plant operating conditions, these oxygenated by-products will be ultimately vented to the atmosphere and/or discharged with wastewater from the process, depending upon the separation and recovery processes employed and their operating conditions.

However, it has been found that these by-products are toxic and result in damage to property, particularly crops and foliage and are probable contributors to photochemical smog and haze, especially when vented as aerosols. It is not only desirable to eliminate or at least reduce this source of air and water pollution, but such control is essential in many locations.

Removal from Water

In a process developed by *T. Hutson, Jr. and R.E. Riter; U.S. Patent 3,646,239; February 29, 1972; assigned to Phillips Petroleum Company*, excess purge water containing oxygenated hydrocarbons resulting from hydrocarbon dehydrogenation processes is rendered nontoxic by the conversion of the oxygenated hydrocarbons to water and carbon oxides in the presence of microorganisms, e.g., saprophytic bacteria, protozoa, yeast and fungi, under aerobic conditions.

Preferably the purge water prior to treatment with microorganisms is reboiled or stripped with a hot gas to reduce the concentration of oxygenated hydrocarbons.

In a process developed by *R.A. Hinton and J.E. Cottle; U.S. Patent 3,679,764; July 25, 1972; assigned to Phillips Petroleum Company,* water containing oxygenated hydrocarbons resulting from hydrocarbon oxidative dehydrogenation processes is rendered substantially nontoxic by stripping with steam to remove the oxygenated hydrocarbons including carbonyls. A portion of the water is returned to the oxidative dehydrogenation process to suppress the formation of additional oxygenated hydrocarbons and another portion is converted to steam for use as stripping medium. The steam stripped water phase can be neutralized with a base prior to reboiling and converted to steam for use as the stripping medium.

A process developed by *E.O. Box, Jr., and F. Farha, Jr.; U.S. Patent 3,823,088; July 9, 1974; assigned to Phillips Petroleum Company* is one in which the aqueous effluent of an oxidative dehydrogenation containing contaminating oxygen-containing organic materials is subjected to oxidizing conditions in the presence of a promoted zinc aluminate catalyst to convert the water to a potable aqueous product.

A process developed by *R.C. Woerner, L.D. Tschopp and C.O. Oelze; U.S. Patent 3,884,650; May 20, 1975; assigned to Petro-Tex Chemical Corporation* involves reducing carbonyl compounds in wastewater from an oxidative dehydrogenation process by stripping an overhead, enriched in carbonyls, condensing an aqueous portion of the overhead which contains the predominate amount of the carbonyls in the overhead, and destroying the carbonyl concentrate. Operating the stripping unit under vacuum essentially eliminates fouling of the stripping unit, caused by aldehyde polymer formation. Such a scheme is shown in Figure 101.

The hydrocarbon stream **1** comprising butene-2 as the major component is dehydrogenated to butadiene-1,3 in reactor **A**. The feed to reactor **A** comprises butene-2, air, and steam. The effluent stream **2** comprises butadiene-1,3, unreacted butene, carbonyl compounds, steam, noncondensable gaseous components such as nitrogen and various dehydrogenation by-products such as CO_2, CO and the like.

The composition entering contacting zone **B** as a stream **2'** exclusive of any water will comprise from 3.5 to 80 mol percent of unsaturated organic compound, from 0.0005 to 2.5 mol percent of carbonyl compounds and from 20 to 93 mol percent of noncondensable gases, all based on total mols of gaseous composition. The oxygen content will generally be less than 10 mol percent of stream **2'**. The stream **2'** may also contain hydrocarbon by-products and unconverted hydrocarbons such as olefins or unsaturated hydrocarbons. The composition of stream **2** exclusive of any water may be within the same ranges as given for stream **2'**.

The aqueous medium will remove the principal amount of carbonyl compounds contained in stream **2'**. The gaseous stream passes out of the contacting zone **B** as stream **3** to be subjected to further processing for recovery of product, for example, butadiene-1,3, recyclable butenes and the like. The aqueous effluent **4** from contacting zone **B** may contain from 0.015 to 0.12 mol percent of carbonyl compound, from 0.004 to 0.05 mol percent of hydrocarbon compounds and from 99.78 to 99.97 mol percent of water. In addition there may be small amounts of carbon dioxide, nitrogen and oxygen dissolved in stream **4**, but this is of no consequence in regard to water purity.

The system is provided with a feed drum **C** where a portion of the hydrocarbons contained in stream **4** are flashed overhead. The flashing is the result of pressure

FIGURE 101: APPARATUS FOR LIQUID WASTE DISPOSAL FROM OXIDATIVE DEHYDROGENATION

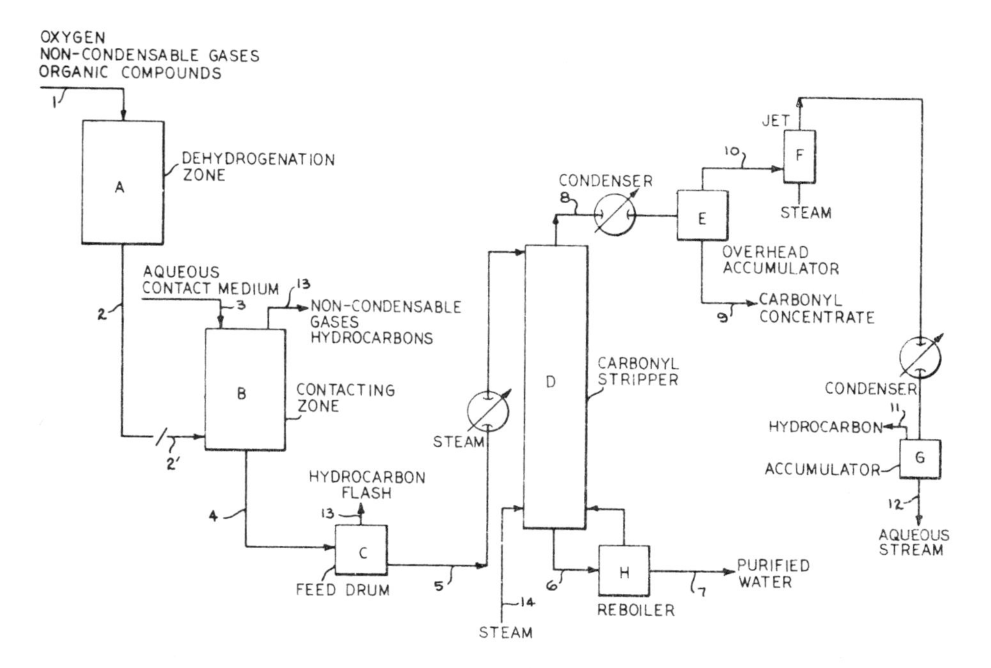

Source: U.S. Patent 3,884,650

reduction in drum **C**. Stream **13** will contain butadiene as well as some carbon dioxide. Only a small portion of stream **4** will be flashed as overhead **13**, however, from 60 to 95 mol percent of the hydrocarbons other than carbonyl compounds may be removed from stream **4**.

The stream **5** from feed drum **C** is passed through a heater where steam is used to adjust the temperature to a temperature in the range of 80° to 220°F. The feed stream entering carbonyl stripper **D** as noted can have a greatly reduced hydrocarbon content from stream **4** or may be substantially the same composition as stream **4** with the hydrocarbons being taken overhead in stream **8** from stripper **D** with the carbonyl compounds.

The carbonyl stripper **D** is any apparatus or equipment which will carry out the stripping as described and can be a plate tray, cap tray, bubble tray, baffle tray or combination arrangements of such components. An internal reflux is maintained by control of conditions within the tower, thus eliminating the need for an external reflux. The retention time of the liquid on the trays will depend to some extent on the nature of the water being treated; however, the holdup time in the stripper **D** will generally be from a few minutes to a few hours, e.g., 30 minutes to 4 to 5 hours. Control of the flow rate of overhead stream **8** in the stripper **D** can be maintained in any conventional manner, however, as shown a reboiler **H** serves quite well in this regard. Steam can also be added directly to tower **D** via line **14**.

The carbonyl compounds are substantially completely removed from the aqueous stream **5**. Similarly, the hydrocarbons are substantially removed to provide a stream **6** which is at least 99.90 mol percent water and more preferably contains less than 40 ppm of combined carbonyl compounds and hydrocarbons, and more preferably less than 20 ppm. A portion of stream **6** coming from the bottom may be employed in reboiler **D**, and the balance discharged as stream **7**, having substantially the same composition as stream **6**. The reboiler may, for example, be a kettle type or the circulating type.

In the case of the former type, where stream **7** is drawn from the kettle there may be a lower concentration of carbonyl compounds than found in stream **6**, since some carbonyl will flash in the kettle. Where stream **7** is drawn off before passing into the reboiler, i.e., circulating reboiler, the composition of stream **6** and **7** will be the same. Stream **7** can be cooled and recycled to contacting zone B or sent to the biological pond for final treatment and returned to the surface water supply.

The overhead stream **8** contains the carbonyl compounds from stream **5**, concentrated several hundred times. The gaseous overhead contains at least 80% (by weight) and more preferably at least 85% of the carbonyl compounds entering stripper **D** in stream **5**. This gaseous stream **8** will normally comprise from 0.2 to 3.8 mol percent carbonyl compounds, from 0.008 to 0.5 mol percent hydrocarbons, and from 95.7 to 99.8 mol percent water.

This gaseous overhead **8** is passed to a condenser and the aqueous portion collected in overhead accumulator **E**. The stream **9** will contain substantially all of the carbonyl compounds from overhead stream **8**. Stream **9** is now in a concentrated form which can be destroyed by incineration.

PACKINGHOUSE EFFLUENTS

Removal from Water

Generally speaking, free fat and oil, i.e., not emulsified fat and oil, present no serious problems in regard to separation from water as they will generally float to the surface and can be skimmed off. Emulsified fats, on the other hand, stay in solution causing severe pollution problems.

In this connection, it has been the usual practice in the past to run the wastewater from a packinghouse to a settling tank or basin having baffles wherein the water could set for a half hour or so and the free fat would rise to the top and be skimmed off. The emulsified fat would, of course, remain in the water and would accompany it to the sewers.

Various means such as aeration and complex apparatus have been employed in attempts to deemulsify the wastewaters. Usually, however, unless the emulsified oil was very valuable, no effort was made to separate it from the water that was eventually passed to the sewers and hence to the streams and rivers.

In processes where water is reused, the oil can be removed from the system by coagulation with aluminum sulfate and alkali, followed by filtration through a nonsiliceous filter medium. The oil is caught in the floc and is filtered out of the system. Periodical backwashings of the filter with hot caustic soda are required.

It should be noted that the processes used to completely remove the oil from the water are clearly uneconomical for use in cleaning up wastewater from packinghouse and edible oil operations. Yet on the other hand, it is desirable to remove these low-grade oils from sewage as they may be used in industrial cutting oils, etc. Further, it is now necessary to clean up the water prior to its discharge into streams, rivers and seas.

In this connection, several systems are known to those skilled in the art. Generally, these are systems wherein either cathodes and anodes are alternately spaced or which use a current with sufficiently high anode voltage that vigorous formation of gas bubbles takes place on both anodes and cathodes and these bubbles rise and tend to carry some of the greasy fatty substances to the surface.

A process developed by *H.T. Anderson; U.S. Patent 3,816,274; June 11, 1974; assigned to Swift & Company* is one in which wastewater containing oil and water emulsions and dissolved or suspended solids is deemulsified and clarified by creating a three-dimensional anolyte stream resulting from the careful placement of anodes and impressing direct or galvanic current through the water.

Wastewater is first contacted with an anode system in a restricted zone so as to give substantially all of the wastewater a rapid pH change of several units and is then conveyed to a second zone wherein a three-dimensional anolyte stream is formed causing the oily particles to float to the surface of the water where they can be skimmed off.

PAINT MANUFACTURING EFFLUENTS

The water-thinned paints, in relatively few decades, provide the major portion of the domestic residential finishing of painted interior and exterior surfaces. The water-thinned paints are commonly identified as latex paints.

It has been found with water-thinned paints that the film-forming resin and other ingredient materials of a proteinaceous nature act as a nutrient to many living organisms. The addition of an agent to retard growth of these living organisms is necessary to preserve the appearance of the paint both before and after application upon the work surface. Various preservatives and fungicides have been employed, but organic mercury compounds are predominately used in domestic paint manufacture. These organic mercury compounds include phenyl mercuric acetate, phenyl mercuric oleate, and other similar types of alkyl-mercuride and organic acid derivatives.

The manufacture of a major portion of the water-thinned paint is by a batch process. At the completion of a batch preparation of the paint, the various mixing and formulation vessels must be cleaned in preparation of receiving the ingredients from which the next succeeding batch of paint is to be prepared.

This cleaning operation is especially critical when the subsequently prepared paint is of a substantially different color than the preceding batch of paint manufactured. Generally, the batch preparation vessels are washed with water, with substantial agitation and possible addition of detergent additives, so that all the residual amounts of the paint are removed in the wastewater generated from this cleaning operation.

Usually, the wastewater contains very large concentrations of the organic mercury compound employed as the preservative and fungicide. An estimate, in average manufacturing procedures, has placed the concentration of the organic mercury compound at about 500 parts per million of the wastewater. In addition, the wastewater contains suspended water-dispersed solids which are the solids residue of the water-thinned paint. The high concentration of mercury and other suspended solids makes unacceptable the discharge of the wastewater into public water courses without severe chemical treatment to reduce the mercury and solids to acceptable levels.

The environmental restrictions require a water to have a mercury content of not more than ten parts per billion before being discharged into public water courses. Additionally, the water must not have a solids content greater than a few hundred parts per million. Otherwise, the discharge of the wastewater with suspended solids produces clouding and coloring of public water courses.

Removal from Water

A process developed by *R.A. Shoberg and P.E. Cravens; U.S. Patent 3,836,459;*

September 17, 1974; assigned to Petrolite Corporation yields satisfactory removal of organic mercury compounds and dispersed solids from wastewater generated in the manufacture of water-thinned paints. The wastewater receives an addition of phosphoric acid until the pH is adjusted to about 3.0.

A major portion of the organic mercury compound (e.g., phenyl mercuric acetate) and water-dispersed solids precipitate from the wastewater. The precipitate is separated from the wastewater. Then, the wastewater is adjusted to a pH of at least about 7.5 by the addition of calcium hydroxide so that a substantially complete precipitation of the residual organic mercury compound and water-dispersed solids is obtained.

The precipitated solids are removed from the relatively mercury- and solids-free water phase. Then, the water phase is passed to a subsequent utilization. The precipitated mercury and solids may be treated to provide a substantially dry material suitable for landfill. A wastewater with zero mercury content can also be produced. Figure 102 shows a suitable form of apparatus for the conduct of this process.

A supply of wastewater derived from the manufacture of water-thinned paints is moved through an inlet conduit **11** and suitable mixer, such as mixing valve **12**, into the decanter **13**. The wastewater usually has a pH of about 7, and if not neutral, then it is slightly basic. A stream of phosphoric acid is introduced through an inlet conduit **14** into the stream of wastewater in conduit **11** upstream of the mixing valve **12**.

The amount of phosphoric acid passing through the conduit **14** is regulated by a motor control valve **16** which valve is actuated by a pH sensor **17** providing a control signal represented by a chain line **18**. The sensor **17** and valve **16** are adjusted so that phosphoric acid and water phase mixture has a pH of about 3.0 upon entering the decanter **13**. A pH of about 3.0 is preferred in the operation of the decanter **13**.

However, an intermixing of the wastewater and phosphoric acid to produce an aqueous mixture with a pH in the range of from about 2.5 to about 4.5 provides for satisfactory operation of the process. The mixing valve **12** pressure differential is regulated by a downstream sensor generating a control signal indicated by chain line **19**. The phosphoric acid and wastewater are thoroughly mixed by passing through the valve **12** before being introduced into the settling conditions with the decanter **13**.

The controlled addition of phosphoric acid causes a major portion of the organic mercury compound and water-dispersed solids to precipitate from the wastewater. These water-dispersed solids include the filmogen, binders, fillers, pigments and other like water-dispersed solid materials. The described pH adjustment by acidification of phosphate salts with mineral acid does not produce consistently the results provided by direct addition of phosphoric acid.

The decanter **13** separates a treated wastewater with a highly reduced mercury and solids content from the precipitated solids. For this purpose, the decanter **13** has a construction such as batch settling tanks, unit settlers, or sedimentation tanks.

FIGURE 102: PROCESS FOR MERCURY REMOVAL FROM LATEX PAINT WASTEWATER

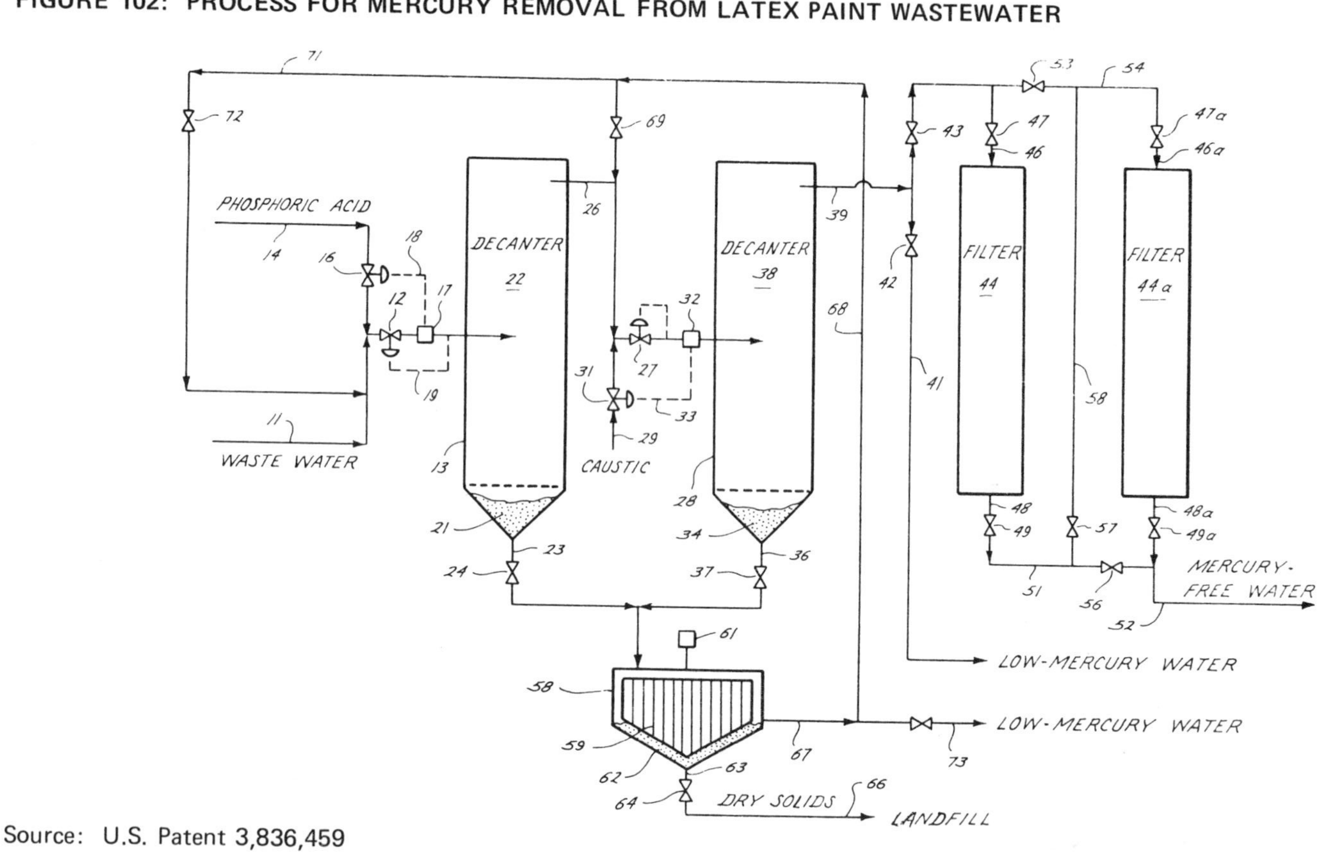

Source: U.S. Patent 3,836,459

Preferably, the decanter **13** has an inverted conical bottom in which the precipitated solids **21** accumulate. A relatively mercury- and solids-free wastewater accumulates in the upper portion **22** of the decanter **13**. For example, the wastewater in the upper portion **22** may have a mercury content less than 10% of the incoming wastewater in conduit **11**.

The solids **21** are removed through a sludge outlet **23** under the controlling action of a valve **24**. These solids **21** may be removed on a discontinuous or continuous basis, as desired. The solids removed from the decanter usually have a residual water content of between 35 and 50% by weight depending upon the efficiency in solids separation of the decanter.

The treated wastewater is removed from the upper portion **22** of the decanter **13** by an outlet **26** which conducts the wastewater through a mixing valve **27** into the decanter **28**. The mixing valve **27** is adjusted to provide a suitable pressure differential to the fluids flowing therethrough for a complete mixing of the treated wastewater with a metal hydroxide base.

The metal hydroxide base introduced through the conduit **29** in the decanter **28** can be an alkali metal or alkaline earth metal hydroxide. Preferably, these metal hydroxide bases are sodium hydroxide and calcium hydroxide. Calcium hydroxide is the preferred base to be employed in the process. This base is introduced through the caustic inlet conduit **29** and through a motor control valve **31** into the stream of treated wastewater upstream of the mixing valve **27**.

A pH sensor **32** is placed downstream of the mixing valve **27** and produces a signal, indicated by chain line **33**, to operate the motor valve **31**. The pH sensor **32** functions to regulate the flow of the metal hydroxide base so that the resultant base and water mixture has a pH of at least about 7.5 upon entry into the decanter **28**. However, the pH of the base and water mixture may reside from slightly above 7.0 to about 9.0 for satisfactory operation of the steps performed in the decanter **28**. The wastewater and metal hydroxide base mixture within the decanter undergoes a reaction which precipitates as solids **34** substantially all of the remaining water-dispersed solids and residual organic mercury compound.

The decanter **28** has a construction which may be identical to the decanter **13**, although other constructions may be employed. The precipitated solids **34** accumulate within the conical lower portion of the decanter **28**. If desired, sludge raker arms (not shown) may be employed for substantially reducing the amount of occluded water within the solids **34**.

The solids **34** are removed from the decanter **28** in regulated amounts through an outlet **36** by the controlled actuation of the valve **37**. The relatively mercury- and solids-free purified water phase accumulates in the upper portion **38** of the decanter **28**. This purified water phase is removed through an outlet **39**. Then the water phase can be passed through a first outlet conduit **41** to any suitable utilization.

Generally, the purified water phase has a mercury content in amounts of less then ten parts per billion. In many cases, careful operation of the decanter **28** produces a purified water phase having a residual mercury content of less than five parts per billion. If a completely mercury-free water is desired, an additional step is practiced. The block valve **42** in the conduit **41** is closed so that the water phase passes through a valve **43**. The water phase can now pass from an inlet **46** controlled by valve **47** through a bed of adsorbent contained in filter **44**. The filter **44** contains a suitable adsorbent such as Anthrafil or activated charcoal.

These adsorbents, and like materials, adsorb any residual mercury compound contained in the wastewater. The mercury-free water is removed through an outlet **48** from the filter **44**, passes through a control valve **49** into a treated water outlet **51**, and lastly, it is directed to a subsequent utilization through a treated water line **52**. The mercury-free water in the line **52** can be reintroduced into the manufacture of the water-thinned paint or can be delivered for disposal into public water courses. The mercury-free water in the line **52** has for practical purposes a zero solids content (filterable).

If desired, a second filter **44a** may be employed in series or parallel flow to the filter **44**. The second filter **44a** and its associate components are denoted by the subscript a following the numerical designation for like parts in the filter **44**. The filters **44** and **44a** are operated in parallel water flow by opening a control valve **53** in connecting line **54** to place the filters in parallel operation. If series water flow is desired through the filters **44** and **44a**, the valve **53** is closed, the block valve **56** in the water outlet **51** is closed and valve **57** in bypass conduit **58** is opened. As a result, the mercury-free water from the filter **44** passes through conduit **58** and into inlet **46a** of the filter **44a**, and exits therefrom through the outlet **48a** into the line **52** for subsequent utilization.

The solids **21** and **34** removed from the decanters **22** and **23** could be treated separately to reduce the occluded water contents. Preferably, these solids are combined and treated in a simultaneous operation. For this purpose, the solids are placed into a clarifier or thickener **58** which preferably includes a set of rotating sludge raker arms **59** above the conical bottom.

The raker arms are actuated by a prime mover **61** mounted above the thickener **58**. The operation of the thickener is conventional and separates the solids into relatively dewatered solids **62** which accumulate in the lower portion of the thickener. The solids are removed through a sludge outlet **63** in regulated amounts by operation of the valve **64** and sent through a conduit **66** to a suitable disposal. For example, the thickener **58** produces solids **62** with a water content of less than 50%. In many cases, these solids have a water content of less than 35% by weight. The dewatered solids are substantially dry and can be disposed in a landfill or other nonpolluting containment. If desired, the solids could be processed for recovering the mercury and other materials which they contain.

The thickener **58** produces a decant water phase which is removed through a water outlet **67**. The decant water may be either basic or acidic depending upon the composition of the water phases occluded in the solids **21** and **34**, and their relative amounts. If the decant water is acidic, it is preferably passed through conduit **68** and block valve **69** into admixture with the treated water removed by the outlet **26** from the decanter **13**. If the decant water is basic, it is preferred to send the water through the conduits **68** and **71** and through the block valve **72** to be intermixed with the wastewater passing through the inlet conduit **11** upstream of the mixing valve **12**. If neither of these recycle functions is desired, the block valves **69** and **72** may be closed, and the decant water is delivered from the conduit **67** into an outflow conduit **73** for subsequent usages where a residual mercury content in the range of 10 to 50 parts per billion can be tolerated. However, for purposes of the process, the decant water in the conduit **67** should be recycled into one of the incoming streams to the decanters **22** and **38**.

The steps in the operation of the process destroy the organic mercury compound and produce precipitable mercury compounds through the addition of phosphoric acid and the metal hydroxide base. In addition, these two materials intermix with the wastewater and serve to remove substantially all of the water-dispersed solids including residual filmogen, fillers, pigments and binder materials.

As a result, the purified water phase removed from the outlet conduit **41** has an extremely low content of residual mercury material, and also water-dispersed solids. In many instances, this purified wastewater phase can be disposed directly in public water courses. If a zero mercury level water is desired, the filters **44** and **44a** are employed for this result.

PAINT SPRAY BOOTH EFFLUENTS

One of the important problems in the use of sprayed paints is the collection of the overspray paint in an economical and efficient manner. The quantity of overspray paint may represent a substantial portion of the original material. For example, in some paint operations 70% of the paint results in overspray. Irrespective of whether this overspray paint is to be reprocessed so it can be used again, it must be collected and prevented from affecting other operations or environments.

Removal from Air

In general, relatively efficient air washing systems have been used in which the paint particles are captured in a water solution. This solution ordinarily contains surface active agents provided to kill, coagulate, and float the paint material so as to prepare it for easy disposition, for example, by scooping the solid components out of one section of the spray booth.

Such a method works quite satisfactorily with many types of paint materials provided efficient formulations of paint killing materials are used. Thus, lacquer-type materials and most paints can be handled effectively by such processes. However, other types of paints are not readily handled in this manner, and merely form gummy masses which adhere to the surfaces of the spray booth and to the moving parts of the air washer.

One example of such a paint is the material referred to as chassis-black paint which is one used to paint, for example, the understructures of automobiles. There are many different formulations of chassis-black paint, but ordinarily it is formed of gilsonite or asphalt dissolved or dispersed in a petroleum-base solvent such as mineral spirits. A low-boiling solvent of the aliphatic type is generally used so that the parts will dry quickly.

A process developed by *O.M. Arnold et al; U.S. Patent 2,982,723; May 2, 1961; assigned to Ajem Laboratories, Inc.* involves the recovery of overspray paint in liquid or semiliquid form and is particularly applicable to chassis-black paints.

A process developed by *L.C. Hardison; U.S. Patent 3,395,972; August 6, 1968; assigned to Universal Oil Products Company* utilizes a multistage cleaning system which includes the treating of the air stream so as to permit it to be returned to the processing zone, as for example, a paint spray booth in a clean, warm and purified state.

Various systems have been utilized for handling and treating an air stream containing particulates and volatiles from a paint spray booth, or from operations providing an equivalent type of problem. For example, there has been used a method of water-washing the exhaust stream from a spray booth to remove entrained paint particles. Such a system results in a disposal problem for the paint sludge and further causes saturation of the air stream so that it cannot satisfactorily be reused in the spray booth.

For proper paint spraying and drying conditions, there is a need for a dry air environment within the spraying zone. Also, there is a particular cost advantage which can be obtained by eliminating the heat requirements for drying and preheating all of the air that is being introduced into the spray booth. The water-wash system or any other system which precludes reuse of the air and necessarily results in the total discharge of the air stream from the system is thus costly and undesirable.

Briefly, this improved purifying system includes at least one moving metal screen type of belt for collecting particulates, a burn-off zone for the continuous oxidation and removal of the deposited material on the belt and at least one fume elimination zone usable for the further incineration and purification of combustion gases from the burn-off zone, and/or for use in treating the air stream to be returned to the processing zone which evolves the particulates and combustible materials.

PAINT WASTES

The washwater which is discharged from painting apparatus and especially from those which operate according to the electrophoresis method generally contains a suspension of approximately 0.1 to 0.3% of paint in the water. Since such a paint concentration is approximately ten times as high as is legally allowed in wastewater which is to be passed through sewers or channels, for example, lakes, rivers or the like, it is necessary to reduce the paint content of the water considerably.

This has so far been done by mixing the washwater with a so-called coagulant, for example, $CaCl_2$, which causes small paint particles to agglomerate into larger particles. According to Stokes' equation, the separating out of these particles occurs the more quickly the larger the particles are. However, the period which this known method requires for effecting the floating of the paint particles to the surface of the water is always relatively long.

Removal from Water

A process developed by *R. Eisenmann; U.S. Patent 3,764,013; October 9, 1973; assigned to Otto Durr, KG, Germany* involves separating paint or the like from water, especially for purifying the water before passing it into a lake, river or the like, by intimately mixing the paint-contaminated water with a coagulant and compressed air so as to form a foamy mixture, allowing the foam to float quickly to the surface of the water, wiping the foam off the surface and filtering the water before discharging it from the apparatus.

The apparatus is preferably designed as a circular unit requiring relatively little ground space in which the various separating containers are of an annular shape surrounding each other about a common axis, while the coagulator is mounted above the central cylindrical space within the innermost annular container and extends likewise coaxially thereto. Such a process is shown in Figure 103.

FIGURE 103: APPARATUS FOR SEPARATING PAINT FROM WATER

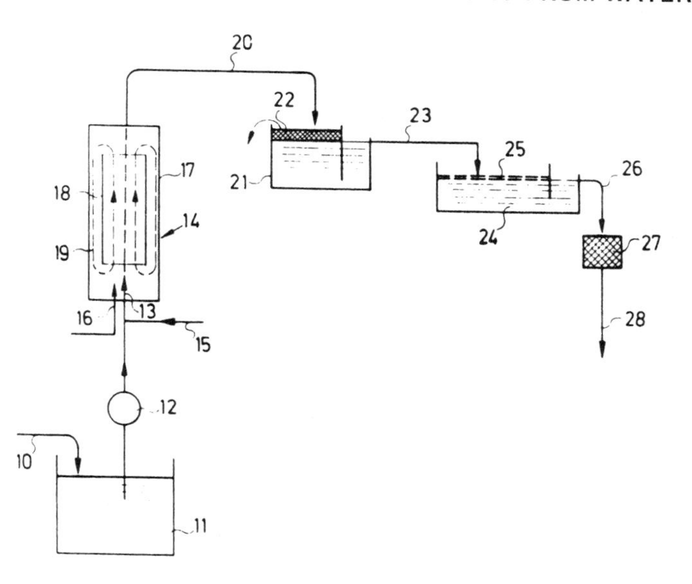

Source: U.S. Patent 3,764,013

The wash water which is mixed with paint particles and comes, for example, from an electrophoresis painting apparatus, passes through a line **10** into a storage or booster tank **11** from which it is then drawn out by a pump **12** and passed under pressure through an injection nozzle **13** into a coagulator which comprises a circulating chamber **14**.

Before the water and paint mixture enters this chamber **14**, it is supplied through a line **15** with a coagulant, for example, calcium chloride. At the same time when the mixture of water, paint and coagulant is injected into the circulating chamber **14**, air is also supplied thereto through a line **16**. Although the air may also be drawn into the circulating chamber **14** by the injection effect of the water mixture passing through the nozzle **13**, it is more advisable to blow the air into the chamber under pressure.

The circulating chamber **14** essentially consists of an outer housing **17** in which an inner pipe **18** is mounted at such a distance from the wall of housing **17** that the liquid which is injected through the nozzle **13** substantially in the axial direction of the inner pipe **18** and fills the circulating chamber **14** may circulate through the inner pipe and the annular chamber **19** between this pipe and the housing **17** and around the upper and lower ends of the inner pipe **18**.

Due to the injection effect of the water mixture passing into chamber **14** through the nozzle **13**, a forced circulation of this mixture together with the air and coagulant is produced which results in an intimate mixture of the different components within the circulating chamber **14** by circulating several or many times around the inner pipe **18** under the effect of the injected current of liquid.

This injected current draws from the annular chamber **19** a certain amount of water which is mixed with the injected current. A corresponding amount of the mixture then flows mostly in the form of a liquid foam from the circulating chamber **14** through a line **20** into the separating container **21** in which the foam rises as a foam layer **22** to the surface of the water. This foam layer may then flow off via an overflow to the outside or it may be skimmed off the surface of the water by simple means, while the water passes through the foam and leaves the separating container through an overflow line **23**.

The water and paint mixture needs to remain within the separating container only for a very short time, for example, for a few minutes, since the floating of the paint particles to the surface of the liquid within this container occurs almost instantaneously. From the overflow line **23** the liquid passes into a clearing basin **24** in which the paint particles which still remain in the liquid will separate therefrom and form a more or less solid cake of paint **25** on the surface of the liquid. This cake may then be easily removed by a stripper, wiper or the like. The water which then contains only very small amounts of paint particles may then be passed through a filter **27**, for example, a coke filter or a paper filter, before it is discharged through a line **28** to the sewer.

A similar process developed by *R. Eisenmann; U.S. Patent 3,772,190; November 13, 1973; assigned to Otto Durr KG, Germany* involves purifying paint-polluted water by adding compressed air and intimately mixing it by circulation in a first operating stage so as to form a foamy mixture, then allowing the foam to float to the surface of the water and removing the foam and thereby partly cleaning the water, then subjecting the partly clean water to a second operating stage in which compressed air is again added to the water and the mixture is again thoroughly circulated and intimately mixed to produce a foam, and the foam is again allowed to float to the surface of the water, and then removing all of the foam from the cleaned water.

A process developed by *J.F. Wallace et al; U.S. Patent 3,528,901; September 15, 1970; assigned to Pressed Steel Fisher, Ltd.* applies to a painting plant in which articles are coated with a water-borne paint and then water-rinsed. The rinse water is conveyed back to the coating tank, the contents of which are conveyed under pressure to a reverse osmosis unit to separate water and low concentrate solutions of pigments and resins, the high concentrate solution of resins and pigments being conveyed to a filter bank, the output from which is passed back to the reverse osmosis unit or under pressure to a second reverse osmosis unit to separate still further relatively high and low concentration of pigments and resins, the circulation of the effluent continuing until separation is substantially complete.

Paint waste generated in painting work has been disposed of by dumping on vacant ground, by reclamation, by open incineration, etc. However, such conventional disposal causes secondary pollution such as the diffusion of solvents and heavy metals into the ground, and the generation of black smoke. Therefore, it is most preferable to dispose of paint waste by complete combustion. However, paint waste is in the form of semisolid or paste and has high adhesiveness. For this reason, it is very difficult to handle paint waste, and it is impossible to burn it continuously and completely by conventional means.

A process developed by *M. Nakayama and Y. Yamahata; U.S. Patent 3,951,581; April 20, 1976; assigned to Mitsui Shipbuilding & Engineering Company, Ltd., Japan* is a process for disposing of solid or paste-like paint waste in which the paint waste is finely comminuted and mixed with waste oil and water, preferably with detergent present, to form a low-sedimentation-rate slurry and the slurry then incinerated to provide substantially complete combustion of the paint waste.

Figure 104 shows the essential features of a batch process for such disposal of paint wastes. The paint waste, waste oil and water are thrown in the shear-type mixing tank **1** in the proper ratio, and a cutting, dispersing and mixing operation in the tank is performed for a predetermined period. The resulting slurry is fed to a storage tank **2** and wherefrom it is fed continuously to an incinerator **3** to be burned.

FIGURE 104: APPARATUS FOR INCINERATION OF PAINT WASTES

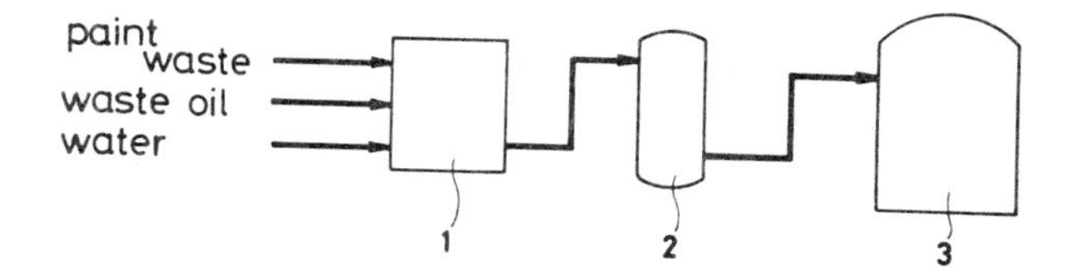

Source: U.S. Patent 3,951,581

PAPER BOX PLANT EFFLUENTS

Removal from Water

In some paper box plants, the flexographic ink wastes are treated separately with inorganic chemicals such as ferrous sulfate and lime or sodium hypochlorite and alum to flocculate and settle the colored material thus producing a clear supernatant. The starch wastes are also usually treated separately by enzymatic hydrolysis which alleviates the pollution problem somewhat by hydrolyzing the long chain, high molecular weight polymers of which starch is composed into glucose which is more readily consumed by bacteria.

Separate treatments such as these are expensive when one considers that a medium sized paper box plant will generate about 400 pounds of starch solids and 40 pounds of flexographic ink solids per day. Thus, an efficient method of treating paper box plant effluents in a single operation would be highly desirable from both an economical and ecological viewpoint.

A process developed by *S.A Hider, J.K. Rogers and C.W. Wilkins; U.S. Patent 3,868,320; February 25, 1975; assigned to Owens-Illinois Inc.* is one in which waste streams from paper box plants consisting primarily of flexographic ink wastes and starch wastes, which are presently treated separately, can be combined and treated as a single system.

Thus, the starch effluent after mixing with the ink effluent can be flocculated more readily by treatment with multivalent metal compounds such as the alkaline earth oxides or salts such as alum or ferrous sulfate. The clarified ink-starch effluent can then be recirculated if desired with a concomitant savings in chemical costs.

PAPERMILL WHITE WATER

In the pulp and paper industry the term white water describes the water which drains off during the formation on a Fourdrinier wire or the like of the pulp web which will constitute a paper or board product after further processing. This white water, sometimes referred to as "back" water, contains finely divided solid material in suspension, including so-called noil fibers and other components of the wood which was used to make up the pulp slurry.

Removal from Water

Numerous methods are known for separating noil fibers from white water. These known methods are based on the use of various substances to cause precipitation, sedimentation or flotation. However, prior art efforts have been directed to the removal, or in some cases, recovery, of noil fibers only, leaving other dissolved and partially dissolved wood material in the white water.

Accordingly, white water treated by prior art methods cannot be repeatedly recycled as fabrication water, and valuable raw material is lost with the wastewater even in those cases where the noil fibers are recovered and not simply discarded. The resulting water pollution problems are notorious.

In a process developed by *E.K. Brax; U.S. Patent 3,873,418; March 25, 1975; assigned to Savo Oy, Finland* noil fibers and wood material which has dissolved or is becoming dissolved in white water from wood processing industries are recovered and used as additional raw material, and resulting clarified water is recycled for use as fabrication water. The pH of the water is controlled during the recovery process and the material to be recovered is caused to agglomerate by the addition of agents whose presence in the final product is beneficial. A polyelectrolyte or the like can be added to enhance flock formation.

A process developed by *G.R. Carta; U.S. Patent 3,778,349; December 11, 1973; assigned to Research Corp.* is one in which a single-cell protein material useful as an animal feed or human food supplement is prepared by culturing the microorganism *Cellulomonas cartalyticum* (ATCC No. 21681). A suitable substrate for culturing this microorganism for growth and production of the cell protein material is an aqueous cellulose-containing medium, such as papermill white water.

A process developed by *A. Boniface; U.S. Patent 3,833,468; September 3, 1974; assigned to Dorr-Oliver Inc.* is one whereby reusable long fibers are recoverable from mill waste effluent, wherein the fibers are entrapped in contaminating impurities comprising coarse particles along with fibrous and nonfibrous fines. The improvement provides a fiber recovery section whereby a suspension of the entrapped fibers is supplied under pressure to a sieve bend type of screen, effecting the decontamination of the fibers, and delivery of recovered fibers suitable for papermaking. Figure 105 shows the essential elements of one variant of the process.

FIGURE 105: SYSTEM FOR RECOVERY OF FIBER FROM PAPERMILL EFFLUENT, INCLUDING A SIEVE BEND SCREEN

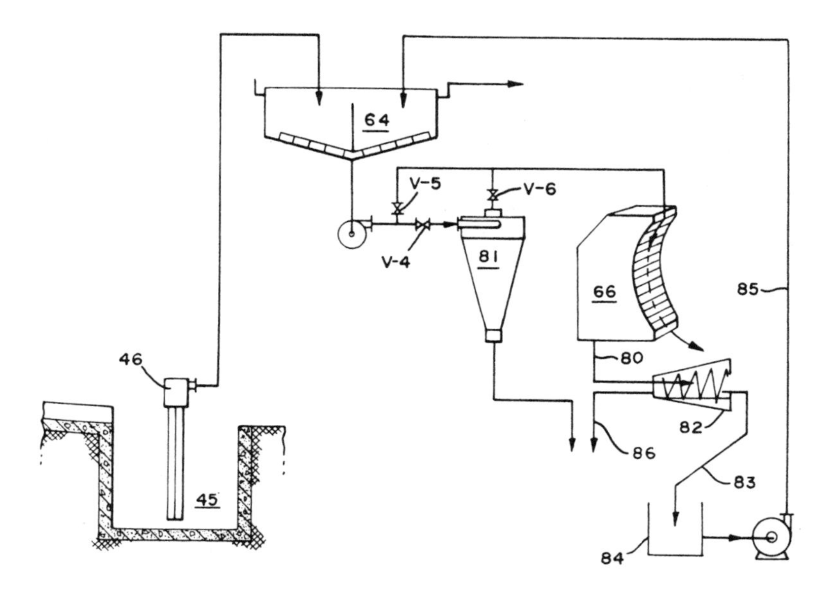

Source: U.S. Patent 3,833,468

The mill waste effluent collects in the sump or pit **45** when a pump **46** delivers it to a clarifier **64**. There is a hydrocyclonic cleaner interposed between the clarifier and the sieve bend screen. That is to say, the underflow from the clarifier is pumped through the cyclone at a pressure sufficient not only to effect therein the desired separation of the coarse impurity fraction from the reusable fibers and fines, but also to provide the feed pressure required by the sieve bend screen.

The underflow **80** from the sieve bend screen is subjected to a dewatering operation. The alternate modes of operation, that is with or without the use of the hydrocyclone cleaner **81**, is indicated by valves **V4, V5** and **V6,** operable to include or exclude the hydrocyclone in the operation.

The example shows a bowl-type dewatering centrifuge **82** which at the wide end delivers overflow **83** into a reservoir **84** from which it may be pumped via line **85** to the clarifier. The highly concentrated separated waste product solids **86** are delivered from the narrow end of the centrifuge.

A process developed by *M. Nakajima and K. Kuwabara; U.S. Patent 3,960,648; June 1, 1976; assigned to Nippon Carbide KK, Japan* is a process for treating a pulp-containing waste liquor which comprises adding a flocculating agent to the pulp-containing waste liquor and separating and removing the resulting flocculated matter, characterized in that about 1 to about 50 parts by weight, per 1,000 parts by weight of the turbid components in the waste liquor, of a dicyandiamide-formaldehyde resin is admixed with the waste liquor, and then aluminum sulfate is added to the mixture.

PARTIAL OXIDATION PROCESS EFFLUENTS

Removal from Water

A process developed by *P. Visser and L.W. Ter Haar; U.S. Patent 3,694,355; September 26, 1972; assigned to Shell Oil Company* is one in which solid particles such as soot are removed from aqueous suspensions by a two-step process, the first step being to agglomerate the particles by contacting them with gentle agitation with a water-immiscible liquid, and then to contact the agglomerate-containing aqueous phase with a continuous phase that is water-immiscible under conditions such that the agglomerates enter the nonaqueous phase.

PARTICULATES

Removal from Air

In the processing of dirty gases evolving from basic oxygen furnaces, electric arc furnaces, rotary kilns or other chemical processes, several methods have been suggested for cleaning or removing the fine particulates from the dirty gas stream. In many instances electrostatic precipitators or wet scrubbers have been utilized, or combinations of these two types of devices have been used for cleaning dirty gas streams. In the case of wet scrubbers, it has been necessary to utilize other clarification equipment in combination with the wet scrubbing equipment. On the other hand, in the case of electrostatic precipitators, expense as well as further clarification of the dirty gas stream have presented several problems.

A system developed by *R.G. Huntington; U.S. Patent 3,984,217; October 5, 1976; assigned to American Air Filter Company, Inc.* is a wet gas cleaning system for removing particulates from a dirty process air stream. This system includes means for introducing the dirty process gas into a drying chamber and simultaneously therewith spraying or otherwise reintroducing the removed particulates or solids in finely divided form as a slurry into the chamber, collecting agglomerated particulates resulting from drying of the mixture of process gas and slurry in a separator while passing the resulting gas stream containing the remaining fine particulates into a high energy wet scrubber. Simultaneously therewith a recycled scrubber solution is introduced thereto and the remaining fine particulates are removed from the scrubber in the form of a slurry. Such a system is shown in Figure 106.

FIGURE 106: WET GAS CLEANING SYSTEM FOR REMOVAL OF PARTICULATES FROM AIR

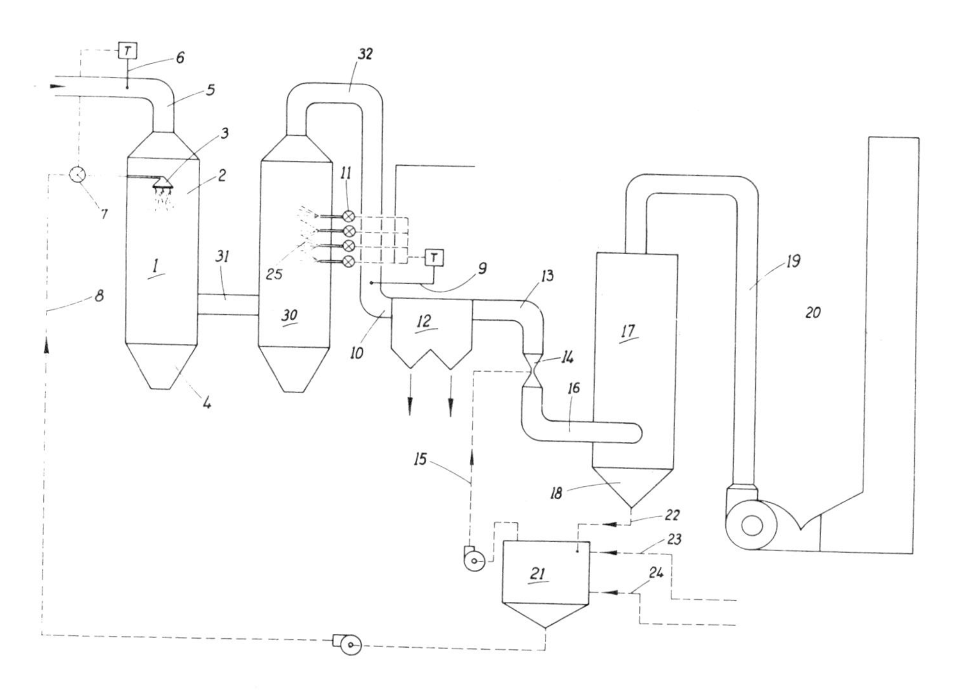

Source: U.S. Patent 3,984,217

The dirty process gas emanating from a source (not shown), such as a metallurgical furnace, is introduced into a spray dryer **1** through conduit **5**, spray dryer **1** being any one of several types known in the art. The hot process gases are introduced at the top portion **2** of the spray dryer **1** and brought into contact with a first slurry sprayed into the spray dryer through slurry spray nozzles **3**.

Upon mixing the particulate materials in the process gases with the sprayed slurry, particulate materials from the gas stream combine with the solid particles in the slurry as the liquid portion of the slurry vaporizes, the combined particulates forming an agglomeration and the vaporized liquid becoming a part of the hot gas stream. The large or heavier agglomerated particles fall to the bottom portion **4** of the spray dryer and are removed therefrom by any known means. The spray dryer may also be provided with a fresh water cooling spray such as the one identified by the numeral **25** which is utilized to cool the process gases even more, if necessary.

As shown, the cooling of the process gases is carried out in a separate cooling chamber **30**. In this embodiment, hot gases exit through conduit **31** and flow upwardly through chamber **30** exiting therefrom through conduit **32**. Cooling liquid is sprayed downwardly through the upcoming gases through nozzles **25**. The spray through nozzles **25** is controlled by a temperature-sensing device **9** disposed within conduit **32**, sensing device **9** actuating control valves **11** in response to the preselected temperature.

A first temperature control means is provided with a temperature-sensing device **6** in the inlet stream to the spray dryer, the temperature-sensing device being in contact with a flow valve **7** which is disposed in the slurry line **8**, slurry line **8** providing the slurry to the spray dryer **1**. The slurry utilized is the underflow from a recycle tank **21**, the slurry being a thickened agglomerated recycle solution which is obtained from the system at a point downstream of the spray dryer, the recycle solution being discussed in more detail hereinafter.

A second temperature control means is also provided with a temperature-sensing device **9** in a duct **10**, duct **10** being the exiting gas duct from the spray dryer **1**, the duct carrying the processed gas and vaporized liquid from the spray dryer to a centrifugal collector **12** where the spray-dried dust and other particulate matter are removed from the gas stream. The second temperature-sensing device **9** is in communication with fresh water control valves **11** which provide cooling water for the spray dryer thereby maintaining a preselected temperature of the exiting gases from the spray dryer.

Gases leaving the centrifugal collector **12** containing fine particulate matter which was not removed by the centrifugal collector **12** then pass to a venturi scrubber **14** through duct **13** wherein a slurry containing processed dust is brought in contact with the gas by way of feed line **15**, the slurry being the overflow from the recycle tank **21**.

From the venturi scrubber **14** the wetted gas stream passes into a centrifugal separator **17** by means of duct **16** with the wetted particles being separated from the wetted gas stream in the form of a slurry and caught in the hopper **18** of the separator **17**, the resulting cleaned gases being removed from the top of the separator **17** through duct **19** and out to the atmosphere through stack **20**.

It is realized that a single high-energy wet scrubber may be utilized in place of the venturi scrubber **14** in combination with the centrifugal separator **17**, but this arrangement has proved to be a preferred apparatus for carrying out the process.

A recycle tank **21** is provided to receive the slurry coming from the separator **17** by means of line **22**. The slurry is generally allowed to settle in the recycle tank wherein the recycle tank is provided with the underflow feed line **8** for the spray dryer **1** and the overflow feed line **15** for the venturi scrubber **14**. Fresh water make-up for evaporative losses in the process are provided for the recycle tank **21** through line **23** and in the case of scrubbing in chemical processing, a neutralizing feed line **24** is provided for adjusting the pH of the solution, as necessary, at the recycle tank **21**.

A process developed by *A.J. Teller; U.S. Patent 3,957,464; May 18, 1976; assigned to Teller Environmental Systems, Inc.* is an essentially adiabatic process for the crossflow scrubbing of particulate material from a gas. The scrubbing liquid is recirculated without cooling. By a combination of a minimum gas temperature of about 150°F, saturation conditions, and a certain amount of turbulence, it has been found possible to achieve particulate nucleation prior to passage of the gas through the scrubber units which, in turn, makes it possible to recycle the scrubbing liquid while maintaining it at a substantially constant temperature. Such an apparatus is shown in Figure 107.

The hot inlet gas stream S_1 at a temperature of about 350°F and a dew point of about 140°F, having already been treated for preliminary particle removal is directed at a velocity of about 50 fps into a venturi **101**. The gas is subjected to a water quench **102** prior to reaching the venturi throat **103**. A plug **104** having an essentially diamond-shaped cross-section may be inserted in the venturi throat and has been found to improve the efficiency of recovery.

In particular, the use of a venturi with a diamond-shaped plug has been found to facilitate the removal of intermediate-sized particles of about 1 to 4 microns at this stage of the process. The removal of particles in the size range of 1 to 4 microns at this stage is believed to be the result of beginning nucleation, and such particles drop out of the gas stream either by action of gravity or by impinging on the diamond-shaped plug and being washed therefrom by the water spray.

These particles are separated from the excess liquid introduced via the spray quench to suitable means such as filter screen **107** located under the venturi from which they are continuously or periodically removed. The excess liquid drains through filter screen **107** to collection chamber **108** for recirculation as hereinafter described.

The turbulent gas stream S, cooled but still at a temperature above 150°F and moisturized to near saturation is next channeled through a set of baffles **105** which are continuously washed by a set of water jets **106**. At this point nucleation occurs among particles below about 1 micron in size, and some of these are removed from the gas stream at this stage either by action of gravity or by impinging on the baffles and being washed therefrom by the water sprays. The wash water is drained to the bottom of the apparatus through filter screen **107** for effecting the separation of particles and liquid and into collection chamber **108**.

FIGURE 107: APPARATUS FOR CROSSFLOW SCRUBBING OF PARTICULATE FROM GASES

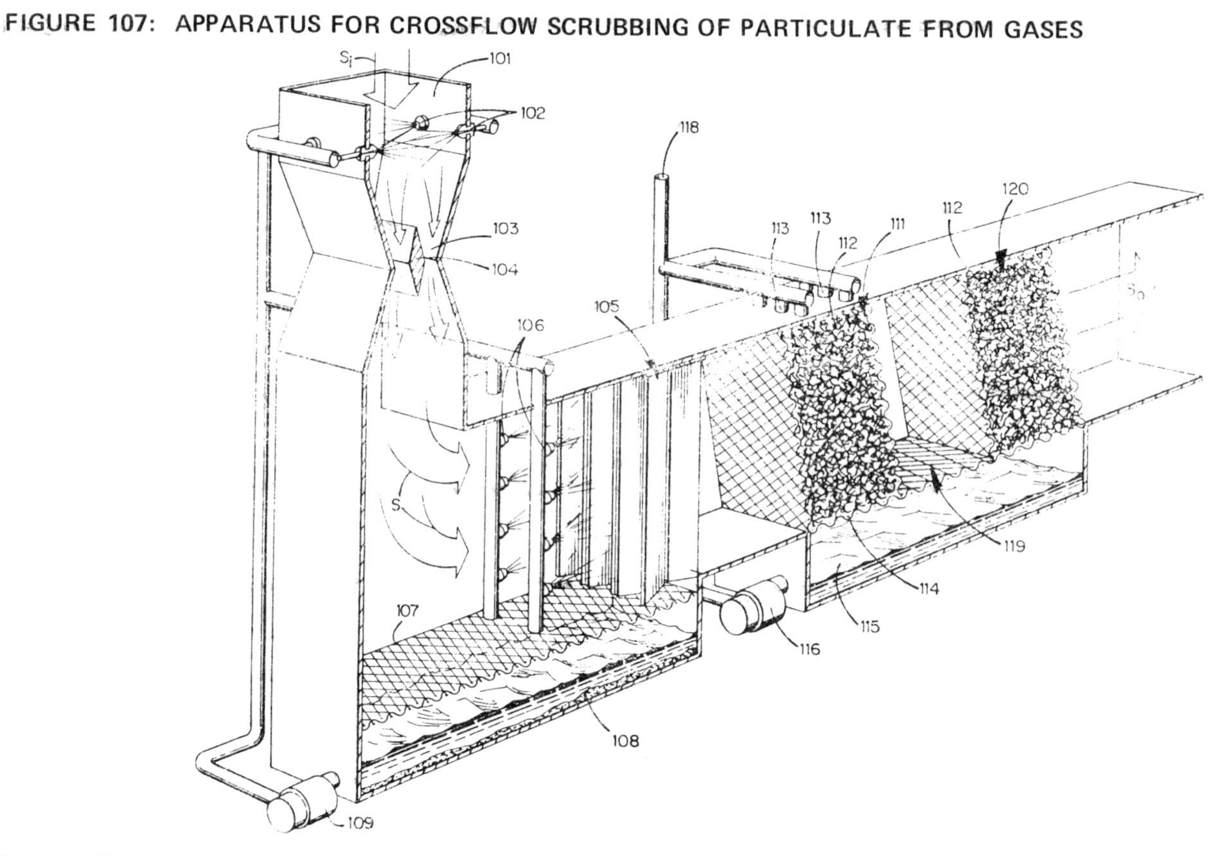

Source: U.S. Patent 3,957,464

A pump **109** is used to recirculate the wash water, and the particles of filter screen **107** are continuously or periodically removed therefrom by suitable means.

Emerging from the baffle system, the turbulent gas is saturated with water vapor at a temperature of at least about 150°F. Nucleation of the particles is complete by the time the gas stream reaches the scrubber. The gas together with the entrained, nucleated particles is then passed in an essentially horizontal path through scrubber bed **111** packed with a suitable packing material **112** where it is brought into crossflow contact with a scrubbing liquid, such as water, which is continuously sprayed into the scrubber bed by water jets **113**.

Although the figure shows a scrubber having a single scrubber bed, the number of beds and the size of the beds are not critical and may be varied to suit individual process requirements. For example, in the case of a gas stream containing both fluoride and sulfur oxide contaminants, two scrubber beds in series may be used to individually remove the contaminants in the manner hereinafter described.

The bed shown is inclined at an angle of about 15° from the vertical in the direction in which the gas is moving. Such a construction is not critical but helps to prevent "channeling" of the gas through the packing and thus insures thorough crossflow contact. The scrubbing liquid together with particulates is drained to the bottom of the scrubbing bed through a support screen **114** which is of such a mesh size that the packing material **112** is supported while the particulates pass through and into a collection and settling chamber **115**. Pump **116** is used to recirculate the scrubbing liquid, and particulates are periodically or continuously removed from the bottom of chamber **115**. In some applications, a single collection chamber may replace **108** and **115** and this single chamber may be served by one pump.

Instead of water, the scrubbing liquid may be brine solution such as a mildly alkaline solution or slurry of about 0.05 to 10 weight percent aqueous solution of sodium hydroxide or sodium carbonate or a lime slurry. The alkaline solutions and slurries are especially useful in reducing the concentration of acid gases such as hydrogen halides and sulfur oxide contaminants to below the allowable limits for release into the atmosphere.

By employing two scrubber beds in series, the first washed with a mild alkaline solution such that the pH is reduced below about 6 as acid is absorbed and the second washed with a stronger alkaline solution such that the final pH is about 8, hydrogen fluoride may be preferentially removed in the first scrubber bed and sulfur dioxide in the second.

Because some of the scrubbing liquid is invariably lost by vaporization or entrainment in the scrubber units, it is usually necessary to supplement the liquid recirculation system with a feed of make-up scrubbing liquid **118**. The treated outlet gas stream S_o leaving the apparatus is saturated at a temperature of about 150°F or higher and is substantially free of particulate matter larger than about 0.10 micron.

Advantageously, after leaving the scrubbing bed **111** the gas stream is passed through an open drainage zone **119** to allow drippage of entrained liquid droplets followed by a demisting chamber **120**. The demisting chamber is packed with any suitable packing material, preferably the same material **112** used to pack the scrubber, and acts to reduce the liquid content of the gas stream to minimize fogging at the outlet point. It is often convenient, as shown, to connect the recirculated water for the baffle wash system **106** and the venturi water quench **102**. In place of water, both the venturi quench and the baffle wash may employ a mild alkaline solution as suggested above for the scrubber liquid to reduce acid gas contaminants.

The following is one specific example of the operation of this process. 100,000 actual cubic feet per minute of kraft recovery boiler gas was fed to the nucleation unit. The gas at 300°F dry bulb temperature and 170°F wet bulb temperature contained particulates (particle size range 0.2 micron to 1.1 microns) at a level of 0.20 gr/scf and 50 ppm total reduced sulfur. The liquid recirculation rate in the scrubber was 5,000 gpm with 1,000 gpm recirculating through the humidifying venturi. The exhaust gas at 170°F saturated has a particulate level of 0.012 gr/scf. The particle size cut point was 0.5 micron, and the system pressure drop was 12 inches water gauge.

See Aluminum Refining Effluents (U.S. Patent 3,900,298)
See Catalytic Cracking Process Effluents (U.S. Patent 3,817,872)
See Glass Industry Effluents (U.S. Patent 3,944,650)
See Lime Kiln Effluents (U.S. Patent 3,793,809)
See Mining Effluents (U.S. Patent 3,792,568)
See Nickel Smelter Effluents (U.S. Patent 3,775,095)
See Seafood Processing Effluents (U.S. Patent 3,721,068)
See Steam Electric Industry Effluents (U.S. Patent 3,726,239)
See Steam Electric Industry Effluents (U.S. Patent 3,890,207)
See Steel Converter Effluents (U.S. Patent 3,820,510)
See Steel Converter Effluents (U.S. Patent 3,844,745)

PENTACHLOROPHENOL

The use of pentachlorophenol and other antimicrobial agents to treat wood and using steam to remove impregnating solvents or using water as the carrier has resulted in aqueous wastewater containing the pentachlorophenol and antimicrobials. It is necessary to recover the major portion of the antimicrobial for economic reasons and now necessary to recover the remainder because of ecological reasons.

The removal of the pentachlorophenol residue and other antimicrobials in the treating water permits discharge of the water into streams without detrimental effect to the environment. However, more advantageously it permits reuse of the waste in chemical processing, removing any ecological problems and producing an economy in water usage permitting plant site locations off riparian ways.

Removal from Water

A process developed by *W.D. Winn; U.S. Patent 3,931,001; January 6, 1976; assigned to The Dow Chemical Company* involves removing antimicrobials such as pentachlorophenol from wastewater by liquid-liquid extraction with a water-immiscible organic liquid in a packed column; recovering the antimicrobial from the organic phase by distillation if desired; and, treating of the aqueous phase to remove residual antimicrobial and organic liquid. The process recovers about 99% of the antimicrobial pentachlorophenol in the wastewater producing a wastewater containing less than 30 ppm pentachlorophenol and less than about 2% organic liquid, both of which may be removed by conventional carbon absorption or soil percolation to acceptable ecological levels for introduction into a stream or reuse as process water. The elements of such a process are shown in Figure 108.

FIGURE 108: PROCESS FOR PENTACHLOROPHENOL RECOVERY FROM WASTEWATER USING METHYLENE CHLORIDE AS A SOLVENT

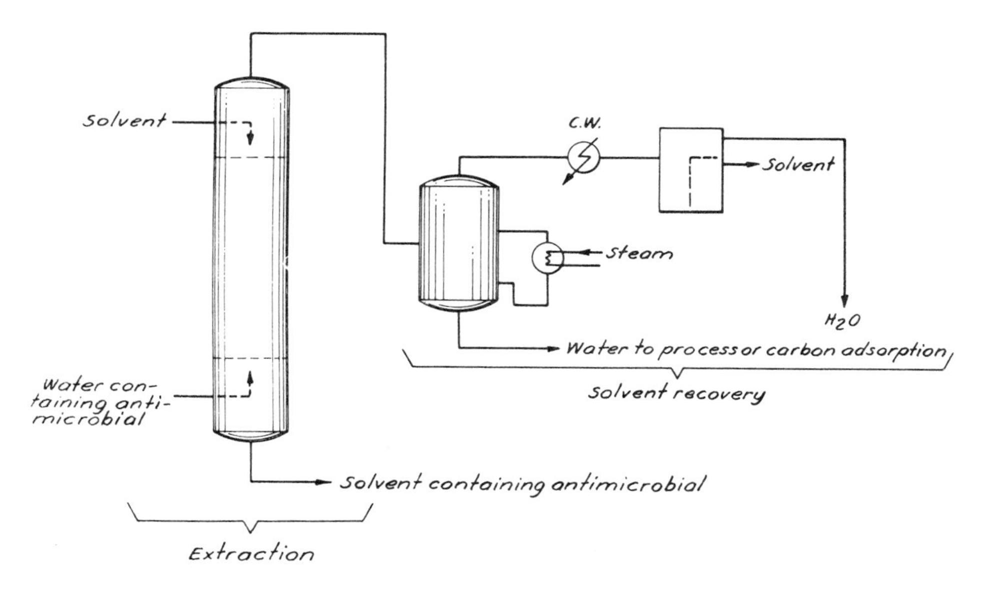

Source: U.S. Patent 3,931,001

In a typical example of its operation, a wastewater from a pentachlorophenol-methylene chloride wood treating process containing 0.02 pound of pentachlorophenol per ten pounds of water was contacted in a countercurrent packed column at a rate about 0.4 gallon per minute per square foot with one-half pound of methylene chloride containing 0.001 pound of water.

The wastewater was flowed upwardly and the methylene chloride flowed downwardly. The temperature of the incoming streams and liquids in the contacting column were about 20°C. An overhead was taken from the column above the methylene chloride inlet. This overhead contained 0.2 pound methylene chloride, 10 pounds water and 15 ppm pentachlorophenol.

A bottom stream taken from the tower below the wastewater inlet had a composition: 0.3 pound methylene chloride, 0.0006 pound water and 0.02 pound pentachlorophenol. The overhead, water containing 0.2 pound of methylene chloride and 15 ppm pentachlorophenol was heated and the methylene chloride flashed off. The vapors were condensed, sent to a water separator and 0.15 pound of methylene chloride was recovered.

The water from the flash evaporator was employed as steaming water in a methylene chloride-pentachlorophenol wood impregnating process after suitable treatment to adjust the pH. The water had 15 ppm pentachlorophenol and a trace of methylene chloride. If the water was not to be reused it could be cleaned up of both pentachlorophenol and the trace of methylene chloride by passing it through a carbon adsorption column, by soil percolation or by biodegradation in ponds.

The methylene chloride from the bottom of the column and that from the water separator, a total of 0.4506 pound (90%) and the pentachlorophenol dissolved therein, 0.02 to 0.00015 lb (99%), are available for use in the wood treating process or can be separated by simple distillation.

The process may be operated under superatmospheric pressures and temperatures upwards of 200°F, particularly when the wastewater is of these higher temperatures as it comes from the primary process. Such temperatures with appropriate pressures enable flash distillations to be accomplished without heating as employed in the example.

PERCHLORATES

Removal from Water

A process developed by *V.N. Korenkov, V.I. Romanenko, S.I. Kuznetsov and J.V. Voronov; U.S. Patent 3,943,055; March 9, 1976* is a process for the purification of industrial wastewaters from perchlorates and chlorates by way of intermixing industrial wastewaters with household communal wastewaters, biochemical reduction of the perchlorates and chlorates under anaerobic conditions, followed by separation of purified waters.

The biochemical reduction of perchlorates and chlorates is effected by means of a strain of the microorganism *Vibrio dechloraticans Cuznesove* B-1168 produced by way of successive inoculations on a liquid nutrient medium containing sources of carbon, nitrogen, and phosphorus under anaerobic conditions in the presence of a perchlorate as a donor of oxygen.

PESTICIDE MANUFACTURING EFFLUENTS

As is known to those skilled in the art, many processes for preparing pesticidal and herbicidal compositions are accompanied by the evolution of odors which may be highly objectionable to persons residing in the area near their source. It is also known that the odorous materials are usually present in relatively small amounts in the gas stream. Although the odor of the gas is unpleasant, the odorous materials are present in such small quantities that they are usually not toxic to plant or animal life.

Removal from Air

Many different types of odor control systems have been used including such devices as catalytic burners and liquid scrubbers to eliminate or control these odors. In some instances, the incorporation of masking agents or the addition of ozone to the gas stream has also been attempted. The small amounts of odorous material which contaminate large quantities of gases make the treatment of the gas streams difficult on a commercial scale. However, in certain processes, notably the preparation or manufacture of certain pesticides and herbicides none of the various approaches utilized has proven to be completely successful.

A process developed by *N.C. Lamb; U.S. Patent 3,798,877; March 26, 1974; assigned to Stauffer Chemical Company* involves passage of the effluent stream or exhaust gases associated with the manufacturing and handling of pesticide compositions through a zone into which finely divided activated carbon is introduced so as to make contact with the exhaust gases and to become entrained therein prior to the venting of the gases into the atmosphere.

This system is particularly useful for controlling the odors which accompany the manufacture and preparation of synthetic organic pesticide and herbicide compositions containing sulfur and particularly phosphorothioate, phosphorodithioate and thiocarbamate compositions. Figure 109 shows the essential elements of the process.

FIGURE 109: ODOR CONTROL SYSTEM FOR EFFLUENT GASES FROM PLANT PRODUCING SULFUR-CONTAINING PESTICIDES

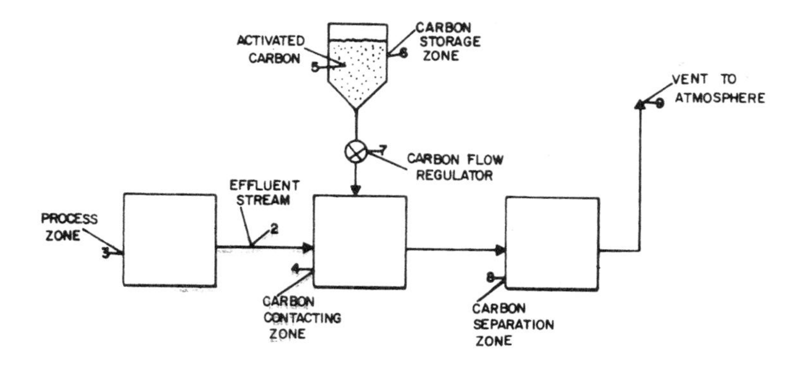

Source: U.S. Patent 3,798,877

The effluent stream of gases **2** emerging from the process zone **3** of the manufacturing process enters a carbon contacting zone **4** into which there is introduced powdered activated carbon **5** which is contained within a carbon storage zone **6** having a suitable means **7** for regulating its flow into the stream of exhaust gases within the carbon contacting zone.

Upon entering the carbon contacting zone, the particles of the finely divided activated carbon make contact with and are entrained in the flow of the effluent stream of exhaust gases thereby adsorbing the various gaseous, vaporous and colloidal solid odor-forming contaminants which are present therein.

The effluent, having the finely divided, activated carbon particles entrained in its moving stream, then proceeds to pass out from the carbon contacting zone **4** and into a carbon separation zone **8** which contains suitable means for removing the particles of finely divided, activated carbon, as well as any other particulate contaminants which may be present, from the effluent stream of exhaust gases prior to their being vented into the atmosphere **9**. The following example illustrates the operation of this process.

In manufacturing O,O-diethyl O-(2-isopropyl-4-methyl-6-pyrimidinyl) phosphorothioate, the latter compound being widely utilized as an insecticide, various odor-forming compounds, including alkyl mercaptans, mercaptoacetates, thiophosphates, and mercaptophosphates are believed formed as by-products of the reaction and decomposition products of the compound.

The preparation of agricultural formulations using this compound involves blending it with an inert carrier which is usually finely divided, hydrated synthetic calcium silicate (Microcel). The phosphorothioate is mixed with the synthetic calcium silicate in a high intensity mixer and the product is air classified.

The large particles are returned to the mixing means to be reduced in size. During the blending and classification a significant quantity of the particulate material is entrained in the air stream utilized in the apparatus. About 8,000 cubic feet per minute of air enters the odor control apparatus from the mixer and area near the mixer. The exhaust gases are passed to a filtration-type dust collector in which about 2,213 pounds per day of Microcel containing the phosphorothioate is separated from the gas stream. The gas stream leaving has an obnoxious and unpleasant odor and carries about 1.65 pounds per day of the phosphorothioate adsorbed on the Microcel out of the dust collector. In addition to the odorous components, the exhaust gas stream contains unadsorbed phosphorothioate vapor.

In an attempt to control the odor problem, a catalytic burning unit which utilized a platinum catalyst to thermally degrade the odorous materials at 800°F was installed in the effluent stream subsequent to its exit from the bag-type, fabric filter dust collector. The catalytic combustion system did not markedly reduce the amount of obnoxious and disagreeable odor being vented into the atmosphere.

In another attempt to solve the odor problem, the gas stream was contacted with 2% sulfuric acid solution as a scrubbing medium in a liquid scrubber subsequent to its exit from the bag-type, fabric filter. Liquid scrubbing also failed to reduce to acceptable levels the highly objectionable odor in the effluent stream which was being vented into the atmosphere. The liquid scrubbing also created a liquid waste disposal problem.

A system incorporating the carbon adsorption process was then installed. This system involved the passage of the effluent stream from the final mixing unit used in preparing the above-described phosphorothioate through a primary dust collector which comprised a bag-type, fabric filter apparatus having a capacity of 8,000 cubic feet per minute (cfm) and an air-to-fabric ratio of 7 cfm per square foot of fabric area. This device was used with a 12-ounce Dacron sock as a filtering medium.

As noted above, the gas stream entering this collector contained 2,213 pounds per day of particulate matter. The effluent stream emerging from this primary dust collector entered a conduit having an inner diameter of 20 inches and a length of 57 feet which led into a secondary dust collector. At a point 15 feet from the primary dust collector outlet, minus 325 mesh (U.S. Sieve Series) activated carbon (Darco S-51) was fed into the above-described conduit at a rate of about 6 pounds per day. At the end of the conduit the gas stream was devoid of the unpleasant odor associated with the phosphorothioate. The gas stream entered a secondary dust collector.

Upon emerging from the secondary dust collector, which was a bag-type, fabric filter comprising a 12-ounce, felted cotton sock having an air-to-fabric ratio of 1.5 cfm per square foot of fabric, the effluent stream which was being vented into the atmosphere was found to be devoid of the objectionable mercaptan-type odor with only a light naphtha odor, which was readily dispersible within only about 30 feet from the vent, being emitted into the atmosphere so that it was completely undetectable at a distance of 200 feet from the vent.

By contrast, when using either the catalytic burning unit or the liquid scrubber, as described hereinabove, at a 1:1 dilution, the effluent being vented into the atmosphere could be detected at a distance of a one-fourth mile using a Model 1-3 Barnaby Cheney Scentometer.

Removal from Water

A process developed by *L.A. Pradt and J.A. Meidl; U.S. Patent 3,977,966; August 31, 1976; assigned to Sterling Drug Inc.* involves treating nonbiodegradable industrial wastes by wet oxidation, followed by biological oxidation of the liquid phase by aeration in the presence of biomass and powdered activated carbon, and wet oxidation of the spent carbon and excess biomass to regenerate the activated carbon.

A unique feature of the wet oxidation is that complex organic molecules are broken into simpler, lower molecular weight molecules prior to complete oxidation to the ultimate products, carbon dioxide and water. The proportion of partially oxidized materials can be regulated by the reaction time and temperature. These principles can be utilized for nonbiodegradable wastewater treatment in that the nonbiodegradable components are converted to less complex molecules, and the wastewater is thereby rendered amenable to biological treatment.

For instance, a wastewater containing 1,1-bis(p-chlorophenyl)-2,2,2-trichloroethane (DDT), which would be regarded as toxic or very difficult to treat at anything but very dilute levels in a biological system was reduced to 1% of its initial concentration after an 80% wet air oxidation (chemical oxygen demand reduction as used herein) and to less than 0.1% after a 90% wet oxidation.

There are basic reasons why treatment of a wastewater is not usually effected totally by wet oxidation alone. The capital cost of a wet oxidation facility increases rapidly as the degree of oxidation gets closer to 100%. Even the effluent after a 99% wet oxidation of a concentrated raw waste will seldom be of acceptable quality. The wastewater is however quite amenable to biological treatment even after a considerably less severe wet oxidation, as little as 30% reduction in chemical oxygen demand.

Accordingly, the wet oxidation is supplemented by a biological oxidation step which serves to remove substantially all of the remaining chemical oxygen demand of the wastewater. The addition of powdered activated carbon to the biological oxidation step serves to adsorb any nonbiodegradable materials which had survived the wet oxidation and to improve the removal of chemical oxygen demand (COD) and biochemical oxygen demand (BOD) from the wastewater. Furthermore, the addition of activated powdered carbon improves the stability of operation of the biological system and provides an aeration system which requires less dilution with water.

An apparatus developed by *M.N. Bernadiner, A.A. Dobrovolsky, B.S. Esilevich, N. Rubinshtein, V.G. Gubarev, E.I. Shipov, P.M. Sharov, B.I. Lurie, A.D. Vodnev, A.B. Moshkovich and P.A. Lupanov; U.S. Patent 3,974,021; August 10, 1976;* for disposal of wastewater is a cyclone combustion reactor as shown in Figure 110.

The reactor comprises a vertical cylindrical chamber **1**. The top portion of chamber **1** accommodates burner apparatus **2** of a known design for feeding therethrough fuel and air into the chamber. The burner apparatus **2** is directed tangentially to the chamber to create a whirling flow of combustion products. Injection nozzles **3** are disposed below the burner apparatus and are radially directed in the transverse plane of the chamber for feeding wastewater containing organic and fusible mineral impurities.

Injection nozzles **4** are located below the injection nozzles **3** in a transverse plane of the reactor chamber **1** and directed at an angle with respect to a tangent to the inner periphery thereof for feeding wastewater containing refractory mineral impurities in combination with organic impurities. The angle α is selected such as to ensure the penetration of the wastewater in the whirling flow of combustion products.

The bottom portion of the chamber **1** has a base **5** provided with an opening **6** communicating the chamber **1** with a gas duct **7** for discharging combustion products. The bottom portion of the gas duct is provided with a tap **8** for discharging the melt of fusible mineral impurities along with the refractory impurities. The following is one specific example of the operation of this apparatus.

Fuel and air required for combustion of both fuel and combustible impurities of wastewater are fed into the reactor in the tangential direction. Sprayed wastewater containing organic and fusible mineral impurities, such as waste liquor from weed killers production containing in grams per liter: sodium chloride, 230; methyl alcohol, 10; formaldehyde, 1; aniline, 13; polyamines, 7 is introduced into a flow of combustion products at a temperature of 1400° to 1550°C.

FIGURE 110: INCINERATOR FOR REFRACTORY WASTEWATERS FROM PESTICIDE MANUFACTURE

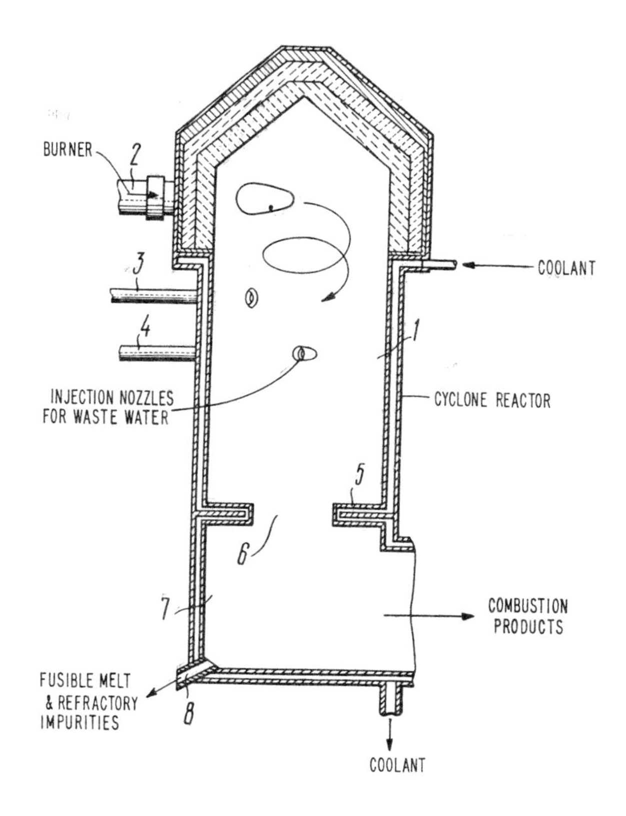

Source: U.S. Patent 3,974,021

High-temperature combustion products cause the evaporation of water drops, oxidize the combustible impurities and melt the particles of fusible mineral impurities (a melting temperature of fusible mineral impurities is of about 810°C), this wastewater being fed into the reactor in such a manner that during the burning out of organic impurities a film of a melt of the fusible mineral impurities is formed on the reactor walls which flows down therealong.

The formation of the film of melt of the fusible mineral impurities may be effected by different methods known per se, e.g., by introducing wastewater tangentially to the reactor walls, whereby the fusible mineral impurities are projected to the reactor walls to form a film of melt, or by introducing wastewater into just whirled flow of combustion products of fuel and air. This film can be also formed by uniformly spraying the reactor walls with wastewater using swingable injection nozzles. The film of melt of the fusible mineral impurities moves down under gravity and by force of the kinetic energy of the flow of combustion products.

At a temperature of fume gases of 1100° to 1200°C the drops of wastewater are evaporated, and the combustible impurities are oxidized with an excess of oxygen into harmless gaseous products (CO_2 and H_2O) to be discharged from the reactor along with fume gases. Dried solid particles of refractory mineral impurities are discharged from the reactor along with the film of melt of the fusible mineral impurities.

See Dithiocarbamates (U.S. Patent 3,966,601)
See Nitroanilines (U.S. Patent 3,458,435)

PESTICIDE RESIDUES

Removal from Water

A process developed by *G.K. Kohn; U.S. Patent 3,852,490; December 3, 1974; assigned to Chevron Research Company* is one in which organic pesticide residues in aqueous solutions can be removed by treating the aqueous solution with a lower alkene-vinyl lower alkanoate polymer, e.g., an ethylene-vinyl acetate polymer.

The process is generally applicable to the purification of any aqueous solution which contains detectable minor amounts of an organic pesticide, e.g., from about 0.01 ppm to 1,000 ppm of pesticide, based on weight of solution, preferably from about 0.01 ppm to 50 ppm of pesticide.

PETROLEUM PRODUCTION EFFLUENTS

Many oil field waters contain up to about 100 to 500 ppm of oil. This should be reduced to essentially 0 ppm of oil if the water is to be dumped into surface streams. If this water is to be injected into an underground reservoir to aid in driving out the oil, as is common in secondary recovery operations, the oil content should be less than about 10 ppm. An economical way of removing this dispersed oil from such water would be a big help in reducing this phase of water pollution.

Removal from Water

A system developed by *L.W. Jones; U.S. Patent 3,957,647; May 18, 1976; assigned to Amoco Production Company* is a system for removing dispersed oil from water by contacting the oily water with sulfur to cause the oil to coalesce or agglomerate. In a preferred embodiment, the water containing the dispersed oil is passed through a bed of granular media presenting a surface area of solid phase sulfur to coalesce the dispersed oil. The coalesced oil is then separated from the water.

A process developed by *L.L. Wilkerson; U.S. Patent 3,547,190; December 15, 1970; assigned to Shell Oil Company* involves treating wastewater associated with hydrocarbon production by steam or hot water secondary recovery methods wherein residual hydrogen sulfide gas concentrations in the water are substantially removed therefrom by spraying the water vertically into the air and the temperature of the water is lowered to a point whereby it will not upset the ecological balance of surrounding bodies of water.

See Fireflood Operation Effluents (U.S. Patent 3,845,196)

PETROLEUM REFINERY EFFLUENTS

In a refinery operation processing approximately 40,000 barrels of crude oil per stream day, water is an essential commodity for effecting cooling, condensing and stripping in relatively large quantities approximating of the order of about 26 million gallons per day of water. Thus it is imperative that such large volumes of water be treated so that unnecessary contamination of the streams to which the water is returned will be minimized if not completely avoided.

In such a refinery operation the quantity of water used at the crude unit may approximate as much as 11 million gallons per day. It has been found in such an operation that the major pollutants entering the water system from the crude distillation operation include oil, phenols, chlorides, sulfides and solids.

In the reforming unit approximately 4 million gallons of water are used primarily as a coolant in the reforming process heat exchange equipment with additional minor quantities of water being used for pump cooling. The oil process water comprising steam condensate from the catalytic cracking unit is identified as the largest source of phenols, ammonia and sulfides in the refinery water system. These pollutants are formed in the cracking operation and thus a substantial portion thereof concentrates in the process steam condensate.

Removal from Water

In a process developed by *T.R. Morrow; U.S. Patent 3,671,422; June 20, 1972; assigned to Mobil Oil Corporation* the refinery process water streams recovered from the separate processing units containing phenols, ammonia and sulfides are combined and charged to existing processing equipment for effecting removal of accumulated phenols in addition to effecting stripping of water for the removal of ammonia and sulfides. Thereafter the cleaned water may be reused or discharged from the refinery operation and returned to the source from which obtained. Such a process is shown in generalized form in Figure 111.

Referring to the figure, a crude oil charge is introduced to the process by conduit **2** for passage to desalter vessel **4**. Water recovered in the process as hereinafter described at a temperature of about 105°F and a pressure of about 250 psi is combined with crude oil feed in substantially two separate increments by conduits **6** and **8**.

Initially from about 1 to about 2 volume percent water is combined with the crude by way of conduit **6** with from about 5 to about 7 volume percent water thereafter combined with the crude charge by way of conduit **8**. The water oil mixture thus formed is then passed to desalter **4** for removal of impurities as discussed above.

FIGURE 111: OVERALL SCHEME FOR WATER POLLUTION ABATEMENT IN A PETROLEUM REFINERY

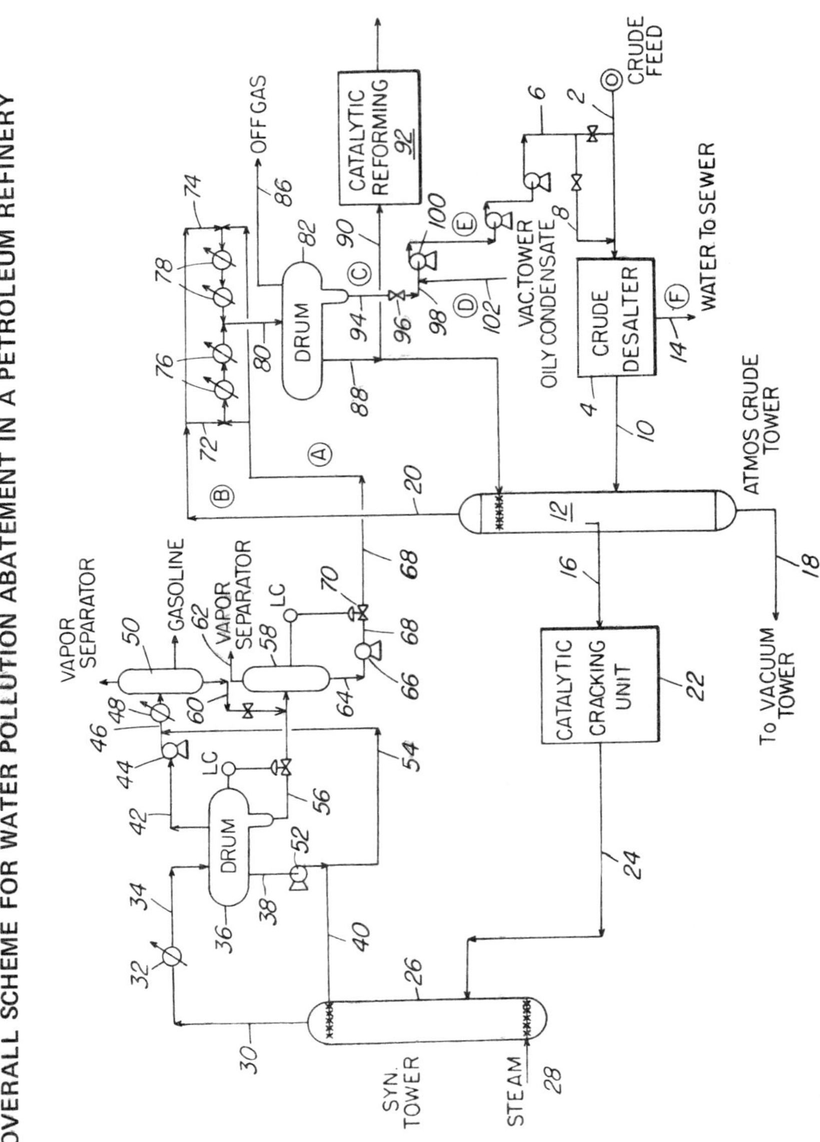

Source: U.S. Patent 3,671,422

The crude oil freed of contaminants in desalter **4** is then passed by conduit **10** to atmospheric fractionating tower **12**. Water containing less than 1 ppm phenol is removed from desalter **4** by conduit **14**.

In atmospheric tower **12**, a number of fractionations are made using a pressure of about 12 psi and a bottom temperature of about 640°F. A gas oil charge for catalytic cracking is removed by conduit **16**. A residual fraction is withdrawn from the bottom of fractionator **12** by conduit **18** for passage to a vacuum tower not shown. An overhead fraction containing steam is withdrawn by conduit **20** amounting to approximately 600 gpm (gallons per minute) of hydrocarbon and 20 gpm of water as process steam condensate at a pressure of about 10 psi at a temperature of about 240°F.

The gas oil charge in conduit **16** is passed to cracking unit **22** wherein it is converted to gasoline and other products typical of catalytic cracking operations. The catalyst employed in the cracking operation may be an amorphous silica-alumina cracking catalyst or one of the zeolitic cracking catalysts known in the art.

The products of the cracking operation are passed by conduit **24** to fractionator tower **26**. In fractionator tower **26** known in the art as a Syn tower, the products of cracking are separated in a manner familiar to the art. Steam is usually introduced to the lower portion of the synthetic crude tower as by conduit **28** in combination with operating the tower to maintain a top temperature of about 270°F and a tower pressure of about 10 psi.

In the processing sequence herein described, an overhead stream **30** is withdrawn from the top of Syn tower **26** comprising steam and light hydrocarbons. This overhead stream **30** is cooled and partially condensed by coolers represented by cooler **32** and then passed by conduit **34** to drum **36**. Drum **36** is maintained at a temperature of about 100°F and a pressure of about 2 psi.

A reflux hydrocarbon stream is withdrawn from drum 36 by conduit 38 with a portion thereof recycled by conduit **40** as reflux to the upper portion of tower **26**. A vapor phase is withdrawn from drum **36** by conduit **42** for passage to a compressor **44** which raises the pressure of the vapor up to about 50 psi. The compressed vapors are passed by conduit **46** to cooler **48** and then to separator drum **50**. The portion of the hydrocarbon condensate in conduit **38** not returned to the tower as reflux is passed to pump **52** and then conduit **54** for admixture with compressed vaporous material in conduit **46** described above.

A water phase separated in drum **36** is withdrawn by conduit **56** in response to a level control valve and system shown. The withdrawn water phase is passed to a second separator **58** maintained at a pressure of about 2 psi and a temperature of about 100°F. In separator **50**, maintained at a pressure of about 50 psi and 100°F, a vapor phase is separated from a gasoline phase and a water phase which are separately withdrawn from the separator.

The water phase is removed by conduit **60** from separator **50** and combined with the water phase in conduit **56** passed to separator **58**. Vaporous material is removed from the upper portion of separator **58** by conduit **62**. Water condensate accumulated in separator **58** is withdrawn from the bottom thereof by conduit **64** and passed to a pump **66**. Pump **66** is used to raise the pressure of the water phase up to about 130 psi. The compressed water phase is passed by conduit **68** containing level control valve **70** for admixture with the overhead vaporous phase in conduit **20** withdrawn from the upper portion of atmospheric fractionation tower **12** as hereinafter described.

The vaporous phase in conduit **20** comprising from about 600 gpm hydrocarbon and 10 to 20 gallons per minute of water is separated into two separate streams **72** and **74** for passage through a plurality cooler **76** and **78** before being combined and passed by conduit **80** to drum **82**. Water condensate in conduit **68** is caused to be separated into two separate streams as shown for admixture with vaporous material in conduits **72** and **74** before being cooled.

Maintained at a temperature of about 100°F and 6 psi pressure in drum **82**, a gas phase is separated and removed therefrom by conduit **86**. A hydrocarbon phase is also separated and withdrawn by conduit **88**. A portion of the hydrocarbon phase comprising naphtha charge material for catalytic reforming is recycled in part as reflux to the upper portion of tower **12** with the remaining portion being passed by conduit **90** to catalytic reforming unit **92**. As discussed and shown herein, the naphtha phase passed to catalytic reforming comprises adsorbed phenolic material collected from the water phases as herein discussed.

The water phase separated in drum **82** is removed therefrom by conduit **94** and passed through check valve **96** and conduit **98** to pump **100**. A water phase recovered from a vacuum tower not shown and known as an oily condensate is added by conduit **102** at a rate in the range of from 5 to 10 gallons per minute to the water phase in conduit **98** passed to pump **100**.

The combined water phase comprising from about 55 to about 80 gpm is then passed through one or more sequentially arranged pumps to raise the pressure of this water phase up to about 350 psi for admixture with the hydrocarbon charge being passed to desalter **4** as hereinbefore discussed. The oily condensate in conduit **102** is added primarily as a source of water make up to the process since it contains little if any phenols. On the other hand, it is important to recover the oily constituent of this water stream and such is accomplished by adding it to the crude oil passing to the desalter.

The simplified flow arrangement shown and described above has been intentionally so limited to emphasize the flow of phenolic waters in the system and the recovery of phenols in the hydrocarbon phase passed to catalytic reforming. It will be apparent to those skilled in the art of petroleum processing, however, that the flow arrangement of the drawing represents only a small part of such an operation and further details of such a refinery combination may be derived from the prior art.

A process developed by *A.W. Liles and R.D. Schwartz; U.S. Patent 3,968,036; July 6, 1976; assigned to Exxon Research and Engineering Company* involves treating refinery wastewater by contacting the wastewater with an activated sludge at conditions at which biological oxidation takes place. In this process various water-insoluble inorganic oxides selected from the group consisting of silica, alumina, and silica-alumina are combined with the sludge whereby increased rates of biological oxidation are obtained and the sludge shows increased settleability.

In a preferred embodiment of this method, the water-insoluble inorganic oxide is a spent cracking catalyst, that is, a zeolite which has been used in fluid cracking of hydrocarbon feeds and thus contains vanadium, iron, nickel, copper, and/or carbon. This catalyst may be recovered from aqueous scrubber solutions which are utilized to reduce stack losses in fluidized cracking processes and which present a solids waste disposal problem. Figure 112 shows such a scheme.

FIGURE 112: APPARATUS SEQUENCE FOR TREATMENT OF OIL REFINING WASTEWATERS

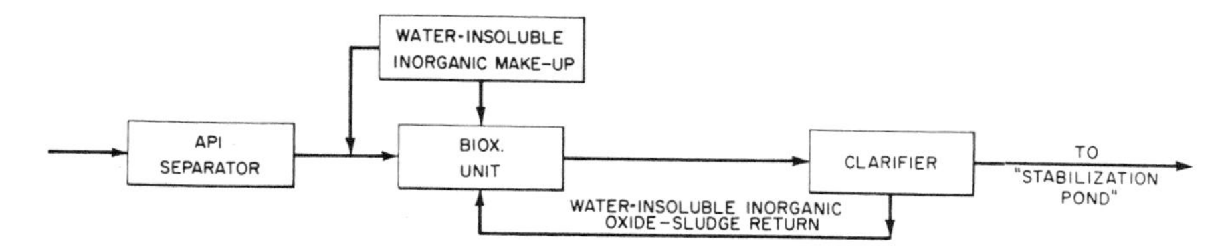

Source: U.S. Patent 3,968,036

The effluent from a petroleum refinery is brought into an API separator where oil is separated from the water. This water is then brought into a Biox unit wherein the activated sludge is contained. In this Biox unit the water-insoluble inorganic oxide may be mixed with the sludge, or the water-insoluble inorganic oxide may be premixed with the water at some point prior to the Biox unit. The wastewater may be contacted with the sludge and the water-insoluble inorganic oxide in a batch or continuous manner.

The effluent from the API separator may have a total organic carbon (TOC) greater than 300 ppm. The residence time in the Biox unit may vary from 30 minutes to 24 hours, for example, 16 hours. The mixture of inorganic oxide, sludge and wastewater is pumped from the Biox unit into the clarifier where a residence time of 30 minutes to 7 hours may be obtained, for example, 4.5 hours.

The temperature maintained in the Biox unit and the clarifier may vary from 10° to 60°C, for example, 25°C. The effluent from the clarifier may show a total organic carbon of 50 or less. The sludge is recycled from the clarifier to the Biox unit. As the process continues, it may become necessary to separate a portion of the sludge at this stage for incineration or other disposable since the sludge volume has a tendency to continuously increase. After incineration the residual ash may be recycled to the Biox unit where it may be used as make-up. This recycled material may be more active than the starting material due to the deposition of metals, oxides, carbon, etc. on its surface.

A process developed by *W. Dardenne-Ankringa, Jr.; U.S. Patent 3,963,611; June 15, 1976; assigned to Chevron Research Company* is one in which a wastewater stream is improved by oxidizing sulfur-containing impurities in the stream to sulfate. In the method, the stream is contacted with molecular oxygen under particular conditions which include an elevated temperature, a substantial oxygen gas partial pressure, a pH of at least 9.6 and the substantial absence of a heavy metal oxidation catalyst. For each gram atom of sub-six sulfur, the contact mixture must contain at least one equivalent of a strong inorganic base such as sodium hydroxide. Figure 113 illutrates such a process.

FIGURE 113: APPARATUS FOR OXIDIZING SULFUR-CONTAINING IMPURITIES IN REFINERY WASTEWATERS TO SULFATES

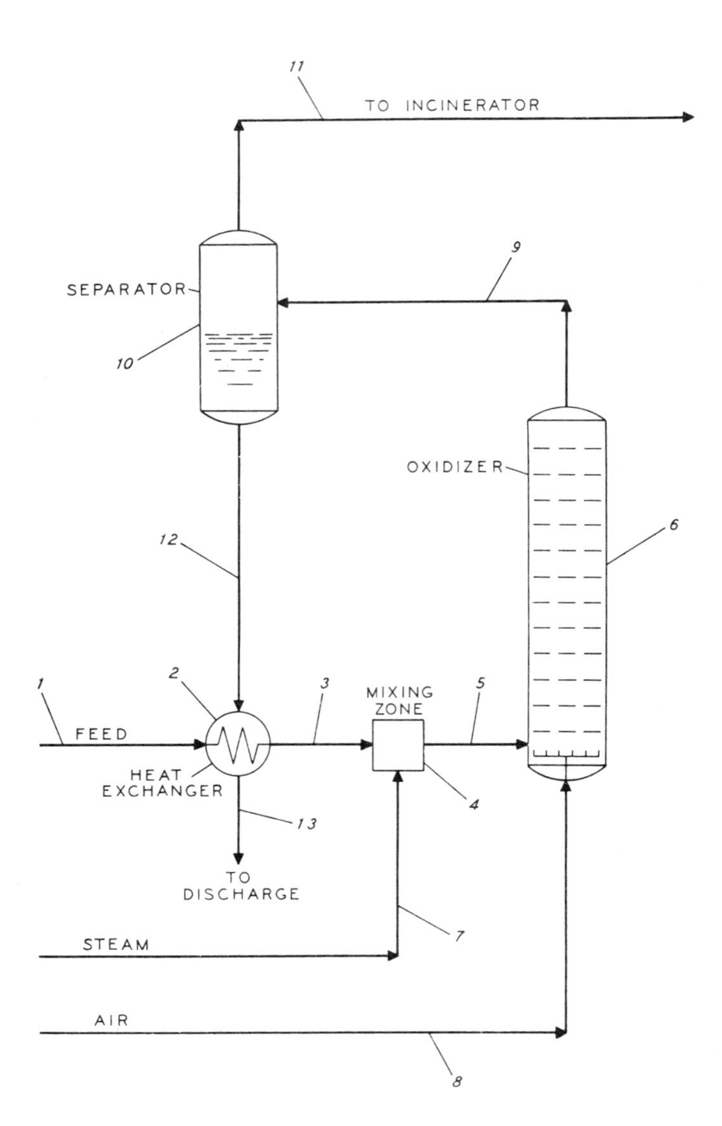

Source: U.S. Patent 3,963,611

As shown in the figure, a typical aqueous spent caustic (3 weight percent free NaOH) scrubbing stream, a refinery effluent stream, is introduced to the process via line **1**. This stream is substantially free of hydrocarbons and has a highly obnoxious odor, is very toxic, and has a high biological oxygen demand as well as a high chemical oxygen demand.

The principal sulfur-containing impurity in the stream is sulfide ion; there is some hydrosulfide ion. The counter-ions are mainly sodium, although ammonium and other alkali ions may also be present. Based upon the water, about 9,000 ppm of sulfide-sulfur is present in the stream. The pH of the spent stream is above 13. The process loop comprises oxidizer **6**, liquid-gas separator **10**, indirect heat exchanger **2**, and associated connecting piping, pumps, relief valves, and the like.

Via line **1** spent caustic feed is passed into heat exchanger **2**, in which it is heated indirectly by the hot liquid effluent stream from separator **10**. The heated fresh feed is withdrawn via line **3** from heat exchanger **2** and is passed into mixing zone **4**, in which it is further heated by steam which is passed into the mixing zone via line **7**. Sufficient steam is introduced into mixing zone **4** to adjust the temperature of the effluent stream from the mixing zone to about 135°C. This stream is passed via line **5** to oxidizer **6**.

Oxidizer **6** is a multitrayed bubble-cap tower capable of operation at a pressure of about **11** atmospheres and is fitted with a suitable gas dispersion unit. On the basis of a complete oxidation of the sulfide-sulfur values of the spent caustic stream to sulfate for a feed rate of 238.5 cubic meters per operating day, oxidizer **6** has a total volume of 34 cubic meters and a diameter of 1.8 meters. The process conditions in the oxidizer are:

Liquid rate, l/min	178
Vapor velocity, m/sec	0.046
Temperature, °C	135
Pressure, atm	11
Residence time, hr	2.5

The effluent liquid-gas mixture is withdrawn from oxidizer **6** via line **9** and is passed to separator (knockout drum) **10**, in which the liquid portion of the process stream is separated from the vaporous portion of the stream. The separated gas is vented from unit **10** via line **11** for further processing, if desirable, for example to an incinerator (not shown).

The separated and hot liquid portion of the process stream is withdrawn from separator **10** and is passed via line **12** to heat exchanger **2**, and then to discharge via line **13**. The discharged aqueous stream contains little or no sub-six-sulfur-containing inorganic compounds and has little or no odor. Optionally, if the discharge stream is off-color, it may be passed through a bed of absorbent carbon. Since the sulfur content is essentially in the sulfate form, it exerts no oxygen demand upon the environment.

See Hydrogen Sulfide (U.S. Patent 3,903,250)
See Sulfides (U.S. Patent 3,725,270)
See Sour Water (U.S. Patent 3,984,316)

PETROLEUM STORAGE EFFLUENTS

In the storage of petroleum in large storage tanks, vapors are expelled due to changing temperatures of the atmosphere surrounding the storage tanks and the vapors are also displaced when the tank is filled.

Removal from Air

In the past, it has been customary to provide a floating roof in each storage tank so that the amount of vapors was reduced. Such floating roof construction has been very expensive, and with older tanks, such construction has been generally impossible since they are usually out of round and/or they have steel framing which interferes with an adequate seal between the floating roof and the side wall of the tank.

In other instances, vapor recovery has been effected from storage tanks by passing the hydrocarbon vapors through a carbon bed, then regenerating the bed and cooling it down for reuse. Such vapor recovery system is also very expensive and as a consequence is not used extensively. Refrigeration heat exchange has also been employed for vapor recovery, but it suffers from the disadvantage that the water which is usually present with the hydrocarbon vapors freezes in the heat exchanger tubes, causing heat transfer and flow problems.

A system developed by *G. Parker, Sr.; U.S. Patent 3,778,968; December 18, 1973* overcomes the problems of the prior art and is far less costly. The system provides for the flow of petroleum vapors under controlled pressure conditions to a brine or glycol solution where the vapors are passed directly into contact with such solution, the temperature range of which is controlled so as to condense the hydrocarbon vapors and allow the water therewith to go into solution while the air bubbles on through the solution for discharge to atmosphere.

Thus, the hydrocarbon vapors are not only controlled to prevent air pollution if they escaped to the atmosphere, but they are also condensed and recovered with a reduced cost compared to the known prior art systems.

A system developed by *R.F. Battey; U.S. Patent 3,714,790; February 6, 1973; assigned to FMC Corporation* is one in which hydrocarbon vapors normally released to the atmosphere during the filling of tanks with a relatively volatile liquid are condensed out and recovered as liquid by simultaneously compressing the vapors and saturating the same with vapors of the liquid, followed by a step where the compressor effluent is contacted under elevated pressures in a condensation column with a refrigerated stream of the volatile liquid.

It is a feature of this process to maintain within the system a relatively small reservoir of refrigerated liquid of substantially the same composition as that being loaded into the receiving tank. This permits the use of refrigeration equipment and of a cold liquid holding reservoir of modest size and cost. The capacity of this reservoir is such that vapors collected during an extended period of peak tank-loading operation can readily be processed and recovered using the reserve of refrigerated liquid maintained in the reservoir. This method of operation proves to be far less costly than is that wherein a large vapor-holding tank is used to receive vapors during peak periods. Such a system is shown in Figure 114.

FIGURE 114: SYSTEM FOR VAPOR RECOVERY IN PETROLEUM STORAGE OPERATIONS

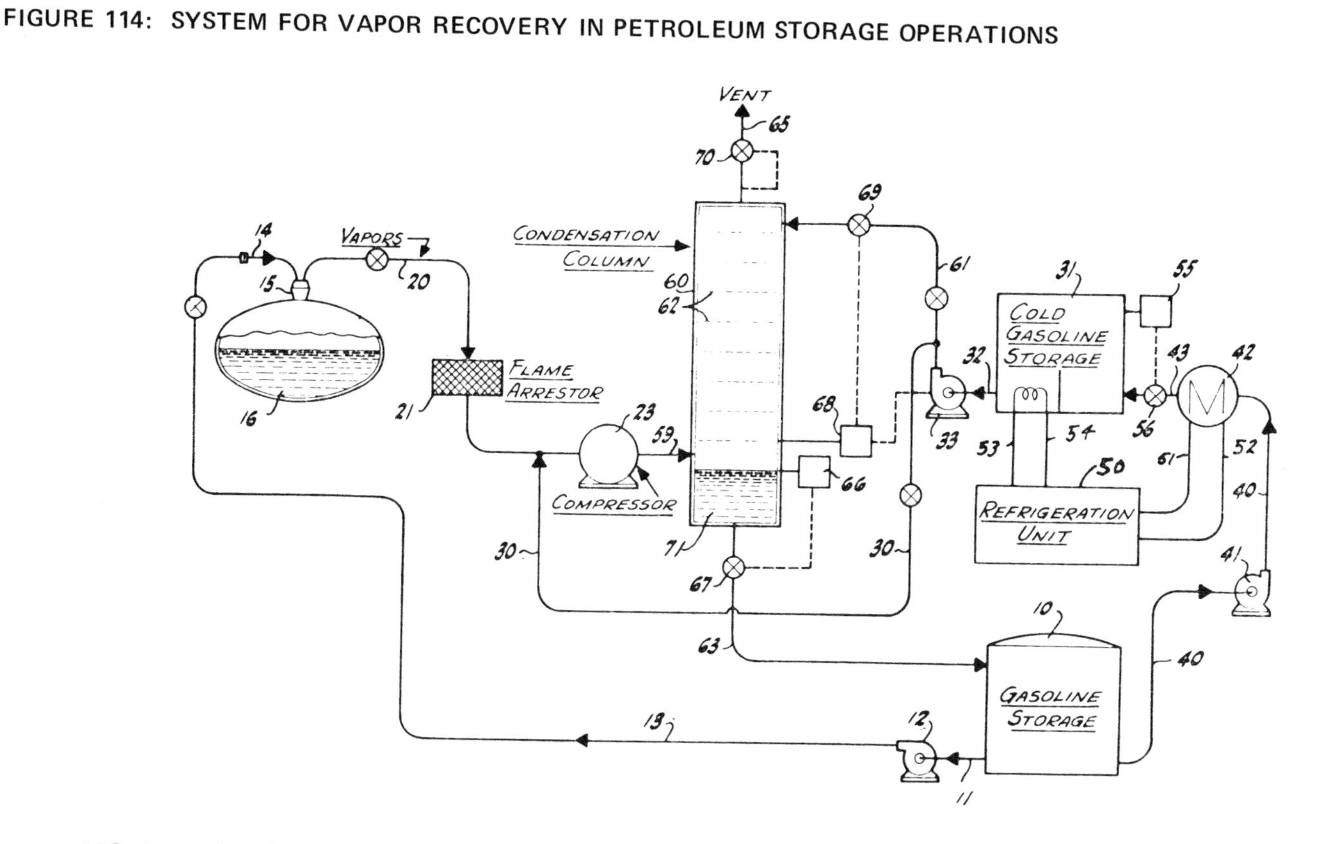

Source: U.S. Patent 3,714,790

Referring to the figure, gasoline from a primary storage tank **10** is withdrawn through line **11** and pumped by pump **12** through line **13** to a filler hose **14**. The hose extends through a vapor collecting hood **15** shaped to be fitted into the gasoline receiving opening of a truck or other receiving tank **16**. The hood **15** has a vapor receiving line **20** which extends from the hood to a flame arrestor **21** and then to a compressor **23** which operates only when gasoline is being loaded into tank **16**.

Along with the vapors, the compressor is supplied with a stream of cold gasoline through line **30**, the gasoline being withdrawn from a cold gasoline storage reservoir **31** and supplied to line **30** through an outlet line **32** and a pump **33** which runs when the compressor **23** is actuated. As the vapors pass through the compressor, they also become saturated with the vapors of the liquid, thus, producing a nonexplosive vapor mixture.

The compressor **23**, which can be one of the liquid piston type or a liquid-driven ejector compressor, is sized to receive vapors at peak rates of loading tank **16**. It can be provided with a speed control (not shown) or other means to permit operation at lower than maximum rates.

The gasoline in reservoir **31** is supplied from the storage tank **10** through line **40**, the gasoline being pumped by means of pump **41** through a heat exchanger **42** which cools the gasoline before it is passed through line **43** into the reservoir. A refrigeration unit **50** is connected with the exchanger and withdraws heat therefrom as the refrigerant is passed through lines **51** and **52**.

Similarly, coolant from the refrigeration unit is passed through lines **53** and **54** into the body of cold gasoline in reservoir **31**, thereby providing further cooling of the gasoline therein. The reserve of cold gasoline in the reservoir is kept at a predetermined level by means of a level control **55** which acts to regulate the flow of gasoline in line **43** by controlling the setting of valve **56** in the line.

Generally speaking, pump **41** will operate in most instances only when pump **33** and compressor **23** are actuated to process incoming vapors inasmuch as gasoline is pumped into reservoir **31** at about the same rate as it is withdrawn therefrom for passage to the compressor and to a condensation column **60** described below.

The temperature of the refrigerated gasoline in reservoir **31** is maintained at the desired level, usually about 5° to 15°F, by means of a thermostat (not shown) which actuates the refrigeration unit **50** when further cooling of the gasoline in the reservoir is required. Thus, it will be seen that the refrigeration portion of the system operates independently of the others, it being actuated when the temperature of the gasoline in reservoir **31** rises above the proper level either upon long standing or as a result of introducing fresh gasoline to replace that which is pumped to the compressor and to the condensation column.

Returning to the processing of the recovered vapors, the effluent from compressor **23** passes via line **59** to a lower portion of a condensation column **60** operated under moderately elevated conditions or pressure, e.g., about 25 to 85 psig. On entering the column, the liquid component of the compressor effluent falls into a pool **71** of gasoline maintained in the bottom of the column while the compressed, saturated vapors pass upwardly in countercurrent flow to a supply of refrigerated gasoline from reservoir **31** which is pumped by pump **33** through line **61** into the top of the column. The interior of column **60** is fitted with trays **62** which can be of the bubble-cap or other variety whereby good contact is achieved between the rising vapors and the downwardly cascading, refrigerated gasoline stream.

The interrelated conditions of temperature and pressure within column **60** are maintained at such levels that vapors exiting from the column to the atmosphere through a vent line **65** will contain less than 10% by weight of hydrocarbon or other vapors of the liquid being loaded.

Such vapor exit occurs as pressures build up above the level to be maintained in the column, thereby opening a back pressure or pressure relief valve **70** in the vent line. For example, when loading gasoline this method will reduce the hydrocarbon content of the vented stream to a level of approximately 7% by weight when the column is operated at a vapor exit temperature of 10°F and a pressure of 30 psig.

The same results can be obtained by increasing the vapor exit temperature to 20°F while maintaining a column pressure of about 60 to 65 psig, or by using a vapor exit temperature of about 30°F and pressures of approximately 80 to 85 psig. The column is preferably operated at pressures of about 25 to 40 psig and at the corresponding vapor exit temperatures ranging from about 5° to 15°F.

As indicated above, a pool **71** of gasoline is maintained in the bottom of column **60**, and liquid is continuously withdrawn from this pool through line **63** for return to the storage tank **10** during loading periods, when gasoline is entering the column through line **61**. A liquid level control **66** maintains this pool at a generally constant level through activation of valve **67** in line **63**.

The unit is preferably sized so that at peak vapor-receiving rates the temperature of the gasoline leaving the absorber column approaches ambient; for example, gasoline which is supplied at the top of the column at 10°F is warmed to about 60° to 65°F in its passage downwardly through the column. At less than maximum vapor-receiving rates, the flow of gasoline to the top of the absorber column is cut back by means of a temperature control instrument **68** which senses the presence of unduly low temperature and acts to partially close valve **69** in line **61**, thereby cutting down the flow of refrigerated gasoline as well as effecting any necessary change in the speed of pump **33**. This reduces the operating costs of the system.

It has been found that by starting with a supply of approximately 3,000 gallons of gasoline in the cold storage reservoir at 10°F, and by pumping refrigerated gasoline therefrom to the condensation column (operated at 30 psig) at a rate of 85 gallons per minute (gpm) and to the compressor at a rate of 25 gpm, it is possible to load tank **16** with gasoline for 50 minutes at a peak rate of 5,000 gpm while still maintaining the hydrocarbon content of the vapors discharged through vent **65** at a level of approximately 7% by weight.

In such an operation, liquid from column **60** at 60°F would be discharged through line **63** at a rate of approximately 118 gpm. This reflects an increment of 8 gpm of gasoline recovered from the vapors as a result of vapor condensation and absorption taking place in column **60**. Possible difficulties due to the formation of ice crystals in the portion of the system subjected to refrigeration can be overcome by the addition to the gasoline passing through line **40** of an effective amount of an antifreeze agent.

PHARMACEUTICAL INDUSTRY EFFLUENTS

In the biosynthesis of cephalosporin for example, large amounts of air are passed through fermenters and in the process become contaminated with sulfur-containing, volatile, obnoxious smelling organic substances. Several fermenters are connected to a common duct and give off varying quantities of malodorous air, depending on the prevailing state of the discontinuously operating fermentation processes.

Removal from Air

A process for deodorizing such waste gases has been described by *V. Fattinger; U.S. Patent 3,923,955; December 2, 1975; assigned to Ciba-Geigy Corporation.* It is a process for deodorizing waste or exhaust gases, preferably containing CO_2, in at least two washing stages by introduction of active chlorine, in which process a stream of the waste gas to be deodorized is subjected in an initial chlorine washing stage to the action of a washing liquid containing active chlorine, and in a subsequent chlorine washing operation to the action of a washing liquid, likewise containing active chlorine, of alkaline pH.

Even since the development of the submerged culture fermentation of penicillin, the disposal of the waste products from the growth of the microorganism producing the antibiotic and the spent broth from the fermentation process has posed a problem.

From this early development came the process wherein the antibiotic activity in the aqueous fermentation broth is partitioned into a water-immiscible solvent for further processing. However, few, if any of such processes have ever achieved a complete partitioning of any antibiotic from the fermentation broth to the solvent phase. Consequently, the disposal of the spent broth, from which a substantial portion of the antibiotic activity is removed, has always presented a problem because the residual antibiotic activity is organic in character and a large amount of oxygen is required to degrade such material in the water.

Removal from Water

A classic procedure for reducing the BOD and COD of spent antibiotic fermentation broth so it can be discharged into the environment involves (1) adjusting the pH to 6.5 to 7.0; (2) adding aerobic or anaerobic activated sludge (microbial mass) and vigorously agitating the mixture for an extended time; (3) clarifying the mixture by settling or filtration; and (4) continuously introducing oxygen into the broth until the BOD and COD is reduced to a level suitable for discharge into the environment. The EPA has set standards for discharge of such material of no more than a BOD of 0.04 g/l.

Such a process has been developed by *R.H.L. Howe; U.S. Patent 3,923,650; December 2, 1975; assigned to Eli Lilly & Company.* It is shown schematically

FIGURE 115: EQUIPMENT SEQUENCE FOR REDUCING BOD & COD OF SPENT FERMENTATION BROTH FROM ANTIBIOTIC MANUFACTURE

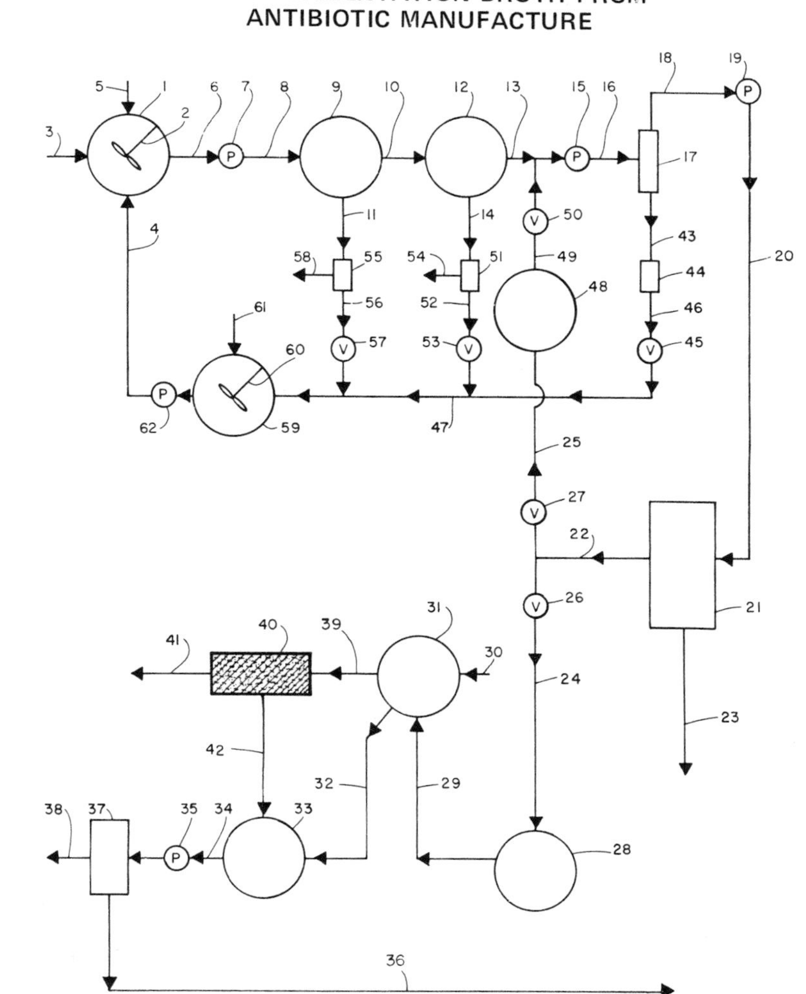

Source: U.S. Patent 3,923,650

in Figure 115. Untreated fermentation broth is fed through a conduit **3** into holding vessel **1** which can be a container of any configuration and is equipped with an agitator **2**. A cylindrical tank with a dished bottom and top is preferred because better mixing conditions can be achieved in such a tank. The agitator **2**, can be a turbine, propeller or any other suitable design effective for mixing solids in a liquid into a homogeneous dispersion. In the figure the agitator appears to be side-entering, but the required mixing can be provided just as well by either a top- or bottom-entering agitator. Coagulants, pH, and viscosity-adjusting agents can be fed into holding vessel **1** through conduit **5**. Mycelia wash water can be introduced into holding vessel through conduit **4**.

In holding vessel **1** the required coagulants and viscosity-adjusting agents are thoroughly mixed into the fermentation broth and the mycelia are homogeneously dispersed therein by the vigorous action of agitator **2**. From holding vessel **1** the treated fermentation broth is conveyed through conduit **6** to the suction side of pump **7** and discharged from pump **7** to a first means **9** for separating the undissolved material, principally mycelia, from such broth.

The solids removing means **9** can be a centrifuge, a continuously revolving drum filter, a string filter or any of several other designs known to those skilled in the art, preferably a centrifuge. The solids removed by the means are continuously taken away through conduit **11**.

The fermentation broth, minus a substantial quantity of the solids is carried on to a second solids removing means **12** through conduit **10** if desired. As with solids removing means **9** the solids removing means **12** can be a centrifuge or any design of a filtration means that is appropriate to the process and such will be known to those skilled in the art.

It is important that a substantially complete removal of the dispersed solids can be achieved by the time the liquid effluent is discharged from the second solids removing means **12**. The solids which are removed by the means are conveyed away through conduit **14**. The undissolved solids-free fermentation broth is conveyed from the second solids removing means **12** through conduit **13** to the suction side of pump **15** and discharged from the pump **15** through conduit **16** into a coalescing strainer **17**.

The coalescing strainer **17** serves the purpose of removing coalesced colloidal and gelled particles which are not removed in the solids removing means **9** and **12**. From coalescing strainer **17** the fermentation broth is fed through conduit **18** into the suction side of a high-pressure pump **19** capable of generating discharge pressures in the range of from about 200 to about 500 psi. The coalesced material removed in the coalescing strainer are conveyed away through conduit **43** and can eventually be returned to the holding vessel through a series of operations which will be described hereinafter.

From the discharge side of high-pressure pump **19** the fermentation broth is carried through conduit **20** to the reverse osmosis element **21**. In the reverse osmosis element **21** there is contained a semipermeable membrane which is capable of stopping the flow therethrough of compounds having a molecular weight greater than 100 or in the alternative, molecules having a molecular weight slightly below the molecular weight of the antibiotic being processed.

In the reverse osmosis element **21**, the fermentation broth under a pressure as indicated heretofore is contacted with the semipermeable membrane and such contact maintained until from about 80 to 90% of the volume of the broth penetrates the semipermeable membrane and the concentration of the antibiotic is increased from five to tenfold.

By utilizing a semipermeable membrane with a molecular weight cut-off of about 400 and above, the high pressure applied to the fermentation broth will cause a flow of the solvent from the concentrated side to the side opposite without permitting any of the dissolved material having such a molecular weight to pass therethrough.

Consequently, the concentration of the dissolved material is greatly increased in the solvent, in this case water and the solvent which passes through the semipermeable membrane is essentially free of all materials which can be labeled as either biologically or chemically in demand of oxygen for either metabolism or degradation. Inasmuch as a fair number of common antibiotics have molecular weights in excess of 400 such as tetracycline, streptomycin, Ilotycin, monensin, tylosin, and the like, such a molecular weight cut-off can be used effectively when these antibiotics are processed.

On the other hand, certain antibiotics have molecular weights below 400 such as Aureomycin, neomycin and penicillin. It is necessary that the reverse osmosis element be equipped with a semipermeable membrane having the molecular weight cut-off of about 100 when these antibiotics are processed. It will be apparent to those skilled in the art that it is advantageous to utilize that semipermeable membrane having the highest molecular weight cut-off consistent with the process in which materials having a molecular weight slightly above that retained by the membrane can be restrained from passing therethrough in order that the flow rates may be maintained at the highest level consistent with the principle of the reverse osmosis.

The conditions which make this process operate are the steps that are taken to remove the undissolved solids prior to the application of the reverse osmosis step and the pressure which is applied to the broth in the latter operation.

The wastewater containing minimum organic material, essentially no antibiotic activity and exhibiting a low BOD or COD is conveyed from the reverse osmosis element **21** through conduit **23**. This clean permeate, after analysis, can either be reused in the process or treated further if needed by such a simple procedure as a deionizing operation to render such permeate suitable for discharge into the environment, or for use in any suitable process.

The fermentation broth concentrated from 5 to 10 times in the reverse osmosis operation is conveyed away from such operation through conduit **22**. From conduit **22** such concentrated broth can be directed to a holding tank **48** through conduit **25**, or, when the concentrated fermentation broth is to be continued through the process for the isolation of the antibiotic activity contained therein, valve **26** is opened, valve **27** is closed and the concentrated antibiotic broth is conveyed through conduit **24** to a receiver **28** from which broth is fed to a solvent extraction process through conduit **29** or, alternatively, to a lyophilization operation before the solvent extraction procedure.

In the solvent extraction process identified as **31** in the accompanying drawing a water-immiscible extraction solvent is introduced through conduit **30**. With the concentrated broth and extraction solvent coming together in the extraction operation, the antibiotic activity is partitioned into the water-immiscible solvent by one of several extraction processes which are known to those skilled in the art.

Generally, such an extraction procedure will be carried out in a countercurrent flow which can be effected in mechanical apparatus designed for the purpose or in a simple countercurrent extraction column wherein the heavier of the two immiscible liquids is introduced into the top and the lighter into the bottom.

In any event the antibiotic activity is essentially transferred from the fermentation broth to the water-immiscible solvent. The spent broth is carried away through conduit **32** to a receiver **33**. The solvent containing the antibiotic activity is conveyed from the extraction process through conduit **39** to a means **40** for removing the last vestiges of the fermentation broth from the solvent stream.

This can be done either by a centrifuge, filtration or absorption operation. The broth-free solvent containing the antibiotic activity is conveyed to a crystallization procedure and subsequent solvent recovery through conduit **41**. The spent broth removed in the scavenging operation at **40** is carried through conduit **42** to the spent broth storage receiver **33**.

The spent broth accumulated in receiver **33** is conveyed through conduit **34** to the suction side of pump **35** and from there to a second reverse osmosis element operating as was described for the first such element. The residual antibiotic activity contained in the spent broth entering the second stage reverse osmosis element **37** is concentrated to from 5 to 10 times the dissolved material content and such is discharged therefrom through conduit **38** and returned to the extraction receiver **31** from whence it is again cycled through the extraction process.

The clear permeate from the reverse osmosis element **37** is carried through conduit **36** and as with the clear permeate from the first reverse osmosis operation is of a quality that can be reused or alternatively can be passed over a deionizing bed and meet the requirement for both biological and chemical oxygen demand for waste materials to be discharged into the environment.

If it is preferred, this wastewater, as with the wastewater from the first reverse osmosis element, can be reused in the fermentation process or as mycelia wash water. As was indicated earlier, the concentrated fermentation broth from the first stage reverse osmosis can be conveyed through conduit **25** to the holding tank **48**. This can be accomplished by opening valve **27** and closing valve **26**.

From the holding tank the concentrated fermentation broth can be conveyed through conduit **49** to the suction side of pump **15** and recycled through the coalescing strainer **17** and the first stage reverse osmosis element **21**. A valve **50** is shown in the conduit **49**.

The mycelia removed from the fermentation broth by the solids removing means **9** and **12** respectively and carried away therefrom through conduits **11** and **14** respectively can be diverted at the elements designated as **55** and **51** which are conveying means, respectively to either a drying operation through conduits **58** and **54**, respectively, or through conduits **56** and **52**, respectively, to a mycelia wash tank **59** as the antibiotic activity and concentration justifies.

Valves **57** and **53** are shown in the conduits **56** and **52**, respectively, and are utilized in the diversion of the mycelia to the mycelia wash tank **59**, both diversions being shown as tying into conduit **47** which enters the mycelia wash tank **59**.

The wash tank is equipped with an agitator **60** and either fresh make-up water or the clear permeate being conveyed from the process through conduits **23** and **36**, respectively can be introduced into the mycelia wash tank **50** through conduit **61**.

Inasmuch as certain of the antibiotics are retained in the mycelia in economically important amounts, the mycelia wash tank serves the purpose of recovering a significant amount of this antibiotic activity through a washing, and perhaps even a mascerating operation, which releases the antibiotic substance into the wash water.

Such wash water containing an economic quantity of antibiotic activity can be conveyed through conduit **4** to the holding vessel **1** with pump **62** being utilized to make the transfer. When this is done the wash water containing the mycelia can either be added to the fermentation broth in holding vessel **1** or it can be conveyed through the entire process beginning with the steps in the operation which begin with the presence of material in holding vessel **1** having antibiotic activity contained therein. The application of the process is illustrated by the following specific example.

Four liters of tylosin fermentation broth were placed in a suitable vessel and about 5 grams of aluminum chloride added thereto as a coagulant. This mixture was vigorously agitated for about 5 minutes and fed to a first-stage solids removal centrifuge at the rate of about 2.5 liters per minute.

The centrifuge was generating about 6,000 *g*. About 2.9 liters of effluent containing about 0.13% suspended solids were delivered from the centrifuge. The centrifuge discharged approximately 1.1 liters of concentrated suspended solids which analyzed 20.5% dry solids. The 2.9 liters of effluent from the first-stage solids removal centrifuge were fed to a similarly operating second-stage solids removal centrifuge (also at 6,000 *g*) at the rate of approximately 0.4 liter per minute.

The second-stage solids removal centrifuge delivered 1.88 liters of substantially clear effluent containing approximately 50 ppm suspended solids. The second-stage solids removal centrifuge discharged 1.01 liters of concentrated suspended material which analyzed 20.8% dry solids. The clarified effluent from the second-stage solids removal centrifuge was conveyed to a coalescing strainer where the last of the suspended solids and coalesced colloids and gels were removed. The 2.11 liters of concentrated solids from the first and second solids removal centrifuges were set aside for recycling.

Approximately 1.7 liters of tylosin containing effluent from the coalescing strainer were pumped under a pressure of 200 psig to a reverse osmosis element which utilized a semipermeable membrane of cellulose acetate having the capacity to restrict the flow of compounds with a molecular weight greater than 200. Approximately 1.42 liters of permeate water was collected on the reverse side of the semipermeable membrane. About 0.28 liter of concentrated tylosin containing fermentation broth was taken from the contact side of the reverse osmosis element. This represented a concentration ratio of approximately 5:1.

The semipermeable membrane utilized in this experiment had a surface area of approximately 50 square feet. The rate of flow through the element was approximately 4.0 liters per minute maximum. The permeate water was crystal clear and was found on analysis to have a biological and chemical oxygen demand of approximately 0.23 gram per liter. The tylosin activity in the original fermentation spent broth was about 875 mcg per milliliter. The concentrate removed from the reverse osmosis element analyzed about 4,400 mcg per milliliter.

The reduction of the biological and chemical oxygen demand to the 0.23 gram of oxygen per liter represented approximately a twenty-fold lowering of the oxygen requirements over that experienced in the conventional process for treating spent fermentation broth through a pH adjustment, cycling with activated sludge, and subsequent clarification.

During the manufacturing process, the antibiotic can accumulate within the cells of the microorganism or be excreted into the aqueous fermentation broth, or both. Recovery, of the antibiotic typically first involves a filtration step to give a filter cake, commonly referred to be the term "mycelial cake," and filtrate, commonly referred to as "beer" or "fermentation beer."

The mycelial cake consists essentially of the microorganism, i.e., mycelia, and excess insoluble nutrients. In some cases, a filter aid also may be present. Next, the mycelial cake and/or fermentation beer are further processed by such means as solvent extraction, selective ion exchange chromatography, precipitation techniques, or some combination thereof. After processing, the fermentation beer is referred to as "spent beer."

Alternatively, the antibiotic can be recovered from the aqueous fermentation broth by an azeotropic distillation procedure. Briefly, an organic solvent which forms an appropriate azeotrope with water is added to the aqueous fermentation broth. The resulting mixture is subjected to azeotropic distillation to remove all or part of the water from the broth, leaving a mixture comprising organic solvent and insoluble material, wherein the antibiotic is in solution in the organic solvent.

The insoluble portion can be separated by suitable means, such as filtration or centrifugation. Such insoluble portion, i.e., azeotropic distillation process solids, consists essentially of the microorganism, i.e., mycelia, excess water-insoluble nutrients, excess water-soluble nutrients, and inorganic salts.

The term "mycelial waste," as used herein, is meant to include both mycelial cake and azeotropic distillation process solids. Furthermore, such a term is meant to include such cake and solids without regard to physical form. That is, such term includes such cake and solids, per se, or slurries of such cake and solids in water, spent beer, or other aqueous media. In its broadest sense, the term mycelial waste is meant to include any mycelia-containing by-product, in whatever form, from antibiotic production.

It is well known to convert aqueous fermentation broth to a solid product by the removal of water and to employ the dried product as an animal feed additive. However, such an additive is employed primarily for the purpose of utilizing the antibiotic activity of the product, typically as a growth regulator or growth promotor.

In those instances where it is desired to obtain the antibiotic separate from the fermentation broth, disposal problems relative to mycelial waste and spent beer arise. The spent beer usually can be treated in the same manner as other plant effluent. However, the mycelial waste is not readily disposed of. Often, such waste is either burned or buried in landfills. Such procedures obviously add to the overall manufacturing cost of an antibiotic, not to mention the adverse environmental impact of such procedures.

Mycelial waste also can be processed readily to a dried product, useful as an animal feed supplement, in order to take advantage of the inherent nutrient value to animals of such waste. However, such product contains some antibiotic activity which presents marketing problems.

It is difficult to thermally decompose the antibiotic in such product because of handling and scorching problems. While the scorching problems can be avoided by heating the product by means of steam injection, water is added which then must be removed, thereby adding to processing costs.

Frequently, it is desired to market mycelial waste as a flowable or pumpable slurry, again for use as an animal feed supplement. In addition to the problem of antibiotic activity already mentioned, substantial handling problems are present. In general, a slurry of mycelial waste having a solids content in excess of about 10% cannot be prepared without gelling.

Obviously, marketing such a slurry will greatly increase shipping and handling costs per unit weight of dry mycelial waste. The economic production and marketing of a mycelial waste slurry requires a slurry solids content in excess of about 20%, and preferably of the order of about 30% or higher. Such levels of solids have not heretofore been attainable.

It is known that lysing microorganisms often can result in an ability to obtain a slurry having increased solids content. However, many of the microorganisms employed to produce antibiotics are resistant to typical lysing techniques, such as shearing, sonic cavitation, strong acid at 100°C, strong base at 100°C, and enzymatic methods.

Specific examples of some attempts to utilize these known lysing techniques include, by way of illustration, heating the organism at 90°C in aqueous solution at a pH of 1.0 for 2 hours; heating at the same temperature under similar conditions but at a pH of 10.0; heating at the same temperature in 5% aqueous sodium hydroxide; heating at the same temperature in 5% aqueous sulfuric acid; mechanical shearing in a blender; and treatment with such enzymes as amylases and proteases.

A process developed by *A.W. Hubert and R.T. Russell; U.S. Patent 3,928,642; December 23, 1975; assigned to Eli Lilly & Company* is one in which antibiotic-producing microorganisms are lysed by the process which comprises heating the mycelial waste in the presence of water at a temperature of from about 140° to 200°C or higher and at a pressure from the equilibrium pressure to about 2,000 psig and at a pH of from about 2 to about 10, until substantially no antibiotic activity remains. Treatment times can vary from about 10 minutes to about 5 hours or longer. The treated waste then is concentrated and/or dried by known procedures.

The lysis procedure permits the preparation of a slurry which still is flowable or pumpable, yet which has a solids content of up to about 30%. Furthermore, the process destroys antibiotic activity, which destruction is necessary in order to market the lysed mycelial waste solely for its nutrient value in animal feeds.

PHENOL MANUFACTURING EFFLUENTS

Effluent streams from phenol plants in which phenol is produced by the oxidation of cumene followed by decomposition of the oxidation product, cumene hydroperoxide, to phenol and acetone, normally contain water, traces of phenol and acetone and other impurities.

Such effluent streams give very low results when tested for biological oxygen demand (BOD) despite their high chemical oxygen demand (COD). It is believed that this is due to the presence of certain biocides, other than phenol itself which sterilize the microorganism responsible for biodegradation in, for example, the standard BOD determination. It is very desirable that the effluent stream from a phenol plant should be biodegradable.

Removal from Water

A process developed by *T. Bewley, M.D. Cooke and M.M. Wirth; U.S. Patent 3,746,639; July 17, 1973; assigned to BP Chemicals Ltd., England* for the biodegradation of a phenol plant effluent stream comprises treating the phenol plant effluent stream containing water, phenol, acetone and other impurities, with sulfur dioxide, or a soluble metal sulfite, bisulfite, or sulfide and then contacting the effluent with an activated sludge biodegradation unit. The following is a specific example of the conduct of such a process.

The chemical oxygen demand (COD) and the biological oxygen demand (BOD) of a sample of a phenol plant effluent were measured before and after treatment of the effluent with 0.4% of sulfur dioxide, a quantity such that no excess sulfite ion could be detected after treatment. The results were as follows:

	COD	BOD
Before SO_2 treatment	13,850	<1
After SO_2 treatment	13,850	700

Following the treatment with sulfur dioxide the effluent stream was contacted with activated sludge, slurried and allowed to settle. The example shows that treatment with sulfur dioxide greatly increased the test result for BOD, that is to say the biodegradability of the effluent stream.

PHENOLIC RESIN PROCESS EMISSIONS

Wastewaters contaminated with phenol occur in the production of phenol and phenolic resins and in other industrial processes. The wastewaters, which generally contain 0.01 to 5 weight percent of phenol, must be subjected to a purification because of the high toxicity of phenol.

Removal from Water

It is known that the separation of phenol from wastewaters by distillation presents considerable difficulties in practice, and this hinders satisfactory separation. However, the separation of phenol from water by distillation can be facilitated by prior chemical blocking. Chemical blocking is a term used for the conversion of phenol into products having low volatilities by reaction with a suitable reagent. However, processes of this type must satisfy the following four conditions.

1. The cost of the blocking agent should be low.
2. No expensive operations should be necessary for the blocking of the phenol and subsequent separation of the water.
3. The quantities of phenol and blocking agent recovered should be very high.
4. The products remaining after the separation of the water should be of industrial value. Their value should as far as possible, correspond to or exceed the recovery costs.

A process developed by *S. Vargiu, S.S. Giovanni, G. Mazzoleni and S. Pezzoli; U.S. Patent 3,869,387; March 4, 1975; assigned to Societa Italiana Resine S.I.R. SpA, Italy* utilizes urea and formaldehyde for such chemical blocking. The phenol is extracted from wastewaters by reacting formaldehyde, urea and phenol containing wastewaters at a basic pH and high ratios of formaldehyde to urea, keeping the product at relatively acidic pH-values for short reaction times, reacting the obtained product with additional urea and wastewater at slightly acidic pH values and low ratios of formaldehyde to urea and adjusting the obtained product to a basic pH. The condensates obtained are suitable for use as adhesives and binders for wood.

A process developed by *J. Ackermann, P. Radici and P. Erini; U.S. Patent 3,911,046; October 7, 1975; assigned to Societa Italiana Resine S.I.R. SpA, Italy* is one in which both formaldehyde and phenol are recovered from waste-liquors as useful products, with purification of the liquors.

The process involves blending of the relative dilute solutions under conditions such as will firstly bring about virtually complete elimination of the phenol in the form of resol phenolic resin and in the subsequent preparation of a mixed resin from phenol, formaldehyde and urea, by reaction of a further quantity of formaldehyde with urea, the reaction taking place in the presence of the previously prepared resol phenolic resin.

A process developed by *M. Adegeest; U.S. Patent 3,741,392; June 26, 1973; assigned to Corodex NV, Netherlands* involves the removal of impurities from wastewaters resulting from the manufacture of phenol/formaldehyde resins. The apparatus includes a mixing tank with lines to feed it with contaminated water, acid, and phenol.

The tank discharges to a heatable reservoir which has a gravity drain to a settling tank as well as an off-take for gaseous products and a bleed-off line from an intermediate zone for conveying fluids to an overflow tank. That tank has a return line to the reservoir and a gravity drain for liquid resin. It is connected also to an expansion boiler which is heated by fluid from heat exchange units.

A stack removes vapor upwardly and a gravity drain conducts liquid resin downwardly to a storage vessel. A bleed-off line from an intermediate zone in the last-mentioned drain to an acid recovery vessel may be added, if desired. Air injectors are installed as needed to maintain circulation.

Such an apparatus is shown in Figure 116. A is the reaction vessel in which the resin is prepared. The apparatus comprises a mixing tank **1** which is supplied through a pipe **1'** with wastewater to be purified. Vacuum created in the tank through vacuum line **22** draws in phenol through the suction line **8**, and acid, preferably mixed with water, through the suction line **21**.

FIGURE 116: PROCESS FOR REMOVAL AND RECOVERY OF POLLUTANTS FROM PHENOLIC RESIN PLANT WASTEWATERS

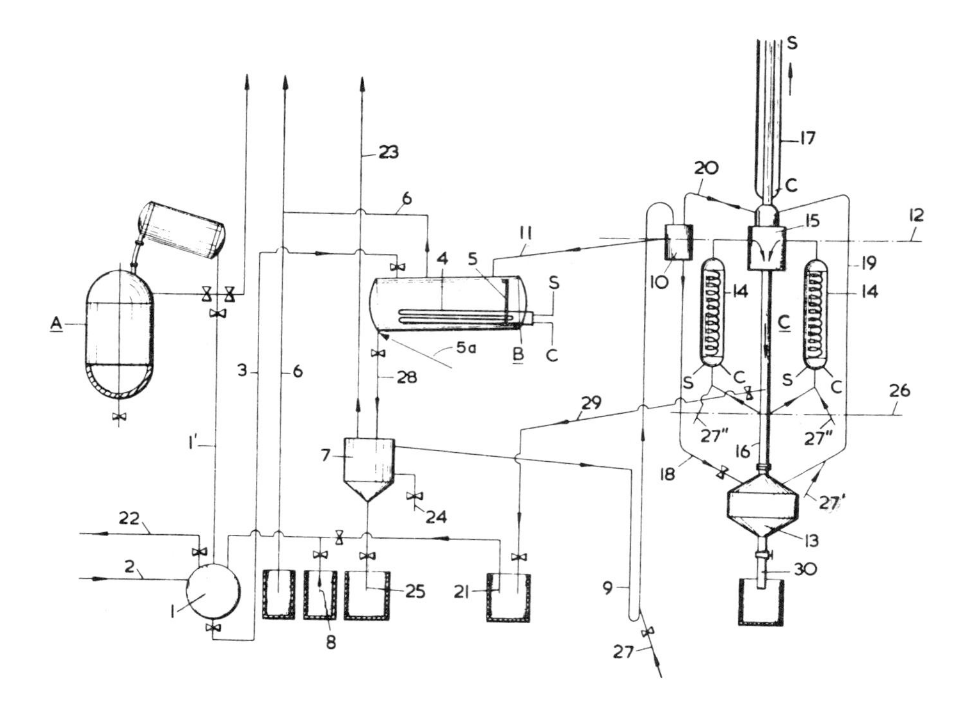

Source: U.S. Patent 3,741,392

Appropriate valving is supplied for closing these suction lines when the mixing tank is to be emptied by means of pressure applied through pipe **2**, which forces the water, phenol and acid through the pipe **3** to the reservoir **B**. **5** is a gauge-glass; **S** and **C** are, respectively, steam feed and steam discharge lines.

The reservoir **B** is installed at an inclination. It is equipped with a heating member **4** and it is connected at an upper point to the discharge pipe **6** for the removal of methanol vapor and other volatile compounds. Finely divided air may be forced into the tank through an air injector **5a**. The pipe **28** carries fluid from the lower end of the reservoir **B** to a resin settling tank **7**, which is connected to a pressure safety valve by way of line **23**, and also has a gravity drain line **25** for removing the resin which has settled out in the lower levels of the tank.

In line **28** near vessel B, means may be provided to introduce finely divided air into B. A drain **24** may be provided for removing fluid from the tank 7. A pipe **9** is connected to an upper portion of the settling tank **7** for removing lighter material from that tank.

An air injector **27** is connected into the pipe **9**, and carries the fluid from the tank **7** to an overflow tank **10**, from which fluid may be led back to the tank **B** by way of the overflow line **11**. Heavier constituents of the material in the tank **10** may flow through the line **18** into the tank **13**, where separated resin is collected. Lighter constituents in the tank **10** are carried to an expansion boiler or evaporator **15** by way of a line **20** equipped with a pressure equalizer. **C** is the after-reaction unit.

Between the expansion tank **15** and the separating tank **13** there is a primary circulating pipe **16** which interconnects these two vessels. At or about the level indicated by the broken line **26**, indicating the interface of hot and cold zones, connections are provided from the line **16** to heat exchangers **14, 14**, these being supplied with fluid from the line **16** by the aid of air injectors **27", 27"**.

The point at which the connections to the heat exchangers come off is just slightly below the zone at which the hot and the lukewarm water meet in the pipe **16**. Slightly above this level is connected the drain pipe **29**, which withdraws water mixed with acid and conveys it back to the tank from which the line **21** extends.

A heated discharge pipe **17** leads from the top of the expansion tank **15** for the removal of volatile compounds. There is a gauge **19** mounted between the tank **13** and the tank **15** for the purpose of checking on the fluid level. This gauge may be provided with an air injector **27'**. At the bottom of the tank **13** there is a drain **30** for the removal and recovery of the liquid resin. The following is one specific example of the conduct of the process.

624 liters (683 kg) of wastewater, remaining after the preparation of a phenol-formaldehyde resin which had been condensed in a basic medium, was admitted to the mixing tank together with 312 liters (341 kg) of wastewater resulting from the condensation of a phenol-formaldehyde resin in an acidic medium. 84 liters (90 kg) of water containing acid was mixed with these wastewater portions, this water-containing acid having been drawn through the continuous drain **29** from the after-reactor of the evaporating plant. 109 kg of phenol was added to this mixture.

Before the addition of phenol, the phenol:formaldehyde molar ratio was 0.43 gram mol per liter to 1.8 gram mol per liter, i.e., 1:4.18. This is 40.46 grams of phenol and 54 grams of formaldehyde per liter. After the addition of phenol the phenol:formaldehyde ratio was 0.57 gram mol per liter to 0.61 gram mol per liter, i.e., 1:1.07. This is 53.64 grams of phenol and 18.38 grams of formaldehyde per liter.

The normality of the total acid amounted to 0.015. After the addition of phenol the mixture was held at a temperature of 80° to 85°C for 48 hours. The water was subsequently removed by being heated to boiling temperature. There was obtained 200 kg of a thin, liquid resin mixture which was separated off by settling.

Inasmuch as 109 kg of phenol was added, there have thus been recovered 91 kg of resin product, i.e., ± 8.8% by weight from the acid wastewater. Subsequently, another 84 liters of water containing acid was obtained from the continuous drain **29** from the reaction-evaporation zone **C**. This water is required for maintaining the acidity in this zone at a constant level, while it also serves to adjust the acidity in the mixing tank.

A process developed by *J.H. Thayer and E.C.Y. Fan; U.S. Patent 3,736,292; May 29, 1973; assigned to General Electric Company* is one whereby the phenol content is markedly reduced in waste discharged from apparatus used for producing phenol-aldehyde resins. Aqueous solutions produced by the reaction between the phenol and aldehyde at different stages of the reaction and containing different proportions of phenol are mixed together at a temperature in the range of about 30°C to about 45°C to provide rapid separation into two solution layers, of which one contains low phenol concentration (8 to 12%) and may be discharged as waste, and the other contains high phenol concentration (70 to 75%) which is returned to the reaction vessel for further use in resin production. The low phenol waste solution thus discharged enables a reduction in pollution of streams into which products of the resin process are usually discharged.

The "low phenol content" stream is claimed to be sufficiently low in phenol content to be disposed of through usual sewage treatment facilities, where the remaining phenol is readily removed by suitable treatment, such as biological oxidation reactions commonly employed in such facilities.

PHENOLS

Removal from Air

See Fiber Glass Manufacturing Effluents (U.S. Patent 3,528,220)

Phenolated residual water is found in the effluent of gasworks, coking plants, refineries, chemical synthesis works (manufacture and conversion of phenols), works manufacturing plastics materials (phenoplasts) and, in general, works in which there are processed coal, tars and their derivatives, pesticides, dyestuffs, etc.

The residual phenol concentrations vary widely, depending on the specific industry concerned. They may attain several grams per liter. It is known that the phenols are toxic to fish at levels as low as 0.1 mg/l. Furthermore, in water which is to be rendered drinkable by the addition of potassium hypochlorite, a phenol content as low as 0.01 mg/l will suffice to impart an extremely disagreeable taste thereto, due to the formation of chlorophenols; it is thus necessary to process such water, in order to eliminate the phenols therefrom.

Removal from Water

Hitherto, the processes for the purification of phenolated water have not been very numerous and, above all, they have not been totally effective. These known processes may be classified in accordance with two large categories.

First of all, there are recovery processes involving operation either by liquid-liquid extraction by means of a solvent, steam distillation, absorption on active charcoal or on an ion-exchanger resin, or by a foaming process whereby a surface-active agent is added to the water and the phenol accumulates in the foam.

Then, secondly, there are processes involving chemical and biological destruction. There may be mentioned treatment by activated sludges and bacterial beds. Depending on the nature of the biological bed employed, it is possible to purify effluents of initially 50 to 100 mg/l of phenol (such as may issue from an installation for physical recovery such as those enumerated hereinabove) and to expel after several hours water which still contains some milligrams per liter of phenol.

Some extremely elaborate plants make it possible to go down as far as 0.2 to 0.5 mg/l, but only by dint of extremely strict monitoring of the composition of the water to be treated (pH, concentration in respect of phenol and foreign bodies, poisons of the bacterial bed).

It has also been proposed to effect oxidation treatment by ozone or permanganate; these two products are extremely costly; the second results in the production of large quantities of sludges. Treatment by chlorine may be considered to be interesting, but it frequently produces chlorophenols and this almost always opposes the achievement of the desired aim, which is precisely to avoid the formation of such by-products. Treatment by means of catalyzed hydrogen peroxide is also known.

Apart from the chemical oxidation processes, all the other processes are never quantitative and always leave a certain residual phenol content ranging between 0.2 and several milligrams per liter.

A process developed by *J.P. Zumbrunn and F. Crommelynck; U.S. Patent 3,711,402; January 16, 1973; assigned to L'Air Liquide, Société Anonyme pour L'Étude et L'Exploitation des Procedes Georges Claude, France* is a process for the complete purification of industrial effluent polluted by phenolated impurities wherein the degradation of the pollutant is obtained by the action of an oxidizing reagent containing the HSO_5^- anion.

The oxidizing reagent is a member of the group constituted by monopersulfuric acid and its salts. The oxidizing reagent is an aqueous solution containing HSO_5^- anion and NH_4^+, HSO_4^- ions and, optionally, alkaline cations and hydrogen peroxide.

A process developed by *B. Hauschulz, H. von Barneveld, W. Jordan, J. Mertmann, G. Rasner and H. Brenienek; U.S. Patent 3,963,610; June 15, 1976; assigned to Phenolchemie GmbH, Germany* is one in which phenol is removed from wastewaters that occur in the manufacture of phenol by the cumene process wherein cumene is used as an extractant.

The cumene which is used for extracting the phenol from the wastewater and which is thereafter washed with sodium hydroxide to remove the phenol is treated prior to this washing with an aqueous solution of 1 to 20 weight percent preferably 5 to 10 weight percent sodium carbonate.

If desired the aqueous solution can also contain sodium sulfate in amounts of 1 to 15% by weight. The sodium carbonate solution may also contain a non-ionogenic, surface-active substance such as aryl polyglycol ethers in amounts of from 0.001 to 1% by weight.

A process developed by *C.E. Hamilton; U.S. Patent 3,931,000; January 6, 1976; assigned to The Dow Chemical Company* is one in which an aqueous waste stream containing polysubstituted phenols, such as bromo-, nitro-, amino-, sulfo-, or chlorophenols, is fed into one end of a container, passed around the outside of a bundle of hollow fibers and removed from the other end of the container.

A reactant for the phenols, e.g., NaOH solution, is fed into the hollow fine fibers. The substituted phenols coalesce on the fiber outside surface, and being soluble pass through the fiber to react with the sodium hydroxide to concentrate sodium phenate salts which are insoluble in the membrane. The product is swept out of the system as the NaOH stream is removed from the fiber tube sheet.

A process developed by *R.A. Wiley; U.S. Patent 3,673,070; June 27, 1972; assigned to Petrolite Corp.* is a process for removing and concentrating such acidic organic material from a water stream. The water stream is dispersed within an organic liquid solvent for removing substantial amounts of acidic organic material, such as phenol, mercaptans and thiophenols etc., from the water.

The enriched organic solvent phase is separated from the purified water stream phase. Next, the enriched solvent is intimately contacted with substantially stoichiometric amounts of an immiscible concentrated caustic solution forming a three-phase mixture in a second dispersion. This three-phase liquid mixture is separated into a regenerated solvent phase, a second liquid phase of the alkali-metal salts of extracted acidic organic material, and a third phase of excess caustic solution. Preferably, both phase separations are undertaken in the presence of an electric field.

The regenerated solvent is recycled into contact with the water stream; the high-purity alkali-metal salts of extracted acidic organic material are passed to some suitable utilization, and the excess caustic solution is recycled for regenerating further amounts of the enriched solvent. Only small amounts of caustic need to be added to maintain a circulating inventory of the caustic solution.

A process developed by *M. Tarjanyi, M.P. Strier and H.D. Siegerman; U.S. Patent 3,730,864; May 1, 1973; assigned to Hooker Chemical Corp.* is a process for decreasing the phenolic content of a solution which comprises passing an electric current through a solution containing phenolic material, which solution is contained as the electrolyte in a cell, the cell having at least one positive and one negative electrode, between which the current is passed, and where the electrolyte also contains a bed of particles, distributed therein, such that the porosity of the bed is from about 40 to 80%, porosity being defined as

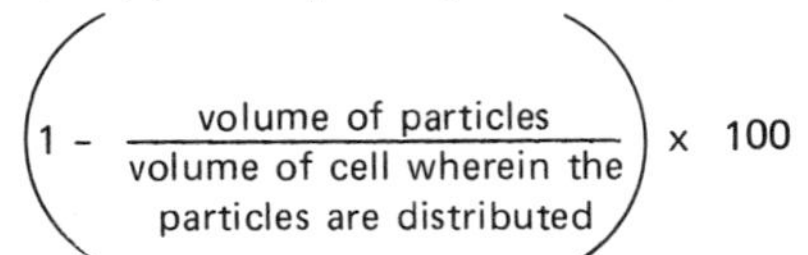

$$\left(1 - \frac{\text{volume of particles}}{\text{volume of cell wherein the particles are distributed}}\right) \times 100$$

The electrolysis of the electrolyte is continued until the desired reduction in the phenolic content thereof is obtained.

The solutions which are electrolyzed to effect the reduction in the phenolic content thereof, i.e., the electrolyte solutions in the cell, may be various solutions which contain phenolic materials although, preferably, these are aqueous solutions. These solutions may contain varying amounts of the phenolic materials, solutions containing as much as 10% by weight and as little as one part per million of the phenolic material being suitable for treatment to effect a reduction of the phenolic content.

In referring to the phenolic material in the solutions, it is intended to include not only phenol itself, i.e., C_6H_5OH, but also chlorinated phenols, such as mono-, di-, and trichlorophenol, as well as various alkyl-substituted phenols, such as 3,4,5-trimethylphenol, and other chemical compounds in which there is present a phenyl ring with a hydroxyl group attached thereto.

Additionally, since it is believed that the phenolic materials are removed from the solutions treated by the process by means of oxidation, phenol going through various oxidation states and resulting ultimately in carbon dioxide, the solutions treated may also contain various oxidized states of phenol and other phenolic materials, such as maleic acid, quinone, and in its reduced form, hydroquinone and the like.

See Coke Oven Emissions (U.S. Patent 3,855,076)
See Coal Gasification Process Emissions (U.S. Patent 3,972,693)
See Petroleum Refining Effluents (U.S. Patent 3,671,422)
See Phenolic Resin Process Emissions (U.S. Patent 3,741,392)
See Phenolic Resin Process Emissions (U.S. Patent 3,911,046)

PHENYLENE DIAMINE

See Photographic Processing Effluents (U.S. Patent 3,721,624)

PHOSGENE

In the production of isocyanates by phosgenating amines, large quantities of waste gases which contain principally hydrogen chloride, phosgene and inert gases are formed. It is necessary to decompose the phosgene because of its highly poisonous property.

Removal from Air

In the old process for decomposing phosgene, large amounts of wastewater and low-concentration acid are produced and seriously add to the expense of the process. Moreover, the dilute hydrochloric acid produced is valueless.

A process developed by *H. Richert, et al; U.S. Patent 3,314,763; April 18, 1967; assigned to Farbenfabriken Bayer AG, Germany* involves working up waste gases which contain hydrogen chloride and which may also contain other impurities in addition to phosgene and inert gases, and comprises adiabatically absorbing the hydrogen chloride and hydrolyzing the phosgene on active carbon wherein possibly after an active carbon absorption, the hydrogen chloride is first of all removed from the waste gas by adiabatic absorption with water and/or dilute hydrochloric acid, concentrated hydrochloric acid being formed.

Thereafter, the gas mixture still containing substantially phosgene and steam (possible after adding more steam) is conducted at temperatures above the dew point of the mixture over active carbon, whereby a molar ratio of water:phosgene of at least 1:1, preferably of between 5:1 and 20:1 is adjusted, the hydrogen chloride thereby formed is condensed and the hydrochloric acid is recovered by separation from the residual constituents of the gas mixture.

A process developed by *R.W. Beech et al; U.S. Patent 3,411,867; November 19, 1968; assigned to Allied Chemical Corp.* involves removing phosgene from gas streams (especially those containing relatively small quantities of phosgene) and comprises contacting the gas streams with crystalline alumina in the presence of sufficient water to dissolve the HCl formed by the resulting conversion of phosgene.

A process developed by *A.A. Allemang and H.J. Bachtel; U.S. Patent 3,789,580; February 5, 1974; assigned to The Dow Chemical Company* provides a method for removing phosgene from an essentially anhydrous gas stream which involves contacting such a stream with activated alumina at a temperature of from 110° to 200°C under essentially anhydrous conditions. This method is particularly useful in removing phosgene from chlorinated hydrocarbon gas streams also containing acid gas impurities.

PHOSPHATES

Phosphates occur in rivers and wastewaters from a variety of sources such as the following: detergent laden wastes where phosphate builders were present in the detergent formulations; agricultural runoff from lands fertilized with phosphate-containing fertilizers; and miscellaneous industrial and municipal wastewaters.

Removal from Water

A process developed by *L.E. Lancy; U.S. Patent 3,562,015; February 9, 1971; assigned to Lancy Laboratories, Inc.* involves removing and neutralizing phosphate type waste or carry-over on workpieces from a metal finishing bath. The carry-over usually consists of large concentrations of iron or zinc phosphates or both and some free phosphoric acid which has been used for phosphatizing, pickling, or metal surface preparation.

An aqueous chemical treatment wash solution is used having a pH of less than about 8 and is applied to surfaces of the workpieces during their movement; the solution as thus contaminated is circulated in a system having a treatment solution reservoir or tank, and is subjected to the introduction of hydrated lime, slaked lime, or powdered limestone, either immediately before its introduction into the reservoir or at the time of its introduction, in an amount determined to be sufficient to precipitate and settle-out iron and zinc as well as calcium phosphates in the reservoir.

Thereafter, the reconditioned solution is moved from the treatment reservoir back to a workpiece treating tank to provide a continuous washing off of the surfaces of the metal workpieces. Care is taken to assure that the solution as returned to the treating tank is, for all practical purposes, free of dissolved calcium compounds.

An alternative is to employ a caustic soda addition to the solution for precipitating the metal phosphates and to employ a small quantity of calcium ion added as a secondary treatment to the solution for the purpose of removing the minor constituent of the carry-over, namely, free phosphoric acid.

A process developed by *J. Block; U.S. Patent 3,583,909; June 8, 1971; assigned to W.R. Grace & Company* is one in which phosphate ions in wastewaters can be removed from the water by first adding an ion which, when added in sufficient amounts, would form an insoluble phosphate precipitate, but here, being added in relatively small amounts, does not per se form such precipitate; then adding an anionic surfactant to form a phosphate-containing precipitate, and floating the precipitate to the surface with bubbles. The precipitate which has been floated to the surface can then be removed in the resultant froth, leaving the remainder of the solution relatively free of phosphate ions.

A process developed by *D.F. Bishop et al; U.S. Patent 3,617,540; November 2, 1941; assigned to the Secretary of the Interior* is one in which nitrogen and phosphorus are removed from wastewaters by a process including the steps of biological nitrification, chemical precipitation of phosphorus and biological denitrification.

Buffering capacity of the water is substantially reduced and in some cases nearly eliminated by reaction of acid, produced in the nitrification step, with bicarbonate ion contained in the wastewater. A precipitate containing phosphate in high concentration is recovered at low chemical cost.

A process developed by *R.B. Hudson et al; U.S. Patent 3,650,686; March 21, 1972; assigned to Monsanto Company* involves recovering phosphorus values from plant effluents by precipitating the phosphorus values with lime. The sludge obtained which is difficult to handle is improved by treating the sludge with phosphoric acid so as to convert phosphorus values in the sludge from basic calcium phosphate to crystalline calcium phosphates which are readily filterable and useful as an animal feed supplement.

A process developed by *G.V. Levin et al; U.S. Patent 3,654,146; April 4, 1972; assigned to Biospherics Inc.* is an activated sludge sewage treatment process in which phosphates are removed from phosphate-enriched sludge by aerating the phosphate-enriched sludge with an oxygen-containing gas. During aeration, the organisms in the sludge, after consuming the available food substrate, go into endogenous respiration, consuming much of their own cellular material.

A process developed by *H.G. Flock, Jr. et al, U.S. Patent 3,655,552; April 11, 1972; assigned to Calgon Corp.* is one in which phosphate is removed from municipal and industrial wastewater by treating the water with a synergistic admixture, of a water-soluble high molecular weight nonionic polymer, preferably polyacrylamide, and a water-soluble salt containing ferric ions, preferably ferric chloride.

A process developed by *M.A. Kuehner; U.S. Patent 3,697,332; October 10, 1972; assigned to Amchem Products, Inc.* is one in which aluminum surfaces with good corrosion resistance and paint adhesion properties, together with a waste liquid substantially free of objectionable ions are produced when the aluminum is treated with a solution of phosphate, molybdate and fluoride, and when the waste stream is rendered basic with lime (pH 11) and then neutralized with sulfuric acid (pH 7).

A process developed by *C. de Latour; U.S. Patent 3,983,033; September 28, 1976; assigned to Massachusetts Institute of Technology* is one in which an aqueous solution containing dissolved phosphorus is seeded with iron oxide and a clay additive, if necessary, and with the electrolyte cation Al^{+3} the latter acting to associate the dissolved phosphate with the iron oxide and clay to form a coagulum. The coagulum is then removed from the solution magnetically.

See Sewage Treatment Effluents (U.S. Patent 3,984,313)

PHOSPHORIC ACID PROCESS EFFLUENTS

In the manufacture of concentrated phosphate fertilizer, the phosphate ore is initially subjected to extraction with sulfuric acid by the "wet process." The resulting dilute phosphoric acid is concentrated by boiling and then used for the production of concentrated liquid and granular phosphate fertilizers. During the extraction of phosphoric acid from the phosphate ore by sulfuric acid, the reaction produces hydrogen fluoride. In the presence of silica, the hydrogen fluoride is converted to fluosilicic acid.

A large quantity of the fluorine remains in the dilute phosphoric acid as hydrogen fluoride and fluosilicic acid. When the dilute phosphoric acid is concentrated by boiling, much of the fluorine is evolved with the steam as hydrogen fluoride and silicon tetrafluoride vapors in variable amounts depending upon the variety of the crude phosphate ore.

The toxicity of both compounds in itself demands careful purification of the waste gases. Furthermore, the waste gases form a valuable raw material source for obtaining fluorine values in the form of inorganic fluorine compounds. From some phosphate plants, as much as 10,000 to 30,000 tons per year of fluorine compounds may be liberated in gaseous plant effluents, which should be captured to prevent air and water contamination.

Removal from Air

A process developed by *A.W. Petersen and J.M. Stewart; U.S. Patent 3,893,830; July 8, 1975; assigned to Stauffer Chemical Company* employing a cocurrent absorber utilizing multiple cyclonic entrainment separators, cocurrent flow and renewed droplet surfaces for recovering inorganic compounds from plant effluents particularly for recovery of fluorine values as low phosphorus fluosilicic acid from phosphate plant effluents. Figure 117 illustrates such a recovery process.

Referring to the flow diagram at the start-up of the system, the heater **15**, the evaporator **16** and circulation pipe **21** are filled with phosphoric acid to a designated level. The level of the acid in the evaporator **16** is kept constant and is regulated by overflow from a barometric leg, pipe **20.**

FIGURE 117: COCURRENT ABSORBER FOR RECOVERING FLUORINE COMPOUNDS FROM PHOSPHORIC ACID PLANT EFFLUENTS

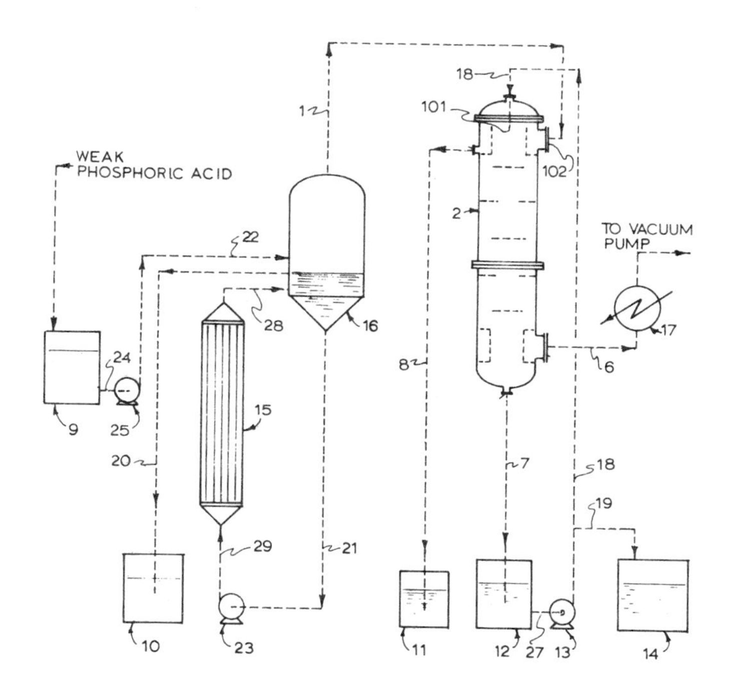

Source: U.S. Patent 3,893,830

Then, water is introduced into the condenser **17**, and with a vacuum pump (not shown), a vacuum is created in the system. The phosphoric acid circulation pump **23** is started and steam is then introduced into the heater **15**. With this, the acid in the tubes of the heater overheats and changes into a vapor-acid mixture, which flows in a continuous stream from the heater into the evaporator. In the evaporator the acid loses vapors and drains into the circulation pipe **21.**

In this way, the system has an uninterrupted circulation of acid from the circulation pipe **21** into the heater **15** through pump **23** and line **29**, then into the evaporator **16** via line **28** and from the evaporator **16** again into the circulation pipe **21** to have the process repeated.

When the acid in the circulation pipe **21** reaches a given concentration, dilute acid is introduced continuously into the system from the feed vessel **9** for the weak phosphoric acid source (not shown) through pipe **24**, pump **25** and pipe **22.**

The concentrated acid, having reached the overflow level, continuously drains through pipe **20** from the evaporator **16** into the collecting vessel **10** for concentrated acid. In the process used prior to this process, the vapor separating in the evaporator **16** continuously flowed into the condenser **17** with the aid of the vacuum pump (not shown). In this manner, the system worked continuously but resulted in fluosilicic acid recovered which contained varying amounts of entrained phosphoric acid.

To recover fluorine from vapor from plant effluent such as a phosphoric acid plant, a vessel **2**, hereinafter called the cocurrent absorber, is added to the concentration flow diagram and apparatus described above. It is preferably placed between the evaporator **16** and the condenser **17**. The vapor separating in the evaporator **16** goes from the evaporator through pipe **1** to the top of the cocurrent absorber **2**, through entrance port **102**, and is contacted with a recirculating liquor.

The vapor is contacted with a recirculating liquor in the cocurrent absorber **2** and exits through pipe **6** from the cocurrent absorber. The vapor enters the condenser **17** where it is condensed. The condenser is maintained at a temperature low enough to condense all the vapors. Thus, the possibility of contamination of air has been effectively removed.

At the start-up of the recovery system, the circulating liquor vessel **12** is filled with water and the centrifugal pump **13** is activated. Then the cocurrent absorber **2** is irrigated with liquid brought into the cocurrent absorber **2** by the centrifugal pump **13** through pipe **18** from the circulating liquor vessel **12**. The vapor entering the upper section of the cocurrent absorber from the evaporator is drawn through the cocurrent absorber and irrigated with liquor from pipe **18** by outlet **101** with the aid of the vacuum pump. The water-soluble components of the vapor are absorbed by the recirculating liquor.

In the case of plant effluents from a wet process phosphoric acid plant, the vapors are absorbed to make fluosilicic acid according to the following reactions. Silicon tetrafluoride reacts with water as shown by the equation:

$$3SiF_4 + 2H_2O \rightleftharpoons 2H_2SiF_6 + SiO_2$$

Hydrogen fluoride reacts with silica to form fluosilicic acid:

$$6HF + SiO_2 \rightleftharpoons H_2SiF_6 + 2H_2O$$

The liquid in the cocurrent absorber **2** is transferred by pipe **7** to the circulating liquor vessel **12**, from where it is fed into the cocurrent absorber for irrigation by the centrifugal pump **13**. As it attains a concentration of up to about 30% H_2SiF_6, the liquid is transferred from the circulating liquor vessel **12** via lines **27** and **19** into the fluosilicic acid product vessel **14**.

The temperature of the unit for recovering fluorine is the equilibrium temperature of the vapor at a given pressure. The cocurrent absorber operates at a vacuum of about 100 to about 740 mm Hg, and the temperature of the vapor under these conditions is about 20° to about 95°C.

A process developed by *W.R. Erickson; U.S. Patent 3,811,246; May 21, 1974;* involves recovering fluorine compounds from the evaporative concentration of aqueous phosphoric acid, without atmospheric pollution. The vapors produced by concentrating aqueous phosphoric acid solution are condensed by contacting same in a system closed to the atmosphere with an aqueous fluorine-compound-absorbing liquid, with the aqueous liquid being cooled by indirect heat exchange and then recycled to condense additional vapors.

The prior art required cooling water ponds or the like and such ponds are eliminated by the process, with the consequent elimination of fluorine pollution of the atmosphere.

A process developed by *R.G. Hartig; U.S. Patent 3,859,423; January 7, 1975* is a process designed to eliminate all contamination by gaseous fluorine liberated in phosphate rock processing from the waters used to slurry gypsum, in wet process phosphoric acid complexes, and to remove fluorine from process gas streams. In addition, various other benefits are obtained including fluorine recovery as a saleable product, elimination of liming costs, higher P_2O_5 recovery, and the like.

In current practice, by-product gypsum from wet process phosphoric acid complexes is slurried with water and pumped to a pond of vast acreage in which the gypsum settles out and the supernatant water is cooled in ponds and recycled to process for use as scrubbing water to remove fluorine from process gas streams, as condensing streams in barometric condensers, and to slurry gypsum.

The fluorine content of this water builds up to appreciable concentrations presenting undesirable problems, such as surface water and ground water contamination by fluorine. Addition of lime to these waters is frequently resorted to in order to control pollution of surface waters into which the plant waters may become intermingled.

By the process, it is possible to eliminate this type of fluorine contamination by completely eliminating the use of gypsum slurry water in scrubbing and processing. All fluorine-contaminated scrubbing liquor is handled and stored in impervious, lined equipment separate from the gypsum slurry system.

The liquid solution circulated through this process and the complex has a pH of 4.5 to 4.7 and therefore has no fluorine vapor pressure because the fluorine therein is present as the two stable salts, Na_2SiF_6 and NaF. This condition precludes the presence of any fluorine in the air exits of cooling towers, and in other equipment. The total gaseous fluorine emission from the entire closed pond system will be less than 15 pounds per day, for a phosphate complex of the size wherein the volume of liquid effluent is in the 16,000 gallons per minute range.

Removal from Water

Phosphate rock is mined by stripping the overburden from the underlying phosphate bearing matrix, slurrying the matrix, consisting of a mixture of clay, sand and phosphate pebble, with water and pumping the slurry from the mine site to a beneficiation plant where the matrix is washed, scrubbed and beneficiated to produce an upgraded pebble, phosphate rock concentrates, a slimes slurry and sand tailings. The slimes slurry consists mainly of clays, sand and very fine particles of phosphate rock. Generally, about 0.75 to 1 part by weight of slimes slurry is produced per part by weight of upgraded pebble and rock concentrate. In practice, the slimes, a dilute aqueous suspension of about 2 to 5 weight percent solids containing about 30 to 40 weight percent of the phosphate matrix, are disposed of by pumping into settling ponds built over previously mined areas. Over a period of time, the entrained solids slowly settle.

As practiced, slimes disposal procedures have many clearly defined and long-recognized deficiencies. Among them is the need for large land areas set aside for slimes ponds. For example, a plant producing 2 million tons of phosphate rock per year will, over a period of 15 years, require approximately 4,500 acres of land for slimes ponds and because of the slow settling rate of the entrained solids, such land is unusable for decades.

Further, since the volume of the hydrated slimes and overburden is approximately 1.5 times greater than the volume of the mine pits, unsightly retaining dams often 35 feet high are necessary around the perimeter of each disposal area. The possibility of dam breakage always exists with resulting river, lake and stream pollution. Also of great importance is the use and retention of large quantities of water.

Another important deficiency in the percent procedures is the poor recovery of phosphate values from the matrix since the 30 to 40% of the phosphate values retained in the slimes are never recovered. The recognition of these deficiencies has led experimenters to propose various alternatives over the years. Flotation techniques for ultrafine particles to separate clays and phosphate values have been unsuccessful in terms of both cost and recovery.

Acid leaching of slimes has not been practical due to extreme difficulty in filtering the clays and other gangue materials from the slurries and the relatively low concentration of phosphate in the filtrate. Further, since the clays and other gangue in both matrix and slimes contain relatively high levels of acid-soluble iron, aluminum and magnesium compounds compared to beneficiated rock, acid leaching of such materials results in phosphoric acid of inferior quality.

Other methods of slimes disposal have been proposed, including using the slimes as aggregates for highway construction or as bricks for home construction or thickening the slimes to provide for improved water conservation and more rapid land reclamation. These methods have not proved practical.

One other method has been proposed to remove the slimes disposal problem; the use of dry unbeneficiated matrix. Previous attempts in this area have failed because of process difficulties and the poor quality of the phosphoric acid produced. These problems arise from the fact that the presence of a significant amount of clay results in poor filtration and from the further fact that the high iron and aluminum content of the clays affects calcium sulfate crystal size and form and product purity.

Thus, there presently exists a need for a slimes disposal process which will accomplish practicably the following design objectives; (1) recovery of discarded phosphate values; (2) elimination of the pond system for slimes disposal; (3) conservation of process water and; (4) provision of potentially closed loop operation of the phosphate mining and phosphoric acid production facility.

A process developed by *R.S. Ribas and J.D. Nickerson; U.S. Patent 3,932,591; January 13, 1976; assigned to United States Steel Corporation* involves preconditioning a phosphate source, prior to digestion, by calcination at a temperature in the range of 1600° to 2200°F, preferably 1750° to 2000°F. This preconditioning allows the use of dry, as-mined matrix in the digestion process and obviates the necessity of a beneficiation step.

This in turn means that no slimes are produced, no slurry water is necessary, mined land can be reclaimed immediately, and presently discarded phosphate values are recovered. Not only will this process provide for immediate land reclamation and water conservation in the future, but it can also be used to recover the phosphate values from previously discarded slimes, thus clearing previously used land for productive use.

In chemical processes for the manufacture of phosphoric acids, there are produced as undesirable by-products aqueous filter aid sludges. These aqueous filter aid sludges normally contain undesirably high levels of arsenic, e.g., up to 5 mg/l, as dissolved arsenic compounds, such as arsenic pentasulfide along with other contaminants such as hydrogen sulfide and other sulfur compounds dissolved in the liquid component.

The aqueous sludges further normally contain about 30 to about 40 volume percent solids content which includes a wide variety of mineral compounds, such as diatomaceous earth filter aid, as well as insoluble arsenic, sulfur, iron, phosphoric, etc., compounds.

A process developed by *S.R. Thompson; U.S. Patent 3,980,558; September 14, 1976; assigned to Browning-Ferris Industries, Inc.* involves contacting such filter aid sludges with a solidifying agent consisting essentially of a hydraulic cement in amounts sufficient to provide a fluid mass that will set to a contiguous rock-like solid upon standing, and then allowing the admixture to set to a contiguous rock-like solid mass which is insoluble in water.

By this process, the soluble and insoluble toxic materials of the waste are wholly entrapped in the contiguous rock-like solid mass which thereby prevents them from being leached into the surrounding environment when exposed to ambient moisture.

PHOSPHORUS

Elemental phosphorus may be formed by electric reduction of phosphate rock. In a particular procedure, ground and pelletized phosphate rock ore is mixed with coke and silica, and the mixture is electrically smelted. The phosphorus is driven off in gaseous form, condensed and recovered as a liquid. The liquid phosphorus is collected under water since it is combustible when exposed to air.

Liquid waste effluents are provided by this procedure including condenser process water called "phossy" water. The phossy water contains small quantities of elemental phosphorus, and, after smelting together with other aqueous effluents in a first pond, the resulting aqueous liquid contains about 3 to 10 ppm. After further lowering of the elemental phosphorus content in scrubbing operations the aqueous liquid is treated with lime to a pH of 5.5 in a second pond. The resulting effluent, No. 2 pond water, contains about 5 to 10 ppb of elemental phosphorus. Since phosphorus is toxic to aquatic life, especially marine fish, discharge of phossy water to aquatic-life-bearing waters should be avoided.

Removal from Water

Due to the colloidal nature of the suspension of the elemental phosphorus in the No. 2 pond water, it is extremely difficult to separate out the solid elemental phosphorus by physical methods, such as filtration and centrifuging. Simple dilution of this waste effluent is also ineffective since the elemental phosphorus tends to accumulate at the discharge point of the wastewater rather than being dispersed.

A process developed by *A.K. Deshpande; U.S. Patent 3,971,707; July 27, 1976; assigned to Erco Industries Ltd.,* is one in which elemental phosphorus in electrothermal phosphorus plant effluents is oxidized either by anodically produced atomic oxygen or by an anodically produced mixture of atomic chlorine and atomic oxygen. Phosphorus content may be lowered by this process from 5 to 10 ppb to an undetectable level below 0.1 ppb. Such a process is shown schematically in Figure 118.

FIGURE 118: PROCESS FOR ELECTROLYTIC OXIDATION OF ELEMENTAL PHOSPHOROUS IN ELECTROTHERMAL PHOSPHORUS PLANT WASTEWATERS

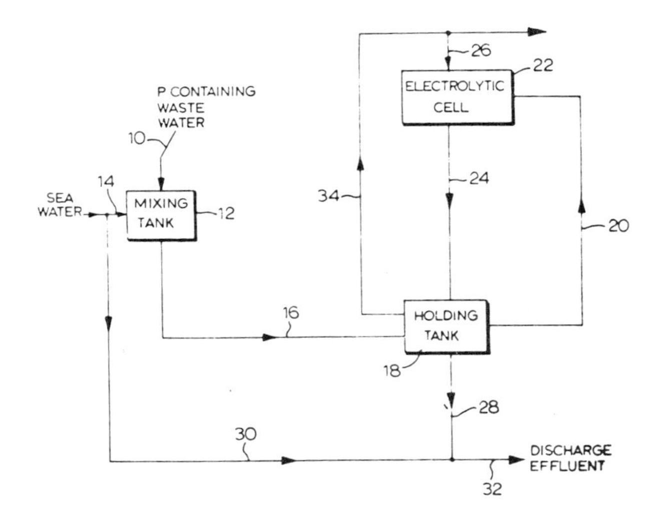

Source: U.S. Patent 3,971,707

Elemental-phosphorus-containing wastewater, typically No. 2 pond water, having an elemental phosphorus content of 5 to 10 ppb (although wastewater with up to 50 ppb has been treated) is fed by line **10** to a mixing tank **12**. In the mixing tank **12**, the wastewater is mixed with seawater fed by line **14**. Any other source of brine solution may be used, but it is convenient to use seawater in coastal regions.

The proportions of mixing of the wastewater and the seawater are not critical although higher rates of oxidation are observed with increasing proportions of seawater, and typically about 1:1 mixture of the components corresponding to a sodium chloride concentration of about ¾ to 1% is employed. The mixture is fed by line **16** to a holding tank **18**. The holding tank **18**, typically a multibaffled tank, has a large capacity as compared to the volume of liquid fed by line **16**, and provides a feed line **20** of electrolyte, typically having a pH of about 3 to 6.5 for an electrolytic cell **22** wherein the electrolyte is subjected to electrolysis.

The holding tank **18** may be any convenient multibaffled type, including a plurality of upright baffles dividing the tank **18** into at least two, preferably a plurality of compartments or zones. The baffles have openings therein allowing passage of liquid from one zone to another through the tank.

The quantity of electrolyte in feed line **20** typically is a portion only of the quantity of liquid in the tank **18**, and hence at any given time there is a substantial pool of liquid present in the holding tank **18**.

The pool of liquid in the holding tank acts as a buffer to smooth out wide variations in the elemental phosphorus content of the wastewater in line **10**, typically between 1 and 50 ppb, so that the liquid feed to the cell **22** in line **20** contains a substantially constant elemental phosphorus content which is about 1 to 3 ppb.

The mixture fed to the tank **18** by line **16** is recycled through the cell **22** a number of times before discharge, as is described in more detail below, and the tank **18** serves to hold the liquid for recycle. Any convenient cell construction may be employed, typically a diaphragm, monopolar or multipolar cell. Preferably, a multipolar cell structure is employed containing a plurality of closely spaced electrodes between which the electrolyte flows upwardly, the liquid product being separated from gaseous products in a gas-liquid separator associated with the cell **22**.

Any convenient electrode material may be employed in the cell **22**, typically graphite. As the electrolyte passes over the electrode faces, it is subjected to electrolysis, forming a mixture of atomic chlorine and atomic oxygen at the anodes which acts as oxidizing agent on the elemental phosphorus.

The current density on the electrodes may vary widely, typically in the range about 0.0125 amp/in^2 to about 0.187 amp/in^2, although values outside this range may be employed, typically up to about 0.25 to 0.33 amp/in^2. The higher values generally are avoided, however, in order to minimize side reactions forming chlorine gas and phosphorus-chlorine compounds. Typically, a current density of about 0.125 amp/in^2 is employed.

The electrode gap within the cell may be varied widely. However, the power requirement of the cell **22** increases with increasing electrode gap and, generally, therefore, the electrode gap is maintained about ¼ to ½ inch. The electrolysis temperature is generally the ambient temperature of the electrolyte in line **20**, typically about 20° to 25°C and little rise in temperature occurs in a multipolar cell. The temperatures above about 40°C are avoided when graphite electrodes are used in the cell due to rapid erosion thereof above 40°C.

The electrolyzed solution is returned from the cell **22** to the holding tank **18** by line **24**. Any gaseous products formed in the cell, mainly in the form of hydrogen with small amounts of chlorine, are vented by line **26** and may be diluted with air prior to discharge to the atmosphere. Some chlorine gas may escape from the solution in the holding tank **18**. This chlorine is vented by lines **34** and **26** to the atmosphere, after dilution, if required.

An effluent which is substantially elemental phosphorus free discharges from the holding tank **18** by line **28**. The feed of the electrolyzed solution by line **24** generally is one compartment of a multicompartment reaction tank **18** from which the elemental phosphorus-free effluent discharges by line **28**. The feed line **16** in this case discharges into a second compartment adjacent the one compartment.

The quantity of liquid removed by line **28** preferably is substantially the same as the quantity fed to the holding tank **18** by line **16**. Therefore, the liquid returned by line **24** to the one compartment of the holding tank overflows the compartment into the remainder of the pool of liquid in the holding tank.

The quantity of liquid cycled from the holding tank through the cell **22** is generally in excess of the quantity of liquid fed to the holding tank by line **16** and is but a fraction of the total liquid in the tank. The ratio of feed of cycled liquid in line **20** to feed of liquid in line **16** may vary from about 2:1 up to about 6:1, with a typical ratio being about 4:1. In addition, the ratio of quantity of liquid cycled in line **20** to the quantity of liquid present in the tank may vary generally up to about 10:1, typically about 5:1.

The mixture of seawater and elemental-phosphorus-containing wastewater, in this way, is periodically recycled between the reaction tank **18** and the electrolytic cell **22** until discharged by line **28**, with a substantial body of liquid being maintained in the holding tank.

The elemental phosphorus contained in the feed liquor in line **16** is oxidized by the mixture of anodically produced atomic chlorine and atomic oxygen in the cell **22** to phosphorus pentoxide which is hydrolyzed to soluble phosphorus oxide materials, which are nontoxic to aquatic life. The recycling is carried out to ensure substantially complete oxidation of the elemental phosphorus to nontoxic harmless materials even in the event of a high level of elemental phosphorus in the feed mixture in line **16**.

The residence time of the liquid within the cell **22** and the holding tank **18** may vary widely depending on the individual flow rates chosen. Generally, however, the residence time of the liquid in the holding tank is considerably in excess of that in the cell, in order to minimize side reactions of the electrolysis while still satisfactorily oxidizing the elemental phosphorus by the mixture of atomic chlorine and atomic oxygen. Typically, an average residence time in the holding tank is 30 to 60, preferably 40 minutes; in the cell, 5 to 15, preferably 12 seconds.

The liquid effluent from the holding tank in line **28** has been found to have no detectable elemental phosphorus content over long periods of operation with widely varying levels of elemental phosphorus in the wastewater in line **10**. There may be present in the effluent in line **28** small quantities of dissolved chlorine, typically about 1 to 5 ppm. In addition, since a slightly acid pH of 6.0 to 6.5 usually is used for the electrolyte in line **20**, the effluent in line **28** usually has a pH of 5.8 to 6.2.

Therefore, while it is possible by this process to overcome the major problem of the presence of toxic elemental phosphorus in wastewater, the presence of the dissolved chlorine in the effluent in line **28** and its acid pH may represent alternative hazards to aquatic life. Therefore, prior to discharge of the effluent in line **28** it is preferred to mix the same with seawater fed by line **30**.

The quantity of seawater employed in this dilution usually should be sufficient to provide an effluent for discharge in line **32** having a pH approximately that of the seawater to which the effluent is to be discharged and a dissolved chlorine concentration less than 0.05 ppm. Typically, the effluent in line **28** is mixed with from 50 to 100 volumes of seawater fed by line **30** to provide the discharge effluent in line **32**.

Where the installation is remote from a coastal region, the effluent in line **28** may be mixed with a considerable volume of water from the body to which the discharge effluent ultimately is to be fed in order to provide the dilution discussed above.

A process developed by *F. Muller et al; U.S. Patent 3,684,461; August 15, 1972* involves the continual workup of wastewater having phosphorus sludge therein, such as that obtained in the electrothermal production of phosphorus. The wastewater is first filtered in a filtration zone, the resulting filter cake is predried in a preliminary drying zone, the predried filter cake is conveyed through at least two additional drying zones maintained at temperatures between about 100° and 380°C, gaseous and vaporous matter issuing from the additional drying zones is delivered to a condensation zone, the phosphorus and water are condensed therein and separately removed therefrom.

PHOTOGRAPHIC PROCESSING EFFLUENTS

Removal from Water

In the past gelatin-bound silver has been recovered from waste washing waters from photographic processing by centrifuging. However, a typical centrifuging operation removes only about 77% of the silver. "Floccing" has been used in an attempt to recover the remainder, or at least a significant part thereof, but known floccing methods often make it necessary to handle uneconomically large volumes of liquids.

Furthermore, the quantity of flocculent must be substantially commensurate with the gelatin content of the wastes for best silver retention. Under these conditions the "floc" will contain a considerable proportion of alumina waste that makes its presence felt during the smelting operation.

A process developed by *M. Korosi; U.S. Patent 3,982,932; September 28, 1976; assigned to Eastman Kodak Company* is one in which silver is recovered from photographic waste liquids containing emulsified silver, silver compounds, and gelatin by reacting therewith a proteolytic enzyme while the reactants are maintained in an alkaline condition.

Then the reactants are acidified to a pH of 4.2 or less by introducing an acid such as HCl or H_2SO_4. Precipitation of gelatin-bound silver and silver compounds occurs upon acidification, and the precipitate is allowed to settle out in a settling tank. Supernatant liquid is removed from the top, neutralized and sent to sewer. Settled sludge is removed from the bottom and silver recovered, as by incineration. The operation can be conducted either batch-wise or continuously. Such a scheme is shown in Figure 119.

FIGURE 119: PROCESS FOR SILVER RECOVERY FROM PHOTOGRAPHIC WASTE LIQUIDS

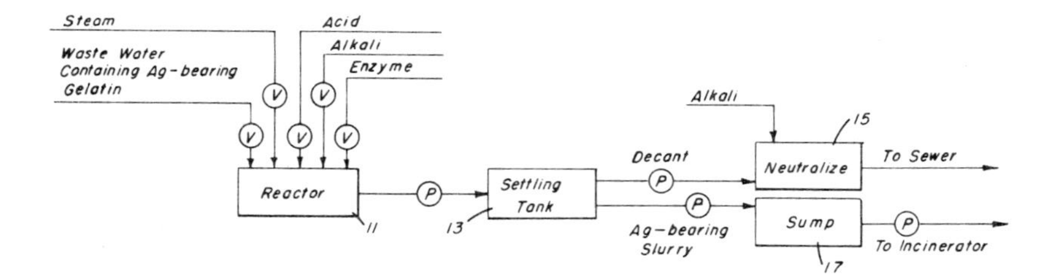

Source: U.S. Patent 3,982,932

The following is an example of the conduct of this process. Reactor tank **11** is filled with 2,000 gallons of waste wash waters from photographic emulsion manufacture containing 130 ppm of silver and steam is injected until the temperature is 50°C. Then 46% sodium hydroxide is added until the pH is raised to 8, at which point 5 ppm of Bioprase PN-10 is introduced in aqueous solution and digestion proceeds for 15 minutes.

Sulfuric acid (98%) is then added in tank **11** until pH is reduced to 3.5, the isoelectric point, and a fine precipitation begins within 1 minute. After 10 minutes the tank contents are pumped into a settling or sedimentation tank **13**. After settling for 24 hours the silver content of the supernatant liquid is less than 1 ppm. The supernatant liquid is pumped to tank **15** where it is neutralized with aqueous NaOH to pH 7 and then discarded.

The slurry in the bottom of tank **13** contains, on a dry weight basis, 23% silver (in metal and halide form) and 60% gelatin, and is pumped to sump **17**, and then injected into an incinerator where the water is vaporized and silver-bearing ash is formed.

The disposal of spent photographic processing baths can introduce harmful and polluting chemicals into the lakes and streams. In recent years color photographic processing has used bleach-fix, hypo or fix baths containing large quantities of thiosulfate (a reducing agent) and color developers containing phenylenediamine compounds as well as hydroquinone, ascorbic acid, hydroxylamine, hydrazine, benzyl alcohol and/or other organic compounds.

Stabilizer baths may contain formaldehyde, acetic or citric acid, and other compounds. Wash and other processing baths also tend to include such materials, depending on their location in the photographic processing sequence.

A process developed by *R. Fisch and N. Newman; U.S. Patent 3,721,624; March 20, 1973; assigned to Minnesota Mining and Manufacturing Company* is a process for reducing the pollution level of spent color photographic processing solutions from a process including a thiosulfate-containing fix solution and a phenylenediamine-containing color developer solution.

The process comprises adding an oxidizing agent to the spent fix solution to oxidize thiosulfate to sulfate, then adding the oxidized fix solution to the spent color developer solution, thereby insolubilizing the phenylenediamine and forming a sludge, and removing the sludge from the solution.

A process developed by *B.A. Hutchins et al; U.S. Patent 3,502,577; March 24, 1970; assigned to Eastman Kodak Company* is one in which potential water pollutants including primary aromatic amine color developing agents are recovered, alone, or simultaneously with color-forming couplers or benzyl alcohol from developer solutions in a separable, second phase by admixture with a water-soluble salt under limited pH conditions.

See Ethylenediaminetetraacetic Acid EDTA (U.S. Patent 3,767,572)
See Iron Cyanides (U.S. Patent 3,772,194)
See Silver (U.S. Patent 3,901,977)

PHOTORESIST PROCESS EFFLUENTS

In various photoresist-making processes, after image-wise-exposure of a solid photosensitive layer comprising a macromolecular organic polymer, the resist-forming element is placed in an organic solvent developer and the areas of the element which are soluble after exposure are dissolved in the solvent, leaving a polymeric image upon the element.

Frequently, a volatile halogenated hydrocarbon solvent, e.g., methyl chloroform or methylene chloride, is used as the developing agent or solvent. After development the solvent contains dissolved polymer and monomer. If no means are available to purify the spent solvent, it must be disposed of. It is thus economically desirable to have a way of recovering the solvent so that it may be used over again. The problem incurred by conventional distillation recovery systems when applied to solvents containing dissolved polymer and/or monomer is caking or solidification and buildup of the scrap polymer on the working parts and particularly the heating surfaces as the solvents are distilled. This fouling precludes efficient recovery, since a cleaning step or steps must be employed to ready the recovery apparatus for further use.

Furthermore, high temperatures within the cake can produce products corrosive to the heating surfaces and a formation of toxic products. The problem results from the use of solid heating surfaces; that is, the necessary heat imparted to the system to effect distillation of the solvent from the dissolved polymer and monomer impurities is supplied directly through solid surfaces in contact with the solution to be purified. As the solvent is driven off, the residue solidifies and congeals on the heated surfaces. If monomeric materials are present, heating might cause crosslinking of them and they would not then be removable by introduction of new solvent.

Removal from Water

A process developed by *A.M. Essex and R.B. Heiart; U.S. Patent 3,666,633; May 30, 1972; assigned to E.I. duPont de Nemours and Company* is a process for the recovery of polymer- and/or monomer-laden organic solvents which permits maximum recovery of the solvent without the buildup of the impurities on the distillation equipment. The process comprises passing steam directly through a batch of spent solvent, condensing the mixture of steam and pure solvent which is given off and separating the water and solvent in a decanter.

The cycle is then repeated by drawing off the polymer- and/or monomer-laden residue together with condensed steam and injecting another batch of spent solvent which in turn dissolves any solid residue left over from the first batch.

The apparatus for carrying out this process comprises a container for holding the spent solvent, a steam source, a condenser, a separation decanter, means for drawing off the polymer- and monomer-laden residue after each batch of solvent is processed, means for injecting additional spent solvent into the container, and electrical means for automating the entire apparatus.

The process and apparatus eliminate fouling of the working parts due to the buildup of residue in two ways. First, no solid heating surfaces are used, which minimizes the tendency of the polymer and monomer to solidify on those surfaces. A slight buildup occurs initially on the pipe conveying the steam into the container; but this is limited since the residue itself gradually becomes a heat insulator and no more buildup can occur. Second, the use of a batch process determines that any residue left after a given batch of solvent is purified will be dissolved by the next batch of spent solvent introduced into the container. This process allows recovery of 80 to 98% of the spent solvent. Figure 120 is a diagram of the apparatus used in this process.

As shown, the process is begun with the activation of a positive displacement gear pump **1** which draws in the spent solvent with dissolved polymer and monomer from an outside source through pipe **2**. Pump **1** forces the solvent into the container **3** for a preset time based on the amount of solvent which vessel or container **3** is designed to hold.

FIGURE 120: APPARATUS FOR PREVENTION OF WATER POLLUTION BY SOLVENT RECOVERY IN PHOTORESIST MANUFACTURE

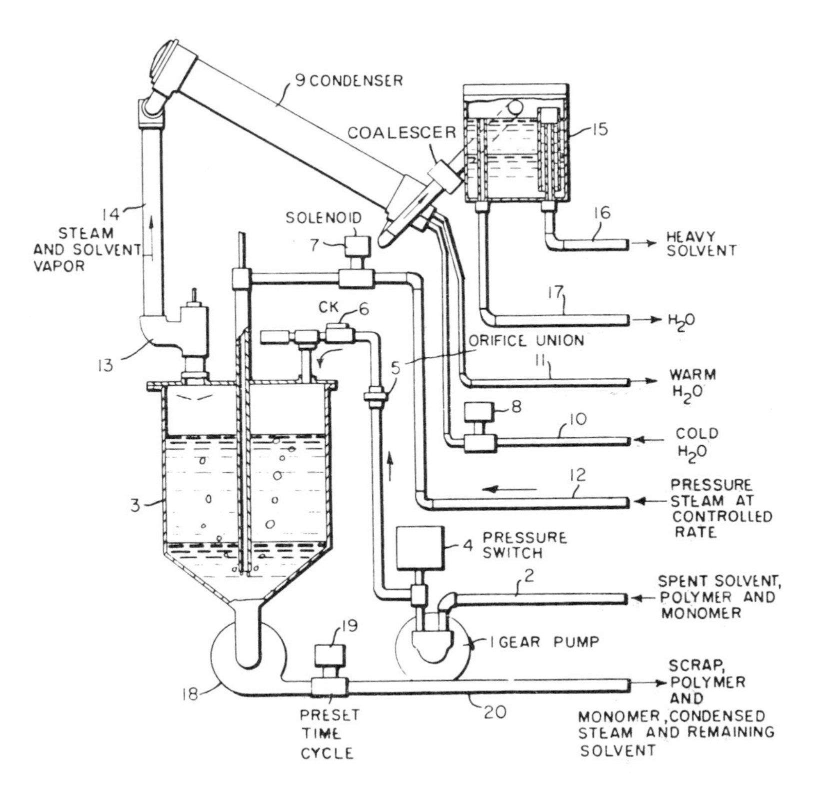

Source: U.S. Patent 3,666,633

For example, if the container **3** is to be charged to five gallons, pump **1** could introduce the solvent at a rate of 2½ gallons per minute for two minutes. The entrance line **2** is equipped with a pressure switch **4** and an orifice union **5** which provides for automatic stopping of the system if no liquid pressure is detected for 5 seconds during the "fill cycle." The solvent passes through a check valve **6** which prevents back flow of the solvent when steam is entering the container **3**.

At the end of the filling time, pump **1** automatically stops and the fill cycle is over. At that time, solenoid valves **7** and **8** open. Water flows through valve **8** from pipe **10** through the shell- and tube-type condenser **9** and out of pipe **11** in order to cool the condenser and liquefy the vapor that will be entering it. Valve **7** admits steam from pipe **12** under pressure at a controlled flow rate.

The steam is bubbled through the solvent and after a few minutes distillation begins. A thermoswitch **13** senses the rise in temperature as the distillation proceeds. The distillation temperature is less than 100°C and gradually climbs as the steam to solvent ratio increases during distillation. The nonvolatile impurities are left behind in the container **3**. The mixture of steam and solvent vapor travels up pipe **14** and into condenser **9**. There it liquefies. The liquefied mixture then travels to the decanter **15** where the heavier liquid is separated by gravity from the lighter liquid and is drained through and collected from pipe **16**.

The water is drained through pipe **17**. It can be determined experimentally at what temperature, detected by the temperature switch **13**, optimum recovery of the solvent is obtained. For example, using methyl chloroform as the solvent to be recovered, 90% of the solvent will have been distilled and recovered when the vapor temperature at the thermoswitch **13** reaches 70°C.

At this point, the temperature switch **13** closes valves **7** and **8** thus ending the steam or distillation cycle, and activates a centrifugal pump **18** and opens solenoid valve **19**. Pump **18** sucks out the scrap polymer and monomer left behind in the container **3** after distillation along with the condensed steam and the remaining amount of solvent. The waste leaves through pipe **20**. The pump **18** operates on a preset time cycle, then stops.

Solenoid valve **19** closes. At this point, gear pump **1** is activated and another batch of spent solvent is introduced into container **3** and the process begins again as just described. The new batch of solvent dissolves any scrap polymer and monomer which the pump **18** was not able to remove. Thus all working parts remain clean through recovery and the apparatus may be used continuously without fear of gumming or fouling by residue buildup. When the parameters of operation have been settled upon, the apparatus works on its own normal cycle without manual intervention.

A specific example of how this process is used follows. A photoresist coating was laminated to a copper-clad, epoxy-Fiberglas board. The laminate was then imagewise-exposed, producing soluble and insoluble areas in the photoresist film coating. The board was then developed, using methyl chloroform as a solvent.

The unexposed, soluble areas of the photoresist film were dissolved in the methyl chloroform and washed from the board leaving the insoluble resist image on it. The methyl chloroform containing the dissolved photoresist was then purified and recovered in the following manner. The gear pump drew in the spent methyl chloroform and charged the container at a rate of 2.5 gallons per minute.

The desired amount of solvent in container **3** was 5 gallons, thus the time delay relay was set to time out in two minutes, ending the fill cycle. As described above, the steam cycle automatically begins. Steam was introduced into the solvent in container **3** at the rate of about 30 pounds per hour. After approximately 23 minutes, the vapor temperature as detected by the thermoswitch **13** had risen from 67° to 70°C, the point at which optimum recovery of solvent was obtained.

The switch automatically turned off the steam and ended the steam cycle. The pump out cycle began. Pump **18** drew out the residual polymer-monomer mixture and steam condensate. The time delay relay had been set for 35 seconds.

When it timed out, the pump stopped, ending the pump out cycle. The next batch of spent solvent was then introduced by pump **1** into container **3**, dissolving any scrap that pump **18** had not emptied and thus cleaning the apparatus as well. The process went on as before. The recovery and purification of the methyl chloroform was 88%, i.e., 88% of the spent solvent was purified.

PHTHALIC ANHYDRIDE PROCESS EFFLUENTS

Removal from Air

A process developed by *D.C. Ferrari et al; U.S. Patent 3,624,984; December 7, 1971; assigned to The Badger Company, Inc.*, is one in which chemical effluent waste gases from chemical plants, particularly effluent waste gases from phthalic anhydride and maleic anhydride plants, are effectively water-washed of residual organic matter (98 to 99% removal) in a wet scrubber using recycled water to concentrate the organic pollutants in the scrubber liquor. A concentrated liquid purge (blowdown) from the scrubber recycle circulating loop is directed to a thermal incinerator where the purge is vaporized and the organic pollutants are oxidized to nonpollutant products.

See Naphthoquinone (U.S. Patent 3,370,400)

Removal from Water

A process developed by *E.L. Cole et al; U.S. Patent 3,642,620; February 15, 1972; assigned to Texaco, Inc.* involves treating wastewater containing very stable and resistant contaminants such as maleic acid, fumaric acid, phthalic acid, terephthalic acid and the like. Currently, these are handled by bacterial oxidation in ponds but this requires considerable land to hold the waste stream and is slow.

In this process, a waste feed stream containing essentially water-soluble organic wastes is continuously subjected to noncatalytic air oxidation at a temperature in the range of 400° to 700°F at a pressure within the range of 300 to 3,100 psi under turbulent conditions for a contact time ranging from 0.1 minute to 2 hours whereby substantially all the organic wastes are oxidized to carbon dioxide and water.

PHTHALIC ESTER MANUFACTURING EFFLUENTS

One of the processes which is most widely used on an industrial scale for the production of dialkyl phthalates is esterification of the corresponding alcohol by phthalic anhydride or acid in the presence of sulfuric acid acting as a catalyst.

In this reaction, besides the dialkyl phthalate, there is also formed a certain number of neutral or acid esters resulting from the reactions as between the alcohol used and the phthalic and sulfuric acids. In order to purify the dialkyl phthalate, the neutral or acid esters are subjected to a series of operations: careful hydrolysis of the esters, destruction of the neutral alkyl sulfate, neutralization of the acid functions by a base, washing operations, or decantation operations; this results in wastewaters comprising all the water-soluble impurities from the process and particularly the salts of organic acid, and a small amount of the unreacted starting materials for the esterification operation.

Before discharging such wastewaters, it is essential that they be subjected to a treatment for removing the major part of the organic molecules. It is common for the wastewater from a phthalate-producing factory to have a BOD of 20,000 milligrams per liter of oxygen, whereas often the standards set for protection of the environment will accept discharge only of water having a very low BOD. Moreover, most of the substances present in the wastewater are due to the excess amounts of the starting materials.

Removal from Water

Most of the conventional treatment processes applied to wastewater (flocculation, absorption on activated carbon, chemical oxidation, microbial digestion) are of low efficiency, expensive, and do not permit recovery of the organic materials.

In order to facilitate separation and recovery of the organic compounds, it has been proposed that the wastewater should be subjected to an operation to hydrolyze the alkyl phthalates and sulfates in acid medium, followed by a neutralization operation and an operation for recovery of the alcohol by decantation.

This process, which makes it possible to recover the alcohol, is however not sufficiently efficient for depollution, as it does not separate the phthalates and as it also requires fairly strict hydrolysis conditions, namely 200°C at a pressure of 20 bars, thus resulting in substantial capital investment and operating costs.

A process developed by *J. Helgorsky and M. Auroy; U.S. Patent 3,933,630; January 20, 1976; assigned to Rhone-Progil, France* is a process for the purification of wastewater containing one or more phthalic esters of alcohols having more than 4 carbon atoms, which comprises acidification of the wastewater by a strong acid to attain a free acidity of the wastewater greater than 0.05 N and extraction of the acidified water with an alcohol having more than 4 carbon atoms; preferably the same alcohol is used as that used to form the phthalate ester. The flow diagram for ester production and integrated wastewater cleanup is shown in Figure 121.

In the figure, a rectangle **1** is used to denote the conventional production apparatus, with the feeds for the various raw materials, the intake or feed **2** for alcohol, feed **3** for phthalic acid, feed **4** for sulfuric acid, feed **5** for the basic material serving for the neutralization of the reaction mixture before separation of the dioctyl orthophthalate and the discharges or outlets, namely **6** for the dialkyl phthalate, **7** for the wastewater and **8** for the various purges.

The conduit **7** leads into an acidification vessel **9** provided with stirrers and provided with a feed **10** for the sulfuric acid, and two discharges, one discharge **11** for gases and the other discharge **12** for the acidified wastewater. The conduit **12** leads into a liquid-liquid extraction contactor vessel **13** (a vessel for extraction by liquid-liquid contact).

The contactor vessel **13** is a known multistage apparatus. In the figure the apparatus is indicated diagrammatically by a rectangle with the conduit **12** for the inlet of the acidified wastewater, conduit **14** for the alcohol feed, outlet **15** for the alcohol charged with organic extract, which is returned to the apparatus **1**, and the discharge **16** for the wastewater which has been treated.

FIGURE 121: SCHEME FOR PURIFICATION OF WASTEWATER CONTAINING PHTHALIC ESTERS

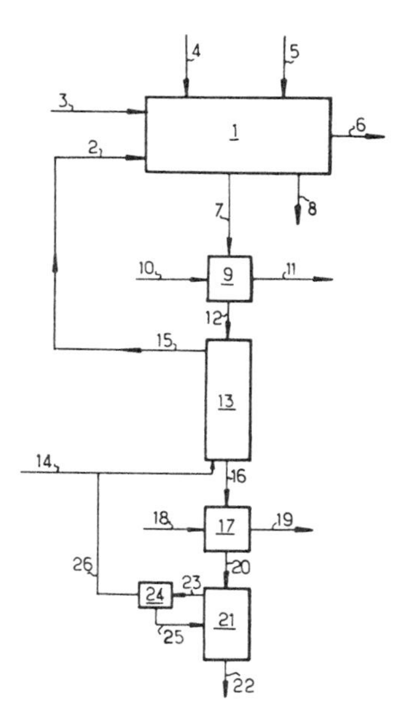

Source: U.S. Patent 3,933,630

The discharge **16** leads into a neutralizer **17** with a feed **18** for neutralizing material, a discharge **19** for the outlet of gas, an outlet **20** by means of which the neutralized water passes into a distillation apparatus **21**, the final discharge **22** for the treated wastewater, and an outlet **23** for the alcohol-water azeotrope.

The conduit **23** leads into a decanter **24** which carries out the separation into water which is returned by way of conduit **25** to the distillation apparatus, and alcohol, which is returned by way of conduit **26** to the conduit **14** leading to the contacting vessel **13**.

In a typical example of the operation of this process, the distillation bottom product which is discharged by way of the conduit **22** has the following characteristics: sodium sulfate, 40 grams per liter; BOD, 50 milligrams of oxygen per liter. It can therefore be discharged into a river.

PICKLE LIQUORS

Removal from Water

A process developed by *C.B. Myers; U.S. Patent 3,468,797; September 23, 1969; assigned to Diamond Shamrock Corp.* is one in which spent hydrochloric acid pickle liquors, especially those liquors which contain iron, may be disposed of in an efficient and economical manner by reacting these liquors with a waste material, which waste material is formed primarily from the discharge of an ammonia-soda plant. In this manner the iron is substantially completely removed from the liquor and therefore does not enter the surrounding watershed.

See Iron and Steel Pickle Liquors for additional process details.

PIGMENT MANUFACTURING EFFLUENTS

Removal from Water

An industrial wastewater from a paint pigment plant at Glens Falls, New York, was treated in a typical example of the application of a process developed by *W.H. Gardner et al; U.S. Patent 3,637,490; January 25, 1972; assigned to Hercules, Inc.*

This particular waste is a plant effluent which is acid (pH $\approx$ 3.0) and contains significant amounts of inorganic paint pigment solids. This waste effluent was treated by adding Bakelite microballoons as the flotation agent and then adding Hercofloc as the flocculant in the proportion of about 0.1 gram microballoons and about 0.001 gram Hercofloc to 1,000 cc of the waste effluent. Complete visual clarification was obtained within about 30 seconds upon subjecting the mixture to flotation.

PLASTIC FILM MANUFACTURE EMISSIONS

The preparation of polyvinyl chloride film by the blown tube method is well known. Polyvinyl chloride is mixed with ingredients such as plasticizers, lubricants and stabilizers in a high intensity mixer. The mixer raises the temperature of the formulation to effect plasticizer absorption and the material is then gravity discharged into cooling blenders.

After cooling, the compound is transferred to extruders. The extruders accept the compound through a feed opening, melt the powder-like material and pump the melt through an annular die at approximately 350° to 410°F. It is at this point upon leaving the die that the misting of the plasticizer occurs. The relatively high temperature of extrusion coupled with the sudden pressure drop upon emerging from the die causes the plasticizer to flash and atomize from the surface of the molten polymer. The plasticizers normally used in this type of operation are dioctyl phthalate or dioctyl adipate.

After the melt emerges from the die, it is inflated with air and trapped between a pair of squeeze rolls and the die to form a sausage-like balloon of inflated plastic. Simultaneously, upon emerging from the die, the mist is impinged with cooling air to set further expansion. It is in this tower area which houses the upper nip roll system and the bubble guide mechanisms where the existing plasticizer mist exhaust system is normally located.

After the inflated bubble has been collapsed to a web by the nip rolls, the web is returned to floor level where it is slit, separated and eventually wound into finished customer size rolls. The rolls of film are then boxed, palletized and placed in warehouse racks for eventual distribution to supermarkets and other establishments where the film is used for the wrapping of food products such as meat and produce.

The die where the misting occurs is usually exposed and some of the mist escapes to the work area. Most of the mist is drawn into the tower area and discharged through a mist exhaust system to the atmosphere. The mist is oily, has an unpleasant odor and causes an unpleasant environment both within the work area and in the neighborhood surrounding the polyvinyl chloride film manufacturing facilities.

When the mist is drawn into the tower area by the exhaust system, large quantities of air are drawn into the tower with the mist. There are systems available for removing the mist from the air such as scrubbers, burners and filters. Due to the large quantity of air drawn into the tower area by the exhaust system, the cost of the removal systems is either very high or the efficiency of the removal system is very low.

Removal from Air

A process developed by *W.P. Davis, J.J. Golner and S.S. Feinstein; U.S. Patent 3,852,392; December 3, 1974; assigned to Borden, Inc.* is one in which hot polyvinyl chloride film is blown in a confined area so as to eliminate escape of plasticizer mist. The plasticizer mist generated is passed through a suitable scrubber, filter or other such known device to remove the plasticizer. The heated air component of the mist, free of plasticizer, is then returned to the process building to recover the heat content, or discharged to the atmosphere as desired.

A process developed by *W.R. Evans, Jr.; U.S. Patent 3,232,029; February 1, 1966; assigned to Celanese Corporation of America* involves solvent recovery from cellulose acetate drying operations. During many operations such as the formation of synthetic fibers and films, e.g., cellulose ester fibers, sheet, and films or the application of organic coatings, considerable amounts of organic solvents are evaporated into the atmosphere or other gaseous media. In order to make such processes economically profitable, it is necessary that such evaporated solvents be recovered from the air or other gaseous media.

In this process the condensation of the vaporized solvent is achieved by bringing the coolant into direct contact with the gaseous medium. This may be advantageously accomplished by spraying the refrigerated coolant into the gaseous medium to condense and carry off or entrain the vaporized solvent as a coolant-solvent admixture.

When this spray condensation process is used, the coolant preferably comprises the organic solvent composition in the liquid state. This avoids the need to completely separate the condensed solvent from coolant prior to the rerefrigeration of the coolant since the condensed solvent forms part of the coolant. If desired, the sprayed coolant may contain a liquid absorbent for the vaporized solvent such as light mineral oil.

PLUTONIUM

See Nuclear Industry Effluents (U.S. Patent 3,862,296)

POLYCHLORINATED BIPHENYLS (PCBs)

Polychlorinated biphenyls (hereinafter referred to as PCBs) are industrial chemicals that are widely used as plasticizers, fire retardant paint ingredients, hydraulic fluids and heat exchange fluids. They have recently been detected in various water sources, in human tissue and in many species of birds and fish.

Being relatively heavy organic molecules, PCBs persist indefinitely under natural conditions. Moreover, they tend to accumulate in the food chain, and once ingested, they are stored in the tissues of birds, mammals and fish used as human food. As with many other chlorinated hydrocarbons, such as DDT, even small dosages of PCBs can be toxic.

Most of the release of PCBs into the environment probably occurs from industrial effluent discharges and dumping the leakage of lubricants, hydraulic fluids and heat transfer fluids into the waterways and soil, from materials containing PCBs as a plasticizer into the atmosphere, and as a result of leaching from dumps and landfills.

Although there is some evidence that PCBs can be dechlorinated by ultraviolet light and, thus, lead to the formation of hydroxyl derivatives and other polar compounds that are more readily degradable, PCBs are very persistent and are extremely difficult to degrade once they are present in the environment.

Removal from Water

A process developed by *E.N. Azarowicz; U.S. Patent 3,779,866; December 18, 1973; assigned to Bioteknike International, Inc.* is a process for the microbial degradation of polychlorinated biphenyls (PCBs) which comprises treating the PCBs with certain nonpathogenic, hydrocarbon-utilizing strains of *Cladosporium cladosporioides, Candida lipolytica, Nocardia globerula, Nocardia rubra* and/or *Saccharomyces cerevisiae* until the PCBs have been substantially degraded.

The process is applicable to degrading PCBs as they may be present as pollutants or contaminants in water, in industrial effluents, in various land areas such as industrial sites and the like or in varied laboratory or commercial installations. The process may also be used to clean up and degrade mixtures of PCBs and various hydrocarbon oils or petrochemicals whenever their presence constitutes a deleterious pollution.

A process developed by *T. Nakamura et al; U.S. Patent 3,841,239; October 15, 1974; assigned to Shinmeiwa Kogyo KK, Japan* is an incinerator process. The incinerator contains a molten salt bath and it is claimed to be capable of decomposing PCBs into harmless materials.

POLYESTER MANUFACTURING EFFLUENTS

In the fractional distillation of glycol from reaction products recovered from polycondensation plants, wherein terephthalic acid is esterified with ethylene glycol and the product is subsequently condensed, wastewater containing diisopropyl amine is produced because diisopropyl amine is often utilized as an inhibitor of ether formation and/or catalyst deterioration. The wastewater must be purified before it is discharged to the environment due to the highly poisonous properties of this substance.

Removal from Water

A process developed by *H. Jakob; U.S. Patent 3,867,287; February 18, 1975; assigned to Zimmer Aktiengesellschaft, Germany* is one in which diisopropyl amine contained in wastewater can be removed by contacting the wastewater in finely divided form with an air stream to produce a wastewater stream containing a lesser amount of diisopropyl amine. A suitable form of apparatus for use in the conduct of the process is shown in Figure 122.

FIGURE 122: PROCESS FOR PURIFYING INDUSTRIAL WASTEWATERS CONTAINING DIISOPROPYL AMINE

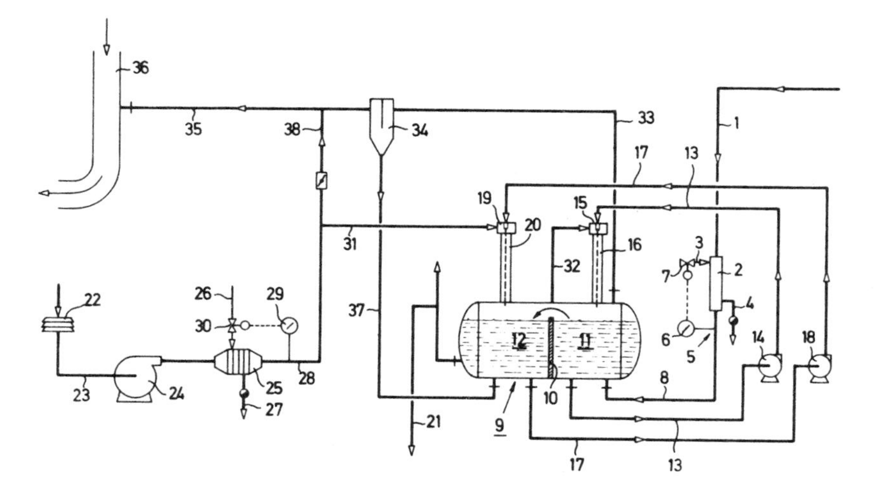

Source: U.S. Patent 3,867,287

The line **1** supplies the head product wastewater from a water column for the distillation of glycol. As a rule, this head product contains between 0.4 and 0.5%, by weight, diisopropyl amine. The wastewater flows through a heat exchanger **2** in a continuous manner to heat the waste. Heating steam is supplied to heat exchanger **2** via line **3** and discharged therefrom via line **4**.

A temperature sensor **5** is located in the line, which in turn controls valve **7** through control circuit **6** thereby controlling the heat output of the heat exchanger in response to wastewater temperature. The wastewater is fed to a container **9** via line **8** which is subdivided into two separate chambers **11** and **12** by a separating wall **10**. The separating wall **10** leaves one section of the cross section open in the upper area of the container **9** so that the wastewater initially entering into the chamber **11** may spill over the separating wall.

A circulation line **13** passes from chamber **11** via a wastewater circulation pump **14** to a liquid jet gas washer **15**, which is attached to container **9** by a tube **16** with its liquid jet nozzle, which is conventional and therefore not specified in more detail, being aligned in a vertical direction. The inside of container **9** and tube **16** communicate with each other so that a jet of droplets dropping down through the tube **16** are collected by a chamber **11** of container **9**.

A similar arrangement is illustrated for chamber **12** by circulation line **17**, circulation pump **18**, a liquid jet gas washer **19** and a tube **20**. The cleaned, purified wastewater containing about 0.04 to 0.05% of diisopropyl amine is taken from container **9** via line **21**. It can be seen that chambers **11** and **12** are arranged in series and that the wastewater within the various chambers is circulated by parallel circulations **13** and **14**, and **17** and **18**.

The gas flow required for the cleaning process enters the apparatus as air via an induction filter **22** and flows into a gas line **23**. A compressor **24** controls the transport of the gas stream first to a gas heater **25**, which in turn is supplied with heating steam via line **26** which steam is discharged via line **27**. A temperature sensor **28** together with a control instrument **29** and control valve **30** provide control for the heat output of the heat exchanger and the resultant temperature of the gas.

The gas flow is fed to the liquid jet gas washer **19** via line **31** and is intensely mixed with the liquid jet in the form of droplets in the gas washer **19**. The gas and liquid together then flow into chamber **12** of container **9** via tube **20** and are then separated into separate gas and liquid phases in the chamber.

The gas is then passed from container **9** via line **32** to the liquid jet gas washer **15**, where the same purification process is repeated. The gas from washer **15**, after separation in chamber **11**, leaves container **9** via line **33** and is fed to discharge air line **35** after passing through a droplet separator **34**. Air discharge line **35** ultimately discharges into an induction air line **36** of the oil furnace of a steam generating unit.

It is clear that the flow of the gas through container **9** with regard to the concentration gradient of diisopropyl amine therein is in the opposite direction or countercurrent to the wastewater. This is not changed by the fact that there are flows in the same direction within the individual stages, and/or within previous lines **20** and **16**.

The wastewater collected in the droplet separator **34** is passed back to container **9** via line **37**. A connecting line **38** permits a control of the gas flow through the liquid jet gas washer or directly into the induction air line **36**.

PRINTED CIRCUIT BOARD MANUFACTURE EFFLUENTS

In the metal processing area, many techniques presently being utilized require wet processing. In particular, wet processing is found to be necessary in the manufacture of printed circuit boards, multilayer circuits, flexible circuits, chassis plating, and even the production of memory cores.

Associated with the wet processing is the problem of handling the wastewater resulting therefrom. This problem has become more severe with stringent local and federal legislation, which regulate the dumping of wastewater thereby requiring highly efficient systems for effluent treatment.

The question of how to handle such a problem becomes even more difficult when quantities of water used in such wet processing systems for metal processing run in magnitudes in the area of 100 gallons of water per minute, roughly involving handling approximately 50,000 gallons of water each day. In attempting to solve this problem, many techniques have been devised within the last few years which are directed to the reusing of the water.

Removal from Water

Most of these techniques in the metal processing area involve recirculating the water obtained through reverse osmosis techniques, where all the contaminated water is routed through a central stream in which, through evaporation techniques, the contaminants are separated from the main stream and reduced to sludge solids which are easily disposed of. In a few exceptional instances, because of the hazards of some contaminants, e.g., chrome or cyanide wastes, specific treatments for reclaiming these specific contaminants are separately effected.

Such systems, in general, have been found to be quite inefficient in the handling of feeding concentrated streams directly to the evaporated stage. The inefficiencies involved include intermediate stages used prior to the evaporated stage for handling sludge due to the heavy concentration of contaminants. In addition, a heavy burden is placed on reverse osmosis membranes and filters, resulting in their frequent replacement and/or cleaning.

A process developed by *D.E. Hewitt and T.J. Dando; U.S. Patent 3,973,987; August 10, 1976; assigned to Data General Corporation* provides a more efficient system for use in metal processing systems which not only provides for more efficient handling of the contaminated water, but in addition, increases the life of reverse osmosis membranes and reduces maintenance efforts on the membranes and replacement of conventional filter units.

In providing such advantages, an economic benefit is gained by significantly reducing the costs in running such an improved system. This is attained while still achieving recycling of better than 90% of the water being used in the metal processing operation. Such a system is shown in Figure 123. With reference to the drawing, there is shown the demand pumps **11** for pumping clean water through two separate streams into process solution tanks **12** and rinse tanks **13**, which together make up a metal processing operation for which the water recycle treatment system is designed.

Although in the embodiment shown, the flow volumes through the process solution tanks are about 2% and flow volumes through the rinse tanks are about 98%, it should be understood a high degree of efficiency may still be attained within the framework of the process if the flow volumes are less than 10% and greater than 90% respectively.

In the stream leading to the process solution tanks **12**, there is included an ion exchange unit containing ion exchange resins for polishing the critical process solution make-up water for critical baths for maximum bath life. This polished water would be used, for example, in gold plating, palladium catalyst, or copper deposition baths. Otherwise, the water could be directly routed to process solution tanks **12** for noncritical baths.

Due to contamination build-up, metal build-up or chemical degradation, the process solutions are periodically dumped, for example, on a daily or weekly basis. In the embodiment, the average daily dump, amounting to about 2 gallons per minute, is routed to a concentrate-neutralizing tank **14** in which the pH is adjusted by use of, for example, caustic soda, sulfuric acid, etc.

By adjustment of the pH, this helps to minimize corrosion of subsequent stainless steel equipment utilized. In the embodiment, the subsequent equipment includes an evaporator **15** known as a Rototherm evaporator. The concentrated chemical waste from neutralizing tank **14** is fed into the evaporator **15** where water is vaporized for discharge into the atmosphere or reclamation, as desired, and where the balance components are ejected as sludge.

The sludge output which contains, in a concentrated form, all the manufacture generated chemical waste may be disposed of in a variety of ways, including sanitary landfill, reclamation, or conversion to building materials.

A major portion of the flow from the demand pumps **11**, roughly about 98%, is routed through a second stream which passes the clean water to the rinse tanks **13**. With the delivery of fresh water, the rinse tanks will overflow the contaminated rinse overflow being directed through troughs to a neutralizing tank **16** where the contaminated rinse overflow is neutralized by means of a pH controller and solution mixer by the use of acid or alkali.

This neutralization allows for constant pH feed to subsequent treatment units. In neutralizing the water in tank **16**, fresh water makeup will be required to balance water lost through the process solution dumps and reverse osmosis waste stream. This water is provided through local water supply when a suitable level controller requires it.

The neutralized rinse water from the neutralizing tank **16** is then pumped through a filter bed **17** for removal of particulate matter. The filter bed material would be of the 30 micron and 10 micron size consisting, for example, of spiral wound cartridges of cotton fiber.

FIGURE 123: WATER RECYCLE TREATMENT SYSTEM FOR PRINTED CIRCUIT BOARD MANUFACTURE

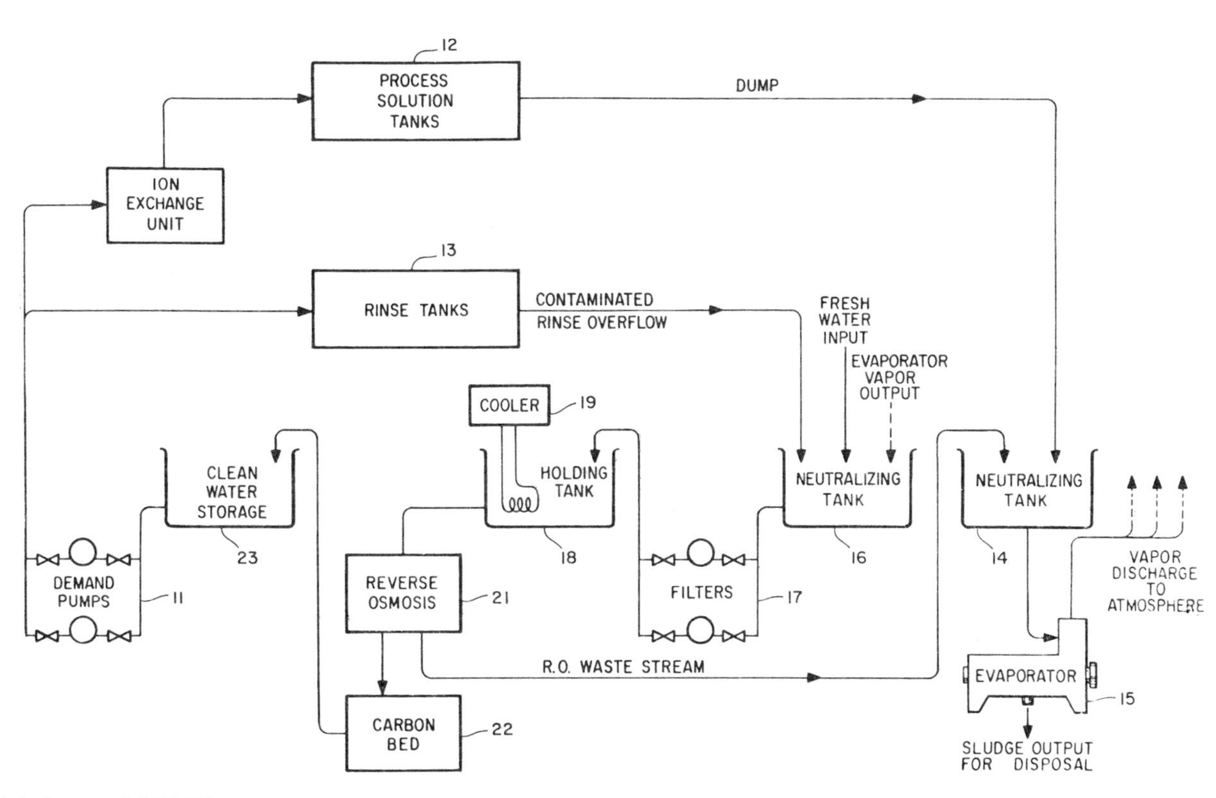

Source: U.S. Patent 3,973,987

Other types would include sand filters, disposable fabric, etc. After passing through the filter bed **17**, the filtered water is collected in a holding tank **18** which serves as a reservoir for reverse osmosis processing and which includes a temperature controlling mechanism to maintain a temperature of the filtered water at less than 85°F to protect the membranes of the reverse osmosis unit. If desired, another filter of the one micron type might be inserted in the water flow path between the holding tank and the reverse osmosis unit **21**.

The reverse osmosis unit **21** is of a conventional type which pumps water through semipermeable membranes which physically separate clean water from contaminants. Contaminants will be removed in a stream consisting approximately of 5% of the feed volume, with approximately 95% clean water passed through the membranes.

In the embodiment where approximately 100 gallons per minute pass through the RO unit **21**, a 5-gallon-per-minute waste stream is routed from the reverse osmosis unit **21** into the neutralizing tank **14** where it is treated in the same manner as the contaminated dump from the process solution tanks **12**.

The 95-gallon-per-minute clean water flow is passed through a carbon bed **22** which will retain organics of small physical size that may have passed through the reverse osmosis membrane. Alternate examples of organic removal include hydrogen peroxide treatment, ultraviolet light and pasteurizing.

The organic-free cleaned water from the carbon bed **22** is discharged into a clean water storage tank **23** which serves as a reservoir for the demand pumps **11** to complete the recycling process providing for 95% recovery. The evaporator **15** discharge to the atmosphere may otherwise be reclaimed for recycling into the system by use of a cooling tower. The flow starts at the demand pumps, as earlier discussed, where constant pressure is maintained for water supply by means of a suitable pressure bypass back to the clean water storage tank **23**.

PROPYLENE OXIDE PROCESS EFFLUENTS

The Halcon process for propylene oxide manufacture is based on the reaction of hydroperoxides with olefins in the presence of selected catalysts leading to high yields of mixtures of alcohol and epoxide as in the following case where the olefin is propylene:

$$ROOH + CH_3{-}CH{=}CH_2 \longrightarrow ROH + CH_3{-}\underset{\diagdown O \diagup}{CH{-}CH_2}$$

When according to this process propylene is reacted with hydroperoxides of ethylbenzene, cumene or isobutane propylene oxide is obtained with transformation levels up to 95% by using catalysts selected from the basic organic salts of molybdenum, tungsten, titanium, vanadium, niobium or tantalum, as for example the naphthenates and acetylacetonates of the aforesaid metals.

The propylene, propylene oxide, hydrocarbon and alcohol are separated by distillation. The residue of this distillation is an organic effluent. This effluent comprising by-products of the Halcon process and containing the metal utilized as catalyst poses a problem with respect to the elimination of the metal. It is impossible to burn the effluent without risking pollution of the atmosphere by the metal oxides in the fumes unless certain precautions requiring the use of rather burdensome apparatus are employed. This problem is especially acute when molybdenum is used because the elimination of molybdenum oxide from the fumes is particularly required by the health authorities of industrial countries.

Removal from Water

A process developed by *H. Lemke; U.S. Patent 3,887,361; June 3, 1975; assigned to Produits Chimiques Ugine Kuhlmann, France* provides nearly complete elimination of molybdenum from these effluents. The molybdenum or other catalyst metal which is present in the effluents can be precipitated almost quantitatively in an unobvious and unexpected fashion by a treatment carried out at temperatures from about 100° to 300°C in an enclosed vessel or under reflux.

This process comprises adding to the effluents or the distillation residues of these effluents from about 5 to 50% by weight of tertiary-butyl alcohol if the effluents or distillation residues do not already contain such an amount of this alcohol, heating the effluents or the mixture to a temperature between about 100° to 300°C in a closed vessel or under reflux, and separating the resulting precipitate containing the metals present in the effluent.

PROTEINS

Proteinaceous wastewaters from operations such as production of vital gluten, starch, beer, and the like, were heretofore deposited in our nation's waterways. Since these waste effluents contain up to 5% solids (mostly protein), they are a major source of water pollution.

Legislation now prevents deposition of these waste liquors into lakes, rivers, municipal sewage treatment systems, and so forth. Consequently, manufacturers are being forced to clean up their wastewaters at great expense or to shut down operations completely.

Removal from Water

A process developed by *J.W. Finley; U.S. Patent 3,898,160; August 5, 1975; assigned to The U.S. Secretary of Agriculture* is one in which there is added to the wastewater a molecularly dehydrated phosphate and a water-soluble salt which provides polyvalent metal ions, preferably ferric or calcium ions. Also, the pH of the mixture is adjusted to 7. This results in precipitation of a polyvalent metal-protein-phosphate complex.

Separation of this complex yields an effluent which may be discharged directly into a sewage treatment system or a water course. The precipitated complex can be utilized in various ways. For example, it may be treated to recover the protein therefrom. Alternatively, where a ferric salt is used in the precipitation, the resulting ferric-protein-phosphate complex can be used per se as a source of nutritional iron in, for example, animal feeds.

A process developed by *R.A. Grant; U.S. Patent 3,697,419; October 10, 1972; assigned to Tasman Vaccine Laboratory, Ltd.* involves the use of a particulate ion exchange material for the purification of waste effluents, such as washings obtained from slaughterhouses, which contain protein or fat, or both. The use of the material can provide effluent with a sufficiently low contamination level for it to be readily disposed of, or even reused for further cleaning purposes. By suitable elution of the material, the protein or fat can be released and isolated for use, for example, as animal food. The ion exchange material can be regenerated for reuse.

A process developed by *B.A. Wennerblom and S.E. Jorgensen; U.S. Patent 3,862,901; January 28, 1975; assigned to Svenska Cellulosa Aktiebolaget, Sweden,* involves removing proteins from solutions containing them, by the steps of bringing the solution into contact with a lignocellulosic material which has been sulfonated to such an extent that its ion exchange capacity reaches to at least 0.15 milliequivalent per gram, and separating the solution from the sulfonated lignocellulosic material. The following is one specific example of the application of this process.

Effluent wastewater from meat processing plants having a permanganate number of 800 mg/l and a BOD value of 1,460 mg/l was precipitated with 0.1 g/l glucose trisulfate. The precipitate was then permitted to settle. The top solution was decanted off and was found to have a permanganate number of 320 mg/l and a BOD value of 500 mg/l. Two liters of the solution were passed through a column bed having 5 grams of sodium bisulfite pulp from pine (chlorine No. 20.5, yield 68%, ion exchange capacity 0.25 meq/g). The outgoing water had a permanganate number of 200 mg/l and a BOD value of 290 mg/l.

See Meat Processing Wastes (U.S. Patent 3,969,203)

PULP MILL EFFLUENTS

In the making of semichemical sulfate pulp by digestion of disintegrated, lignocellulose-containing material such as wood, bagasse, straw and similar materials, by means of alkaline digesting liquid, also called "green liquor," difficult and intricate problems with the obnoxious odor resulting from sulfur-containing organic substances which are produced as by-products, must be solved.

Such obnoxious odors make themselves noticed both in the blowing-off and in the emptying of digesters or cookers. Part of these sulfur-containing compounds remain perceptible in the pulp after the washing thereof and also in the paper manufactured from the pulp. It has thus developed that it is impossible to employ the pulp for the manufacture of "corrugating medium" because the foul-smelling substances in the paper escape in vapor phase during the corrugating process and cause very difficult problems of air pollution.

Removal from Air

A process developed by *K.N. Cederquist; U.S. Patent 3,969,184; July 13, 1976; assigned to Defibrator AB, Sweden* provides odor control for a continuous method of making cellulosic pulp from wood chips within a range of yield from 65 to 90% by digestion with a digestion liquor containing Na_2S and Na_2CO_3. The digestion is effected at temperatures between 150° and 190°C and malodorous sulfur compounds are removed from the pulp, the liquor and the vapors prior to their withdrawal from the digesting process.

In order to eliminate their capacity of emitting obnoxious odors, the sulfur compounds are oxidized by addition of molecular oxygen to the digester prior to discharge of pulp and spent liquor from the digester after which the spent liquor is removed from the digested pulp, concentrated and burned for renewed use in preparation of fresh digesting liquor. A flow diagram of the improved pulping process with provision for odor control is shown in Figure 124.

FIGURE 124: SEMICHEMICAL PULPING OPERATION WITH AIR POLLUTION CONTROL

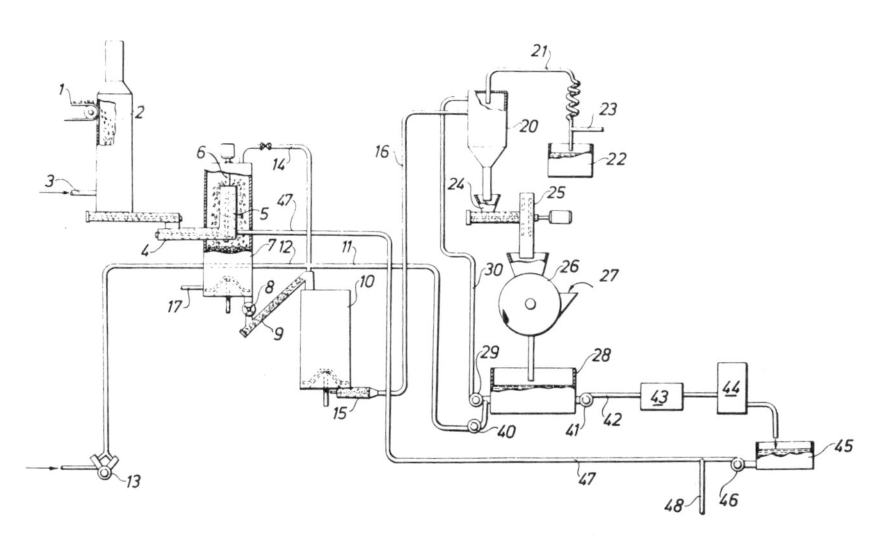

Source: U.S. Patent 3,969,184

From a conveyor **1**, chips fall down into a steaming vessel **2**, to which steam is supplied through a pipe **3**. Thereupon the steamed chips are fed by means of a feeder **4** adapted to form a sealing plug into an impregnating vessel **5**, which is subjected to the pressure prevailing in a digester **7** forming part of the system and is by means of a helical feeder **6** carried upwards through the digesting liquid and falls down into the digester **7**.

The alkaline digesting liquid is then supplied through a pipe **47**. A quantity of sodium monoxide in an amount sufficient to compensate for the losses in chemicals is supplied in the form of Na_2SO_3 through a conduit **48**. Living steam is supplied to the digester through a pipe **17** so that the temperature in the digester is maintained at a constant level, such as 160°C. The duration of stay of the chips in the digester at that temperature may be 10 minutes depending on the desired yield. The digested chips together with their content of liquor are fed through a sluicing feeder **8** and a screw conveyor **9** into an oxidation vessel **10**.

Simultaneously, cold liquor is supplied through a pipe **11** so that incoming material is cooled to 130°C for example and at the same time air is supplied from a compressor **13** through a conduit **12** so that the total vapor-gas pressure at the oxidation station **10** is maintained just below the pressure in the digester **7** in order to enable the sluice feeder to operate.

The duration of stay at the oxidation station is most suitably about 10 minutes. Pulp and accompanying liquor, vapor and gas are continuously blown out through a valve **15**. Air under pressure is introduced all the time so that the pressure remains constant. The pulp is blown through a conduit **16** into a centricleaner **20** from which vapor and gases escape through a conduit **21** and are cooled.

Condensate is collected in a receiver **22** and noncondensable gases are exhausted through a pipe **23**. From the centricleaner **20**, the pulp falls down into a hopper **24** and is fed into a defibrator or other grinding apparatus **25**. At the same time oxidized liquor is supplied to the centricleaner through a conduit **30** to facilitate the flushing of pulp out of the centricleaner.

From the defibrator **25** pulp and liquor are discharged to at least one press **26** for recovery of liquor. Recovered liquor is collected in a tank **28** and the pulp is conveyed further through an outlet **27** in the press. Liquor is circulated by means of a pump **29** and through a pipe **30** into the centricleaner **20**, and by means of a pump **40** and the pipe **11**, cold liquor is supplied to the oxidation station.

The recovered liquor is pumped by means of a pump **41** and through a pipe **42** into evaporator **43** and thickened liquor is fed further to a soda furnace **44**. Resultant black ash or crude sodium carbonate is dissolved in a tank **45** to prepare new digesting liquid which by means of a pump **46** through a pipe **47** is returned to the impregnation vessel **5**. Noncondensable gases possibly formed during the digesting operation can be exhausted through a pipe **14** and fed to the oxidation station.

Air pollution caused by objectionable odoriferous sulfur compounds, for example, hydrogen sulfide, mercaptans and mercaptan ethers, for example, dimethyl sulfide are particularly objectionable to communities located in the vicinity of industrial establishments such as, for example, kraft pulp mills. This is because even extremely small quantities, typically 5 to 50 parts per billion of these compounds present in the atmosphere can be detected by the human nose, and higher levels can cause great discomfort to those who are exposed to these chemicals, since they have most unpleasant odors. Because of the low odor thresholds which are characteristic of these sulfur compounds, a treatment process for their removal, when this is called for, has to be highly efficient.

A process developed by *S. Prahacs and S.P. Bhatia; U.S. Patent 3,701,824; October 31, 1972; assigned to Pulp and Paper Research Institute of Canada, Canada* is one in which objectionable odoriferous sulfurous gases, e.g., H_2S and

CH_3SH, are removed from a gaseous stream by intimately contacting the stream with an aqueous alkaline suspension of activated carbon having a pH of at least about 8 in the presence of at least about 12% by volume of CO_2 and not more than about 6% by volume of O_2, the activated carbon thereby promoting the oxidation of sulfide ions and CH_3SH.

A process developed by *S.P. Bhatia, S. Prahacs, T.L. de Souza and H.G. Jones; U.S. Patent 3,794,711; February 26, 1974; assigned to Pulp and Paper Research Institute of Canada, Canada* is one in which objectionable odoriferous sulfur compounds are removed from gaseous emissions, e.g., methyl mercaptan and dimethyl sulfide or mixtures thereof, either alone or together with hydrogen sulfide, from kraft-recovery furnace stack gases, in industrial operations.

The removal is achieved by the use of alkaline or near neutral scrubbing solutions containing defined minimum amounts of certain phenolic compounds, e.g., substituted phenols and derivatives thereof as additional scurbbing agents, in the presence of gaseous oxygen. The absorption efficiency of the alkaline aqueous solution thus is increased and the hydrosulfide ions are substantially instantaneously converted to soluble sulfate and thiosulfate ions.

A process developed by *R.C.A. Brannland, B.G. Hultman and B.V. Hubert; U.S. Patent 3,842,160; October 15, 1974; assigned to Mo Och Domsjo Aktiebolag, Sweden* is a method which provides for reducing the emission of hydrogen sulfide when scrubbing sulfur dioxide-containing waste gases obtained from burning waste liquor from cellulose digestion processes with aqueous washing liquor derived from an alkaline waste liquor containing sulfide ion. It has been found that hydrogen sulfide emission is reduced considerably when such aqueous washing liquor has a pH within the range from about 6 to about 7.

A process developed by *B. Sen; U.S. Patent 3,796,628; March 12, 1974; assigned to Cariboo Pulp and Paper Company, Canada* is a process for the reduction of malodors emitted from kraft mill lime kiln stacks in the manufacture of pulp by the kraft or sulfate cellulose process. The cooking of wood chips with white liquor is followed by the steps of separating the black liquor from the pulp, evaporating excess water from the liquor, burning the evaporated liquor to form green liquor and then treating the green liquor with lime to form white liquor and a lime mud slurry.

The white liquor is recycled for the cooking of further chips, and the lime mud slurry is calcined in a kiln to produce lime and carbon dioxide. The slurry is intimately mixed with an oxidizing agent to reduce the sulfite content of the slurry before the slurry is burned in the lime kiln. The oxidizing agent may be a gas such as air, chlorine or chlorine dioxide. The gaseous oxidizing agent may suitably be intimately mixed with the slurry by the use of an oxidation tube or any suitable contact apparatus.

A process developed by *K.A. Zetterström; U.S. Patent 3,940,253; February 24, 1976; assigned to Volvo Flygmotor Aktiebolag, Sweden* involves the purification of process waste gases using a combustion device and conditions used in gas turbine technology. The process gases to be purified are fed into the combustion chamber through the axial central inlet thereof, and the combustion chamber is provided with a special inlet for further additional air. A flame tube is disposed inside the combustion chamber wall and spaced therefrom. The flame tube is arranged to be cooled by means of a film of the main part of the further additional air entering through the special inlet.

Such a combustion chamber device has been used to treat 800 Nm^3 waste gases from the cooking and the evaporation steps in a sulfate cellulose plant. The process waste gases contained carbon monoxide and odorous sulfur compounds while the purfied gases leaving the chamber outlet had a temperature of 1400°C and contained only harmless compounds, carbon monoxide and water and a minor amount of sulfur compounds which were not odorous.

One of the virtues of the ammonia-base pulping process is that it presumably enables the recovery of a major amount of the chemicals involved. A process developed by *M. Toivonen, J. Kettunen and E. Salunen; U.S. Patent 3,819,812; June 25, 1974; assigned to Savon Sellu Oy, Finland* is one in which ammonia, furfural, sulfur dioxide and volatile organic acids such as acetic and formic acids are recovered from ammonia-based sulfite waste liquor by volatilizing them, drying the liquor to a powder, and separating them from the volatilized gases.

A process developed by *C.S. Gaillard; U.S. Patent 3,753,851; August 21, 1973; assigned to Westvaco Corp.* is a process for treating hot, odoriferous blow gases from a pulp digester whereby a significant portion of the available heat is recovered, turpentine is recovered, a portion of the odoriferous gases is removed and recovered and the remaining gases are burned resulting in further heat recovery.

The digester blow gases containing water vapor, turpentine vapor and gaseous odoriferous sulfur compounds are passed through a series of direct condensers to a gas storage means before being conducted to an indirect condenser. The uncondensed gases leaving the indirect condenser are scrubbed while the condensate mixture leaving the indirect condenser is conducted to a turpentine separating means. Such a process is shown schematically in Figure 125.

As shown in the figure, a conduit **11** supplies blow gases to the system from blow tanks which are not shown. As heretofore pointed out, pulp digesting is generally a batch process. Therefore, it will be appreciated that the flow rate of blow gases is subject to wide variation. For example, it is common to encounter digesting operations wherein the flow rate of blow gases will vary from as low as a hundred cfm to as high as 200,000 cfm.

Therefore, if one wishes to avoid air entrainment within the blow gas recovery system, one may wish to use a steam purging system as shown wherein the pressure within the blow gas supply line **11** is measured, as at **6**, and the pressure signal is supplied to a pressure controller **7** whose output manipulates a valve **8** that regulates the steam injected into the blow gas line **11** as at **9**.

The blow gases supplied to the system through the conduit **11** are first conducted to a primary or first direct condenser **12**. In a typical installation, the blow gases entering the primary condenser, for example at **13**, are at a temperature of 212°F. Within the upper portion of the primary condenser **12**, the blow gases are directly contacted by a coolant, for example, water, wherein the coolant is supplied to the top of the primary condenser **12** as at **14**.

FIGURE 125: METHOD FOR TREATING PULP DIGESTER BLOW GASES

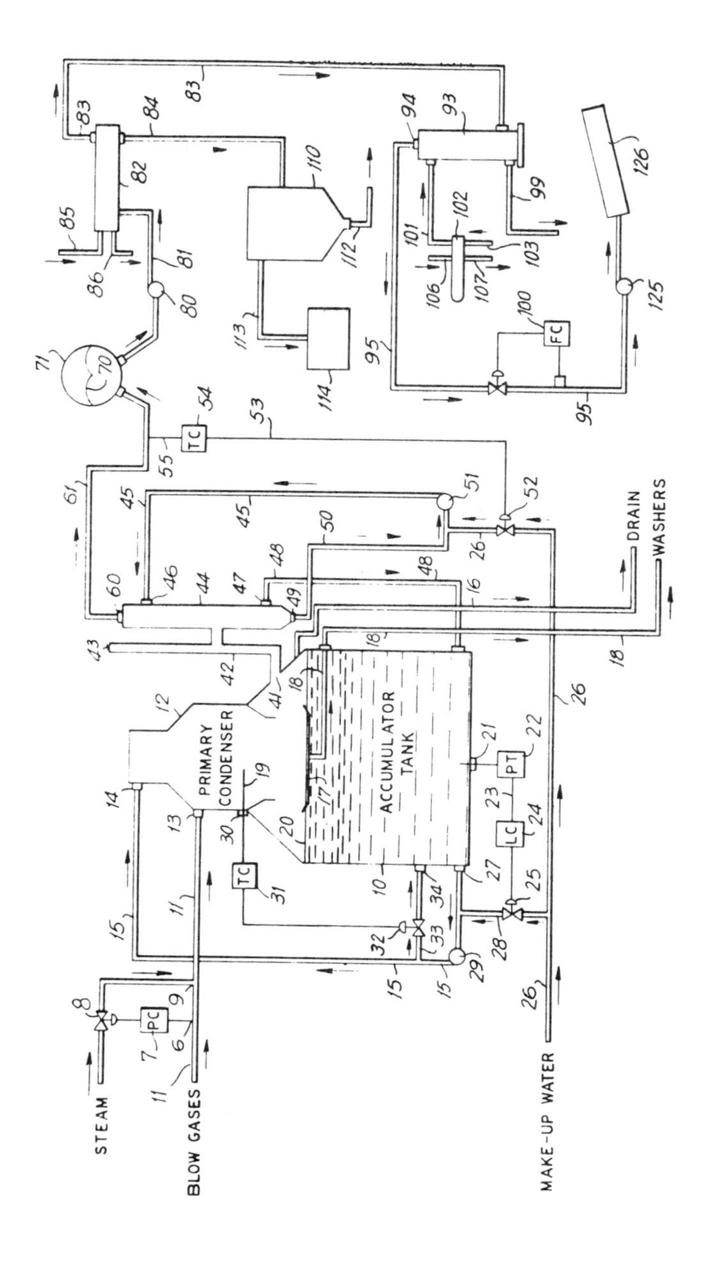

Source: U.S. Patent 3,753,851

As the coolant passes down through the direct condenser **12** and contacts the blow gases, there is, of course, a heat exchange between the coolant and the blow gases. Within the lower portion of the primary condenser **12**, heated coolant may be collected by the trough **17** and conducted from the primary condenser by the conduit **18** to insure a supply of hot water for use in other unit processes, e.g., washers or blow heat evaporators.

In the primary condenser the design of the trough **17** is such as to allow a portion of the heated coolant to pass into an accumulator tank **10** where it may accumulate to a predetermined level, for example as indicated at **20**.

Those skilled in the art will perceive that the temperature of the heated coolant exiting the primary condenser **12** provides an indication of the heat transfer operation which is occurring in the upper part of the primary condenser. Thus, in order to satisfy objectives which will be hereinafter described in more detail, a temperature sensor **19**, e.g., a thermocouple, is mounted in the wall of the primary condenser as at point **30** and provides to temperature controller **31** a signal representative of the temperature of the heated coolant as it leaves the primary condenser **12**.

The temperature controller **31** manipulates a flow control valve **32** so as to control the flow of coolant to the primary condenser. More specifically, a primary condenser pump **29** continuously withdraws accumulated and cooled coolant from the base of the accumulator tank **10** as at **27**.

The primary condenser pump **29** discharges through conduit **15** thus supplying a flow of coolant to the top of the primary condenser **12** as at **14**. However, downstream of the discharge side of the primary condenser pump **29**, the line **15** is tapped by line **33** which is essentially, a recirculation line back to the accumulator tank **10**, as at **34**.

The recirculation flow rate, and thus the flow of coolant through the line **15**, is controlled by manipulation of the flow control valve **32** in response to control signals from the temperature controller **31**. As such, control of the heat transfer or condensation process within the primary condenser **12** may be achieved. Still more specifically, the temperature set point provided to the temperature controller **31** is maintained above the condensation temperature of the turpentine fraction in the blow gases supplied to the primary condenser **12**.

Further, since the aforementioned temperature set point will essentially determine the temperature of the heated coolant collected in the trough **17**, external process conditions (e.g., blowheat evaporator, or washer operating conditions) associated with the use of the heated coolant collected in the trough **17** will generally determine the temperature set point subject, of course, to the limitation that it is above the condensation temperature of the turpentine fraction in the blow gas. In a representative installation, the temperature of 205°F is the temperature set point to the temperature controller **31**.

Since a portion of the heated coolant is collected in the trough **17** and conducted outside of the primary condenser **12** through the line **18**, it becomes necessary to provide make-up coolant. Thus, as shown a pressure tap **21** is provided in the base of the accumulator tank **10**, whereby a pressure signal indicative of the level of the accumulated coolant is provided to a pressure transmitter **22**. Pressure transmitter **22** supplies a pressure signal **23** to a level controller **24** where there is generated a level control signal that is applied to a flow control valve **25**. Manipulation of the flow control valve **25** controls the flow of make-up water through line **28** thus insuring that the desired level **20** is maintained.

Returning to a consideration of the heat transfer or condensation process which occurs in the primary condenser **12**, it will be appreciated that if the temperature of the heated coolant is maintained above the condensation temperature of the turpentine fraction within the blow gas, but below the boiling point of water, for example, if the temperature of the heated coolant is maintained at approximately 205°F, a major portion of the blow gases supplied to the primary condenser **12** will be condensed thus providing a high volume source of hot water for other unit processes and simultaneously minimizing the volume of uncondensed blow gases which must be subjected to subsequent treatment.

That fraction of the blow gases which is not condensed within the primary condenser **12**, exits therefrom as at **41** into a standpipe **42** which may be provided with a pressure relief flapper **43**. In any event, the aforementioned uncondensed portion of the blow gases are passed to a second direct condenser **44**. Within the second, direct condenser **44**, the uncondensed portion of the blow gases from the primary condenser may pass countercurrently with respect to a coolant supplied to the secondary condenser as at **46**.

While the configuration of the second direct condenser **44** may take any of a number of forms, it is critical to the process that the condenser be operated such that the temperature of the remaining uncondensed fraction of the blow gases leaving the secondary condenser **44** is maintained at a temperature just slightly above the condensation temperature of the turpentine fraction in the remaining uncondensed fraction.

Thus, the remaining uncondensed fraction of the blow gases exits from the secondary condenser **44**, as at **60**, into line **61** where the temperature of the remaining uncondensed fraction is measured at **55**. A signal representative of the temperature of the remaining uncondensed fraction as measured at point **55** is supplied to a temperature controller **54** where it is compared to a temperature set point which is slightly greater than the condensation temperature of the turpentine fraction in the remaining uncondensed fraction.

Based upon this comparison, the temperature controller **54** provides a control signal **53** to a flow control valve **52**. Flow control valve **52** controls the flow of make-up water through the line **26** to the secondary condenser pump **51**. As may be clearly seen, the secondary condenser pump **51** has its suction side connected to both make-up line **26** and return line **50**.

Thus, the temperature of the coolant discharged into the coolant supply line **45** from the secondary condenser pump **51** will be determined by the position of the flow control valve **52**. Recognizing that there is provided means for controlling the temperature of the coolant supply in line **45** in response to the temperature of the remaining uncondensed fraction of the blow gases in line **61**, it will be perceived that the condensation process which occurs within the secondary condenser **44** is controlled in response to the temperature of the remaining uncondensed fraction of the blow gases discharged from the secondary condenser **44**.

Since some condensation will occur in the secondary direct contact condenser **44**, an overflow, as at **47**, is provided whereby excess coolant is discharged to the line **48** and thereafter to the accumulator tank **10** which is equipped with an overflow line **16** while the remainder of the coolant from the secondary direct condenser **44** is discharged as at point **49** to the conduit **50** whereby it is returned to the suction side of the secondary condenser pump **51**.

As hereinbefore observed, most pulp digester operations function in a manner which creates wide swings in the flow rate of blow gases. Thus, prior art blow gas treatment systems which have been designed to handle the maximum flow rate which may be encountered are such that one or more pieces of equipment are seldom operated at their design capacity.

To avoid this inefficiency, this process contemplates the provision of a gas storage means which essentially has the capacity for absorbing or flattening out wide variations in the flow rate of blow gases. For example, the remaining uncondensed portion of the blow gases passing from the secondary condenser **44** through line **61** are stored in a vapor storage means **71**.

It is most economical to use a vapor sphere wherein the stored gas is maintained at, essentially, atmospheric pressure by expanding against a flexible diaphragm **70**. The gas within the sphere **71** may be withdrawn therefrom by a gas blower **80** which discharges the gas through line **81** to an indirect condenser **82**. The condensation process within the indirect condenser **82** operates subject to the criterion that the gases supplied to the indirect condenser are cooled below the condensation temperature of the turpentine in such gases.

Therefore, the cooling medium supplied through the line **85** to the indirect condenser **82** is accordingly temperature controlled. As a result of operating in accordance with the aforementioned criterion, there will be formed within the indirect condenser **82** a turpentine-condensate mixture which may be discharged through the line **84**. Gases which are not condensed within the primary condenser **82** are substantially comprised of noncondensable gases and, in most cases, such gases will be highly odoriferous, e.g., hydrogen sulfide, methyl mercaptan, dimethyl sulfide and dimethyl disulfide.

Considering the turpentine-condensate mixture formed within the indirect condenser, this mixture may be conducted by a conduit **84** to a turpentine separating process. In the embodiment shown, line **84** conducts the turpentine-condensate mixture to a decanter **110**, of a type commonly used, where the turpentine is withdrawn from the top through a line **113** to a turpentine storage means **114** and the condensate is withdrawn from the lower portion through a line **112**. Thus recovery of saleable turpentine is achieved.

Returning to the operation of the indirect condenser **82**, it was previously indicated that uncondensed and odoriferous gases were conducted therefrom through a line **83**. In the past, prior art systems had discharged such gases directly to the atmosphere. However, through the process, atmospheric discharge of such gases is avoided.

In a typical installation, the concentration of the noncondensable gases in the line **83** may be quite high. At such concentrations, if the gases within the line were mixed with air, there would be produced a gaseous mixture which was far too flammable for direct, safe incineration. Additionally, if all these gases were burned, chemical recovery of the compounds which form such gases would be precluded. Therefore, as shown, the remaining, uncondensed blow gases in line **83** are passed to a scrubbing tower **93** where they are scrubbed with a caustic scrubbing agent, e.g., white liquor.

In the scrubbing process step, some of the odoriferous gases are removed, e.g., hydrogen sulfide and methyl mercaptan, while the remainder of the noncondensable, odoriferous gases passes from the tower **93**. In this manner, the excess flammability problem is solved by reducing the concentration of the noncondensable gases.

Moreover, if a scrubbing agent such as white liquor is employed, the white liquor leaving the tower **93** will be enriched by the sulfur compounds scrubbed from the blow gases and the enriched white liquor may be returned to the digesters (not shown). Through this procedure, less sulfur compounds need to be added to the white liquor and a chemical recovery and monetary saving are thus achieved.

A further beneficial method of operating the scrubbing process would involve the step of maintaining the temperature of the gas discharged from the tower at a temperature not greater than the temperature of the gases feeding the tower. In this manner, the specific volume of the gases discharged from the tower **93** is not excessive, thus reducing the size of equipment downstream from the tower.

Thus, a caustic scrubbing agent, such as white liquor, is supplied to the scrubbing tower through line **101** and descends through the tower scrubbing the uncondensed gases rising through the tower and recovering the chemicals in those gases. The scrubbing agent leaves the tower **93** through line **99**.

The scrubbed, uncondensed gases pass from the tower, as at **94**, into a line **95**. In a typical installation, the gases leaving the scrubbing tower have had their concentration of noncondensable gases reduced by more than 90%. Therefore, the remaining odoriferous, noncondensable, flammable gases (largely dimethyl sulfide and dimethyl disulfide) can be incinerated thus extracting their heat of combustion while converting them into nonoffensive gases.

A convenient way to achieve this objective in a paper mill is to burn the gases in a lime kiln. Thus, the gases are passed, subject to a flow control loop **100**, to the suction side of an air fan **125** and are discharged therefrom to a lime kiln **126** wherein they are incinerated.

With further regard to the scrubbing process, as was hereinbefore pointed out, the volume of the gas discharged from the tower **93** can be minimized if the temperature of such gases is approximately equal to the temperature of the gases fed to the tower.

One method of achieving this objective is to pass the scrubbing agent through a heat exchanger before it enters the tower. Thus, there is shown a heat exchanger **102** through which a coolant is passed from **106** to **107**. The scrubbing agent if hot, is supplied through line **103** to the heat exchanger **102** and is discharged therefrom through line **101**. If needed, the operation of the heat exchanger **102** may be automatically controlled by measuring the temperature of the gas in line **94** and, in response thereto, regulating the flow of coolant through the line **106**.

In a kraft mill equipped with a continuous digester with internal washing, with external diffusion washing as a final washing step, with a recovery furnace which avoids direct contact evaporation of black liquor, the significant sources of odor and foul turpentine are limited to the following four sources:

(1) Vent gases from the digester condenser.
(2) Vent gases from the evaporator ejector.
(3) Foul condensate from the digester condenser.
(4) Foul condensate from the evaporator condensers.

In the past there have been various schemes developed by others for turpentine recovery and also schemes for collection and destruction of odorous gases. The processes used to date have often been inefficient or complex. Seldom has high percentage recovery of turpentine been achieved where continuous chip digesters have been employed.

A process developed by *J.H. Fisher; U.S. Patent 3,745,063; July 10, 1973; assigned to British Columbia Forest Products, Ltd., Canada* is one in which vaporous relief gases from the flashers and the chip preheaters used in the standard kraft pulping process are directed to a digester condenser; vaporous relief gases from the evaporators of the process are directed to condensing apparatus; the former gases are partially condensed in the digester condenser and the condensate therefrom is combined with part or all of the condensate from the evaporator condensing apparatus and both of these are directed to a steam stripper.

The uncondensed gases from the digester condenser are conducted to the condensing system of the stripper where these gases are condensed together with vaporous gases arising from the stripper. Following this simultaneous condensation of the two sources of gas, the turpentine is recovered and the odorous foul gases are directed to equipment for their destruction together with uncondensed gases from the evaporator condensing apparatus.

Removal from Water

See Kraft Paper Mill Effluents for details of several processes.

RADIOACTIVE MATERIALS

During the operation of a nuclear reactor, stable and radioactive gaseous substances are produced. In nuclear power systems with direct circulation, such as boiling water reactors, these substances leave the reactor with the steam and reach the turbine system (turbine, preheater, condenser) where they are sucked out of the condenser, together with the air which has entered through poor seals and connections and other unavoidable minor leaks in the waste gas system which produces and maintains the vacuum in the condenser.

Thus these gases are continuously being removed from the primary circuit via the waste gas system. During the operation of a water cooled and moderated reactor, substantially three types of gases are produced in the reactor system: (a) nonradioactive gases from the coolant, such as H_2 and O_2, (b) radioactive gases from the coolant, primarily isotopes of N, O, F, and (c) gases from the fuel, isotopes of Kr, Xe.

Hydrogen and oxygen are produced by radiolysis. Radioactive gases from the coolant are formed during nuclear reactions of neutrons and protons with the oxygen of water. During the fission of the fuel elements, radioactive isotopes of krypton and xenon are produced. The production rate for all gases is dependent on the output of the reactor.

The gases formed by the activation, i.e., the isotopes of N, O, F, hereinafter called the activation gases, as well as the radiolytically formed hydrogen and oxygen, are continuously removed from the nuclear system in the form of steam. In the ideal case (absolutely tightly sealed fuel shells) the fission gases remain in the fuel elements.

In practice, however, such perfection is not feasible so that a small portion of the fission gases therein leaks out into the reactor water through this shell and is also removed with the steam. This means that, during normal operation, some radioactive fission gases are present without there being any real defects in the shells. Of course, should a substantial leakage occur in one or more of the fuel elements, other and more extensive measures must be employed to contain and remove the radioactive materials.

These radioactive fission and activation gases ultimately mix in the condenser and must be treated before their release to the atmosphere. The purpose of this treatment is to prevent there occurring an unacceptably high contamination of the environmental air around the nuclear reactors, with the unacceptable radiation load associated with such contamination.

Removal from Air

A process developed by *H. Queiser, H. Schwarz and H.-J. Schroter; U.S. Patent 3,871,841; March 18, 1975; assigned to Licentia-Patent-Verwaltungs-GmbH and*

Bergwerksverband GmbH, Germany is one in which the radiation level of waste gases from nuclear power plants containing both activation and fission gases is controlled at or below limits permitted by applicable standards by passing such gases, prior to release to the atmosphere, through an adsorptive delay path including a body of activated carbon having the relation to the throughput and character of such gases determined by the formula

$$t_v = (K \cdot E/\nu)$$

where t_v = dwell period in minutes
E = quantity of activated carbon in gases
ν = gas throughput in cm^3/min
$K \doteq$ dynamic adsorption coefficient of the gas in cm^3/g

A process developed by *H. Queiser and H. Schwarz; U.S. Patent 3,871,842; March 18, 1975; assigned to Licentia-Patent-Verwaltungs GmbH, Germany* embodies an exhaust gas cleaning system utilizing the principle of delaying radioactive gases to permit their radioactive decay to a level acceptable for release to the atmosphere.

The equipment comprises an adsorbent for adsorbing radioactive gas and a container for containing the adsorbent and for constraining gas to flow through the adsorbent, the adsorbent and the container forming simultaneously an adsorptive delay section and a mechanical delay section. By means of a predetermined ratio of volume of voids in the adsorbent to total volume of the container containing the adsorbent, provision is made for delaying radioactive gas to permit its radioactive decay to a level acceptable for release to the atmosphere. Such a scheme is shown in Figure 126.

FIGURE 126: EXHAUST GAS CLEANING SYSTEM FOR HANDLING RADIOACTIVE FISSION AND ACTIVATION GASES

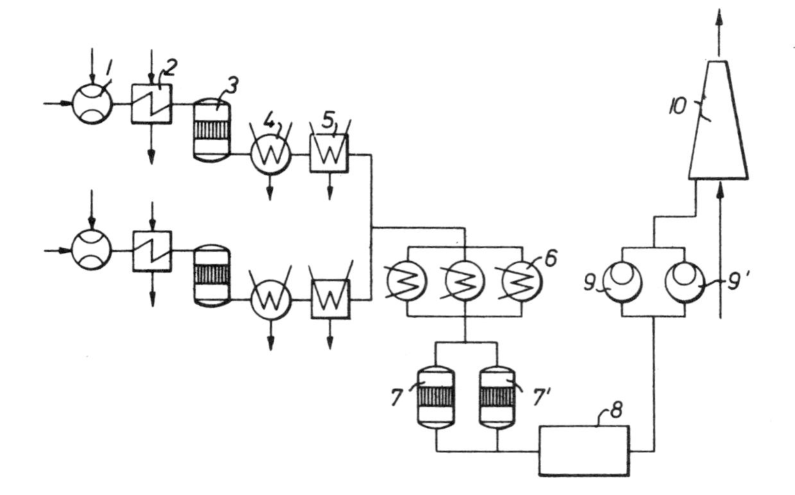

Source: U.S. Patent 3,871,842

The radioactive fission and activation gas to be treated is sucked out of a turbine condenser (not illustrated) by evacuation pump **1** and, in mixture with the driving steam, fed to the exhaust gas treating plant. After a slight superheating of the steam/gas mixture in a heater **2**, the mixture moves on to recombiner **3**, where free hydrogen and oxygen in the mixture are burned to water on the surface of a catalyst, for example, of platinum.

The heat given off by the exothermic reaction causes a large temperature increase. Temperatures of 400°C in the mixture are comprehended. The thus strongly superheated steam/gas mixture goes next into a condenser **4** and then a cooler **5**. The driving steam and the steam from the radiolysis gas being burned to water are largely condensed and withdrawn, so that only an inert gas mixture, composed primarily of air with some remaining steam, comes from the cooler.

The flow diagram shows a second, parallel line of devices **1** to **5**. Following the condenser and cooler is a gas cooling and drying plant **6**. The gas leaving the cooler and dryer, usually will be at a temperature in the range of 0° to 30°C, with a humidity corresponding to a dew point in that range. Then comes preliminary adsorber **7**, and a second preliminary adsorber **7'**, which is used as needed. Following directly on preliminary adsorber **7** is the main adsorber **8**. The main adsorber can include one or more adsorption columns. There follows vacuum pumps **9** and **9'**, by which the treated gas is led to an exhaust air and exhaust gas chimney **10**.

The essential components of the exhaust gas cleaning system are the preliminary adsorber **7**, and perhaps **7'**, and the main adsorber **8**. The preliminary adsorber, which usually represents the smaller part of the adsorption plant, serves essentially for the decay of the short-lived fission products, present as the vastly greater portion of the gaseous nuclide mixture. In conformance with the high proportion at which the short-lived fission products are present, there is a correspondingly large heat and daughter product production.

The forming solid daughter products are simultaneously filtered off. The preliminary adsorber is constructed in such a manner that the adsorbent, and the solid daughter products, which deposit in the adsorbent and are thus filtered off simultaneously, can be exchanged easily and without danger.

The main adsorber, which usually represents the larger part of the system, serves both for mechanically and adsorptively delaying the remaining, longer-lived isotopes and for the filtering of the resulting solid daughter products. The container of the main adsorber **8** is, therefore, of larger volume than the preliminary adsorber and is provided with simple installations for an improbable, yet conceivable, exchange of the adsorbent.

By choosing a suitable ratio of volume of voids in the adsorbent to the total volume of container containing adsorbent, which ratio may be controlled for example by the particular shape or size of adsorbent particles used, it is possible to provide in main adsorber **8** for the further decay of poorly adsorbable gas components in the voids between the adsorbent packing bodies.

See Nuclear Industry Effluents (U.S. Patent 3,963,460)
See Nuclear Industry Effluents (U.S. Patent 3,964,887)

Removal from Water

A process developed by *K.J.A. Peeters and N.L.C. Van De Voorde; U.S. Patent 3,896,045; July 22, 1975; assigned to Belgonucleaire SA, Belgium* is one in which radioactive liquids and the products thereof are decontaminated by contacting the radioactive liquid with a sorbent in a sulfate-containing medium selected from the group consisting of a barium salt and a barium salt mixed with up to 50% of a metal ferrocyanide.

A technique developed by *W.J. Mecham; U.S. Patent 3,983,050; Sept. 28, 1976; assigned to U.S. Energy Research and Development Administration* is one in which metal canisters for long-term storage of calcined high-level radioactive wastes can be made self-sealing against a breach in the canister wall by the addition of powdered cement to the canister with the calcine before it is sealed for storage. Any breach in the canister wall will permit entry of water which will mix with the cement and harden to form a concrete patch, thus sealing the opening in the wall of the canister and preventing the release of radioactive material to the cooling water or atmosphere.

See Nuclear Industry Effluents (U.S. Patent 3,862,296)

RADON

Radon is a chemically inert gas which diffuses from uranium rock surfaces into the mine atmosphere or is released when ore is crushed as during processing.

Radon gas diffuses into the mine atmosphere after emanating from rock surfaces and can be inhaled as can radon gas released during ore processing; however, because it is a gas, radon is exhaled before it is able to emit appreciable amounts of alpha particles. Radon daughters formed in air can also be inhaled. Where there is inhalation of both radon and its daughters, it has been estimated that radon contributes only about 5% of the alpha radiation dosage received by the lungs. Later estimates of the alpha dose percentage from radon in normal mine atmospheres indicate a smaller percentage.

The daughter nuclides are small enough that their principal means of transport is by diffusion and turbulent mixing. Few unattached radon daughter atoms are found in mine air because of their kinetic interaction with the natural aerosols present. When the mine air is breathed, a portion of the dust is trapped in the respiratory system where attached radon daughters decay, and the soft lung tissue is irradiated by the alpha particles emitted.

The few daughters which are not attached to dust or other condensation nuclei when inhaled tend to be deposited and concentrated in the upper respiratory tract before reaching the lungs. Some investigators have suggested that this part of the overall alpha dose received by the respiratory system may be of special importance so far as the health hazard is concerned. Considerable evidence indicates that excessive exposure to radon daughter products is associated with a high incidence rate of lung cancer.

Removal from Air

A process developed by *C.W. Stringer; U.S. Patent 3,853,501; Dec. 10, 1974; assigned to Radon Development Corporation* utilizes a filter for water contaminated with radon daughters which is used in conjunction with air cleaning systems which employ water as a trap for radon daughters. The filter medium involves use of granular substrate which has been dried and oil-wetted with a normally-liquid hydrocarbon material.

A process developed by *J.W. Thomas; U.S. Patent 3,890,121; June 17, 1975; assigned to U.S. Energy Research and Development Administration* involves removing a noble gas such as krypton, xenon or radon from air comprising the use of activated carbon filters in stages in which absorption and desorption steps in succession are conducted in order to increase the capacity of the filters.

RAILROAD EQUIPMENT SERVICING EFFLUENTS

Wastewater is produced by the operation of railroad equipment servicing facilities. Such a waste stream includes the run-off from engine washing, fueling and sanding aprons, diesel shops, round house pit drainage and miscellaneous other water streams released in servicing and maintenance of railroad locomotives and related equipment. These waste streams contain free oil, emulsified oil, sand, silt, grease, carbon and various other solids. Dissolved substances such as lye and other water-soluble alkaline, acidic and salt materials may be present as a result of cleaning operations for servicing railroad rolling stock. The resultant wastewater stream has a pH of above 10 and cannot be merely settled in a hope of providing a sufficiently clear water stream acceptable for disposal in public water systems such as streams and rivers.

Removal from Water

A process developed by *W.F. Burns and R.B. Martin; U.S. Patent 3,707,464; December 26, 1972; assigned to Petrolite Corporation* is a process for clarifying an oil-solids contaminated aqueous stream such as may arise from the operation of a railroad equipment servicing facility. A primary clarifier gravitationally separates the stream into settled solids, a merchantable oil product and a clarified water stream. The clarified water stream is treated by creating a flocculation product for removing solids and oil therefrom to produce a clear water stream delivered to a subsequent utilization. Settled solids may be returned to the primary clarifier. The flocculation product is accumulated in a sludge cleaner to be periodically converted into water and oil phases. The oil phase is returned into the primary clarifier. The water phase is reformed into an oil-free flocculation product and separated from a water filtrate returned into the primary clarifier.

Solids from the primary clarifier are accumulated in a sand cleaner to be periodically cleaned to separate oil-free solids from a solids-free water-continuous filtrate returned to the primary clarifier. The reformed flocculation product and oil-free solids are separated in a solids dewatering vessel from a water filtrate returned into the primary clarifier. The relatively water-free solids are then delivered to a subsequent utilization such as landfill.

RARE EARTH IONS

The trivalent rare earth ions present in wastewaters, may be unreacted ions which have been used as precipitant for the removal of phosphorus-containing ions from wastewater. The phosphorus-containing ions may be in the form of simple phosphates such as orthophosphate or complex phosphates such as pyrophosphate or tripolyphosphate. The wastewater may be, for example, municipal wastewaters such as the tertiary wastes from a secondary processing stage.

In a typical treatment of municipal wastewaters, there is a primary settling step for solids removal and a secondary treatment step for aeration, biological oxidation of organic matter and further solids removal. The combined primary and secondary treatments generally remove no more than about 20% of the phosphorus present in sewage, the remainder being passed on to the tertiary treatment stage.

The trivalent rare earth ions may also be unreacted ions which have been used as precipitant for the removal of coloring matter from wastewater. The coloring matter may be natural or synthetic in origin and the wastewater may be, for example, the effluent from pulp and paper mills, tanneries or textile mills. Illustrative pulp and paper mill effluents are the acid sulfite waste from sulfite pulping and the alkaline or acid bleach waste from the kraft process. The acid sulfite waste is a dark brown-colored liquor having a pH of about 2 to 4 when it is discharged from the pulp mill. The acid bleach waste in the kraft process results from the chlorine and hypochlorite bleaching stages and the alkaline bleach waste results from a caustic extraction stage which follows the chlorine bleaching.

Removal from Water

A process developed by *H.L. Recht and M. Ghassemi; U.S. Patent 3,692,671; Sept. 19, 1972; assigned to North American Rockwell Corporation* provides a method for chemically removing trivalent rare earth ions from wastewater by treating the trivalent rare earth ion-containing water with a carbonate salt, such as an alkali metal carbonate, to form an insoluble rare earth carbonate precipitate and then separating the precipitate from the water. The rare earth ions may be regenerated in the form of a soluble rare earth salt, for example, by treating the separated rare earth carbonate precipitate with acid, such as hydrochloric or sulfuric acid.

RENDERING PLANT EFFLUENTS

A rendering plant is an example of an industrial operation producing a discharge of contaminated air. In the rendering process, animal parts and various organic waste products are cooked to render protein and fat materials for feed supplements and other uses. During the rendering process, organic matter from the raw materials is heated and cooked in water, producing a by-product including water vapor and a mixture of organic matter, gases, and other materials borne in the escaping water vapor effluent.

The organic matter, such as protein and fat, usually consist of lighter compounds and tends to be in a liquid particulate state. Gases such as ammonia, hydrogen sulfide, and various organic gases resulting from decomposition during the cooking operation, are also frequently found in the vapor discharge. The atmospheric contamination produced by such rendering operations is in part liquid organic compounds which are in colloidal or aerosol solution within the liquid together with gases. The organic matter in aerosol solution is not susceptible to removal by filtration. Furthermore, this organic matter is not highly soluble and therefore can be only partially removed from the aerosol by conventional water scrubbing techniques.

The above described airborne discharge of a rendering plant diffuses through the atmosphere and contains odors which are notoriously nauseating and obnoxious. Moreover, this vapor discharge contains obnoxious and corrosive compounds such as hydrogen sulfide and acidic gases. The organic matter in the discharge oxidizes slowly in the atmosphere and thus may produce noticeably unpleasant odors at long distances from the rendering plant.

Removal from Air

A process developed by *J.W. Gooch; U.S. Patent 3,803,290; April 9, 1974; assigned to Chlortrol, Inc.* is a waste oxidation and waste extraction process for deodorizing, coalescing, agglomerating, coagulating, and extracting organic and inorganic waste materials such as found in the particulate and molecular waste effluent streams of packing and rendering plants, paper and pulp mills, food processing plants, wood preserving plants, rice processing plants, and similar facilities.

The process comprises introducing into the waste stream an activated gas such as chlorine which is adsorbed on the surfaces of suspended organic and inorganic particulate and molecular waste materials to react with the surfaces of the waste materials and produce surface characteristics which eliminate malodorous properties and induce agglomeration and coagulation of the solids and materials. Such a process is shown schematically in Figure 127.

Referring to the figure, a discharge stream containing pollutant is admitted through the line **10** into the reactor **11**. At the same time, a quantity of chlorine gas from the gas source **12** is admitted through the valve **13** for mixture with a quantity of steam from the source **14** and admitted through the valve **15**. The active principal (chlorine gas, in the embodiment depicted in the figure) is merely mixed with the carrier medium (steam), and is not dissolved in the carrier medium.

The mixture of chlorine gas and steam passes through the line **16** and is introduced into the reactor **11** through the nozzle **17** disposed therein causing introduction of molecular chlorine into the reactor. The nozzle **17**, in preferred embodiment, is essentially an atomizing nozzle which sparges and sprays the individual particles of the active gas-steam mixture into the flow stream within the reactor. In the case of a very hot discharge flow stream entering the reactor **11**, a relatively cold medium can be substituted for steam from the source **14**, with the resultant injection of the relatively cool mixture into the reactor accomplishing a rapid dispersion of the active principal caused by the immediate superheating of the relatively cooler medium and active principal.

After the reaction which follows introduction of the active principal into the reactor **11**, it is often necessary to mix or agitate the mixture of contaminants

FIGURE 127: PROCESS FOR DESTROYING ODOROUS POLLUTANTS IN RENDERING PLANT EFFLUENT GASES

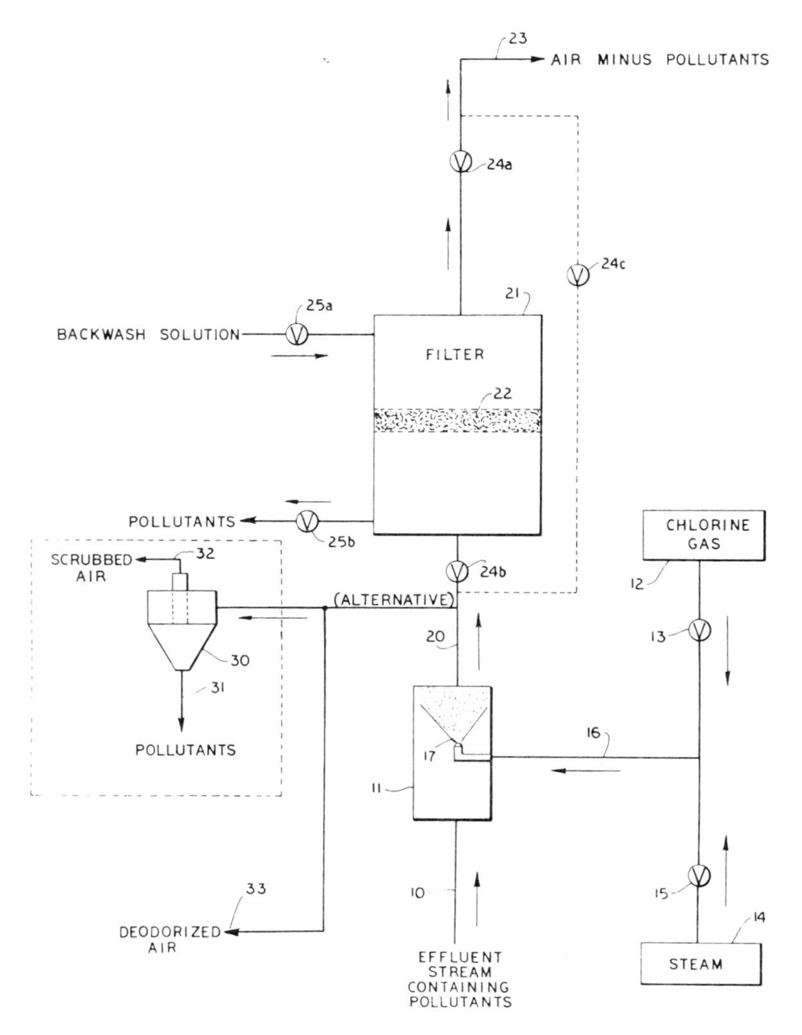

Source: U.S. Patent 3,803,290

plus active principal with a moderate degree of turbulence. This mixing or agitating may be accomplished through the use of a reactor **11** containing spiralling, baffling, or other internal structural features which produce mixing, agitation, and turbulence of the flow stream therein.

The reactions which take place within the reactor **11** between the discharge stream and the active principal are of the type described above, so that the flow leaving the reactor **11** along the line **20** includes contaminants or pollutants which have become coalesced, coagulated, and/or agglomerated for subsequent separation.

The flow through the line **20** may, in one alternative, be applied to a filter **21** having a carbon particle filter bed **22** which removes the treated contaminants entering the filter. The vapor of the stream, minus the contaminants which remain in the particle bed **22**, exits the filter **21** and may be vented to the atmosphere by the line **23**.

As is known to those skilled in the art, the filtering capabilities of the carbon particle bed **22** periodically become exhausted and it is necessary to backwash the filter **21** with an appropriate solution to remove the contaminants extracted by the particle bed, and thus to rejuvenate the filter. This is accomplished by first closing the valves **24a** and **24b** while opening the valve **24c**, to provide a discharge stream by-pass around the filter **21** (assuming, of course, that the process producing the polluted discharge stream remains operative during the filter backwashing operation).

The backwash valves **25a** and **25b** are then opened to admit a suitable backwash solution, and the liquid and organic polluting matter previously retained by the particle bed **22** is removed from the filter. The valves are returned to their original positions upon the termination of the backwashing operation, and the filter **21** again functions in the foregoing manner.

Since the filter **21** is necessarily subjected to intermittent shut-down to permit backwashing, it can be seen that continuing operation of the industrial process during periods of shut-down will release pollution-containing effluent to the atmosphere unless a second filter (not shown) is provided to receive the treated stream emerging from the reactor **11** along the line **20**. The alternatives are, accordingly, either (1) periodically shutting down the industrial process producing the contaminated effluent stream, (2) discharging contaminants to the atmosphere, or (3) incurring the additional expense of duplicate filters **21**. While one or more of these alternatives may be acceptable in many circumstances, the problem of intermittent filter shut-down is avoided through the alternative use of a continuously-operating filtration system such as a cyclone separator **30** connected to receive the flow of treated pollutants discharged from the reactor **11**.

The cyclone separator **30** functions in the conventional manner to produce a discharge **31** of pollutants and a discharge **32** of scrubbed air suitable for venting to the atmosphere. Where it is desired to accomplish only deodorization, the oxidized flow can be discharged as **33** into the atmosphere. The following is one specific example of the operation of this process.

A rendering plant which processes animal parts to produce a feed supplement includes cookers and drying kilns which develop by-product vapors and gases. The by-product gases and vapors form an aerosol which has a very rancid odor and which has corrosive characteristics which are deleterious to equipment. The gases and vapors from the rendering plant form an aerosol which is unfilterable and only partially removable by scrubbing. Chemical analysis indicates that the untreated aerosol from the meat cooker contains 58.8 parts per million organic matter and 4.0 parts per million of ammonia gas.

This contaminated aerosol is flowed through a reactor at the rate of 100.0 cubic feet per minute. Chlorine gas flowing at the rate of 10.0 cubic feet per minute is mixed with steam flowing at the rate of 110.5 cubic centimeters per minute, and this mixture is injected into the reactor through a nozzle which disperses the chlorine-steam mixture throughout the contaminated aerosol stream to introduce molecular chlorine to the aerosol. The treated aerosol is then flowed through a particle bed of carbon 1.5 inches thick. Analysis of the solution after flowing through the particle bed shows that the organic matter is reduced to 5.2 parts per million and the ammonia gas is reduced to 0.4 part per million. Residual chlorine was 0.0 part per million. The odor of the treated aerosol is reduced to a level which is only slightly perceptible to the olfactory system of humans.

The odor units or odor threshold of the untreated air is 25,000 odor units and the treated air rates 10 odor units. Odor units are expressed as a unitless number which indicates the number of dilutions of the odorous air with fresh air. The odoriferous units test is described in Standard Methods, American Health Association, also known as the ASTM method adopted by the Environmental Protection Agency.

The foregoing results are produced with chlorine flow rates varying between 9.0 to 12.0 cubic centimeters per minute, and with steam flow rates varying between 90.0 to 150.0 cubic centimeters per minute. The temperature of the chlorine-steam mixture is varied between 220° to 270°F with no appreciable change in results. There was no residual chlorine or free chlorine in the air discharge after treatment. The residual chlorine test was conducted as in Standard Methods, American Health Association.

RESTAURANT EFFLUENTS

Pollution problems are caused by grease, smoke and other pollutants from the exhaust coming from a restaurant or the like as a result of cooking and broiling operations. Such operations are often accompanied by the discharge of clouds of smoke and grease laden air. The grease tends to clog up removal equipment and poses both a fire hazard and a maintenance problem.

Removal from Air

Devices utilizing water spray nozzles are fairly effective in separating the grease from the air but difficulty has been encountered in efficiently collecting and removing the grease from the water so that the water can be recirculated. Moreover, water spray systems are not particularly effective in separating smoke and similar small particles.

This is better accomplished by electrostatic precipitators which electrically charge the particles and cause them to collect upon oppositely charged plates of the precipitator. However, a combination of an electrostatic precipitator with a water wash system presents a problem in that the moisture must be substantially completely removed from the air before it enters the precipitator or the precipitator will not operate efficiently. Moreover, this must be done without unduly blocking the air flow or a very large exhaust fan would be required.

A process developed by *H.F. Ohle; U.S. Patent 3,802,158; April 9, 1974* provides a pollution control apparatus for eliminating grease, smoke and other pollutants from the air discharged by restaurants and the like as a result of food preparation. Water spray nozzles wash grease and large particles from the restaurant exhaust, and the separated material falls into a collection tank having a skim dam at one end and spray nozzles at the other end. The water spray skims the grease off the water so that the water can be circulated. Baffles disposed across the path of the washed air remove most of the water, and a downstream filter system removes most of the remaining moisture. The generally nonturbulent air leaving the filter system is slowed in a stall chamber, and electrostatic precipitators next act on the air as it leaves the stall chamber to remove smaller particle pollutants not removed by the water wash section.

See Grease (U.S. Patent 3,628,311)

RHENIUM

Removal from Water

A process developed by *C.N. Wright and K.J. Richards; U.S. Patent 3,870,779; March 11, 1975; assigned to Kennecott Copper Corporation* is one in which rhenium values are recovered from dilute solutions thereof by introducing a soluble sulfide into the solution and reacting the rhenium values with a base metal in the presence of the sulfide to form a precipitate which may be easily separated from the solution. The rhenium is recovered from the precipitate by conventional procedures.

ROCK DRILLING DUST

As is well known, difficulties have been encountered in boring holes into rock-like materials containing high percentages of quartz or silicates since dust formed from these materials cannot be breathed continuously without causing a disease of the lungs known as silicosis.

Removal from Air

While many devices have been proposed for settling drill dust both prior to and after leaving the hole being bored, such devices have been unsatisfactory. The introduction of water through the drill bit reduces the life of the bit due to excessive wear to the bearings of the bit. The introduction of water exteriorally of the hole being bored by prior art devices has been unsatisfactory due to the fact that the bailing velocity of the moving stream of dust is such that it outruns the velocity of the water spray. Accordingly, only a small percentage of the dust particles are actually contacted by the water spray. That is, the velocity of the stream of moving chips, dust and air ranges from approximately 4,000 to 5,000 feet per minute.

A process developed by *J.B. Loftis, D.L. Moody and J.H. Phillips; U.S. Patent 3,716,108; February 13, 1973; assigned to Robbins Machinery Company* utilizes a first nozzle assembly in position to discharge an atomized mixture of water and air under high pressure toward the hole being bored with the mixture moving in a direction opposite the direction of movement of the dust from the hole.

A second nozzle assembly discharges atomized water and air under high pressure in the same direction as the moving stream of dust after the stream passes the first nozzle assembly. To introduce the mixture of water and air under extremely high pressure, a relatively small tank is provided which communicates with the lower end of a large tank or reservoir for water. Air under high pressure is introduced into the small tank while air under low pressure is introduced into the larger tank and the smaller tank is exhausted at intervals permitting water to flow from the larger tank to the small tank.

Accordingly, there is provided a continuous, fine particle size vapor which moves at an extremely high velocity to thus assure intimate mixing of the vapor with all the dust particles discharged from the hole being bored.

In prior art devices for collecting drill dust produced by rock drilling equipment and transported from the drill bore by means of an air stream, from which the drill dust is separated in a cyclone and filter apparatus, the drill dust is usually collected in a container, a bag or the like. When the bag or container is to be emptied or shifted the drilling apparatus must be stopped to avoid free discharge of dust from the drill. In connection with the removal of the bag a certain amount of drill dust may whirl up therefrom and thereby to some extent counteract the aim of the device which is to prevent dust from escaping into the atmosphere and causing unhealthy dust concentrations within the working area.

A technique developed by *G. Jysky, I. Mardla and B. Eriksson; U.S. Patent 3,895,929; July 22, 1975; assigned to Ilmeg A/B, Sweden* provides for a stock of flexible tube material in the shape of a folded flexible hose which is placed around the lower portion of the separating apparatus and arranged to be successively withdrawn in desired lengths to form successive bags, means being provided for constantly securing a bag thus formed tightly around the downwardly opening outlet of the separating apparatus.

ROLLING MILL EFFLUENTS

Removal from Air

A process developed by *T. Eklund; U.S. Patent 3,204,393; Sept. 7. 1965; assigned to A/B Svenska Flaktfabriken* relates to apparatus for the collecting and transporting away of dust, fumes and gases in rolling mills. As a part of the general tendency to obtain better and more hygienic working conditions in the industry, arrangements of the abovementioned type have come into use. It has, however, proved difficult to obtain an effective and suitable exhaust of dust, fumes and gases without the use of exhaust hoods of such dimensions and such placement that they jeopardize the attending and supervising of the mills.

A primary object of the process is to eliminate the abovementioned difficulties and to make possible a more widespread use of such exhaust devices. The principal form is characterized in that suction boxes are located just above the path of the rolled stock and provided with a suction slot facing the path. Further, a throttling valve is arranged in the branch duct from each suction box, the valves being adapted to be automatically adjusted by means of impulses from the driving means of the mill in such a manner that the throttling valve located at the discharge side of the roll stand always is open, while the other valve simultaneously is closed.

By arranging the exhaustion point close to the points where the development of fumes primarily takes place the fumes can effectively be removed by using relatively small quantities of air and in this manner the suction boxes can be given dimensions so that they do not disturb the running process. Because in reversing type mills the exhaust need be employed at only one side at a time the required fan may be relatively small and this also works well for the dust separator used in the exhaust line.

Therefore the plant as a whole is simple and inexpensive while simultaneously taking but little space and working effectively. According to a suitable form of the process the suction boxes are combined with the strippers of the roll stand and are connected to the exhaust ducts by means of flexible tubes disposed at the pivot center of the strippers.

During the cold rolling of steel to reduce the thickness thereof for producing strapping and sheeting of various gauges, rolling oil is used to facilitate the cold rolling operation. In industry, a considerable amount of oil is used during the cold rolling operation and must be disposed of periodically. The dimensions of the problem are understood when it is known that in excess of 100,000 gallons of waste rolling oil and water emulsion may be discarded each month in a medium size steel producing facility. This large quantity of rolling oil and water emulsion is costly to dispose of and also presents an environmental hazard if dumped untreated into the sewer system or into lakes or streams.

An additional operation in the production of steel strapping or sheeting is the pickling of the steel with a hydrochloric acid bath. The pickling operation removes mill scale and the like, and it is necessary for the production of good quality unstained sheets and strips. The waste pickle liquor, which contains iron ions therein, also presents a disposal problem, the treatment of which is discussed in U.S. Patent 3,434,797. The disposal problem is aggravated by the fact that perhaps thirty times more waste pickle liquor is produced per month than waste rolling oil and water emulsion.

Removal from Water

A process developed by *V.D. Beaucaire; U.S. Patent 3,986,953; October 19, 1976; assigned to Interlake, Inc.* is one in which the oil and water emulsion used during the cold rolling of steel to reduce the thickness thereof is treated by breaking the emulsion with waste pickling acid solution and thereafter converting iron ions present in the waste pickling acid solution to magnetite particles which absorb the oil. The magnetite particles and the oil absorbed thereby are separated from the solution leaving a clarified solution.

A process developed by *G.L. Kovacs; U.S. Patent 3,301,779; January 31, 1967; assigned to New Canadian Processes, Ltd., Canada* involves treating the oil-bearing cooling water from industrial plants such as cold rolling mills with a view to separating the oil and water in such a manner that the oil is suitable for further processing into reusable or saleable form rather than fit only for disposal, and the water is in an acceptable state for discharging into adjacent sewers or water courses.

The process involves forming in the effluent cooling liquid an insoluble precipitate which coagulates and concentrates the oil, placing the total effluent under pressure and dissolving pressurized air therein, releasing the pressure on the effluent liquid so that the air returns to the gaseous state and removes the coagulum and thus the oil upwardly by flotation, and skimming off the resultant layer of scum. By "insoluble precipitate" is meant a material which is insoluble in the effluent cooling liquid. It is mandatory that this material be capable of dissolution in some subsequent process for oil recovery.

The precipitate is that formed by mixing an alkaline waste with acid pickle liquor. In steel mills there are "cleaning lines" which are alkaline washers for steel strip. These produce the alkaline waste which usually contains free caustic soda, soap, and oil in an emulsified form which will not settle.

ROOFING FACTORY WASTES

Removal from Air

Roofing material factories using paper, fiber, bituminous products, sand, gravel, talc, etc., have as a necessary by-product solid wastes in the form of combustible material, effluent gas from the oxidizer, and exhaust fumes from the saturators that are used in the production of the roofing. The disposal of such solid wastes, as well as that of the saturator exhaust fumes, and oxidizer effluent gas has, in the past, been troublesome from an economical as well as an ecological standpoint, due to the noxious nature of such fumes and effluent gas, as well as the products of combustion involved in the simple burning of the solid wastes that are combustible. In the past such gas products have simply been discharged from stacks high enough to be blown away by air currents.

A process developed by *H.B. Johnson; U.S. Patent 3,662,695; May 16, 1972; assigned to GAF Corporation* is one in which the solid wastes of combustible and noncombustible roofing material are comminuted in an attrition mill, fed to a surge bin, and then used as fuel in a solid waste and fume incinerator which also receives and ecologically incinerates the exhaust fumes from a saturator. A fume incinerator also receives saturator exhaust fumes along with effluent gas from the oxidizer of the factory, and ecologically incinerates the same with a minimum consumption of conventional fuel.

The fume ducts from the saturator exhausts to the incinerators include a cross duct and switching valves for selectively controlling the flow of the saturator exhaust fumes from one or both saturators to one or both incinerators for optimum operational efficiency of the system in accordance with current production of the roofing factory. Granules resulting from the solid scrap incineration are washed and recovered for reuse. Figure 128 illustrates the essential elements of the process.

Referring to the drawing, saturators **10** and **12**, and oxidizer **14** constitute components of a roofing factory which also includes a felt section **16**. A by-product of the factory is scrap, including tabs which may be termed solid waste that contains combustible material used in making the roofing products. The scrap from the factory is loaded on a platform **18** for movement by a conveyor

FIGURE 128: ROOFING FACTORY FUME AND SOLID WASTE DISPOSAL SYSTEM

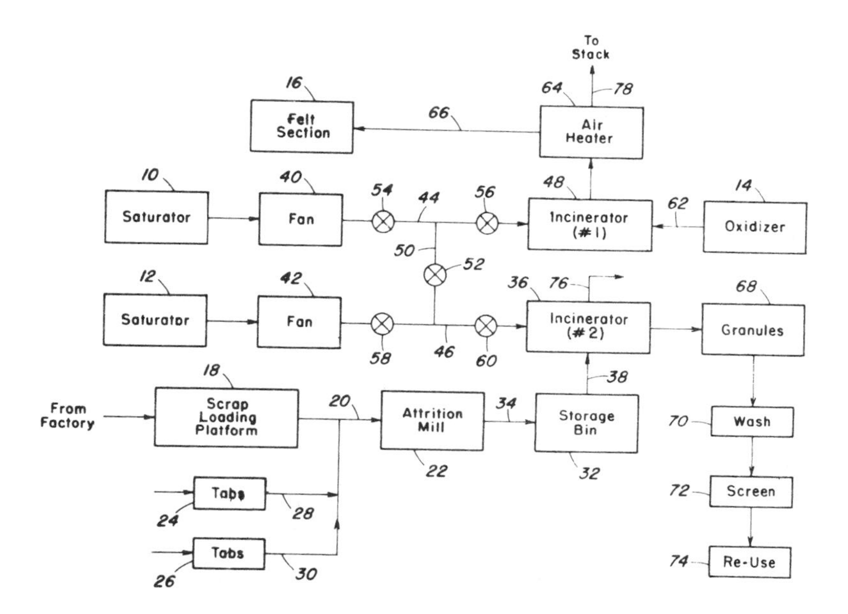

Source: U.S. Patent 3,662,695

20 to an attrition mill **22**. The tabs **24** and **26** are also moved by conveyors **28** and **30** onto the conveyor which, in turn, carries them to the mill. The output of the mill goes to a surge or storage bin **32** on a conveyor **34**, and from the bin to a fuel input of incinerator **36** by a suitable conveyor **38**.

Fume exhaust blowers or fans **40** and **42** are associated with the saturators **10** and **12**, respectively, and are connected by ducts **44** and **46** to incinerators **48** and **36**, respectively. A cross duct **50** having a damper or valve **52**, is connected to the ducts **44** and **46** for selectively separating, or putting the latter ducts in communication with each other. The ducts are also provided with switching valves **54** and **56** and **58** and **60** at opposite sides of the cross duct **52** for selectively controlling the flow of exhaust fumes from either one or both saturators **10, 12** to either one or both of the incinerators **36, 48.**

The effluent gas from the oxidizer **14** flows to the incinerator **46** through a duct **62**. The effluent of the incinerator **48** goes to a suitable stack or chimney by way of an air heater **64** from which the heated air flows by way of a duct **66** to the felt section **16** of the factory, providing auxiliary heat for the vapor absorption system of such felt section **16**.

Residual noncombustible granules produced as a by-product of the incinerator **36**, are collected at station **68**, then washed and screened at stations **70** and **72**, respectively, for reuse from a station **74** where they are finally collected for such purpose.

Effluent from the incinerator **38** goes directly to an individual stack from effluent outlet **76**. Such effluent as well as that of incinerator **48** going to its stack through outlet **78**, is ecologically clean due to complete combustion of the exhaust fumes, and the oxidizer effluent gas, as well as all of the combustible material contained in the solid wastes that are fed to the incinerators **36** and **42**. The residual granules of incinerator **36** are also ecologically clean even before washing by virtue of the operation of the complete incinerator **36**.

A technique developed by *E.A. Busse and G.W. Raker; U.S. Patent 3,718,131; February 27, 1973* provides means for drawing off the smoke and fumes given off by bituminous material when heated and feeding such fumes directly into the heating flame for substantially complete combustion with reduced danger of a blow-back occurring.

This technique reduces the smoke, fumes and vapors given off in kettle operation. It is suitable for retrofitting existing kettles as well as for embodiment in new kettle installations being built.

A device developed by *W.L. Hart, G.D. Vevang and H.J. Walkowicz; U.S. Patent 3,880,143; April 29, 1975; assigned to U.I.P. Engineered Products Corporation* is a combined fume oxidizer and asphalt heater which substantially reduces fumes normally associated with the production of asphalt roofing materials. These fumes are utilized to heat additional asphalt and are substantially combusted. A method is described wherein the fumes from an asphalt shingle saturation machine are directed through a burner of the gas or oil type and the combined streams of burned gases are utilized to heat asphalt material.

The heat of combustion is controlled by a thermocouple and the temperature in the heating chamber by by-passing a portion of the combined combusted streams away from the heating chamber and into an exhaust duct to which the other portion of the combusted gases is directed after heating the asphalt material. A damper control means directs the portions of the combusted gases to the heating chamber or through the by-pass ducts depending upon the temperature desired in the heating chamber for heating the asphalt material.

A process developed by *L.A. Fry, Sr.; U.S. Patent 3,724,173; April 3, 1973; assigned to Lloyd A. Fry Roofing Company* is a process for substantially reducing the particulate emission, opacity and odor of an asphalt saturator effluent. More particularly, this is a process for substantially reducing the particulate emission, opacity and odor of an asphalt saturator effluent to form a substantially odor-free gaseous stream that is substantially free of hydrocarbonaceous particles and suspended particulates utilizing an exhaust stack stream that has previously been substantially purified for recycle to the process. Figure 129 illustrates the essential elements of this process.

Referring to the figure, asphalt saturator **101** represents the portion of the roofing preparation process utilizing felt saturators wherein an asphalt saturator effluent containing particulates, and having a high opacity and an objectionable odor are generated. The effluent from asphalt saturator **101** passes via line **102** into a first contacting zone **103**, labeled contacting zone **I** wherein the effluent

FIGURE 129: APPARATUS FOR REDUCING PARTICULATE CONTENT, OPACITY AND ODOR OF ASPHALT SATURATOR EXIT GASES

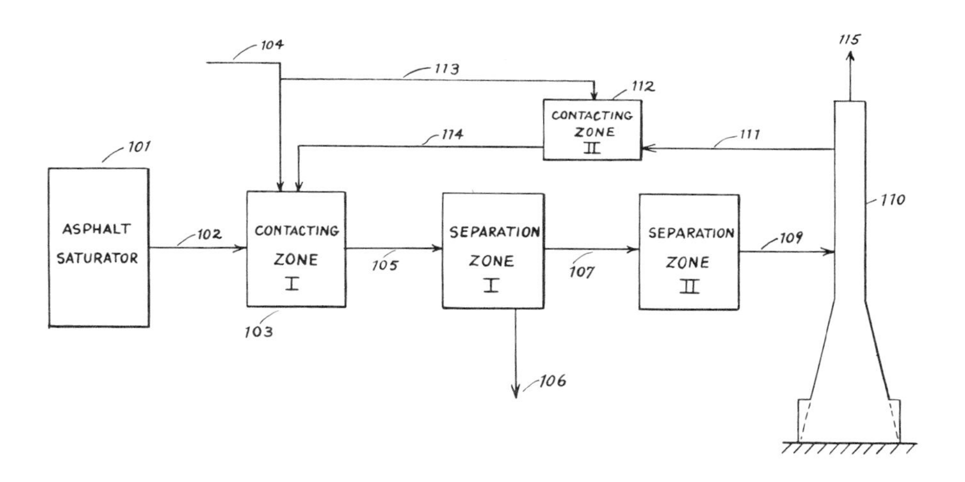

Source: U.S. Patent 3,724,173

is contacted, at contacting conditions, with a contacting fluid passing into contacting zone **I** via line **104**.

A fluid stream comprising hydrocarbonaceous particles and finely divided suspended particulates is withdrawn from contacting zone **I** via line **105** and introduced into a first separation zone, labeled separation zone **I** wherein the hydrocarbonaceous particles and suspended particulates are separated from the fluid stream at separation conditions so as to produce a hydrocarbonaceous particle-suspended particulate stream exiting from the process via line **106** for further treatment, not shown, and a hydrocarbonaceous particle-suspended particulate-free gaseous stream which is withdrawn from separation zone **I** via line **107** and passed to separation zone **II**.

Separation zone **II** is maintained at separation conditions so as to produce a substantially odor-free gaseous stream. The substantially odor-free gaseous stream is removed from second separation zone **II** via line **109** and introduced into exhaust stack means **110** which may be an exhaust stack. At least a portion of the substantially odor-free gaseous stream is now recycled via line **111** to a second contacting zone **112**, labeled contacting zone **II**.

The substantially odor-free gaseous stream is again contacted with contacting fluid via lines **104** and **113** at contacting conditions and the stream is then passed via line **114** to contacting zone **103**, labeled contacting zone **I** for admixture with the untreated asphalt saturator effluent from asphalt saturator **101**. The remaining portion of the substantially odor-free, low opacity and unobjectionable gaseous stream is removed from the exhaust stack **110** via line **115** for passage to the atmosphere.

RUBBER CHEMICAL MANUFACTURING EFFLUENTS

See Dithiocarbamates (U.S. Patent 3,966,601)

RUG INDUSTRY EFFLUENTS

See Latex (U.S. Patent 3,726,794)

SALICYLALDEHYDE PRODUCTION WASTES

See Glycolic Acid (U.S. Patent 3,819,516)

SCRAP MELTING FURNACE EFFLUENTS

The melting of scrap metals for reuse is a well-known procedure but, quite often, the scrap metals are contaminated with oils, such as cutting oils, and the contaminating material will be volatilized when the metal is heated and will pass off from the metal as an atmospheric pollutant. Such emissions, due to the fact that the principal contaminant of scrap metal is oil, are essentially hydrocarbon and are quite objectionable when discharged directly into the atmosphere.

Removal from Air

A number of different schemes have been proposed for collecting such emissions to prevent them from entering the atmosphere, but heretofore, arrangements for accomplishing this have been expensive and imperfect and have represented no economy whatsoever in respect of the melting of the scrap metal for reuse.

A process developed by *K.M. Habayeb; U.S. Patent 3,869,112; March 4, 1975; assigned to Wabash Alloys Inc.* is one in which an enclosing structure is erected about the charging well having a hood portion at the top to collect emissions that are given off by metal placed into the charging well to be melted. The hood portion of the enclosure erected about the charging well has, at the top, at least one duct leading therefrom which is connected to the suction side of a blower.

The end wall of the furnace opposite the end from which the charging well projects is provided with burners, preferably gas burners, and the discharge side of the aforementioned blower is connected to these burners to supply the combustion air thereto, or leads to other burners, or to the base of the stack of the respective furnace.

The enclosure erected about the charging well has an operator's opening, or window, through which an operator can observe the conditions of the metal in the melting chamber and also has an opening through which the metal to be melted is introduced into the charging well. This last mentioned opening may be large enough to admit a powered lift truck or the like which conveys batches of metal to the charging well but, alternatively, the metal could be supplied to the charging well by a conveyer arrangement thereby permitting the opening in the enclosure through which metal is conveyed to be smaller.

It is proposed to shield both openings provided in the enclosure by air curtains so that no emissions escape from the enclosure whereby all of the emissions are withdrawn from the enclosure by the duct or ducts connected to the upper hood portion thereof. Under normal circumstances, all of the emissions can be conveyed to the aforementioned gas burners as combustion air for the gaseous fuel supplied thereto but, if the emissions from the metal supplied to the charging well exceed the amount that can be delivered to the gas burners, at least a portion of the emissions can be conveyed either to other gas burners, or are injected into the base of the stack leading from the furnace, or are injected into the furnace downstream from the burners. In every case, except where the emissions are injected into the stack, the heat of combustion of the emissions is realized in respect of heating metal.

It has been found that, by supplying the emissions to a place of combustion as referred to above, the total emissions from a furnace of the nature referred to are reduced to a fraction of the maximum allowed by pollution codes while, at the same time, a substantial fuel economy is also realized.

SEAFOOD PROCESSING EFFLUENTS

It has been common practice to process fish meal by cooking such meal and then draining moisture therefrom. The meal is then further run through a press to remove additional moisture therefrom and is then advanced to a dryer for final drying prior to storage. It has been common practice to exhaust the vapors released during the drying stage directly to the atmosphere thus substantially contributing to the already acute air pollution problems. Further, such direct exhaust to the atmosphere necessitated careful control of the dryer temperature to avoid burning of the meal and emission of consequent air polluting smoke.

Removal from Air

An apparatus developed by *V.J. Evich; U.S. Patent 3,837,272; September 24, 1974; assigned to Star-Kist Foods, Inc.* consists of a dryer having a recirculation conduit leading from the outlet end thereof to return gases from such dryer to an incinerator generating the drying heat. The recirculation conduit splits into a reheat conduit and a residual heat conduit with the reheat conduit connecting to the upstream end of the incinerator for incineration of recirculating gases directed through such reheat conduit.

A chimney rises from the central portion of the incinerator and the temperature in such incinerator is maintained sufficiently high to assume incineration of any particles in the gas returned through the reheat conduit prior to exhausting out the chimney. Valves are provided in the reheat and residual heat conduits for balancing flow therethrough to provide the necessary flow through the reheat conduit and incinerator to maintain a selected drying temperature in the dryer for proper drying of the meal.

A system developed by *D.B. Vincent; U.S. Patent 3,721,068; March 20, 1973* is a countercurrent system, for removing particles, such as fish solids and other dust particles from the exhaust gases from a dryer or the like. The system includes a vessel provided with coaxially interdigitated upper and lower baffles which control a countercurrent flow of the washing liquid and gas stream. The baffles are of graduating heights and the upper set floats at a level determined by the pressure within the vessel. The washing liquid with entrained gas assists in cleansing the baffles. Cleaning ports may also be provided.

Removal from Water

The problem of treating emulsified wastewaters has been most particularly troublesome in the rendering or reducing of whole fish to produce fish meal and oil. In North America species of fish commonly utilized are menhaden, alewives, herring and mackerel, with menhaden being the most abundant. Fish meal is utilized as a feed supplement for poultry and swine, while fish oil is for a large part exported for use in margarines, shortenings and cosmetics.

Previously, liquid wastes from plants producing fish meal and oil have been discharged without any treatment, under the assumption that the wastes consisted only of material which had come from the aquatic environment and that therefore the wastes could not harm this environment. This assumption by segments of the fish industry has led to much difficulty with both state and federal regulatory agencies, especially in recent years when pollution control has been emphasized. This is because the primary waste material consists of bits and pieces of whole fish which are in a partially decomposed state. Since the pieces of fish are organic in nature, they create an oxygen demand on the receiving waters, thus depleting the amount of oxygen available to fish life inhabiting the receiving waters.

In the wet rendering process, the primary process for producing fish meal and oil from whole fish, the major source of pollution in terms of pounds per day of pollutants is the transport or bailing water stream which is utilized in unloading the fish from the boats and transporting them by pumping to the process plant. On reaching the processing unit the whole fish are removed from the transport water stream by a dewatering screen or other similar means.

The catching, storing and pumping of the fish all result in damage to the whole fish, and consequently the transport water after the removal of the whole fish contains fish particles, fish oil and products of decomposition. It has been found that this stream will at times achieve a 5-day biochemical oxygen demand (BOD_5) in excess of 40,000 milligrams per liter and a chemical oxygen demand (COD) in excess of 400,000 milligrams per liter.

Obviously effluents from fish processing plants containing BOD concentrations in thousands of milligrams per liter are of considerable concern to the regulatory agencies, considering that most effluent standards for BOD_5 are less than 100 milligrams per liter. The fact that large volumes of water are utilized in these plants further magnifies the problem. For example, a typical plant processing 5,000,000 menhaden fish per day would utilize 1,500,000 gallons of water to transport the fish from the boats to the dewatering screens.

As the result of fish kills in streams receiving wastewaters from menhaden and other similar fish processing plants, the regulatory agencies have become even more concerned about the early development of some method for effectively treating the effluents from these plants. Despite this pressure on the fish processing industry by the regulatory agencies, no feasible process or method has been developed for effectively treating the wastewaters from these fish processing plants.

Some fish processing plants in order to reduce the volume of wastewater being discharged have recycled or reused the transport water in the unloading of fish from more than one boat. However, the presence of hydrogen sulfide, a highly toxic and soluble gas, in the transport water stream resulting from the rapid decomposition of the bits and pieces of fish in the water has restricted this practice. This is because of several deaths that have occurred in the industry which are believed to have resulted from asphyxiation due to the presence of the hydrogen sulfide.

In considering possible methods of treating transport water discharged from menhaden treatment plants and other similar plants, the use of normal biological or secondary treatment methods for oxidizing the waste constituents of the water was found not to be feasible or practicable because of the high BOD strength and large volume of this wastewater stream.

Also investigated was the use of conventional methods and techniques for the removal or the recovery of the waste constituents, the fish solids and oil, from the wastewater. For instance, sedimentation, flotation, filtration, centrifuging, adsorption, reverse osmosis and ultrafiltration were investigated but all were found to be unsatisfactory.

The primary problem in developing a feasible process for treating the transport water stream by recovery or removal of the waste constituents was the tenacious emulsion of fish oil and water which inhibited normal reaction to conventional methods of treatment. Although refrigeration was found to break the emulsion, the utilization of a process employing refrigeration would not be economically feasible in view of the approximately 1.5 million gallons of wastewater to be treated per day in a typical plant operation.

A process developed by *E.H. Pavia and A.D. Tyagi; U.S. Patent 3,787,596; January 22, 1974; assigned to Pavia-Byrne Engineering Corporation* involves treating emulsified wastewater from fish processing plants by (1) degassifying the wastewater to break the emulsion and to remove any toxic gases and (2) separating the fish solids and oil from the deemulsified wastewater. This method allows the safe reuse of transport water used in the unloading of fish from boats. Such a scheme is shown diagrammatically in Figure 130.

From the unloading operation **10**, the fish in the transport water are transferred via conduit **11** to dewatering screens **12**. The fish to be processed are removed through conduit **13**. The wastewater stream from dewatering screens **12** is transported via conduit **14** to liquid removal screens **15** where fish solids of significant size are removed through conduit **16** to be processed to produce fish meal and oil.

The emulsified wastewater stream from screens **15** is conducted via conduit **17** through degassifier **18** where the stream is subjected to a vacuum of greater than 10 inches of Hg. From the degassifier the deemulsified stream is transported to sedimentation tank **19** where the oil and the fish solids covered or saturated with oil are separated from the water. The oil is removed at the top of sedimentation tank **19** through conduit **21** and is combined with the fish solids which are removed at the bottom of the tank through conduit **20**. The combined oil and fish solids mixture is then processed to produce fish meal and oil. The supernate or the water from which the fish solids and oil have been removed is conveyed

FIGURE 130: PROCESS FOR TREATMENT OF WASTEWATERS FROM THE PROCESSING OF FISH

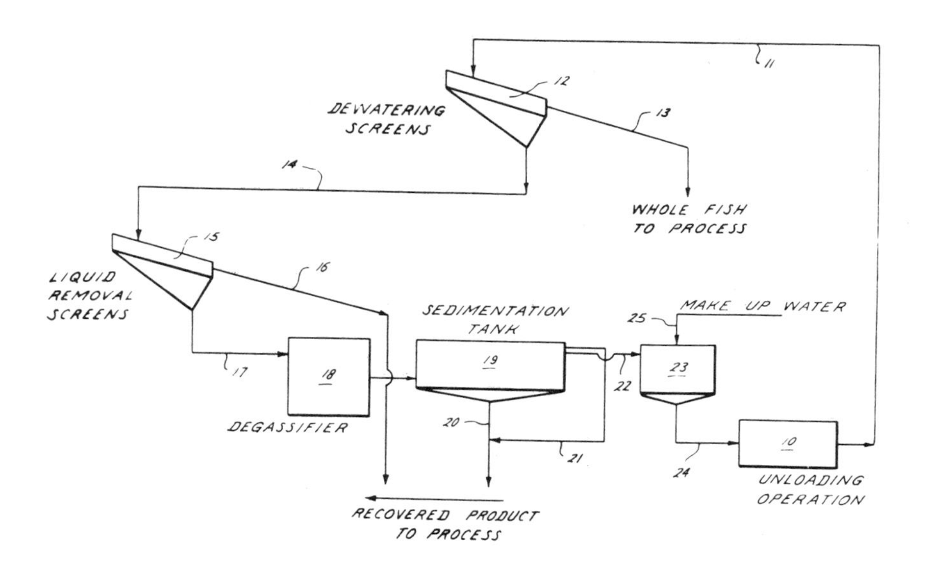

Source: U.S. Patent 3,787,596

from tank **19** through conduit **22** to tank **23**. The water from this tank can then be reused for other unloading operations through conduit **24**. Make-up water as needed may be added to tank **23** through conduit **25**.

A process developed by *M. Furuta; U.S. Patent 3,892,640; July 1, 1975; assigned to Zenkoku Reitogyoniku Kyokai, Japan* purifies the contaminated wastewater that is produced in the processing of aquatic animals, the method comprising adding to the wastewater at least one electrolyte selected from the group consisting of the halides, halogen acid salts and nitrates of the alkali metals, alkaline earth metals and ammonium in an amount such that the concentration of the electrolyte in the wastewater is 0.03 to 1.7% by weight, passing a direct-current electricity through the electrolyte-added wastewater to effect its electrolysis, and thereafter separating the precipitated solid matter from the wastewater.

SELENIUM

A large percentage of zinc is obtained industrially from sulfide ores such as sphalerite, ZnS. The ores are initially concentrated to remove gangue and other metals such as lead, iron, and copper. The concentrates are then smelter-roasted to convert the zinc sulfide to oxide according to the equation:

$$2ZnS + 3O_2 \longrightarrow 2ZnO + 2SO_2$$

Offgases from the smelter operation are water-scrubbed for cooling, humidification and removal of particulates. In the scrubbing process SO_2 and other soluble materials are absorbed by the scrubber water. When the concentrate contains selenium, typically in concentrations of about 0.1 to 1%, the scrub solution will also contain selenium in the form of selenious acid, H_2SeO_3, and generally in concentrations of about 5 to 30 ppm. Typically, these solutions also contain SO_2 in concentrations of about 0.3 to 2 grams per liter, and have pH values of about 2.1 to 2.6.

This scrub solution, which constitutes effluent from the smelter operation, contains concentrations of selenium that are too high to meet regulations pertaining to the quality of water acceptable for discharge into public waterways. Accordingly, removal of a substantial proportion of the selenium from the effluent is essential.

Removal from Water

Commonly employed water treatment operations, such as lime coagulation and sand filtration, are not sufficiently effective for removal of selenium from solution. Ion exchange processes are more effective, but are complex and, therefore, expensive.

A process developed by *W.N. Marchant; U.S. Patent 3,933,635; January 20, 1976; assigned to the U.S. Secretary of the Interior* is one in which selenium is removed from solution in acidic wastewater by treatment of the water with a metallic reducing agent. The process is particularly effective for removal of selenium from zinc smelter effluent by reaction of the effluent with powdered zinc.

SELENIUM COMPOUNDS

See Glass Production Effluents (U.S. Patent 3,966,889)

SEWAGE TREATMENT EFFLUENTS

The exhaust gases issuing from oxidation of sewage sludge contain, in addition to carbon dioxide and water vapor, undesirable contaminants such as volatile hydrocarbons and other low molecular weight organic compounds, as well as inorganic substances, especially ammonia. These substances are toxic and impart an unpleasant odor to the waste gases, and it is necessary that a substantial proportion of these contaminants be removed in order to avoid pollution of the atmosphere.

Removal from Air

Conventional gas scrubbers containing water are not effective in removing the contaminating substances because of the limited solubility of the latter in water. Solutions of sodium hydroxide and ferric chloride are also ineffective. Aqueous acid removes the ammonia but does not affect the hydrocarbon and other organic components.

A buffered solution of potassium permanganate (pH 8.0 in sodium borate) improves odor characteristics and effects a substantial reduction in hydrocarbon content by oxidation, but this method has drawbacks in that insoluble manganese dioxide is precipitated from the solution, and it is necessary to replenish the permanganate as it is used up.

A process developed by *W.M. Copa and L.A. Pradt; U.S. Patent 3,828,525; August 13, 1974; assigned to Sterling Drug Inc.* is one in which contaminants and odors are removed from the waste gases emanating from wet air oxidation reactors by passing the gases through an aqueous suspension of activated sewage sludge or the mixed liquor obtained by suspending activated sewage sludge in fresh sewage.

The ammonia nitrogen contained in drainage such as the effluent from the wastewater treatment plant often causes environmental pollution such as, for example, eutrophy of lakes, marshes, rivers, and bays less influenced by current, and inland seas. These phenomena bring about harm to the coastal and culturing fishery industries and destruction of the ecological system in the inland areas.

Removal from Water

Some processes have been proposed for removing the ammoniacal nitrogen in the drainage: (1) ammonia stripping, (2) break point chlorination and (3) ion-exchange method. In the ammonia stripping method, ammonia is less effectively removed because of a decrease of stripping ability when it is cold. Furthermore, some procedures are necessary to prevent the secondary pollution due to ammonia gas liberated from ammonium ion in the drainage.

In the method of break point chlorination, adjustment of pH decreased by the direct addition of chlorine to the drainage and treatment of chloramines, formed as a by-product, with active carbon and of remaining residual chlorine in the drainage are necessary. In addition the treating plant cannot help being a complicated one because liquid chlorine is used in proportion to the concentration of the ammonia nitrogen in the drainage and some devices are needed for handling of liquid chlorine. Moreover, the inevitable effect of the break point chlorination is to increase the chloride concentration of the effluent.

In the ion-exchange method, the ammonium ion contained in drainage is adsorbed on an ion-exchange substance such as natural or synthetic zeolite. And then the ammonium ion is eluted in the alkali salt or alkali hydroxide solution from the ion-exchange substance and is discharged to the atmosphere by a stripping method. Accordingly, in the ion-exchange method, some procedures are necessary to prevent the secondary pollution and the decrease of stripping ability when it is cold which is common in the ammonia stripping method.

In another ion-exchange method, the ammonium ion is directly decomposed by the oxidizing action of chlorine in the ion-exchange column. Accordingly the ammonium ion eluting is obstructed by generating gas.

A process developed by *T. Hiasa and H. Ishimard; U.S. Patent 3,929,600; Dec. 30, 1975; Iwao Engineering Co., Inc., Japan* is a process for removing ammonia nitrogen by recycling a treating liquid as a medium in a closed system including an ion-exchange process, an adsorption process, an elution process, an electrolysis process, an adjusting process and a cleansing process to prevent secondary pollution and improve removing ability. Such a system is shown in Figure 131.

As shown there, drain pit **1** is connected with ion-exchange column **3** via drainage-supplying path **2**. Natural or synthetic zeolite is usually packed as the ion-exchange substance in the column **3**, with which drainage-discharging path **4** is joined. Another ion-exchange column **5** which is interchangeable, as occasion demands, with the column **3** is provided and is joined with electrolytic cell **7** via path **6**. Natural or synthetic zeolite is packed in the column **5** too.

The electrolytic cell **7** is joined with adjusting bath **9** through path **8** for adjusting bath and the adjusting bath **9** is united with active carbon column **11** via path **10** for active carbon column. The active carbon column **11** is united with the aforementioned zeolite column **5** through circulating path **12** to form a circulating circuit. The path **6** for electrolytic cell and the circulating path **12** are installed with pumps **13** and **14**, respectively. The adjusting bath **9** is joined, via path **15** for adjusting liquid, with adjusting liquid bath **16** which is united with supplying path **17** of adjusting liquid.

Furthermore, the electrolytic cell **7**, the adjusting bath **9**, and the active carbon column **11** are united, through path **18** for generating gas, with the adjusting liquid bath **16** which is united with active carbon column **20** via path **19** for clean gas. With gas-discharging path **21** is connected the active carbon column **20**.

When a more simplified system, rather than more effective removal, is desired, the adjusting bath **9** can be joined with the circulating path **12** through pump **14** and the path **18** for generating gas can be directly united with the gas-discharging path **21**.

FIGURE 131: PROCESS FOR REMOVING AMMONIACAL NITROGEN FROM WASTEWATER

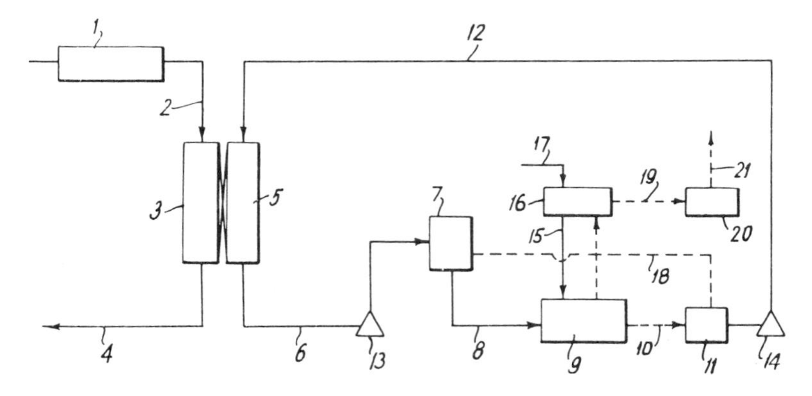

Source: U.S. Patent 3,929,600

A process developed by *M.J. Stankewich, Jr.; U.S. Patent 3,764,524; October 9, 1973; assigned to Union Carbide Corporation* involves removing both carbon food and phosphorus pollutants by biochemical oxidation and chemical precipitation using oxygen gas in the presence of activated sludge, where most of the carbon food and pollutants are removed in a first covered zone with the addition of phosphorus-precipitating compound and under high food-to-biomass ratio, and the effluent water is further purified in a second covered zone under low food-to-biomass ratio.

A process developed by *I.R. Higgins; U.S. Patent 3,984,313; October 5, 1976; assigned to Chemical Separations Corporation* is one in which sewage water containing many pollutants including ammonia and phosphates is treated so that the ammonia and phosphates are preferentially removed through ion exchange while the other contaminants are allowed to remain in the water. Ferric hydroxide is deposited within the matrix of strong acid type cation exchange resin and the resin is rejuvenated with sodium hydroxide. Figure 132 illustrates the essential features of such a process.

In the drawing a treatment vessel **10** is shown with an influent line **12** for bringing in water to be treated and an effluent line **14** for carrying away treated water. The water which contains additives common to any sewage water including calcium, magnesium and sodium salts, as well as sulfates and chlorides, also contains phosphates which are undesirable, as explained above. The water is treated by contacting it with a specially treated cation exchange resin which is present in the treatment vessel.

The resin is treated so that a metallic hydroxide is deposited inside the matrix of each of the resin beads. This is done by sorbing iron in the resin and placing the resin in ammonium or sodium hydroxide. In the case where it is desirable to sorb iron, the iron can be provided by a ferric salt such as ferric nitrate, ferric sulfate or ferric chloride. When ferric chloride is contacted with the resin, it separates into ferric and chloride ions.

The ferric ions settle on the ion exchange site present in the matrix of each resin bead. The chloride can be washed away. Thereafter, the resin is thrown into a reactant hydroxide which may be sodium hydroxide or ammonium hydroxide. As a result, ferric hydroxide is implanted inside of the matrix of the resin beads at locations other than the ion exchange sites so that the ion exchange sites are free to sorb the ammonium ions which are present in the sewage water or other solutions being treated.

Phosphate ions are attracted by the ferric hydroxide so that as the sewage water passes through the treatment vessel both phosphate and ammonium ions are sorbed on virtually every bead of the ion exchange resin and thus the solution is free of these pollutants.

The resin is prepared from any standard cation resin of the strong acid type, such as the resin Dowex HCR-W, IR-120, or Duolite C-20. The resin is regenerated and intermittently fed to the treatment vessel by a regenerating circuit indicated generally at **16**. The regenerating circuit includes the treatment vessel, as well as a conduit **18**. Conduit **18** has a straight section **20** which extends vertically and parallel to the longitudinal axis of the treatment vessel. The straight section is connected at its bottom to a U-shaped section **22** which connects the straight section with a vertically extending resin feed section **24** which extends upward to the bottom of the treatment vessel. Extending from the top of the treatment vessel to the straight section is a curved resin exhaust section **26** which connects with the vertical straight section at a location adjacent to the top of the section.

FIGURE 132: PROCESS FOR AMMONIA AND PHOSPHATE REMOVAL FROM WASTEWATERS BY CONTINUOUS ION EXCHANGE

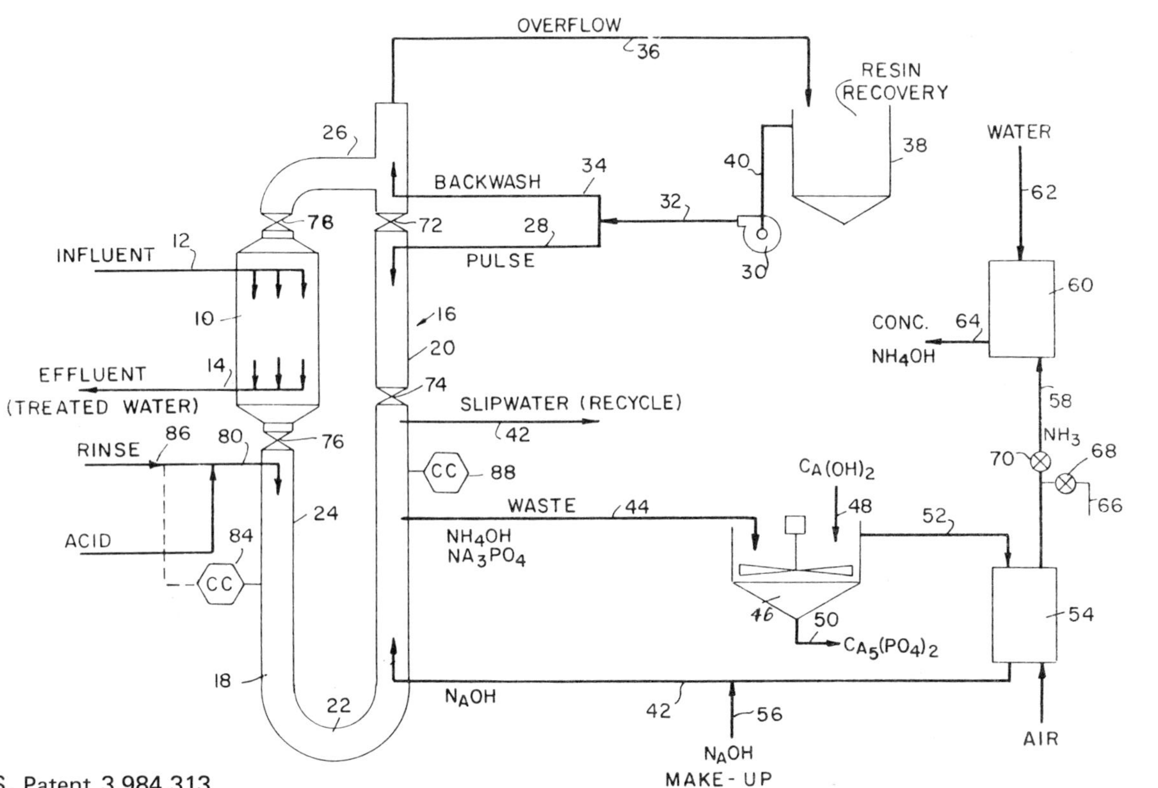

Source: U.S. Patent 3,984,313

In order to provide for resin regeneration, resin is intermittently moved through the regenerative circuit so that regenerated resin moves upwardly through the resin feed section **24** and into the treatment vessel and then out through the resin exhaust section **26** and downward through the straight section. This movement of resin occurs only after the influent of sewage water is shut off and continues only long enough to replenish the contaminated resin within the treatment vessel **10** with regenerated resin.

The movement of resin takes place because of a pulse of water and resin mixture which is fed through a pulse line **28** under pressure impressed by a circulation pump **30**. The pulse line is connected to the pump **30** by means of a pump exhaust line **32**. Since movement of fluid through the pulse line will move entrained resin downward through the straight section and then upward through the resin feed section and into the treatment vessel and since resin within the vessel will be moved upward through the resin exhaust section, it is possible that there will be an overflow of resin at the upper portion of the straight section.

Some of the fluid being pumped into the system by the pump bypasses the pulse line through a back wash line **34** which is connected to the straight section at a location slightly above that at which the pulse line is connected and adjacent to the connection between the resin exhaust section and the straight section.

Excess resin washed upwardly by the backwash water coming through the backwash line **34** moves upwardly and out of the conduit **18** and out through an overflow line **36** and into a resin recovery reservoir **38**. A supply of water is always available and is fed by pump through a pump supply line **40** extending between the pump and the resin reservoir. Movement of the resin out of the treatment vessel and movement of regenerated resin into the treatment vessel can be provided without pressure buildups and in spite of temporary changes in the volume of the total amount of liquid within the system because differences in volume can be accommodated within the resin reservoir **38**.

Since the energy to move the resin through the system is supplied by flowing water, in order to prevent the resin from being diluted in its aqueous transportation medium, water is withdrawn from the straight section through a slip water line **42** located below the pulse line. The slip water can be recycled at the influent line **12**.

The resin is regenerated when it is in the lower portion of the straight section by sodium hydroxide which is fed through a regenerative supply line **42** which feeds sodium hydroxide upwardly in the straight section. The sodium hydroxide removes the ammonia and phosphates from the resin by reacting with the ammonium and phosphate ions to yield ammonium hydroxide and sodium phosphate. These products of reaction move out from the straight section through a waste line **44** which leads to a reaction vessel **46**.

The purpose of the reaction vessel **46** is to facilitate the conversion of the sodium phosphate and ammonium hydroxide into marketable and useful by-products. To this end, calcium hydroxide (lime) is added to the reaction vessel **46** through line **48** and calcium phosphate is removed from the bottom of the vessel through line **50**. Calcium phosphate is a solid and thus the addition of the lime affords a convenient way of eliminating the phosphates in a condensed form.

Sodium hydroxide and ammonium hydroxide are run off of the reaction vessel **46** and pass through a line **52** into an air sparger **54**. The sodium hydroxide will pass through the air sparger **54** and out of it through the regenerative supply line **42** to be fed to the conduit and serve as regenerative. Since losses will occur, it is necessary to constantly supply a small amount of sodium hydroxide through a sodium hydroxide makeup line **56** which feeds into the regenerative supply line **42**. Ammonia gas passes up through an ammonia supply line **58** to a water scrubber **60** where water is fed into it through a water line **62**. Concentrated ammonium hydroxide leaves the water scrubber through a line **64**.

If desired, ammonia gas can be taken off of the ammonia supply line through an ammonia takeoff line **66** by opening a valve **68** in the line **66** and closing a valve **70** in the line **58** between the line **66** and the scrubber **60**. If ammonia gas is not desired, the valve **68** is closed and the valve **70** in the line **58** is opened to allow the gas to flow into the water scrubber **60** to produce concentrated ammonium hydroxide. The choice is really one of by-products and marketability will determine the choice.

In operation, sewage is run through the influent line **12** to pass through the resin within the treatment vessel which has been specially prepared, as described above, and out through the effluent line **14** for a predetermined time. The time is selected so that the flow of fluid will not be longer than the resin is effective. During this time, sodium hydroxide is being added to the lower portion of the straight section through regenerative supply line **42** and ammonium hydroxide and sodium phosphate are taken off through the waste line **44**.

After sewage water is passed through the treatment vessel for the predetermined time, a valve **72** automatically closes. The valve **72** is in the straight section between the pulse line and the back wash line **34** so that the pulse water will move the resin downward when the valve **72** is closed. The movement of the resin is facilitated because as the valve **72** opens a valve **74** which is below the valve **72** in the straight section and above the slip water line **42** opens. Valves **76** and **78** which are placed in the conduit **18** below and above the treatment vessel **10** respectively, also open simultaneously with the valve **74** to allow an upward flow of regenerated resin into the treatment vessel.

The resin flows for a time which is long enough to allow a change of resin within the vessel. Thereafter, valves **74, 76** and **78** close while valve **72** opens. This prevents movement of the resin and allows excess water to pass upward through the overflow line **36** to the resin reservoir **38**. While the valves **74, 76** and **78** are closed, sewage water passes through the treatment vessel.

In order to assure effective operation of the resin within the treatment vessel, it is necessary to rinse it to remove any salts which have been retained by the regenerating process and also to maintain the proper acidity. To this end, rinse water is added through a rinse line **80** which connects with the resin feed section **24** below the valve **76** and acid is added through an acid supply line **82** which connects with the rinse line **80** and which supplies an acid such as hydrochloric or sulfuric to the rinse water when necessary to lower the pH. The rinse water is added automatically when needed. This is so because the conductivity of NaOH is higher than water and a conductivity control **84** senses when the conductivity of the upward moving NaOH in the resin rinse section **24** is high enough to indicate that there is an absence of rinse water in it. When this occurs, the conductivity control **84** opens a valve **86** in the rinse line **80**. As soon as the conductivity becomes low enough to indicate that the resin has been sufficiently rinsed with water, the conductivity control **84** closes the valve **86**.

Similarly, a conductivity control **88** determines when there is too much water in the resin suspension passing down the vertical section **20** and when too much water is present automatically opens a valve **90** in the slip water line **42** to pass excess water out of the conduit **18**.

A process developed by *R.J. White; U.S. Patent 3,698,881; October 17, 1972; assigned to Chevron Research Company* involves separating solid material and an aqueous stream containing inorganic nutrients from sewage, using the aqueous stream as a source of nutrients to aid in growing plants, and reacting at least a portion of the plants and at least a portion of the solid material separated from the sewage with steam in a reaction zone to produce synthesis gas. According to a preferred embodiment, a portion of the plants which are grown are fed to animals and solid wastes from the animals are used as feed for synthesis gas production.

A process developed by *R.L. Meyers; U.S. Patent 3,917,564; November 4, 1975; assigned to Mobil Oil Corporation* is one in which sewage sludges can be put to useful purpose with concurrent recovery, in whole or part, of inherent fuel values. Upon injection of such by-products diluted with added water to a delayed coker as aqueous quench medium, water content of the sludge is utilized to cool the hot coke. The dispersed solids and any organic liquids present are apparently dispersed through the coke mass, contrary to the expectation that these contaminants would be filtered out on the surface of the coke face first contacted thereby. The combustible solid portions of the by-product become a part of the primary fuel (coke). Noncombustible solids (e.g., sand, rust, silt) are distributed throughout the mass of coke in such manner that increase in ash content is acceptable as being within commercial specifications.

Municipal and industrial wastewater treatment processes usually produce not only sedimentary wastes, such as sedimentary sewage sludge, but also floating wastes which usually comprise, for example, grease, oil, scum and other materials which float to the water surface. Such floatage is particularly found in primary treatment tanks.

Because such materials are capable of being removed from the water surface by a skimming process, they are hereinafter referred to as skimmings. A blade, chain-driven scoop or other devices can be used to skim the water surface; sometimes a weir is used and the water and/or weir height controlled so the skimmings run off over the weir.

The skimmings comprise a large percentage of water (usually over 50%), together with the aforementioned grease, scum, wood chips and other floatage. Usually the skimmings comprise organic materials, but sometimes plastic pieces and synthetic filaments are also included. Such very nonhomogeneous skimmings present difficult handling and disposal problems.

Known methods of disposal of the skimmings include burial, chemical digestion and burning. Equipment for burning the skimmings generally comprises a combustor or burner designed to handle a liquid waste through a steam, air, or mechanical atomizing nozzle. The liquid feed is atomized as finely as possible to present the greatest surface area for mixing with combustion air. Sometimes a secondary incineration chamber is provided and that may comprise, for example, a vertically arranged cylinder which acts as its own stack.

It is also known to feed skimmings directly to sludge incinerators such as multiple-hearth furnaces, sometimes obviating the necessity of using auxiliary fuels to burn the solid sludge and at the same time disposing of the skimmings. However, due to their high thermal values, the skimmings must be fed slowly to the incinerator and then must be burned immediately or else there is the possibility of a subsequent uncontrollable flare-up. Such flare-ups can exceed the temperature capacity of the incinerator and, consequently, can destroy the furnace exhaust ducting and the like or even create an explosion.

Because the skimmings are not an evenly-flowing medium and often include large solid pieces or fibrous and synthetic strands, it is difficult to feed the skimmings to an incinerator while achieving therein an even distribution. To avoid clogging the supply system, the skimmings generally are introduced to the incinerator through a single large pipe. However, the flow from the single large pipe does not evenly distribute the skimming and, consequently, there is uneven burning inside the incinerator. As a result, ash stalactites sometimes form in the furnace and eventually interfere with its operation. Furthermore, grease may soak into the refractory brick of the furnace so that, after a period of time, the bricks are destroyed.

An apparatus developed by *U.F. Rinecker; U.S. Patent 3,894,833; July 15, 1975; assigned to Envirotech Corporation* for incinerating grease-laden aqueous mixtures includes a burner assembly having a swirl chamber whereinto the liquid is tangentially fed and swirled. From a nozzle connected to the chamber, a minor fraction of the feed mixture is emitted as a rotating, atomized spray for subsequent ignition. A major fraction of the feed is continuously recycled thereby maintaining a high liquid flow rate through the burner assembly.

In modern efforts to reduce pollution of the environment, various waste treatment systems have been employed. One system which has enjoyed increasing attention employs biological oxidation to consume organic contaminants in wastewater. In such operations a population of microorganisms, generally called activated sludge, is selected or adapted to consume the organic contaminants while the microorganisms are actively growing in the presence of an abundant supply of oxygen or air.

However, such activated sludge processes continually produce large quantities of waste bacterial sludge which is difficult to dewater and which must itself be disposed of. Recently it has been shown in U.S. Patent 3,335,798 that under certain conditions activated sludge disposal may be accomplished by pumping the sludge into porous subterranean formations.

In any event, however, such sludge may present a water pollution problem. A process developed by *C.E. Hamilton; U.S. Patent 3,724,542; April 3, 1973; assigned to The Dow Chemical Company* is based on the discovery that fresh activated organisms can utilize hydrocarbons as a food source under substantially anaerobic conditions with the production of gases consisting of substantial quantities of methane and having appreciable utility as fuel gas. Thus, the process provides a method for disposing of excess activated sewage sludge with a minimum of contamination of potable water sources and under aesthetically pleasing conditions while providing fuel gas for future use.

SILICON TETRAFLUORIDE

When phosphate rock is decomposed with strong mineral acid to produce phosphoric acid or a phosphate such as superphosphate or triple superphosphate, or when phosphoric acid is concentrated, or when glass is etched with hydrofluoric acid, silicon tetrafluoride and fluosilicic acid are formed as by-products, and appear in the waste gases together with small amounts of hydrogen fluoride. Also when hydrogen fluoride is manufactured for silica containing fluorspar, silicon tetrafluoride and hydrofluosilicic acid are formed.

These fluorine- and silicon-containing compounds have very little economic value and have usually been disposed of by venting the waste gases to the atmosphere. The small amounts of hydrogen fluoride that may appear in such gases are usually not sufficiently great to warrant recovery. However, these waste gases are highly corrosive, because silicon tetrafluoride is reacted with water to form silica and hydrogen fluoride. With the emphasis today on avoiding pollution of the environment, it has become necessary to avoid the discharge of such materials to the atmosphere, and to develop an economic process for doing so.

Removal from Air

A process developed by *G.L. Flemmert; U.S. Patent 3,969,485; July 13, 1976;* is a process for converting silicon and fluorine-containing waste gases into silicon dioxide and hydrogen fluoride, absorbing the waste gases in water to form hydrofluosilicic acid, decomposing the hydrofluosilicic acid in the presence of concentrated sulfuric acid to form silicon tetrafluoride and hydrogen fluoride, converting the silicon tetrafluoride in the vapor phase to silica and hydrogen fluoride, and recovering the hydrogen fluoride. The elements of such a process are shown in Figure 133.

In the process illustrated, water and silicon- and fluorine-containing gases are passed into an absorber **1**, where silicon tetrafluoride, hydrogen fluoride and any silica which may also be present are absorbed, and dispersed in the water. The mixture of hydrofluosilicic acid, high silica fluosilicic acid, $H_2SiF_6{\cdot}SiF_4$ and mixtures of either of these with solid silicon dioxide are passed to the combined decomposition unit **2b** and scrubber **2a**, where concentrated sulfuric acid is added to decompose the hydrofluosilicic acid to SiF_4, HF and H_2O. At least part of the sulfuric acid first passes through the scrubber, in countercurrent flow to the gaseous effluent from the decomposition unit, where it absorbs steam and hydrogen fluoride therefrom. If the proportion of hydrogen fluoride is more than stoichiometrically equivalent to the silica present in the decomposition unit, sand is added in sufficient amount to take up the excess hydrogen fluoride.

The remaining gaseous reaction product, silicon tetrafluoride, is recovered and sent on to the hydrolysis unit **3** where it is blended with additional silicon tetrafluoride from the scrubber **6a** in a subsequent stage and with combustible gas and with air, and burned, using the process of U.S. Patent 2,819,151 to form silica and hydrogen fluoride.

The residual sulfuric acid in the decomposition unit, now diluted by water, is withdrawn from the decomposition unit **2b**, and can be concentrated and recycled. The silica formed in the hydrolysis unit **3** is carried by the effluent gases to the separator **4**, where it is recovered in a highly active form. The effluent gases, containing hydrogen fluoride, water vapor, unreacted silicon tetrafluoride and combustion products, such as carbon dioxide, nitrogen and oxygen, are passed to a second absorber **5**, where the silicon tetrafluoride and hydrogen fluoride are absorbed in water or aqueous hydrofluoric acid solution, while the exhaust gases, now relatively free from fluorine-containing materials, are exhausted to the atmosphere.

FIGURE 133: PROCESS FOR CONVERTING SILICON- AND FLUORINE-CONTAINING WASTE GASES INTO SILICON DIOXIDE AND HYDROGEN FLUORIDE

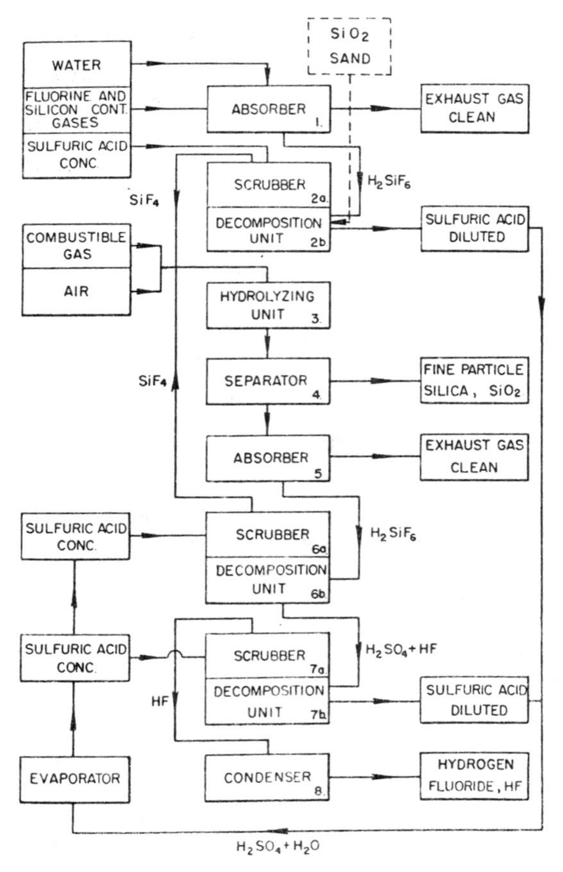

Source: U.S. Patent 3,969,485

In order to maintain a low concentration of hydrogen fluoride in the exhaust gases, it may be suitable to introduce water or hydrofluosilicic acid into the absorber **5**.

The aqueous absorbent liquor is mixed with concentrated sulfuric acid in decomposition unit **6b**. At least part of the sulfuric acid is introduced into scrubber **6a**, through which it flows into decomposition unit **6b** into countercurrent flow to the effluent gases from the decomposition unit. Silicon tetrafluoride is liberated and is returned to the hydrolysis unit **3** for hydrolysis into hydrogen fluoride and silica.

The liquid effluent from decomposition unit **6b**, composed of a mixture of sulfuric acid, hydrogen fluoride and water, is passed to decomposition unit **7b**, where it is heated to a somewhat more elevated temperature, and combined with additional concentrated sulfuric acid.

The gaseous effluent from this reaction mixture, after scrubbing by at least part of the entering sulfuric acid in scrubber **7a**, is composed substantially completely of hydrogen fluoride which can then be recovered in a substantially pure dry form in the condenser **8**.

Dilute sulfuric acid is recovered from decomposition units **2b** and **7b** and can be passed to the evaporator, concentrated, and recycled, or used for reaction with phosphate rock. Thus the reaction products from the process shown in the flow sheet are substantially silica and hydrogen fluoride, both in a relatively pure condition, while all the other products recovered can be recycled, including the dilute sulfuric acid from decomposition units **2b** and **7b**.

SILICONE POLYMER PROCESS EFFLUENTS

A wide variety of different products are usually produced in the same silicone-polymer producing plant, involving the use of enormous quantities of water particularly for the hydrolysis of monomeric chlorosilanes to produce the corresponding siloxane polymers. A major problem facing the silicone industry at the present time is the treatment of the aqueous wastewater so as to convert materials therein which adversely affect the ecology into harmless compounds in order to safely dispose of such waste streams. The problem of purifying these aqueous waste streams is greatly magnified due to the usual production of a vast number of different products in a silicone-polymer producing plant which in turn results in the contamination of the wastewater stream with a large variety of different types of harmful impurities.

The effect of the many varieties of different types of impurities in the wastewater stream is more than being merely accumulative since the presence of one type of impurity may greatly increase the difficulty of finding a suitable means to remove another type of impurity from the aqueous waste stream.

Another major problem present in attempting to purify aqueous wastestreams from silicone plants is that the types and concentrations of impurities vary considerably depending upon which of the many processes in the plant is responsible for the specific aqueous waste stream to be purified at any particular time.

Accordingly, the task of designing a relatively inexpensive purification system which can handle all the different waste streams is extremely difficult to accomplish. Moreover, because of the specific types of impurities present in the aqueous waste stream, many conventional types of purification methods are unsuitable or impractical.

Removal from Water

The design of a suitable purification process is rather difficult since the aqueous stream contains various inorganic salts and since the concentration of the salts drastically changes from batch to batch. Biological methods are not very practical for purifying aqueous waste streams from silicone producing plants because of the chlorine ion content in the waste stream.

In addition, sorption of impurities by a sorbent such as activated carbon is not a very satisfactory purification step because of the presence of silicones in the aqueous waste stream. The silicones are extremely harmful to the normal regeneration processes for such sorbents as activated carbon. Accordingly, the inability to successfully regenerate the carbon to any appreciable extent would render such a process nonfeasible from an economic viewpoint.

In addition, if the sorbent cannot be regenerated, then the additional problem of disposal of the sorbent must be solved. Various oxidizers such as permanganates have been suggested but such are not entirely satisfactory for the purification purposes of this process.

One particular disadvantage of such oxidizers is the necessity of employing extremely high temperatures to effect oxidation. Another suggested means of purification which also has not attained general commercial acceptance is electrolysis. Some of the difficulties associated with electrolysis are high equipment cost, slow reaction rates, and extremely high power requirements.

In addition, it has previously been suggested to treat certain industrial wastewater streams with ozone under certain conditions. Ozone treatments however have been too expensive and too inefficient for the treatment of the large quantities of wastewater which are usually produced in silicone plants. Accordingly, ozone treatment of wastewater streams from silicone plants has not heretofore been used to any appreciable extent in large scale commercial operations. Ozone treatments suggested heretofore require relatively large pieces of equipment as compared to the amount of the material being treated. Also such treatments are extremely expensive due to the poor ozone utilization and high cost of ozone production.

A process developed by *H. Lapidot; U.S. Patent 3,855,124; December 17, 1974; assigned to General Electric Company* is a process for the purification of an aqueous waste stream from a silicone-polymer producing plant which includes conducting the aqueous waste stream to a flotation and sedimentation zone; adjusting the pH of the wastewater stream to at least about 12; conducting the stream to a clarification zone; conducting the waste stream to an ozonation zone; conducting the waste stream to at least one holding zone prior to the ozonation zone and subsequent to the flotation and sedimentation zone; and obtaining a purified wastewater stream. This process is shown in Figure 134.

FIGURE 134: PROCESS AND APPARATUS FOR THE PURIFICATION OF AN AQUEOUS WASTE STREAM FROM A SILICONE-POLYMER PRODUCING PLANT

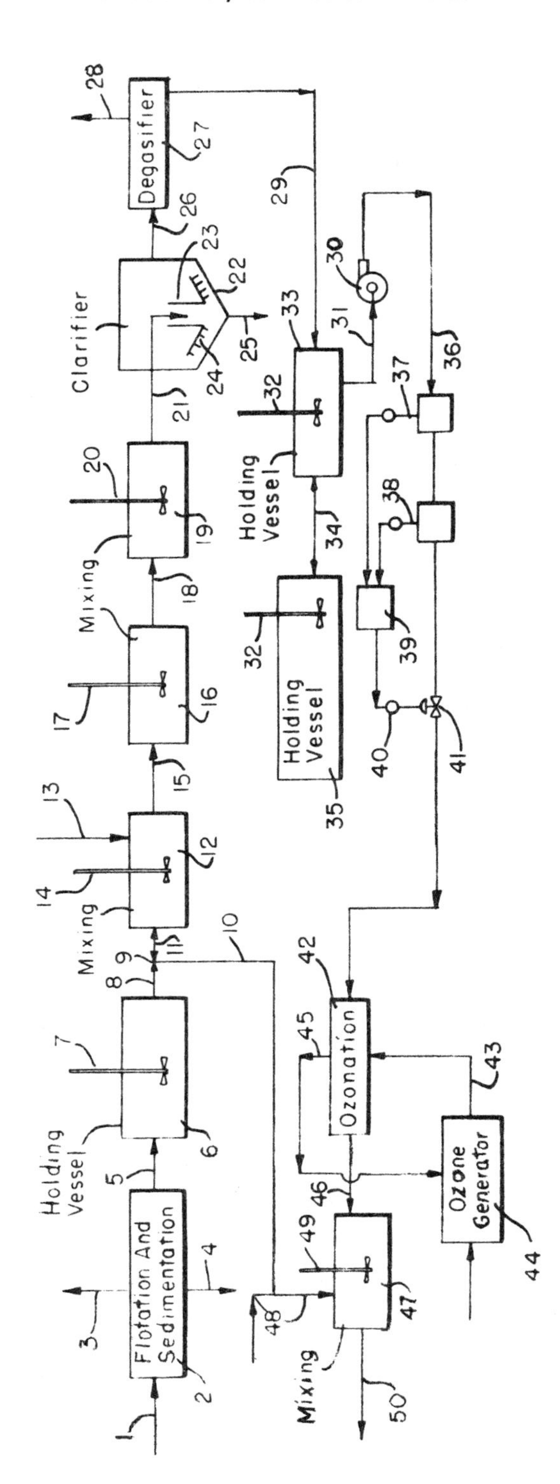

Source: U.S. Patent 3,855,124

The waste stream line **1** connects to a flotation and sedimentation vessel **2**. Sedimentation vessel **2** contains removal means **3** and **4**, and outlet line **5** which connects to holding tank **6**. Holding tank **6** is provided with stirring means **7** and outlet line **8**. Outlet line **8** contains a valve **9** at which point it splits up into lines **10** and **11**. Line **11** connects to a mixing vessel **12**. Mixing vessel **12** contains an inlet line **13**, mixing means **14** and outlet line **15**. Outlet line **15** connects to mixing vessel **16** which is provided with mixing means **17** and outlet line **18**. Outlet line **18** connects to mixing vessel **19** which is provided with mixing means **20** and outlet line **21**.

Outlet line **21** connects to clarifier **22** which has a material introducing means **23**, scraping means **24**, and outlet means **25** and **26**. Outlet means **26** connects to a vacuum degasifier **27** which is provided with outlet lines **28** and **29**. Outlet line **29** connects to storage vessel **33** which has an outlet line **34** which connects to storage vessel **35** and outlet line **31**. Storage vessels **33** and **35** contain stirrers **32**. Outlet line **31** connects to pump **30** which has an outlet line **36**. Outlet line **36** contains a flow measuring device **37**, an organics concentration analyzer **38**, a multiplier element **39**, a valve **41**, and a valve control means **40**. Line **36** connects to ozonation reactor **42** which contains liquid outlet line **46**, gas outlet line **45**, and gas inlet line **43**. Outlet line **45** connects to ozone generator **44**. Outlet line **46** connects to mixing vessel **47**. Mixing vessel **47** contains inlet line **48** and outlet line **50**. Line **10** from line **8** connects to line **48** before line **48** enters mixing vessel **47**.

In operation, the aqueous waste stream from a silicone producing plant is conducted through line **1** through sedimentation tank **2** wherein water-insoluble materials having densities lower than water are skimmed off the top of the aqueous waste material and wherein water-insoluble materials having densities greater than water are removed from the bottom of the sedimentation tank.

An aqueous waste stream is removed from this sedimentation tank through line **5** and is conducted to holding tank **6**. The aqueous stream is then removed from the holding tank through line **8**. A portion of the aqueous stream is fed from line **10** to line **48** to mixing vessel **47**, depending upon the requirements discussed above for use of this stream as a back neutralization source of material. The rest of the aqueous stream is conducted to a mixing vessel **12** through line **11**. In addition, an alkaline material is added to the mixing vessel **12** via line **13**. The stream is removed from vessel **12** via line **15** and conducted to vessel **16** whereafter it is removed via line **18** and conducted to vessel **19**. A stream having a pH of at least about **12** is removed from vessel **19** by line **21** and is conducted to the clarifier **22** whereby the material flows through conduit **23**.

A slurry is removed from clarifier **22** through line **25**. An aqueous waste stream is removed from clarifier **22** via line **26** whereafter it is conducted to degasifier **27**. Gases are removed through line **28** and an aqueous waste stream is removed from the degasifier via line **29** whereby it is conducted to small holding tank **33**. The stream is pumped from tank **33** via line **31** by pump **30** and is analyzed for amount of material and for concentration of organics by flow measurer **37** and organics analyzer **38** respectively.

These measurements are then multiplied in multiplier **39** whereby a flow controller **40** can regulate the opening or closing of valve **41** in accordance with the above measurements to provide a flow which introduces a predetermined amount of organics per unit time into the ozonation zone. Material will flow from small tank **33** to large tank **35** which is at the same level or will flow from the large to the small tank depending upon the relative flow rates of material into the small tank from line **29** and out of the tank through line **31**.

The material entering the ozonation reactor through line **36** is contacted with ozone-containing gas entering through line **43**. A gaseous stream is removed from ozonation reactor **42** via line **45** and is conducted to an ozone generator **44** whereby the ozone-containing gas is generated and removed therefrom and conducted back to the ozonation reactor. An ozone-treated aqueous product is removed from ozone reactor **42** and is conducted to a mixing vessel **47** via line **46**. An acid such as hydrochloric acid is added to mixing vessel **47** through line **48**. A product having a pH between the range of about 6 and 9 is removed therefrom through line **50**.

SILVER

Removal from Water

A process developed by *J.S. Bentley; U.S. Patent 3,901,777; August 26, 1975; assigned to Photographic Silver Recovery Limited, England* provides an electrolytic process for the recovery of silver from used photographic solutions.

See Photographic Processing Effluents (U.S. Patent 3,982,932)

SINTERING PLANT EFFLUENTS

The sintering process involves exposing a particulate metallic oxide, e.g., iron ore or iron oxide (as from basic oxygen furnace fume), in a burden on a moving grate to transversely moving hot oxidizing gases to fuse particles of the ore together to form an agglomerated mass. The agglomerated mass may then be used in a known manner in iron-producing equipment, e.g., a blast furnace. Frequently, the ores contained naturally or were made to contain a combustible material, e.g., coke or coal, and a flux, e.g., limestone and/or dolomite, and a burden of a granular mixture with or without added moisture deposited upon a traveling grate.

In the initial stages of the movement of the burden along its predetermined path on the traveling grate, the burden was ignited by passing under gas torches or oil burners whereby ignition of combustibles and a flame front were established in the burden. Thereafter, by means of fans, ambient air was drawn downwardly through the burden as it moved along the path causing the flame front to move downwardly through the burden toward the grates.

The temperature of the burden by this process was raised along the flame front to approximately 2500°F (lower or higher depending upon the ore), whereby the individual particles became fused into a solid foraminous mass. As the hot sinter was discharged from the end of the traveling grate, it was broken into large chunks. The residence time on the traveling grate was sufficient to cause the fusion to occur substantially through the depth of the burden. The gases which were exhausted from the lower side of the burden when downdraft sintering was employed were generally exhausted to the atmosphere. These gases in addition to containing entrained and fumed solid particulate material also contained unburned hydrocarbons, sulfur compounds, and oxides of carbon.

Removal from Air

Recent efforts to control atmospheric pollution have necessitated reconsideration of the methods by which sinter exhaust gases are handled. To remove hydrocarbons and lower oxides of carbon, afterburner or incinerator means have been utilized to ignite the combustible components of the exhaust sinter gases and convert them to harmless gases, e.g., carbon dioxide and water vapor. However, afterburner and incinerator means typically employ hydrocarbon fuels which can result in an increase in the concentration of hydrocarbons in the exhaust gases. Entrained sulfur or lower sulfur oxides are also converted to higher oxides which may be removed by scrubbing means in a known manner. The large quantity of solid particulate material entrained in such exhaust sinter gases typically is passed through electrostatic means which effect precipitation of such particulate material in a known manner.

Unfortunately, conventional sintering practice is such that the moisture content and temperature of the exhaust gases are not optimal for cleaning by electrostatic means. Thus, in order to improve the combustion and formation of sinter, particularly in the case of iron ore, the mixture of iron ore and carbonaceous material and optionally a fluxing material, e.g., limestone and/or dolomite, has usually been moistened with up to 6 to 14% by weight of water to form wet agglomerated masses of approximately the size of rice. This enables the use of finely divided carbonaceous material and facilitates the distribution thereof with respect to the iron oxide in a wet agglomerated mass.

The use of water as a binder aid results, during sintering, in considerable moisture in the gases traversing the burden. If the burden is too deep for the draft system employed as the gases encounter the lower regions of the burden which are at a much lower temperature, the moisture picked up at or near the flame front is condensed by and saturates the lower regions of the burden.

When the problem becomes acute, a condition known as sogging out occurs and the passage of the gases through the burden is greatly impeded. Hence it has been desired to severely limit the depth of the burden or to minimize the amount of moisture in the burden and in the gases traversing therethrough in order to approach the capacity of the apparatus. However, when the moisture content of the gases is low, the resistivity of the gases and particules is high and the effectiveness of electrostatic precipitation means is greatly reduced. Also, the temperature of the gases normally passing through the burden in conventional procedures and exiting from the lower side thereof is low (225° to 300°F). At both low moisture content and such relatively low temperature, the electrostatic precipitation characteristics are generally inferior.

A process developed by *T.E. Ban; U.S. Patent 3,909,189; September 30, 1975; assigned to McDowell-Wellman Engineering Company* is an improved traveling grate sintering process for treating solid mineral matter such as iron ore and utilizing a sinter draft system which recycles a part of the total draft, minimizes the exhaust draft effluent, and reduces the costs of draft treatment to meet environmental air quality standards. The system contemplated herein recycles relatively cool draft from the initial windboxes adjacent the traveling grate to the hood or hoods downstream of the path of grate travel to significantly decrease the concentration of hydrocarbons in the sinter exhaust gases and to produce a hotter, more humid exhaust for subsequent processing to remove pollutants, particularly with respect to the electrostatic treatment of the exhaust draft to remove solid particulate material. At the same time a sinter product comparable in quality to that produced by the basic downdraft sintering system is obtained without substantial change in rates of production.

SMOKEHOUSE EFFLUENTS

Food products, such as ham, bacon, sausage, and the like, are often pumped or infused with a liquid to facilitate cooking and to improve flavor and appearance, as is well known in the art. Some food products, especially meat, fish, and the like, also may be smoked or treated with a smoke derivative, known in the art as liquid smoke, so that there will be imparted a desirable smokey flavor to the cooked product. The product may be immersed in the liquid smoke, or the liquid smoke may be sprayed onto the product. Alternatively, the food processing chamber can be supplied with a separate source of smoke either from ordinary wood burning or by generating smoke from a smoke derivative product, for direct contact with the food product during a portion of the cooking cycle.

However, whether or not smoking is used during a part of the cooking cycle, the liquid added to the product must be substantially removed and the weight of the food product restored to approximately the original weight before it is considered suitable for consumption. Thus one major problem is efficiently removing the moisture from the food product treated in a processing chamber.

Simple recycling of the moist exhaust air would be ineffective because of the relatively high humidity thereof and the resultant inability of the air to absorb additional moisture from the food product.

Accordingly it has been suggested in the past to treat the air emitted from a food processing chamber by various systems to remove particulate matter therefrom, to discharge the treated air to the atmosphere, and to supply fresh relatively dry air into the processing chamber for absorption of additional moisture from the food product.

There are, however, severe problems attendant the discharge of such air into the atmosphere. The emitted air from the processing chamber not only contains moisture, but often contains excessive particulate matter, grease aerosols, and various tars and other organic substances picked up during the processing cycle. There may also be emitted toxic gases, and odors which may be considered nontoxic but which nevertheless are often considered offensive.

At the present time, federal and local governmental agencies are concerned with environmental protection, and particularly air pollution control, so that the various emissions traditionally associated with food processing chambers and especially food smokehouse operations are, or will soon be, subject to strict regulations.

Removal from Air

One method of treating the contaminated air emitted from a food processing chamber in the past has been by direct incineration, which requires the use of a separate source of energy such as natural gas or fuel oil. However such sources of fuel require afterburners or related apparatus, adding considerable costs and even providing for further pollutants from combustible fuel, in effect, exchanging one pollutant for another.

It has also been suggested to utilize wet scrubber devices to remove water-soluble and oxidizible contaminant matter from the emitted air, but these devices have proven to be ineffective and inefficient especially against submicron smoke particulates. Electrostatic precipitator devices also have been suggested, but at an initially very high cost along with the continuous problem of accumulation of grease aerosols and particulate matter requiring considerable maintenance during operation.

Separate boiler incineration may also be used for treating the emitted air from a food processing chamber but this required considerable maintenance of the burner and all of the connecting duct work which may well extend to some distance from the processing chamber, along with the necessity for insuring continuous efficient boiler operation during the cooking cycle to avoid discharging untreated air to the atmosphere.

A process developed by *C.W. West; U.S. Patent 3,805,686; April 23, 1974;* provides an air pollution control system for treating contaminated air generated by smokehouse processing apparatus. Moist air laden with impurities and odors during food treatment is withdrawn from a processing chamber, cooled to condense moisture and particulate matter out of the air, and returned to the processing chamber, without discharge into the atmosphere. The conditioned air is reheated, absorbing moisture from the food product and again fed into the system. Such a system is shown in Figure 135.

Referring now to the drawing, there is illustrated a food processing chamber **10**, in which food products, such as ham, sausage, and the like, are supported on movable racks or other devices (not shown) and a suitable heat source **11** is coupled to the processing chamber **10** for cooking the food products, in a manner well known in the art. During the cooking cycle, which is dependent upon the selected food product being processed, a smoke generator **12** may be actuated to provide a supply of smoke for delivery by way of a conduit or duct into the smokehouse processing chamber for direct contact with the food product contained therein.

Alternatively, the food product may be infused with liquid smoke to provide the desirable smokey flavor during the cooking process and without the need for generating a separate supply of smoke to contact the food product.

Referring again to the drawing, the food product is processed in the chamber **10** for a predetermined cycle which may extend from a period of only a few hours up to about 12 to 14 hours, depending on the selected product. In the illustrated embodiment, a portion of the food cooking cycle includes treatment of the food product by smoke introduced by way of the smoke generator **12**. Moist air containing various contaminants is emitted from the smokehouse chamber **10**

FIGURE 135: AIR POLLUTION CONTROL SYSTEM FOR FOOD PROCESSING APPARATUS

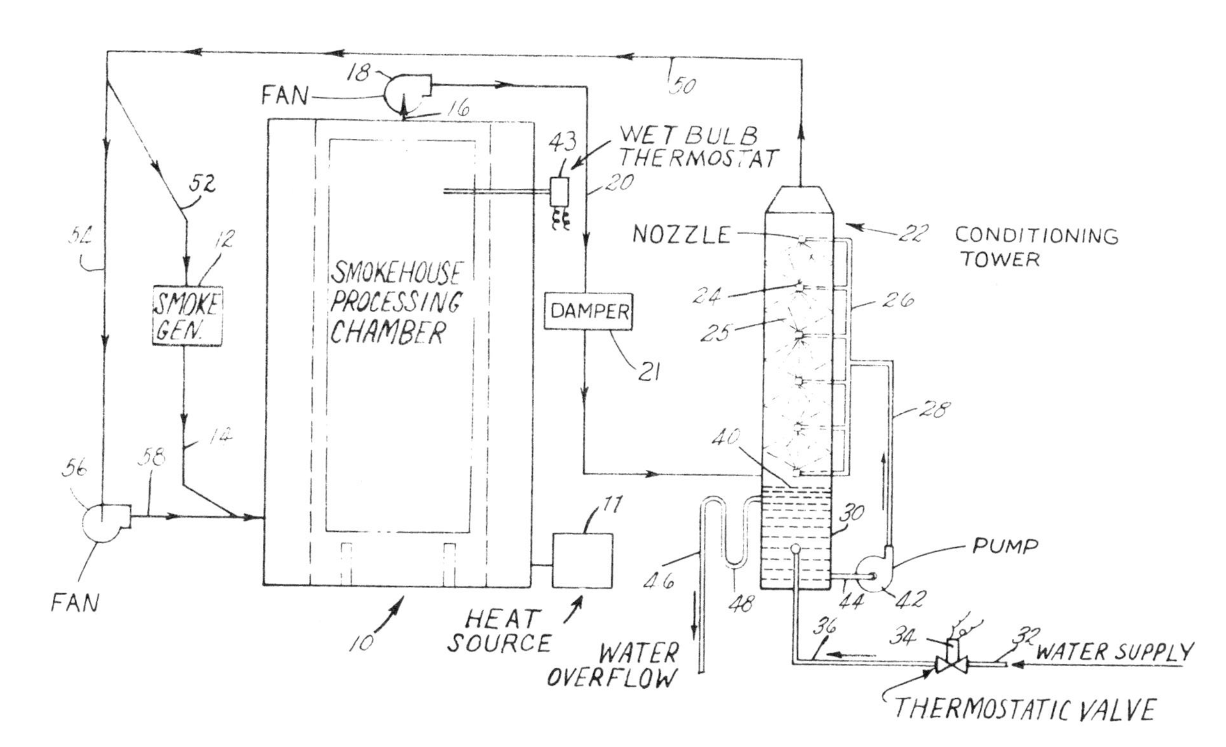

Source: U.S. Patent 3,805,686

through a duct **16** and is circulated by means of a booster fan **18** to a duct **20** for subsequent conditioning. A damper **21** may be provided in the duct **20** for modulating the flow of air to be conditioned, as will be described below.

For the purpose of conditioning the contaminated air there is provided a conditioning tower **22**, preferably in the form of an upstanding chamber, through which the contaminated air is directed. The tower is provided with a plurality of water spray nozzles **24** extending into the chamber and mounted in a vertical array, preferably with individual control valves (not shown). The nozzles are connected by suitable conduits **26, 28** to a supply of water. A water reservoir **30** is provided in the lower part of the tower **22** and is adapted to supply water to the conduits **26, 28** and thus to the nozzles **24** by means of a suitable circulating pump **42**, as will be described below.

A source of fresh water (not shown) is fed through a supply pipe **32** to a conventional control valve **34** which is connected to a suitable thermostatic control (not shown). The thermostatically controlled valve is arranged for regulating the supply of fresh water, which may be at a temperature of about 55°F, for example, through a supply pipe **36** into the reservoir **30**, in order to maintain the water in the reservoir at a preferred operating temperature. The water in the reservoir is maintained at a preselected level **40** by means of an overflow device described hereinafter.

Water from the reservoir **30** is drawn by the pump **42** through a conduit **44** connected to the reservoir **30**, the pump circulating the water through conduits **26, 28** to the spray nozzles **24** within the conditioning tower **22**. The cool water is sprayed through the nozzles **24** providing a fine shower, shown by dotted lines **25**, and collected in the reservoir **30** located therebelow. The valve **34** is selectively operated to furnish enough fresh water to maintain the water in the reservoir at a preferred operating temperature, which may be from about 80° to 90°F, for example.

The contaminated air in the duct **20** is fed into the tower **22** and passes upwardly through the water tower **25**, while the water is free to drain downwardly into the reservoir **30**. By selectively operating the pump **42**, a given supply of water can be recirculated and replenished as needed to chill the warmer air which passes through the tower to such an extent that a given amount of moisture and at least some of the particulate matter will be removed therefrom.

To this end, the pump **42** is coupled to a suitable thermostatic control such as a wet bulb thermostat **43** which is arranged to be responsive to variations in the humidity or moisture content of the air in the processing chamber **10**. Thus, the cooling effect of the water sprayed onto the warm air passing through the tower **22** can be varied and controlled in accordance with predetermined variations in the moisture content or humidity of the air within the processing chamber.

When the humidity of the air in the processing chamber exceeds a desirable amount, as measured for example by variations in the wet bulb air temperature, the pump **42** is activated to supply water to the conditioning tower for the purposes of cooling the contaminated air passing through the tower, thus reducing the temperature of that air, and consequently, lowering the humidity and removing some particulate matter.

The cooking temperatures and moisture content of the air within the processing chamber **10** will be understood to be dependent upon the selected food product being treated. A given amount of a fish product, for example, can be treated so as to emerge from the processing apparatus in a somewhat drier condition than a similar amount of ham, for example, by adjusting the temperature responsive control for the pump **42**. Accordingly, the pump **42** can be arranged to operate at any selected humidity level of the air in the chamber.

Connected to the tower **22** is a return duct **50** for recycling the conditioned air back to the smokehouse processing chamber **10**. By use of a suitable damper (not shown) the air can be selectively directed back to the smoke generator **12** by a branch duct **52** connected to return duct **50**, while at other times all or a portion of the conditioned air is conducted from duct **50** through another branch duct **54** to a booster fan **56** and a supply duct **58** which is connected to the processing chamber **10**.

It will be appreciated that the warmer air passing through the tower **22** will raise the temperature of the water used for cooling, and thus it will be necessary for the water supply control valve **34** to be opened at certain times to admit fresh water, for example at about 55°F, into the cooling system to maintain the reservoir water temperature within the preferred range of about 80° to 90°F.

A water discharge pipe **46** is connected to the reservoir **30** for overflow, particularly where fresh, cooler water is being supplied to the reservoir. The water level **40** can be controlled by a suitable valve (not shown), or may drop to the level of the discharge pipe **46**. Preferably the discharge pipe is provided with a water seal **48**, which may take the form of a conventional trap having a U-shaped configuration, to prevent discharge of gases from the tower.

The particulate matter which is removed from the air passing through the tower is collected in the water reservoir **30** and safely treated and disposed of by known methods such as by chemical treatment and/or holding tanks, and the like.

When pump **42** is operating, the warm air passing through the conditioning tower **22** is cooled to a temperature below the dew point thereof by direct contact with the fine sprays **25** of cool water, whereby moisture is condensed or wrung out of the air. The air is therefore returned at a lower temperature and with a substantially reduced capacity for holding moisture. Upon being recycled into the processing chamber, the relatively dry air is reheated and again becomes capable of absorbing additional moisture from the food product contained in the chamber. Thus the same air can be recycled over and over again, and a supply of dry outside air is unnecessary.

It will also be understood that there is no need to remove all of the particulate matter or smoke from the recycled air since it is contained in a closed system without discharge to the atmosphere. Thus since the returned air may contain smoke there is less need for generating and supplying additional new smoke into the chamber during the smoking portion of the cooking cycle. By reconditioning the air over and over, after the smoke generator is shut off, the smoke content of the air will progressively decrease until substantially no particulate matter will remain in the air at the end of the cooking cycle.

In treating bacon a typical cooking and smoking cycle would be about 7 hours total time elapsed, whereas for treating ham, the complete cycle could extend up to about 10 or 12 hours. However, whether treating bacon, ham, or other food products, the smoking portion of the processing cycle would be intermittent, and of relatively short duration, for example about 1 to 1½ hours.

When the smoking portion of the cycle is completed, and the smoke generator is shut off, the air continues to be recycled and conditioned during at least part of the remainder of the cooking cycle. The food product thus continues to absorb smoke from the returning dry air, while additional moisture is removed from the food product, so that substantially no smoke particles or residual cooking products remain in the air at the end of the processing cycle. The smokehouse chamber then can be safely opened to remove the treated food product at the end of the processing cycle without any danger of exhausting fouled air to the environment.

In one example of the conduct of this process, 600 pounds of hams were pumped with about 60 pounds or about 10% by weight of moisture known in the art as the pickle, a solution of brine, brown sugar and other additives. The hams thus treated were cooked and smoked in a smokehouse and the moisture was removed from the hams.

To indicate the efficiency of moisture removal from the air during the cooking cycle, the average dry bulb air temperature in the smokehouse was maintained at about 179°F while the average wet bulb temperature was about 153°F, the humidity in the air being about 48%. This was equivalent to a moisture content of about 0.225 pound of water per pound of air. The moist contaminated air was withdrawn from the smokehouse through the duct **20**, directed into the dehumidifying tower **22**, and subjected to the cooling water spray **25** provided by the nozzles **24**. The air was thereby chilled to below its dew point and moisture was condensed therefrom.

The conditioned air directed back to the processing chamber was found to have an average dry bulb temperature of 102°F, while the average wet bulb temperature was 101°F, or nearly 100% humidity, but containing only about 0.044 pound of moisture per pound of air. Comparing the moisture content of the air within the processing chamber before conditioning and the moisture content of the air returning to the processing chamber, it was found that a total of about 0.181 pound of water per pound of air had been removed.

The cooler and relatively drier air was again heated upon reentering the processing chamber, and due to its lower moisture content such air was capable of absorbing substantial additional moisture driven off by the food product being cooked therein. The reheated air having additional moisture added to it, or rehumidified, is then withdrawn into the treatment system for again cooling and condensing moisture therefrom and recycling.

This process was continued until the air recycled in the system was substantially smoke free. By the end of the cooking cycle the processing chamber was safely opened to remove the cooked and smoked hams, without undesirable discharge of any consequence into the environment.

SNG (SUBSTITUTE NATURAL GAS) PROCESS EFFLUENTS

Like many industrial processes, the conversion of crude oil into SNG (substitute natural gas) presents a number of environmental problems.

Removal from Air and Water

A process developed by *N.L. Carr, W.A. Roe, Jr and H.C. Stauffer; U.S. Patent 3,732,085; May 8, 1973; assigned to Gulf Research & Development Company* is one in which no sulfur (in the form of either SO_2 or hydrogen sulfide) is released by the process to the atmosphere or to a public waterway, and essentially no sulfur is present in the pipeline gas product. Accordingly, neither the process nor the product of this process contribute to air or water pollution. The only by-products of the process are relatively pure elemental sulfur, carbon dioxide and ammonium hydroxide each of which have commercial value.

For example, the ammonium hydroxide may be converted to ammonium sulfate fertilizer. The only feedstock to the process are crude oil, water and air, and there is but a single hydrocarbon product of the process, namely, a nonpolluting pipeline gas. Such a process is shown in Figure 136.

The sulfur-containing crude oil is introduced into the system by means of a line **10** into an atmospheric and vacuum distillation unit **12** wherein the crude oil is separated into a light gas fraction containing hydrogen sulfide and C_1 to C_4 gases, which light stream is discharged from the distillation unit by means of a line **14** to a gas treatment plant hereinafter described.

Likewise, a light stream boiling in the C_5-375°F range (e.g., a naphtha fraction) is withdrawn by means of line **16**, while a heavy stream boiling in the 375° to 1040°F range (e.g., the gas oil-furnace oil range) is withdrawn by means of a line **18**. The bottoms fraction (e.g., 1040+°F) is discharged from the bottom of tower **12** by means of line **20** and is treated as will be hereinafter described.

The 375° to 1040°F boiling range fraction is introduced by means of line **18** along with hydrogen from line **22** into hydrocracker unit **24**. Hydrocracker **24** converts the gas oil fraction to a lighter, naphtha fraction, for example, in the C_5-350°F boiling range and also converts the sulfur present to hydrogen sulfide, while converting organically-bound nitrogen to ammonia. The hydrogen sulfide is withdrawn along with C_1 to C_4 hydrocarbons by means of line **26** and the ammonia is withdrawn in the form of ammonium hydroxide from line **27**, while the product naphtha fraction is discharged from the hydrocracker unit **24** by means of the line **28**.

The naphtha fraction from hydrocracker **24** is passed by means of line **28** along with hydrogen gas introduced by means of the line **30** to a naphtha desulfurizer unit **32** wherein the naphtha and hydrogen circulate over a desulfurizing catalyst in order to desulfurize the naphtha and, in addition, saturate any olefins present in the naphtha stream. The naphtha can be fed directly to the gasification plant if its sulfur (organic) content is only about 1 ppm. An advantage of this process is that only a very small amount of olefins are produced. Significant amounts of olefins cannot be tolerated in the pipeline gas product.

FIGURE 136: THERMALLY EFFICIENT NONPOLLUTING SYSTEM FOR PRODUCTION OF SUBSTITUTE NATURAL GAS

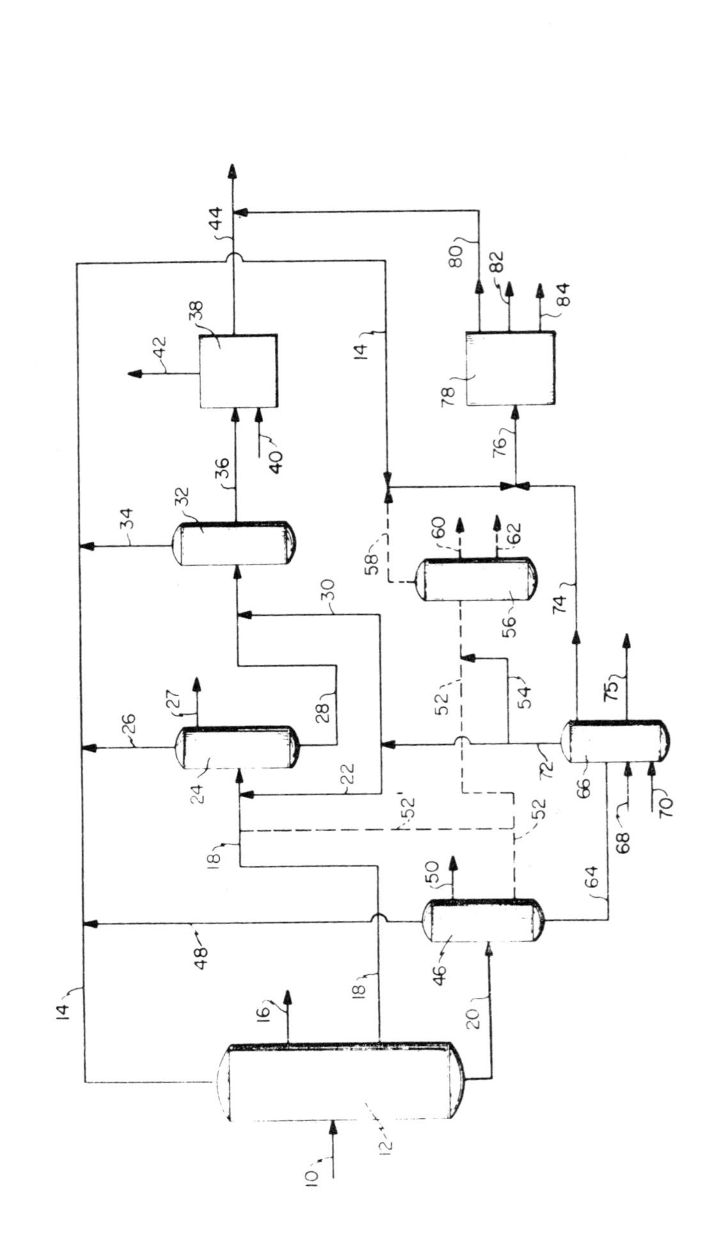

Source: U.S. Patent 3,732,085

The purified naphtha stream is discharged from the desulfurizer **32** by means of line **36** and is introduced into a substitute natural gas plant **38** wherein the naphtha is admixed with steam introduced by means of line **40** and converted along with some C_3 and C_4 gases obtained from another part of the system to a sulfur-free, methane-rich fuel gas and carbon dioxide. The pure by-product carbon dioxide is discharged from the substitute natural gas plant **38** by means of the line **42**, while the product methane-rich gas is discharged by means of the line **44**.

The bottoms fraction **20** is passed to a visbreaker unit **46**. The 1940+°F bottoms fraction line **20** is subjected to thermal cracking in the visbreaker **46** wherein the oil passes through heated coils and a portion thereof is cracked therein to lower molecular weight hydrocarbons. In this manner, additional light gases and naphtha suitable for conversion to additional methane are produced. The use of a visbreaker in the manner described is significant, since the volume of tar passed to the partial oxidation unit (hereinafter described) is reduced thereby reducing the oxygen requirements for the oxidizer.

Light gases including hydrogen sulfide and C_1 to C_4 hydrocarbons are withdrawn from the visbreaker **46** by means of line **48**, while a naphtha fraction boiling in the C_5-400°F range is withdrawn by means of a line **50** and is passed to naphtha desulfurizer **32** (by a means not shown).

A fraction boiling in the range of 400° to 950°F is withdrawn from the visbreaker unit **46** by means of a line **52** and is introduced along with hydrogen introduced by means of line **54** to a desulfurizer unit **56**.

The desulfurizer **56** may employ a noncracking catalyst similar to that employed in the naphtha desulfurizer **32**. Suitable conditions which may be employed in the desulfurizer **56** include temperatures in the range of between about 600° and 850°F, preferably between about 675° and 775°F, while suitable pressures include, for example, between about 750 and 2,000 psig, preferably between about 800 and 1,200 psig. The feed to the unit **56** may be passed therethrough at a space velocity of between 0.5 and 5 LHSV, preferably between 1 and 3 LHSV. Hydrogen is employed at a rate of between about 1,000 and 10,000 scf/bbl. Preferably between about 2,000 and 5,000 scf/bbl is used.

A light gas stream **58** comprising hydrogen sulfide and C_4 hydrocarbons is withdrawn from the unit **56**, while a fraction boiling in the range of C_5-400°F is withdrawn by means of line **60** and may be passed to the naphtha desulfurizer unit **32** by a means not shown. A stream boiling in the range of 400+°F is discharged from the desulfurizer **56** by line **62** and this stream may be suitably employed as refinery fuel. This stream may constitute, for example, up to about 10% by volume of the crude oil feedstock.

Alternatively, at least a portion of the 400° to 950°F stream that is passed by means of line **52** to the desulfurizer unit **56** may be passed instead directly to the hydrocracker unit **24** by means of the line **52′** and may be cracked therein to form additional naphtha and to increase the thermal efficiency of the process.

The desulfurizer **56** may be omitted altogether and all of the material in the stream **52** may be diverted to the hydrocracker unit **24** by means of the line **52′**. In this instance, refinery fuel may be drawn from the substitute natural gas product stream in line **44** or may be obtained as a portion of the heavy effluent from the hydrocracker.

According to another important aspect of this process, a residue pitch or coke fraction **64** is discharged from the bottom of the visbreaker unit **46** and is passed to a partial oxidizer unit **66** along with air and water which are introduced by means of the lines **68** and **70**. Thus, portions of the crude oil which are finally provided in the form of either pitch or coke, which materials have the lowest economic value of any process fraction, are converted to useful hydrogen and to carbon dioxide.

The sulfur present in this high boiling fraction is converted to hydrogen sulfide, while any organic nitrogen present is converted to ammonia. In this manner, the hydrogen requirements for the desulfurization, denitrification, and hydrocracking of the lower boiling portions of the crude oil are provided from the portions of the crude having the lowest value, thus rendering the process autogenous in hydrogen requirements and economically desirable. No extraneous hydrogen source is required to supply the large amount of hydrogen needed to convert the various streams to the ultimate product, viz, a methane-rich pipeline gas.

The hydrogen in line **72**, which is withdrawn from the oxidizer **66** is passed to process lines **22**, **30** and **54** to supply the various hydrogen requirements in the units **24**, **32** and **56**. The partial oxidizer unit **66** may be operated at temperatures, for example, in the range of between about 2000° and 3000°F, while employing pressures in the range of between about 400 and 1,500 psig at suitable residence times.

Hydrogen sulfide is withdrawn by means of line **74** and is passed to a line **76** wherein it is joined by other light gas streams **14**, **26**, **34**, and **58** from other units in the system. The combined light gas streams are introduced into the gas treatment plant **78** for removal of hydrogen sulfide by a conventional scrubbing unit therein. Sulfur-free hydrogen, and C_1 and C_2 gases are withdrawn by means of the line **80** and are passed to the line **44**, while sulfur-free C_3 and C_4 gases are withdrawn by means of the line **82** and are introduced into the substitute natural gas plant **38** along with the feed-in line **36** (by a means not shown). Elemental sulfur is recovered from the plant **78** and is withdrawn by means of the line **84**.

Substantially pure carbon dioxide is withdrawn from the oxidizer **66** by means of the line **75** and combined with carbon dioxide which is withdrawn from the SNG plant by means of the line **42** and recovered therefrom.

SODIUM CARBONATE

Removal from Air

A process developed by *C.J. Howard et al; U.S. Patent 3,634,999; January 18, 1972; assigned to Allied Chemical Corp.* is one in which dust and fines issuing from trona processing systems in the manufacture of sodium carbonate are removed by scrubbing the dust laden gases with an aqueous scrubbing solution under conditions sufficient to remove the fines from the gases and preferably effect particle growth of the dust and fines in the scrubbing solution. The solids of larger particle size are separated and returned to the trona processing system.

SODIUM HYDROSULFITE PROCESS WASTES

In the process for manufacturing sodium hydrosulfite via the sodium formate process, sulfur dioxide and sodium hydroxide are used as raw materials in addition to sodium formate. The reaction is effected in a medium consisting predominantly of methyl alcohol but containing substantial quantities of water. As a result of intermediate reactions, side reactions, and the incomplete utilization of raw materials, the reaction medium at the completion of the synthesis contains, in addition to methyl alcohol and water, substantial quantities of dissolved sodium formate, methyl formate, sodium bisulfite, sodium sulfite, sodium thiosulfate and possibly some other salts. The product, sodium hydrosulfite, is completely insoluble in this medium and thus forms a slurry with it. The separation of the two, the sodium hydrosulfite and the reaction medium is accomplished by filtration.

Economic considerations dictate that the filtrate resulting from this separation be subjected to a distillation process for the recovery of the very substantial amount of methyl alcohol which it contains. In this distillation, the methyl formate is recovered along with the methyl alcohol since the methyl formate is equally as volatile as the methyl alcohol.

The bottoms, or residue, remaining when the distillation is completed consists of the water originally contained in the reaction medium plus the previously mentioned soluble salts, sodium formate, sodium bisulfite, sodium sulfite and sodium thiosulfate.

Disposal of this residue presents a very serious pollution problem in that all of the dissolved substances are reducing agents and would thus create a chemical oxygen demand in any body of water into which they might be discharged. Chemical oxidation prior to disposal is difficult and extremely expensive. More attractive is an economical method of recovery of the chemical values in this residue, and subsequently reusing or selling the valuable compounds thus recovered.

Removal from Water

A process developed by *C.E. Winslow, Jr., J. Plentovich and M.A. Kise; U.S. Patent 3,718,732; February 27, 1973; assigned to Virginia Chemicals Inc.* is one in which chemical values from a formate-sodium hydrosulfite reaction medium may be recovered after filtration and removal of the sodium hydrosulfite by introducing sulfuric acid into the reaction medium, boiling gently so as to expel sulfur dioxide, continuing boiling and distilling with reflux of methyl alcohol and methyl formate, filtering the reaction medium residue thereby removing sulfur. The reaction medium residue may then be evaporated, so as to separate sodium sulfate from water. The recovered chemical values may then be recycled or sold on the open market.

SODIUM MONOXIDE

When diaphragms are being replaced in cells for the electrolysis of fused mixtures of sodium chloride and calcium chloride for the manufacture of sodium and chlorine a fume is obtained containing fine particles of sodium monoxide together with a little calcium chloride and sodium chloride and a little chlorine.

Again a fume containing fine particles consisting almost entirely of sodium oxide is obtained when sodium residues are being disposed of in burner bays. Such fumes are very objectionable and numerous attempts have been made to try to remove such fume effectively. Thus the fume has been passed through a filter composed of slag wool fibers but it was found that such a filter quickly blocked with collected solids. This filter was also attacked by small quantities of chlorine which were present in the fume.

Removal from Air

A fiber filter composed of very fine untreated glass fibers (wettable fibers) will not effectively remove fumes of sodium and potassium oxide particles entrained in air since high pressure drops are required and in any case the filter soon blocks. Moreover such filters cannot be irrigated with a liquid such as water to remove deposited solids as it very quickly logs and becomes useless as a filter medium.

A process developed by *G.L. Fairs et al; U.S. Patent 3,135,592; June 2, 1964; assigned to Imperial Chemical Industries, Ltd.* is based on the discovery that when using a nonwettable fiber filter, suitably one composed of Terylene polyester fibers or one composed of glass fibers having an adherent silicone surface, which is continuously irrigated with water, these filters were completely effective in removing sodium monoxide fumes.

Moreover when using such filters the process of removing the particles may be operated at relatively low pressure drops. While such filters can be used to remove extremely fine liquid and solid particles of extremely fine particle size, for instance of size less than 5 μ, they may also be used for the removal of particles of considerably greater size, for instance 25 μ or more.

SODIUM SULFUR OXIDE WASTES

Removal from Water

In preparing water for use as boiler feedwater, nuclear reactor coolant water, spent fuel storage water, or sump wastewater are decontaminated or demineralized to remove various components, including borates, calcium, sodium, silica, radioactive components, various anionic components including sulfite and sulfate ions and the like, since presence of these components cause problems in the operation of the boilers or in waste disposal.

Ambient water supplies have normally been treated by a series of demineralizers or ion exchange resins. These resins are periodically backwashed or cleaned and result in a backwash containing large amounts of sodium sulfur oxide components normally assayed as sodium sulfate. For example, such demineralizer wastes can contain from 15 to 25,000 parts per million of sodium sulfate. The wastes also may contain other ionic components, heavy metals, radioactive components and the like which pose disposal problems. Typical heavy metals include iron, cobalt, copper, manganese, and radioactive components include radioactive potassium, ^{137}Cs and ^{131}I.

Still another source of sodium sulfate is that contained in the blowdown from nuclear or fossil fuel fired power plant cooling towers. A 2,000 megawatt nuclear station employing cooling towers typically will produce cooling tower blowdown sludge or magma in the amount of approximately 3.4 tons per hour of 90% solids material which, prior to concentration, assays approximately 8,880 ppm of Na^+ and $SO_4^=$ in the ratio of 2:1. The vastness of the quantity of this sludge material is evident in the fact that approximately 15 tons per megawatt are produced each year.

Far more serious is the potential for pollution from the use of sodium alkali materials, such as sodium hydroxide, sodium carbonate, sodium bicarbonate, sodium sulfite, and the like for SO_2 control in wet or dry scrubbing of SO_2 from tail or flue gases from industrial plants, smelters, paper plants, glass plants, power plants and the like which burn sulfur-containing fossil fuels such as coal or oil. The magnitude of the problem can be seen from the fact that it is estimated that from 20 to 35 million tons of sulfur dioxide were vented to the atmosphere in the United States in 1972 from industrial plants and power plants by burning fossil fuels containing bound sulfur.

While a principal approach to SO_2 control has been wet scrubbing of tail gases with lime or limestone, these processes involve several disadvantages. The energy required to pump the water through the scrubber is relatively high and the scrubbers are prone to scale formation. Normally, there must be recycle of unreacted alkaline calcium compounds and the liquid/gas ratio must be relatively high. The inverse solubility and scaling problems lead to high capital and operating costs due to complex piping, demisters, surface contact areas and the like.

Still further, the calcium scrubber sludge containing calcium sulfite and sulfate poses very expensive and serious disposal problems. Normally, it is produced at the rate of approximately one acre foot per megawatt per year. The sludge material is the "fourth" state of matter, being a thixotropic mixture of finely divided crystalline particles and water. Normally, the material contains from 40 to 60% water and will not settle or completely dry out at ambient conditions.

Its disposal involves the cost of transporting the water into arid areas where water is precious or unavailable. In addition, sludge cannot be piled up above ground since the pile will flow under its own weight. This has led to the use of flocculants and scrubber ponds, which are typically clay-lined ponds, or pits scooped out of clay formations.

However, calcium sulfate has a tendency to break down clay and render it relatively permeable. This then permits the soluble constituents of the scrubber sludge to leach from the pond. Normally, the starting limestone contains from 10 to 50% magnesium carbonate which, in the scrubber SO_2 removal reaction, produces highly soluble (700 g/l) magnesium sulfate (epsom salt) pollutant which can leach from the pond due to calcium sulfate induced permeability. Calcium sulfate/sulfite sludge disposal costs were running about $20/ton in the early 70's.

In addition, the tail gas clean-up processes also collect fly ash and heavy metals which are originally present in the coal. The heavy metals, among them beryllium, boron, cadmium, strontium, magnesium, arsenic, chromium, barium, cobalt, fluorine, mercury, manganese, nickel, tin, selenium, vanadium, lead, radioactive elements such as ^{226}Ra, ^{228}Ra, ^{228}Th, ^{230}Th, and ^{232}Th, are present in various quantities in the coal. These are collected in the scrubber sludge and may be leached from the ponds.

To overcome the problems of using calcium in scrubbers, and the high capital and operating costs, there have been proposed both wet and dry systems for use of sodium alkaline compounds. The sodium compounds have the advantage of increased reactivity and a lower liquid to gas ratio due to the fact that both the alkalis and end-product sodium sulfur oxide compounds are highly soluble in the scrubber water. The scrubbers employing sodium alkaline compounds are less complex primarily because no substantial attention need be paid to the problem of scaling in the scrubber.

Dry sodium processes, principally involving the use of sodium bicarbonate injected as a dry powder upstream of a baghouse have been successfully tested on full-scale sized pilot baghouse operations for removal of SO_x and particulates from oil and coal-fired power plants. The resultant baghouse filter cake material is a mixture of fly ash containing residual heavy metals, and the sulfur oxide compounds, principally sodium sulfite and sodium sulfate. The dry baghouse cake is periodically removed from the baghouse hoppers.

A hybrid process involves spraying a solution of sodium bicarbonate or carbonate at a rate which permits drying after reaction and collection of fly ash and sodium sulfite/sulfate in a cyclone and/or baghouse and/or electrostatic precipitator. Mixtures of molten alkali metal carbonates are used as SO_x sorbents, producing Na, K, Li sulfites/sulfates.

However, while it would be of great advantage to adopt the use of sodium alkalis, either wet or dry for SO_x air pollution control in order to take advantage of the increased efficiency as compared to calcium systems, the major problem is the fact that the resultant sodium wet scrubber liquor or the dry sodium sulfur oxide-containing waste material is water-soluble. The net result is that utilities and industrial users are left with a water pollution problem after having solved their air pollution emissions problem.

Although the wet or dry sodium alkali SO_2 emissions control processes are generally cheaper than calcium or MgO systems, the disposal of the resultant sodium sulfur oxide wastes has been a major barrier to the adoption of such systems as a solution to SO_x and particulate emissions problems.

Therefore, there is a very great need for a process which will result in reducing the solubility of the SO_2 emissions control wastes or the solubility of waste sodium sulfite and sulfate from various types of industrial processes. There is also great need for a process which will simultaneously dispose of two or more of the fly or bottom ash residuals from the burning of fossil fuels such as coal or oil, potentially toxic heavy metals, or radioactive elements, which are normally present as the initial components of fossil fuels such as coal or oil or are produced in the energy-generating process (nuclear reactor power plants). There is also the need for a process which will reduce the leaching or result in the fixing or partial fixing of heavy metals, radioactive elements, and sodium sulfur oxide components permitting the thus-fixed product to be cheaply disposed of, by landfill or useful as an aggregate product (where radioactivity is nil or low).

A process developed by *J.M. Dulin, E.C. Rosar, R.B. Bennett, H.S. Rosenberg and J.M. Genco; U.S. Patent 3,962,080; June 8, 1976; assigned to Industrial Resources, Inc.* involves mixing sodium sulfur oxide wastes with an alumina and silica-containing fly ash and/or bottom ash, forming an agglomerate, such as a pellet or briquette, and sintering in the range of about 1000° to 2300°F.

The resultant sintered particle shows lower solubility than the current standard of calcium sulfate, being 10^{-2} to 10^{-4} as soluble as the starting sodium sulfur oxides, has increased density (as high as 2.3 g/cc), and reduced volume as compared to the dry fly ash or sodium sulfur oxide wastes, and fixes heavy metals. Fly ash leaching and dusting problems are substantially overcome. Sintered particle may be disposed of by known landfill techniques, or used as an aggregate for mulch, road beds, concrete, asphalt or the like.

See Cooling Tower Blowdown Effluents (U.S. Patent 3,876,537)

SOLVENTS

Removal from Air

It is quite customary to effect either thermal or catalytic incineration of fume laden waste gas streams prior to their discharge into the atmosphere. For example, in connection with ovens used for the drying of protective lacquer coatings on containers, metal sheets, etc., there is a high production of vaporized solvents which, if allowed to escape to the atmosphere, would cause a high degree of contamination in the area of the drying oven. In effecting a catalytic incineration of the combustible fumes there is a release of useful heat energy which, of course, should be recovered where possible.

A process developed by *E.C. Betz; U.S. Patent 3,486,841; December 30, 1969; assigned to Universal Oil Products Company* is one in which a liquid heat transfer medium is used in a closed and pressured heat exchanger section of a heat recovery system for integration with catalytic oxidation of an oven discharge stream in order to provide a form of heat-sink and a more uniform heat release to a circulating gaseous stream. The system is of particular advantage for use with a discharge stream having a cyclical release (or varying quantity) of oxidizable volatile materials.

Equipment developed by *C.R. Wilt, Jr. and F.L. Schauermann; U.S. Patent 3,868,779; March 4, 1975; assigned to Salem Corporation* includes apparatus for incinerating combustible solvents evaporated in an oven of a paint drying conveyor system to thereby effectively eliminate emission of obnoxious pollutants to the atmosphere, as well as to dilution control apparatus for regulating solvent concentration in the oven to a safe value.

See Linoleum Manufacturing Effluents (U.S. Patent 3,618,301)
See Lithographic Printing Process Effluents (U.S. Patent 3,768,232)

Removal from Water

A device developed by *J.D. Hummell; U.S. Patent 3,748,081; July 24, 1973; assigned to PPG Industries, Inc.* is adapted to efficiently dispose of combustible liquid wastes of various types, particularly liquid organic wastes containing solvents, resins, paint sludges, chemical residues, or the like. It is useful with essentially any liquid containing appreciable proportions of combustible organic materials, including aqueous wastes containing up to 70% or more of water. Such a device is shown in both plan and elevation in Figure 137.

FIGURE 137: INCINERATOR FOR DISPOSAL OF AQUEOUS SOLVENT-CONTAINING WASTES

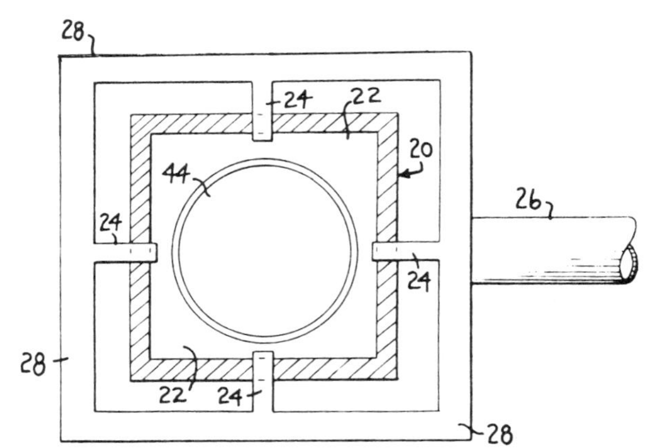

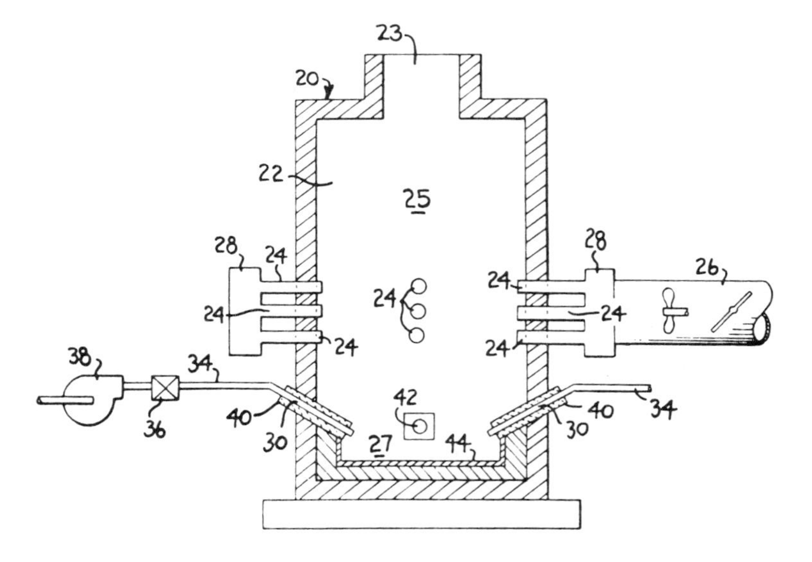

Source: U.S. Patent 3,748,081

Referring specifically to the drawings, the apparatus includes a housing, indicated generally at **20**, which forms a chamber **22** having an upper combustion zone **25** generally defined by the upper portion of housing **20**.

A plurality of air inlet ports **24** are provided in the side walls of housing **20**, in the combustion zone, and are communicated via air duct **28** to a regulated source of combustion air, such as a conventional blower diagrammatically illustrated at **26**.

A pair of fuel inlet passage means **30** are provided in the lower portion of housing **20** and are communicated via pipe lines **34** and valve means **36** to a conventional pump **38** which in turn is communicated with a storage tank or source of liquid wastes, not shown.

Pump **38** and valve means **36** mainly function to meter the flow of liquid wastes into the incinerator as the liquid wastes are not required to be preheated nor are high pumping pressures required.

The dimensions of inlet passages **30** although not critical are relatively large as compared to restricted orifices or the like which function to divide the inlet fluid into a spray as employed in prior liquid waste disposal apparatus. Generally substantially all liquid wastes commonly incurred could function with merely a gravity feed and pump **38** is used primarily to control the flow rate of liquid entering the housing.

Inlet passages **30** are preferably protected from the heat generated in the combustion chamber by a surrounding insulating material **40** to maintain the temperature of passages **30** low enough to prevent reaction of any of the materials of premature vaporization of the volatile constituents in the liquid waste being fed into housing **20**. Suitable insulating materials include water, air, ceramics, etc.

Fuel ignition means, such as auxiliary burner **42**, is provided adjacent to the vaporizing surface **44**, which is in the bottom portion of chamber **22**. The vaporizing surface is usually the bottom of the housing along with the lowermost portion of the sidewalls. This area of the chamber **20**, having the fuel inlets and the vaporizing surface, forms vaporizing zone **27**.

Vaporizing surface **44** is preferably heated initially by burner **42** to cause an increased rate of vaporization of the volatile constituents present in the thin liquid layer of fuel on the vaporizing surface. The combustible vapors are then ignited by the burner **42** as they become mixed with the combustion air entering inlet ports **24**. This permits rapid and smoke-free start-up even with liquid waste fuels having relatively low volatility.

After the start of the operation, the heat generated in the upper portion of housing **20** by the burning fuels and radiated down to the surface of the liquid waste and the vaporizing surface is generally sufficient to provide rapid enough vaporization of the volatile constituents to maintain efficient smoke-free combustion. Then auxiliary burner **42** is no longer needed.

See Nonhalogenated Solvent Wastes (U.S. Patent 3,980,417)

SOOT

See Partial Oxidation Process Effluents (U.S. Patent 3,694,355)

SOUR WATER

As is well-known, water which contains hydrogen sulfide and other intermediate sulfur compounds such as thiosulfate, tetrathionate, polythionate and polysulfide originates from many sources. A number of processes used in refining petroleum produce water effluents which contain high concentrations of hydrogen sulfide and ammonia. These waters are typically called foul or sour water. Typical operations producing sour water include such widely practised processes as crude distillation, hydrotreating, catalytic cracking, thermal cracking, delayed coking and hydrocracking.

Ammonia also is usually present in sour water streams either because it has been added to neutralize the H_2S for corrosion control or as the result of hydrogenation of nitrogen during the refining process. The H_2S and ammonia react to form ammonium sulfide and ammonium hydrosulfide, depending on the pH of the water and, if free sulfur is present, polysulfides. Normally the pH of the refinery sour water is approximately 9.0 and the sulfides are present as hydrosulfide ion HS^-.

Removal from Water

A process developed by *D.E. McCoy, R.M. McEachern and R.M. Dille; U.S. Patent 3,761,409; September 25, 1973; assigned to Texaco Inc.* is a continuous process for the liquid phase, noncatalytic, air oxidation of sour water. The sulfur compounds in the water are converted to nonpolluting, nonoxygen demanding compounds by heating the water and oxidizing the heated water with countercurrent or cocurrent flow of the water and oxidizing medium. The process rapidly converts objectionable sulfides and intermediate sulfur compounds such as thiosulfates, tetrathionates, polythionates, sulfites and polysulfides to sulfates.

A process developed by *R.D. Kent; U.S. Patent 3,754,376; August 28, 1973; assigned to Texaco Inc.* is one in which water containing contaminating gases such as hydrogen sulfide, carbon dioxide and ammonia is freed of these contaminants by a closed system stripping process which employs an inert gas and steam.

As shown in Figure 138 the sour water **10** is introduced near the top tray of a conventional stripping tower **11**. Inert gas enters the stripping tower at or near the bottom at **12**. Steam is also introduced at or near the bottom of the stripping tower at **13**. Alternately, heat may be applied to the bottoms by use of a reboiler or the like to generate steam from the bottoms liquid. The water stripped of contaminants leaves the bottom of the stripping tower at **14**.

The inert gas, steam, and stripped contaminants are routed through a condenser and reflux system **15** to increase the efficiency of the stripping tower. The inert gas and stripped contaminants (including hydrogen sulfide, ammonia and carbon dioxide) leave the stripping tower and the reflux accumulator drum and are

FIGURE 138: CLOSED CYCLE INERT GAS STRIPPING WITH SINGLE ABSORBER FOR AMMONIA AND HYDROGEN SULFIDE

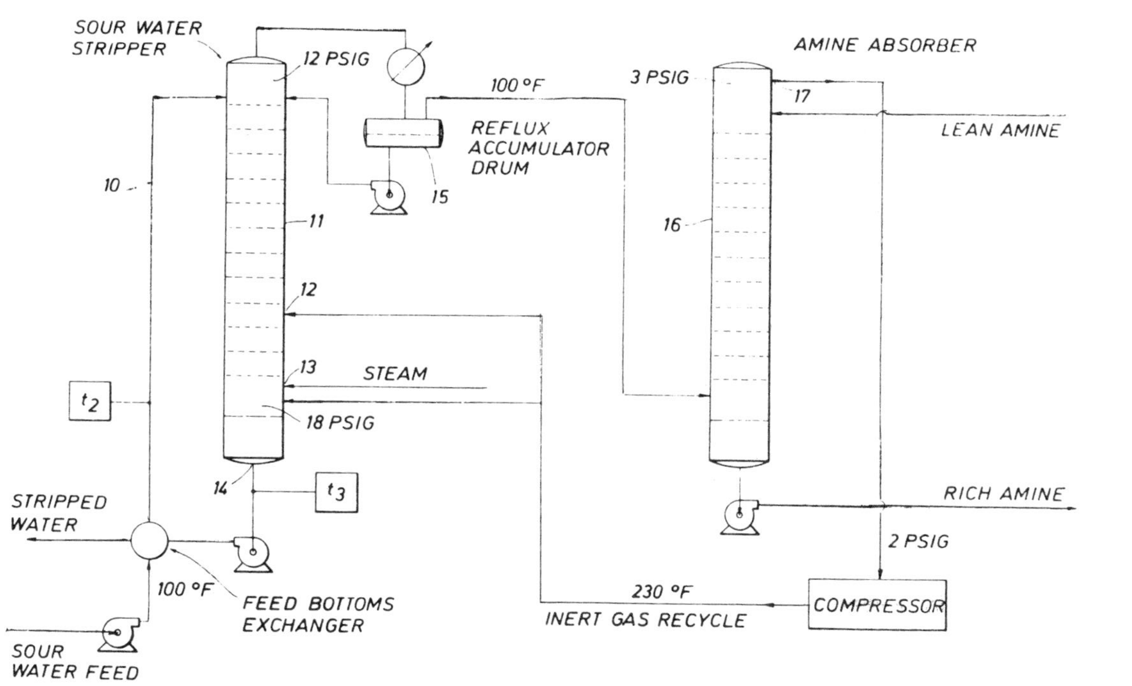

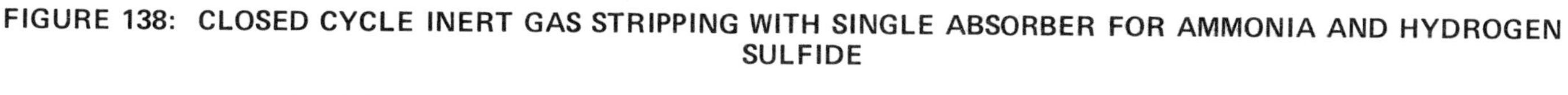

Source: U.S. Patent 3,754,376

passed through an absorber **16** wherein the hydrogen sulfide and ammonia are removed by a suitable solvent. The inert gas now relatively free of contaminants leaves the absorber at **17** and is reintroduced by means of a compressor to the original stripping tower at **12** and the process is repeated.

Most of the current processes for removing ammonia and hydrogen sulfide from foul refinery wastewaters require the use of steam as a stripping source. A process which will reduce the need for a steam stripping source is thus highly desirable because of energy shortages and resultant high steam costs. Petroleum refineries usually produce a quantity of residual gas which is available at pressures of between 100 and 220 psig. Ordinarily the energy contained in this residual gas is not utilized because the gas is released to a header operating at 40 to 80 psig. This gas subsequently is burned in various furnaces.

A process developed by *G.I. Worrall, D.A. Strege and G.D. Myers; U.S. Patent 3,984,316; October 5, 1976; assigned to Ashland Oil, Inc.* is one in which refinery wastewaters containing dissolved hydrogen sulfide and ammonia are stripped of hydrogen sulfide by a countercurrent stream of refinery absorber gas which removes substantially all of the hydrogen sulfide and none of the ammonia. The dissolved ammonia is subsequently removed by steam stripping, leaving a wastewater sufficiently clean as to be nonpolluting.

A process developed by *E. Ruschenburg; U.S. Patent 3,804,757; April 16, 1974; assigned to Deutsche Texaco AG, Germany* is a process for treating sour water which contains for example, hydrogen sulfide, ammonia and phenol in the presence of nonvolatile and strongly alkaline ions. An ammonium salt of a strong mineral acid such as ammonium sulfate is added to the contaminated wastewater and the ammonia and hydrogen sulfide are stripped away. The phenol separates as an oily layer and may be removed. The ammonium sulfate needed for the process may be generated during the process by treating the stripped ammonia with sulfuric acid.

A process developed by *G.C. Lahn; U.S. Patent 3,853,744; December 10, 1974; assigned to Exxon Research and Engineering Company* is one in which sour water from fluid solids hydrocarbon conversion processes, such as catalytic cracking and fluid coking, is disposed of by vaporization in a fluid solids bed. Heat for the vaporization is supplied by circulating hot solids between the steam generation vessel and the coker burner or heating unit in a fluid coking process or the regenerator in a catalytic cracking process. This process is particularly useful in the residuum conversion process which integrates coke gasification with fluid coking.

See Hydrogen Sulfide (U.S. Patent 3,754,376)
See Sulfides (U.S. Patent 3,725,270)

SPENT SULFURIC ACID

Removal from Water

A process developed by *S.J. Shirley; U.S. Patent 3,888,653; June 10, 1975; assigned to Shamrock Chemicals, Ltd., Canada* for a fertilizer (potassium sulfate) utilizes "spent" sulfuric acid (40 to 98% H_2SO_4) and potash (potassium chloride) according to the following chemical equation:

$$H_2SO_4 + 2KCl \longrightarrow K_2SO_4 + 2HCl\uparrow$$

SPRAY BOOTH EFFLUENTS

Spray paint booths ordinarily have large exhaust fans to remove hydrocarbon solvents as they evaporate from the sprayed paint. Heretofore the mixture of air and hydrocarbons, referred to herein as dirty gas, was simply exhausted to the atmosphere.

Removal from Air

More recently afterburners have been required in paint booth installations to meet pollution code requirements. These afterburners must be capable of heating the dirty gas to the auto-ignition temperature of the hydrocarbons and maintaining this temperature for a sufficient dwell or residence time to insure that all of the hydrocarbons will burn to water vapor and carbon dioxide which are commonly considered as clean gases and are not detrimental to the environment.

Pollution codes establish auto-ignition temperatures and dwell times at levels sufficient to insure thorough hydrocarbon combustion and clean gas. Typical codes for afterburners applied to paint spray booths require that the gases be heated to approximately 1400°F for a residence time of at least 0.2 to 0.5 sec. These operating limits require afterburners having high capacity gas burners which use relatively large amounts of gas.

Heretofore, many afterburner installations have simply exhausted the heated clean gases out a stack to the atmosphere without any attempt to recover the heat expended. Such installations must use burners requiring a constant gas flow input to provide the necessary rise to 1400°F.

Some afterburner installations have placed heat exchangers in the stack after the gas has passed through the burner to recover some heat from the clean gases which is then used to preheat the dirty gas before passing through the burner. However, these installations are generally inefficient, recovering at best up to only 50% of the heat discharged.

A somewhat more efficient heat recovery system circulates the dirty gases through ductwork in the stack beyond the burner, thereby preheating the dirty gas prior to passing it through the burner. However, these heat exchangers tend to plug from the impingement of carbonaceous materials against hot surfaces making them difficult to clean and expensive to operate.

An apparatus developed by *H.E. Shular and J. Sellors, Jr.; U.S. Patent 3,881,874; May 6, 1975; assigned to Pyronics, Inc.* is one in which control of air pollution from organic solvents issuing from paint spray booths is achieved by a device for burning air-entrained hydrocarbons. The device comprises a housing, an elongated passageway, a burner in the center of the passageway dividing the passageway into two alternating combustion chambers, honeycombed firebrick at the remote end of each chamber for absorbing heat from the exiting gases and means for periodically reversing the flow of air-entrained hydrocarbons through the passageway whereby heat absorbed by the firebrick at one end of the passageway preheats the entering gases when the flow is reversed.

See Surface Coating Effluents (U.S. Patent 3,861,887)

STARCH WASTES

In many industries it is conventional practice to employ various starch compositions for adhesive purposes, and among these may be mentioned the corrugated board and box industries where starch compositions are employed for sealing or laminating two or more pieces or piles of corrugated board or box components together during the course of their manufacture.

These components may be paper, cardboard, plywood, etc. The waste starch residues that result from the manufacturing operations in these industries pose serious problems of waste control and disposal due to the physical and chemical characteristics of the starch adhesive refuse materials.

Removal from Water

Normally this starch refuse material is directed into a pit or tank from the corrugators or other starch composition utilizing apparatus, together with an indeterminate amount of flushing water, grease, oil and/or other liquid debris. On standing, this material becomes compacted and doughy or sludge-like in consistency with concomitant microbial and bacterial action, resulting in odors of decay rendering its removal problem even more difficult.

Since many municipalities have imposed severe restrictions upon the dumping of such malodorous and otherwise objectionable waste materials into streams, it is frequently impossible to dispose of this waste residue in such fashion. Moreover, even where the plant may be in a relatively isolated location with perhaps few, if any, legal restrictions upon the method of disposal of the waste, nevertheless to attempt to dispose of the waste in such manner still gives rise to serious problems.

Heretofore, removal of this waste residue was effected periodically such as approximately every month by means of manual excavation from the pit or tank, transfer of the refuse to an isolated area and burial of same. This, however, entails considerable expense in terms of labor costs, the procurement of permits from municipal authorities and general inconvenience to all concerned.

A process developed by *J.T. Gayhardt; U.S. Patent 3,037,931; June 5, 1962* involves the disposal of wastes derived from starch compositions employed as adhesives in certain industries, with particular reference to the corrugated board and box industries.

The process comprises mixing the residues with a treating composition comprising cyclohexylamine, an alkyl phenyl polyethylene glycol ether and tetrakis (2-hydroxypropyl) ethylene diamine in order to disperse components of the sludge-like residues therein and running the resulting mixture to waste in the normal sewage system of an industrial plant.

See Paper Box Plant Effluents (U.S. Patent 3,868,320)

STEAM-ELECTRIC INDUSTRY EFFLUENTS

One of the many sources of air pollution is the flue gases emitted from fuel burning equipment such as steam generating units. The sulfur oxides, SO_2 and SO_3, are of major concern as air pollutants in such flue gases. The particulate matter such as fly ash and other dust particles also contribute to the pollution problem if not completely removed.

Removal from Air

One system for cleansing flue gases of their sulfurous and particulate impurities utilizes a wet scrubber in which intimate contact of the flue gas with wash water effects the purification of the gas. Such a process has been described by *H.E. Burbach; U.S. Patent 3,726,239; April 10, 1973; assigned to Combustion Engineering, Inc.* and is shown schematically in Figure 139.

FIGURE 139: STEAM GENERATOR WITH WET SCRUBBER FOR FLUE GAS PURIFICATION

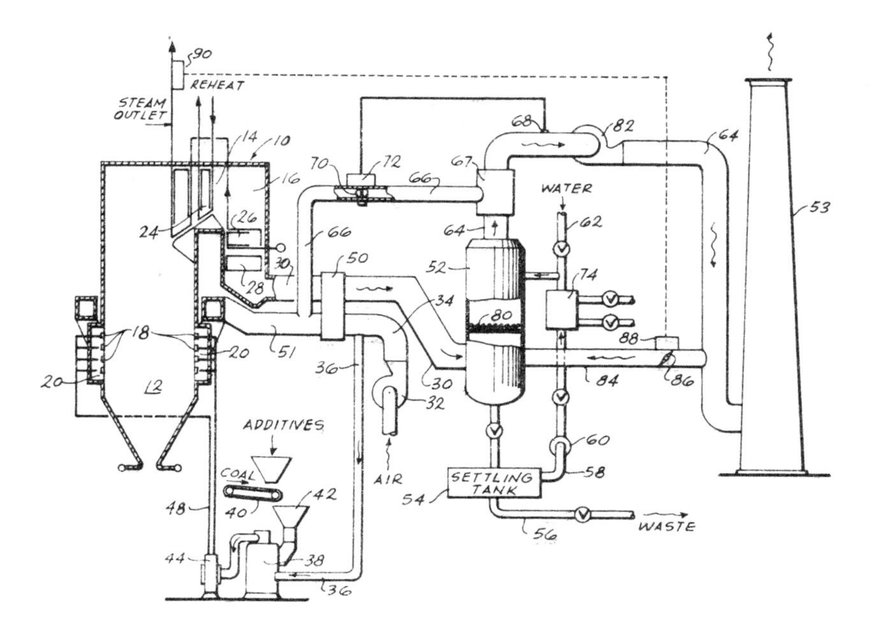

Source: U.S. Patent 3,726,239

In the drawing there is shown a steam generating unit **10** including a furnace portion **12**, a horizontal gas pass **14** and a rear gas pass **16**. The furnace portion **12** contains a plurality of burners **18** which feed a mixture of pulverized coal and primary combustion air into the furnace chamber. Secondary combustion air is fed to the furnace through the wind boxes **20**.

The products of combustion produced in the furnace rise through the furnace and enter the horizontal gas pass **14** in which they contact the finishing superheater **22** and the reheater **24**. The products of combustion or flue gases then enter the rear gas pass **16** and contact, in turn, the primary superheater **26** and the economizer **28** after which the flue gases enter the duct **30**.

The combustion air is supplied to the furnace by means of the forced draft fan **32**. A portion of the air from the fan discharge duct **34** is withdrawn through duct **36** and fed to the pulverizer **38**. Coal is also fed to the pulverizer by means of the conveyor **40** and the chute **42**.

The pulverizer pulverizes the coal and mixes it with the primary combustion air from duct **36**. This mixture of pulverized coal and primary combustion air is conveyed to the inlet of the exhauster fan **44** via the conduit **46**. From the exhauster fan **44** the coal-air mixture is conveyed through duct **48** to the burners **18**.

The remaining combustion air from the forced draft fan **32** passes through the conventional regenerative air preheater **50** and then through duct **51** to the wind boxes **20** from which the air enters the furnace. The air preheater **50** serves to extract the heat from the flue gases in duct **30** and transfer the heat to the secondary combustion air in duct **51**.

During the course of the combustion process the sulfur contained in the fuel is converted to SO_2 and SO_3 and water vapor whose serious corrosive effects are avoided by the addition of oxides, hydroxides and carbonates of alkali and alkaline earth metals to the combustion gas stream as described in U.S. Patent 3,320,906.

In addition a wet scrubber **52** is interposed between the steam generating unit and the stack **53** in order to scrub the combustion gases prior to their discharge to the atmosphere. The wet scrubber **52** not only removes the sulfate and sulfite particles which have been formed in the steam generator by the reaction of the alkaline additives with sulfur compounds but also effects an additional reaction of the additives with the sulfur compounds to remove most of the latter from the combustion gases.

One form of wet scrubber that is most effective for purifying the combustion gases prior to their discharge from the stack **53** is one incorporating a bed **80** comprised of discrete spheroidal marbles through which the combustion gases are caused to flow and to be intimately mixed with wash water admitted thereto. In practice the flow velocity of the gas through the marble bed **80** must be maintained within certain prescribed limits. For example, gas flow velocity is ideally maintained between 500 and 550 feet per minute in order that the marbles in the bed will be agitated to such a degree as to effect best results. Adequate results will be obtained at gas flow velocities between 350 to 500 ft/min. If the gas velocity falls below 350 ft/min the marble bed cannot be agitated to any significant degree such that adequate gas-wash water contact is not usually attainable.

The scrubber operates in a continuous manner with the liquid effluent from the scrubber being fed to a settling tank **54**. The reaction products settle rather rapidly in the settling tank and the sludge is discharged through line **56** and disposed of. The supernatant water is drawn off the top of the settling tank through line **58** by the pump **60** and recirculated through the scrubber **52**. Make-up water is added to the scrubbing system through line **62**. Considerable quantities of water may be necessary in excess of that required for sulfur removal in order to keep the scrubber from clogging. The scrubbed gases are conducted from the wet scrubber through duct **64** to the stack **53**.

To reduce the effects of a water vapor plume the gases emanating from the scrubber **52** are preheated by extracting air from duct **51** via duct **66** and introducing it into duct **64** in direct mixing relation with the combustion gases passing therethrough. The preheated air is introduced into duct **64** at the enlarged portion or chamber **67**. As illustrated in the figure, the preheated air duct **66** enters the chamber **67** tangentially so as to create a swirling motion of the gases to promote mixing.

The temperature and amount of combustion gases leaving the scrubber **52** will of course vary under certain conditions such as a change in load on the steam generator. Therefore the amount of preheated air necessary to reheat the stack gas will also vary. Means may be provided to control the reheating air such as the temperature measuring device **68** which controls the damper **70** in duct **66**. This control is accomplished by means of suitable conventional control apparatus **72**.

The amount of water vapor being carried over to the stack **53** can be reduced by maintaining the scrubbing water at a low temperature so as to cause the flue gases to leave the scrubber at a low temperature. This results in a lower weight of water vapor in the gases prior to reheating thus lowering the dew point temperature of the gases leaving the stack. If sufficient scrub water cooling does not take place in the settling tanks or by the addition of make-up water, a cooling heat exchanger **74** may be inserted into the scrub water circuit to provide the desired cooling.

The need for a plurality of scrubbers **52** in the system is obviated by provision being made for recirculating part of the processed combustion gas under certain conditions again through the wet scrubber **52** in order to maintain the gas flow velocity substantially uniform through the marble bed **80** and, concomitantly, its effectiveness to contact wash water with the combustion gas over the full operational range of the steam generator **10**.

To accomplish this result a gas recirculation duct **84** is connected between the combustion gas discharge duct **64** and the wet scrubber **52**. The duct **84** has its inlet end connected to the duct **64** downstream of an induced draft fan **82** and its outlet end communicating with the scrubber **52** upstream of the marble bed **80**. By means of this arrangement the recirculated part of the processed combustion gas is admitted to the scrubber and caused to flow through the bed **80** in mixed relation with the raw combustion gas fed to the scrubber through duct **30**.

The operation of the gas recirculation system is such as to controllably pass part of the purified combustion gas that flows through duct **64** toward the stack **53** back through the wet scrubber **52** when the flow velocity of the raw gas entering the scrubber through duct **30** is reduced below about 500 feet per minute. The controls provided to effect such controlled gas recirculation include a damper **86** disposed in the gas recirculation duct **84** having a controller **88** that is operated in response to the operation of element **90** arranged to sense the load conditions of the steam generator **10**.

In the illustrated embodiment the sensing element **90** is one operative to detect the steam output delivered by the unit which parameter will vary in direct proportion to the combustion gases exiting the steam generator. The controller **88** and the sensor **90** cooperate to open the damper **86** whenever the load conditions on the steam generator **10** fall below full rated value. The damper **86** is regulated in response to the signal emitted by the sensor **90** to admit processed gas in such amounts as to ideally maintain total gas flow through the bed at about 500 ft/min. Moreover, at extremely low load conditions where flow velocities of 500 ft/min are not possible the flow permitted by the damper **86** prevents the flow velocity from falling below 350 ft/min, the minimum required for scrubber operation.

It will be recognized that by employing this gas recirculation system, the use of a single wet scrubber can be extended into the lower load ranges of steam generator operation. Thus, the need for incorporating a plurality of scrubbers, each being designed for reduced gas flow capacity, is avoided together with the expense attendant therewith. Alternatively, in installations where a plurality of scrubbers are employed, gas recirculation obviates the need to remove scrubbers from service over the full load range of steam generator operation.

Additionally, retreatment of a portion of the processed gas, as is characteristic of this system, becomes effective up to the capability of the scrubber for additional gas purification. This feature is becoming increasingly more important to meet regulations which limit the amount of pollutants that can be discharged to the atmosphere.

A process developed by *W.H. Chapman and J.F. Eichelmann, Jr.; U.S. Patent 3,890,207; June 17, 1975; assigned to El Paso Southern Company* involves contacting the effluent gases with water to give a purified gaseous stream, which is evacuated to the atmosphere, and a water effluent containing suspended and dissolved contaminants. The suspended contaminants are separated and the resultant water effluent is freed of dissolved noxious gases and fumes and is substantially distilled to give a pure water condensate and a concentrated effluent containing dissolved solids. The concentrated water effluent is mixed with a circulating, relatively nonvolatile carrier liquid, the water is evaporated therefrom and recovered as an industrial water condensate, and the precipitated solids are removed from the carrier liquid. Such a process is shown in Figure 140.

The chimney **10** of an industrial power plant is fitted with a removable stack seal **12**. The stack effluent gases, comprising particulate matter (fly-ash) and gases such as CO, CO_2, SO_2, nitrogen oxides, H_2S and various mercaptans, pass through pipe **14** into an inductive blower **16**. The latter eliminates back pressure due to the pressure drop and provides the necessary pressure to force the gases through the wet cleaning system. The stack gases, which in this instance have

FIGURE 140: ALTERNATIVE WET SCRUBBER SYSTEM FOR FLUE GAS PURIFICATION

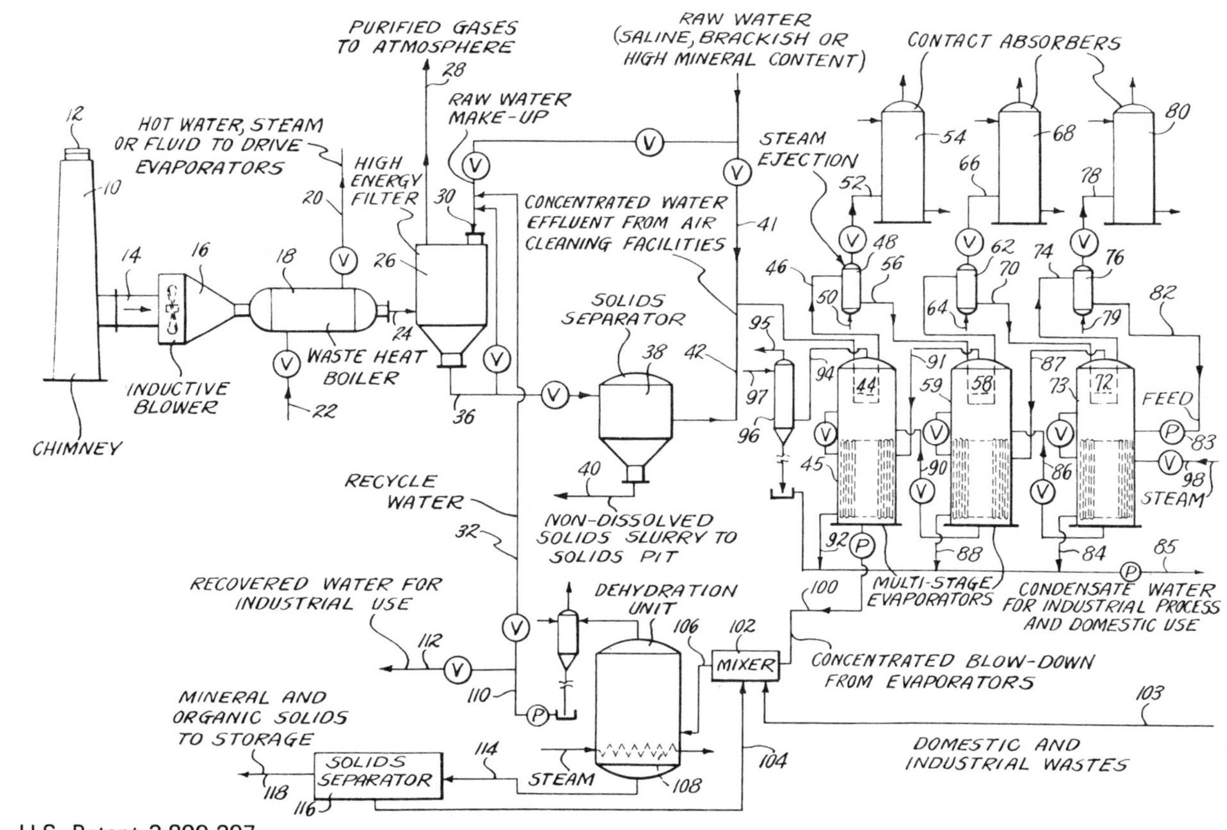

Source: U.S. Patent 3,890,207

a temperature above about 500°F, generally about 700° to 800°F, are cooled by passing them into the waste heat boiler **18.** Water, or other cooling fluid or heat transfer agent, passes into boiler **18** through line **22,** and the resultant hot water, steam or heated fluid is exited at line **20** and may be used to heat the dehydration unit or evaporators used in the system.

The cooled effluent gases, which are now below about 200°F, are passed through the exit pipe **24** and into a high energy water filter or scrubber **26.** Purified gases are vented to the atmosphere through vent pipe **28.** The water used in the scrubber enters at water inlet **30.** Either potable water, recycle water, or raw water obtainable from a typical source of saline or brackish water can be used. Typically, the raw water can contain up to about 1,000 parts per million dissolved solids. If the source of raw water contains a mineral content higher than this level, it can be blended by mixing it with recycled industrial water from recycle line **32** which is of relatively low mineral content. Water obtained from this source generally contains no more than about 15 to 20 parts per million dissolved solids.

The effluent water from unit **26** containing suspended particulate matter, absorbed gases and dissolved solids is passed through the pipe **36** to the solids separator or centrifuge **38** where the suspended particulate portion (mainly fly-ash) is withdrawn as a thick slurry by exit pipe **40.** As mentioned above, this fly-ash may find use for various applications such as in the preparation of cinder blocks. The water effluent from separator **38,** containing dissolved solids and absorbed gases such as SO_2, is delivered through line **42** to preheating zones of multiple effect evaporators **45, 59** and **73.** Any conventional multiple effect evaporators may be modified to include such preheating zones. The wastewater in line **42** can also be optionally mixed with additional saline, brackish water or water of high mineral content by adding such water through line **41.**

The water from line **42** is passed in series flow through heat exchanger or preheat zones **44, 58** and **72** contained within evaporator units **45, 59** and **73,** respectively, of the multiple effect evaporator system, and through steam ejection degassers **48, 62** and **76** wherein selective flashing of the absorbed gases occurs.

In a representative operation, the effluent water from the solids separator **38** is transmitted by line **42,** optionally mixed with raw water from line **41,** into heat exchange zone **44** in the evaporator **45.** The water is heated to about 90° to 100°F and directed by line **46** to the steam ejection unit **48.** Steam is directed into the unit by means of inlet **50** and is passed through the heated effluent mixture whereupon various absorbed gases, such as SO_2, are driven out of the water mixture and carried by line **52** to the contact absorption tower **54,** where they are absorbed. The expelled gases can be collected by contacting them with suitable absorbent liquids appropriately selected for the particular constituent to be absorbed.

The effluent water then passes by way of line **56** into the heat exchange zone **58** contained in the distillation unit **59** where it is heated to a higher temperature, e.g., about 100° to 150°F. The heated effluent water is then passed to an additional steam ejection unit **62.** Again steam is passed into the unit by means of inlet **64** and another fraction of dissolved gases are displaced and expelled from the effluent water. The expelled gases pass through exit line **66** into the contact absorption tower **68.**

The effluent water is heated to a still higher temperature (150° to 200°F) by directing it by means of line **70** into the heat exchange zone **72** in the distillation unit **73**, which is the first distillation unit of the forward feed system shown. The heated effluent water is conducted by line **74** into the steam ejection unit **76** and once again steam is conducted into the effluent water through inlet **79**. The remaining dissolved gases and fumes are driven through line **78** into the contact absorption tower where they are removed.

It should be understood that while a three stage evaporation and heat exchange system is shown, this number is not critical and a greater or lesser number of units can be employed depending upon the particular pollution problem under consideration. Typically, the temperature gradients over the three units of the evaporator system shown may range from 250°F in unit **73** to 215°F in unit **45**. However, the temperature gradient over both preheat zones and distillation zones can be adjusted to best suit the volume and nature of the impure input water.

The heated effluent water in line **82**, now freed of particulate matter and dissolved gases and noxious fumes, is pumped into the distillation units of the multiple effect evaporation system and distilled. The condensate water from each unit is available for domestic and industrial purposes, and the concentrated blowdown is passed into a Carver-Greenfield dehydration system or other treating system.

Referring to Figure 140, the heated effluent water passes, by way of line **82** and pump **83**, into the distillation unit **73** and is partially evaporated. Heat for this unit is supplied by steam through line **98**. Steam-condensate water is collected and delivered by means of line **84** to the main condensate collection line **85**.

The concentrated effluent is passed through line **86** into the distillation unit **59** where it is again subjected to distillation, heat being supplied by vapors from line **87** and condensate water being collected and directed to the main collection line **85** by means of line **88**. The further concentrated effluent from unit **59** is passed through line **90** to final distillation unit **45**. The distillation is continued with heat supplied by vapors from line **91** and the condensate water being collected and directed by line **92** to the collection line **85**.

Vapors from final unit **45** pass through line **94** where moisture is condensed by contact with pure water entering condenser **96** through line **97**. Pure water effluent from the condenser **96** passes through line **99** into condensate water line **85**. Uncondensed gases, substantially free of contaminants, leave the system through line **95**. Again, it should be pointed out that while three distillation units are shown, more or less than this number can be used depending upon the particular industrial circumstances.

The highly concentrated blow-down water from unit **45** is now collected and directed by line **100** into a mixing tank **102** comprising part of a so-called Carver-Greenfield system (such as that illustrated in U.S. Patent 3,323,575).

The mixing tank **102** may also receive domestic and industrial wastes such as raw sewage, garbage and the like, through the line **103**. The combined wastewaters are mixed with reclaimed fluidizing oil, which is recirculated through line **104** into the mixing tank **102**. The resultant fluidized slurry mixture is introduced by means of line **106** into the dehydration unit **108** of the Carver-Greenfield system. In this unit substantially all the water is removed leaving the solids suspended in the nonvolatile fluidizing oil.

The condensate water is withdrawn through line **110** and made available for various uses. Thus, it can be recycled through line **32** and reused in the high-energy filter **26**. Alternatively, it can be diverted through line **112** for other industrial uses. Because of its low mineral content (about 15 to 20 parts per million or less dissolved solids) it can be used for irrigation purposes.

The remaining dehydrated material, consisting of solids suspended in oil, is withdrawn through line **114** to a solids separator **116** which may be a centrifuge or the like. The reclaimed oil is recycled back into the system through line **104** to the mixing tank as described above, and the solids are withdrawn through line **118**. The solids can be stored for commercial use, used for fuel or fertilizer, or sent to a waste pit.

Removal from Water

See Cooling Tower Blowdown Effluents (U.S. Patent 3,810,542)
See Cooling Tower Blowdown Effluents (U.S. Patent 3,901,805)
See Sodium Sulfur Oxide Wastes (U.S. Patent 3,962,080)

STEEL CONVERTER EFFLUENTS

In the bottom-blown oxygen steelmaking process, the refractory lined converter vessel has tuyeres in its bottom so that oxygen, other gases such as hydrocarbon cooling fluids, and powdered fluxes and other additives needed for the conversion process may be introduced through the bottom of the vessel and diffused through the melt. This contrasts with the well-established top-blown oxygen conversion process where oxygen is injected into the melt by means of a lance which extends through the mouth of the vessel to near the surface of the melt. In the top-blown process, the fluxes and other materials are also admitted through the mount of the vessel to the top surface of the melt.

In both the top and bottom-blown processes, oxygen (O_2) reacts with silicon, manganese, carbon and phosphorus in the melt during the blowing period. In the bottom-blown process, oxygen (O_2) also reacts with hydrogen resulting as a decomposition product from the hydrocarbon fluids injected in the bottom.

The duration of blowing or end point of the heat is indicated by the reduction of carbon oxidation products (CO and CO_2) in the evolved gases which means that the carbon content of the melt has been reduced to the desired level and that the melt may be poured provided that the temperature is correct.

In both processes gases are evolved during the blow period. The primary gases in the bottom-blown processes are carbon monoxide (CO), carbon dioxide (CO_2), hydrogen (H_2) and water vapor (H_2O) which may vary in proportion as the melt proceeds. In general these primary gases are directed into a hood over the vessel. The hood is spaced from the vessel so air enters and burns the CO and H_2 in the gas and the products of combustion are conducted through an off-gas cleaning system with an exhaust fan, and then led to a chimney and discharged to the atmosphere as waste.

In such a combustion system, iron vaporized from the bath in localized high temperature zones, and such iron that is discharged as oxide, is completely oxidized to red fume (Fe_2O_3) and diluted by combustion air and excess air. The gas cleaning system required to clean such fine red fume and high volumes of resulting off-gases is voluminous and costly due to high fan capacity and power consumption resulting from high pressure drop required to achieve air pollution standards.

The bottom-blown oxygen process in comparison to the top-blown, is characterized by a smaller percentage of iron vaporized but in an extremely fine dispersion of the iron particles. Therefore, a noncombustion system has to be used in order to avoid oxidation and further dilution with air, and to take advantage of the increased agglomeration and wetting properties of dust particles comprised largely of non- or semioxidized iron (Fe, FeO and Fe_3O_4) prevailing under reducing conditions. Besides the economic advantages of such a system, it provides the release of the lowest possible gas volumes with minimum concentration of dust in the exhausted stack gases. This ensures that all applicable pollution code standards can be economically met.

Removal from Air

A scheme developed by *K. Baum, J.P. Baum, J.K. Pearce and D.L. Schroeder; U.S. Patent 3,908,969; September 30, 1975; assigned to Pennsylvania Engineering Corporation* provides economic and efficient pollution control on a bottom-blown oxygen converter vessel, with simultaneous provisions for safe capture and collection of uncombusted high calorific value waste gases. The benefits derived from use of such collected gases will significantly enhance economics by reducing the costs required to meet air pollution standards. Such a process is shown in Figure 141.

The bottom-blown converter vessel is generally designated by the reference **10**. This comprises a metal shell **11** lined with refractory **12** except for its top mouth **13**. At the bottom of the vessel is a housing **14** in which there is a gas and powdered material distributor box **15**. These substances are injected under pressure into the molten metal within vessel **10** by means of several nozzles **16** which extend into tuyeres **17** in the bottom of the vessel. All gases and powdered solid materials which are injected in the melt diffuse upwardly through it. Intimate and extensive contact between active gases such as oxygen and the powdered flux materials results in practically stoichiometric reactions. Inert gases, of course, diffuse through the melt and arrive at the interior top of the vessel.

The vessel **10** is shown mounted conventionally on a tiltable trunnion ring **18** which has laterally extending trunnion shafts **19** and **20** that are normally journaled in supports which are not shown. Thus, the vessel **10** may be inverted on trunnion shafts **19** and **20** that are normally journaled in supports which are not shown. Thus, the vessel **10** may be inverted on trunnion shafts **19** and **20** to discharge slag or it may be tilted substantially horizontally to discharge the molten metal through a tapping side spout **21**.

FIGURE 141: APPARATUS FOR AIR POLLUTION CONTROL COMBINED WITH SAFE RECOVERY AND CONTROL OF GASES FROM A BOTTOM-BLOWN STEEL CONVERTER VESSEL

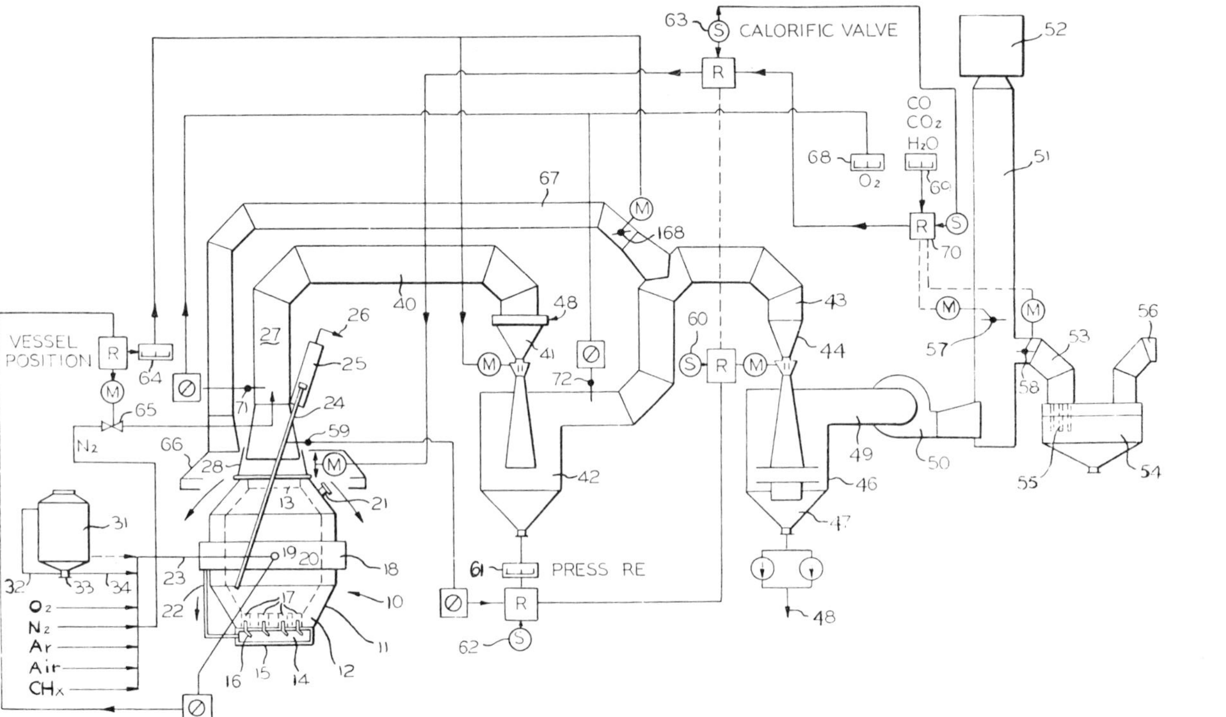

Source: U.S. Patent 3,908,969

Selected gases, liquids and powdered solids are delivered for injection to distributor chamber **15** at the bottom of vessel **10** through a pipe **22** that is connected to trunnion shaft **19** which is hollow. A feed pipe **23** connects to hollow trunnion shaft **19** by means of a swivel joint which is not shown. Thus, gases and materials may be injected when the vessel **10** is upright and gases only may be injected when the vessel is tilted.

A thermocouple probe **24** for sensing temperature at the top center of the melt projects through the vessel **10** at an angle from a reversibly moveable thermocouple support **25.** The lead conductors from the thermocouple are symbolized by a line marked **26.**

Above the top of the mouth of vessel **10** is a water-cooled hood **27**. The hood **27** is provided with a moveable skirt **28** which can be elevated and lowered with respect to the vessel **10** so as to allow an annular gap between the vessel mouth and the hood skirt for reasons which will be explained later.

Typical gases which may be injected in the bottom of vessel **10** are designated O_2, N_2, Ar (Argon), air and CH_x for a hydrocarbon gas or vapor. The sources for the gases are shown connected into main feed pipe **23.** Hydrocarbon fluid is always injected around the oxygen stream through tuyeres in the bottom of vessel **10.** The hydrocarbon prevents premature deterioration of the bottom lining.

A typical pressurized container for storing powdered material that is to be injected into the bottom of the melt within the vessel **10** is marked with the numeral **31.** In an actual installation there are several such containers for storing such powdered flux materials as lime, limestone and fluorspar as well as iron oxide, a desulfurizing agent and other additives.

Oxygen is usually used to entrain and convey the powdered materials at a controlled rate. Nitrogen and other gases are used as required as well. An oxygen pipe **32** feeds into vessel **31** and a mixing device **33** from which the oxygen and the powdered material is delivered through a pipe **34** to main feed pipe **23** and ultimately to vessel **10.**

The water-cooled hood **27** over vessel **10** is connected with a water-cooled hood stack **40** leading to a saturator-venturi **41** in which the evolved gases are quenched and cooled to saturation temperature at the prevailing pressure. Water is drained from a separator **42.** The gases are then conducted through a pipe **43** to a second venturi scrubber **44** for final cleaning which is subject to variable flow control. The gas then goes to a droplet separator **46.** The water is drained to a sump **47** and recycled to the overflow trough **48** at the inlet of saturator-venturi **41.** The gas then flows through a pipe **49** to an exhaust fan **50** from where it is delivered to a stack **51** and burnt at a torch **52.**

In case of gas recovery stack **51** is connected to a pipe **53** which leads to a water seal check valve **54** in which there are a plurality of pipes **55** whose ends are immersed in water so that gas cannot back up into the system. The gas which comes through pipe **53** and trap **54** in that case is the high calorific value gas comprised largely of CO and H_2 and some CO_2, H_2O and N_2. The useful gas is conducted away through a pipe **56** to a pump and storage pressure vessel, the last two items not being shown.

Stack **51** has a damper or valve **57**. The damper is controlled so that the inert gas plug, of the low calorific gas, may be selectively discharged to the atmosphere through the stack rather than pipe **53** for storage. Pipe **53** also has a damper **58** which is opened to pass the high calorific gas when such is being evolved from the vessel **10**. When damper **57** is closed, damper **58** is opened and vice versa. Switching of these valves is controlled on the basis of a preset calorific valve related to the actual readings from the combined outputs of the partial pressure sensor **68** and infra-red sensor **69**.

The point of gas sampling for pO_2 (oxygen partial pressure) analyzer **68** and infra-red analyzer **69** (CO, CO_2, H_2O) will be either at the hood at point **71** or after the venturi scrubber **41** at point **72**. Pressure sensor and its location is designated by numeral **59**. Signals from the pressure sensor **59** converted to electric signals is used to control the motor that operates movements of the variable throat of secondary venturi **44**.

During start of a blow the throat opening of secondary venturi scrubber **44** is preset by a set value indicator **60**. As the $CO + H_2$ level increases and combustion takes place with aspirated air through the air gap between the hood and vessel mouth with hood skirt **28** in a raised position, the pO_2 analyzer **68** monitors changes in combustion conditions.

When the actual point of stoichiometric combustion is reached and the presence of an inert gas plug is identified by the pronounced signals from the pO_2 analyzer **68**, this signal is used to initiate closure of hood skirt **28**. Closure of hood skirt **28** is continued through signals from the pO_2 analyzer **68** till such time that the pressure in the hood as measured by sensor **59** and indicated on pressure indicator **61**, is equal to a preset negative pressure on a set value indicator **62**. At this position of hood skirt **28**, control of the negative pressure in the collection and cleaning system is then taken over by the variable throat of the secondary venturi **44**.

The pressure signal from sensor **59** to the variable throat of secondary venturi scrubber **44** is supervised by signals from the pO_2 analyzer **68** and infra-red analyzer **69** giving the calorific value of the gas to maintain a preset calorific value set in set value indicator **63**.

When the preset calorific value of the gases is reached, as set on set value indicator **63**, the controller **70** initiates switching of valves **57** and **58** from gas exhaust to gas collection.

In the case of high carbon heats and for emergency turn down of vessel **10** at levels when high $CO + H_2$ exist, as the vessel is turned down, this condition is monitored by vessel position indicator **64** which will open nitrogen injection valve **65**. Such opening will provide for a nitrogen purge for inertization of the gases in the system. Simultaneously the decreasing calorific value of the gases monitored by instruments **68** and **69** and compared with set value indicator **63**, will cause controller **70** to switch valves **57** and **58** such that gases are exhausted through stack **51**.

Additionally, the nitrogen injection valve **65** will be triggered by signals from the pO_2 analyzer **68** and infra-red analyzer **69** indicating the approach of a potential explosive emergency condition through the identified presence of oxygen with a high calorific value gas containing high $CO + H_2$ levels.

Fumes emitted during tilting or charging and tapping are collected jointly through the main hood and auxiliary hood **66** connected through duct **67** via shut off valve **168** to inlet duct **43** of the secondary venturi scrubber **44**. As the vessel is tilted from the vertical, position indicator **64** causes progressive opening of the shut-off valve in the auxiliary hood duct and the progressive closure of primary venturi scrubber throat **41** to a minimum preset opening accomplished through limit switches. Fumes from the auxiliary hood are thus cleaned through the secondary venturi scrubber **44** while those collected through the main hood are cleaned through both venturi scrubbers **41** and **44**, utilizing the main gas cleaning exhaust fan **50**.

A steel plant exhaust system described by *G. Hausberg and K.-R. Hegemann; U.S. Patent 3,976,454; August 24, 1976; assigned to Gottfried Bischoff Bau kompl. Gasreinigungsund Wasserruckkuhlanlagen KG, Germany* is one in which dust laden waste gases from a converter and from ancillary equipment are exhausted by respective blowers through a first and a second duct to a main and a secondary scrubbing station where the purified gases escape through respective chimneys. The first duct is provided with two cascaded washing stages, the upstream stage being operable to throttle or block the flow of gases therethrough.

An annular cowl for the interception of peripherally escaping converter gases is connected to a third duct which joins the first duct at a location between the two washing stages but is also connectable through a switching valve to the second duct. The switching valve is open in the blowing phase of the converter while a movable insert at the upstream washing stage unblocks the first duct; in the charging phase, the switching valve is closed and the insert at the upstream stage obstructs the first duct as the gases picked up by the cowl are drawn through the third duct and the downstream washing stage of the first duct for discharge through the chimney of the main scrubbing station. Such a process is shown in Figure 142.

The exhaust system shown comprises a converter **1** tiltable into an off-normal (charging) position; this converter forming part of a steel-making plant further including such ancillary equipment as a transfer pit **17** and a mixer **19**. Molten pig iron, transported in a railroad car **43**, is poured into the pit **17** where it is received by a transfer ladle **18** serving to carry it to the mixer **19** for the admixture of additives therewith; the contents of the mixer are then emptied into a charging ladle **20** which dumps them into the converter. These steps occur, as is well-known, at different times in the course of an operating cycle of the plant.

The mouth of the converter **1** in its normal upright position is overlain by a hood **3** opening into a first duct **2** which has a vertical pipe section in line with the converter axis; this vertical pipe section is surrounded by a downwardly open annular cowl **26** positioned to intercept gases rising around the duct **2**.

A second duct **24** has branches **24a, 24b, 24c** whose entrance ends form hoods **25a, 25b, 25c** overlying the pit **17**, the loading side of mixer **19** and the unloading side of that mixer, respectively, to collect the rising gases. Cowl **26** opens at **21** into a third duct **22** which merges with duct **2** at a junction **6** within a main scrubbing station **I**. A fourth branch **24d** of duct **24** is connected to duct **22**, branches **24a-24d** being individually closable by slide valves **23a-23d** which are open only when the equipment concerned is in use, e.g., during loading and unloading of the mixer **19** in the case of valves **23b** and **23c**, respectively. A

FIGURE 142: EXHAUST SYSTEM FOR STEEL-MAKING PLANT

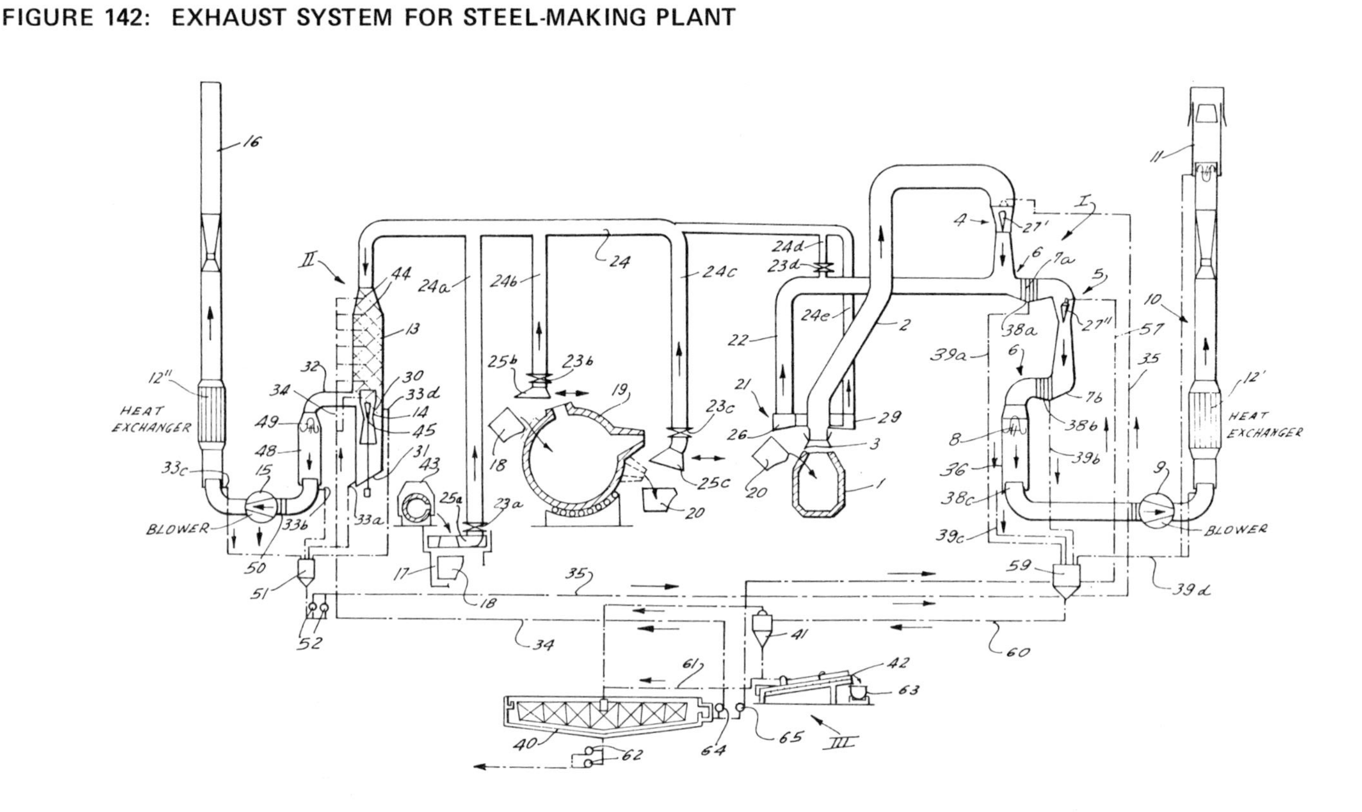

Source: U.S. Patent 3,976,454

fifth branch **24e** of duct **24** communicates at **29** with the cowl **26**. Duct **24** terminates at a secondary scrubbing station **II** including a blower **15** which draws the gases from its several branches through a wash tower **13** provided with a nozzle **14** downstream of a set of spray heads **44**.

An insert **45** in nozzle **14** is vertically shiftable by a mechanism not shown under the control of a sensor responsive to the pressure differential between an upstream compartment and a downstream compartment separated by an inclined partition **30**. The downstream compartment has a sloping bottom **31**, provided at its lowest point with a drain **33a**, and opens well above the lower end of nozzle **14** into a lateral outlet **32** whereby the wetted gases descending through the nozzle are sharply deflected to shed most of their entrained particles before escaping into a vortex chamber **48** provided with vanes **49** in which further separation between liquid and solids takes place.

More liquid is drained off at **33b** and the purified gas goes through a valve **50** and the blower **15** into a chimney or flue **16** which may open into the atmosphere and is shown provided with a heat exchanger **12"** for the recovery of residual thermal energy from the gas. Residual moisture is collected at a drain **33c** whereas spent wash water from spray heads **44** runs off at **33d**. All the drains **33a-33d** empty into a vessel **51** whose contents are fed by pumps **52** to a conduit **35** leading to an upstream washing stage **4** in duct **2**.

Stage **4**, located in a descending vertical portion of that duct, comprises a spray head above a constriction which forms a nozzle adapted to be selectively blocked or unblocked by a pear-shaped insert **27'** vertically displaceable therein. The insert is controlled, e.g., manually, via a stem and in its unblocking position defines a relatively wide annular passage for the wetted gases.

A downstream washing stage **5** comprises a spray head just above a similar insert **27"** which is disposed in a nozzle formed by another constriction of duct **2**. Insert **27"** is vertically adjustable by a mechanism under the control of a sensor responsive to the pressure differential across the constricted passage surrounding that insert.

In the blowing or refining phase of converter **1**, nozzle **5** is opened wide by the insert **27'** and switching valve **23d** is open to interconnect the two ducts **22** and **24**. The first blower **9** then aspirates the major part of the converter gases via duct **2** through the cascaded washing stages **4** and **5** while the second blower **15** draws the remaining gases, escaping laterally around the inlet of duct **2**, via cowl **26** and ducts **22** and **24** through wash tower **3** on their way to flue **16**.

In the changing phase, with the converter tilted into its off-normal position, insert **27'** is lowered to throttle the duct **2** at nozzle **4** and switching valve **23d** in branch **24d** is closed to disconnect the duct **22** from the duct **24**; now, the suction of blower **9** draws the evolving gases from cowl **26** by way of duct **22** through washing stage **5** supplied with fresh wash water from a conduit **57**.

Duct **2** further includes baffle-type water separators **7a** and **7b** provided with drains **38a** and **38b** leading via conduits **39a** and **39b** to a collecting vessel **59**. A further such conduit **39c** extends from a drain **38c** at the bottom of a vortex chamber **36** provided with vanes **8**, this chamber serving as a water separator common to ducts **2** and **22**. The gases freed from solids and from most of the liquid pass from water separator **36** through blower **9** into a flue **10** provided with a heat exchanger **12'** and with an additional vortex chamber **11**; the latter chamber and other parts of the flue **10** are drained into vessel **59** by a conduit **39d**.

The effluent from scrubbing station **I**, collected in vessel **59**, exits therefrom via a conduit **60** terminating at a hydrocyclone **41** in a regenerating station **III**. The sludge precipitated in the cyclone is dried on a screen **42** from which the residual water is filtered out at **61** for delivery to a settling tank **40**. The residual sludge collected in the tank is carried off by pumps **62** whereas solids retained by the screen can be removed by a bucket conveyor **63**.

The supernatant liquid from the tank is recirculated by a pump **64** via a conduit **34** to the spray heads **44** of station **II** and by a pump **65** via conduit **57** to the spray head **54** of station **I**. The circulation of the wash water through scrubbing stations **I** and **II** in tandem, coupled with the recovery of a large portion of the spent water in settling tank **40**, ensures a particularly economic mode of operation of the system.

With junction **6** located upstream of water separator **7a**, the gases arriving at that separator via duct **22** in the charging phase are prewetted by water from spray head **53**, flowing through the narrowed passage of nozzle **28'**, and are freed from some of their entrained solids before reaching the washing stage **5**.

A process developed by *A.T. Dortenzo; U.S. Patent 3,788,619; January 29, 1974; assigned to Pennsylvania Engineering Corporation* is one in which the hot evolved gases from a bottom blown steel converter vessel are delivered to a spray chamber near the vessel in which the gases are cooled by sufficient amount to make them acceptable to a remotely situated electrostatic precipitator in which finely divided solids are removed from the gases. The clean gas is withdrawn from the precipitator with a suction fan and delivered to a stack which discharges the gases to the atmosphere.

STEEL MILL EFFLUENTS

Removal from Air

See Electric Arc Melting Furnace Effluents (U.S. Patent 3,979,551)
See Rolling Mill Effluents (U.S. Patent 3,204,393)

Removal from Water

A process developed by *L.A. Duval; U.S. Patent 3,844,943; October 29, 1974* has as its object the treatment of material coming largely from settling pits and settling lagoons in a steel mill so as to: separate the valuable heavier components, mostly particulates of iron oxide from the insoluble oils and lubricants and the water so as to recover the valuable raw material in the oxides; conserve the waste volume; render the waste material easily handled; reduce sewer maintenance; and provide a cleaner waste discharge.

The equipment arrangement suitable for the conduct of such a process is shown in Figure 143.

The input indicated at **10** is preferably through a flexible hose **30** which is preferably arranged to permit adjustment of the input hose at different positions with respect to the side of the tank **11** and at different angles to the horizontal from about 12° to 45°.

The stream circulates generally in a spiral inside of the tank **11** in a nonturbulent manner giving the heavy particles, mostly iron oxide, an opportunity to settle out downwardly while the liquids, mostly water and oil, float upwardly and pass out through an overflow pipe **12**. Preferably, this pipe is cut away for about one quarter of the circumference on the upper side of the pipe to insure a proper skimming action.

Means are provided for controlling the discharge of the heavier components separated in the tank **11** through the converging bottom **14** thereof. The overflow from the separator tank (mostly oil, slime and water) passes through the outlet **12** and passes to a thickener **13**.

FIGURE 143: APPARATUS FOR PROCESSING WASTEWATER SLIMES OF STEEL MILL WATER TREATMENT SYSTEMS

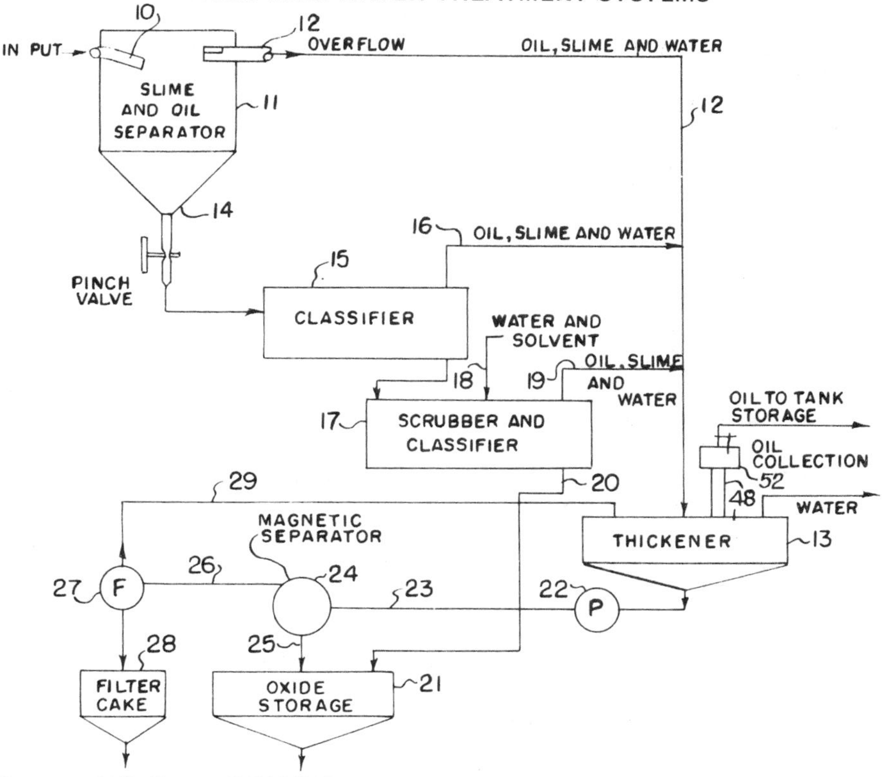

Source: U.S. Patent 3,844,943

The heavier components discharge from the tank, pass through line **46** to a classifier **15**. This classifier works in the usual manner to separate the heavier components of the stream from the lighter components thereof, such as oil, slime and water. The latter passes through the line **16** and line **12** to the thickener **13**. The heavier material is discharged and fed to the intake end of a combined scrubber and classifier indicated at **17**. As the material passes through the driving screw of the scrubber and classifier, additional water and solvent are added through line **18**. The water may be any reasonably clear water and the solvent is suitable for dissolving the heavier oil and lubricant particles which are still contained in the stream when it passes through the apparatus at **17**.

The heavier components discharge from the classifier **17** at **20** consisting mostly of oxides which are transferred directly to an outside storage bin. The lighter components discharging from the classifier are discharged through line **19** and the line **12** to the thickener.

The lighter components discharged from the vertical generally cylindrical vessel **11** through line **12**, the lighter components discharged from the classifier **15** through line **16**, and the lighter components discharged from the combined scrubber and classifier **17** through line **19** are all fed to the thickener **13**.

The heavy components which settle to the bottom of the thickener are driven by pump **22** through line **23** to a magnetic separator **24**. The magnetic particulates separated at **24** pass directly through line **25** to the oxide storage. These particulates are substantially mill scale. The overflow from the separator passes through line **26** to cake filter **27** from which the filter cake passes through a means to the filter cake storage **28**. The overflow from the filter passes through line **29** back to the thickener.

See Rolling Mill Effluents (U.S. Patent 3,986,953)

STRONTIUM

Removal from Water

A process developed by *D.W. Rhodes et al; U.S. Patent 3,032,497; May 1, 1962; assigned to U.S. Atomic Energy Commission* involves the removal from solutions of strontium ions including those of the radioactive isotope Sr^{90} or radiostrontium, particularly when such ions are present in very small or trace amounts not economically removable by known methods.

This process is based on the discovery that while either calcium carbonate or calcium phosphate alone make but indifferent ion exchange materials for the removal of Sr^{90} ions from solutions, calcium and other alkaline earth phosphates, as well as other metal phosphates, in the process of being created through the reaction of carbonates or other salts with phosphate ions, make highly efficient ion exchange materials for this purpose. A typical, but of course not the only reaction whereby a metal phosphate is created is the following.

$$5CaCO_3 + NaOH + 3Na_3PO_4 \longrightarrow Ca_5(PO_4)_3(OH) + 5Na_2CO_3$$

Chemists are familiar with a number of similar reactions whereby phosphate salts are created. If strontium is present during the course of these main reactions it will be found to be removed from the solution of the reaction, even if present in only trace, or residual amounts.

SUGAR PROCESSING EFFLUENTS

Removal from Water

A process developed by *G. Pascarella and F. Salvemini; U.S. Patent 3,962,077; June 8, 1976; assigned to Tecneco SpA, Italy* involves purifying wastewaters which come from the regeneration of the anionic and cationic resins used in the treatment of sugar juices. The method includes subjecting the wastewaters to one or more zones which include an inverse osmosis section and one or more electrolysis cells.

The waters coming from the regeneration treatments of the resins present a high content of either salts or acids or bases and a remarkable amount of organic substances such as group B vitamins, simple proteins (such as lysine, arginine, tyrosine), more complex compounds (such as betaine, glutamic acid, aspartic acid) and a fraction of sugars.

The volume of the wastes is very high (for a medium to big industry there are from 2,000 to 3,000 m^3/day for about 100 working days). Its discharge brings forth remarkable problems since its organic content which can ferment and its saline content obviously do not permit it to be sent to open basins.

It is not only impossible to send the wastes to sewers primarily because of their polluting content but also because in almost all cases the location of a sugar mill is somewhat far from highly crowded zones or from high industrial density zones.

The methods heretofore used for solving the problem have not given satisfactory results. The biological treatment has been unsatisfactory both because of the low obtained yields and because of the complication due to the necessity of starting the plant with synthetic waters before the start of the production period. The chemical treatment is also unsatisfactory since the clear appearance of the wastewaters due to the very low content of colloids, does not permit the use of coagulants for the flocculation.

Other methods have been proposed, among which are:

(1) Lagooning which presents drawbacks due to the uneconomical utilization of a large ground area and to the development of disagreeable smells;

(2) Reverse osmosis which does not allow separation of the organic compounds from the salts in the polluting wastewaters having a reduced volume obtained after the treatment; and

(3) Finally electrodialysis which presents the drawback of the high cost and of the difficulty in separating the organic compounds from amino acids which migrate toward the electrodes with a high speed.

This purification process yields a complete purification of the wastewaters with relatively low costs, if one considers also the possibility of recycle or recovery. Figure 144 is a flow diagram of the process.

The waters, coming from the regeneration of the decoloring resins, rich in sodium chloride and containing organic substances are introduced into a reverse osmosis section consisting of two stages (**1** and **2**). From the first of these stages, through **3**, a solution is obtained which is rich in organic substances and has a low NaCl content. Through **4** is obtained a diluted NaCl solution free from organic substances. In the second stage there is effected the concentration of sodium chloride up to the desired value. This is recycled through **5** to the resins and practically deionized water is separated (**6**). Through **7** the solution recycled to the resins is restored.

The waters coming from the regeneration of the anionic resins (containing ammonium hydroxide and organic substances) are fed to the anodic compartment of an electrolytic cell (**8**) having two compartments and a cation-selective septum. There they are subjected to electrolysis in order to obtain in the cathodic compartment ammonium hydroxide at the necessary concentration for being recycled through **9** to the regeneration, and hydrogen **10**. **11** indicates the make-up to the recycle. A diluted organic solution leaves the anodic compartment through **12**, and is further demineralized to the desired value by coupling a cell having cation-selective septums to another cell having anion-selective septums.

For the waters coming from the regeneration of the cationic resins the treatment is analogous, the only difference being that these waters are fed to the cathodic compartment of an electrolytic cell (**13**) having two compartments with anion-selective membranes obtaining sulfuric acid at the concentration necessary for the recycle (**14**), oxygen (**15**) and a diluted solution of the organic substances (**16**). **22** indicates the make-up. The deionization limit depends on the possible sequence of anion-selective and cation-selective membranes.

FIGURE 144: PURIFICATION OF SUGAR PROCESSING WASTEWATERS

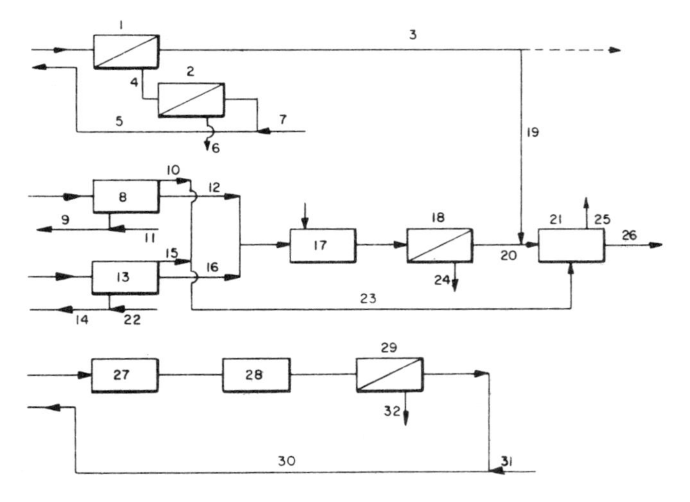

Source: U.S. Patent 3,962,077

The organic waters, if not completely demineralized, pass to a neutralization and mixing stage (**17**) from which they are sent to a reverse osmosis treatment (**18**). In this way there are obtained concentrated organic waters which, mixed with the ones coming, through **19**, from the osmotic processes of the waters from the decoloring resins, are fed, through **20**, to an evaporator (**21**) for recovering the dry organic substance, making use of, as nonpolluting fuel, the hydrogen produced in the electrolytic cells for the anionic waters, possibly with addition of methane, fed to **21** through **23**. Purified water is discharged through **24**; **25** is the vent for steam and the organic products are recovered through **26**.

The waters coming from the regeneration of decalcifying resins are sent to a treatment stage (**27**) wherein magnesium is precipitated in the form of hydroxide by adding sodium hydroxide and calcium is precipitated by addition of carbon dioxide, possibly recovered from the lime furnaces present in the plant. A neutralization phase (**28**) with hydrochloric acid, at last, a reverse osmosis treatment (**29**) for the concentration of the solution to the values requested for the recycle to the resins, through **30**, follow.

31 indicates the make-up. At the same time high purity water (**32**) is obtained. Alternatively the organic effluent stream, containing sodium chloride, leaving the first stage of the osmosis treatment of the waters from the regeneration of the decoloring resins, can be sent to a roasting furnace from which the sodium chloride to be recycled is recovered.

SULFIDES

Waste liquor discharged from plants in the field of petroleum refining, pulping and food processing industries contains hydrogen sulfide and other sulfides. Such sulfidic contaminants cause offensive odor in rivers and corrosion of steel structures. Accordingly, elimination of the sulfidic contaminants has been a goal in the prevention of environmental pollution.

Removal from Water

A method known hitherto for eliminating hydrogen sulfide from waste liquor comprises blowing air thereinto. However, this method has drawbacks such as that acidification and oxidation with an oxidizing agent are needed as preliminary treatment and that it is difficult to eliminate hydrogen sulfide so entirely from the waste liquor as to permit complete destruction of its foul odor. Since waste liquor discharged from plants especially in the field of petrochemical and papermaking industries contains a considerable amount of various sulfidic contaminants in addition to hydrogen sulfide, there is a large demand for an effective method for eliminating these sulfidic contaminants from waste liquor.

A process developed by *M. Ohta; U.S. Patent 3,806,435; April 23, 1974* is a process for treating waste liquor which comprises subjecting a waste liquor containing one or more sulfidic contaminants selected from hydrogen sulfide, alkaline sulfides and inorganic and organic hydrosulfides to an electrolytic treatment using an anode of iron/aluminum or zinc/aluminum and a cathode of iron or zinc, thereby eliminating the sulfidic contaminants from the waste liquor.

A process developed by *J.P. Tassoney and R.M. Dille; U.S. Patent 3,725,270; April 3, 1973; assigned to Texaco Inc.* is a process for the abatement of pollution from sulfide-bearing wastewaters. Waste aqueous solutions, e.g., sour water, treated sour water, and other refinery wastewaters containing sulfides are admixed with a hydrocarbonaceous fuel and reacted by partial oxidation with an oxygen-rich gas in a free-flow noncatalytic gas generator at an autogenous temperature in the range of about 1500 to 3500°F and a pressure in the range of about 1 to 250 atmospheres to produce synthesis gas from which hydrogen sulfide may be subsequently recovered and converted into by-product sulfur.

SULFUR

Removal from Air

A process developed by *M.W. MacAfee; U.S. Patent 2,780,307; February 5, 1957; assigned to Pacific Foundry Co., Ltd.* involves the removal of elemental sulfur from a gas stream in which the sulfur may exist as a vapor or an entrained liquid or solid. Sulfur vapor may be recovered from hot gases containing it by scrubbing it in water, but the resulting product is not suitable for market and requires dewatering and other operations before it can be sold. Consequently, there is an advantage to absorbing the sulfur from a gas stream in molten sulfur, but heretofore customary practices along this line have not been entirely successful and tend to produce a sulfur aerosol which passes through the system and escapes.

In accordance with this process, sulfur vapor is recovered from a gas stream by cooling the gas stream to a temperature below the boiling point of sulfur but not below about 507°F by indirect heat exchange and thereafter the cooled gas is brought into direct contact with molten sulfur, preferably in a scrubbing tower. For optimum results the gas stream and the molten sulfur stream should flow concurrently with each other through the scrubber so that the hottest gas comes into contact with the coolest molten sulfur.

After the contact between the gas and the molten sulfur, some sulfur may remain entrained in the gas stream, but the great bulk of this can be removed by swirling the gas to throw entrained sulfur out of it by centrifugal action. Small residual amounts of sulfur, if any, may be removed thereafter by passing the gas stream through a porous solid medium such as a tower packed with Raschig rings or berl saddles, or by electrostatic precipitation.

SULFURIC ACID

Removal from Air

A process developed by *R.G. Hartig et al; U.S. Patent 2,901,061; August 25, 1959; assigned to International Minerals & Chemicals Corp.* is based on the discovery that chemical mists such as sulfur trioxide can be eliminated by passage of mist-bearing gas through a packing of inert materials having a critical interstitial pore size, which can be controlled within limits through control of the volume of liquid used to wet the packing.

In utilizing the principle herein referred to, the mists and gases are passed through a confined space such as a tower. This confined space is filled in its entirety or in part, depending upon use conditions, with a packing inert with respect to the gases and mists such as Orlon filaments or woven Orlon cloth, asbestos, sawdust, coke, rock, ceramic material and the like. When using cloth or filaments and asbestos, interstitial space must be provided so as to be equivalent to that provided by a size graded bed of coke, rock, sawdust, ceramic material, or the like, having particles of a size to pass through standard screens in the range between about 10 mesh, 1,650 microns and about 80 mesh, 175 microns (Tyler Standard screens).

Packing in the confined space may be wetted with liquid which is capable of absorbing the nuisance gases, i.e., water, dilute acids, aqueous solutions such as sodium carbonate, sodium bicarbonate and the like. The greater the volume of liquid passing through the confined space and over the packing, up to the point of flooding the combined space, the greater the effectiveness of the absorption. A factor limiting the volume of liquid flowing through the confined space is the back pressure on the gas as the fluid reduces the interstitial space free to pass the gases.

A process developed by *C.L. Leonard et al; U.S. Patent 2,906,372; Sept. 29, 1959; assigned to Union Carbide Corp.* involves efficient recovery of mist from the exhaust gases from the concentration of sulfuric acid. Sulfuric acid mist is removed from a gaseous mixture containing the same by passing the mist containing gaseous mixture upward through porous plates covered with a free level of sulfuric acid of 60 to 72% concentration.

A process developed by *W. Plaut et al; U.S. Patent 3,107,986; October 22, 1963; assigned to Imperial Chemical Industries, Ltd.* involves sulfuric acid mist removal using a fiber filter made up of a class of fibers which, for convenience, are termed nonwettable fibers. An outstanding and most surprising increase in filtration efficiency is obtained; these are fibers whereon the mist is deposited in separate and independent droplets not connected (as in the case of the wettable fibers hitherto used) by a bridging film of liquid. The following is a specific example of the application of this process.

Example: In a plant for the manufacture of sulfuric acid the gases leaving the contact chamber were first cooled, then absorbed in strong sulfuric acid and subsequently passed through alkali-containing absorption towers. The sulfuric acid content of the mist-containing exit gas varied between 0.05 and 0.1 gram H_2SO_4 per cubic meter of gas and the mist particles were all of size less than 2 μ, 10% of them by weight being less than 1 μ.

Glass fiber of diameters in the range of 5 to 50 μ was treated with the Silicone Fluid M441 and was packed and compressed to a density of 10 pounds per cubic foot (160 kg/m^3) to form a layer 5 cm deep and the filter was held between confining gauzes made of resin coated stainless steel. The surface area of the filter presented to the gas stream was approximately 0.46 square meter. The mist-containing exit gas was passed downwardly through the filter at the rate of 300 to 350 m^3/hr per square meter of filter surface and the pressure drop was 19 cm water gauge.

While the filter was in continuous operation for over 900 hours there was no visible fume in the exit gas and a weak acid varying in strength between about 2½ and 10% H_2SO_4 was collected by drainage from the filter. The sulfuric acid content of the exit gas as measured by means of an electrostatic sampler was less than 0.0007 to 0.0008 g/m³. By way of comparison the example was repeated except that the mist-containing gas was passed through a similar filter of untreated glass fiber but in this case a persistently visible gas left the filter, the sulfuric acid content of the tail gases being 0.007 to 0.012 g/m³.

A process developed by *C.A. Sumner; U.S. Patent 3,927,975; December 23, 1975; assigned to Stauffer Chemical Co.* is one which provides a seal over the surface of liquid sulfur trioxide so as to prevent reaction with moisture vapor in the air to form sulfuric acid aerosol commonly referred to as smoke. This is brought about by adding an inert immiscible oil slurried with glass bubbles to reduce the specific gravity of the oil-slurry so that it will float on and seal the surface of the liquid sulfur trioxide.

See Battery Charging Effluents (U.S. Patent 3,926,598)

Removal from Water

See Iron and Steel Pickle Liquors (U.S. Patent 3,969,207)
See Iron and Steel Pickle Liquors (U.S. Patent 3,743,844)
See Iron and Steel Pickle Liquors (U.S. Patent 3,745,207)
See Mining Effluents (U.S. Patent 3,823,081)
See Mining Effluents (U.S. Patent 3,795,609)
See Mining Effluents (U.S. Patent 3,717,703)
See Spent Sulfuric Acid (U.S. Patent 3,888,653)

SULFURIC ACID PROCESS EFFLUENTS

Removal from Air

A process developed by *K.-H. Dörr, H. Grimm, U. Sander, R. Peichl and M. Tacke; U.S. Patent 3,944,401; March 16, 1976; assigned to Metallgesellschaft AG, and Sud-Chemie AG, Germany* is one in which sulfur compounds oxidizable to form sulfuric acid and organic compounds to form CO_2 and H_2O are removed from an exhaust gas of a contact-process plant for producing sulfuric acid by treating the exhaust gas with a scrubbing solution consisting of dilute sulfuric acid and peroxydisulfuric acid. The peroxydisulfuric acid is produced electrolytically from fresh dilute sulfuric acid and the resulting electrolyte is continuously introduced into the scrubbing acid cycle. The exhaust gas is treated in a vertical venturi with uniflow, i.e., codirectional flow of the gas and the scrubbing solution, and the gas is then passed through a horizontal venturi and subsequently upwardly through a packing layer. Such a process is shown in Figure 145.

A mixing vessel **3** is supplied with sulfuric acid through conduit **1** and with diluent water through conduit **2**. The dilute sulfuric acid is used as electrolyte acid and is supplied through conduit **4** into the cathode space **5** of the electrolytic unit. The acid electrolyte flows through conduit **6** over filter **7** for an intermediate purification and through conduit **8** into the anode space **9** of the

FIGURE 145: THE REMOVAL OF IMPURITIES FROM CONTACT PROCESS SULFURIC ACID

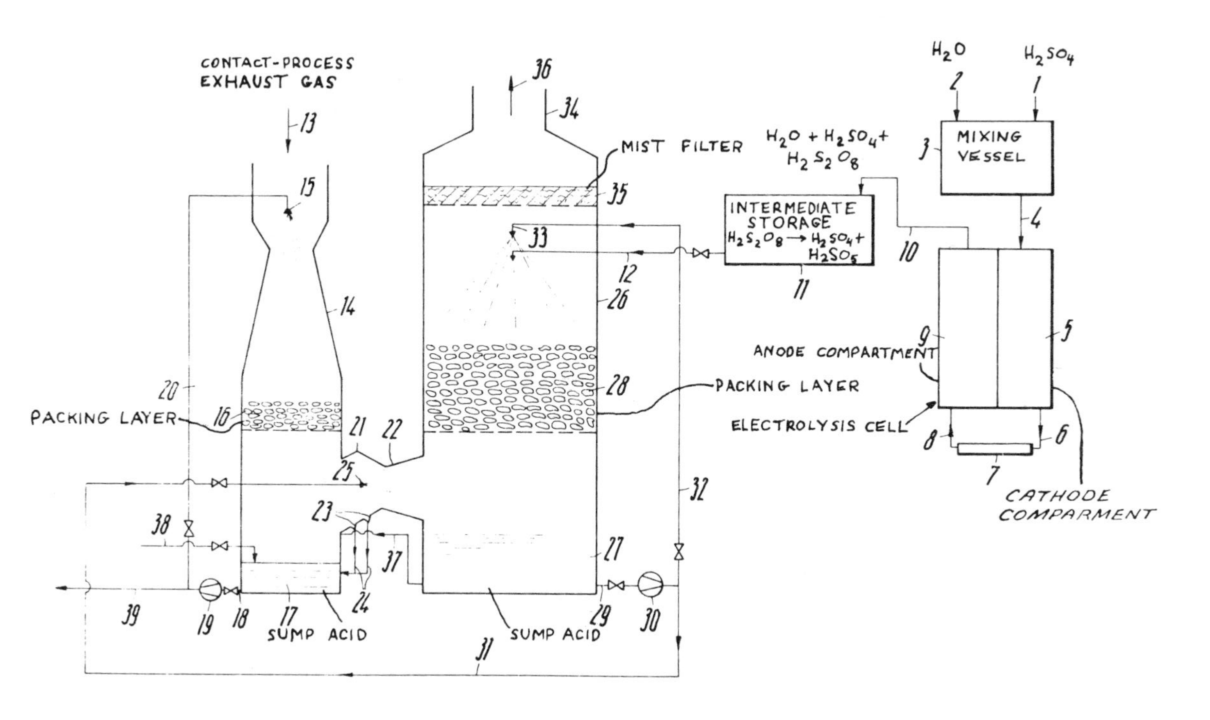

Source: U.S. Patent 3,944,401

electrolytic unit. The electrolyte acid which contains peroxydisulfuric acid flows through conduit **10** into a supply container **11** for an intermediate storage to increase the degree of hydrolysis. Scrubbing acid which contains active oxygen is supplied through conduit **12** into the scrubbing acid cycle at the required rate.

The exhaust gas **13** from the final absorber of the contact process plant for producing sulfuric acid is fed into the head of the vertical venturi tube **14**. Through the nozzle **15**, circulating scrubbing acid is injected into the head of the venturi tube and is mixed with the exhaust gas. A packing layer **16** is disposed below the outlet opening of the venturi tube. In this layer, the gas and scrubbing acid are further mixed and the injected scrubbing acid is separated.

A large part of the injected scrubbing acid is collected in the sump **17** of the venturi tube. The circulating scrubbing acid which contains active oxygen is pumped from the sump through conduit **18**, pump **19** and conduit **20** to the nozzle **15**. The gas leaves the vertical venturi tube and enters the inlet pipe **21** of the substantially horizontal venturi tube **22**.

The inlet pipe **21** is downwardly inclined, and drain openings **23** are installed in the lower part of the inlet pipe and of the inlet portion of the venturi tube **22** and are connected to the sump **17** by return conduits **24**. As a result, a large part of the scrubbing acid entrained from the venturi tube **14** into the venturi tube **22** is separated and flows into the sump **17**. Through the nozzle **25**, circulating scrubbing acid is injected into the venturi tube **22** and mixed with the gas. The mixture of gas and scrubbing acid enters the tower **26**.

A large part of the scrubbing acid which has been injected into the venturi tube **22** is separated already in the lower part of the tower **26** and enters the sump **27**. The gas rises through the packing layer **28**.

Circulating scrubbing acid which contains active oxygen is pumped from the sump **27** through conduit **29**, pump **30** and conduit **31** to the nozzle **25** and through conduit **32** to the nozzle **33** in the upper part of the tower **26**. Electrolyte acid which contains active oxygen is also supplied from the container **11** through conduit **12** to the nozzle **33**. The circulating scrubbing acid and acid electrolyte supplied into the tower trickle through the packing layer **28** into the sump **27**.

The gas outlet **34** is preceded by a wire mesh filter **35**, in which residual acid is separated from the gas and drips onto the packing layer **28**. The purified gas **36** is discharged into the atmosphere. Scrubbing acid from the sump **27** is transferred over the overflow **37** into the sump **17**. The scrubbing acid flows over at a rate which corresponds to the rate at which electrolyte acid containing active oxygen is supplied to the tower plus the rate at which sulfuric acid is produced by oxidation in the tower **26** and the venturi tube **22**. Water is supplied through **38** into the sump **17** of the venturi tube **14**. The rate is controlled so that a steady-state concentration of sulfuric acid in the sump **17** remains constant.

Scrubbing acid which contains the oxidation product is withdrawn through conduit **39**. The rate is controlled so that the level in the sump remains constant. The withdrawn acid is supplied to the final absorber of the contact process plant where the residual active oxygen is utilized for oxidation.

The advantages afforded by the process reside mainly in that a conversion of 95% and more of SO_2, depending on the rate at which active oxygen is added, may be obtained so that the purified gas contains less than 50 ppm SO_2, and that the 70 to 90% of the sulfuric acid mist contained in the exhaust gas are absorbed. This purification of the exhaust gas is accomplished with small equipment and with a relatively high economy and results in a true increase of the total production of the sulfuric acid plant.

Only electric energy is consumed as an operating supply; all other components are supplied from and returned to the cycle of the contact process plant for producing sulfuric acid. The water which is supplied is used to adjust the concentration in the acid cycles of the contact process plant. The residual active oxygen contained in the product which is withdrawn, is utilized in the final absorber so that virtually no active oxygen is lost. The pressure loss is low.

Because the cooling water for the electrolytic unit is only slightly heated, it may be used in the contact process plant, e.g., in the coolers for the acid cycles. In the purification of exhaust gases that have been used for reconcentration of acid, organic components and nitrous gases are oxidized in addition to the sulfur compounds. The compact and simple installation requires only a small space and may be made to a large extent from plastic material. It may well be integrated in contact process plants.

A process developed by *E. Jenniges; U.S. Patent 3,907,979; September 23, 1975; assigned to Chemiebau Dr. A. Zieren GmbH & Co. KG, Germany* is a low SO_2-emission sulfuric acid process using sulfur and pure oxygen as feed materials. Such a process is shown in Figure 146.

In a sulfur combustion furnace **3** with waste heat boiler **4**, liquid sulfur fed via conduit **1** is combusted to sulfur dioxide with technical oxygen supplied via conduit **2**. The sulfur combustion gas gives off the largest portion of its heat in the waste heat boiler and the feed water preheater **5** and is recycled through conduit **6** into the sulfur combustion furnace.

SO_2 gas is withdrawn from the SO_2 gas cycle through the sulfur combustion furnace via conduits **7** and/or **7a** and fed, together with technical oxygen supplied via conduit **8** and cycle gas returned via conduit **15**, to the first hurdle **9a** of the contact vessel **9**. By selecting the ratio of the amounts of gas withdrawn through the conduits from the SO_2 gas cycle, the gas temperature required for the contact process can be set at the inlet of the contact vessel.

After partial conversion in hurdle **9a** and recooling in heat exchanger **10**, the gas is further reacted in hurdle **9b**, recooled in heat exchanger **11**, and, after further conversion in the third hurdle **9c**, cooling in heat exchanger **12** and feed water preheater **13**, is introduced into the absorption tower **14**.

In the latter, the thus-formed SO_3 is removed from the gas with concentrated sulfuric acid. The waste gas of the absorption tower is fed via conduit **15** to the heat exchanger **12**; then, it is conducted successively through the heat exchanger **11** and **10**, and, after combining the gas with the SO_2/O_2 mixture from conduits **7** and **8**, it is introduced into the contact vessel **9**. The cycle gas is to be heated in the heat exchangers to such an extent that the mixture of cycle gas and SO_2/O_2 mixture has the start-up temperature of the first contact hurdle.

FIGURE 146: LOW SO_2 EMISSION SULFURIC ACID PROCESS USING SULFUR AND OXYGEN AS FEED MATERIALS

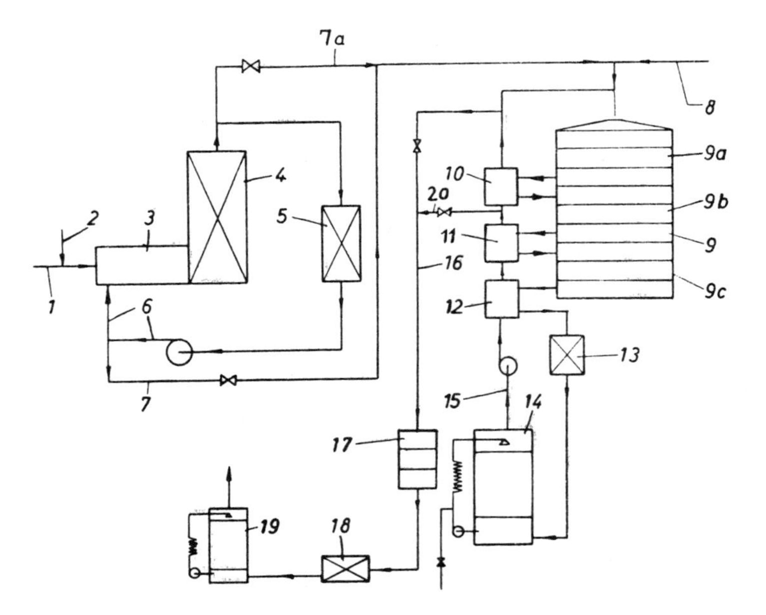

Source: U.S. Patent 3,907,979

In order to maintain a steady state inert gas concentration in the contact gas cycle, a portion of the gas is withdrawn via conduit **16** between the first heat exchanger **10** and the contact vessel **9**, and is further reacted in the separate, small post-contact stage **17**. The thus-formed minor SO_3 content is removed from the gas, after cooling the latter in heat exchanger **18**, in a small final absorption tower **19**. The final gas leaving this tower consists essentially of nitrogen, noble gas, some oxygen, and traces of sulfur dioxide.

Since the start-up temperature of the post-contact stage is lower than that of hurdle **9a** of the main contact stage, a bypass line **20** is provided which connects the cycle conduit **15** between heat exchangers **10** and **11** with the conduit **16**, by means of which it is possible to feed also less strongly heated cycle gas to the post-contact stage. By correspondingly adjusting the proportion of the amounts of cycle gas withdrawn upstream and downstream of the heat exchanger **10**, respectively, with the aid of valves provided in conduits **20** and **16**, it is possible to adjust the gas temperature to that required for the operation of the final contact stage **17** which is preferably about 400° to 420°C.

SULFUR OXIDES

Sulfur dioxide has become a major pollutant of the atmosphere, particularly in urban areas. The presence of sulfur dioxide in the atmosphere is due primarily to the combustion of fossil fuels, i.e., coal and oil, which contain sulfur. Electric power plants constitute a major source of sulfur dioxide pollution of the atmosphere.

Removal from Air

Various processes have been suggested for removal of sulfur dioxide from flue gas, although none has gained a general industry acceptance to date. These processes may be grouped generally as wet processes and dry processes. Wet processes are those which employ an absorbent solution, usually aqueous, for removal of sulfur dioxide from a gas stream.

A flue gas desulfurization process has several requirements. First, it must be capable of removing most of the sulfur dioxide content of the flue gas, preferably 90% or more of the SO_2 present, under widely varying load conditions. Second, it should not create any air or water pollution problems. Third, the process should be easy to operate and maintain. The process should have a low net cost.

In many instances this would require the production of a salable by-product. The process should be capable of incorporation into existing power plants if it is to achieve maximum application. This requirement favors wet processes, which operate at a low temperature and therefore can be placed after the conventional air preheater in which incoming air for combustion is heated by the hot flue gas. Dry processes usually require a much higher operating temperature, and therefore must be inserted ahead of the preheater and integrated with the power plant.

A process developed by *W. Groenendaal, F.C. Taubert, J.E. Naber and G.A. Bekker; U.S. Patent 3,966,879; June 29, 1976; assigned to Shell Oil Company* for the removal of particulate matter and sulfur oxides from waste gases comprises crosscurrent contacting of the waste gas stream with a moving bed of supported, copper-containing acceptor in a first zone, thereby accepting the sulfur oxides and filtering out the particulate matter, removing in subsequent separate zones the particulate matter and the sulfur oxides from the acceptor and, optionally, reactivating the acceptor in a subsequent zone before introducing it back into the first zone for further removal of sulfur oxides and particulate matter. A suitable form of apparatus for conduct of this process is shown in Figure 147.

Flue gas is introduced via line **1** into crosscurrent contacting vessel **2**. The acceptor is fed to this vessel through line **3**. The acceptor is passed through the vessel and removed therefrom via line **4**. The waste gas, which was in crosscurrent contact with the acceptor in the vessel is removed therefrom via line **5**, and may be led to the stack. In general, it may be partly cooled by heat transfer to air to be used for combustion in the furnace from which the off-gas emerges.

The acceptor removed from the crosscurrent contacting vessel via line **4** is led to a vessel **6**, which is provided with perforated plates **7** at a slight angle with the horizontal. The acceptor passes over these plates by gravitational force, and the fly ash present thereon is removed with the aid of steam introduced via line **8**. The fly ash is transported by the steam via line **9** to a cyclone **10**. From this cyclone dry fly ash is removed through line **11**, and a slurry of fly ash in water via line **12**. The acceptor is removed via line **13** from vessel **6** to a riser. Via

FIGURE 147: PROCESS FOR REMOVAL OF SULFUR OXIDES AND PARTICULATE MATTER FROM WASTE GAS STREAMS

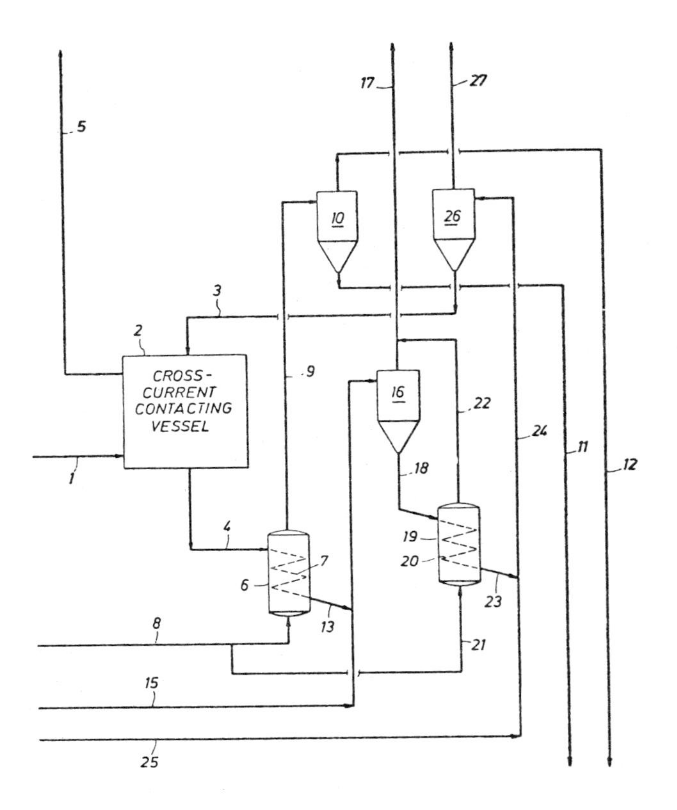

Source: U.S. Patent 3,966,879

line **15** a reducing gas (for example a mixture of steam and hydrogen) is introduced into the riser, and the acceptor may be transported to cyclone **16**. In the riser the SO_2 is removed from the acceptor. From cyclone **16** the gas, which now contains appreciable amounts of SO_2, is removed via line **17**. The acceptor is removed from cyclone **16** via line **18** to vessel **19** equipped with perforated plates **20**, which are at a slight angle with the horizontal.

Steam is introduced into this vessel via line **21**, and this steam, after having been in contact with the acceptor (which may pass over plates **20** by gravitational force) is removed via line **22**. Line **22** may be connected with line **17**. The acceptor leaves vessel **19** through line **23** to a riser. Air is introduced into line **24** via line **25**, and the acceptor is led to a cyclone **26** via line **24**. In this line copper on the acceptor is oxidized to copper oxide. From cyclone **26** the gas is removed via line **27**. It may be used as feed air for the furnace from which the waste gas emerges, or it is passed to stack. The acceptor is recycled from cyclone **26** to reactor **2** via line **3**.

A process developed by *E.D. Tolles; U.S. Patent 3,862,295; January 21, 1975; assigned to Westvaco Corporation* is a process of sorbing noxious sulfur-oxide-containing gases onto activated carbon by increasing the moisture content of the gas being treated by injecting water either prior to or concurrent with the sorption reaction which substantially increases the rate of sorption and permits more accurate temperature control of the process. A bed of fluidized carbon moving substantially countercurrent to the gas being treated is one system found to be especially efficient and effective.

Figure 148 shows a suitable form of apparatus for the conduct of this process consisting of a multistage fluidized bed system contained in a single tower **10.** This tower includes a lower sorbent bed **12** supported on a first perforated plate **14,** a second sorbent bed **16** supported on a second perforated plate **18,** and a third or upper sorbent bed **17** supported on perforated plate **19.** Preferably, the sorbent used in the beds is activated carbon. It should be appreciated that the number of sorption beds will vary depending upon the process conditions.

In operation, flue gas leaves the power plant or other plant effluent stream (not shown) through line **11,** passes upwardly through the beds into cyclone **20,** and is discharged from the cyclone through a chimney into the atmosphere substantially free of sulfur compounds. Temperature sensing means (not shown) are disposed in the beds and control the amount of water introduced into the sorbent beds by sprays through controls **24.** Temperature controller **25** can be utilized for faster response.

Where the SO_3 concentration in the flue gas is substantially zero, or, where corrosion effects can be negated, the sorbent beds can be combined as a single tower. Activated carbon is the preferred sorbent, however, any carbonaceous sorbent can be used, if desired, but with a lesser economic advantage.

FIGURE 148: ABSORBER FOR SULFUR OXIDE REMOVAL FROM WASTE GASES

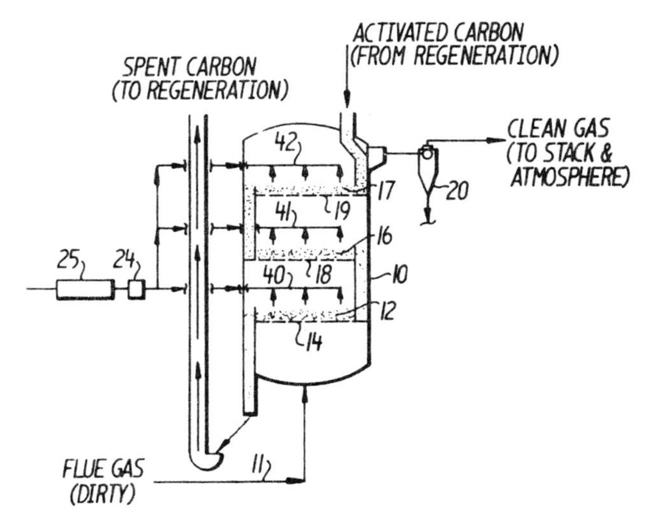

Source: U.S. Patent 3,862,295

The following is a specific example of the conduct of the process. An 18 inch diameter adsorber tower was used to treat flue gas from a gas-fired steam plant employing the fluidized bed system shown. The flue gas flow rate was 15,250 standard cubic feet per minute (measured at 70°F), having a composition comprising 995 ppm SO_2, 3.1% by volume O_2 and 12.7% by volume H_2O. The inlet gas temperature measured 310°F. Activated carbon at the rate of 23.2 lb/hr was passed downwardly through the tower countercurrent to the upwardly-flowing flue gas giving a carbon bed height per stage of about 3 to 4 inches. The temperature at the second stage was kept at about 175°F by the injection of about 29.8 pounds of water per hour. This addition raised the moisture content of the inlet flue gas from 12.7 to 16.4%.

The thus-treated flue gas had an outlet SO_2 concentration of about 69 ppm, that is, more than 90% of the SO_2 present in the inlet flue gas was removed. The bulk of the SO_2 removal occurred in the last three stages after the additional moisture was injected. Leaving the first stage, the gas SO_2 content measured 925 ppm, a decrease of about 70 ppm. This contrasted with a SO_2 reduction in the other stages which ranged from 194 to 384 ppm per stage.

A process developed by *Y. Hishinuma and Z. Tamura; U.S. Patent 3,981,972; September 21, 1976; assigned to Hitachi, Ltd., Japan* is one in which an active carbon suspension is prepared by mixing particles of active carbon with water, and flue gas containing sulfur dioxide is introduced into a desulfurization column, wherein the flue gas is contacted with the active carbon suspension. Sulfur dioxide is absorbed into the active carbon suspension, thereby purifying the flue gas. The absorbed sulfur dioxide is converted to sulfuric acid by a catalytic action of the active carbon in the desulfurization column, and the resulting sulfuric acid is separated from the active carbon suspension.

A process developed by *D.F. Greene, R.J. Lang and A.B. Welty, Jr.; U.S. Patent 3,989,798; November 2, 1976; assigned to Exxon Research and Engineering Company* is one in which flue gas is desulfurized by absorbing it onto a selective absorbent-catalytic material such as vanadia on K_2O. The absorbed SO_2 is recovered by contacting it in a desorption cycle with a reducing desorption gas.

In one embodiment, multiple absorbent beds are employed and alternate absorption-desorption cycles are used in each of the multiple beds with the flue gas and the desorption gas used alternately. The sulfur oxide is recovered in concentrated form in the desorption gas stream and is then utilized as feed in a conventional sulfuric acid plant yielding concentrated sulfuric acid.

See Aluminum Cell Exit Gases (U.S. Patent 3,919,392)
See Fly Ash (U.S. Patent 3,969,094)
See Steam-Electric Industry Effluents (U.S. Patent 3,726,239)
See Glass Industry Effluents (U.S. Patent 3,944,650)
See Cooling Tower Blowdown Effluents (U.S. Patent 3,810,542)

SURFACE ACTIVE AGENTS

See Detergents (U.S. Patent 3,898,159)

SURFACE COATING EFFLUENTS

Spray-type coating chambers, or hoods, are widely used in manufacturing plants, such as in auto plants for painting articles with liquid and powdered paints or coating them with oil. Such chambers normally have entrance and exit openings through which conveyed articles pass. When spray-type coating chambers are used, however, pollution problems often arise. That is, coating-material contaminants escape from the chambers and contaminate surrounding atmospheres, thereby creating unfavorable working conditions in such manufacturing plants.

Removal from Air

One prior art solution to this contamination problem is to employ a vacuum ventilation system for continually sucking excess sprayed coating material from a coating chamber and expelling it into outside atmosphere. Such a ventilation system usually sucks large amounts of air into the coating chamber so as to dilute the coating material which is expelled.

Although this method cuts down on indoor air pollution it increases outdoor air pollution. Further this method is somewhat wasteful in that large amounts of coating material are thrown away. In addition, this method creates excessive air currents through the coating chamber which may cause undesirable concentrations of spray coating material if counter measures are not taken.

A system developed by *A.B. Repp and W.B. Berk, Jr.; U.S. Patent 3,750,622; August 7, 1973* is one in which a fluid such as air, is circulated in a closed-loop path outside of a coating chamber. The circulating fluid is directed so as to form a fluid curtain across entrance and exit openings of the coating chamber. The air curtain picks up waste passing out of the openings. Because there is very little outside air added to the circulating fluid, coating material and contaminants are at a high concentration level in the circulating fluid and can therefore be efficiently separated or cleaned out of the circulating fluid.

A process developed by *S.W. Forney; U.S. Patent 3,861,887; January 21, 1975* is one in which water used to wash the air in paint or lacquer spray booths in order to remove over-sprayed paint or lacquers is treated with a blend of polycationic water-dispersible polymer and a water-soluble salt of an amphoteric metal. This acts to reduce the tackiness of paint and lacquer solids and thereby reduce the tendency of over-sprayed paints and lacquers to adhere to walls, ceilings and floors of the spray booths and also to condition the paint and lacquer solids removed with the water so that they can be separated and the water recycled for further use in washing the air in the spray booth.

See Spray Booth Effluents (U.S. Patent 3,881,874)

In the coating of articles with a water-borne paint by an electrophoretic process, the article after coating is rinsed to remove any residual coating medium and any resin or pigment which is only partially coagulated. The rinse water contains pigments and resins which are most desirable to recover for further use.

Removal from Water

It is known to treat effluents from the process, i.e., the pigment and resin contaminated rinse water, by the reverse osmosis process. In one such process effluents from a painting plant are brought into contact with a semipermeable membrane which is porous to water molecules but impermeable to large molecules, the effluent being pressurized to cause the water in the effluent to pass through the membrane. This treatment provides for the disposal of the effluent and enables a concentrated solution of resins and pigments to be recovered for further use.

In this prior treatment process, where the rinse operation is performed over the coating tank, the effect of the rinse water on the concentration of pigment and resin is very small and it is quite a simple matter to reconcentrate the contents of the tank; where however, the rinsing operation is carried out at a position remote from the coating tank, as is frequently the case, it is essential, in order to effect treatment of the tank contents to maintain the pigment and resin concentation and to hold the pH value at the desired level, that the rinse water containing resins and pigments be conveyed quickly back to the coating tank, thereby avoiding the danger of instability of the pigment and resin content so that the contents of the tank can then be treated effectively by the reverse osmosis process.

A process developed by *J.F. Wallace and J.G. Ransome; U.S. Patent 3,556,970; January 19, 1971; assigned to Pressed Steel Fisher Limited, England* in which the contaminated rinse water is conveyed from a rinsing station directly back to a treatment tank and the thus diluted contents of the treatment tank are passed as an effluent under pressure through a reverse osmosis unit to separate the effluent into parts of high and low concentration of the original contents of the treatment tank.

One example of a reverse osmosis unit suitable for this purpose comprises a semipermeable member of copper ferrocyanide formed in the walls of an unglazed reinforced porous porcelain pot. A second example comprises a membrane of cellulose acetate and magnesium perchlorate dissolved in dioxan, the membrane being formed into a tube and inserted into a porous tube of glass fiber, the whole being encased in a perforated stainless steel tube. Effluent having a high concentration of resins and pigments is returned directly to the tank while water or effluent having a low concentration is conveyed to the rinse station or to waste or for use in a further process.

TALL OIL PROCESSING EFFLUENTS

The fractionation of tall oil is conventionally carried out under vacuum using stripping steam which carries from the top of the fractionation column the most volatile and therefore the most odorous of the unsaponifiable material in the crude. A condenser communicates with the top of the column to maintain the vacuum and transform most of the stripping steam and low-boiling odorous compounds into liquid.

In one type of system being used, the condenser is of the barometric type and the cooling water is recirculated foul water from that condenser. After the condenser, the water is usually passed through a settling pond to remove an oil layer after which the water is circulated over a cooling tower and returned to the condenser. The use of a settling pond is advisable since it minimizes fouling of the cooling tower.

A disadvantage to such a system is that it creates an air pollution problem since the cooling tower air strips odors from the oily water. This is true whether or not a settling pond is used. The odor produced is not wholly unpleasant but is extremely strong particularly when redistilling the heads produced to make the prime products. While objectionable, the odor is not considered hazardous and presents no health problem. This type of system reduces to a minimum any problems associated with stream pollution because under most weather conditions the evaporative loss from the cooling tower exceeds the stripping steam added to the system in the barometric condenser. Thus there is no foul wastewater except in extended periods of rainy weather.

A two-stage jet with intercondenser usually follows the first condenser. The first condenser removes essentially all the organic contaminants so that a relatively pure stream of noncondensable gases and water enters the two-stage system and the condensate produced by the two-stage system is therefore not objectionable.

Another type of system which is used is one where vacuum surface condensers are used for the stripping steam. This is slightly more expensive than the barometric type even though the latter requires a separate clean water cooling tower. The surface condenser does eliminate the odor from the cooling tower but a foul stream of condensate equivalent to the stripping steam is produced.

Still another system in use employs a closed barometric condensing system with a cooler to cool the circulating foul water indirectly with air or water. A purge from the water circulating, equivalent to the stripping steam, is then produced as a polluted stream which requires further treatment.

It can be seen that the tall oil distillation plants in use create both air and stream pollution problems. The trend has been toward the closed condensing systems. The stream of polluted water resulting from such systems is often treated, but complete purification has been too difficult and expensive to be practicable. Mechanical and chemical treatment systems have been used, but only with limited success. In addition, biological treatments in conventional aerated activated sludge plants have been performed on tall oil wastes.

The polluted water stream even after settling will contain as much as 1% of dissolved and emulsified organic chemicals. These are mainly unsaponifiable materials as contrasted to the rosin and fatty acid which make up the bulk of the crude tall oil in the distillation system. These include sterols, hydrocarbons, terpenes, alcohols as well as decomposition products resulting from the cracking of fatty acid rosin in the distillation process.

Removal from Air and Water

A process developed by *D.F. Bress; U.S. Patent 3,709,793; January 9, 1973; assigned to Foster Wheeler Corporation* is one in which tall oil is fractionated with little or no environmental pollution. The stripping steam and the odorous compounds it carries with it are condensed by foul water being recycled through the condenser, an amount of foul water equal to the stripping steam coming into the condenser being purged, revaporized and used again as stripping steam in the fractionation process.

Such a process is shown in Figure 149. In the drawing a stream **1** of stripping steam plus contaminants from a fractionation tower (not shown) leads to a condenser **2** which may be of the barometric type. Foul water from the condenser **2** flows through line **3** to a hot well **4** from which a pump **5** circulates it by line **6** to a cooler **7**. The cooler **7** is fed coolant through a line **8** which cools the foul water which is then returned to the condenser **2** through line **9**.

Noncondensable gases from the hot well **4** are vented by line **4a**. Preferably, the gases are not vented directly to the atmosphere but are disposed of by incineration, an elevated stack or by other means which would eliminate any odor.

Noncondensable gases coming out of the condenser **2** through line **10** have water vapor with little if any organic content. The noncondensable gases flowing through line **10** are compressed in a multistaging steam evaporator system **11** having an interstaging condenser **12** from which the condensate is fed to a hot well **13** from which it overflows to a sewer or other water system through line **14**. It has already been pointed out that the condensate from the interstaging condenser is not objectionable. Gases from the hot well **13** are disposed of in the same manner as the gases vented through line **4a**.

A portion of the circulating water coming out of the pump **5** is purged by line **16**. The amount of water which is purged is such that in any given time, it will be approximately equal to the amount of stripping steam condensed in the condenser **2**. The water in the line **16** is heated in the heat exchanger **17** which is fed heating medium through a line **18**. The heating medium may be steam, flue gas or a hot process stream. Such heating promotes the separation of emulsified organic material when the water is led to a coalescing filter **19** through line **20**.

The liquid then flows through line **21** to a decanter **22** which is suitably baffled to separate the organic or oil phase from the water phase. The organic phase leaves through line **23** and may then be added to low-grade fractions produced in the tall oil plant or it may be burned. Gases are vented through line **24** and may be treated in the same manner as the gases which are vented through lines **4a** and **15**.

FIGURE 149: SCHEME FOR CONTROL OF ENVIRONMENTAL POLLUTION IN TALL OIL FRACTIONATION

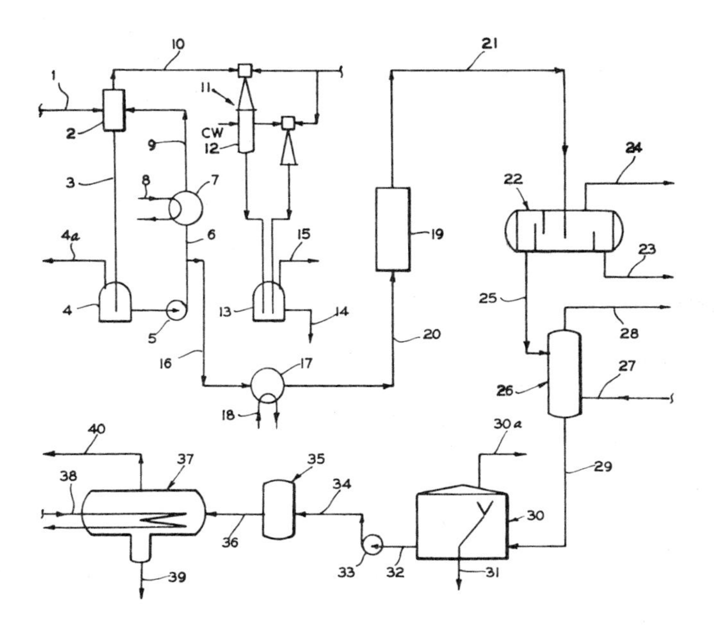

Source: U.S. Patent 3,709,793

The water phase passes through line **25** to a stripper vessel **26** in which a small amount of steam is injected through line **27** to strip odors from the water. The vapor from the stripper **26** passes through line **28** and is preferably treated in the same manner as the gases vented through lines **4a**, **15** and **24**. Stripped water from the stripper **26** passes through a line **29** to a surge tank **30** having a floating suction drain **31** which is used to remove any remaining organic matter which is separating from the water phase.

The organic material exiting through line **31** is added to the material leaving through line **23** to be treated in the same manner as the material flowing in that line. Gases leave surge tank **30** through line **30a** and are treated as the gases flowing out through lines **4a, 15, 24** and **28**.

After passing through the surge tank **30**, the water passes through line **32** to a pump **33** which forces it through line **34** to an activated carbon treating vessel **35** which enhances the quality of the water. After passing through the vessel **35**, the water is led through a line **36** to a vaporizer **37** which employs a heating coil **38** to vaporize the water. The heating coil **38** uses a heating medium such as hot process stream or low-pressure steam. Organic material which is separated from the water in the vaporizer **37** is withdrawn through line **39** and added to that withdrawn through lines **23** and **31**. The vaporized water passes through line **40** to be used as stripping steam in the fractionation process.

The process eliminates air and water pollution. It should also be appreciated that it obviates the treatment of boiler feed water. In conventional systems, the stripping steam is generated in a boiler and then used in the fractionating tower. Subsequently, it is condensed and the condensates are purged. By utilizing boiler steam, hot process fluid or other means to revaporize the condensate for reuse, the need for fresh boiler feed water is eliminated.

TAR SAND PROCESSING EFFLUENTS

Tar sands which are principally located in the Athabasca region of Alberta, Canada, are used as a supplement source of synthetic crude oils, and it is hoped that in the future, their importance as a source of oil would increase. The tar sands are contained in beds in mixture with sand and other inorganic minerals. Typically, these sands contain from about 6 to 20% bitumen (also referred to herein as oil), from about 1 to 10% of water, and from about 70 to 90% of mineral solids. The major portion by weight of the mineral solids in bituminous sand is quartz sand having a particle size greater than about 45 microns and less than 2,000 microns.

Various methods are known for separating bitumen from bituminous sand. Some of these methods involve the use of water for preparing a slurry at a temperature above about 75°F. Most of the coarse sand and portions of the fines are separated from the slurry by various means, such as settling, to recover an emulsion or froth which contains some of the fines and quantities of coarse sand. Such an emulsion or froth is simply referred to herein as froth. One well known method for preparing such froth is often referred to as the hot water process.

In the hot water method the bituminous sand is slurried with steam and hot water at about 180°F and the pulp is then agitated with a stream of circulating hot water and carried to a separation cell maintained at an elevated temperature of about 180°F. In the separation cell, entrained air causes the bitumen to rise to the top of the cell in the form of a froth. The froth contains air with the emulsion of bitumen, water and mineral solids. The mineral solids are extremely difficult to separate from the bitumen and unless the froth is further treated it will generally contain at least 3% of mineral solids. The solids and water remaining may also be processed.

The hot water process not only results in the separation of the bitumen from the tar sands, but also in the classification and separation of the minerals contained therein from the water, called middlings water. Classification of middlings water, both to recover valuable mineral by-products and to provide a water which may be returned to native waters is an extremely costly and time-consuming process.

Generally, the waters are introduced into sludge settling ponds where aprroximately three months are required for untreated material to settle to a contamination level where clarified water may be returned to the source. Organic flocculating agents have been utilized to remove the clays and fine solids from the water prior to its discharge into the native water, but have proven far too costly and inefficient an accelerator of the clarification process. Inorganic agents such as calcium chloride leave residues in the waters which are harmful to the overall clarification process. What is required is an economical method to clarify middlings water, and in some instances tailings water, to provide a clarified water suitable for return to the native water source.

Removal from Water

A process developed by *R. Schutte; U.S. Patent 3,816,305; June 11, 1974; assigned to Gulf Oil Canada Limited, Canada-Cities Service, Ltd., Imperial Oil, Limited and Atlantic Richfield Canada Ltd., Canada* for the clarification of middlings waters comprises the introduction of sulfuric acid into the water to reduce the pH to within the range 5.5 to 7.0 to cause flocculation and coagulation of the clays and small solids contained therein. The product is allowed to stand for several days to allow the pH to return to neutral. In addition to solids removal, a decrease in surface active compounds and organic residue is then observed.

The process may further comprise stirring the treated tailings or middlings water until a pH equal to about 7.0 is obtained wherein the water may be returned to the native water source subsequent to clarification and neutralization.

TELLURIUM HEXAFLUORIDE

A continuous fluoride volatility process has been studied at the Argonne National Laboratory to separate uranium and plutonium from spent nuclear fuel elements. During the process, in which fuel elements are oxidized and later fluorinated to produce volatile fluorides, tellurium hexafluoride as well as other fission product fluorides must be separated for storage and eventual disposal. Because the gas stream which contains the tellurium hexafluoride will also contain large amounts of elemental fluorine, systems designed to eliminate the tellurium hexafluoride must also be able to handle the fluorine present.

Removal from Air

Sorption of fission product gases or other radioactive gases onto solids which have a relatively small volume and are easily stored for long periods of time is one answer to the waste disposal problem. A process developed by *D.R. Vissers, M.J. Steindler and J.T. Holmes; U.S. Patent 3,491,513; January 27, 1970; assigned to U.S. Atomic Energy Commission* is one in which tellurium hexafluoride and fluorine are removed from a gas by passing the gas through a fluidized bed of activated alumina to remove the greater portion of fluorine and thereafter passing the effluent from the fluidized bed through a packed bed of activated alumina to remove the tellurium hexafluoride.

TEREPHTHALIC ACID PROCESS EFFLUENTS

Waste streams from a plant which produces terephthalic acid by hydrolysis of terephthalonitrile will contain varying amounts of ammonia, terephthalonitrile, tolunitrile, benzonitrile, terephthalic acid, toluic acid, benzoic acid, and ammonium salts of these acids.

Removal from Water

A process developed by *R.T. Whitehead, B.J. Luberoff and M.C.-Y. Sze; U.S. Patent 3,836,461; September 17, 1974; assigned to The Lummus Company* is one in which an aqueous waste stream from an industrial chemical or other processing plant can be utilized to furnish cooling water for use in the same or another plant by partially vaporizing it in a cooling tower, after filtering off any suspended solids, utilizing the cooled liquid as cooling water for heat exchangers, and returning the heat-exchanged water to the cooling tower.

Chemicals precipitated from the waste stream during vaporization can be recovered or burned as fuel in the plant. Small amounts of volatile wastes can be removed by vaporization in the cooling tower. Total recycle or elimination of all materials in the waste stream results in the reduction or elimination of pollution of local waters. Such a process is shown in Figure 150.

FIGURE 150: PROCESS FOR PURIFYING WASTEWATERS FROM PLANT FOR PRODUCING TEREPHTHALIC ACID FROM TEREPHTHALONITRILE

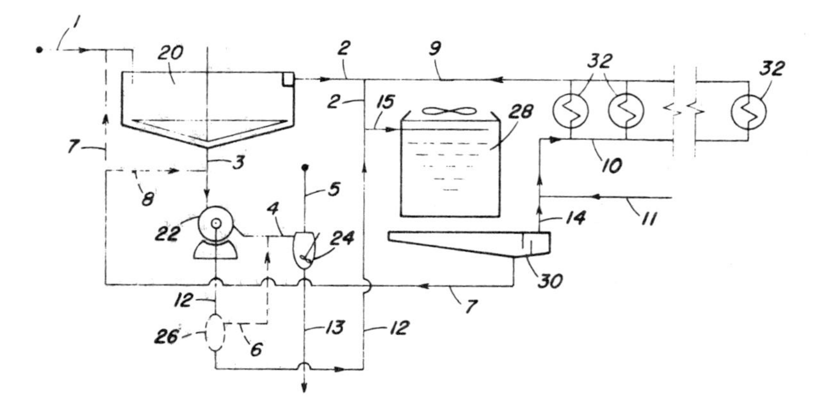

Source: U.S. Patent 3,836,461

The aqueous waste stream is conducted in line **1** to a settling basin **20**, in which the contained solids are permitted to settle to the bottom. Flocculating and/or antifoam agents may be added to facilitate settling. The heavy sludge is removed in line **3** to filtering or centrifuging apparatus **22** (shown in the figure as a rotary filter). Light immiscible liquids can be skimmed off the top of settling basin **20**.

The filtrate discharged from apparatus **22** is then introduced via line **12** into the top of cooling tower **28**. If necessary, the filtrate can first be introduced into another settling basin **26** for removal of heavy immiscible liquids. Alternatively the filtrate can be returned to the settling basin **20**.

The filtrate in line **12** is introduced into the cooling tower through line **15**, together with the clear liquid overflow from settling basin **20**, transmitted via line **2**, and recycled cooling water in line **9**. In the cooling tower, the aqueous waste stream becomes concentrated due to evaporation of part of the water, forming an aqueous saturated solution of the contained chemicals. Concentration of the effluent in this manner also will result in precipitation of some waste components.

Cooling water is removed via line **14** from settling basin **30** and fed through heat exchangers **32** via line **10**, cooling one or more process streams. The heated coolant is then returned in line **9** to the cooling tower. Sludge from settling basin **30** is removed in line **7** and either passed to settling basin **20** or, via line **8**, to filter **22**. The waste stream thus serves essentially to furnish makeup water to the system. If necessary, additional makeup water can be added in line **11**.

In some cases the sludge from refineries or organic synthesis processes may contain unreacted or partially reacted starting materials which may be recycled to the process for production of additional product. Such is also the case with some waste streams from mining operations. The inorganic minerals and chemicals recovered have little or no fuel value, but they can be recycled to ore processing steps.

For example, in the process shown, sludge from filter **22** is mixed in tank **24** with fuel oil introduced via line **5**, the mixture is removed in line **13** and used for fuel in the plant as needed. Immiscible liquids in the waste stream can be removed by passing the filtrate from filter **22** into another settling basin **26**. The immiscible liquid layer is removed in line **6** and incorporated in the sludge fuel oil mixture. The choice of whether to utilize the sludge for fuel, or fertilizer, or to recover chemicals contained in it, is essentially based on two factors: the technical feasibility of recovering the chemical or chemicals and the economics of doing so, as opposed to utilization for fuel.

In the application of the process illustrated to the effluents from a plant producing terephthalic acid by the hydrolysis of terephthalonitrile, the ammonium salts are dissociated in the cooling tower; the remaining materials can be precipitated and mixed with fuel oil. The stream in line **11** will contain additional ammonia to prevent the precipitation of terephthalic acid in the heat exchangers. For example, from a plant designed to produce 300,000,000 pounds per year of terephthalic acid by hydrolysis of terephthalonitrile, aqueous waste streams from several processing steps can be combined to produce a stream comprising about 84,648 pounds of material per hour, containing:

	Pounds per Hour
Tolunitrile	11
Terephthalonitrile	672
Benzonitrile	156
Terephthalic acid	9
Ammonium terephthalates	165
Water	83,635

The combined stream is introduced via line **1** into settling basin **20**. The sludge, comprising tolunitrile (approximately 1 lb/hr), benzonitrile (approximately 6 lb/hr), terephthalonitrile (672 lb/hr), terephthalic acid (149 lb/hr) mainly transferred from the sludge in basin **30** and water (821 lb/hr) is recovered and slurried in tank **24** with fuel oil from line **5**.

Filtrate, comprising the remainder of the water, tolunitrile, ammonium terephthalates and benzonitrile, is introduced via lines **2, 12** and **15** into cooling tower **28**, along with recycled cooling water in line **9**. Air is passed through the cooling tower at a rate of about 5,500,000 lb/hr, and the air leaving the tower would carry with it about 287,200 lb/hr water, plus all the benzonitrile and tolunitrile stripped from the circulating cooling water (150 and 10 lb/hr respectively), and about 58 lb/hr ammonia formed by decomposition of ammonium terephthalates.

The liquid removed from settling basin **30** under the cooling tower, is passed via lines **10** and **9** through heat exchangers **32**. Ammonia still bottoms, comprising about 205,000 lb/hr water and 33 lb/hr ammonia are added in line **11** to maintain the terephthalic acid in solution. The total stream circulating through the heat exchangers comprises more than 12,500,000 lb/hr of material.

Thus, the benzonitrile and tolunitrile have been effectively disposed of by vaporization in the cooling tower, the terephthalonitrile and terephthalic acid burned for fuel, and ammonium salts removed in the cooling tower by dissociation into ammonia and terephthalic acid.

TETRAALKYLLEAD COMBUSTION PRODUCTS

Removal from Air

Air pollution may be reduced by the complete removal of alkyllead compounds from gasoline. The reduction may occur in two ways: first, by removing tetraalkyllead poisons for automotive pollution converter catalysts; and second, by eliminating a source of particulate lead in the atmosphere.

A process developed by *D.D. Whitehurst, S.A. Butter and P.G. Rodewald; U.S. Patent 3,944,501; March 16, 1976; assigned to Mobil Oil Corporation* is based on the discovery of sorbents which selectively remove alkyllead compounds from gasoline. The sorbents are comprised of metal halides (preferably tin tetrachloride or antimony pentachloride) bonded to a suitable substrate through at least one amine or alkyl halide functional group. The sorbents can be effectively regenerated.

TETRAALKYLLEAD MANUFACTURING EFFLUENTS

Removal from Air

A process developed by *W.C. Jaasma; U.S. Patent 3,403,495; October 1, 1968; assigned to Ethyl Corporation* involves the recovery of tetraalkyllead compounds from an inert gas stream by contacting the stream at least once with an inert organic scrubbing liquid.

Figure 151 shows the configuration of apparatus for the process as applied to the production of tetramethyllead in the presence of toluene. Referring to the drawing, inert gas vapor from a steam still or other product recovery or production means containing quantities of tetramethyllead and toluene or other volatile condensables such as aluminum compounds enters the process by line **10** and is fed to fume scrubber **11** where it is contacted with aqueous caustic solution entering fume scrubber **11** by line **12** and line **5** to produce a scrubbed gas phase indicated in first sump vessel **13** at **14** and an aqueous liquid phase **15**.

FIGURE 151: APPARATUS FOR RECOVERY OF TETRAALKYLLEAD FROM AIR BY SOLVENT SCRUBBING

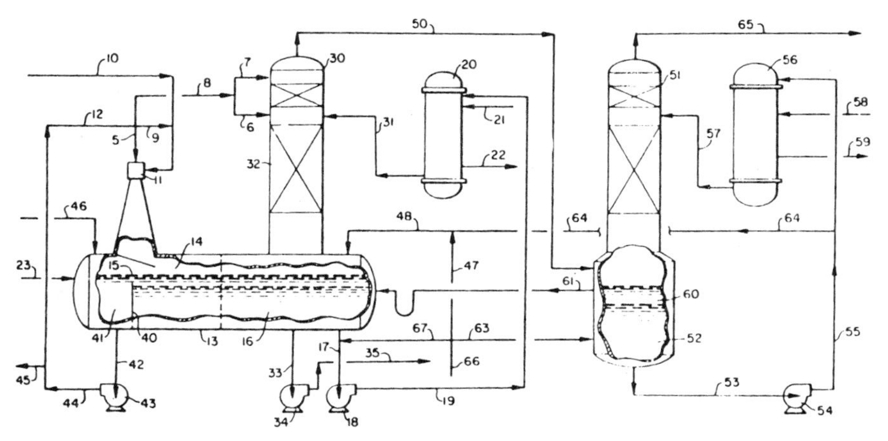

Source: U.S. Patent 3,403,495

Should the quantity or quality of volatile aluminum compounds, or other material, in the inert gas stream of line **10** be such that condensation thereof may cause plugging of fume scrubber **11**, additional quantities of aqueous caustic solution may be added to or blended with the entering inert gas stream by line **9** to prevent such plugging. It is clear also that when the process is used for the recovery of tetraalkyllead product from an inert gas stream containing no or substantially no volatile aluminum compound or like material, fume scrubber **11** may be bypassed or eliminated, and the product containing inert gas fed directly to vapor space **14** without any pretreatment.

A first sump is vessel **13** containing the aqueous liquid phase **15** and a first organic liquid **16** from this first sump is removed by line **17** through pump **18** and line **19** to cooler **20** where the first organic liquid phase may be cooled to a temperature of from -10° to 50°F by brine entering cooler **20** by line **21** and leaving by line **22**. Cooled first organic liquid phase from the cooler enters tower **30** by line **31** where it contacts the scrubbed gas from vessel **13** countercurrently.

Suitable packing section **32** or other contacting means such as distillation plates may be provided in the tower to permit good contact of the liquid and gas phases. The first organic liquid phase after contact returns to the vessel and settles to the organic liquid phase with extracted tetramethyllead and toluene contained therein, and tetramethyllead product and toluene recovered from the scrubbed gas phase, along with first organic liquid are removed from the system as product by line **33** to pump **34** and then to process product recovery by line **35**.

By the use of weir **40** located within the vessel, first organic liquid phase is retained within the vessel, except for that removed to product recovery, and aqueous liquid phase is permitted to overflow into aqueous sump **41** where it may be removed by line **42** through pump **43** and recycled to fume scrubber **11** by line **44** joining line **12**. As necessary, portions of contaminated aqueous caustic solution may be removed from the system to waste by line **45** and fresh aqueous caustic solution added to the system by line **46**.

In operation of the process over long periods, it may be necessary or desirable to provide a continuing or intermittent aqueous caustic wash for the tower to prevent buildup of deposits in the contacting section **32** or the demisting section, if any, above the contacting section. When this is necessary or desirable, aqueous caustic solution from line **12** may be supplied in a desired quantity by line **8** through line **6** or **7** or both. As necessary, and generally in a volume approximating that volume removed to product recovery, substantially pure organic scrubbing liquid may be added to the system from a supply not shown by line **66** to line **47** and then to line **48**.

In most process applications, the concentration of desirable products in the second scrubbed gaseous phase exiting the tower by line **50** is reduced to a minimum and the venting gas phase either may be recovered for reuse as inert gas or vented to the atmosphere without pollution. In operation in this manner as a single tower system, it is desirable to supply substantially pure organic scrubbing liquid in a necessary volume to that portion of the scrubbing liquid being pumped to cooler **20** and this may be accomplished as shown by line **67** from supply line **66**.

Under certain conditions of production and high volumes of vent gas, it is sometimes desirable to remove only a major portion of product from the inert gas system in the tower and to remove the final minor portion of the product in tower **51** which may be substantially identical to tower **30**. When operating the process in this manner, second scrubbed gaseous phase exiting tower **30** by line **50** is fed to the bottom portion of tower **51** where it is contacted countercurrently with organic scrubbing liquid from sump **52** in the bottom portion of tower **51**.

The organic liquid is removed by line **53** and pumped by pump **54** through line **55** to cooler **56**, then to the top portion of tower **51** by line **57**. As in the case with cooler **20**, chilled brine enters cooler **56** by line **58** and exits by line **59** to cool the organic liquid to a temperature of from -10° to 50°F.

The second contacting organic liquid phase flows down tower **51** to sump **52** which is connected by line **61** to first sump **16** so that product containing organic liquid does not build up within sump **52** beyond the level of connecting line **61** and any excess product containing organic liquid flows to sump **16** within vessel **13** for use in the first contacting in tower **30** and subsequent removal to product recovery. Any aqueous phase which may be present in second sump **52** as indicated at **60** also may flow to the first sump in vessel **13** for joining with the aqueous phase therein as indicated at **15**.

When the system is operated using two contacting towers, substantially pure organic scrubbing liquid may be charged to the system of both sumps; however, it is preferred to charge it by line **63** only in a quantity substantially equal to that being removed to product recovery. In this manner, substantially pure organic contacting liquid is used to remove the remaining trace quantities of product from gaseous phase entering tower **51** to obtain a more efficient operation. Connecting line **64** joining line **48** may be provided to pump quantities of organic liquid to first sump **16** to maintain the desired level, if needed, and the substantially product-free inert gas exiting tower **51** by line **65** may be vented to the atmosphere without air pollution problems or recovered for reuse in an inert gas system, if desired.

In a typical example of the operation of the process for the recovery of tetramethyllead and toluene from a nitrogen gas stream, a vapor feed composition having a concentration of 57% nitrogen, 40.2% tetramethyllead-toluene mixture and 2.8% water, all percentages by weight, and having a temperature of 100° to 110°F, was fed to a fume scrubber where it was scrubbed with a 4% by weight aqueous solution of sodium hydroxide. Scrubbed vapors continued to a first tower where they were contacted countercurrently with ethylene dichloride at a temperature of approximately 35°F in a quantity sufficient to cause the gaseous phase leaving the first tower to have a temperature of approximately 50°F.

The contacted vapors from the overhead of the first tower were fed to the bottom of a second tower where they were contacted countercurrently with ethylene dichloride entering the tower at a temperature of approximately 32°F. Contacted vapors entered the second tower at a temperature of approximately 50°F and exited at a temperature of approximately 35°F. Both towers were substantially identical and contained 2-inch steel Pall rings in a depth in the tower sufficient to provide approximately three theoretical distillation plates in each tower and each tower was operated at substantially atmospheric pressure.

The concentration of the sodium hydroxide in the aqueous caustic solution was maintained at approximately 4% by weight by the addition of a 25% by weight aqueous sodium hydroxide solution as necessary and fresh ethylene dichloride was added to the sump of the second tower as necessary to compensate for the volume of ethylene dichloride-tetramethyllead-toluene mixture removed from the sump below the first tower.

Approximately 90% of the tetramethyllead-toluene mixture in the vapor feed stream was recovered in the first tower and a total of approximately 99% of the tetramethyllead-toluene mixture in the original vapor feed stream was recovered in the two-tower system. Based upon total quantity of tetramethyllead produced from a reaction method not utilizing the process, sufficient tetramethyllead-toluene mixture was recovered using the process to increase the overall yield of the tetramethyllead produced approximately 4 to 5%.

Removal from Water

Aqueous effluents from tetraalkyllead manufacturing processes usually contain dissolved organic and inorganic lead compounds. The dissolved organic lead content can amount to 100 to 100,000 ppm, the dissolved inorganic lead content, to as much as 4,000 ppm.

It is known that water-soluble organic lead compounds can be formed from tetraalkyllead compounds under acidic or oxidizing conditions. For example, the aeration or oxygen purification of tetraalkyllead compounds in the presence of water leads to the formation of some soluble trialkyllead compounds. Moreover, the transalkylation of a mixture of tetraalkyllead compounds in contact with a Lewis acid to form tetra(mixed alkyl)lead compounds usually is accompanied by the formation of water-soluble alkyllead compounds. Disposal of the aqueous effluents may provide pollution problems because of their high lead contents and, in addition, may result in undesirable loss of lead.

The prior art has disclosed the decomposition of hydroxides, halides and carbonates of trialkyllead compounds by steam distillation to form tetraalkyllead compounds, inorganic lead compounds and alkyl halides or hydrocarbons. The decompositions do not go to completion and are relatively slow, requiring 2 to 8 hours to achieve substantial decomposition.

U.S. Patent 3,308,061 discloses the removal of soluble organolead compounds from aqueous effluents produced in the manufacture of alkyllead compounds by treating the effluents with ozone at a pH of 8.0 to 9.5. Such a treatment, using a high molar ozone to lead ratio, may be expensive and may preclude recovery of alkyllead from the effluent.

Processes for lowering the concentration of dissolved inorganic lead in aqueous effluents are known. Canadian Patent 572,192 discloses the removal of inorganic lead as its carbonate at a pH of 8.0 to 9.5. Although this method can reduce inorganic lead to 2 to 4 ppm, it does not materially affect the level of lead present as alkyllead cations.

A process developed by *C. Lores and R.B. Moore; U.S. Patent 3,770,423; Nov. 6, 1973; assigned to E.I. du Pont de Nemours and Company* involves treating an aqueous medium containing dissolved organic and inorganic lead compounds, for example, aqueous effluent from the manufacture of tetraalkyllead, with an alkali metal borohydride to substantially reduce the level of dissolved lead in the aqueous medium.

In the process, an aqueous medium containing 2 to 300 ppm of lead in the form of soluble organolead compounds and up to 10 ppm of lead in the form of soluble inorganic lead compounds is treated with an alkali metal borohydride to produce insoluble lead compounds and an aqueous medium containing less than 2 ppm of lead in the form of soluble compounds.

A process developed by *E.R. Taylor, Jr.; U.S. Patent 3,697,567; October 10, 1972; assigned to E.I. du Pont de Nemours and Company* involves the removal of dissolved organic lead from an aqueous effluent produced in the manufacture of alkyllead compounds by contacting a metal more electropositive than lead but essentially nonreactive with water with the effluent until the dissolved organic lead is converted to an insoluble lead-containing product, leaving a reduced dissolved organic lead content in the effluent. The following is one specific example of the conduct of this process.

250 ml of effluent water from the tetraethyllead process with a pH of 11.2 and containing 9 ppm lead [90% as organic lead, tri(mixed C_1 and C_2 alkyl)lead, and 10% inorganic lead as Pb^{++}] were passed at 10 ml per minute through a column of No. 40 mesh iron filings 3 cm in diameter and 13 cm deep. The effluent water had a contact time of 9.2 minutes with the column of iron. The effluent water after the iron contact contained no organic phase and a maximum lead

content of 0.2 ppm, thereby showing a greater than 97% reduction of lead content.

A process developed by *H.E. Collier, Jr.; U.S. Patent 3,308,061; March 7, 1967; assigned to E.I. du Pont de Nemours and Company* is a process for treating an aqueous effluent from the manufacture of alkyllead compounds and containing about 5 to 5,000 ppm of lead as dissolved organic lead not precipitable by pH adjustment to 8 to 9.5, which process comprises the steps of:

(1) Adjusting the pH of the effluent to between 8.0 to 9.5;
(2) Intimately contacting the aqueous effluent with an ozone-containing gas;
(3) Precipitating the converted lead compounds; and
(4) Separating the precipitated lead-containing compounds from the aqueous effluent.

This process is based on the surprising discovery that ozone is highly effective in removing water-soluble organic lead compounds from wastewater by producing water-insoluble products. Indeed, the discovery that ozone successfully converted the soluble products to water-insoluble materials was most surprising since air and other oxidizing agents have been found unsatisfactory to effect this removal.

By this process the dissolved alkyllead can be removed from the aqueous effluent to such an extent that the aqueous effluent from tetraalkyllead manufacture can be safely discharged into lakes and streams. The process is easy to operate, and the quantity of materials, the time and intimacy of contact are easily coordinated to reduce the dissolved organic lead content to an acceptable level, below 5 ppm, or substantially nil if desired.

TETRABROMOMETHANE

Processes are known for manufacturing copies and stencils using formation of free radicals from halogen compounds to develop colored or polymer reaction products. Preferred radical formers are short chain polyhalogen compounds particularly tetrabromomethane. An explicit description of such a process is given, e.g., in U.S. Patent 3,042,519.

The high volatility and toxicity of the tetrabromomethane is a hindrance to the practical use of this process which is to produce higher quality copies. At any one place during the manufacture of the copying material or its processing, tetrabromomethane must be passed on to or into the copying material before exposure to light. In this way a certain part of the tetrabromomethane is constantly being evaporated in the surrounding gas phase. Due to the toxicity of the tetrabromomethane this quantity must be kept low. If this is not directly successful, then care must be taken that the operator does not inhale any air containing tetrabromomethane.

Removal from Air

A process developed by *R. Dietrich; U.S. Patent 3,437,429; April 8, 1969; assigned to Kalle AG, Germany* is characterized by the fact that the gas containing the tetrabromomethane is washed with a liquid which contains at least one aliphatic or araliphatic amine or an amine closed to form a ring and/or an amino alcohol with up to 20 carbon atoms in such a way that, in a conventional manner, an extensive surface of contact between the washing liquid and the gas phase is provided.

The process is based on a very fast-acting chemical reaction between tetrabromomethane and the amine group, the mechanism and reaction products of which are not yet exactly known. The fact that the reaction concerned is a chemical reaction means that the washing liquid cannot be regenerated, as is possible in an adsorption process. The process is not limited to air purification, but can also be applied to other inert gases which contain tetrabromomethane.

The process provides for the removal of gaseous tetrabromomethane from gases, operates effectively both with high and low concentrations, and provides purification of contaminated air to a level of less than about 0.3 ppm of tetrabromomethane.

TEXTILE FIBERS

Various textile-fiber processing operations, such as the carding operation, customarily generate large amounts of particulate matter of fly such as lint, fibers, dust and other particulate matter. Air-suction devices are now frequently employed to entrain such matter and remove it from the vicinity of the fiber-processing operation. The suction devices may be associated with entire processing operations or areas, or with individual fiber-processing machines therein.

In the case of the carding operation, for example, various fly-generating sections of the individual cards of a series of carding machines may be enclosed by hoods or shrouds, through which air is continuously withdrawn while the machines are in operation. Such a system may illustratively encompass 100 carding machines each requiring 1,000 cfm of air at a negative pressure, at the card suction orifice, equivalent to 4 inches of water.

Removal from Air

A system of the type described above must of course include some means for subsequently removing, from the air employed in the system, the particulate matter entrained therein. Such removal should be accomplished with maximum efficiency and minimum pressure-head losses, since otherwise the cost of operating the system could easily become excessive. Rotating-drum filters, by themselves or sometimes in association with other filtering means, have heretofore been proposed for the aforesaid purpose.

However, when the air contains large quantities of entrained particulate matter, as is frequently the case, filters of the aforesaid type quickly become loaded in operation, even when equipped with automatic stripping means, and as a result undergo severe drops in efficiency. In an effort to alleviate the aforesaid problem, it has also been heretofore proposed to first remove part of the particulate matter from the air by passing the same through a conventional cyclone device disposed upstream from the rotating-drum filter.

This, however, presents additional problems. Conventional cyclone devices are relatively expensive both from the viewpoint of fabrication and the viewpoint of installation. Additionally, the considerable size of a conventional cyclone device does not permit its use at all in some installations, or at best requires that the device be located some distance from the rotating-drum filter and the remaining components of the system. In the latter case, the requisite additional duct-work increases installation costs and also duct pressure losses.

A further and perhaps most important disadvantage, which is present irrespective of the foregoing considerations, resides in the fact that a pressure loss of some three to four inches is inherent in the operation of a conventional cyclone device, which normally conducts the air and entrained particulate matter introduced therein at high velocity through at least one and usually several complete revolutions about the device's central axis.

In a system such as previously mentioned wherein 100 cards are each supplied air at the rate of 1,000 cfm and at a negative pressure of four inches of water, pressure losses in the duct-work and in the rotating drum and/or other conventional filtering means normally total approximately three inches. An additional pressure loss of some three to four inches entailed by the inclusion of a conventional cyclone device would therefore require that the system realize a suction-pressure of some 10 to 11 inches.

As will be apparent to those skilled in the art, the realization of such a pressure in a system handling some 100,000 cfm of air would be exceedingly expensive, in terms of the power required and expended.

A process developed by *J.F. Baigas, Jr.; U.S. Patent 3,864,107; February 4, 1975* is one which includes a compact unitary housing enclosing a partial-cyclone device, a rotating drum filter and other filtering means arranged in sequence for progressively removing particulate matter from the air. Automatic stripping means is associated with the rotating drum filter, and the additional filtering means disposed downstream of the drum filter preferably includes a bank of V-cell filters. The partial-cyclone device is of generally C-shaped configuration, and extends through an arcuate distance of less than 360° and preferably no more than approximately 180°.

Following its introduction into the housing, air containing entrained particulate matter is conducted through an inlet duct containing adjustable velocity-regulating valve means and is directed thereby substantially tangentially against the concave inner surface of the aforesaid partial-cyclone device adjacent the upper end portion thereof.

Under the impetus of centrifugal force and gravity, a major portion of the larger-size particulate matter passes along the concave inner surface of the device to a purge-outlet adjacent its lower end. The particulate matter and a minor part of the air introduced into the housing are withdrawn through such outlet and from the housing. Adjacent its axial center portion and intermediate the aforesaid inlet and outlet associated therewith, the partial-cyclone device is open so as to permit free passage therefrom of the major portion of the air introduced therein and partially cleaned thereby. Such air passes directly to and through the rotating-drum filter disposed closely adjacent the partial-cyclone device and downstream therefrom within the housing. The minor amount of air purged with the particulate matter from the housing by the partial-cyclone device is, following condensation of the particulate matter therefrom exteriorly of the housing, preferably reintroduced into the housing so as to also pass with the aforesaid main flow of air first through the rotating-drum filter and thereafter through the additional filtering means provided within the housing downstream from the rotating-drum filter.

Due to the precleaning of the air by the partial-cyclone device, the rotating drum-filter and the other filtering means disposed downstream therefrom do not become overloaded during operation, and effectively perform their respective filtering functions at all times. In its passage through the housing of the apparatus, the air is therefore thoroughly cleaned, yet undergoes only a minimal pressure loss as compared to that loss in pressure which would ensue if the air were first conducted through a conventional cyclone.

A device developed by *J.C. Neitzel; U.S. Patent 3,727,383; April 17, 1973; assigned to Hardwicke-Etter Company* is a separator for textile fibers from a stream of carrier air. The stream of carrier air is directed substantially tangentially to and at substantially the same speed as a cylindrical drum surface rotatively mounted in the housing. The cylindrical drum wall closely approaches opposed housing walls for restricting the tendency of the carrier air to pass around the drum.

The cylindrical wall of the drum has relatively fine perforations, but the total area of perforations in any portion of the cylinder wall which spans the casing interior between the inlet and the first of the opposed housing walls is substantially greater than the cross-sectional area of the inlet. The resulting reduction in velocity of the carrier air together with movement of the drum surface substantially at the speed of and in the same direction as the transported fibers results in continuation of the fibers around the drum due to inertia, while the carrier air tends to pass into the drum.

TEXTILE INDUSTRY WASTES

In many textile processing operations, it is necessary to apply a finish to the fiber, particularly synthetic fibers, in order to improve the fastness characteristics during the processing in subsequent applications. Frequently, in the processing of synthetic fibers, it is necessary to subject the fibers to a heat treatment in order to obtain or improve upon the desirable properties of the fiber. Such heat treatments tend to volatilize the finish, which may be an organic material dispersed in water or may be a combination of volatile oils and waxes, which escapes as a fume commonly referred to as smoking.

Such fumes escaping into the atmosphere not only may be irritating to the workers in the area but frequently condense on operating machinery causing problems in cleanliness, housekeeping, potential flammable conditions and even degradation of the quality of the textile product being produced. Many attempts have been made to eliminate the potential problems by the choice of formulation of the finish to avoid smoking components. However, it is not always possible to arrive at a suitable compromise between adequate finish requirements and reduction of smoking components.

Other attempts to minimize the problems have included: extraordinary measures to make process equipment air tight to prevent escape of fumes; more sophisticated ventilation systems with increased capacity to exhaust fumes which may then require scrubbing and taller stacks or other conventional means for minimizing the potential for air pollution. Frequently, increased ventilation reduces the heating efficiency of the heat treatment process thereby requiring excess amounts of energy in order to provide the required degree of heat treatment.

Removal from Air

The use of high-temperature catalytic systems known in the prior art would require extensive cooling systems and an additional cooling step in a yarn heat-treating process which is sensitive to temperature change and in which it is difficult to maintain the exact temperature conditions for proper heat treatment of the textile fibers.

A process developed by *H. Fernandes and W.H. Walsh; U.S. Patent 3,725,532; April 3, 1973; assigned to E.I. du Pont de Nemours and Company* involves removing at least a portion of undesirably vaporized components by directing the components through a gas-permeable structure containing an oxidation catalyst for the textile finish in a finely dispersed form to oxidize the components to a more desirable state at about the same temperature range used to treat the textile structure. Specific oxidation catalysts disclosed are palladium and platinum. The process can operate at extremely high space velocities making the process feasible from an economic viewpoint.

A process developed by *R.J. Tarves, Jr.; U.S. Patent 3,800,505; April 2, 1974; assigned to Air Pollution Specialties, Inc.* is one in which textile oil is removed from hot effluent gas by cooling the gas, scrubbing the gas in two stages, removing entrained droplets in a demister and introducing the gas into an electrostatic precipitator. After the gas leaves the electrostatic precipitator, approximately 99%, by volume, of the oil is removed from the gas and the gas is rendered substantially invisible.

The gas is thus in a condition to be returned to the atmosphere. The apparatus consumes little energy and is relatively maintenance free in operation. A portion of the water used in a heat exchanger for cooling the gas may be used in the two scrubbing stages and recycled within the system.

Removal from Water

A relatively large number of organic wastes are heavily diluted with water during manufacturing operations and present a serious problem of waste disposal. Such wastes cannot be easily burned, due to their high water content, nor can they be conveniently disposed of by chemical reaction without considerable expense and, sometimes, without formation of other products which, themselves, are objectionable if discharged to the sewer.

An example of such substances are the finish oils utilized in textile yarn manufacture, which are relatively high-boiling materials and, also, very stable and resistant, so that they survive even severe chemical treatments aimed at their destruction. Inevitably, finish oils find their way into waste collection sewers, heavily diluted with washdown water, so that they cannot be restored to the process and constitute a waste which has to be disposed of.

Attempts have been made to incinerate wastes such as dilute finish oils by introducing them into combustion flames in regulated amounts, but these have not always been successful, the chilling effect frequently extinguishing the flame, or the organics being only partially burned to heavy soots or being incompletely vaporized, or burned, so that there is objectionable deposition on furnace walls, and other problems. Moreover, such combustion disposition processes have been exceedingly expensive, since approximately 5,000 Btu of fuel have hitherto been required for each pound of aqueous waste.

A process developed by *E.S. Monroe, Jr.; U.S. Patent 3,611,954; Ocotber 12, 1971; assigned to E.I. du Pont de Nemours and Company* involves the oxidative disposal of organic wastes heavily diluted with water by (1) vaporizing the dilute organic waste stream by directly heating waste sprayed counter to a combustion flame in a general envelope pattern and (2) catalytically oxidizing the hot vaporized effluent from the combustion flame to equilibrium products. Analysis of effluent products has shown that complete waste disposal is effected and the heat consumption required has been as low as 1,300 Btu/lb of finish oil waste having a water dilution as great as 90%.

A process developed by *F.S. Whitfield; U.S. Patent 3,734,035; May 22, 1973; assigned to Millhaven Fibres Limited, Canada* involves the conversion of liquid textile wastes containing combustible carbonaceous material and water but being substantially free of substances yielding solid residue. The process converts such wastes into useful energy, simultaneously reducing pollution of the environment. A suitable apparatus for the conduct of such a process is shown in Figure 152.

FIGURE 152: APPARATUS FOR COMBUSTION OF LIQUID TEXTILE PROCESSING WASTES

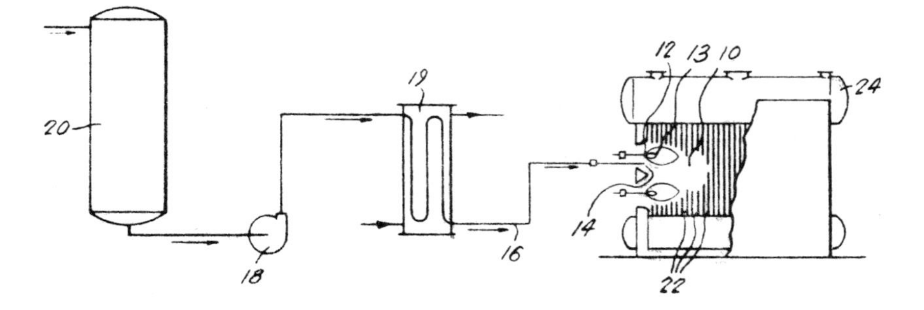

Source: U.S. Patent 3,734,035

The incinerator comprises a furnace partly shown at **10** provided with at least one burner having a nozzle **12** generating an open flame **13**. The burner is coupled with auxiliaries not shown. A spray jet **14** is spaced in relationship to the open flame **13**. The jet **14** sprays a liquid waste received from the line **16** via pump **18**. Generally the liquid waste is stored in tank **20**. Before pumping it into line **16** it is preheated to a temperature above its boiling point at atmospheric pressure, for instance, by means of a heat exchanger **19**. The pressure exerted by pump **18** must be such as to maintain the waste in liquid state until it reaches the spray jet **14**. Generally the pressure exerted on the waste may be in the order of 50 lb/in^2 to 60 lb/in^2.

On leaving the spray jet **14**, the liquid waste is sprayed into the fire or flame and the liquid is flashed out, thereby evaporating the water from the liquid waste while burning the combustible carbonaceous material. Thus the water content is converted into vapor at high temperature. This vapor together with the products of combustion circulate around a plurality of running pipes **22** and boiler **24**, giving up most of their heat contents, then pass through the boiler stack to finally emerge to the atmosphere. These running pipes coupled with boilers are used to supply steam to the required units not shown. The following is one specific example of the operation of such a scheme.

A mixture of liquid carbonaceous waste was preheated to its boiling point at atmospheric pressure and kept under 50 psi. The mixture of liquid carbonaceous waste was composed of approximately: (1) about 7 parts of alcohol still washing containing approximately 80% methanol, 10% of glycol and 10% of water; and (2) 22 parts of spin finish waste containing approximately 2% of oil and 98% of water.

The mixture was fed to a fine spray nozzle or jet located adjacent to the oil jets in the fire box of a boiler. The distances between the jet and nozzle were about one-half foot each respectively, vertically, center to center, and horizontally, face of spray to face of nozzle.

The pressure and delivery rates were arrived at by visual observation. Under these conditions, complete combustion was obtained and confirmed by analysis of the flue gas. The spray jet was provided with an automatic shut-off of the liquid in the event of a flame outage of the boiler.

An approximate value of about 700 Btu per pound of the mixture was obtained. The boiler was rated at 30,000 lb/hr at 250 lb/in^2. When incinerator equipment was in use at a rate of about 137 imperial gallons per hour at 50 lb/in^2 delivery pressure, a minimum of approximately 15,000 lb/hr at 250 lb/in^2 per boiler was obtained.

The incineration temperature was 2000° to 2200°F. The brick stack was 125 feet high above grade, 4 feet 6 inches inside diameter. The flue gas temperature of the boiler breeching was approximately 500°F and at the top of the stack approximately 450°F. Flue gas by test indicated absence of unburned combustibles and approximately 3% excess oxygen.

Synthetic fiber plant wastewaters or textile finishing mill effluents, alone or in admixture with municipal sewage, may be effectively treated by a process described by *D.G. Hutton and F.L. Robertaccio; U.S. Patent 3,904,518; Sept. 9, 1975; assigned to E.I. du Pont de Nemours and Company*. The process involves subjecting the wastewater to a biological treatment process in the presence of activated carbon or fuller's earth. A gas containing molecular oxygen is distributed within the liquid mixture during treatment to provide oxidation means.

Through the use of finely divided activated carbon of fuller's earth, the activated sludge process for the treatment of wastewater is significantly improved upon. Not only does the presence of these additives cause a greater percentage of BOD removal, but BOD removal is accomplished in a much shorter time.

For example, in identical procedures, the activated sludge process with no activated carbon present showed a BOD percent removal of 68% and a TOC percent removal of 60% at an aeration time of 7.6 hours; while when activated carbon was added, BOD removal percentage was 95% and TOC removal percentage was 85%, at an aeration time of only 2.5 hours (a 67% reduction in time of aeration).

See Wool Scouring Effluents (U.S. Patent 3,847,804)
See Wool Scouring Effluents (U.S. Patent 3,909,407)

THERMAL POLLUTION

Removal from Water

A scheme developed by *J.A. Lahoud and D.L. Orphal; U.S. Patent 3,851,495; December 3, 1974; assigned to Computer Sciences Corporation* is one in which thermal pollution from industrial manufacturing and electric generation plants is avoided by incorporating in the water cooling system of such plants an underground water storage and heat dissipation space such as that which may be economically formed by a nuclear explosion.

In a preferred embodiment, the system incorporates a wet-type water cooling system (primarily because of its lower cost) and one or more nuclear chimneys below the level of the local aquifer or water table so that makeup water for water lost through evaporation in the wet-type cooling unit is added by water draining or leaching to the nuclear chimney from the aquifer level.

Such a scheme is shown in Figure 153. Referring to the drawing, the system is shown as applied to a nuclear electric power generating station **10** having a reactor **11** through which cooling water is caused to flow. Cooling water from reactor **11** is conveyed in a conventional manner to cooling tower **12** by way of a pipe system **13**. Cool water from cooling tower **12** is discharged through pipe **14** to the underground storage reservoir system.

Cooling air for the cooling tower enters the lower end of the cooling tower at **15** in a conventional manner and exits through the top. Such cooling towers are typically of two broad categories namely the wet type and the dry type. Water discharged from the cooling tower **12** may be several degrees above the natural temperature of the natural water. Such an elevated temperature of water cannot be discharged directly into the local natural waters because of the problem of thermally polluting such waters and changing the ecological condition thereof.

Accordingly, such water is caused to flow into a nuclear chimney or storage reservoir **17** by means of a pipe system **16**. It will be noted that the water is introduced at the upper level **18** of nuclear chimney **17**. Cooler water as it gravitates through the rubble and rocks and fissures in nuclear well or chimney **17** is tapped or removed from the lower level **19** thereof by means of a pipe system

20 which leads to the surface of the earth. Such cooled water is therefore recirculated through reactor **11** by means of pipe system **21**. Thus, a substantially closed cycle system for cooling water is shown.

FIGURE 153: SCHEME FOR REDUCTION OF THERMAL POLLUTION USING UNDERGROUND NUCLEAR CAVITY FOR WATER ACCUMULATION AND STORAGE

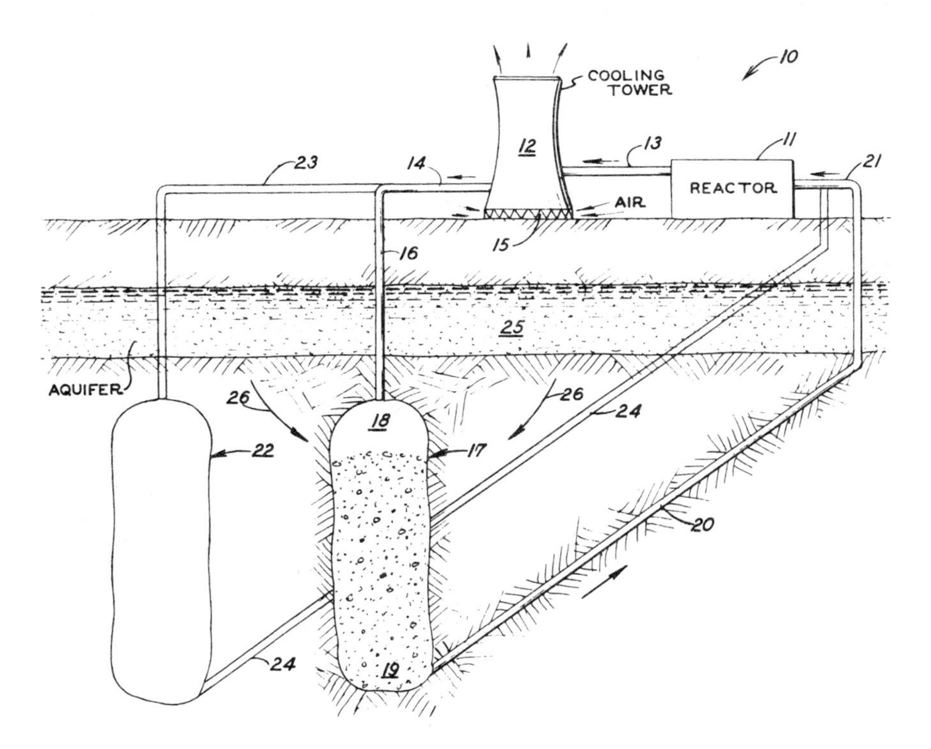

Source: U.S. Patent 3,851,495

In order to make up for lost water due to evaporation of water in cooling tower **12**, the storage and cooling nuclear chimney **17** is located below the aquifer **25** so that water flows from the aquifer to chimney **17**. This avoids heated water being carried to a possibly nearby stream by the flowing water in the aquifer.

Typically, the chimney **17** may be 250 to 500 feet in diameter and 1,500 to 2,500 feet in height. Moreover, when a nuclear explosion is used to form the chimney **17**, observance of Nuclear Test Ban Treaty Safety Rules, of course, is mandatory so the detonation is set off at approximately 3,500 feet below the earth surface and such charge is typically less than a megaton charge. Details of the manner of forming such subterranean reservoirs of various sizes and configurations are well known in the art and by way of example attention is invited to U.S. Patent 3,589,773.

As shown, makeup water generally flows from the aquifer **25** down to the lower level of the chimney **17** as indicated by dotted arrows **26**, it being understood that aquifer **25** may be connected or tapped by pipe or other means (not shown) for supplying makeup water to chimney **17**, to provide the makeup water in case the cooling tower **12** is of the wet type.

There may be one or more nuclear chimneys, one additional chimney being shown and designated as **22** in the drawings. This second nuclear chimney **22** may be connected by pipe **23** to receive water from cooling unit **12** at the top of this chimney and a second pipe **24** connected to the bottom for delivering cooling water to reactor **11** in such a way that chimney **22** is connected in parallel with chimney **17**.

THIOSULFATES

Removal from Water

A process developed by *P. Urban; U.S. Patent 3,709,660; January 9, 1973; assigned to Universal Oil Products Co.* is one in which a water-soluble inorganic thiosulfate compound is reduced to the corresponding sulfide compound by contacting an aqueous solution of the thiosulfate compound and hydrogen with a catalyst, comprising a catalytically effective amount of cobalt sulfide combined with a porous carrier material, at reduction conditions.

Principal utility of this treatment procedure is associated with the cleanup or regeneration of aqueous streams containing undesired thiosulfate compounds so that they can be reused in the process which originally produced them or discharged then into a suitable sewer without causing a pollution problem.

See Photographic Processing Effluents (U.S. Patent 3,721,624)

TIN COMPOUNDS

Removal from Air

See Glass Treatment Effluents (U.S. Patent 3,885,929)

Certain industrial operations generate aqueous waste that contains both tin and fluoride in solution. One of these is electrolytic tinplating from halogen tin lines. In such operations, wash solutions and other aqueous wastes are produced from rinsing sprays, the scrubbing of effluent fumes, the liquid effluent from tin recovery operations, and from other sources.

Both tin and the fluoride must be removed from or substantially reduced in such waste solutions, for two reasons. In the first place the loss of large quantities of recoverable tin is economically unacceptable and in the second place the fluoride is a pollutant that cannot be pumped into streams and rivers in large quantity.

Removal from Water

In the past, it has been conventional practice to remove and recover tin from aqueous solutions of this type by treating the solutions with sodium carbonate so as to precipitate the tetravalent tin solute as stannic hydroxide. However, such treatment was ineffective to precipitate or otherwise remove the fluoride from the solution.

A separate treatment had to be provided for fluoride removal. Even if a single reagent could have been found that would precipitate both the tin and the fluoride, the recoverable stannic hydroxide sludge would be greatly diluted by precipitated fluoride and would be of correspondingly decreased market value.

A process developed by *R.C. Williamson; U.S. Patent 3,284,350; November 8, 1966; assigned to National Steel Corp.* provides for the removal both of tin and of fluoride from aqueous solutions of the same, while at the same time recovering a tin concentrate of desirably high tin content.

The process is based on the discovery that both tin and fluoride can be removed from aqueous solution by precipitation with calcium hydroxide and that, moreover, the precipitation can be carried out stepwise in such a way that in at least one early precipitation stage a greater percentage of the original tin solute can be precipitated at fairly low pH compared to that of the original fluoride solute, while during a later precipitation stage at fairly high pH, a greater proportion of the original fluoride solute can be precipitated than of the original tin solute.

In other words, the process can be used to produce first a tin-rich and second a fluoride-rich precipitate. The tin-rich precipitate is valuable for recovery of its tin values. The fluoride precipitate assures that the liquid can be discharged without giving rise to objectionably high levels of fluoride contamination.

The process is best adapted for the treatment of large volumes of water in which the tin and fluoride are in relatively low concentration. For relatively small volumes of highly concentrated tin solute, it is preferable to rely on tin recovery by precipitation with sodium carbonate if fluoride is present only at such a low level as not to give rise to undesirable fluoride contamination.

TIRE RETREADING PROCESS EFFLUENTS

It is known that tire buffing incident to tire retreading is a very serious cause of air pollution, particularly because the dust and smoke represent a combination of gross particles, even strips of rubber, intermediate particles in the range of 2 to 10 microns, and fine smoke particles in the submicron range resulting from the actual burning of rubber in the buffing operation.

With the emphasis placed by cities on eliminating air pollution, the ability to remove dust and smoke caused by tire buffing in the process of tire retreading may be a limiting factor in the operation of tire retreading plants near cities where, of course, most of the retreading business is located.

Removal from Air

A process developed by *J.J. Cunningham, Jr.; U.S. Patent 3,579,314; May 18, 1971; assigned to Quinn Brothers, Incorporated* is one in which effective removal of dust and smoke from tire buffing can be accomplished by combining a cyclone dust collector with a spray chamber and an absolute filter in that order. The mechanism can be made economical by providing a bypass from a high pressure point in the cyclone dust collector which will return dust and smoke to the hood of the tire buffer so that instead of imposing all the smoke and dust collection load on the mechanism over the short period of a tire buffing cycle, for example, of the order of 30 seconds, the dust collection can be spread over several minutes, and can complete its cycle while the buffed tire is being removed from the buffer and a new tire is being placed on the buffer for buffing.

TITANIUM COMPOUNDS

Removal from Air

A process developed by *W.P. Vosseller; U.S. Patent 3,250,059; May 10, 1966; assigned to National Lead Co.* involves removing aerosols and certain obnoxious gaseous components from the effluent gases discharged from kilns used to calcine TiO_2 hydrate material.

Heretofore it has been the practice to cool these effluent kiln gases, by passing them through a succession of spray towers and/or electrostatic precipitators, so as to remove the sulfur oxide components and in particular H_2SO_4 before exhausting the gases to the atmosphere. However, the removal of the liquid aerosols, including liquid H_2SO_4 and water, formed as a result of cooling the kiln gases, has been quite incomplete.

In this process, the effluent gases from a calciner are piped or otherwise conveyed into a multistage filtering system which, in essence, comprises two filtering areas or chambers arranged in sequence each provided with a filter arranged so that the gases will pass therethrough and with means for spraying a liquid onto the upstream side of the first filter to wash the aerosols and gaseous components therethrough in a direction concurrent with the gas flow. Each filtering area or chamber is provided also with an outlet on the downstream side of its respective filter for draining off the liquids and/or solids-burdened liquids which collect on the downstream side thereof.

The filters used may be in any suitable form, i.e., substantially flat pads, filter candles or a combination thereof, and are formed of a liquid repelling material and in particular hydrophobic fibers such as silicone treated glass wool or preferably a synthetic, hydrophobic polyester fiber such as Dacron (polyethylene terephthalate), Terylene or the like. While the size of the fibers is not critical it is preferred to use fibers having a diameter of less than about 30 microns.

A process developed by *E.O. Kleinfelder and H. Valdsaar; U.S. Patent 3,564,817; February 23, 1971; assigned to E.I. du Pont de Nemours & Co.* involves removing small amounts of titanium tetrachloride and other chloride impurities from waste gas produced during the chlorination of a titaniferous ore by first acid scrubbing the gas with sulfuric acid of 75 to 95 weight percent concentration and then scrubbing with water to produce a clear gas that may be vented to the atmosphere without fuming.

See Glass Treatment Effluents (U.S. Patent 3,885,929)
See Glass Treatment Effluents (U.S. Patent 3,919,391)

TNT EXPLOSIVE WASTES

It has been common practice to dispose of organic nitrogenous waste materials, including explosives such as TNT, and propellants such as nitrocellulose by incineration of the waste material in the open air. Disposal methods include burning in an open field, in a sand pit, and on a concrete pad. Such burning methods result in the formation of considerable quantities of nitrogen oxides (NO_x), and also frequently lead to pollution of the soil with unburned or partially combusted residues.

Removal from Air

A process developed by *C.D. Kalfadelis and A. Skopp; U.S. Patent 3,916,805; November 4, 1975; assigned to Exxon Research & Engineering Co.* is a process for incineration of nitrogenous waste materials in a manner which minimizes NO_x pollution. A nitrogenous waste material, such as TNT, is burned with a fuel and less than a stoichiometric quantity of air in the presence of a catalyst in a fluid bed. Secondary air is added to the gaseous products, and the resulting gas mixture is burned to yield a stack gas which has minimal amounts of NO_x, carbon monoxide and hydrocarbons. Nickel or a compound thereof is preferred as the catalyst.

Figure 154 shows a suitable form of apparatus for the conduct of such a process. Referring to the drawing, **10** is a vertical fluid bed reactor having a perforated distributor grid **11** near its base for supporting a bed **12** of finely divided solid particles of catalyst, inert refractory material, or both. Below the distributor grid **11** is an air plenum chamber **13**. A feed inlet line **14** for admitting an aqueous slurry of TNT or other nitrogenous material, and an auxiliary fuel inlet line **15**, are provided for the fluid bed **12** a short distance above distributor grid **11**. A primary air inlet line **16** communicates with the air plenum chamber **13**.

This air line **16** has a preheater **17**, shown here as an electrical preheater. Above the feed inlet line **14** and auxiliary fuel inlet line **15**, but below the top of bed **12** when the bed is in its expanded or fluidized condition, is a secondary air inlet line **18**. The secondary air line **18** is controlled by a shutoff valve **19**. An overhead outlet line **20** is provided at the top of the reactor **10** for removing gaseous reaction products.

The apparatus also includes a feed tank **21** in which an aqueous slurry of TNT or other explosive, propellant, or waste nitrogenous compound is prepared. This tank is provided with a stirrer **22** driven by a motor **23**. This slurry feed system also includes a centrifugal pump **24**, a recirculation line **25** for returning part of the slurry from the outlet of pump **24** to the top of tank **21**, and a slurry feed line **26** for feeding the aqueous slurry to the inlet of a metering pump **27**. The metering pump **27** provides for feeding the aqueous slurry into the reactor at the desired rate.

The gaseous reaction products and any fines which are carried overhead from the fluid bed **12** pass from overhead outlet line **20** to a cyclone separator **28**. Gases are vented to the atmosphere via vent **29**. Separated solids are conveyed from the cyclone separator **28** to a solids receiver **30**. These solids may be returned to the reactor **10** either periodically or continuously, or may be discarded, as desired. The solids receiver **30** is provided with an outlet line having a shutoff valve **31**.

FIGURE 154: APPARATUS FOR INCINERATION OF TNT EXPLOSIVE WASTES WITHOUT AIR POLLUTION

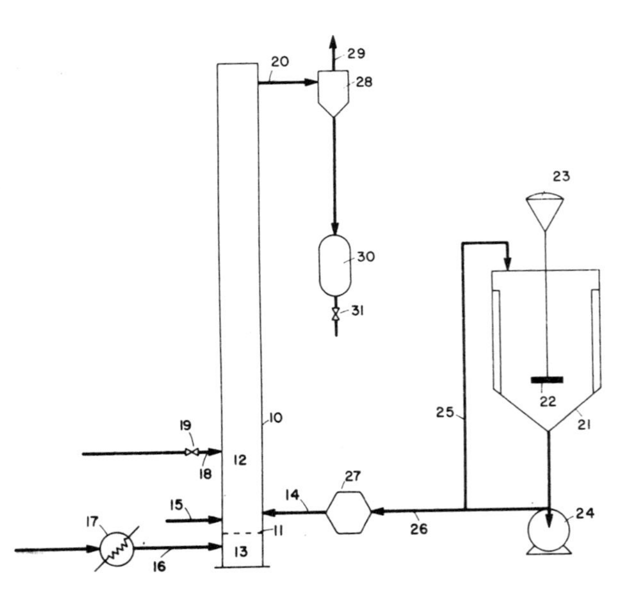

Source: U.S. Patent 3,916,805

TNT PROCESS WASTES

In the production of nitrogen derivatives of benzene, toluene, xylene, phenol, and alkyl-phenols, as in the production of explosives such as TNT (trinitrotoluene), organic compounds are treated with nitric acid. This acid is usually in the form of an aqueous solution containing nitric acid and sulfuric acid. The treatment process produces wastewater containing nitric acid, nitrous acid, sulfuric acid, nitrates, nitrites and sulfates, the nitrogen-containing components being generally described below as nitro compounds. The concentration of these solute impurities is relatively low so that their removal presents a considerable difficulty.

Removal from Water

A process developed by *P. Marécaux; U.S. Patent 3,954,381; May 4, 1976; assigned to Société pour l'Equipement des Industries Chimiques, SPEICHIM, France* is one in which a concentrated aqueous solution containing nitro compounds is incinerated in a chamber to produce a hot dry vapor that is fed through a heat-exchange chamber through which the unconcentrated aqueous solution containing the nitro compounds is dripped so as to evaporate the greater portion of the water in this solution and produce the concentrate that is incinerated.

The incineration is carried out at a temperature above 1000°C and the hot vapors leave the heat exchange chamber in a dry state. Thereafter these hot vapors are scrubbed in a tower to remove any particulate material therefrom and are allowed to escape into the atmosphere as a nonpolluting gas. Figure 155 illustrates such a process.

FIGURE 155: APPARATUS FOR INCINERATING AN AQUEOUS SOLUTION CONTAINING NITRO COMPOUNDS

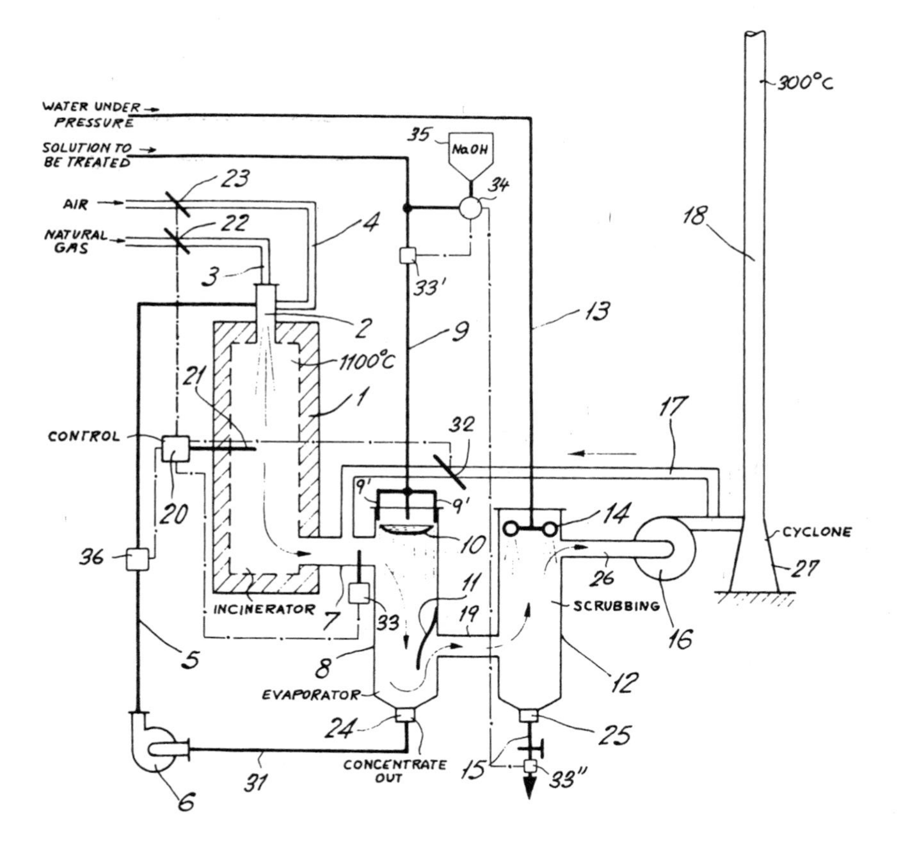

Source: U.S. Patent 3,954,381

As shown in the drawing an incineration chamber **1** is provided with a burner **2** which is fed with natural gas through a conduit **3** and with air through a conduit **4** so as to heat the interior of this chamber **1** to a temperature between 1000° and 1200°C. Vapors produced in chamber **1** pass, as shown by the arrow, through a conduit **7** into an evaporator or trickle tower **8**.

A solution to be treated, here an aqueous solution containing toxic nitro compounds, is fed under low pressure via a conduit **9** into the top of the trickle tower **8**. A perforated vessel **11** in the top of this trickle tower receives most of the flow from the conduit **9**, and branches **9'** from the conduit **9** allow the solution to flow down the side walls of the tower **8**.

Thus the liquid collects in the vessel **10** and falls downwardly in the tower **8** in the form of streams or large drops. The vapors enter the top of tower **8** and leave at the bottom thereof, and pass at a speed approximately the same as the rate at which the solution falls in the chamber, so that the falling streams or drops have virtually zero velocity relating to the vapors and therefore do not tend to pick up particles therefrom.

The conduit **7** opens into the tower **8** well above an outlet conduit **19** which is protected by a deflector **11** so that the gases must pass downwardly through the upwardly elongated tower **8** thereby heating the solution falling from the vessel **10** and trickling down the sidewalls of this evaporator tower **8**. This operation causes much of the water phase of the solution to evaporate, thereby producing a concentrate which passes out of the bottom of the chamber **8** through a particle filter **24** and into a conduit **31**. A pump **6** in this conduit **31** conducts the concentrate back to the chamber **1** and injects it at the burner **2** so that the concentrate is integrally incinerated, that is its temperature is raised to approximately 1100°C to vaporize the water component completely into hot and dry steam, and to transform the toxic solute component into a nontoxic phase as will be described below.

The evaporated water and other vapors leave the evaporator **8** through a conduit **19** opening into the lower region of a scrubbing tower **12**. Water under pressure is fed by a conduit **13** into a spray ring **14** at the upper end of the scrubbing tower **12** which atomizes this water and scrubs the gas rising countercurrent to it from the lower inlet **19** to an upper outlet **26**.

Any particulate material in this gas will be trapped by the relatively rapidly moving atomized water droplets, and the spray will reduce the temperature of the hot dry gas to such a level that condensation will begin, this condensation naturally using any particulate material carried in the gas as nuclei. The scrubbing water is filtered at **25** and leaves the scrubbing tower **12** through an outlet **15**. This water is nontoxic and can be allowed to run into the sewer system.

A blower **16** draws the air out of the outlet **26** of the scrubbing tower **12** and forces it into a cyclone **27** at the bottom of a chimney **18**. This cyclone removes any remaining particulate material, so that only hot air, steam and CO_2 emerge as gases from tower **18**. A shunt conduit **17** extends from the conduit **26** between the blower **16** and the cyclone **27** to the conduit **7**. This conduit **17** is provided with a valve **32** which can be operated to allow a certain amount of cool-vapor feedback to the inlet of the evaporator **8**. Thus it is possible to control the temperature inside the evaporator according to a thermostat **33** which operates the valve **32**. A similar thermostat **21** is connected to a control circuit **20** which serves to operate the valve **32** as well as valve **22** and **23** in the conduits **3** and **4**, respectively. A valve **34** in the inlet line **9** may be opened to allow caustic soda in a hopper **35** to be admixed with the water, this additive serving to neutralize particular acids in the solution.

It has been found that it is possible to reduce the consumption of heating gas in such a system by a factor of approximately 5 over a system which simply incinerates the unconcentrated solution. In addition the apparatus is relatively inexpensive because a furnace **1** only a fifth as large as would hitherto be required is needed, and the evaporator **8** is of relatively simple construction and long service life.

As a general rule in industrial processes when the size of an element such as a furnace is reduced by a factor of 5 its cost is reduced by a factor of at least 3. In addition the gases ultimately escaping from the chimney **18** have a maximum temperature of between 250° and 350°C, so that there is little pollution.

The control circuit **20** is also connected to a sensor **36** in the conduit **31** which ascertains the concentration of the solute in the concentrate passing through this conduit **31**. Thus if the concentration becomes too weak the control circuit **20** opens valves **22** and **23** to increase the amount of combustibles fed to the burner **2**, and thereby raises the temperature inside the oven **1**. This will, of course, cause the temperature of the gas leaving the chamber **1** to increase, thereby increasing the amount of evaporation in the chamber **8** and increasing the concentration.

TRICHLOROETHYLENE

See Degreasing Process Wastes (U.S. Patent 3,803,005)

TURPENTINE

See Pulp Mill Digester Effluents (U.S. Patent 3,745,063)
See Pulp Mill Digester Effluents (U.S. Patent 3,753,851)

TV TUBE MANUFACTURING EFFLUENTS

In the manufacture of color cathode ray tubes, on the inner surface of the face panel of the color cathode ray tube, phosphors emitting red, green and blue lights, respectively, are securely arranged alternately in a specified dots pattern. These phosphor dots are excited by electron beams and produce color pictures.

For application of said phosphor dots to the face plate, usually the so-called photochemical dot-forming method is employed. This dot-forming method consists of first coating the face panel with a slurry made by diffusing phosphor powder in a photosensitive film-forming material containing polyvinyl alcohol and bichromate, then after a photosensitive film has been formed on the face panel, exposing the film in a specified dotted pattern through a multiholed plate called a shadow-mask, and finally developing the film by showering it with hot water to complete the phosphor dots.

During this process, large quantities of waste liquid of slurry, as well as of those of materials unexposed when developing, are produced. For the phosphor, a compound of rare earth elements such as zinc sulfide (ZnS), cadmium sulfide (CdS), yttrium oxide (Y_2O_3), yttrium oxysulfide (Y_2O_2S), etc., is used. These phosphors, when absorbed into human or animal bodies or into agricultural or dairy products, are almost invariably noxious.

Removal from Water

A process developed by *T. Sakagami, S. Kakumoto, K. Okawa and K. Miyashita; U.S. Patent 3,915,691; October 28, 1975; assigned to Matsushita Electric Industrial Company, Ltd., Japan* involves treating such an industrial waste liquid containing powders of metals such as cadmium or yttrium and/or their compounds comprising the steps of:

(1) mixing the waste liquid with a liquid containing a water-soluble high molecular compound such as polyvinyl alcohol,
(2) evaporating the water in the mixture by heating with hot blast for concentration, and
(3) cooling the dehydrated mixture to form a solid film of the high molecular compound enclosing the waste metals and/or their compounds.

A suitable form of apparatus for the conduct of such a process is shown in Figure 156. First, waste liquid is dripped from a container **1** onto a metal belt **2** while rotating it with driving drums **3** and **4**, and leading it into a hot blast stove **5**.

The hot blast stove **5** is designed to blow a hot blast of, for instance, 200°C from its blast intake **6** and to exhaust it away from an exhaust hole **7**. In this stove, water in the waste liquid is evaporated and the polyvinyl alcohol solution is concentrated.

In the abovementioned stage, all the phosphor contents of the waste liquid are deposited in the concentrated polyvinyl alcohol solution. Therefore, as the next step, the metal belt **2** passes through a cold chamber **8**, where the metal belt **2** is gradually cooled and finally, a thin film **9** of the polyvinyl alcohol formed on the metal belt **2** is peeled off by a peeling blade **10** and rolled up, thus reclaiming all residual substances in the waste liquid.

The design of the hot blast stove **5** in this apparatus depends on the quantity of waste liquid to be treated and the concentration of high molecular compound. In the case of treating a waste liquid of slurry prepared with a water solution of polyvinyl alcohol in the concentration of, for instance, 2 to 10% in the abovementioned process of manufacturing color cathode tubes, an operation with the hot blast at about 200°C, running speed of the metal belt **2** at 300 millimeters per minute, and drying surface area of the metal belt at about 0.75 square meter has a treating capacity of about 12 liters per hour.

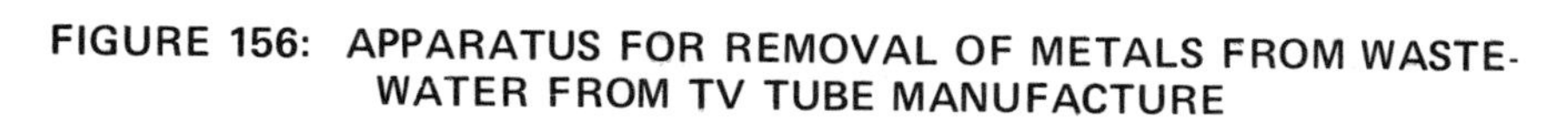

FIGURE 156: APPARATUS FOR REMOVAL OF METALS FROM WASTE-WATER FROM TV TUBE MANUFACTURE

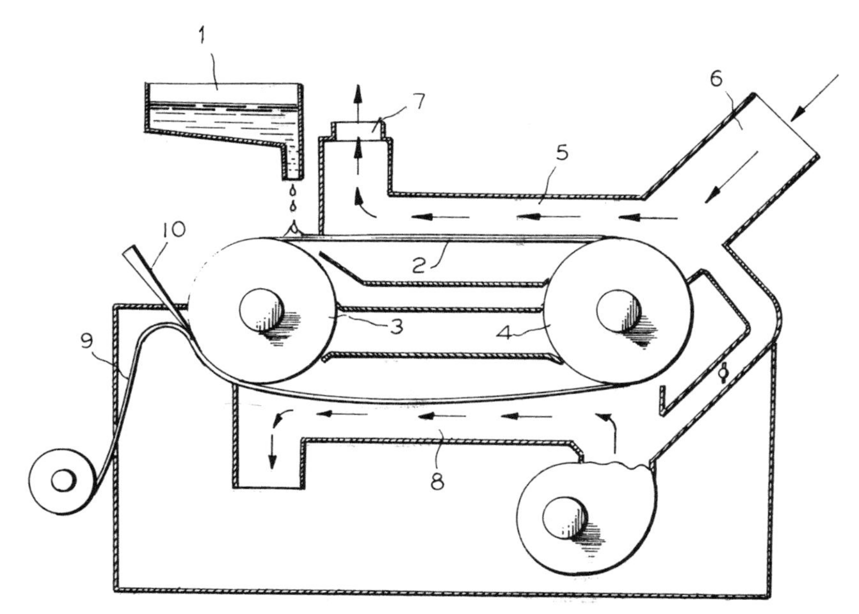

Source: U.S. Patent 3,915,691

URANIUM PRODUCTION EFFLUENTS

Removal from Air

See Radon (U.S. Patent 3,853,501)

Uranium, which principally finds application as fissionable material or nuclear fuel in modern atomic power plants, is obtained from uranium ore which at most contains only a fraction of 1% of uranium.

In order to extract the uranium, the ore is subjected to a chemical treatment after some preliminary filtration and other mechanical steps. During such chemical treatment, which employs fresh water and in which gross quantities of auxiliary chemicals (e.g., acids during the ion exchange or adsorption operations for removing certain radioactive components) are used up extraordinarily large amounts of radioactively contaminated amounts of wastewater are generated. The wastewater emitted from installations of this type is in the form of aqueous solutions of inorganic salts, and also contains residual uranium and elements resulting from the radioactive decay of uranium, such as radium, polonium and lead. The treatment of such wastewater to negate its dangers represents a difficult problem for each installation.

Removal from Air

A process developed by *V. Klicka, J. Mitas and J. Vacek; U.S. Patent 3,988,414; October 26, 1976; assigned to Vyzkumny ustav chemickych zarizeni, Czechoslovakia* is a closed-loop process for treating wastewater resulting from chemical extraction of uranium from ore. The water is evaporated to form a concentrated solution and is then subjected to crystallization of the least soluble salt component thereof via further evaporation, or cooling or simultaneous cooling and a partial vacuum. The crystallized component is then separated from the mother liquor, whereupon the latter is fed back after removal of residual uranium therefrom to the extraction installation to replace the acids used therein. Additionally, the pure condensate produced during evaporation of the wastewaters is employed as a replacement for the fresh water employed in processing of the ore.

In preparing fuel elements for nuclear reactors it is desirable to employ UO_2 for this purpose. A widely used method is to process UF_6 to prepare ammonium diuranate (ADU) which is then calcined and reduced to UO_2. In preparing ADU, enriched UF_6 gas is reacted with water to produce an aqueous uranyl fluoride (UO_2F_2) solution which is then treated with an excess of ammonium hydroxide to cause a precipitate of ADU which is recovered as an aqueous slurry. The precipitate can be filtered out or centrifuged from the aqueous phase of the slurry. The resulting wastewater contains a high concentration of fluorides, primarily as NH_4F, with excess ammonia, and small quantities of dissolved uranium.

The disposal of this wastewater entails considerable problems and difficulty because of the toxicity of the fluorides as well as the excess ammonia and the small but significant quantities of radioactive uranium. On both ecological and health accounts it would be highly desirable to eliminate or greatly reduce this disposal problem, as well as to recover from the wastewater substantially entirely the traces of enriched uranium, which latter is a very costly material as well as being a health hazard.

In a process developed by *T.J. Crossley; U.S. Patent 3,961,027; June 1, 1976; assigned to Westinghouse Electric Corporation* such wastewater is rendered suitable for cyclic reuse by initially treating the wastewater with sufficient lime to precipitate substantially all of the fluorides present in the wastewater to a relatively insoluble CaF_2 precipitate. The treated solution is then subjected to distillation to drive off ammonia for reuse in the ADU precipitation and the CaF_2 precipitate is separated from the aqueous distilland leaving water with dissolved calcium.

The distilland is treated by a cationic ion-exchange material to remove substantially all of the calcium and other cationic metal impurities and the resulting water containing small amounts of uranium, fluoride and ammonia is recycled to react with UF_6 or to be combined with the ammonium hydroxide distillate and then treated with additional concentrated ammonium hydroxide to form a solution of the desired NH_3 content for use in precipitating ADU.

UREA MANUFACTURING EFFLUENTS

Fertilizer dusts such as urea dust and ammonium nitrate dust contained in an air stream used for cooling fertilizer granules are exhausted from prilling towers at large scale fertilizer plants. The prilling of the fertilizer can be conducted by a known process which comprises steps of producing an anhydrous melt by melting the fertilizer or a concentrated solution of over 90 weight percent of the fertilizer by concentrating an aqueous solution thereof, dividing the anhydrous molten urea into liquid droplets by, for example, passing it through nozzles, and cooling the liquid droplets with an air stream to form urea prills.

Removal from Air

A process developed by *T. Takae, T. Kawabe and S. Maeno; U.S. Patent 3,861,889; January 21, 1975; assigned to Mitsui Toatsu Chemicals, Inc., Japan* is one in which such dust is removed from a gas stream containing the same by passing the gas stream through a layer of a foamed material having a noncellular porosity of over 90%, the layer being impregnated with a solvent for the dust. Prior to being passed through the layer of foamed material, the gas stream may be washed with the solvent for the dust.

Figure 157 shows a suitable form of apparatus for the conduct of such a process. The gas stream containing dust from duct **1** is washed with a solvent for the dust sprayed from nozzle **2**. The solvent is fed from a solution reservoir **3** to spray nozzle **2** by means of pump **4** so that dust of relatively large diameter can be dissolved into the solvent. The gas stream which has been washed with the solvent is then filtered through filter layer **5** of a foamed material having a noncellular porosity of over 90% and thereby the fine particles are removed and exhausted through exhaust duct **6**. The solvent is flowed down on the surface at the gas outlet side of the filter layer **5**. The filter layer is suitable for a compact size apparatus wherein it is composed of a filter layer fixed to a frame **10** whose cross section perpendicular to the surface of the filter layer is of an isosceles trapezoid.

FIGURE 157: APPARATUS FOR REMOVAL OF DUST FROM UREA-PRILLING EFFLUENTS

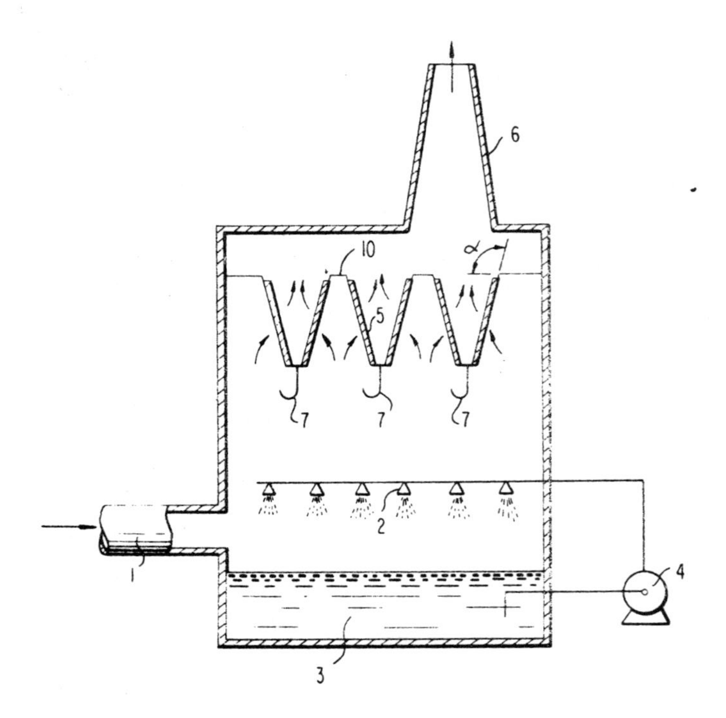

Source: U.S. Patent 3,861,889

An angle α of the surface of the filter layer to the horizontal should be an obtuse angle of less than 120°, preferably within the range of 95° to 110°. In case the angle exceeds the upper limit of the above range, the liquid tends to stagnate on the filter surface, thereby lowering the effectiveness of the filter and resulting in a poor removal ratio of dust. On the other hand, when the angle is too small to fall in this range, then there will be a great pressure loss due to the filter layer. Shown at **7** is a U-shaped tube adapted to exhaust the solvent which has flowed down on the surface of the filter layer. To prevent splashing of the solvent droplets, it is preferable to provide a net or mesh of metal or synthetic fiber on the surface at the gas outlet side of the foamed material. The following is a specific example of the application of this process.

Example: The exhaust air stream from urea prilling towers was introduced through duct **1** into the apparatus. The exhaust air stream contained 500 to 1,000 milligrams per cubic meter of urea dust whose diameter was within the range of 1 to 10 microns. The exhaust air stream was then washed with an aqueous urea solution sprayed from spray nozzle **2** and thereby 90% of the dust was washed into the aqueous urea solution. The aqueous urea solution was stored in a solution reservoir **3** and then fed by means of a pump **4** to the spray nozzles for circulation. A part of the urea solution was discharged therefrom for the recovery of urea dissolved therein and then water was added in an amount sufficient to supplement the amount of the urea solution discharged.

The exhaust air stream which had been washed with the urea solution was filtered through a filter layer at a surface velocity of 1.0 meter per second, thereby removing the urea dust from the air stream and was then withdrawn through exhaust duct **6**. The air stream from exhaust duct **6** contained 10 milligrams per cubic meter of the urea dust. The pressure loss was 29.5 mm H_2O.

The filter layer used was of ether type polyurethane foam resin useful for cushioning and having a thickness of 10 millimeters and an open-cellular porosity of 98%. The angle α of the filtrating surface to the horizontal was 100°. The small droplets of urea solution formed by means of the spray nozzle reached the surface of the filter layer at the gas outlet side, thereby the filter layer was impregnated by the urea solution. The urea solution into which urea dust had been dissolved was exhausted through U-shaped tube **7** into solution reservoir **3**.

Removal from Water

As is well known, urea is conventionally produced by reacting ammonia and carbon dioxide under high temperatures or pressures. The output of the reactor may be directed to a vacuum crystallizer from which the desired products such as, for example, a 74% urea solution are withdrawn. The process also forms large amounts of water which are removed from the vacuum crystallizer as vapor along with relatively small but contaminate amounts of ammonia, carbon dioxide and urea.

Heretofore this water and the contaminants have frequently been condensed and simply dumped. However, the dumping of untreated and contaminated urea plant condensate is a source of water pollution and efforts have been made to treat the condensate so as to remove the contaminants. Heretofore such efforts have not been satisfactory since they have resulted in materially increased operating costs and/or significant additional costs for new equipment.

A process developed by *W.H. Van Moorsel; U.S. Patent 3,922,222; November 25, 1975; assigned to CF Industries, Inc.* is one in which condensate from a urea crystallizer is directed to a low pressure boiler or heat exchanger and evaporated to accomplish thermal decomposition of urea and any biuret contained therein into ammonia and carbon dioxide which are subsequently removed and recycled to the urea plant, if desired, leaving substantially urea-free water of sufficient quality to satisfy environmental regulations or guidelines or for use as boiler feed water.

VANADIUM

See Adipic Acid Process Effluents (U.S. Patent 3,673,068)

VEGETABLE OIL REFINERY WASTES

Removal from Water

In the processing of vegetable sources to obtain edible oils, the wastewater from the various processing steps is pumped to a gravity separation station. This wastewater contains fatty constituents in both dispersed and nondispersed forms. A dispersion of fatty constituents can be formed in the conduit to the separation station where fatty acids combine, under appropriate conditions of concentration and turbulence, to form micelles which act as emulsifying agents for other fatty constituents such as triglyceride oils.

At the separation station, the nondispersed fatty constituents, which have an average specific gravity of about 0.79, rise to the surface and are removed for further commercial processing. The aqueous phase from the separation station, which contains dispersed fatty constituents, has heretofore been discharged as wastewater. Since this water frequently contains from 600 to 1,000 parts per million of fatty constituents, it constitutes a source of environmental pollution.

A process developed by *H.F. Keller, Jr.; U.S. Patent 3,803,031; April 9, 1974; assigned to GBK Enterprises, Inc.* is one in which aqueous systems containing fatty constituents and particulate solids and having a pH from about 1 to about 4.8 are filtered through a finely divided, acid and alkali resistant filter media having a particle mesh size range from about 12 to about 60 at a rate from about 1 to about 50 gallons per minute per square foot of filter media surface area whereby fatty constituents and particulate solids are retained by the filter media and the effluent therefrom is clarified water.

The filter media is periodically regenerated by: (1) agitating the media in the presence of a saponifying alkaline solution to extract fatty constituents therefrom, (b) withdrawing the alkaline solution from the filter media, (c) passing fresh water through the filter media in the same direction of flow as that of the aqueous system to remove residual water-soluble materials, and (d) backwashing the filter media with fresh water to remove insoluble and nondispersible particulate solids.

A process developed by *P. Bradford; U.S. Patent 2,925,383; February 16, 1960; assigned to Swift & Company* applies to the removal of soapy materials from vegetable oil refinery wastewaters. Considerable difficulty has been experienced in the vegetable oil refining industry in separating insoluble soaps from the condensate of the steam deodorizers. It is common practice in the refining of vegetable oils to subject the oil to a steam deodorization operation during which the oil is held in a large enclosed vessel and a vacuum drawn on its surface while steam is bubbled therethrough.

The volatile materials, which include principally free fatty acids, are drawn off with the steam by the vacuum and later condensed by spraying with a cooling liquid in barometric condensers. In those refineries which are located adjacent to the ocean, seawater may be used for condensing the vapors in the barometric condensers. Seawater contains large quantities of calcium and magnesium ions which react with the free fatty acids to form insoluble soaps.

The soaps are widely dispersed and form insoluble flocs, which tend neither to float nor settle and will remain suspended in the liquid for long periods of time. The seawater from the condensers of the steam deodorizers is collected in a basin known as a hot well. It has proven difficult to economically separate these insoluble soaps from the hot well liquor, presumably because of their low concentration and because of their tendency to remain suspended.

The process developed by Bradford involves introducing an aerated second wastewater which carries fine solid particles or a precipitate into the first waste. The treated waste and the introduced aerated stream are intimately mixed and then passed to a quiescent zone where the floc and the precipitate of the two aqueous waste bodies may be floated to the surface of the combined liquors.

VINYL CHLORIDE

When vinyl chloride is polymerized in an aqueous medium by suspension or emulsion polymerization techniques, there is obtained a latex or slurry that contains polyvinyl chloride and up to about 5% by weight of vinyl chloride. Most of the unreacted monomer is usually removed by heating the latex or slurry under reduced pressure to about 65°C. This stripping process produces large volumes of gases that contain low concentrations of vinyl chloride.

In view of developed safety standards that require that the amount of vinyl chloride in the atmosphere that workers breathe be maintained at very low levels, it is necessary that the vinyl chloride in the effluent gas streams be recovered or destroyed so that these requirements can be met.

Removal from Air

A number of procedures have been proposed for the removal of vinyl chloride from gas streams that contain a low concentration of vinyl chloride, but none has proven to be entirely satisfactory. Procedures that involve its adsorption on activated carbon are effective in removing vinyl chloride from gas streams, but carbon has a limited capacity for the adsorption of vinyl chloride, and when the carbon has adsorbed from about 5 to 20% by weight of vinyl chloride it is saturated and must be regenerated. The activated carbon gradually loses its ability to adsorb vinyl chloride and must be replaced by fresh activated carbon at frequent intervals. Vinyl chloride and other compounds that are desorbed from the surface of the carbon must be recycled to the process or destroyed.

Among the chemical methods that have been proposed for the destruction of vinyl chloride in gas streams is reaction with ozone. This method has the disadvantage of being slow and requiring long residence times to reduce the vinyl chloride content of the gas stream to 1 ppm or less. In addition, it is difficult to meter ozone into the gas streams in amounts that will destroy substantially all of the vinyl chloride without leaving an appreciable amount of ozone in the effluent gas. There are also environmental problems arising from the presence in the effluent gas of ozonides formed by the reaction of vinyl chloride with ozone.

A process developed by *J.R. Sudduth and D.A. Keyworth; U.S. Patent 3,983,216; September 28, 1976; assigned to Tenneco Chemicals, Inc.* is one in which vinyl chloride is removed from gas streams that contain from about 10 ppm to 1,000 ppm of vinyl chloride by contacting the vinyl chloride in the gas stream with ozone in the presence of activated carbon. The gas streams treated in this way contain less than about 1 ppm of vinyl chloride and no detectable amount of ozone or ozonides.

A process developed by *L.A. Smalheiser; U.S. Patent 3,933,980; January 20, 1976; assigned to Stauffer Chemical Company* is one in which the amount of vinyl chloride in a gas stream can be reduced by contacting the stream with ozone.

The process can be used to treat gas streams which arise in ethylene oxychlorination processes, ethylene dichloride cracking operations, vinyl chloride polymerization processes in which vinyl chloride is a monomer or comonomer, ventilation streams from areas in which vinyl chloride monomer is or may be present, processes for preparing vinylidene chloride, polymerization processes in which vinylidene chloride is utilized as a monomer or comonomer.

After reaction with ozone, the gas stream contains hydrogen chloride, oxygenated compounds such as carbon dioxide and water, and can contain phosgene and partially oxygenated hydrocarbons such as methanol and the like.

Contacting the treated gas stream with an aqueous medium is advantageous in that products of the reaction are removed from the gas stream and partially oxidized hydrocarbons can further react with any unreacted ozone present in the gas stream. The aqueous medium also aids in hydrolysis of reaction products of ozone and the chlorinated hydrocarbons present.

A process developed by *P.J. Patel, C.G. Thompson, E.J. Hourihan and C.S. Stutts; U.S. Patent 3,984,218; October 5, 1976; assigned to Tenneco Chemicals, Inc.* is one in which vinyl chloride is removed from gas streams that contain from 10 ppm to 100 mol percent of vinyl chloride by passing the gas stream through a bed of coconut shell-derived or petroleum-derived activated carbon. The exit gas stream contains less than 5 ppm of vinyl chloride.

The vinyl chloride-saturated carbon is treated with steam at 100° to 150°C to desorb the vinyl chloride, which is then recovered. The wet carbon is contacted with an inert gas at 90° to 150°C until its water content is less than 1% by weight; the hot dry carbon is cooled to ambient temperature by contacting it with a cold inert gas. The regenerated activated carbon is then used to remove additional vinyl chloride from the gas stream.

The vinyl chloride adsorption-carbon regeneration cycle can be repeated for long periods of time without loss of adsorptive capacity of the activated carbon or formation of polyvinyl chloride on the surface of the carbon.

In a typical polyvinyl chloride plant, unreacted vinyl chloride vapors are present in the reactor void space, e.g., about 20% of the reactor capacity, the vapors being commonly at a pressure of about 100 psig and temperatures on the order of 120° to 140°F. As a typical plant will have a number of such polyvinyl chloride reactors, each coming off service on a regular schedule, such as every 8 hours, a relatively large number of reactor changes per day will occur, commonly at uniform intervals of time. As each reactor comes off service, the unreacted vinyl chloride vapors therein are vented from the reactor, as to an atmospheric gas-holder by release of the gas pressure in the reactor. The reactor is then purged with either nitrogen or some other inert gas, e.g., flue gas, to assure that all of the vinyl chloride vapors are removed from the reactor. The purge gas is also passed to the gas-holder.

The gaseous mixture thus obtained in the gas-holder comprises vinyl chloride vapors, together with the inert purge gas and minor amounts of various gaseous impurities vented from the reactor. The impurities present in the gaseous mixture in the gas-holder typically include acetylene, butadiene, methyl chloride, acetaldehyde, hydrogen chloride, sulfur and nonvolatile material. The recovery and purification of the vinyl chloride present in the gaseous mixture thus obtained is, of course, highly desirable in the overall economy of such plants.

While liquid solvents for the various components of the gaseous mixture can be determined, the recovery and purification of the vinyl chloride from such gaseous mixtures are particularly difficult in the absence of a selective solvent facilitating the desired recovery and purification of the vinyl chloride present in such gaseous mixtures. For example, the volatility of vinyl chloride is intermediate between that of methyl chloride as a more volatile material, and butadiene as a less volatile material, with the desired purification of vinyl chloride therefrom being made particularly difficult in that the methyl chloride, butadiene and other impurities are all present in the gaseous mixture in only trace quantities. A selective solvent permitting the desired separation of vinyl chloride from the inert purge gas and associated gaseous impurities removed from polyvinyl chloride reactors is, therefore, highly desirable in the art.

A process developed by *A.A. Bellisio; U.S. Patent 3,807,138; April 30, 1974; assigned to GAF Corporation* is one in which vented vapors and inert purge gas from a polyvinyl chloride reactor are contacted with an N-alkyl lactam liquid solvent in an absorption zone maintained at a liquid temperature of about 20° to 40°C to remove a substantial portion of the vinyl chloride content of the gaseous mixture of the vented vapors and purge gas. The N-alkyl lactam contains water in an amount of from about 2% to about 10% by weight. The gaseous mixture introduced into the absorption zone is at a pressure of from about 25 to about 100 psig, the temperature of the gas being from about 10° to about 40°C.

Rich lactam solvent having the recovered vinyl chloride absorbed therein is passed to a stripper zone wherein reboiler vinyl chloride vapors strip therefrom vapor impurities removed from the gaseous mixture for recycle to the absorption zone. A stripper bottoms liquid stream is passed to a distillation zone wherein the vinyl chloride is stripped from the stripper bottoms and recovered as an essentially pure vinyl chloride free of the impurities vented therewith from the polyvinyl chloride reactor and the associated purge gas. N-methyl pyrrolidone is a preferred N-alkyl lactam solvent for the recovery and purification of vinyl chloride vapors.

VIRUSES

Removal from Water

A process developed by *C. Wallis and J.L. Melnick; U.S. Patent 3,770,625; November 6, 1973; assigned to The Carborundum Company* involves treating an adsorbing media such as activated carbon with an inorganic hydrolyzing composition of matter containing sodium. Viruses are removed from a fluid containing viruses by bringing the fluid into contact with the treated activated carbon adsorbing media.

Many of the deficiencies of the virus adsorbents of the prior art are overcome by this process. For example, the treated activated carbon media binds viruses so that a high pressure stream of fluid including organics passing through the virus containing treated media will not elute viruses. The exhaustion or any ineffectiveness of the treated media can be easily detected. At that time the adsorbed viruses can be rendered inactive and removed without the risk of environmental contamination.

After the viruses have been removed from the exhausted media, the media removed can be retreated for reuse in adsorbing additional viruses. The use of readily available and inexpensive materials in the practice of the process makes it economically feasible to reclaim large quantities of wastewater. Since high flow rates per treated media can be achieved, the apparatus disclosed is also suited for reclaiming small volumes of water.

An apparatus developed by *C. Wallis and J.L. Melnick; U.S. Patent 3,836,458; September 17, 1974; assigned to The Carborundum Company* is a filter system for the purification of water, whereby particulates and viruses may be economically and reliably removed. The system comprises a filter means for removal of silts and colloidal materials, a treated carbon virus filter, and a specially treated honeycomb cartridge for removal of microbiological flora, e.g., bacteria, fungi, molds and algae one micron or larger. Optionally, the system may include reverse osmosis means, an ion exchange demineralizer and an ultrafine membrane filter.

WASTE WOOD PROCESSING EFFLUENTS

The disposal of hydroscopic combustible material, such as scrap wood and bark, needs to be carried on at a temperature of about 600°F or somewhat lower to prevent flashing off certain volatiles which at or above that temperature level cause the formation of a noxious blue smoke.

Removal from Air

An apparatus developed by *R.M. Williams; U.S. Patent 3,826,208; July 30, 1974; assigned to Williams Patent Crusher and Pulverizer Company* is one which moves a large volume of air through a furnace to provide the needed heat for drying the material to a state where it can be consumed by serving as a fuel. The air volume so moved is dust laden, and the apparatus concentrates the dust and particulate material for burning, and effectively mixes the combustible products with cleansed air to produce a source of heat at a controlled temperature level.

A device developed by *A.B. Baardson; U.S. Patent 3,837,303; September 24, 1974; assigned to Mill Conversion Contractors, Inc.* is a burner by means of which waste wood and the fumes emitted by drying wood can effectively be burned to not only provide additional heat energy but also to reduce atmospheric pollution. The burner basically comprises two chambers. The wood, fed in powdery form, is burned in the first chamber, the product of this combustion being passed to the second chamber where it is mixed with the fumes before being combusted a second time.

WELDING PROCESS EFFLUENTS

It is well-known that welding processes, electric arc welding in particular, produce noxious pollutants in the air of the welding shop, particularly when galvanized iron is to be welded. Small concentrations of these pollutants may be tolerated, but concentrations of appreciable magnitude constitute a recognized health hazard for workmen in the welding trade.

Removal from Air

One attempt to remedy this situation comprises ventilating the entire shop: this has the first disadvantage of requiring enormous ventilating equipment with enormous power requirements; the second disadvantage of transmitting the pollutants into the general outside atmosphere; and the third disadvantage of conducting the warmth of the shop to the outside as well as the pollutants, thus greatly increasing the demand for heating energy in cold weather.

A second attempt to remedy the situation divides the shop into bays or partially enclosed work areas, each vented to the outside. The same disadvantages apply here, on an only slightly smaller scale, and the additional problem arises of how to work on pieces larger than a work bay.

Another factor must be taken into consideration, and this is that welding shops must be served by cranes, which means that the design and installation of suitable duct work for the conduction and discharge of the fumes is often difficult and sometimes impossible because craneways cannot be blocked.

A technique developed by *J.H. El Dorado; U.S. Patent 3,926,104; December 16, 1975; assigned to Midwest Mechanical Services, Inc.* involves the use of a portable hood for positioning over a workpiece to be welded, an air purifier in the form of an electrostatic air cleaner, and means for drawing shop air over the workpiece into the hood and through the purifier and discharging it back into the shop.

WOOD DRYER EFFLUENTS

In typical dryers of the type wherein hot air is passed in contact with the material to be dried, the air circulating through the dryer tends to pick up the finer particulate solids associated with the material as well as gaseous by-products of the drying process. Heretofore these waste gases and gas-borne solids have simply been exhausted to the atmosphere, creating air pollution problems. In veneer and rotary dryers used in the plywood industry, the fine sander dust picked up in the gas stream is highly combustible and tends to be at least partially incinerated by the hot air even while in the dryer itself, intensifying the air pollution problem because of the presence of dense smoke in the emissions.

Removal from Air

A scheme developed by *R.B. Burden, Jr. and E.J. O'Gieblyn; U.S. Patent 3,749,030; July 31, 1973; assigned to Wasteco, Inc.* is one in which heat generated within an incinerator is transferred to oil in a combustion gas-to-oil heat exchanger. The hot oil is circulated through a radiator in a material dryer to heat air used as the drying medium. The hot air passing in contact with the material to be dried picks up particulate solids and gases which heretofore have been exhausted to the atmosphere. These gas-borne waste emissions are conducted to the combustion chamber of the incinerator where they are burned, thus becoming a source of heat for the dryer.

A process developed by *V.J. Tretter, Jr., L.M. Stefensen and R.A. Rydell; U.S. Patent 3,853,505; December 10, 1974; assigned to Georgia-Pacific Corporation* involves the removal of haze-forming constituents from wood veneer dryer effluent by condensing the haze-forming constituents, agglomerating the condensed constituents and removing them from the effluent.

A suitable form of apparatus for the conduct of such a process is shown in Figure 158. As shown there, a cyclone separator **10** is shown used in combination with a tank-type separator **11**. The cyclone separator comprises a cylindrical body **12** having an inlet duct **13**. The inlet duct enters the cylindrical body tangentially at a point near the top **14**. The inlet duct **13** has water spray nozzles **15** positioned in the duct for the dispersion of water into the incoming effluent. The water spray nozzles are positioned in the conduit at a distance from the entry **14** to provide sufficient distance for the agglomeration of the extractives in the effluent prior to entry into the cyclone separator.

FIGURE 158: APPARATUS FOR REMOVAL OF HAZE-FORMING CONSTITUENTS FROM WOOD DRYER EFFLUENT

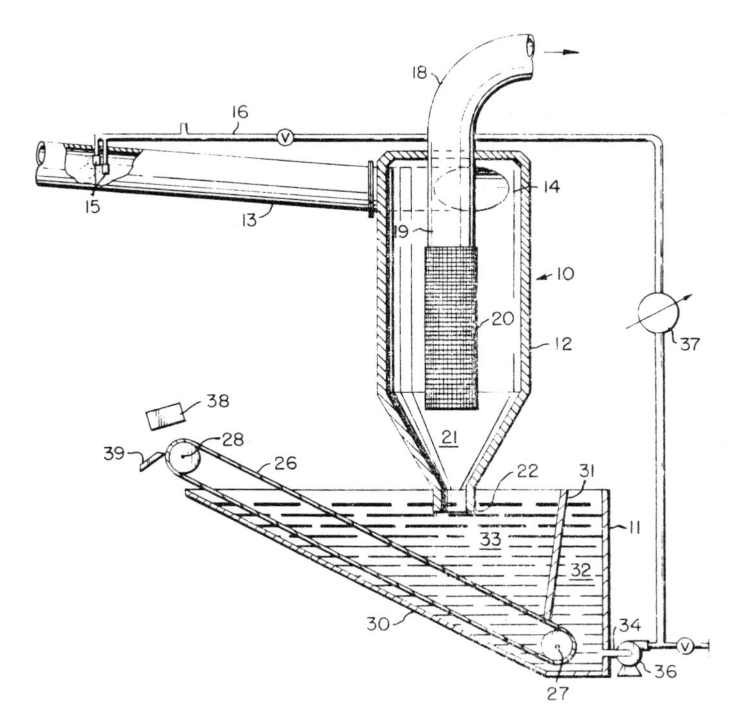

Source: U.S. Patent 3,853,505

Appropriate piping **16** is provided to supply the required water to the nozzles with proper control equipment (not shown) to disperse the required amount of water and to obtain the temperature desired in the conduit. The inlet conduit is positioned usually at an incline of about a degree or two, preferably from 10° to 15°, from the horizontal to provide a downward approach to separator **10**. The angle can be increased to the extent that the inlet conduit is substantially vertical, as long as a 90° bend is provided at the cyclone separator to obtain the horizontal tangential entry into the separator.

The effluent is discharged from the top of the cyclone separator through discharge conduit **18** which extends into the cylindrical body from the top with a portion **19** of the conduit extending below the inlet conduit entry **14**. Preferably the discharge conduit is equipped with an impingement-type separator **20** through which the effluent passes prior to entry into the discharge conduit, even though additional entrainment equipment may be used outside of the cyclone. As shown in the drawing, the impingement separator **20** is a wire-mesh screen extending through the cylindrical body and into the conical bottom portion of the cyclone where the diameter of the cylindrical body is progressively decreased to form discharge opening **22** at the bottom of the cylindrical body from which the liquid constituents, comprising the haze-forming constituents and water removed from the effluent, are discharged.

The wire-mesh screen is a simple, inexpensive type impingement separator which may be used. It has an advantage over many separators in that the wire-mesh has some flexibility resulting in movement of the screen during operation which enhances the removal of the haze-forming constituents agglomerated on the screen. However, in place of the wire-mesh screen, other types of impingement separators which provide surfaces for the impingement of the flowing effluent or change in direction of flow may be used.

For example, a conduit containing a multiplicity of holes, or an assembly of closely-spaced rods, or closely-spaced plates providing a change in direction of flows may be employed. The effluent is drawn through the cyclone by means of a fan or blower (not shown) communicating with effluent discharge conduit **18**. The haze-forming constituents and water removed by the cyclone separator are discharged through discharge opening **22** to be further processed to separate the condensed haze-forming constituents from the water.

The separator used for the separation of the haze-forming constituents from the water is a sloping-side rectangular tank-type vessel **4** in which a continuous belt or conveyor **26** is positioned on rolls **27** and **28** to move at an incline, usually for convenience at an angle of from 15° to 45°. Roll **27** is located near the bottom of the tank at one end and roll **28** is located over the top of the sloping side **30** of the tank at the opposite end of the tank. A plate **31** partitions the upper portion of the tank into two compartments **32** and **33**. Plate **31** extends into the tank from the top to the proximity of belt **26** near the bottom of the tank so that compartment **32** formed by the plate encompasses a portion of the belt at the greatest depth in the tank.

A discharge **34** near the bottom of compartment **32** is provided through which water separated from the haze-forming constituents may be removed or recycled through pipe **16** by means of pump **36**. The water recycled may be cooled if necessary by heat exchanger **37**. The liquid level in the tank is maintained at a predetermined height by means of liquid level control instrumentation (not shown) or other methods obvious to people skilled in the art.

Tank **11** is positioned with respect to cyclone **10** such that discharge opening **22** is above belt **26** and below the liquid level in the tank in compartment **33** or the compartment in which the belt emerges from the tank. A radiant heater **38** placed above belt **26** heats a portion of the belt above the tank before the belt passes by knife-edge scraper **39**.

A system developed by *D.E. Drake and J.J. Nelson; U.S. Patent 3,945,331; March 23, 1976; assigned to Enertherm, Inc.* is a dual-function thermal recovery system for reducing pollutant emissions to the ambience and for utilizing normally wasted heat energy for beneficial purposes. The system is particularly applicable to the lumber mill industry wherein there is a need to incinerate large quantities of waste wood products and a corresponding need for hot, relatively clean gas for lumber drying purposes.

WOOD PRESERVING PLANT EFFLUENTS

Removal from Water

A process developed by *J.W. Gooch; U.S. Patent 3,733,269; May 15, 1973* is a process which comprises introducing a reagent solution into the waste stream in such a manner that the reagent is released and dispersed within the waste stream in the form of a microscopically gaseous phase which is adsorbed on the surfaces of the suspended solids and organic waste materials to react with the surfaces of the solids and materials and produce surface characteristics which induce agglomeration and coagulation of the solids and materials. Figure 159 shows such a process.

FIGURE 159: APPARATUS FOR TREATMENT OF WOOD PRESERVING PLANT WASTEWATERS

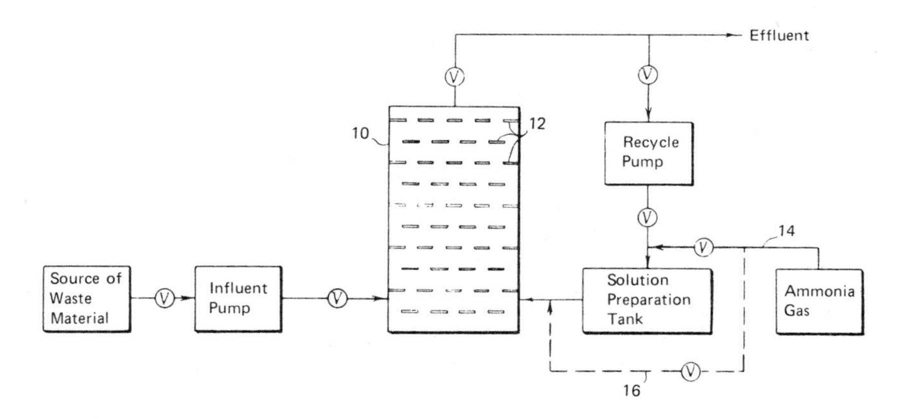

Source: U.S. Patent 3,733,269

In this schematic diagram of the process, liquid influent from a source of waste material is pumped by an influent pump through a valve to the bottom of dispersion tank **10**. Also entering the bottom of the dispersion tank is metastable saturated ammonia water solution. The ammonia water solution is prepared by combining ammonia gas with a heated water solution within a solution preparation tank. The temperature of the ammonia saturated water solution should be maintained at a level sufficient to cause the ammonia to be metastable. A suitable temperature for a water solution is 90°C.

The ammonia gas is admitted to the solution preparation tank under pressure and an increase in pressure over the pressure required to repress release of the ammonia gas is maintained within the tank, suitably 2 atmospheres. As the heated metastable ammonia water solution is introduced into the bottom of the dispersion tank, the ammonia gas is released from the solution in the form of microbubbles. Foraminous baffles **12** within the dispersion tank assist gentle dispersion of the microbubbles throughout the waste material influent. In addition, the baffles cause the waste material influent together with the generated microbubbles to follow a tortuous path through the dispersion tank and establish a contact period between waste material and bubbles during which the microbubbles can be adsorbed on the particulate surfaces. Other equipment may be used in place of the baffles to assist the dispersion of the microbubbles if so desired.

From the top of the dispersion tank, water solution and coagulated and agglomerated waste material flow as an effluent through a valve to a separation station where the waste material may be easily and efficiently separated from the water solution by conventional mechanical techniques. As shown in the diagram, makeup water solution for use in preparing the saturated ammonia water solution may be withdrawn from the effluent and pumped by an effluent recycle pump to the solution preparation tank. The ammonia gas, as shown by the solid line **14** and the dotted line **16** may be added to the water solution either before or after heating of the water solution within the solution preparation tank.

The following is a specific example of the conduct of this process. A viscous sludge of cresol, phenolic acids, wood fiber and other contaminants of wood origin is taken from the bottom of a settling basin of a wood preserving plant and diluted with water to 1 part sludge per 4 parts water. The pH of this solution is measured at 4.0.

A metastable ammonia water solution is prepared and heated to a temperature of about 90°C in an air-tight solution preparation tank by mixing ammonia gas with distilled water. The ammonia solution is then gradually pumped under 2 atmospheres of pressure into the bottom of a dispersion tank, containing approximately 500 ml of the waste solution at 20°C, until 0.15 gram of ammonia has been added. Microbubbles of ammonia gas form within the dispersion tank and the waste material begins to agglomerate and coagulate as a precipitate soon thereafter. The waste solution is clear and is removed by decanting.

The supernatant liquid is found to have a pH of 6.8. The chemical oxygen demand (COD) is determined for the diluted waste solution and the supernatant liquid for comparison purposes. It is found that the diluted waste solution has a COD of 170,000 mg/l and the treated supernatant liquid has a COD of 600 mg/l.

WOOL SCOURING EFFLUENTS

Raw wool is typically treated by employing a process referred to as emulsion-scouring to wash the wool. In the emulsion-scouring process, the raw wool is passed through a series of scouring tanks, whereby the impurities present in the wool, such as suint, dung, earth, sand and wool grease, are removed from the wool. Wool grease removal is specifically achieved in a tank containing a surfactant solution (nonionic or anionic surfactant). During and after the scouring treatment, the various impurity components are separated for disposal or recovery, depending on their economic value.

Typically, the spent surfactant solution contains emulsified wool grease rich in lanolin and having a high economic value. Further, the removal of this wool grease from the spent scouring liquor is desirable in that it reduces the pollution load on the waterways into which it is typically discharged. Thus, in the emulsion-scouring process, wool is introduced into a cold-water bath so as to remove water-soluble impurities initially present in the fleece (referred to as suint), and, thereafter, introduced into a hot-water surfactant-containing bath for wool grease removal. The wool cleaned in this fashion results in a spent emulsified wool-grease solution typically containing from about 1 to 8% wool grease.

Removal from Water

Although the wool grease in the emulsion-scouring liquor has a high economic value, the cost of recovering the wool grease is often quite high for small companies that do not have the capital assets, and, accordingly, such companies lagoon the emulsion-scouring liquor to provide for its disposal by biodegradation. Others employ one or both of the two main methods now available for recovering wool grease from aqueous-scouring liquors. These methods are the centrifugal process and the acid-cracking process, or combinations thereof.

In the centrifugal process, the emulsion-scouring liquor is passed through one or more primary desludge separators, followed by a desludging centrifuge. The three component streams removed from the desludging centrifuge are a wool-grease concentrate stream, a partially degreased emulsion-scouring liquor stream, and, thirdly, a dirt-rich sludge. The dirt sludge is often discharged to drain or filter-pressed to concentrate the disposal of the sludge. The centrifugal recovery process provides for recovery of only up to about 50% by weight of the wool grease, but typically, 25 to 40% of the grease. The wool-grease concentrate stream is rich in lanolin which is a valuable product useful in the soap, cosmetic and other industries.

A series of centrifuge separators may be employed to compound recovery percentages if desired. If desired, the partially degreased liquor stream exiting the centrifuge may be disposed of or treated by the acid-cracking process to recover additional wool grease. However, the additional grease recovered is a low-quality grease not suitable for lanolin recovery. Recovery of the grease is for pollution-abatement purposes, and not for economic advantages. The centrifugal process, although simple, does not provide for very high overall economic recovery of the total wool grease.

Another method of wool-grease recovery from emulsion-scouring liquors is known as the acid-cracking process. In the acid-cracking process, the emulsion-scouring liquor again is initially desludged by a primary desludger. The liquor is then acidified with an acid to a pH between about 2 and 4, to provide for acid-cracking or destabilization of the wool-grease emulsion. The acid-cracked liquor is then collected in a settling tank where a major portion of the wool grease and dirt are separated as a sludge, with the wool grease accumulating with the dirt particles on the bottom of the settling tank. The sludge recovered is then treated to recover the valuable wool grease, such as by filter-pressing the sludge. Generally, the recovery yield of the wool grease using the acid-cracking process is about 65% by weight, with losses occurring in the unsettled supernatant and filter cake.

In the centrifugal recovery process discussed earlier, the partially degreased liquor is often treated by the acid-cracking process to achieve additional wool grease, but usually, to provide for a cleaner plant aqueous-effluent stream. One disadvantage of the acid-cracking process is that the lanolin is composed of a mixture of alcohol, acid and ester compounds. Acid treatment hydrolyzes the ester and produces acid and alcohol components, which are often not desirable and are of lower economic value.

In summary, the centrifugal process employed for the recovery of wool grease or an aqueous-scouring liquor is characterized by an unsatisfactory level of wool-grease recovery and the need to treat further the partially degreased liquor prior to discharge as an effluent, while the acid-cracking process, although providing for higher recovery levels of wool grease, has the inherent disadvantage of producing a low-quality grease unsuitable for lanolin recovery.

A process developed by *J. Del Pico; U.S. Patent 3,847,804; November 12, 1974; assigned to Abcor, Inc.* involves treating wool-scouring liquor and wool-scouring liquor centrifuge effluent, both containing wool-grease emulsion, by the utilization of membranes to concentrate the liquor before centrifuging, or to eliminate the centrifuging step by concentrating the wool grease in the scouring liquor to a concentrate level through the use of ultrafiltration membranes. The process permits the efficient and economic recovery of the wool grease from the membrane concentrate fraction, and provides for reuse or disposal of the aqueous-permeate fraction.

A process developed by *W.A. Heisey; U.S. Patent 3,909,407; September 30, 1975; assigned to Geo. W. Bollman & Co., Inc.* is a process for continuously treating wool scouring wastes in which the scouring liquor at an elevated temperature is acidified and treated with bentonite while continuously mixing, and the acidulated liquor is in turn treated with lime and flocculated while also continuously mixing at elevated temperatures. Finally the conditioned mixed liquor is transferred to a settling tank where the sludge is removed for further treatment and clear liquid is drawn off for discharge into a sewer system.

YTTRIUM

See TV Tube Manufacturing Effluents (U.S. Patent 3,915,691)

ZINC

See TV Tube Manufacturing Effluents (U.S. Patent 3,915,691)

ZINC SMELTER EFFLUENTS

See Selenium (U.S. Patent 3,933,635)

NOTICE

WASTE TREATMENT WITH POLYELECTROLYTES AND OTHER FLOCCULANTS 1977

by Sidney J. Gutcho

Pollution Technology Review No. 31

The recycling of wastewater to a purity acceptable and useful in population centers and industry can be accomplished through the judicious use of polyelectrolytes and other flocculants, such as natural latexes.

Polyelectrolytes are high molecular weight polymers and are classified into three types—anionic, cationic and nonionic. They promote the process whereby suspended solids and colloidal materials in the water are agglomerated into masses sufficiently large to settle. Being polyelectrolytes, they neutralize any static charges on dirt particles, allowing them to come together.

Correctly applied, organic flocculants are practically nontoxic, biodegradable, small in volume, easily incinerated, and effective under varied pH and temperature conditions. This makes them ideal for clarification of water from reservoirs, lagoons, etc. They are being used at an increasing rate for the removal of solids from various industrial wastewaters including the effluents of the mining and papermaking industries.

This book presents a broad spectrum of the manufacturing technologies and practical applications, as they have been elaborated by many industries and municipalities. A partial and condensed table of contents follows here. Numbers in parentheses indicate the number of processes per topic.

ISBN 0-8155-0648-1 **274 pages**

TOXIC METALS 1976
Pollution Control and Worker Protection

by Marshall Sittig

Pollution Technology Review No. 30

This is a practical book about those metals and their compounds that are most likely to cause poisoning—in industry while being processed, and in the general environment while the disposal of their process-effluents is being implemented.

Each toxic metal survey gives leads to available information on

Toxicity
Extent of Exposure
Detection & Determination
Environmental Standards
Handling Procedures & Precautions
Removal from Air
Removal from Water
Solid Waste Disposal
Economic Impact of Controls

Specific data are given on permissible intake, acute and chronic toxicities, also on sensitization and the resulting hypersensitivities. In addition some mention is made of possible carcinogenicity and other immunologic pathogenesis as seen in beryllium poisoning.

What is most important and effective in the prevention of toxic metals poisoning is thorough removal of the toxic agents in a manner that positively precludes ingestion or contact by all potential victims, and therein lies the emphasis and main endeavor of this book.

Where it is necessary to remove toxic metals from the air, it is frequently a question of particulate removal. General techniques for their removal from water include

1. **Chemical Precipitation**
   a. **Lime**
   b. **Alum**
   c. **Iron Salts**
2. **Chemical Treatment**
   a. **Oxidation**
   b. **Reduction**
3. **Ion Exchange**
4. **Filtration & Ultrafiltration**
5. **Electrochemical Treatment**
6. **Evaporative Recovery**

Detailed monographs are furnished for the following metals:

ANTIMONY
ARSENIC
BARIUM
BERYLLIUM
BORON
CADMIUM
CHROMIUM
COPPER
INDIUM
LEAD
MANGANESE
MERCURY
MOLYBDENUM
NICKEL
SELENIUM
TIN
VANADIUM
ZINC

Each survey monograph, including the rather comprehensive introduction, is followed by a detailed list of pertinent references. In the case of government publications complete bibliographical data and report numbers are given in order to facilitate acquisition.

Not discussed in this book are the harmful effects—acute, delayed, or chronic—that may be produced in body tissues by exposure to ionizing radiations from radioactive isotopes of metals. Specialized uses in the atomic energy field require specialized antipollution measures and worker protection far beyond the scope of this book.

ISBN 0-8155-0636-8 **349 pages**

CORROSION RESISTANT MATERIALS HANDBOOK 1976

Third Edition

by Ibert Mellan

This famous book, first published in 1966 and now already in its third, greatly enlarged and completely revised edition, will help you cut losses due to corrosion by enabling you to choose the proper *commercially available* corrosion resistant materials for your particular purpose.

The great value of this outstanding reference work lies in the extensive cross-indexing of thousands of substances. The 151 tables are arranged by types of **corrosion resistant materials.** The **Corrosive Materials Index** is organized by *corrosive chemicals* and other *corrosive substances:* it refers you to specific recommendations in the tables. New in this edition are a separate **Trade Names Index** and a listing of **Company Names and Addresses.**

For the first time there appears also a group of 13 tables comparing the respective anticorrosive merits of commercial engineering and construction materials essential to industry.

The tables in this book represent selections from manufacturers' literature made by the author at no cost to, nor influence from, the makers or distributors of these materials.

Contents:

ISBN 0-8155-0628-7 **665 pages**